Principles of Electromagnetic Waves and Materials

Second Edition

Principles of Electromagnetic Waves and Materials

Second Edition

Dikshitulu K. Kalluri

CRC Press
Taylor & Francis Group
Boca Raton London New York

CRC Press is an imprint of the
Taylor & Francis Group, an **informa** business

CRC Press
Taylor & Francis Group
6000 Broken Sound Parkway NW, Suite 300
Boca Raton, FL 33487-2742

First issued in paperback 2019

ISBN-13: 978-1-4987-3329-8 (hbk)
ISBN-13: 978-0-367-87385-1 (pbk)

Library of Congress Cataloging-in-Publication Data

Names: Kalluri, Dikshitulu K., author.
Title: Principles of electromagnetic waves and materials / Dikshitulu K. Kalluri.
Description: Second edition. | Boca Raton : Taylor & Francis, CRC Press, 2018. | Includes bibliographical references and index.
Identifiers: LCCN 2017015930| ISBN 9781498733298 (hardback) | ISBN 9781315272351 (ebook)
Subjects: LCSH: Electromagnetism--Mathematical models. | Electromagnetic waves--Computer simulation.
Classification: LCC QC760 .K363 2018 | DDC 537--dc23
LC record available at https://lccn.loc.gov/2017015930

Visit the Taylor & Francis Web site at
http://www.taylorandfrancis.com

and the CRC Press Web site at
http://www.crcpress.com

Contents

Part II Electromagnetic Equations of Complex Media

Part IV Chapter Problems

Preface

The second edition retains the flow and the flavor of the first edition, which was stated in the preface to the first edition. Item 1 describes this flavor in a few words.

1. The millennium generation of students, immersed in computer culture, responds to teaching techniques that make heavy use of computers. They are very comfortable in writing the code to implement an algorithm. In this connection, I remember a graduate student doing research with me mentioning that his struggle is always in understanding the concept. Thus, my approach in teaching is to definitely include the simplest possible model that brings out the concept to the satisfaction of the student and then show how one can improve on the accuracy of the prediction of the performance of a more realistic model to be solved by using more advanced mathematical and computational techniques.
2. It is important to include material that makes my comprehensive book a resource to be referred to when the student encounters, at some stage, topics that are not covered in one or two courses he/she could take using my book as textbook. Having taken one or two courses, the student should be able to quickly get to the new material in the book as and when needed and proceed in a mode of self-study.
3. In view of the advances in the areas of electromagnetic metamaterials and plasmonics, the discussions of the underlying topics of "electromagnetics and plasmas" are strengthened in the second edition.
4. In addition to correcting the typos and errors of the first edition and making a few minor changes of rearrangement of the material in the chapters, the following are the major additions in the second edition:
    a. Chapter 13 in Volume 1 is entirely new, bringing into focus direct solutions of Maxwell's equations in the time domain (a rather long chapter). Section 13.2 reviews the one-dimensional solutions of bounded ideal transmission lines. Sections 13.3 and 13.4 introduce more advanced techniques to take care of the lossy transmission lines and direct solution based on the Klein–Gordon equation. Section 13.5 uses a nonlinear transmission line model to explain the soliton theory.
    b. Section 13.6 describes the charged particle dynamics based on Newtonian formulation as well as Lagrange and Hamiltonian formulations of equations of motion.
    c. The more advanced topics on charged particle dynamics such as generation of nuclear electromagnetic pulse (Section 13.7) and magnetohydrodynamics (Section 13.8), though not always done in an electromagnetics course unless the instructor wishes to lay the foundation for a course on electromagnetics and plasma science, are included for completeness. So is Section 13.10 on statistical mechanics and the Boltzmann equation.
    d. An emerging research area in electromagnetics is the effective use of space-time modulation of an electromagnetic medium for various applications.

While the space variation is well understood with a long history of the study of inhomogeneous media, the study of the time-varying medium is of more recent origin. With the possibility of synthesizing metamaterials, researchers are exploring the exploitation of the effects that can be obtained by a simultaneous space-time modulation of the electromagnetic parameters to achieve the desired frequency and polarization changes of the waves. Section 13.9 gives the conceptual basis, while Appendix 13C explores an application of these concepts to conceive of a frequency and polarization transformer to convert a 10 GHz signal to a 1000 GHz signal by collapsing the ionization of a magnetoplasma in a cavity. The interesting physics that makes this possible is explained.

e. Appendices 13A and 13B deal with Maxwell's stress tensor and electromagnetic forces.
f. Chapter 14 is an extensively revised Chapter 14 of the first edition and deals with uniform motion of the electromagnetic medium and Lorentz transformations. Its appendices deal with the application of the Lorentz transformations to the bounded moving magnetoplasma medium, including transmission, reflection, critical angle, Brewster angle, and other wave phenomena. Appendix 14 G deals with Lienard–Wiechert potentials due to a point charge moving along a specified trajectory.
g. Appendix 11C is unique to this book. We bring together the effects of anisotropy and time and spatial dispersion by discussing the reflection from a warm magnetoplasma slab. The problem is so formulated so as to highlight the effect of the temperature parameter of the warm magnetoplasma. The resonance of the X wave at the upper hybrid frequency of a cold plasma case is removed by the temperature parameter. In the neighborhood of the upper hybrid frequency, the characteristics of the waves change from basically electromagnetic wave to that of an electron plasma wave.

MATLAB® is a registered trademark of The MathWorks, Inc. For product information, please contact:

The MathWorks, Inc.
3 Apple Hill Drive
Natick, MA, 01760-2098 USA
Tel: 508-647-7000
Fax: 508-647-7001
E-mail: info@mathworks.com
Web: www.mathworks.com

Acknowledgments

In preparing the second edition, I thankfully acknowledge the help I received from Dr. Constantine Taki Markos in editing my figure sketches of Chapter 13 to the standard required by the publisher. Moreover, he is the coauthor of the new Appendix 11C, which is based on his doctoral thesis.

Author

Dikshitulu K. Kalluri, PhD, is a professor emeritus of electrical and computer engineering at the University of Massachusetts Lowell. He received his BE in electrical engineering from Andhra University, India; a DIISc in high-voltage engineering from the Indian Institute of Science, Bangalore; a master's degree in electrical engineering from the University of Wisconsin, Madison; and his doctorate in electrical engineering from the University of Kansas, Lawrence.

Dr. Kalluri began his career at the Birla Institute of Technology, Ranchi, India, advancing to the rank of professor, heading the Electrical Engineering Department, and then serving as (dean) assistant director of the institute.

Since 1984, he had been with the University of Massachusetts Lowell, Lowell, advancing to the rank of full professor in 1987. He coordinated the doctoral program (1986–2010) and codirected/directed the Center for Electromagnetic Materials and Optical Systems (1993–2002/2003–2007). As a part of the center, he established the Electromagnetics and Complex Media Research Laboratory. He had collaborated with research groups at the Lawrence Berkeley Laboratory, the University of California, Los Angeles; the University of Southern California; New York Polytechnic University; and the University of Tennessee and has worked several summers as a faculty research associate at the Air Force Research Laboratory. He retired in May 2010 but continues his association with the university guiding doctoral students. He continues to teach his graduate courses online as professor emeritus. CRC Press published his two recent books: *Electromagnetics of Time Varying Complex Media Second Edition*, in April 2010, and the first edition of this book, in August 2011. He has published extensively on the topic of electromagnetics and plasmas. He supervised doctoral thesis research of 17 students.

Dr. Kalluri is a fellow of the Institute of Electronic and Telecommunication Engineers and a senior member of IEEE.

Selected List of Symbols

A or $\bar{A}$	Vector A
$\bar{\mathbf{A}}$ or $\bar{\bar{A}}$	Tensor A
$\tilde{A}$	Phasor A
$\hat{n}$ or $\hat{a}_n$	Unit vector

List of Book Sources

Balanis, C. A., *Advanced Engineering Electromagnetics*, Wiley, New York, 1989.

Cheng, D. K., *Field and Wave Electromagnetics*, 2nd edn., Addison-Wesley, Reading, MA, 1989.

Harrington, R. F., *Time Harmonic Electromagnetic Fields*, IEEE Press, New York, 2001.

Hayt, Jr., W. H., *Engineering Electromagnetics*, 5th edn., McGraw-Hill, New York, 1989.

Heald, M. A. and Wharton, C. B., *Plasma Diagnostics with Microwaves*, Wiley, New York, 1965.

Inan, S. A. and Inan, S. I., *Electromagnetic Waves*, Prentice-Hall, Englewood Cliffs, NJ, 2000.

Ishimaru, A., *Electromagnetic Wave Propagation, Radiation, and Scattering*, Prentice-Hall, Englewood Cliffs, NJ, 1991.

Jackson, J. D., *Classical Electrodyanamics*, Wiley, New York, 1962.

Kong, J. U., *Electromagnetic Wave Theory*, EMW Publishing, Cambridge, MA, 2000.

Kraus, J. D. and Carver, K. R., *Electromagnetics*, 2nd edn., McGraw-Hill, New York, 1973.

Neelakanta, P. S., *Handbook of Electromagnetic Materials*, CRC Press, Boca Raton, FL, 1995.

Papas, C. H., *Theory of Electromagnetic Wave Propagation*, McGraw-Hill, New York, 1965.

Ramo, S., Whinnery, J. R., and Van Duzer, T., *Fields and Waves in Communication Electronics*, Wiley, New York, 1967.

Stratton, J. A., *Electromagnetic Theory*, McGraw-Hill, New York, 1941.

Ulabi, F. T., *Applied Electromagnetics*, Prentice-Hall, Englewood Cliffs, NJ, 2001.

Van Bladel, J., *Electromagnetic Fields*, McGraw-Hill, New York, 1964.

Yeh, P., *Optical Waves in Layered Media*, Wiley, New York, 1988.

Part I

Electromagnetics of Bounded Simple Media

1

*Electromagnetics of Simple Media**

1.1 Introduction

The classical electromagnetic phenomena are consistently described by Maxwell's equations; these vector (see Appendix 1A) equations in the form of partial differential equations are

$$\nabla \times \mathrm{E}(\mathbf{r}, t) = -\frac{\partial \mathbf{B}(\mathbf{r}, t)}{\partial t}, \tag{1.1}$$

$$\nabla \times \mathbf{H}(\mathbf{r}, t) = \mathbf{J}(\mathbf{r}, t) + \frac{\partial \mathbf{D}(\mathbf{r}, t)}{\partial t}, \tag{1.2}$$

$$\nabla \cdot \mathbf{D} = \rho_V, \tag{1.3}$$

$$\nabla \cdot \mathbf{B} = 0, \tag{1.4}$$

where, in standard (RMKS) units, **E** is the electric field intensity (V/m), **H** the magnetic fieldintensity (A/m), **D** the electric flux density (C/m^2), **B** the magnetic flux density (Wb/m^2), **J** the volume electric current density (A/m^2), and ρ_V the volume electric charge density (C/m^3).

In the above equation, **J** includes the source current $\mathbf{J}_{\text{source}}$.

The continuity equation

$$\nabla \cdot \mathbf{J} + \frac{\partial \rho_V}{\partial t} = 0 \tag{1.5}$$

and the force equation on a point charge q moving with a velocity ν

$$\mathbf{F} = q(\mathbf{E} + v \times \mathbf{B}) \tag{1.6}$$

are often stated explicitly to aid the solution of problems. In the above equation, q is the charge (C) and ν the velocity (m/s).

* For chapter appendices, see 1A through 1D in Appendices section.

Solutions of difficult electromagnetic problems are facilitated through the definitions of electromagnetic potentials (see Appendix 1B):

$$\mathbf{B} = \nabla \times \mathbf{A}, \tag{1.7}$$

$$\mathbf{E} = -\nabla\Phi - \frac{\partial \mathbf{A}}{\partial t}, \tag{1.8}$$

where **A** is the magnetic vector potential (Wb/m) and **Φ** the electric scalar potential (V).

The effect of an electromagnetic material on the electromagnetic fields is incorporated in constitutive relations between the fields **E**, **D**, **B**, **H**, and **J**.

1.2 Simple Medium

A simple medium has the constitutive relations

$$\mathbf{D} = \varepsilon\mathbf{E} = \varepsilon_0\varepsilon_r\mathbf{E}, \tag{1.9}$$

$$\mathbf{B} = \mu\mathbf{H} = \mu_0\mu_r\mathbf{H}, \tag{1.10}$$

$$\mathbf{J} = \sigma\mathbf{E}, \tag{1.11}$$

where ε_0 and μ_0 are the permittivity and permeability of the free space, respectively,

$$\varepsilon_0 = 8.854 \times 10^{-12} = \frac{1}{36\pi} \times 10^{-9}\,(\text{F/m}), \tag{1.12}$$

$$\mu_0 = 4\pi \times 10^{-7}\,(\text{H/m}). \tag{1.13}$$

Different materials have different values for the relative permittivity ε_r (also called the dielectric constant), the relative permeability μ_r, and the conductivity σ (S/m). A simple medium further assumes ε_r, μ_r, and σ to be positive scalar constants (see Figure 1.1).

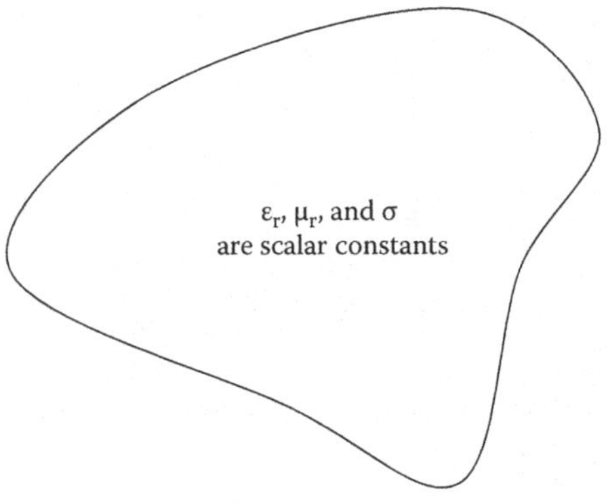

FIGURE 1.1
Idealization of a material as a simple medium.

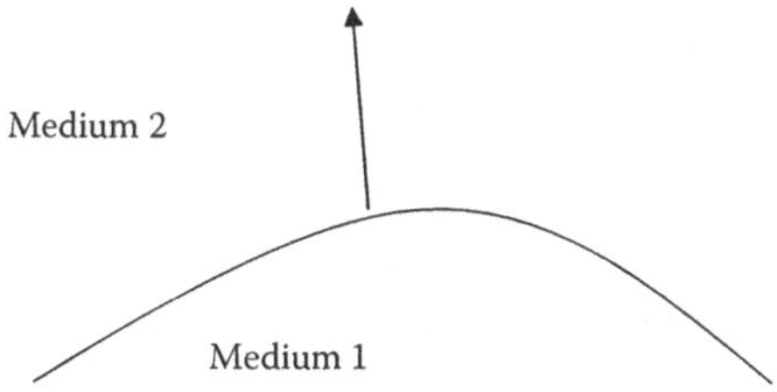

FIGURE 1.2
Boundary conditions.

Such an idealization of material behavior is possible in the solution of some electromagnetic problems in certain frequency bands. In fact, the first course in electromagnetics often deals with such problems only. In spite of such an idealization, the electromagnetic problems may still need the use of heavy mathematics for analytical solutions due to the size, shape, and composition of the materials in a given volume of space. At a spatial interface between two materials, the fields on the two sides of the boundary can be related through boundary conditions (Figure 1.2):

$$\hat{\mathbf{n}}_{12}\cdot(\mathbf{D}_2-\mathbf{D}_1)=\rho_s, \tag{1.14}$$

$$\hat{\mathbf{n}}_{12}\cdot(\mathbf{B}_2-\mathbf{B}_1)=0, \tag{1.15}$$

$$\hat{\mathbf{n}}_{12}\times(\mathbf{E}_2-\mathbf{E}_1)=0, \tag{1.16}$$

$$\hat{\mathbf{n}}_{12}\times(\mathbf{H}_2-\mathbf{H}_1)=\mathbf{K}. \tag{1.17}$$

In the above, ρ_s is the surface charge density (C/m^2), $\mathbf{K}$ is the surface current density (A/m), and $\hat{\mathbf{n}}_{12}$ a unit vector normal to the interface directed from medium 1 to medium 2 as shown in Figure 1.2.

1.3 Time-Domain Electromagnetics

The equations developed so far are the basis for determining the time-domain electromagnetic fields in a simple medium. For a lossless ($\sigma = 0$) simple medium, the solution is often obtained by solving for potentials $\mathbf{A}$ and Φ, which satisfy the simple form of wave equations

$$\nabla^2\mathbf{A}-\frac{1}{v^2}\frac{\partial^2\mathbf{A}}{\partial t^2}=-\mu\mathbf{J}, \tag{1.18}$$

$$\nabla^2\Phi-\frac{1}{v^2}\frac{\partial^2\Phi}{\partial t^2}=-\frac{\rho_V}{\varepsilon}, \tag{1.19}$$

where

$$v = \frac{1}{\sqrt{\mu\varepsilon}}. \tag{1.20}$$

In obtaining the above equations, we used the Lorentz condition

$$\nabla \cdot \mathbf{A} + \mu\varepsilon \frac{\partial \Phi}{\partial t} = 0. \tag{1.21}$$

The terms on the right-hand sides of Equations 1.18 and 1.19 are the sources of the electromagnetic fields. If they are known, Equations 1.18 and 1.19 can be solved using the concept of retarded potentials:

$$\mathbf{A}(\mathbf{r}, t) = \frac{\mu}{4\pi} \iiint_{\text{source}} \frac{[\mathbf{J}]}{|\mathbf{r} - \mathbf{r}'|} \mathrm{d}V', \tag{1.22}$$

$$\Phi(\mathbf{r}, t) = \frac{1}{4\pi\varepsilon} \iiint_{\text{source}} \frac{[\rho_V]}{|\mathbf{r} - \mathbf{r}'|} \mathrm{d}V', \tag{1.23}$$

where the symbols in square brackets are the values at a retarded time, that is,

$$[\mathbf{J}] = \mathbf{J}\left(\mathbf{r}', t - \frac{|\mathbf{r} - \mathbf{r}'|}{v}\right), \tag{1.24}$$

$$[\rho_V] = \rho_V\left(\mathbf{r}', t - \frac{|\mathbf{r} - \mathbf{r}'|}{v}\right), \tag{1.25}$$

and $\mathbf{r}$ and $\mathbf{r}'$ are the position vectors describing the field point and the source point, respectively (see Figure 1.3a). After solving for $\mathbf{A}$ and $\mathbf{\Phi}$, the electromagnetic fields are obtained by using Equations 1.7 and 1.8. We can then obtain the power density $\mathbf{S}$ (W/m^2) on the surface s:

$$\mathbf{S} = \mathbf{E} \times \mathbf{H}. \tag{1.26}$$

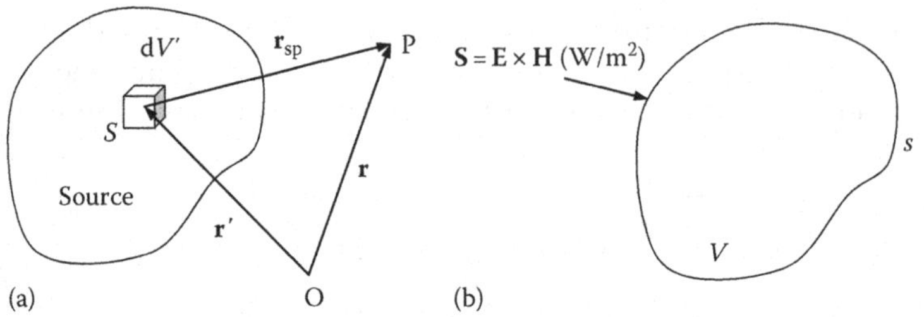

FIGURE 1.3
(a) Source point and field point. (b) Poynting theorem.

The interpretation that **S** is the instantaneous power density follows from the Poynting theorem (see Appendix 1C) applied to a volume V bounded by a closed surface s (Figure 1.3b):

$$\oiint_s \mathbf{E}\times\mathbf{H}d\mathbf{s} + \frac{\partial}{\partial t}\iiint_V \left(\frac{1}{2}\mathbf{E}\cdot\mathbf{D} + \frac{1}{2}\mathbf{B}\cdot\mathbf{H}\right)dV + \iiint_V \mathbf{E}\cdot\mathbf{J}dV = 0. \tag{1.27}$$

Equation 1.27 is obtained from Equations 1.1 and 1.2 and is the mathematical statement of conservation of energy.

1.3.1 Radiation by an Impulse Current Source

The computation of the time-domain electric and magnetic fields radiated into free space due to a time-varying current is illustrated through a simple example. Let the source be an impulse current in a small piece of wire of very small length h. The source can be modeled as a point dipole with

$$I(t) = I_0\delta(t), \tag{1.28a}$$

where $\delta(t)$ is an impulse function defined by

$$\delta(t) = 0, \quad t \neq 0, \tag{1.28b}$$

$$\int_{-\infty}^{\infty}\delta(t) \rightarrow \int_{-\infty}^{\infty}\delta(t)dt. \tag{1.28c}$$

The geometry is shown in Figure 1.4.

Since the source is a filament, Equation 1.22 becomes

$$\mathbf{A}_{\mathrm{P}}(\mathbf{r},t) = \frac{\mu}{4\pi}\int_{\text{source}} \frac{I\left(t - \frac{r_{sp}}{c}\right)}{r_{sp}}\hat{z}dz' \tag{1.29}$$

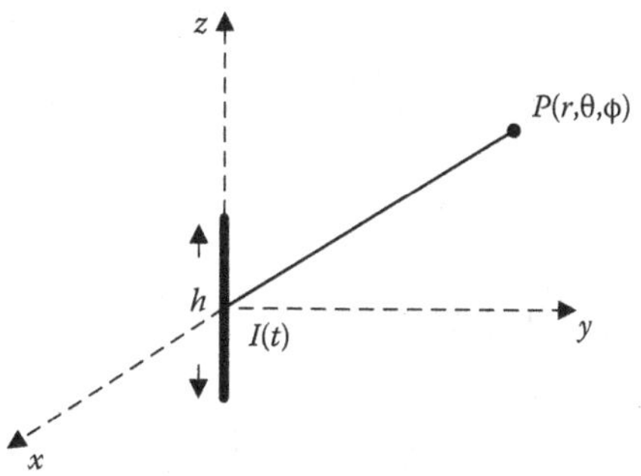

FIGURE 1.4
Impulse current source modeled as a point dipole for very small length h.

and

$$I\left(t-\frac{r_{sp}}{c}\right)=I_0\delta\left(t-\frac{r_{sp}}{c}\right).$$

Since h is very small, we can approximate r_{sp} by r and from Equation 1.29:

$$\begin{aligned}\mathbf{A}_p(\mathbf{r},t)&=\hat{z}\frac{\mu_0 I_0}{4\pi}\frac{\delta\left(t-\frac{r}{c}\right)}{r}\int_{\text{source}}dz' \\ &=\hat{z}\frac{\mu_0 I_0 h}{4\pi}\frac{\delta\left(t-\frac{r}{c}\right)}{r}.\end{aligned} \tag{1.30}$$

Expressing the unit vector $\hat{z}$ in spherical coordinates, we obtain

$$\hat{z}=\hat{r}\cos(\theta)-\hat{\theta}\sin(\theta), \tag{1.31}$$

the vector potential in spherical coordinates is given by

$$\mathbf{A}_p(\mathbf{r},t)=\frac{\mu_0 I_0 h}{4\pi}\frac{\delta\left(t-\frac{r}{c}\right)}{r}\left[\hat{r}\cos(\theta)-\hat{\theta}\sin(\theta)\right]. \tag{1.32}$$

The time-domain magnetic field $\boldsymbol{H}$ is given by

$$\boldsymbol{H}=\frac{1}{\mu_0}\nabla\times\boldsymbol{A}. \tag{1.33}$$

From Equations 1.32 and 1.33, evaluating curl in spherical coordinates, we obtain

$$\boldsymbol{H}=\hat{\phi}\frac{I_0 h}{4\pi}\left[\frac{\delta\left(t-\frac{r}{c}\right)}{r^2}+\frac{1}{cr}\delta'\left(t-\frac{r}{c}\right)\right]\sin(\theta). \tag{1.34}$$

The electric field $\boldsymbol{E}$ can be computed from the Maxwell equation

$$\nabla\times\boldsymbol{H}=\varepsilon_0\frac{\partial\boldsymbol{E}}{\partial t}. \tag{1.35}$$

Such an evaluation gives $E_\Phi = 0$ and

$$E_r=\frac{I_0 h\cos(\theta)}{2\pi\varepsilon_0}\left[\frac{u\left(t-\frac{r}{c}\right)}{r^3}+\frac{\delta\left(t-\frac{r}{c}\right)}{cr^2}\right], \tag{1.36a}$$

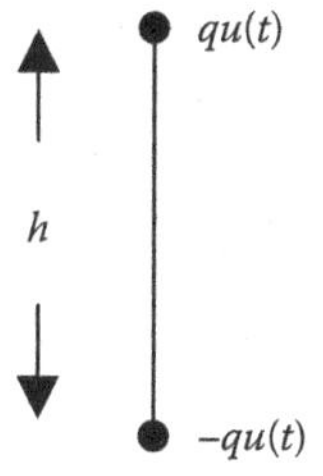

FIGURE 1.5
The impulse current source and its equivalent electric point dipole.

$$E_\theta = \frac{I_0 h \sin(\theta)}{4\pi\varepsilon_0}\left[\frac{u\left(t-\frac{r}{c}\right)}{r^3}+\frac{\delta\left(t-\frac{r}{c}\right)}{cr^2}+\frac{\delta'\left(t-\frac{r}{c}\right)}{cr}\right]. \tag{1.36b}$$

In the above, $u(t)$ is a Heaviside step function defined by

$$u(t)=\begin{cases}0, & t<0^-,\\ 1, & t>0^+.\end{cases} \tag{1.37}$$

The problem has the following physical significance. The impulse current creates at $t = 0$, a dipole as shown in Figure 1.5. Two charges initially at the same location are suddenly separated by a small distance h. Noting that $(\mathrm{d}[u(t)])/\mathrm{d}t = \delta(t)$, q in Figure 1.5 is equal to I_0. Equations 1.34 and 1.36 show that the fields are zero until the instant $t = r/c$ at which the wave front reaches the observation point. At that instant the fields are discontinuous but immediately after that instant, the magnetic field is zero and the electric field has a value corresponding to a static electric dipole of dipole moment

$$p_e = I_0 h\hat{z}. \tag{1.38}$$

1.4 Time-Harmonic Fields

A particular case of time-domain electromagnetics is time-harmonic electromagnetics where we assume that the time variations are harmonic (cosinusoidal). The particular case is analogous to steady-state analysis in circuits and makes use of phasor concepts. A field component is expressed as

$$\mathbf{F}(\mathbf{r},t)=\mathrm{Re}\left[\tilde{\mathbf{F}}(\mathbf{r})e^{j\omega t}\right], \tag{1.39}$$

where $\tilde{\mathbf{F}}(\mathbf{r})$ is the phasor–vector field. The special symbol ~ denotes a phasor distinct from the real-time harmonic field and will be used where it is necessary to make such a distinction. When there is no likelihood of confusion the special symbol will be dropped, the phasor and the real-time harmonic field will be denoted in the same way, and the meaning of the symbol will be understood from the context. Consequently, the time-domain fields are transformed to phasor fields by noting that partial differentiation with respect to time in time-domain transforms as multiplication by jω in the phasor (frequency) domain. Hence, the phasor form of the fields satisfies the following equations:

$$\nabla \times \tilde{\mathbf{E}}(\mathbf{r}) = -\mathrm{j}\omega\tilde{\mathbf{B}}(\mathbf{r}), \tag{1.40}$$

$$\nabla \times \tilde{\mathbf{H}}(\mathbf{r}) = \tilde{\mathbf{J}}(\mathbf{r}) + \mathrm{j}\omega\tilde{\mathbf{D}}(\mathbf{r}), \tag{1.41}$$

$$\nabla \cdot \tilde{\mathbf{D}}(\mathbf{r}) = \tilde{\rho}_V(\mathbf{r}), \tag{1.42}$$

$$\nabla \cdot \tilde{\mathbf{B}}(\mathbf{r}) = 0, \tag{1.43}$$

$$\nabla \cdot \tilde{\mathbf{J}}(\mathbf{r}) + \mathrm{j}\omega\tilde{\rho}_V = 0, \tag{1.44}$$

$$\tilde{\mathbf{B}}(\mathbf{r}) = \nabla \times \tilde{\mathbf{A}}(\mathbf{r}), \tag{1.45}$$

$$\tilde{\mathbf{E}}(\mathbf{r}) = -\nabla\tilde{\Phi}(\mathbf{r}) - \mathrm{j}\omega\tilde{\mathbf{A}}(\mathbf{r}), \tag{1.46}$$

$$\nabla^2\tilde{\mathbf{A}}(\mathbf{r}) + k^2\tilde{\mathbf{A}}(\mathbf{r}) = -\mu\tilde{\mathbf{J}}(\mathbf{r}), \tag{1.47}$$

$$\nabla^2\tilde{\Phi}(\mathbf{r}) + k^2\tilde{\Phi}(\mathbf{r}) = -\frac{\tilde{\rho}_V(\mathbf{r})}{\varepsilon}, \tag{1.48}$$

$$\nabla \cdot \tilde{\mathbf{A}}(\mathbf{r}) + \mathrm{j}\omega\mu\varepsilon\tilde{\Phi}(\mathbf{r}) = 0, \tag{1.49}$$

$$\tilde{\mathbf{A}}(\mathbf{r}) = \frac{\mu}{4\pi}\iiint_{\text{source}} \tilde{\mathbf{J}}(\mathbf{r}')\frac{\mathrm{e}^{-jk|\mathbf{r}-\mathbf{r}'|}}{|\mathbf{r}-\mathbf{r}'|}\mathrm{d}V', \tag{1.50}$$

$$\tilde{\Phi}(\mathbf{r}) = \frac{1}{4\pi\varepsilon}\iiint_{\text{source}} \tilde{\rho}_V(\mathbf{r}')\frac{\mathrm{e}^{-jk|\mathbf{r}-\mathbf{r}'|}}{|\mathbf{r}-\mathbf{r}'|}\mathrm{d}V', \tag{1.51}$$

$$\langle \mathbf{S}_R \rangle = \frac{1}{2}\mathrm{Re}\left[\tilde{\mathbf{E}} \times \tilde{\mathbf{H}}^*\right]. \tag{1.52}$$

In the above

$$k^2 = \omega^2\mu\varepsilon = \omega^2\mu_0\varepsilon_0\mu_r\varepsilon_r = k_0^2\mu_r\varepsilon_r, \tag{1.53}$$

$\langle \mathbf{S}_R \rangle$ is the time-averaged real power density, the symbol * denotes complex conjugate, and k is the wave number. Several remarks are in order. The program of analytical computation of phasor fields given by the harmonic sources is easier than the corresponding computation of time-domain fields indicated in the last section. We can compute $\tilde{\mathbf{A}}(\mathbf{r})$ from Equation 1.50 and obtain $\tilde{\Phi}(\mathbf{r})$ in terms of $\tilde{\mathbf{A}}$ from Equation 1.49. The time-harmonic electric and magnetic fields are then computed using Equations 1.45 and 1.46. The second remark concerns the boundary conditions. For the time harmonic case, only two of the four boundary conditions given in Equations 1.14 through 1.17 are independent.

1.5 Quasistatic and Static Approximations

Quasistatic approximations are obtained by neglecting either the magnetic induction or the displacement current. The former is electrostatic and the latter is magnetostatic. Maxwell's Equations 1.1 through 1.4 take the following forms.

a. Electroquasistatic:

$$\nabla \times \boldsymbol{E} = 0, \tag{1.54a}$$

$$\nabla \times \boldsymbol{H} = \boldsymbol{J} + \frac{\partial \boldsymbol{D}}{\partial t}, \tag{1.55a}$$

$$\nabla \cdot \boldsymbol{D} = \rho_V, \tag{1.56a}$$

$$\nabla \cdot \boldsymbol{B} = 0. \tag{1.57a}$$

b. Magnetoquasistatic:

$$\nabla \times \boldsymbol{E} = -\frac{\partial \boldsymbol{B}}{\partial t}, \tag{1.54b}$$

$$\nabla \times \boldsymbol{H} = \boldsymbol{J}, \tag{1.55b}$$

$$\nabla \cdot \boldsymbol{D} = \rho_V, \tag{1.56b}$$

$$\nabla \cdot \boldsymbol{B} = 0. \tag{1.57b}$$

If both the time derivatives are neglected in Maxwell equations, they decompose into two uncoupled sets describing electrostatics and magnetostatics.

1. Electrostatics:

$$\nabla \times \boldsymbol{E} = 0, \tag{1.58a}$$

$$\nabla \cdot \boldsymbol{D} = \rho_V \tag{1.59a}$$

2. Magnetostatics:

$$\nabla \times \boldsymbol{H} = \boldsymbol{J}, \tag{1.58b}$$

$$\nabla \cdot \boldsymbol{B} = 0. \tag{1.59b}$$

A quantitative discussion of these approximations is given in Appendix 1D, based on a time-rate parameter.

1.6 Maxwell's Equations in Integral Form and Circuit Parameters

Equations 1.1 through 1.4 are partial differential equations. When Equations 1.1 and 1.2 are integrated over an open surface bounded by a closed curve, and Stokes' theorem (1A.66) is used, we get the first two integral forms of the Maxwell's equations. When Equations 1.3 and 1.4 are integrated over a definite volume bounded by a closed surface, and divergence theorem (1A.67) is used, we get the last two of the Maxwell's equations in the integral form. Lumped-parameter circuit theory is a low-frequency approximation of the integral form of the first two Maxwell's equations. One must make a quantitative check of the validity of the low-frequency approximation before applying the circuit theory to the electromagnetic field problems. Some details of the integral forms and circuit concepts derived from them are given next.

Equation 1.1 relates the E field at a point with B field at the same point. When integrated over an open surface bounded by a closed curve and with the application of (1A.66), it will lead to

$$\iint_s \nabla \times \boldsymbol{E} \cdot \mathrm{ds} = \oint_c \boldsymbol{E} \cdot \mathrm{d}l = -\iint_s \frac{\partial \boldsymbol{B}}{\partial t} \cdot \mathrm{ds} \tag{1.60}$$

where c is a closed curve bounding the open surface s. In Equation 1.60, if the surface s is moving, that is, $s = s(t)$ and $c = c(t)$, $\boldsymbol{E}$ and $\boldsymbol{B}$ in Equation 1.60 are those measured by an observer, who is at rest with respect to the moving media. If $\boldsymbol{E}$ and $\boldsymbol{B}$ are measured by an observer at rest in the laboratory, Equation 1.60 takes the form [1,2]

$$V_{\text{induced}} = \underbrace{\oint_c \boldsymbol{E} \cdot \mathrm{d}l}_{\text{term1}} = \underbrace{-\frac{\mathrm{d}\Phi_m}{\partial t}}_{\text{term2}} = -\frac{\mathrm{d}}{\mathrm{d}t}\iint_s \boldsymbol{B} \cdot \mathrm{ds} = \underbrace{-\iint_s \frac{\partial \boldsymbol{B}}{\partial t} \cdot \mathrm{ds}}_{\text{term3}} + \underbrace{\oint_c v \times \boldsymbol{B} \cdot \mathrm{d}l}_{\text{term4}} \tag{1.61}$$

In the above

term1: total induced voltage

term2: total time rate of decrease of the magnetic flux Φ_m

term3: transformer part of the induced voltage

term4: motional part of the induced voltage, where $\boldsymbol{v}$ is the velocity of $c(t)$

Additional discussion from the view point of moving media can be found in Sections 14.3 and 14.4

When the deformation of the moving circuit $c(t)$ is continuous, the total induced voltage is easier to compute by evaluating term2 than evaluating term3 and term4 separately and adding them.

The concept of the circuit element inductance L and its volt–ampere (V–I) relationship

$$V = L\frac{\mathrm{d}I}{\mathrm{d}t} \tag{1.62}$$

can be obtained from Equation 1.61. Let us define inductance L in henries (abbreviated as H) as the magnetic flux Φ_m generated by a current I, that is

$$L = \frac{\Phi_m}{I}(H). \tag{1.63}$$

From Equation 1.61 and noting the induced voltage (also referred to as emf) is opposite in sign to the voltage drop V,

$$V = \frac{\mathrm{d}\Phi_m}{\mathrm{d}t} = \frac{\mathrm{d}}{\mathrm{d}t}(LI) = L\frac{\mathrm{d}I}{\mathrm{d}t}. \tag{1.64}$$

In multiturn systems or situations where all the flux is not external to the current that generates it (e.g., the magnetic flux in a conductor due to the current flowing in the conductor), the concept of flux linkages is used [3].

Next, we will develop the second Maxwell's equation in integral form and use it to explain the capacitance circuit parameter and the concept of displacement current. Equation 1.2 can also be transformed to integral form, using Stokes' theorem again (1A.66):

$$\underbrace{\oint_c \boldsymbol{H}\cdot\mathrm{d}l}_{\text{term1}} = \underbrace{\iint_s \boldsymbol{J}\cdot\mathrm{d}s}_{\text{term2}} + \underbrace{\iint_s \frac{\partial \boldsymbol{D}}{\partial t}\cdot\mathrm{d}s}_{\text{term3}} = I_C + I_D = I_{\text{total}}, \tag{1.65}$$

where, again, the closed curve c is the bound for an open surface s. From Figure 1.6, it is clear that s is not unique. The closed curve c is the bound for all the open surfaces s_1, s_2, and s_3, the implication of which will be used to explain a current jumping through a perfect dielectric in a capacitor. Let us state in words the meaning of various terms marked in Equation 1.65:

term1: circulation of $\boldsymbol{H}$ field over c
term2: conduction current I_C in amperes
term3: displacement current I_D in amperes.

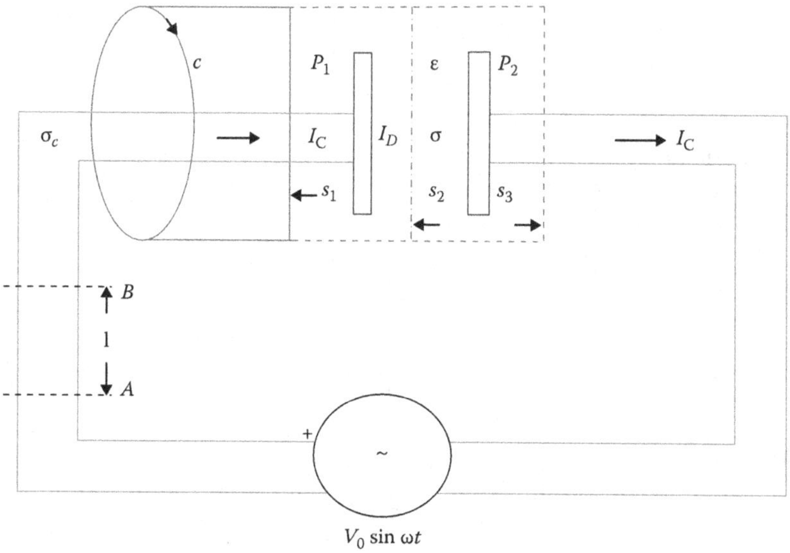

FIGURE 1.6
Integral form of Maxwell–Ampere law (Equation 1.65): Explanation of circuit elements and displacement current.

The capacitance circuit parameter C is defined as

$$C = \frac{Q}{V}, \tag{1.66}$$

where Q is the magnitude of the charge on either of the two perfect conductors embedded in a perfect dielectric medium and V is the potential difference between the two conductors. The V–I relationship of a two terminal capacitive circuit element is given by

$$I = C\frac{dV}{dt}. \tag{1.67}$$

If we identify this current as I_D (term3) in Equation 1.65

$$I_D = \frac{\partial}{\partial t}\iint_s \varepsilon \mathbf{E} \cdot ds. \tag{1.68}$$

Thus, we can write

$$C = \frac{I_D}{dV/dt} = \frac{\iint_s \varepsilon \mathbf{E} \cdot ds}{-\int_{P_2}^{P_1} \mathbf{E} \cdot dl}. \tag{1.69}$$

With reference to Figure 1.6, if $\mathbf{E}$ is in the z direction, independent of the coordinates, and ε is a constant

$$C = \varepsilon \frac{\iint_s ds}{-\int_{P_2}^{P_1} dl} = \frac{\varepsilon A}{d}, \tag{1.70}$$

where A is the plate area.

Suppose the dielectric is a low-loss medium with a small conductivity σ, as shown in Figure 1.6; one can define the conductance G as

$$G = \frac{I_C}{V} = \frac{\iint_s \sigma E \cdot ds}{-\int_{P_2}^{P_1} E \cdot dl}. \tag{1.71}$$

For uniform E and constant σ

$$G = \frac{\sigma A}{d}. \tag{1.72}$$

The parallel plate capacitor with a low-loss dielectric shown in Figure 1.6 can be modeled as C given by Equation 1.70 in parallel with G given by Equation 1.72.

Suppose the connecting wires in Figure 1.6 are not perfect but are good conductors of conductivity σ_c. The familiar form of Ohm's law is

$$V_{AB} = R_{AB}\, I, \tag{1.73}$$

where R_{AB} is the resistance of the conductor between the planes A and B:

$$R_{AB} = \frac{V_{AB}}{I} = \frac{-\int_B^A \boldsymbol{E}\cdot \mathrm{d}\boldsymbol{l}}{\iint \boldsymbol{J}\cdot \mathrm{d}\boldsymbol{s}} = \frac{-\int_B^A \boldsymbol{J}/\sigma_c \cdot \mathrm{d}\boldsymbol{l}}{\iint \boldsymbol{J}\cdot \mathrm{d}\boldsymbol{s}}. \tag{1.74}$$

For the case of constant $\boldsymbol{J}$ and σ_c

$$R_{AB} = \frac{1}{\sigma_c}\frac{\int \mathrm{d}l}{\iint \mathrm{d}s} = \frac{l}{\sigma_c S}, \tag{1.75}$$

where S is the cross-sectional area of the wire. For the current flow governed by the skin effect, S is not the physical cross-sectional area but an effective cross-sectional area. Section 2.6 discusses this case.

Maxwell called term3 as displacement current. Let us explore its role in explaining the flow of the same conduction current through the wires attached to the two plates P_1 and P_2. Let us assume the ideal situation of $\sigma = 0$ and σ_c = infinity. The conduction current flowing through the wires is given by

$$I_C = C\frac{\mathrm{d}V}{\mathrm{d}t} = C\omega V_0 \cos\omega t = \frac{\varepsilon A}{\mathrm{d}}\omega V_0 \cos\omega t. \tag{1.76}$$

The displacement current in the wires is very small compared to I_C. This is term2, where the surface s is either s_1 or s_3 in Figure 1.6. Without the concept of the displacement current, it is difficult to explain the jumping of the current from the left wire to the right wire since the dielectric in the capacitor is a perfect insulator ($\sigma = 0$). However, we show below that the I_D given by term3 in Equation 1.65 will be exactly the same as I_C:

$$I_D = \frac{\partial}{\partial t}\iint_{S_2} \varepsilon \boldsymbol{E}\cdot \mathrm{d}\boldsymbol{s} = \frac{\partial}{\partial t}\varepsilon\frac{V}{\mathrm{d}}\iint \mathrm{d}s = \frac{\varepsilon A}{\mathrm{d}}\frac{\partial}{\partial t}\left(V_0 \sin\omega t\right) = \frac{\varepsilon A}{\mathrm{d}}\omega V_0 \cos\omega t. \tag{1.77}$$

The integral form of Equation 1.3 is obtained by integrating over a definite volume v bounded by a closed surface s:

$$\oiint_s \boldsymbol{D}\cdot \mathrm{d}\boldsymbol{s} = \iiint_v \nabla\cdot \boldsymbol{D}\,\mathrm{d}v = \iiint_v \rho_V \mathrm{d}v = Q. \tag{1.78}$$

In the above, we used divergence theorem (1A.67). The word statement for Equation 1.78 is that the electric flux coming out of a closed surface is the net positive charge enclosed by the surface (Gauss's law).

Similar operations on Equation 1.4 give

$$\oiint_s \boldsymbol{B} \cdot \mathrm{d}\mathbf{s} = \iiint_v \nabla \cdot \boldsymbol{B} \mathrm{d}v = 0, \tag{1.79}$$

leading to the word statement that the magnetic flux coming out of a closed surface is zero. The magnetic flux lines close on themselves, thus ruling out the existence of magnetic monopoles. The familiar source of a small loop of current for magnetic fields can be shown to be equivalent to a magnetic dipole. This aspect is discussed in Section 8.9.

Integration of Equation 1.5 over a fixed surface leads to the mathematical statement of the principle of conservation of charge:

$$\oiint_s \boldsymbol{J} \cdot \mathrm{d}\mathbf{s} = \iiint_v \nabla \cdot \boldsymbol{J} \mathrm{d}v = \iiint_v -\frac{\partial}{\partial t} \rho_V \mathrm{d}v = -\frac{d}{dt} \iiint_v \rho_V \mathrm{d}v. \tag{1.80}$$

The first term in Equation 1.80 is the current going out of the fixed closed surface s and the last term is the time rate of decrease of the charge in the volume v.

References

1. Kalluri, D. K., *Electromagnetic Waves, Materials, and Computation with MATLAB*, 2nd edn., CRC Press, Taylor & Francis Group, Boca Raton, FL, 2016.
2. Rothwell, J. R. and Cloud, M. J., *Electromagnetics*, CRC Press, Taylor & Francis Group, Boca Raton, FL, 2001.
3. Hayt, Jr., W. H., *Engineering Electromagnetics*, 5th edn., McGraw-Hill, New York, 1989.

2

*Electromagnetics of Simple Media: One-Dimensional Solution**

The mathematics needed to solve problems may be simplified if we are able to model the problem as one-dimensional. Sometimes a practical problem with the appropriate symmetries and dimensions will permit us to use one-dimensional models. A choice of the appropriate coordinate system will sometimes allow us to reduce the dimensionality of the problem. A few of such basic solutions will be listed next.

2.1 Uniform Plane Waves in Sourceless Medium ($\rho_V = 0$, $J_{source} = 0$)

A simple lossy medium with the parameters ε_r, μ_r, and σ has the following one-dimensional (say *z*-coordinate) solutions in Cartesian coordinates:

$$E_z = H_z = 0, \tag{2.1}$$

$$\eta \mathbf{H} = \hat{z} \times \mathbf{E}, \tag{2.2}$$

$$\mathbf{E} = \mathbf{E}_t \begin{Bmatrix} e^{+jkz} \\ e^{-jkz} \end{Bmatrix}. \tag{2.3}$$

In the above, the braces {} indicate a linear combination of the functions within parentheses and $\mathbf{E}_t$ is the electric field in the transverse (to *z*) plane. For a lossy medium,

$$k^2 = \omega^2 \mu\varepsilon - j\omega\mu\sigma, \tag{2.4}$$

and hence *k* is complex. The characteristic impedance η is given by

$$\eta = \left(\frac{j\omega\mu}{\sigma + j\omega\varepsilon} \right)^{1/2}, \tag{2.5}$$

which is again complex. If *k* is written as $(\beta - j\alpha)$, then the phase factor $\exp(-jkz) = \exp(-\alpha z - j\beta z)$ gives the solution of an attenuated wave propagating in the positive *z*-direction. Here α is called the attenuation constant (Np/m) and β is called the phase constant (rad/m). Explicit expressions for α and β may be obtained by solving the

* For chapter appendices, see 2A through 2D in Appendices section.

two equations generated by equating the real and imaginary parts of the LHS with the RHS of Equation 2.4, respectively:

$$\alpha = \omega\sqrt{\mu\varepsilon}\left\{\frac{1}{2}\left[\sqrt{1+T^2}-1\right]\right\}^{1/2} \text{(Np/m)} \tag{2.6}$$

$$\beta = \omega\sqrt{\mu\varepsilon}\left\{\frac{1}{2}\left[\sqrt{1+T^2}+1\right]\right\}^{1/2} (\text{rad/m}) \tag{2.7}$$

where

$$T = \frac{\sigma}{\omega\varepsilon} \tag{2.8}$$

is called the loss tangent. For a low-loss dielectric, the loss tangent $T \ll 1$ and

$$\alpha \approx \frac{\sigma}{2}\sqrt{\frac{\mu}{\varepsilon}}, \quad T \ll 1, \tag{2.9}$$

$$\beta \approx \omega\sqrt{\mu\varepsilon}, \quad T \ll 1, \tag{2.10}$$

$$\eta \approx \sqrt{\frac{\mu}{\varepsilon}}\angle\tan^{-1}\left(\frac{\sigma}{2\omega\varepsilon}\right), \quad T \ll 1. \tag{2.11}$$

2.2 Good Conductor Approximation

For the other limit of approximation, called the good-conductor approximation,

$$\alpha \approx \beta \approx \sqrt{\pi f \mu\sigma} = \frac{1}{\delta}, \quad T \gg 1, \tag{2.12}$$

$$\eta \approx \frac{\sqrt{2}}{\sigma\delta}\angle 45°, \quad T \gg 1. \tag{2.13}$$

In the above, δ is called the skin depth. The characteristic impedance in this case is also called the surface impedance Z_s:

$$\eta = Z_s = R_s + jX_s = (1+j)\sqrt{\frac{\omega\mu}{2\sigma}}. \tag{2.14}$$

2.3 Uniform Plane Wave in a Good Conductor: Skin Effect

The attenuated positive-going wave in a good conductor is described by the phasor exp $[-(1 + j)z/\delta]$ and the real fields are of the form

$$E(z,t) = E_0 e^{-z/\delta}\cos\left(\omega t - \frac{z}{\delta}\right), \tag{2.15}$$

TABLE 2.1

Skin Depth for Copper

f	60 Hz	1 kHz	1 MHz	10 GHz
δ	1 cm	2.45 mm	7.75×10^{-2} mm	7.75×10^{-4} mm

$$J(z,t) = J_0 e^{-z/\delta} \cos\left(\omega t - \frac{z}{\delta}\right), \tag{2.16}$$

where

$$J_0 = \sigma E_0. \tag{2.17}$$

From Equation 2.16, we see that the amplitude of the current density drops to e^{-1} (36.8%) of its value at a distance

$$\delta = \frac{1}{\sqrt{\pi f \mu \sigma}}. \tag{2.18}$$

In the case of direct current (DC), $f = 0$ and $\delta = \infty$, indicating that the current density can be uniform. However, for the time-harmonic AC case ($f \neq 0$), δ is finite, indicating that the current density is nonuniform and decays exponentially with distance. For example, at a distance of 4δ, the current density reduces ($e^{-4} = 1.8\%$) practically to zero. In Table 2.1, we list the skin depth δ for a good conductor such as copper at various frequencies.

At high frequencies, the current is practically confined to the skin of the conductor and the phenomenon is hence described as a skin effect.

2.4 Boundary Conditions at the Interface of a Perfect Electric Conductor with a Dielectric

We also note that $\delta = 0$ for $\sigma = \infty$. That is the reason for the statement in Problem P2.1 that the time-harmonic fields in a perfect conductor are zero. If medium 1 is a perfect conductor and medium 2 is a dielectric, then the boundary conditions are

$$\hat{n}_{12} \cdot \mathbf{D}_2 = \rho_s, \tag{2.19}$$

$$\hat{n}_{12} \cdot \mathbf{B}_2 = 0, \tag{2.20}$$

$$\hat{n}_{12} \times \mathbf{E}_2 = 0, \tag{2.21}$$

$$\hat{n}_{12} \times \mathbf{H}_2 = \mathbf{K}. \tag{2.22}$$

Thus, at the interface, the electric field is entirely normal to the interface and is equal to ρ_s/ε. Furthermore, the magnetic field is entirely tangential and is equal in magnitude to the surface current density $\mathbf{K}$.

The PEC boundary condition

$$E_t = 0 \tag{2.23}$$

implies

$$\frac{\partial E_n}{\partial n} = 0, \tag{2.24}$$

where $\hat{t}$ is a tangential unit vector and $\hat{n}$ is a normal unit vector. Also the PEC boundary condition

$$H_n = 0 \tag{2.25}$$

implies

$$\frac{\partial H_t}{\partial n} = 0. \tag{2.26}$$

2.5 AC Resistance

Figure 2.1 shows an infinitely deep good conductor of conductivity σ defined by the half-space $0 < x < \infty$.

The electric, magnetic, and current density phasors may be written as

$$\tilde{\mathbf{E}}(x) = \hat{z}E_0 e^{-x/\delta - jx/\delta}, \tag{2.27}$$

$$\tilde{\mathbf{H}}(x) = -\hat{y}E_0 \frac{\sigma\delta}{\sqrt{2}} e^{-x/\delta} e^{-jx/\delta} e^{-j45°}, \tag{2.28}$$

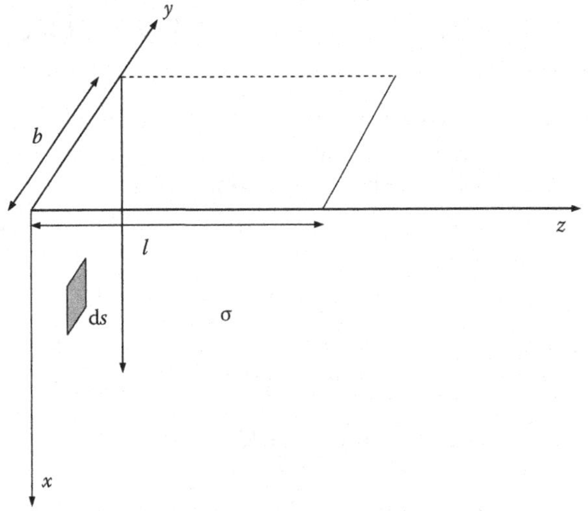

FIGURE 2.1
AC resistance of a semi-infinite good conductor.

and the corresponding real fields are

$$\mathbf{E}(x,t) = \hat{z}E_0 e^{-x/\delta}\cos\left(\omega t - \frac{x}{\delta}\right)(\mathrm{V/m}), \tag{2.29}$$

$$\mathbf{H}(x,t) = -\hat{y}E_0\frac{\sigma\delta}{\sqrt{2}}e^{-x/\delta}\cos\left(\omega t - \frac{x}{\delta} - 45^\circ\right)(\mathrm{A/m}). \tag{2.30}$$

The time-averaged power density is given as

$$\langle \mathbf{S}\rangle = \hat{x}\frac{1}{2}E_0^2\frac{\sigma\delta}{\sqrt{2}}e^{-2x/\delta}\cos 45^\circ = \hat{x}\frac{1}{4}E_0^2\sigma\delta e^{-2x/\delta}. \tag{2.31}$$

The total power entering the conductor of width b and length l is given by

$$P = \langle \mathbf{S}\rangle_{x=0}\, bl = \frac{1}{4}E_0^2\sigma\delta bl. \tag{2.32}$$

The total phasor current entering the conductor of width b is given by

$$\tilde{I} = \iint \tilde{\mathbf{J}}\cdot d\mathbf{s} = \iint \sigma\tilde{\mathbf{E}}\cdot d\mathbf{s} = \int_0^b\int_0^\infty \sigma E_0 e^{-(1+j)x/\delta}dx\,dy. \tag{2.33}$$

After integrating and substituting the limits, we obtain

$$\tilde{I} = \frac{\sigma E_0 b\delta}{\sqrt{2}}\angle -45^\circ, \tag{2.34}$$

$$I(t) = \frac{\sigma E_0 b\delta}{\sqrt{2}}\cos(\omega t - 45^\circ). \tag{2.35}$$

Let us define R_{AC} as an equivalent resistance that consumes the same power, that is,

$$\tilde{I}_{RMS}^2 R_{AC} = \text{Power consumed} = P. \tag{2.36}$$

From Equation 2.35, we obtain

$$\tilde{I}_{RMS} = \frac{\sigma E_0 b\delta}{2}. \tag{2.37}$$

From Equations 2.35 and 2.33, we obtain

$$R_{AC} = \frac{l}{\sigma b\delta}. \tag{2.38}$$

Note that if the current were uniformly distributed (DC case), then we would have got the same resistance as given by Equation 2.38 if the depth of the conductor was δ. Figure 2.2 shows this equivalence.

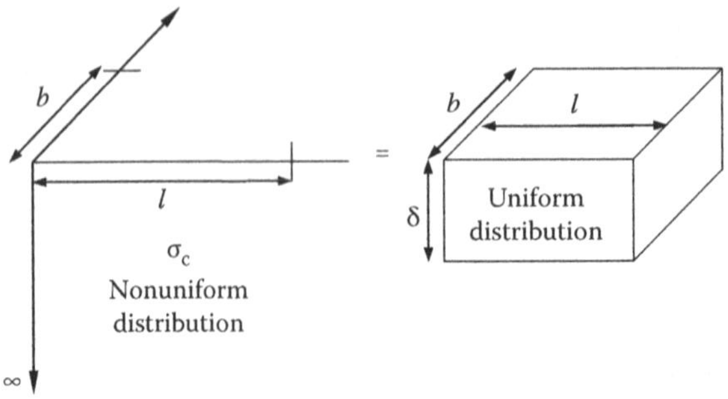

FIGURE 2.2
Equivalent resistance.

2.6 AC Resistance of Round Wires

The AC resistance of a round wire of radius a and length l is easily calculated for the two extreme cases of approximation $\delta \gg a$ and $\delta \ll a$. For the former, the current may be considered as uniformly distributed and the DC resistance formula may be used:

$$R = \frac{l}{\sigma s} = \frac{l}{\sigma \pi a^2}, \quad \delta \gg a. \tag{2.39}$$

For the latter, the analysis of the previous section may be applied. The circular conductor may be considered as a semi-infinite solid because $\delta \ll a$. Therefore, the effective area of cross-section s_{eff} (Figure 2.3) is given by

$$s_{\text{eff}} = \pi\left[a^2 - (a-\delta)^2\right] \approx 2\pi a\delta, \quad \delta \ll a \tag{2.40}$$

and

$$R_{\text{AC}} = \frac{1}{2\pi\sigma a\delta}, \quad \delta \ll a. \tag{2.41}$$

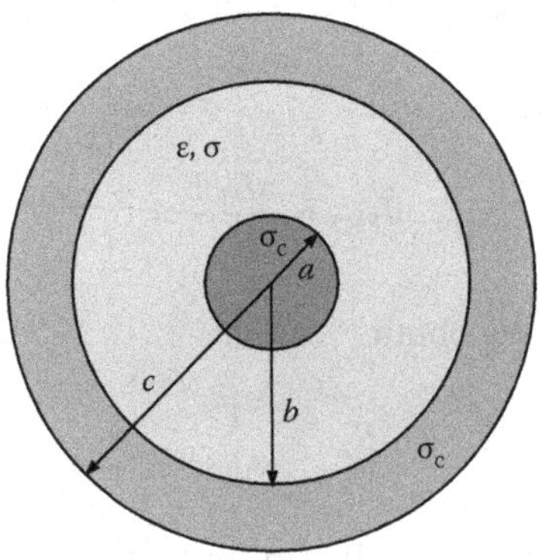

FIGURE 2.3
Cross section of a coaxial cable.

The analysis is more involved for the case where δ is comparable to a. In this case, we have to take a fresh look at the solution based on the one-dimensional solution in cylindrical coordinates (see Appendix 2A).

2.7 Voltage and Current Harmonic Waves: Transmission Lines

A transmission line consists of two conductors that guide an electromagnetic wave from an electromagnetic source to an electromagnetic load. One approach to transmission line theory is through the distributed circuit theory. In this approach, we consider a differential length Δz of a transmission line and describe its properties in terms of its per meter length circuit parameters, R', the resistance due to imperfect conductors, L', the inductance due to magnetic flux generated by the currents in the conductors, G', the conductance due to an imperfect dielectric, and C', the capacitance due to the surface charges on the conductors. We use electromagnetic theory to calculate the circuit parameters that depend on the electromagnetic properties of the materials used and their geometric arrangement. A parallel plate transmission line and a coaxial cable are two simple examples of transmission lines. Strip lines and microstrip lines are extensively used in high-frequency electromagnetic devices. Transmission line theory is often given in terms of voltage and current since in the transverse plane of a transmission line (cross section), the electromagnetic wave may be described as a TEM wave and the transverse fields satisfy the Laplace equation in the transverse plane. The voltage V and the current I can then be described as the appropriate integrals of the electric field and the magnetic field, respectively.

The parameters G', L', and C' are computed as though the fields are static. For a coaxial cable (Figure 2.3), the parameters are given by [1]

$$G' = \frac{2\pi\sigma}{\ln(b/a)}, \tag{2.42}$$

$$L' = \frac{\mu}{2\pi}\ln\left(\frac{b}{a}\right), \tag{2.43}$$

$$C' = \frac{2\pi\varepsilon}{\ln(b/a)}. \tag{2.44}$$

The formulas to be used for the calculation of R' and the corrections to L' due to the internal inductance of the conductor will depend on the frequency. The previous section discussed the calculation of these parameters at high frequencies.

For high frequencies, that is, $\delta \ll a$,

$$R' = \left[\frac{1}{2\pi\sigma_c\delta a} + \frac{1}{2\pi\sigma_c\delta b}\right], \tag{2.45}$$

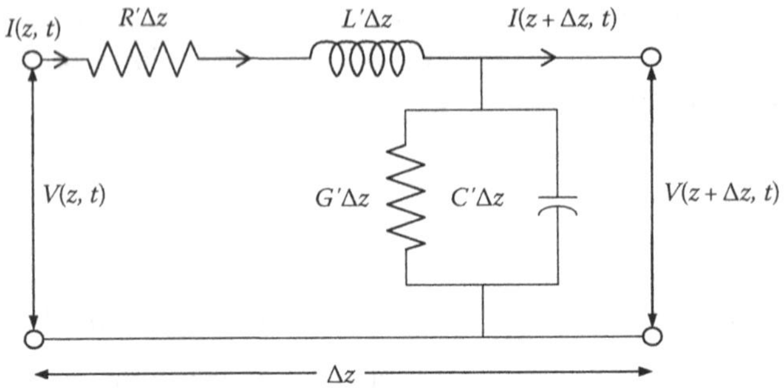

FIGURE 2.4
Circuit representation of a differential length of a transmission line.

and L' including the correction due to the internal inductance (inductance due to the magnetic flux in the conductors) is given by

$$L' = \frac{\mu}{2\pi}\ln\left(\frac{b}{a}\right) + L'', \tag{2.46}$$

where

$$\omega L'' = R'. \tag{2.47}$$

The parameters for other transmission lines may be calculated [1] using the same principles. After obtaining the parameters, the differential length transmission line may be represented as a two-port network shown in Figure 2.4 and expressed

$$V(z,t) - V(z+\Delta z, t) = R'\Delta z I(z,t) + L'\Delta z \frac{\partial I(z,t)}{\partial t}, \tag{2.48}$$

which, in the limit Δz approaches zero reduces to

$$-\frac{\partial V(z,t)}{\partial z} = R'I(z,t) + L'\frac{\partial I(z,t)}{\partial t}. \tag{2.49}$$

Similar analysis for the current shows that

$$I(z,t) - I(z+\Delta z, t) = G'\Delta z V(z+\Delta z, t) + C'\Delta z \frac{\partial V(z+\Delta z, t)}{\partial t}, \tag{2.50}$$

which, in the limit Δz approaches zero reduces to

$$-\frac{\partial I(z,t)}{\partial z} = G'V(z,t) + C'\frac{\partial V(z,t)}{\partial t}. \tag{2.51}$$

Equations 2.49 and 2.51 are the first-order coupled partial differential equations describing the voltage and current waves on a transmission line. The wave nature and the analogy with a simple electromagnetic wave may easily be seen by considering the following justifiable simplifications. As a first approximation, we can often neglect the losses in a transmission line and obtain the following wave equation for the voltage $V(z, t)$:

$$\frac{\partial^2 V}{\partial z^2} - \frac{1}{v^2}\frac{\partial^2 V}{\partial t^2} = 0, \quad R' = G' = 0, \tag{2.52}$$

where

$$v = \frac{1}{\sqrt{L'C'}}. \tag{2.53}$$

If the source is harmonic, then

$$\frac{\partial^2 \tilde{V}}{\partial z^2} + \beta^2 \tilde{V} = 0, \tag{2.54}$$

where

$$\beta = \frac{\omega}{v}. \tag{2.55}$$

The solution may be written as

$$\tilde{V} = \tilde{V}^+ + \tilde{V}^-, \tag{2.56}$$

$$\tilde{V} = \tilde{V}_0^+ e^{-j\beta z} + \tilde{V}_0^- e^{+j\beta z}. \tag{2.57}$$

The superscript + is used for a positive-going wave and the superscript – is used for a negative-going wave. From Equation 2.49, the relationship between the voltages and the current may be obtained:

$$\tilde{V}^+ = Z_0 \tilde{I}^+, \tag{2.58}$$

$$\tilde{V}^- = -Z_0 \tilde{I}^-, \tag{2.59}$$

where

$$Z_0 = \sqrt{\frac{L'}{C'}}. \tag{2.60}$$

The following analogy with a TEM wave may be drawn:

TEM Wave	Transmission Wave
E	V
H	I
ε	C'
μ	L'
η	Z_0
$\beta = \omega\sqrt{\mu\varepsilon}$	$\beta = \omega\sqrt{L'C'}$

Neglecting losses is equivalent to assumptions that the conductors are perfect ($\sigma_c = \infty$) and the dielectric is perfect ($\sigma = 0$). Practical lines have large but finite conductivity for conductors (e.g., copper $\sigma_c = 5.7 \times 10^7$ S/m) and small but nonzero conductivity for dielectrics (e.g., hard rubber, $\sigma = 10^{-15}$ S/m). In such a case, R' and G' are nonzero and the propagation constant γ is

$$\gamma = \mathrm{j}k = \sqrt{(R' + \mathrm{j}\omega L')(G' + \mathrm{j}\omega C')} = \alpha + \mathrm{j}\beta, \tag{2.61}$$

and the characteristic impedance Z_0 is complex:

$$Z_0 = \sqrt{\frac{R' + \mathrm{j}\omega L'}{G' + \mathrm{j}\omega C'}}. \tag{2.62}$$

As expected, the solution to the transmission line equation is a linear combination of positive-going and negative-going attenuated traveling waves. Also, from the field viewpoint, we realize that the electromagnetic wave is no longer a pure TEM wave. To drive the current through the conductors, which are not perfect, a small E_z-component is needed. The integral of this component along the length of the conductors gives the voltage drop. The cross product of this z-component with the magnetic field **H** gives the power density. The time-averaged power density integrated over the surface of the conductor gives the power consumed by the conductors. Problem P2.5 is constructed to illustrate the approximate method of solving a transmission line problem and to understand the power flow in a transmission line.

2.8 Bounded Transmission Line

Figure 2.5 shows a transmission line of length d with a load at $z = 0$, and a source of voltage V_g and internal impedance Z_g connected at the point $z = -d$. From Equations 2.57 through 2.58, the voltage at the terminals AB and the current $\tilde{I}_g = \tilde{I}(d)$ may be written as

$$\tilde{V}(d) = \tilde{V}_0^+ e^{\mathrm{j}\beta d} + \tilde{V}_0^- e^{-\mathrm{j}\beta d}, \tag{2.63}$$

$$\tilde{I}(d) = \frac{1}{Z_0}\left[\tilde{V}_0^+ e^{\mathrm{j}\beta d} - \tilde{V}_0^- e^{-\mathrm{j}\beta d}\right]. \tag{2.64}$$

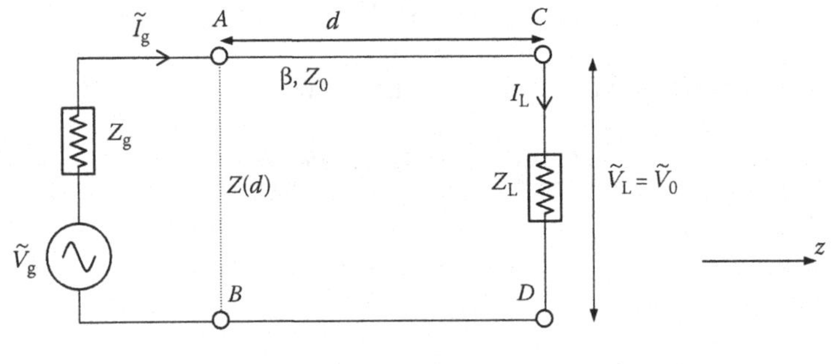

FIGURE 2.5
Bounded transmission line.

The input impedance is given by

$$Z(d) = \frac{\tilde{V}(d)}{\tilde{I}(d)} = Z_0 \left[\frac{1 + \Gamma_0 e^{-j2\beta d}}{1 - \Gamma_0 e^{-j2\beta d}} \right], \tag{2.65}$$

where Γ_0 is the reflection coefficient at the load and is given by

$$\Gamma_0 = \frac{\tilde{V}_0^-}{\tilde{V}_0^+} = \frac{Z_L - Z_0}{Z_L + Z_0}. \tag{2.66}$$

By substituting Equation 2.66 into Equation 2.65, the input impedance may be written in an alternative form as

$$Z(d) = Z_0 \frac{Z_L \cos\beta d + jZ_0 \sin\beta d}{Z_0 \cos\beta d + jZ_L \sin\beta d}. \tag{2.67}$$

From Equation 2.67, several formulas for special lengths may be obtained

$$Z\left(\frac{\lambda}{4}\right) = \frac{Z_0^2}{Z_L}, \tag{2.68}$$

$$Z\left(\frac{\lambda}{2}\right) = Z_L. \tag{2.69}$$

The circuit approximation applies if $d \ll \lambda = \dfrac{2\pi}{\beta}$:

$$Z(d) = Z_L, \quad \frac{d}{\lambda} \ll 1. \tag{2.70}$$

For a matched line, that is, $Z_L = Z_0$, $\Gamma_0 = 0$, and

$$Z(d) = Z_L \quad \text{if } Z_L = Z_0. \tag{2.71}$$

Equation 2.71 shows that for the matched line the length of the line has no effect on the input impedance, the right-going wave travels from the source to the load, there is no reflected wave, and the source wave is absorbed by the load. See Appendix 2B for power calculations.

A graphical method based on "Smith chart" is available to calculate the input impedance [1]. Appendix 2C gives the theory and application of the Smith chart.

The input impedance is reactive, if the load is either short-circuited or open-circuited and the reflection coefficient has a magnitude of 1:

$$Z(d) = jZ_0 \tan\beta d, \quad Z_L = 0, \tag{2.72}$$

$$Z(d) = -jZ_0 \cot\beta d, \quad Z_L = \infty. \tag{2.73}$$

In the above two cases, the waves are standing waves, there are permanent nulls in voltage at certain points on the transmission line, and the source does not supply any real power to the load but supplies only a reactive power.

Transmission line serves as a powerful analog for describing many electromagnetic phenomena since voltage and current variables are easily measurable and electrical engineers have a certain feel for these variables. Appendix 2B may be consulted for details of the transmission line theory. See Appendix 2C for application of the Smith chart to wave reflection problems.

The theory and applications of nonuniform transmission lines is given in Appendix 2D.

2.9 Electromagnetic Wave Polarization

A plane wave solution given by Equations 2.1 through 2.3 allows for x- and y-components of the electric and magnetic fields. Let

$$\tilde{\mathbf{E}}_t(z) = \tilde{\mathbf{E}}(z) = \hat{x}\tilde{E}_x(z) + \hat{y}\tilde{E}_y(z). \tag{2.74}$$

Furthermore, let

$$\tilde{E}_x(z) = E_{x0}e^{-jkz}, \tag{2.75}$$

$$\tilde{E}_y(z) = E_{y0}e^{j\delta}e^{-jkz}, \tag{2.76}$$

where E_{x0}, E_{y0}, and δ are real numbers. The instantaneous vector electric field is given by

$$\mathbf{E}(z,t) = \hat{x}E_{x0}\cos(\omega t - kz) + \hat{y}E_{y0}\cos(\omega t - kz + \delta). \tag{2.77}$$

A wave is said to be linearly polarized if $\delta = 0$ or π. This is because, at a specified value of z, say $z = 0$, the tip of $\mathbf{E}(0, t)$ traces a straight line in the x–y-plane. For $\delta = 0$,

$$\mathbf{E}(0,t) = (\hat{x}E_{x0} + \hat{y}E_{y0})\cos\omega t. \tag{2.78}$$

Figure 2.6a depicts this case. If, furthermore, $E_{y0} = 0$, the wave is said to be x-polarized. On the other hand, if $E_{x0} = 0$, the wave is said to be y-polarized. A wave is said to have left-hand circular polarization (LHC polarization) or L wave as it is called in this book, if $\delta = \pi/2$ and $E_{x0} = E_{y0} = E_0$, that is,

$$E(z,t) = \hat{x}E_0\cos(\omega t - kz) + \hat{y}E_0\cos(\omega t - kz + \pi/2), \tag{2.79}$$

or as a vector phasor

$$\tilde{\mathbf{E}}(z) = (\hat{x} + j\hat{y})E_0e^{-jkz}\,(L\text{ wave}). \tag{2.80}$$

Figure 2.6b shows $\mathbf{E}(0,t)$ for various values of ωt. The tip of the $\mathbf{E}$ vector traces a circle. If the thumb of the left hand is pointed in the direction of wave propagation, the other four fingers point in the direction of the rotation of $\mathbf{E}$.

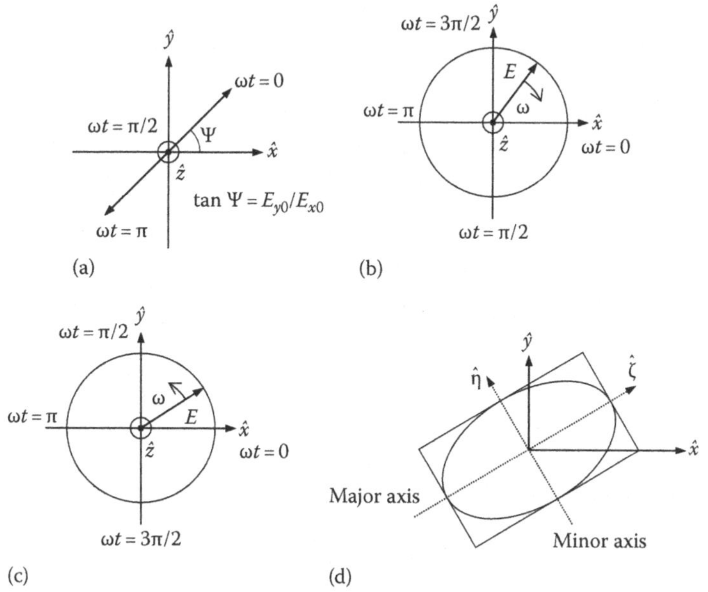

FIGURE 2.6
(a) Linearly polarized wave ($\delta = 0$). (b) L wave. (c) R wave. (d) Elliptic polarization.

A wave is said to have a right-hand circular polarization, if $\delta = -\pi/2$ and $E_{x0} = E_{y0} = E_0$, that is,

$$E(z,t) = \hat{x}E_0\cos(\omega t - kz) + \hat{y}E_0\cos(\omega t - kz - \pi/2), \tag{2.81}$$

or as a vector phasor

$$\tilde{\mathbf{E}}(z) = (\hat{x} - j\hat{y})E_0 e^{-jkz} \quad (R \text{ wave}). \tag{2.82}$$

Figure 2.6c shows $\mathbf{E}(0, t)$ for various values of $\mathbf{E}$. The tip of the $\mathbf{E}$ vector traces a circle. If the thumb of the right hand points in the direction of propagation, the other four fingers point in the direction of rotation of $\mathbf{E}$.

A wave is said to be elliptically polarized if none of the conditions stated above is satisfied. In this most general case, the tip of $\mathbf{E}$ traces an ellipse in the x–y-plane. The shape of the ellipse and its handedness are determined by the values of E_{y0}/E_{x0} and the polarization phase difference δ. Figure 2.6d shows a wave with elliptic polarization. An elliptically polarized wave may be shown to be a linear combination of L and R waves. A linearly polarized wave is a different linear combination of L and R waves.

2.10 Arbitrary Direction of Propagation

Figure 2.7 shows a uniform plane wave propagating at an angle θ in the x–z-plane. The wave is still a one-dimensional wave propagating in the direction of ζ-coordinate, which can be expressed in terms of x and z:

$$\zeta = x\sin\theta + z\cos\theta. \tag{2.83}$$

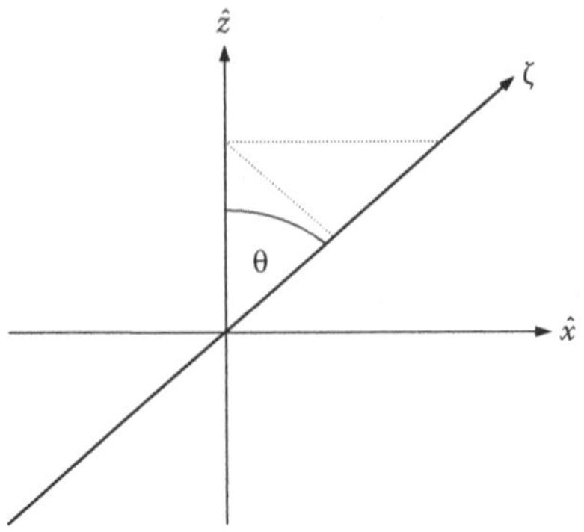

FIGURE 2.7
Arbitrary direction of propagation.

The exponential factor is given by

$$e^{-jk\varsigma} = e^{-jk(Sx+Cz)}, \tag{2.84}$$

where

$$S = \sin\theta, \quad C = \cos\theta. \tag{2.85}$$

We can have two cases of TEM waves, one with the electric field in the plane of incidence and the other with the electric field normal to the plane of incidence. The former is called a *p* wave and the latter an *s* wave. The *p* wave has the magnetic field perpendicular to the plane of incidence. If the *z*-axis is normal to an interface (boundary), then one can define wave impedance Z_w as the ratio of the tangential components of the electric and magnetic fields. For the geometry shown in Figure 2.9, we have

$$Z_w = \frac{|E_x|}{|H_y|} = \frac{E\cos\theta}{H} = \eta\cos\theta. \tag{2.86}$$

Similarly, for an *s* wave (Figure 2.10),

$$Z_w = \frac{|E_y|}{|H_x|} = \frac{E}{H\cos\theta} = \frac{\eta}{\cos\theta}. \tag{2.87}$$

The *p* wave is also referred to as a parallel-polarized wave and the *s* wave as a perpendicular-polarized wave.

2.11 Wave Reflection

At an interface between two media, the wave energy, in general, is partly transmitted and partly reflected. At an interface with a perfect conductor, the wave is totally reflected. The boundary condition $E_t = 0$ at a PEC suggests an analogy with a short-circuited transmission line (Figure 2.8).

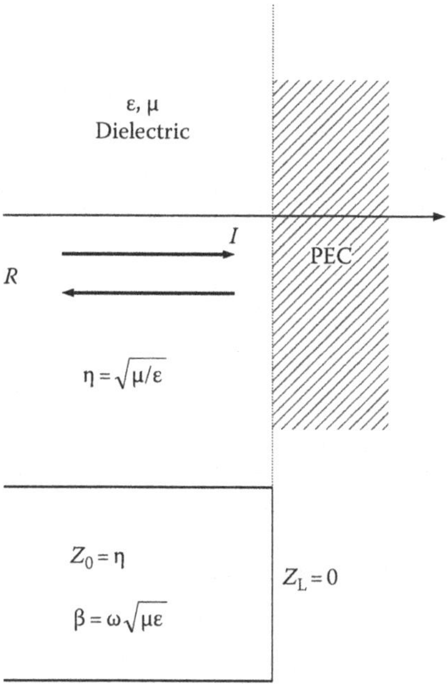

FIGURE 2.8
Dielectric–PEC interface.

Instead of an interface with a PEC, if we have a boundary with a good conductor, then we should use for the load impedance $Z_L = Z_s$, where Z_s is given by Equation 2.14.

2.12 Incidence of p Wave: Parallel-Polarized

Figure 2.9 shows a p wave propagating obliquely in the x–z-plane. The transmission line analogy is also shown. The problem may be solved as an electromagnetic boundary value problem. The unknowns may be reduced to E_x^R and E_x^T, the tangential components of the reflected wave R and the transmitted wave T, respectively. All the other components in either medium may be expressed in terms of these two unknowns and E_x^I, where the superscript I indicates an incident wave. The unknowns may be determined from the boundary conditions on tangential components of E and H at $z = 0$ for all t and x:

$$E_{x1} = E_x^I + E_x^R = E_{x2} = E_x^T, \tag{2.88}$$

$$H_{y1} = H_y^I + H_y^R = H_{y2} = H_y^T. \tag{2.89}$$

The satisfaction of the condition for all t imposes the requirement that the frequency of all the waves is same, and the satisfaction of the conditions, for all x, imposes the Snell's law:

$$\omega_1 = \omega_R = \omega_T, \tag{2.90}$$

$$k_1 \sin\theta_1 = k_R \sin\theta_R = k_2 \sin\theta_2. \tag{2.91}$$

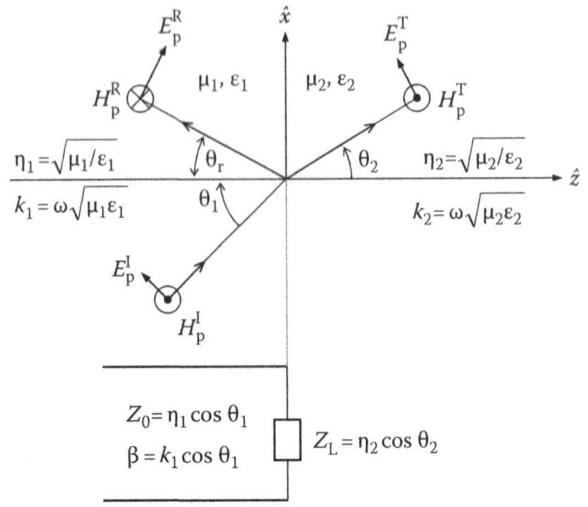

FIGURE 2.9
Oblique incidence of a p wave on a dielectric–dielectric interface. Transmission line analogy.

Sections 9.2 and 9.4 discuss these points in more detail. The solution of the boundary value problem may be completed by imposing Equations 2.88 and 2.89 subject to Equations 2.90 and 2.91. Alternatively, we can use the transmission line analogy given in Figure 2.9 and can also use Equation 2.66 for obtaining the reflection and transmission coefficients Γ_0 and T_0, respectively:

$$\Gamma_0 = \frac{V_0^-}{V_0^+} = \frac{Z_L - Z_0}{Z_L + Z_0} = \frac{\eta_2\cos\theta_2 - \eta_1\cos\theta_1}{\eta_2\cos\theta_2 + \eta_1\cos\theta_1} = \frac{E_{xp}^R}{E_{xp}^I} = \Gamma_{p1}, \tag{2.92}$$

$$T_0 = \frac{V_0}{V_0^+} = 1 + \Gamma = \frac{2Z_L}{Z_L + Z_0} = \frac{2\eta_2\cos\theta_2}{\eta_2\cos\theta_2 + \eta_1\cos\theta_1} = \frac{E_{xp}^T}{E_{xp}^I} = T_{p1}. \tag{2.93}$$

It follows immediately that the reflection coefficient is zero if

$$\eta_2\cos\theta_2 = \eta_1\cos\theta_1. \tag{2.94}$$

Further simplification of Equation 2.94 results if θ_1 and θ_2 are related through Snell's law (Equation 2.91). The angle $\theta_1 = \theta_{Bp}$ for which Equation 2.94 is satisfied is called the Brewster angle and is given by

$$\theta_{Bp} = \tan^{-1}\sqrt{\frac{\varepsilon_2}{\varepsilon_1}}. \tag{2.95}$$

A similar analysis for the s wave shows that θ_{Bs} does not exist for nonmagnetic materials. For the p wave, sometimes, the reflection and transmission coefficients are defined in terms of the ratios of the magnitude of the entire electric fields rather than the tangential components:

$$T_p = \frac{E_p^T}{E_p^I} = \frac{E_{xp}^T/\cos\theta_2}{E_{xp}^I/\cos\theta_1} = \frac{\cos\theta_1}{\cos\theta_2}T_{p1}. \tag{2.96}$$

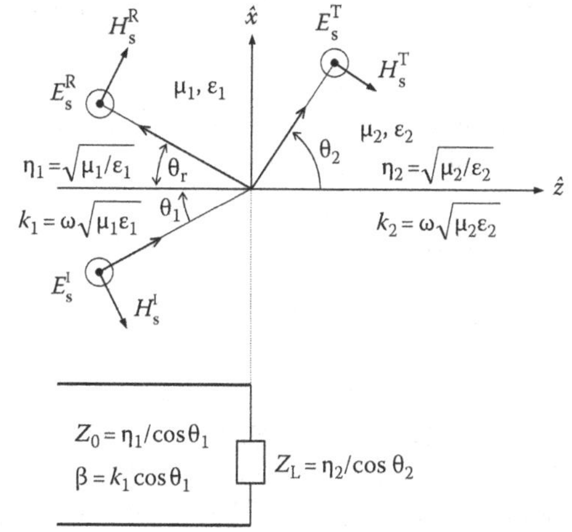

FIGURE 2.10
Oblique incidence of an *s* wave on a dielectric–dielectric interface. Transmission line analogy.

2.13 Incidence of *s* Wave: Perpendicular-Polarized

Figure 2.10 shows an *s* wave propagating obliquely in *x*–*z*-plane. The transmission line analogy is also shown. Equations 2.90 and 2.91 still hold for the reasons mentioned before. The boundary conditions in this case can be stated as, at $z = 0$, for all t and x:

$$E_{y1} = E_y^I + E_y^R = E_{y2} = E_y^T, \tag{2.97}$$

$$H_{x1} = H_x^I + H_x^R = H_{x2} = H_x^T. \tag{2.98}$$

From the transmission line analogy shown in Figure 2.10 and Equation 2.66, the reflection and transmission coefficients are given respectively, by,

$$\Gamma_s = \frac{E_y^R}{E_y^I} = \frac{\eta_2/\cos\theta_2 - \eta_1/\cos\theta_1}{\eta_2/\cos\theta_2 + \eta_1/\cos\theta_1}, \tag{2.99}$$

$$T_s = \frac{E_y^T}{E_y^I} = \frac{2\eta_2/\cos\theta_2}{\eta_2/\cos\theta_2 + \eta_1/\cos\theta_1}. \tag{2.100}$$

2.14 Critical Angle and Surface Wave

Let us next consider wave propagation from an optically dense medium to an "optically rare" medium. In optics, a nonmagnetic medium is described in terms of the refractive index n, which is related to the dielectric constant through

$$n = \sqrt{\varepsilon_r}. \tag{2.101}$$

The problem under consideration assumes that $\varepsilon_1 > \varepsilon_2$ or $n_1 > n_2$. From Snell's law (Equation 2.91),

$$\frac{\sin\theta_2}{\sin\theta_1} = \frac{k_1}{k_2} = \sqrt{\frac{\varepsilon_1}{\varepsilon_2}} = \frac{n_1}{n_2} > 1. \tag{2.102}$$

Note that $\theta_2 > \theta_1$ if $n_1 > n_2$. The angle of transmission $\theta_2 = 90°$ when

$$\theta_1 = \theta_c = \sin^{-1}\frac{n_2}{n_1}, \quad \theta_2 = 90°. \tag{2.103}$$

Such an angle of incidence θ_c is called a critical angle. For this angle, the wave no longer propagates into medium 2 and in fact it propagates along the interface. However, we still have a reflected wave, with the angle of reflection $\theta_r = \theta_c$. The wave is said to be totally reflected since there is no power flow into medium 2. Since $\theta_2 = 90°$, the second medium for p wave propagation behaves like a short-circuited load and for s wave propagation as an open-circuited load. The magnitude of the reflection coefficient in either case is 1. The wave is totally reflected. Let us examine further what happens at the critical angle by considering s wave propagation. The reflection coefficient $\Gamma_s = 1$ and the transmission coefficient $T_s = 2$. Furthermore,

$$\mathbf{E}_s^T = \hat{y}2E_s^I e^{-jk_2x}, \tag{2.104}$$

$$\mathbf{H}_s^T = \hat{z}\frac{2E_s^I}{\eta_2}e^{-jk_2x}, \tag{2.105}$$

which represent a plane wave that travels along the interface in the positive x-direction. The time-averaged power densities of various waves are given by

$$\langle \mathbf{S}^T \rangle = \hat{x}\frac{2\left|E_s^I\right|^2}{\eta_2}, \tag{2.106}$$

$$\langle \mathbf{S}^R \rangle = \frac{2\left|E_s^I\right|^2}{2\eta_1}\left[\hat{x}\sin\theta_1 - \hat{z}\cos\theta_1\right], \tag{2.107}$$

$$\langle \mathbf{S}^I \rangle = \frac{2\left|E_s^I\right|^2}{2\eta_1}\left[\hat{x}\sin\theta_1 + \hat{z}\cos\theta_1\right]. \tag{2.108}$$

Figure 2.11 shows the constant phase planes in medium 2.
When the angle of incidence $\theta_1 > \theta_c$, then

$$\sin\theta_2 = \frac{k_1}{k_2}\sin\theta_1 = \frac{n_1}{n_2}\sin\theta_1 > 1. \tag{2.109}$$

The equation can be satisfied only if θ_2 is complex [2]:

$$\theta_2 = \theta_R + j\theta_X, \quad \theta_1 > \theta_c. \tag{2.110}$$

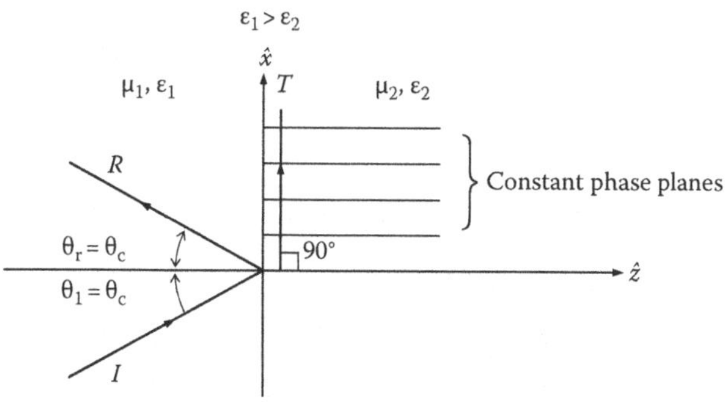

FIGURE 2.11
Constant phase planes in medium 2 for $\theta_1 = \theta_c$.

The transmitted electric field may be written as

$$\mathbf{E}_s^T(z) = \hat{y} T_s E_s^I e^{-\alpha_e z} e^{-j\beta_e x}, \tag{2.111}$$

where

$$\alpha_e = k_2 \sqrt{\frac{n_1^2}{n_2^2}\sin^2\theta_1 - 1}, \quad \theta_1 > \theta_c, \tag{2.112}$$

$$\beta_e = k_2 \frac{n_1}{n_2}\sin\theta_1, \quad \theta_1 > \theta_c, \tag{2.113}$$

and the phase velocity of this wave is given by

$$v_{Pe} = \frac{\omega}{\beta_e} = \frac{v_{P2}}{\left(\frac{n_1}{n_2}\right)\sin\theta_1} < v_{P2} = \frac{\omega}{k_2}. \tag{2.114}$$

This wave is a surface wave. The wave propagates in the positive x-direction but attenuates in the positive z-direction. Figure 2.12 shows the constant-amplitude planes (dotted lines) and the constant-phase planes (solid lines). The wave, while traveling along the interface with phase velocity v_{Pe} less than the phase velocity of an ordinary wave in the second medium v_{P2}, rapidly decays in the z-direction. Such a wave is labeled as a tightly bound slow surface wave. The imaginary part of the complex Poynting vector $\langle S_I \rangle = (1/2)\,\mathrm{Im}[\mathbf{E} \times \mathbf{H}^*]$ represents the time-averaged reactive power density. For $\theta_1 > \theta_c$, it may be shown that [3]

$$\langle S_I \rangle = \frac{1}{2}\hat{z}\sqrt{\left(\frac{n_1}{n_2}\sin\theta_1\right)^2 - 1}\,\frac{|T_s|^2 |E_s^I|^2}{\eta_2} e^{-2\alpha_e z}, \quad \theta_1 > \theta_c. \tag{2.115}$$

The wave penetrates the medium to a depth of $1/\alpha_e$ and the energy in the second medium is a stored energy. The wave in the second medium is called an evanescent wave. The attenuation of the evanescent wave is different from the attenuation of a traveling wave in

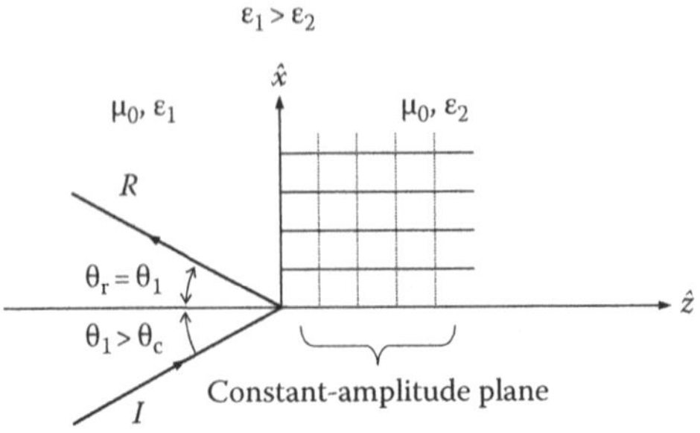

FIGURE 2.12
Constant-phase planes and constant-amplitude planes, for $\theta_1 > \theta_c$.

a conducting medium. The former indicates the localization of wave energy near the interface. Because of this penetration, it may be shown that an optical beam of finite cross section will be displaced laterally relative to the incident beam at the boundary surface. This shift is known as the Goos–Hanschen shift and is discussed in [3].

We will come across the evanescent wave in several other situations. The wave in a waveguide whose frequency is less than that of the cutoff frequency of the wave is an evanescent wave. The wave in a plasma medium, whose frequency is below "plasma frequency," is an evanescent wave. The wave in the cladding of an optical fiber is an evanescent wave. Section 9.5 discusses the tunneling of power through a plasma slab by evanescent waves.

2.15 One-Dimensional Cylindrical Wave and Bessel Functions

Let an infinitely long wire in free space be along the z-axis carrying a time harmonic current $\tilde{I}_0$. From Equation 1.50, it is clear that the vector potential **A** will have only the z-component and from the symmetry it is also clear that the component is at the most a function of the cylindrical radial coordinate ρ:

$$\tilde{\mathbf{A}} = \hat{z}A(\rho). \tag{2.116}$$

From Equation 1.47 we get

$$\nabla^2 A + k_0^2 A = 0, \quad \rho > 0, \tag{2.117}$$

$$k_0 = \omega\sqrt{\mu_0\varepsilon_0} = \frac{\omega}{c}. \tag{2.118}$$

Since A is a function of ρ only, Equation 2.117 becomes an ordinary differential equation of the form

$$\frac{d^2A}{d\rho^2} + \frac{1}{\rho}\frac{dA}{d\rho} + k_0^2 A = 0, \tag{2.119}$$

which is the Bessel equation. It has two independent solutions J_0 and Y_0. These are analogous to cosine and sine functions that are the independent solutions for the Cartesian case. Linear combinations of J_0 and Y_0 give other functions to represent traveling waves in cylindrical coordinates. Table 2.2 gives the power series [4] for these functions and also differential and integral relations involving these functions. Table 2.3 lists these functions and also gives analogous functions in Cartesian coordinates. Since we are on the topic of Bessel functions, J_m and Y_m are the solutions of a more general Bessel equation

$$\frac{d^2 f}{d\rho^2} + \frac{1}{\rho}\frac{df}{d\rho} + \left[k_0^2 - \frac{m^2}{\rho^2}\right] f = 0. \tag{2.120}$$

The sketches of the functions J_m and Y_m for various values of m and real values of the argument are given in Figure 2.13.

Table 2.4 lists the zeros χ_{mn} of the Bessel function J_m and the zeros χ'_{mn} of the derivative of the Bessel function J'_m with respect to its argument. The second subscript n denotes the nth zero of these oscillatory functions. These higher-order Bessel functions will be used in solving the cylindrical waveguide problems.

Getting back to the subject of the electromagnetic fields due to harmonic current $\tilde{I}_0$ in an infinitely long wire, the solution of Equation 2.119 that fits the requirement of an outgoing wave as $k_0\rho$ tends to infinity is expressed in terms of a Hankel function of the second kind (see Table 2.2):

$$\tilde{A}(\rho) = A_0 H_0^{(2)}(k_0\rho), \tag{2.121}$$

where A_0 is a constant to be determined. All the components of the electric and magnetic fields may be obtained from Maxwell's equations and Equation 2.121:

$$\tilde{H}_\phi = -\frac{1}{\mu}\frac{\partial A}{\partial \rho} = -k_0 \frac{A_0}{\mu} H_0'^{(2)}(k_0\rho), \tag{2.122}$$

$$\tilde{H}_\rho = \tilde{H}_z = 0, \tag{2.123}$$

$$\tilde{E}_\rho = \tilde{E}_\phi = 0, \tag{2.124}$$

$$\tilde{E}_z = -j\omega A_0 H_0^{(2)}(k_0\rho). \tag{2.125}$$

Application of Ampere's law for a circular contour C in the x–y-plane of radius ρ, in the limit $\rho \to 0$, relates the constant A_0 to the current $\tilde{I}_0$:

$$A_0 = -j\frac{\mu_0}{4}\tilde{I}_0. \tag{2.126}$$

TABLE 2.2

Bessel Functions: Definitions and Relations

Differential equations

$$\frac{d^2R}{d\rho}+\frac{1}{\rho}\frac{dR}{d\rho}+\left(T^2-\frac{\nu^2}{\rho^2}\right)R=0$$

Solutions

$$J_\nu(T\rho)=\sum_{m=0}^{\infty}\frac{(-1)^m\left(\frac{T\rho}{2}\right)^{\nu+2m}}{m!\Gamma(\nu+m+1)} \qquad J_n(T\rho)=\sum_{m=0}^{\infty}\frac{(-1)^m\left(\frac{T\rho}{2}\right)^{n+2m}}{m!(n+m)!}$$

$$Y_n(T\rho)=\frac{\cos\nu\pi J_n(T\rho)-J_{-\nu}(T\rho)}{\sin\nu\pi}$$

$$H_\nu^{(1)}(T\rho)=J_\nu(T\rho)+jY_\nu(T\rho) \qquad H_\nu^{(2)}(T\rho)=J_\nu(T\rho)-jY_\nu(T\rho)$$

$$R = AJ_\nu(T\rho) + BY_\nu(T\rho)$$

Asymptotic forms

$$J_\nu(x)\underset{x\to\infty}{\to}\sqrt{\frac{2}{\pi x}}\cos\left(x-\frac{\pi}{4}-\frac{\nu\pi}{2}\right) \qquad Y_\nu(x)\underset{x\to\infty}{\to}\sqrt{\frac{2}{\pi x}}\sin\left(x-\frac{\pi}{4}-\frac{\nu\pi}{2}\right)$$

$$H_\nu^{(1)}(x)\underset{x\to\infty}{\to}\sqrt{\frac{2}{\pi x}}e^{j[x-(\pi/4)-(\nu\pi/2)]} \qquad H_\nu^{(2)}(x)\underset{x\to\infty}{\to}\sqrt{\frac{2}{\pi x}}e^{-j[x-(\pi/4)-(\nu\pi/2)]}$$

$$j^{-\nu}J_\nu(jx)\underset{x\to\infty}{=}I_\nu(x)\underset{x\to\infty}{\to}\sqrt{\frac{1}{2\pi x}}e^{x} \qquad j^{\nu+1}J_\nu^{(1)}(jx)\underset{x\to\infty}{=}\frac{2}{\pi}K_\nu(x)\underset{x\to\infty}{\to}\sqrt{\frac{2}{\pi x}}e^{-x}$$

Derivatives

$$R_0'(x)=-R_1(x) \qquad R_1'(x)=R_0(x)-\frac{1}{x}R_1(x)$$

$$xR_\nu'(x)=\nu R_\nu(x)-xR_{\nu+1}(x) \qquad xR_\nu'(x)=-\nu R_\nu(x)+xR_{\nu-1}(x)$$

$$\frac{d}{dx}\left[x^{-\nu}R_\nu(x)\right]=-x^{-\nu}R_{\nu+1}(x) \qquad \frac{d}{dx}\left[x^{\nu}R_\nu(x)\right]=x^{\nu}R_{\nu-1}(x)$$

$$xI_\nu'(x)=xI_\nu(x)+xI_{\nu+1}(x) \qquad xI_\nu'(x)=-xI_\nu(x)+xI_{\nu-1}(x)$$

$$xK_\nu'(x)=xK_\nu(x)-xK_{\nu+1}(x) \qquad xK_\nu'(x)=-xK_\nu(x)-xK_{\nu-1}(x)$$

Recurrence formulas

$$\frac{2\nu}{x}R_\nu(x)=R_{\nu+1}(x)+R_{\nu-1}(x) \qquad \frac{2\nu}{x}I_\nu(x)=I_{\nu-1}(x)-I_{\nu+1}(x)$$

$$\frac{2\nu}{x}K_\nu(x)=K_{\nu+1}(x)-K_{\nu-1}(x)$$

Integrals

$$\int x^{\nu}R_{\nu+1}(x)dx=-x^{-\nu}R_\nu(x) \qquad \int x^{\nu}R_{\nu-1}(x)dx=-x^{\nu}R_\nu(x)$$

$$\int x^{\nu}R_\nu(\alpha x)R_\nu(\beta x)dx=\frac{x}{\alpha^2-\beta^2}\left[\beta R_\nu(\alpha x)R_{\nu-1}(\beta x)-\alpha R_{\nu-1}(\alpha x)R_\nu(\beta x)\right],\alpha\neq\beta$$

$$\int xR_\nu^2(\alpha x)dx=\frac{x^2}{2}\left[R_\nu^2(\alpha x)-R_{\nu-1}(\alpha x)R_{\nu+1}(\alpha x)\right]$$

$$=\frac{x^2}{2}\left[R_\nu'^2(\alpha x)+\left(1-\frac{\nu^2}{\alpha^2x^2}\right)R_\nu^2(\alpha x)\right]$$

Bessel and Hankel functions of imaginary arguments

$$\frac{d^2R}{d\rho^2}+\frac{1}{\rho}\frac{dR}{d\rho}-\left(\tau^2+\frac{\nu^2}{\rho^2}\right)R=0$$

Solution, $x = \tau\rho$

$$I_{\pm\nu}(x)=j^{\mp\nu}J_{\pm\nu}(jx) \qquad K_\nu(x)=\frac{\pi}{2}j^{\nu+1}H_\nu^{(1)}(jx)$$

TABLE 2.3

Wave Functions in Cylindrical and Cartesian Coordinates

	Cylindrical			Cartesian	
Wave Type	**Symbol**	**Name**	**Remarks**	**Symbol**	**Remarks**
Standing waves	$J_0(k_0\rho)$	Bessel function zero order, first kind	Zeros: $k_0\rho = \chi_{0n}$ unevenly spaced. Function: finite at the origin	$\cos(k_0 z)$	Zeros: $k_0 z = (2m+1)\left(\frac{\pi}{2}\right)$ evenly spaced. Even function
	$Y_0(k_0\rho)$	Bessel function zero order, second kind	Zeros: unevenly spaced Function: blows up at the origin	$\sin(k_0 z)$	Zeros: $k_0 z = m\pi$, evenly spaced. Odd function
Traveling waves	$H_0^{(2)}(k_0\rho)$	Hankel function zero order, second kind = $J_0(k_0\rho) - \mathrm{j}\,Y_0(k_0\rho)$	$\lim_{k_0\rho\to\infty} H_0^{(2)}(k_0\rho) = \sqrt{\frac{2\mathrm{j}}{\pi k_0\rho}}\mathrm{e}^{-\mathrm{j}k_0\rho}$ outgoing wave	$\mathrm{e}^{\mathrm{j}k_0 z}$	Positive-going
	$H_0^{(1)}(k_0\rho)$	Hankel function zero order, first kind = $J_0(k_0\rho)$ + $\mathrm{j}\,Y_0(k_0\rho)$	$\lim_{k_0\rho\to\infty} H_0^{(1)}(k_0\rho) = \sqrt{\frac{2\mathrm{j}}{\pi k_0\rho}}\mathrm{e}^{\mathrm{j}k_0\rho}$ incoming wave	$\mathrm{e}^{-\mathrm{j}k_0 z}$	Negative-going
Evanescent waves	$K_0(\alpha\rho)$	Modified Bessel function zero order, second kind	Monotonic function blows up at the origin	$\mathrm{e}^{-\alpha z}$	Monotonic function zero at ∞
	$I_0(\alpha\rho)$	Modified Bessel function zero order, first kind	Monotonic function finite at the origin	$\mathrm{e}^{+\alpha z}$	Monotonic function zero at $-\infty$

In obtaining Equation 2.126, the following small-argument approximation is used:

$$\lim_{k_0\rho\to 0} H_1^{(2)}(k_0\rho) = \mathrm{j}\frac{2}{\pi}\frac{1}{k_0\rho}. \tag{2.127}$$

The fields $\tilde{E}_z$ and $\tilde{H}_\phi$ are given by

$$\tilde{E}_z = -\frac{k_0\eta_0}{4}\tilde{I}_0 H_0^{(2)}(k_0\rho), \tag{2.128}$$

$$\tilde{H}_\phi = -\mathrm{j}\frac{k_0}{4}\tilde{I}_0 H_1^{(2)}(k_0\rho). \tag{2.129}$$

Equations 2.128 and 2.129 are the fields of a one-dimensional cylindrical wave. It can be shown that the wave impedance in the far-zone is

$$Z_w = \lim_{k_0\rho\to\infty} \frac{\tilde{E}_z}{-\tilde{H}_\phi} = \eta_0. \tag{2.130}$$

Since the wave impedance is equal to the intrinsic impedance of the medium, the wave is a TEM as shown in Figure 2.14.

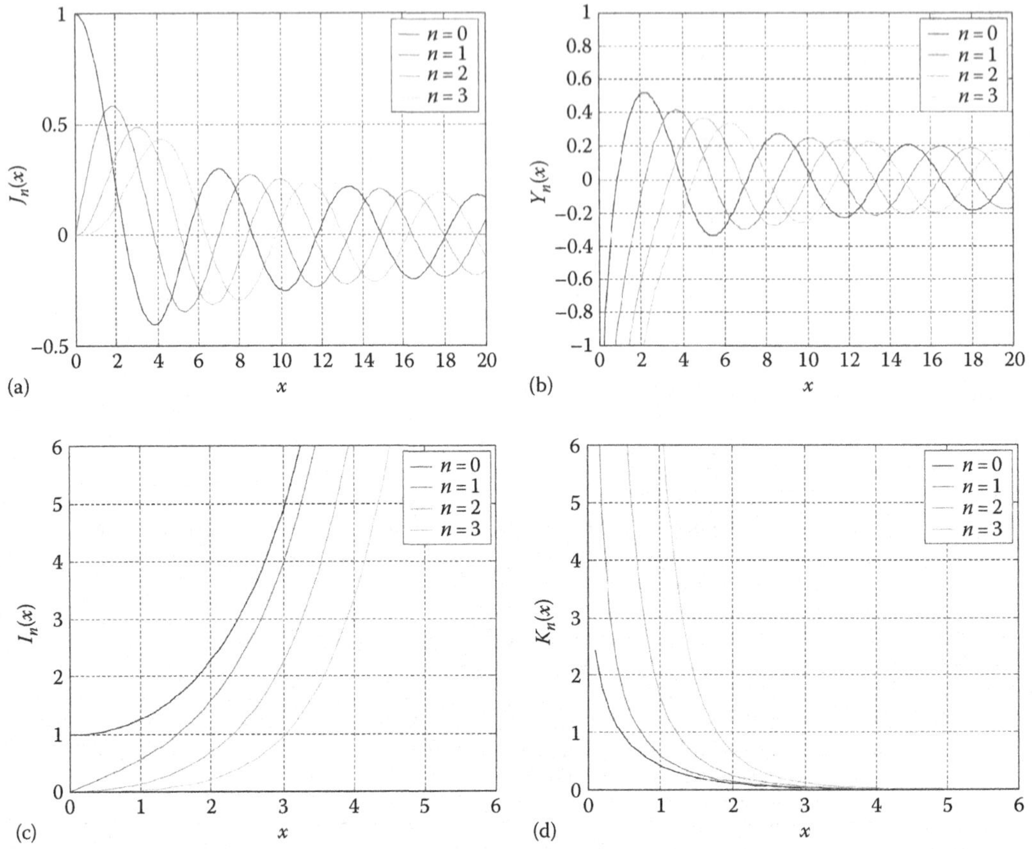

FIGURE 2.13
(a) Bessel functions of the first kind. (b) Bessel functions of the second kind. (c) Modified Bessel functions of the first kind. (d) Modified Bessel functions of the second kind.

TABLE 2.4

Zeros of Bessel Functions

J_0	J_1	J_2	Y_0	Y_1	Y_2
2.405	3.832	5.136	0.894	2.197	3.384
5.520	7.016	8.417	3.958	5.430	6.794
8.654	10.173	11.620	7.086	8.596	10.023

Zeros of derivatives of Bessel functions

J_0'	J_1'	J_2'	Y_0'	Y_1'	Y_2'
3.832	1.841	3.054	2.197	3.683	5.003
7.016	5.331	6.706	5.430	6.942	8.351
10.173	8.536	9.969	8.596	10.123	11.574

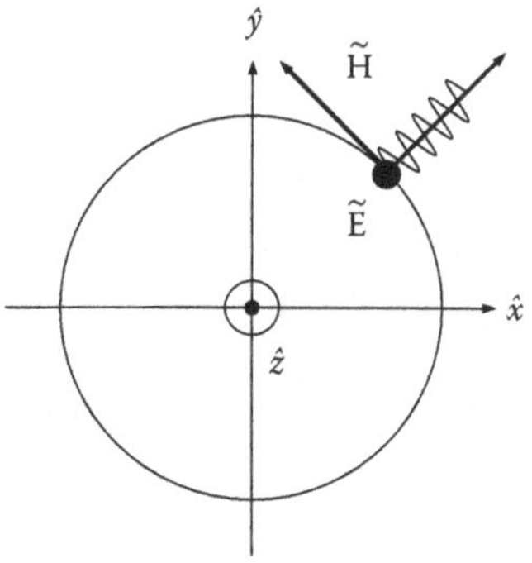

FIGURE 2.14
One-dimensional cylindrical wave.

References

1. Ulaby, F. T., *Applied Electromagnetics*, Prentice Hall, Englewood Cliffs, NJ, 1999.
2. Balanis, C. A., *Advanced Engineering Electromagnetics*, Wiley, New York, 1989.
3. Lekner, J., *Theory of Reflection*, Kluwer Academic Publishers, Boston, MA, 1987.
4. Ramo, S., Whinnery, J. R., and Van Duzer, T. V., *Fields and Waves in Communication Electronics*, 3rd edition, Wiley, New York, 1994.

3

Two-Dimensional Problems and Waveguides

In the previous chapter, we examined one-dimensional solutions. We found the solutions to be plane or cylindrical TEM (transverse electric and magnetic) waves. The TEM waves are the simplest kind of electromagnetic waves.

The next level of simple waves is either transverse magnetic (TM) or transverse electric (TE). Such waves arise when the wave is confined (bounded) in the transverse plane with, say, conducting boundaries, and is traveling in a direction normal to the transverse plane.

3.1 Two-Dimensional Solutions in Cartesian Coordinates

Let us investigate the solution of Maxwell's equations under the following constraints: (1) the wave is traveling in the *z*-direction, (2) it has an electric field component in the *z*-direction, (3) there are no sources in the region of interest, and (4) the medium in the region of interest is lossless.

The $\tilde{E}_z$ component satisfies the equation

$$\nabla^2 \tilde{E}_z + k^2 \tilde{E}_z = 0, \tag{3.1}$$

where

$$k^2 = \omega^2 \mu \varepsilon. \tag{3.2}$$

Let

$$\tilde{E}_z(x, y, z) = F(x, y) e^{-\gamma z}. \tag{3.3}$$

The wave is a positive-going traveling wave when γ is imaginary, that is, $\gamma = j\beta$, where β is real and positive. It is an attenuating wave when γ is purely real, that is, $\gamma = \alpha$, where α is real and positive. Since the medium is assumed lossless, the attenuation in this case is due to the fact that the wave is evanescent. We will confirm this after we complete the solution.

Substituting Equation 3.3 into Equation 3.1, we obtain

$$\frac{\partial^2 F}{\partial x^2} + \frac{\partial^2 F}{\partial y^2} + k_c^2 F = 0, \tag{3.4}$$

where

$$k_c^2 = k^2 + \gamma^2. \tag{3.5}$$

Separation of variable method is a standard technique of solving this partial differential equation (PDE). The technique converts the PDE into ordinary differential equations (ODEs) with a constraint on the separation constants. The meaning becomes clear as we proceed. Let F be expressed as a product of two functions

$$F(x,y) = f_1(x) f_2(y), \tag{3.6}$$

where f_1 is entirely a function of x and f_2 is entirely a function of y. Substituting Equation 3.6 into Equation 3.4, we obtain

$$\frac{1}{f_1}\frac{\partial^2 f_1}{\partial x^2} + \frac{1}{f_2}\frac{\partial^2 f_2}{\partial y^2} + k_c^2 = 0. \tag{3.7}$$

Differentiating partially with respect to x, we obtain

$$\frac{\partial}{\partial x}\left[\frac{1}{f_1}\frac{\partial^2 f_1}{\partial x^2}\right] = 0, \tag{3.8}$$

$$\frac{1}{f_1}\frac{\partial^2 f_1}{\partial x^2} = \text{constant}. \tag{3.9}$$

Let this constant be denoted by $-k_x^2$. Equation 3.9 may be written as

$$\frac{d^2 f_1}{dx^2} + k_x^2 f_1 = 0. \tag{3.10}$$

Following the same argument, the second term in Equation 3.7 can be equated to $-k_y^2$, leading to the ODE

$$\frac{d^2 f_2}{dy^2} + k_y^2 f_2 = 0. \tag{3.11}$$

From Equation 3.7, we can see that the constants k_x^2 and k_y^2 are subject to the constraint

$$k_x^2 + k_y^2 = k_c^2. \tag{3.12}$$

The PDE (Equation 3.4) is converted into the two ODEs given by Equations 3.9 and 3.10 subject to the constraint given by Equation 3.12. Each ODE has two independent solutions. If the constants k_x^2 and k_y^2 are negative, that is,

$$k_x^2 = -K_x^2, \tag{3.13}$$

$$k_y^2 = -K_y^2, \tag{3.14}$$

where K_x^2 and K_y^2 are positive, then the solutions are hyperbolic functions. Thus, the admissible functions are

$$f_1(x) = \begin{Bmatrix} \sin k_x x \\ \cos k_x x \\ e^{+jk_x x} \\ e^{-jk_x x} \\ \sinh K_x x \\ \cosh K_x x \\ e^{+K_x x} \\ e^{-K_x x} \end{Bmatrix}, \tag{3.15}$$

$$f_2(y) = \begin{Bmatrix} \sin k_y y \\ \cos k_y y \\ e^{+jk_y y} \\ e^{-jk_y y} \\ \sinh K_y y \\ \cosh K_y y \\ e^{+K_y y} \\ e^{-K_y y} \end{Bmatrix}. \tag{3.16}$$

The solution to a given problem may be constructed by choosing a linear combination of the admissible functions. The choice is influenced by the boundary conditions. An illustration is given in the following section.

3.2 TM_{mn} Modes in a Rectangular Waveguide

Figure 3.1 shows the cross section of a rectangular waveguide with conducting boundaries (PEC) at $x = 0$ or a and $y = 0$ or b. TM modes have $\tilde{E}_z \neq 0$ and $\tilde{H}_z = 0$. The $\tilde{E}_z$ component satisfies Equation 3.1 inside the guide and is, however, zero on the PEC boundaries. This boundary condition translates into the "Dirichlet boundary condition" $F(x, y) = 0$ on the boundaries given by $x = 0$ or a, or when $y = 0$ or b. The requirement of multiple zeros on the axes, including a zero at $x = 0$, forces us to choose the $\sin k_x x$ function for the x-variation. Moreover,

$$\sin k_x a = 0, \tag{3.17}$$

$$k_x = \frac{m\pi}{a}, \quad m = 1, 2, \ldots, \infty. \tag{3.18}$$

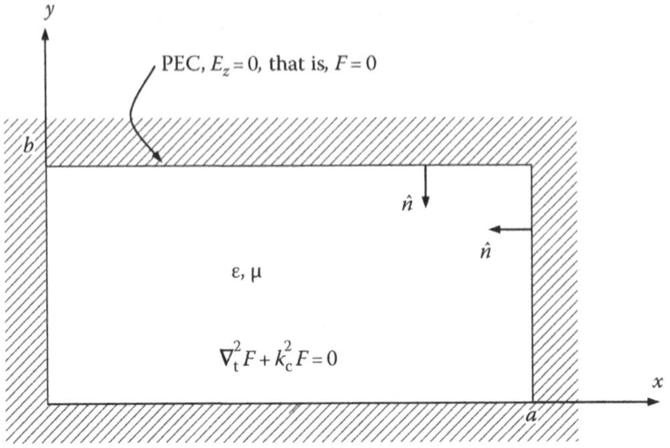

FIGURE 3.1
Cross section of a rectangular waveguide. TM modes. Dirichlet problem.

A similar argument leads to the choice of sine function for the y-variation and also

$$k_y = \frac{n\pi}{b}, \quad n = 1, 2, \ldots, \infty. \tag{3.19}$$

Now, we are able to write the expression for $\tilde{E}_z$ of the mnth TM mode:

$$\tilde{E}_z(x, y, z) = E_{mn} \sin\frac{m\pi x}{a} \sin\frac{n\pi y}{b} e^{-\gamma z}, \tag{3.20}$$

where E_{mn} is the mode constant of the TM_{mn} mode. From Equations 3.5, 3.12, 3.18, and 3.19, we obtain

$$k_c^2 = \gamma^2 + k^2 = \left(\frac{m\pi}{a}\right)^2 + \left(\frac{n\pi}{b}\right)^2, \tag{3.21}$$

$$\gamma = \alpha = \sqrt{k_c^2 - k^2} \quad \text{if } k_c > k, \tag{3.22}$$

$$\gamma = j\beta = j\sqrt{k^2 - k_c^2} \quad \text{if } k_c < k. \tag{3.23}$$

Equation 3.23 has to be satisfied for the wave to be a propagating wave instead of an evanescent wave. Recalling $k^2 = (2\pi f)^2\,\mu\varepsilon$ and defining

$$k_c^2 = (2\pi f_c)^2\,\mu\varepsilon, \tag{3.24}$$

we can obtain

$$\alpha = k_c\sqrt{1 - \frac{f^2}{f_c^2}}, \quad f < f_c, \tag{3.25}$$

$$\beta = k\sqrt{1 - \left(\frac{f_c}{f}\right)^2}, \quad f > f_c, \tag{3.26}$$

where

$$f_c = \frac{1}{2\pi\sqrt{\mu\varepsilon}}\sqrt{\left(\frac{m\pi}{a}\right)^2 + \left(\frac{n\pi}{b}\right)^2}. \tag{3.27}$$

Thus emerges the concept of a mode cutoff frequency f_c of a waveguide. When the signal frequency f is greater than the mode cutoff frequency f_c, then the mode will propagate. When $f < f_c$, the mode will be evanescent. Since the lowest values of m and n are 1 the lowest cutoff frequency of TM modes in a rectangular waveguide is

$$(f_c)_{TM_{11}} = \frac{1}{2\sqrt{\mu\varepsilon}}\sqrt{\left(\frac{1}{a}\right)^2 + \left(\frac{1}{b}\right)^2}. \tag{3.28}$$

Once $\tilde{E}_z$ is determined, we can obtain all the other field components of the TM wave in terms of $\tilde{E}_z$ by applying Maxwell's equations:

$$\tilde{E}_x = \frac{-j\beta k_x}{k_c^2} E_{mn} \cos\frac{m\pi}{a}x \sin\frac{n\pi}{b}y e^{-j\beta z}, \tag{3.29}$$

$$\tilde{E}_y = \frac{-j\beta k_y}{k_c^2} E_{mn} \sin\frac{m\pi}{a}x \cos\frac{n\pi}{b}y e^{-j\beta z}, \tag{3.30}$$

$$\tilde{H}_x = \frac{j\omega\varepsilon k_y}{k_c^2} E_{mn} \sin\frac{m\pi}{a}x \cos\frac{n\pi}{b}y e^{-j\beta z}, \tag{3.31}$$

$$\tilde{H}_y = \frac{-j\omega\varepsilon k_x}{k_c^2} E_{mn} \cos\frac{m\pi}{a}x \sin\frac{n\pi}{b}y e^{-j\beta z}. \tag{3.32}$$

We have chosen $\tilde{E}_z$ appropriately to satisfy the boundary condition $E_{tan} = 0$ on the conducting walls of the guide. However, for a TM wave, there are other tangential components that must also reduce to zero on the walls. For instance (Figure 3.1), $\tilde{E}_x = 0$ on the walls $y = 0$ or b. We note, from Equation 3.29, that this boundary condition is satisfied automatically. We also note, from Equation 3.30, that the boundary condition $E_y = 0$ on $x = 0$ or a is automatically satisfied. Thus, for the TM wave guiding problem, $\tilde{E}_z$ is a potential. The solution is obtained by finding $\tilde{E}_z$ that satisfies the Maxwell's equation and the boundary condition of $\tilde{E}_z = 0$ on the conductor (Dirichlet boundary condition). The other field components are obtained from $\tilde{E}_z$ and the boundary conditions on the other field components are automatically satisfied.

Just as we have defined the wave impedance for the TEM wave, we can also define the wave impedance for a TM wave:

$$Z_w = \frac{\tilde{E}_x}{\tilde{H}_y} = \frac{-\tilde{E}_y}{\tilde{H}_x} = \frac{\beta}{\omega\varepsilon} = \eta\sqrt{1 - \left(\frac{f_c}{f}\right)^2}. \tag{3.33}$$

When the signal frequency is less than the cutoff frequency, Z_w is purely imaginary, showing that the electric fields and the associated magnetic fields in the transverse plane are in time quadrature, the Poynting vector component in the z-direction is imaginary, and the power flow is purely reactive. The wave is evanescent.

The wavelength λ of the signal in the medium without the boundaries is given by

$$\lambda = \frac{2\pi}{k} = \frac{2\pi}{\omega\sqrt{\mu\varepsilon}} = \frac{1}{f\sqrt{\mu\varepsilon}}. \tag{3.34}$$

We can define a guide wavelength λ_g as

$$\lambda_g = \frac{2\pi}{\beta} = \frac{2\pi}{k\sqrt{1-(f_c/f)^2}} = \frac{\lambda}{\sqrt{1-(f_c/f)^2}}. \tag{3.35}$$

Above the cutoff, the wavelength λ_g is greater than the unbounded wavelength λ. It is sometimes more convenient to define a cutoff wavelength λ_c for a waveguide:

$$\lambda_c = \frac{2\pi}{k_c}. \tag{3.36}$$

The waveguide mode propagates if $\lambda < \lambda_c$.

3.3 TE$_{mn}$ Modes in a Rectangular Waveguide

TE modes have $\tilde{E}_z = 0$, but $\tilde{H}_z \neq 0$. The $\tilde{H}_z$ component satisfies

$$\nabla^2\tilde{H}_z + k^2\tilde{H}_z = 0. \tag{3.37}$$

Let

$$\tilde{H}_z(x, y, z) = G(x,y)e^{-\gamma z}. \tag{3.38}$$

The function G satisfies

$$\frac{\partial^2 G}{\partial x^2} + \frac{\partial^2 G}{\partial y^2} + k_c^2 G = 0, \tag{3.39}$$

where k_c^2 is again given by Equation 3.5.

The boundary condition on $\tilde{H}_z$ at the PEC walls, from Equation 2.26, is given by

$$\frac{\partial \tilde{H}_z}{\partial n} = 0, \tag{3.40}$$

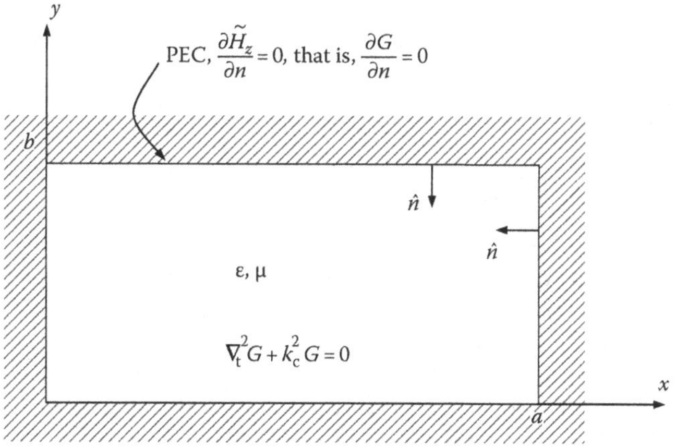

FIGURE 3.2
Cross section of a rectangular waveguide. TE modes. Neuman problem.

where $\hat{n}$ is a normal unit vector as shown in Figure 3.2. This boundary condition translates into the boundary condition on G given by Equation 3.40:

$$\frac{\partial G}{\partial x} = 0, \quad x = 0 \text{ or } a, \tag{3.41}$$

$$\frac{\partial G}{\partial y} = 0, \quad y = 0 \text{ or } b. \tag{3.42}$$

Like in the TM case, it can be shown that all other boundary conditions at PEC are automatically satisfied when Equations 3.41 and 3.42 are satisfied:

$$\tilde{H}_z(x,y,z) = H_{mn} \cos\frac{m\pi x}{a} \cos\frac{n\pi y}{b} e^{-\gamma z}, \tag{3.43}$$

where H_{mn} is the mode constant of the TE_{mn} mode. The lowest allowed value for m or n is zero. The potential for TE modes is $\tilde{H}_z$ and the type of boundary value problem where the normal derivative of the potential is zero on the boundary is called the Neuman boundary value problem. The cutoff frequency f_c for the TE modes is given by Equation 3.27. If $a > b$, then we obtain the lowest cutoff TE modes by choosing $m = 1$ and $n = 0$. The choice of $m = n = 0$ leads to a trivial solution that all the fields are zero in the waveguide.

3.4 Dominant Mode in a Rectangular Waveguide: TE_{10} Mode

The lowest cutoff frequency of all modes of a rectangular cavity is given by Equation 3.49 and such a mode is referred to as the dominant mode. Figure 3.3 shows a sketch of the normalized cutoff frequencies of a rectangular waveguide with $b/a = 1/2$. Normally, a

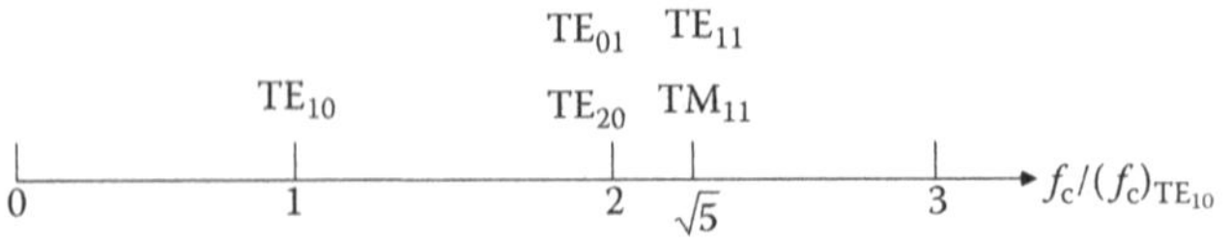

FIGURE 3.3
Normalized cutoff frequencies of a rectangular waveguide with $b/a = 1/2$.

guide is designed so that its cutoff frequency of the dominant mode is about 30% below the operating frequency. Note that from Figure 3.3, the operating frequency is below the next order modes TE_{01} and TE_{20} and therefore, only the dominant mode can propagate. In practice, higher-order modes may be excited at the point of excitation of the guide, but they die away in a short distance from the source since the waves of these higher-order modes are evanescent. In view of the importance of TE_{10} mode, let us study this mode, in more detail. The fields of this mode are given by

$$\tilde{H}_z = H_{10} \cos\frac{\pi x}{a} e^{-\gamma z}, \tag{3.44}$$

$$\tilde{E}_y = \frac{j\omega\mu}{k_c^2}\frac{\partial \tilde{H}_z}{\partial x} = \frac{-j\omega\mu}{\pi/a} H_{10} \sin\frac{\pi x}{a} e^{-\gamma z}, \tag{3.45}$$

$$\tilde{H}_x = \frac{\gamma}{\pi/a} H_{10} \sin\frac{\pi x}{a} e^{-\gamma z}. \tag{3.46}$$

All other components are zero. The wave impedance in this case is given by

$$(Z_w)_{TE10} = -\frac{\tilde{E}_y}{\tilde{H}_x} = \frac{\eta}{\cos\psi}, \tag{3.47}$$

where

$$\cos\Psi = \sqrt{1-\left(\frac{f_c}{f}\right)^2}. \tag{3.48}$$

For this case, the cutoff frequency for the TE_{10} mode is given by

$$(f_c)_{TE10} = \frac{1}{2a\sqrt{\mu\varepsilon}}. \tag{3.49}$$

For $f > f_c$, $\gamma = j\beta$, where β is real, the mode propagates, and the wave impedance given by Equation 3.47 is real. For $f < f_c$, $\gamma = \alpha$, where α is real, the mode does not propagate, and the wave impedance is imaginary. The electric and the magnetic fields are in time quadrature, and there is no real power flow down the guide. The mode is evanescent and the fields decay signifying localization of stored energy.

3.5 Power Flow in a Waveguide: TE_{10} Mode

It is easy enough to study the power flow of the TE_{10} mode. From Equations 3.46 and 3.47, we obtain

$$\langle \mathbf{S}_{ZR} \rangle = \frac{1}{2} \text{Re}\left[\tilde{\mathbf{E}} \times \tilde{\mathbf{H}}^* \right] \cdot \hat{z}, \tag{3.50}$$

$$\left\langle \left(\mathbf{S}_{ZR} \right)_{TE_{10}} \right\rangle = \frac{1}{2} \frac{\eta}{\cos\psi} \left| H_{10} \right|^2 \left(\frac{\beta a}{\pi} \right)^2 \sin^2 \frac{\pi x}{a}, \quad f > f_c, \tag{3.51}$$

where ψ is as given by Equation 3.48. The power flow P_{10} is given by

$$P_{10} = \int_0^b \int_0^a \left\langle \left(\mathbf{S}_{ZR} \right)_{TE_{10}} \right\rangle \cdot \hat{z}\, dx dy = \frac{1}{2} \left| H_{10} \right|^2 \left(\frac{\beta a}{\pi} \right)^2 \frac{\eta}{\cos\psi} \frac{ab}{2}. \tag{3.52}$$

3.6 Attenuation of TE_{10} Mode due to Imperfect Conductors and Dielectric Medium

We have seen that the wave attenuates when the signal frequency is less than the cutoff frequency. The wave becomes evanescent. Even if $f > f_c$, the wave can attenuate due to imperfect materials. The exponential factor of the fields of the wave will have the form

$$e^{-\gamma z} = e^{-(\alpha_c + \alpha_d) z} e^{-j\beta z}. \tag{3.53}$$

In the above, α_c is the attenuation of the fields due to the conversion of the wave energy into heat by imperfect conductors of the guide. α_d is the attenuation due to the conversion of the wave energy into heat by an imperfect dielectric. These are given by

$$\alpha_c = \frac{R_s \left(Z_w \right)_{TE_{10}}}{\eta^2 ba} \left(a + \frac{b\lambda^2}{2a^2} \right) = \frac{R_s}{b\eta \sqrt{1 - \left(\frac{\lambda}{2a} \right)^2}} \left[1 + \frac{2b}{a} \left(\frac{\lambda}{2a} \right)^2 \right], \tag{3.54}$$

$$\alpha_d = \frac{k \left(\frac{\sigma}{\omega\varepsilon} \right)}{2\sqrt{1 - \left(\frac{\lambda}{2a} \right)^2}}. \tag{3.55}$$

In the above, R_s is the surface resistance of the conductor and $\sigma/\omega\varepsilon$ the loss tangent of the dielectric. The derivation of the attenuation constant formulas is given as problems at the end of the book.

3.7 Cylindrical Waveguide: TM Modes

Let

$$\tilde{E}_z(\rho, \phi, z) = F(\rho, \phi)e^{-\gamma z} \tag{3.56}$$

and

$$F(\rho, \phi) = f_1(\rho) f_2(\phi). \tag{3.57}$$

The separation of variable technique applied to the PDE

$$\nabla_t^2 F + k_c^2 F = 0 \tag{3.58}$$

results in two ODEs with separation constants k_c and n:

$$\frac{1}{\rho}\frac{d}{d\rho}\left(\rho \frac{df_1}{d\rho}\right) + \left(k_c^2 - \frac{n^2}{\rho^2}\right) f_1 = 0, \tag{3.59}$$

$$\frac{d^2 f_2}{d\phi^2} + n^2 f_2 = 0. \tag{3.60}$$

Recognizing Equation 3.59 as the Bessel equation (2.120), the solution for F may be written as a linear combination of the product of the functions given below:

$$F = \begin{Bmatrix} J_n(k_c\rho) \\ Y_n(k_c\rho) \end{Bmatrix} \begin{Bmatrix} \sin n\phi \\ \cos n\phi \end{Bmatrix}. \tag{3.61}$$

Since $\rho = 0$ is a part of the field region, we choose J_n function for the ρ variation and from the boundary condition $\tilde{E}_z = F = 0$ when $\rho = a$ (see Figure 3.4):

$$k_c a = \chi_{nl}, \tag{3.62}$$

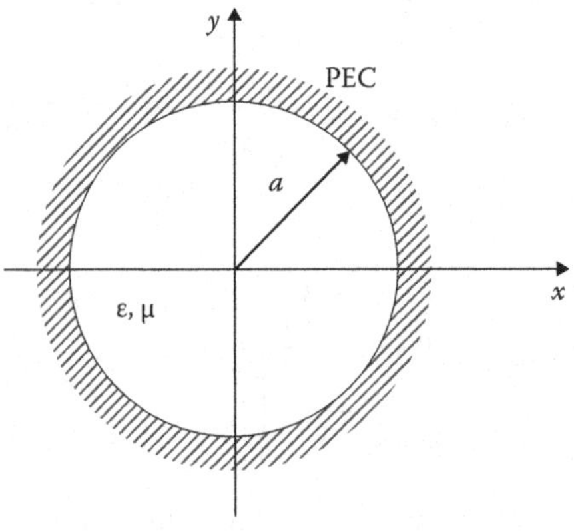

FIGURE 3.4
Circular waveguide with PEC boundary.

where χ_{nl} is the lth root of J_n and a partial list of these is given in Table 2.4.
The cutoff frequency is given by

$$(f_c)_{\mathrm{TM}_{nl}} = \frac{k_c}{2\pi\sqrt{\mu\varepsilon}} = \frac{\chi_{nl}}{2\pi a\sqrt{\mu\varepsilon}}. \tag{3.63}$$

Since the lowest value for χ_{nl} is 2.405 when $n = 0$ and $l = 1$, the lowest cutoff frequency for TM modes is given by

$$(f_c)_{\mathrm{TM}_{01}} = \frac{\chi_{01}}{2\pi a\sqrt{\mu\varepsilon}} = \frac{2.405}{2\pi a\sqrt{\mu\varepsilon}}. \tag{3.64}$$

3.8 Cylindrical Waveguide: TE Modes

Let

$$\tilde{H}_z(\rho, \phi, z) = G(\rho, \phi)e^{-\gamma z}. \tag{3.65}$$

Solution for G is again given by Equation 3.61. However, in the TE case, the Neumann boundary condition

$$\left.\frac{\partial G}{\partial \rho}\right|_{\rho=a} = 0 \tag{3.66}$$

translates into the requirement

$$J_n'(k_c a) = 0, \tag{3.67}$$

where the derivative is with respect to the argument $k_c\rho$. The cutoff wave number is thus given by

$$(k_c)_{\mathrm{TE}} = \frac{\chi_{nl}'}{a}, \tag{3.68}$$

where χ_{nl}' is the lth root of the derivative of nth-order Bessel function of the first kind. The values of χ_{nl}' are listed in Table 2.4. The lowest value is 1.841 and occurs for $n = 1$ and $l = 1$. The lowest cutoff frequency of a circular waveguide is thus given by

$$(f_c)_{\mathrm{TE}_{11}} = \frac{1.841}{2\pi a\sqrt{\mu\varepsilon}}. \tag{3.69}$$

3.9 Sector Waveguide

Figure 3.5 shows the cross section of a sector waveguide with PEC boundaries.

For TM modes, the boundary conditions are the Dirichlet boundary conditions:

$$F = 0, \quad \phi = 0 \quad \text{or} \quad \alpha, \tag{3.70}$$

$$F = 0, \quad \rho = a. \tag{3.71}$$

From Equation 3.71, the ρ variation is given by the Bessel function of the first kind, however, the order of the Bessel function need not be an integer; the field region in this case is limited to $0 < \phi < \alpha$. In the case of a circular waveguide, the field region is given by $0 < \phi < 2\pi$. The ϕ variation in such a case is a linear combination of $\sin n\phi$ and $\cos n\phi$, where n is an integer. Such stipulation is necessary to ensure a unique value for the potential. Note that $\cos n\phi$ and $\cos n(\phi + 2\pi)$ are equal only if n is an integer. For the sector waveguide, the ϕ variation is again a linear combination of $\cos \nu\phi$ and $\sin \nu\phi$ where ν need not be an integer. Bearing in mind these considerations, we write the potential

$$F = J_\nu (k_c \rho) \sin \nu\phi, \tag{3.72}$$

where

$$\nu = \frac{n\pi}{\alpha} \quad \text{and} \quad k_c a = \chi_{(n\pi/\alpha)l}. \tag{3.73}$$

For TE modes

$$G = J_\nu (k_c \rho) \cos \nu\phi, \tag{3.74}$$

where

$$k_c a = \chi'_{(n\pi/\alpha)l}. \tag{3.75}$$

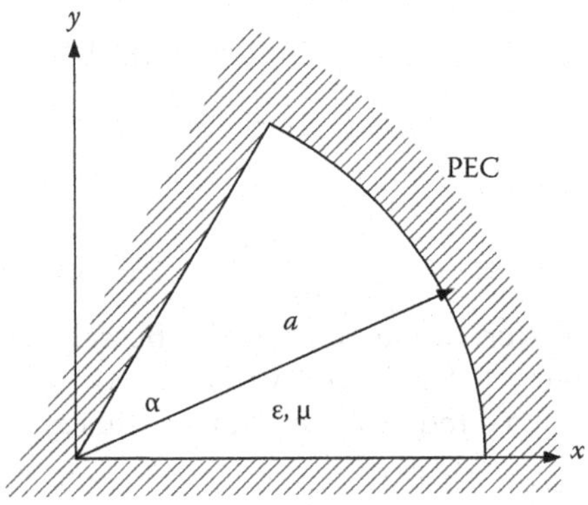

FIGURE 3.5
Sector waveguide with PEC boundaries.

3.10 Dielectric Cylindrical Waveguide: Optical Fiber

We have seen in Sections 3.7 and 3.8 that an electromagnetic wave is guided by a hollow cylindrical tube with a conducting boundary. The fields inside the guide were classified as TM modes ($E_z = 0$ on the boundary) or TE modes ($\partial H_z/\partial n = 0$ on the boundary). In either case, the fields outside the cylinder are zero. Can a cylindrical dielectric rod guide an electromagnetic wave? The answer is yes. A qualitative picture of such wave guiding as a surface wave has been alluded to before in connection with propagation of an electromagnetic wave from an optically dense to optically rare medium when the angle of incidence is greater than the critical angle. In this section, we examine the problem as a boundary value problem. Figure 3.6 shows the variation of the permittivity of a step index optical fiber in its cross section.

In contrast to the conducting guides, one notes here that the boundary condition for this case will be the continuity of the tangential electric and magnetic fields at $\rho = a$ for all ϕ and z:

$$E_{z1} = E_{z2}, \quad \rho = a, \tag{3.76}$$

$$H_{z1} = H_{z2}, \quad \rho = a, \tag{3.77}$$

$$E_{\phi 1} = E_{\phi 2}, \quad \rho = a, \tag{3.78}$$

$$H_{\phi 1} = H_{\phi 2}, \quad \rho = a. \tag{3.79}$$

This suggests to us that, in general, we may not be able to separate the modes into TE or TM modes. A general mode in this case may be a hybrid mode having E_z and H_z

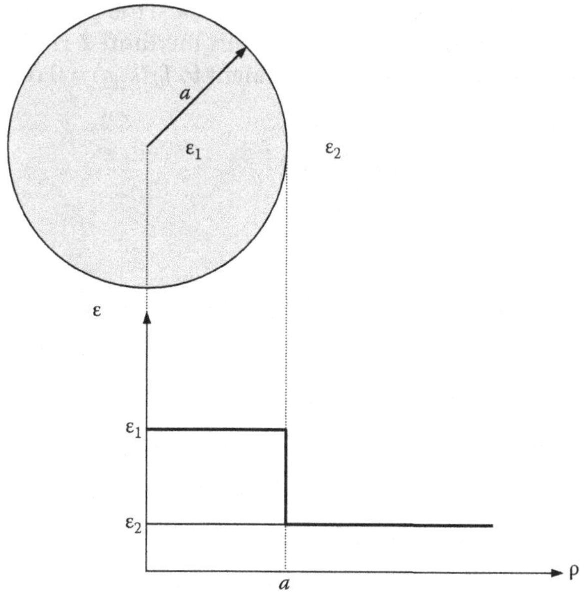

FIGURE 3.6
Step index optical fiber.

components. However, we can show that the optical fiber guide can have a TM_{0l} or TE_{0l} mode of propagation for the special case of $n = 0$. Let us look at the TM case in detail. Let the ρ variation of E_z be chosen as $J_0(k_{c1}\rho)$, for $0 < \rho < a$ and $K_0(k_{c2}\rho)$ for $a < \rho < \infty$. The choice of modified Bessel function K_0 in the latter range is to ensure that the amplitude of electromagnetic fields decay as ρ increases. This choice of the admissible function is in contrast to the choice of $H_0^{(2)}$ in Equation 2.121 where we needed an outgoing cylindrical wave at $\rho = \infty$. The formulation is given as follows:

$$E_{z1} = A_0 J_0\left(k_{c1}\rho\right)e^{-j\beta z}, \quad 0 < \rho < a, \tag{3.80}$$

$$E_{z2} = B_0 K_0\left(k_{c2}\rho\right)e^{-j\beta z}, \quad a < \rho < \infty, \tag{3.81}$$

where

$$k_{c1}^2 + \beta^2 = k_1^2 = \omega^2\mu_0\varepsilon_1, \tag{3.82}$$

$$-k_{c2}^2 + \beta^2 = k_2^2 = \omega^2\mu_0\varepsilon_2. \tag{3.83}$$

It can be shown [1] that the boundary condition (Equations 3.76 and 3.79) can be satisfied if the following transcendental equation is satisfied:

$$\frac{J_1\left(k_{c1}a\right)}{J_0\left(k_{c1}a\right)} = -\frac{\varepsilon_2}{\varepsilon_1}\frac{k_{c1}}{k_{c2}}\frac{K_1\left(k_{c2}a\right)}{K_0\left(k_{c2}a\right)}. \tag{3.84}$$

The transcendental equation has several solutions giving rise to several TM modes. Next, let us consider whether the concept of cutoff frequency holds in the dielectric guiding. The cut off occurs when $k_{c2} = 0$, since for negative values of k_{c2}^2, the modified Bessel function K_0 is no longer a monotonic function and the wave in medium 2 is no longer evanescent. Further consideration shows that $k_{c2} = 0$ is equivalent to $J_0\ (k_{c1}a) = 0$ or

$$\left(k_{c1}\right)_{0l} = \frac{\chi_{0l}}{a} \tag{3.85}$$

and

$$\left(k_{c1}\right)_{01} = \frac{\chi_{01}}{a} = \frac{2.4049}{a}. \tag{3.86}$$

Note that at cutoff

$$k_{c1}^2 = k_1^2 - \beta^2 = k_1^2 - \left(k_2^2 + k_{c2}^2\right) = k_1^2 - k_2^2. \tag{3.87}$$

For TE_{0l} modes, the eigenvalue equation is given by

$$\frac{J_1\left(k_{c1}a\right)}{J_0\left(k_{c1}a\right)} = -\frac{k_{c1}}{k_{c2}}\frac{K_1\left(k_{c2}a\right)}{K_0\left(k_{c2}a\right)}. \tag{3.88}$$

If $n \neq 0$, then the fields do not separate into TE and TM modes. All the fields become coupled through the continuity conditions. The modes are hybrid and have both E_z and H_z nonzero. The fields for the two regions may be written as [1]

$$
\begin{aligned}
&\rho < a, && \rho > a \\
&E_{z1} = AJ_n\left(u\frac{\rho}{a}\right)\begin{Bmatrix}\cos n\phi \\ \sin n\phi\end{Bmatrix}, && E_{z2} = CK_n\left(w\frac{\rho}{a}\right)\begin{Bmatrix}\cos n\phi \\ \sin n\phi\end{Bmatrix} \\
&H_{z1} = BJ_n\left(u\frac{\rho}{a}\right)\begin{Bmatrix}\cos n\phi \\ \sin n\phi\end{Bmatrix}, && H_{z2} = DK_n\left(w\frac{\rho}{a}\right)\begin{Bmatrix}\cos n\phi \\ \sin n\phi\end{Bmatrix} \\
&u^2 = \left(k_1^2 - \beta^2\right)a^2, && w^2 = \left(\beta^2 - k_2^2\right)a^2,
\end{aligned}
\tag{3.89}
$$

The transverse components may be obtained from Maxwell's equations. Applying the boundary conditions (Equations 3.76 through 3.79), one can obtain the eigenvalue equation [1]

$$
\left[\frac{k_1 J_n'(u)}{uJ_n(u)}\right]^2 + \left[\frac{k_2 K_n'(u)}{wK_n(u)}\right]^2 + \frac{k_1^2 + k_2^2}{uw}\left[\frac{J_n'(u)K_n'(w)}{J_n(u)K_n(w)}\right] = \frac{\beta^2 n^2 v^4}{u^4 w^4}, \tag{3.90}
$$

where

$$
v^2 = u^2 + w^2 = \left(k_1^2 - k_2^2\right)a^2. \tag{3.91}
$$

The solutions of the transcendental equation (3.90) may be designated as χ_{nl}. For $n \neq 0$, we get hybrid modes designated as HEM (hybrid electric and magnetic). The *HEM* modes with odd values of the second subscript are called *HE* modes, whereas *HEM* modes with even values for the second subscript are called *EH* modes. HE_{11} mode has no cutoff frequency and so is often called the dominant mode. From [2] and [3], a plot of β/k_0 versus $2a/\lambda_0$ for HE_{11} mode, TE_{01} mode, and TM_{01} mode for the case of $\varepsilon_{r1} = 2.56$, $\varepsilon_{r2} = 1$, $k_2 = k_0$, $\lambda_0 = 2\pi/k_0$ is obtained (see Figure 3.7). For $2a/\lambda_0 < 2.4049/\pi\sqrt{\varepsilon_r - 1}$, only the dominant HE_{11} mode exists.

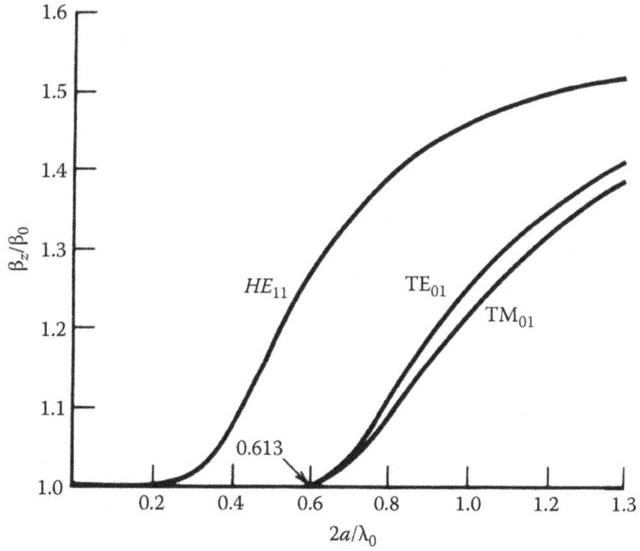

FIGURE 3.7
Ratio of β/k_0 for the first three surface-wave modes on a polystyrene rod ($\varepsilon_r = 2.26$). (Adapted from Collin, R.E., *Field Theory of Guided Waves*, McGraw-Hill, New York, 1960 [3].)

References

1. Ramo, S., Whinnery, J. R., and Van Duzer, T. V., *Fields and Waves in Communication Electronics*, 3rd edn., Wiley, New York, 1994.
2. Balanis, C. A., *Advanced Engineering Electromagnetics*, Wiley, New York, 1989.
3. Collin, R. E., *Field Theory of Guided Waves*, McGraw-Hill, New York, 1960.

4

Three-Dimensional Solutions*

Three-dimensional solutions involve all three spatial coordinates. The discussion of three-dimensional waves in Cartesian coordinates is fairly straightforward and will be discussed next through a cavity example.

4.1 Rectangular Cavity with PEC Boundaries: TM Modes

Consider a rectangular box of $a \times b \times h$ as shown in Figure 4.1.

This cavity can be constructed by taking a piece of a rectangular waveguide of length h and closing the box with end PEC plates at the ends $z = 0$ and h. For TM modes, the additional boundary conditions are

$$\tilde{E}_{\tan} = 0 \quad \text{at } z = 0 \text{ or } h. \tag{4.1}$$

This boundary condition is not on $\tilde{E}_z$, but Equation 4.1 is equivalent to (see Equation 2.24)

$$\frac{\partial \tilde{E}_z}{\partial z} = 0 \quad \text{at } z = 0 \text{ or } h. \tag{4.2}$$

It follows that

$$\begin{gathered} \tilde{E}_z(x, y, z) = E_{mnl} \sin\frac{m\pi x}{a} \sin\frac{n\pi y}{b} \cos\frac{l\pi z}{h}, \\ m = 1, 2, \ldots, \infty, \quad n = 1, 2, \ldots, \infty, \quad l = 0, 1, \ldots, \infty. \end{gathered} \tag{4.3}$$

In the above, the separation constants are $k_x = m\pi/a$, $k_y = n\pi/b$, and $k_z = l\pi/h$. Thus, we obtain

$$k^2 = \omega^2\mu\varepsilon = \left(\frac{m\pi}{a}\right)^2 + \left(\frac{n\pi}{b}\right)^2 + \left(\frac{l\pi}{h}\right)^2. \tag{4.4}$$

* For chapter appendices, see 4A in Appendices section.

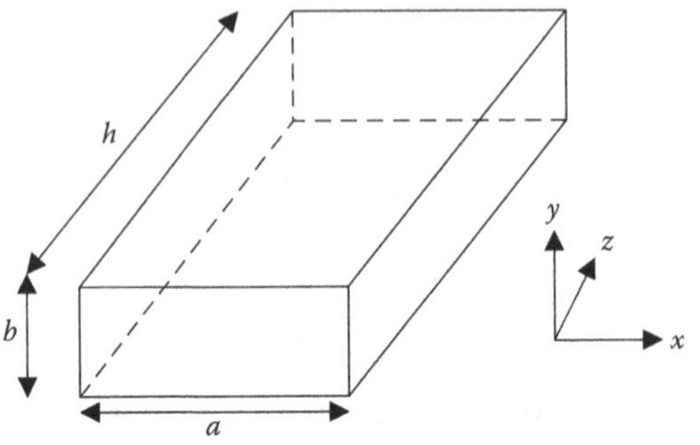

FIGURE 4.1
Rectangular cavity with PEC boundaries. Positive z-direction is opposite to the direction shown in the figure for the right-handed Cartesian coordinate system.

When the size of the cavity is fixed, the frequency ω in Equation 4.4 is dependent on m, n, l, μ, and ε. For a given mode and a specified medium, the frequency ω is a definite value, labeled as the resonant frequency of the cavity. Its value (in Hz) is given by

$$(f_r)_{\mathrm{TM}mnl} = \frac{1}{2\pi\sqrt{\mu\varepsilon}}\left[\left(\frac{m\pi}{a}\right)^2 + \left(\frac{n\pi}{b}\right)^2 + \left(\frac{l\pi}{h}\right)^2\right]^{1/2} \tag{4.5}$$

$$m = 1,2,\ldots,\infty, \quad n = 1,2,\ldots,\infty, \quad l = 0,1,\ldots,\infty.$$

The lowest TM resonant frequency is given by

$$(f_r)_{\mathrm{TM}110} = \frac{1}{2\pi\sqrt{\mu\varepsilon}}\left[\left(\frac{\pi}{a}\right)^2 + \left(\frac{\pi}{b}\right)^2\right]^{1/2}. \tag{4.6}$$

4.2 Rectangular Cavity with PEC Boundaries: TE Modes

The additional boundary conditions to be satisfied in this case are

$$\tilde{H}_z = 0, \quad z = 0 \text{ or } h. \tag{4.7}$$

The expression for $\tilde{H}_z$ is now easily obtained:

$$\tilde{H}_z = H_{mnl}\cos\frac{m\pi x}{a}\cos\frac{n\pi y}{b}\sin\frac{l\pi z}{h} \tag{4.8}$$

$$m = 0,1,2,\ldots,\infty, \quad n = 0,1,2,\ldots,\infty, \quad l = 1,2,\ldots,\infty,$$

but $m = n = 0$ is excluded.

The resonant frequency can now be obtained as

$$(f_r)_{\mathrm{TE}mnl} = \frac{1}{2\pi\sqrt{\mu\varepsilon}}\left[\left(\frac{m\pi}{a}\right)^2 + \left(\frac{n\pi}{b}\right)^2 + \left(\frac{l\pi}{h}\right)^2\right]^{1/2} \tag{4.9}$$

$$m = 0,1,2,\ldots,\infty, \quad n = 0,1,2,\ldots,\infty, \quad l = 1,2,\ldots,\infty,$$

but $m = n = 0$ is excluded.

If $h > a > b$, then the lowest TE resonant frequency is given by

$$(f_\mathrm{r})_{\mathrm{TE}_{101}} = \frac{1}{2\pi\sqrt{\mu\varepsilon}}\left[\left(\frac{\pi}{a}\right)^2 + \left(\frac{\pi}{h}\right)^2\right]^{1/2}, \quad h > a > b. \tag{4.10}$$

The resonant frequency given by Equation 4.10 is lower than that given by Equation 4.6 if $h > a > b$.

4.3 Q of a Cavity

The waveguide and cavity problems assumed PEC boundary conditions. After obtaining the fields, we can relax the ideal assumptions by calculating the losses when the walls are not perfect [1]. We illustrate the technique for TE_{101} mode. It is convenient to write the fields in the following forms:

$$\tilde{E}_y = E_0 \sin\frac{\pi x}{a}\sin\frac{\pi z}{h}, \tag{4.11}$$

$$\tilde{H}_x = -\mathrm{j}\frac{E_0}{\eta}\frac{\lambda}{2h}\sin\frac{\pi x}{a}\cos\frac{\pi z}{h}, \tag{4.12}$$

$$\tilde{H}_z = \mathrm{j}\frac{E_0}{\eta}\frac{\lambda}{2a}\cos\frac{\pi x}{a}\sin\frac{\pi z}{h}. \tag{4.13}$$

We obtain the power loss in the walls by calculating the surface currents. The surface currents are obtained from the magnetic fields:

Front:	$\tilde{K}_y = -\tilde{H}_x\big\rvert_{z=h}$,	Back:	$\tilde{K}_y = \tilde{H}_x\big\rvert_{z=0}$,
Left side:	$\tilde{K}_y = -\tilde{H}_z\big\rvert_{x=0}$,	Right side:	$\tilde{K}_y = \tilde{H}_z\big\rvert_{x=a}$,
Top:	$\tilde{K}_x = -\tilde{H}_z\big\rvert_{y=b}$, $\tilde{K}_z = -\tilde{H}_x\big\rvert_{y=b}$,	Bottom:	$\tilde{K}_x = \tilde{H}_z\big\rvert_{y=0}$, $\tilde{K}_z = -\tilde{H}_x\big\rvert_{y=0}$,

If the conducting walls have surface resistance R_s, then the losses are given by

$$W_\mathrm{L} = \frac{R_\mathrm{s}}{2}\left\{2\int_0^b\!\!\int_0^a \left|\tilde{H}_x\right|^2_{z=0} \mathrm{d}x\mathrm{d}y + 2\int_0^h\!\!\int_0^b \left|\tilde{H}_z\right|^2_{x=0} \mathrm{d}y\mathrm{d}z + 2\int_0^h\!\!\int_0^a \left[\left|\tilde{H}_x\right|^2 + \left|\tilde{H}_z\right|^2\right]\mathrm{d}x\mathrm{d}z\right\}, \tag{4.14}$$

$$W_\mathrm{L} = \frac{R_\mathrm{s}\lambda^2}{8\eta^2}E_0^2\left[\frac{ab}{h^2} + \frac{bh}{a^2} + \frac{1}{2}\left(\frac{a}{h} + \frac{h}{a}\right)\right]. \tag{4.15}$$

If the losses are neglected, then the energy in the cavity passes between electric and magnetic fields, and we may calculate the total energy in the cavity by finding the energy storage at the instant when it is maximum:

$$U = (U_E)_{MAX} = \frac{\varepsilon}{2}\int_0^h\int_0^b\int_0^a \left|\tilde{E}_y\right|^2 dxdydz = \frac{\varepsilon abh}{8}E_0^2. \quad (4.16)$$

The quality factor Q of a cavity is a quantitative measure of how well the cavity is acting as a resonator. It is defined by

$$Q = \frac{\omega_r U}{W_L}. \quad (4.17)$$

From Equations 4.14, 4.16, and 4.17, we obtain

$$Q = \frac{\pi\eta}{4R_s}\left[\frac{2b\left(a^2+h^2\right)^{3/2}}{ah\left(a^2+h^2\right)+2b\left(a^3+h^3\right)}\right], \quad (4.18)$$

which for a cube reduces to

$$Q = 0.742\frac{\eta}{R_s}. \quad (4.19)$$

For an air dielectric $\eta = 377$, and for a copper conductor at 10 GHz, $R_s \approx 0.0261$, giving a quality factor $Q = 10{,}730$. Such a large value of Q cannot be obtained by lumped circuits or even with resonant lines.

Reference

1. Ramo, S., Whinnery, J. R., and Van Duzer, T. V., *Fields and Waves in Communication Electronics*, 3rd edn., Wiley, New York, 1994.

5

Spherical Waves and Applications

So far, we have discussed plane waves and cylindrical waves. In both of these cases, we could develop with relative ease the solutions based on TM^z and TE^z modes. We illustrated these solutions by solving some waveguides and cavity problems. The superscript z is added here to indicate that the transverse plane is perpendicular to the z-coordinates. The structures under consideration had a uniform cross section for all values of z. In spherical coordinates, we do not have any such transverse *planes*, and thus spherical wave problems are mathematically more involved. It is still possible to have mode classification as TM^r and TE^r.

The solution in spherical coordinates involves special functions: spherical Bessel functions and associated Legendre functions. Spherical Bessel functions are related to half-integral Bessel functions.

5.1 Half-Integral Bessel Functions

In connection with the solution for a scalar Helmholtz equation in spherical coordinates, we come across the differential equation

$$u\frac{d}{du}\left(u\frac{df}{du}\right)+\left[u^2-\left(n+\frac{1}{2}\right)^2\right]f=0, \tag{5.1}$$

which, when expanded, gives

$$\frac{d^2f}{du^2}+\frac{1}{u}\frac{df}{du}+\left[1-\frac{(n+1/2)^2}{u^2}\right]f=0. \tag{5.2}$$

By comparing with Equation 2.120, we immediately realize that f can be written as a linear combination of half-integral Bessel functions:

$$f=\begin{Bmatrix} J_{n+1/2}(u) \\ Y_{n+1/2}(u) \\ H^{(1)}_{n+1/2}(u) \\ H^{(2)}_{n+1/2}(u) \\ I_{n+1/2}(u) \\ K_{n+1/2}(u) \end{Bmatrix}. \tag{5.3}$$

TABLE 5.1

Zeros of Half-Integral Bessel Functions $\chi_{n+1/2,l}$

$J_{1/2}$	$J_{3/2}$	$J_{5/2}$
3.1416	4.4934	5.7635
6.2832	7.7253	9.0950

TABLE 5.2

Zeros of Derivatives of Half-Integral Bessel Functions $\chi'_{n+1/2,l}$

$J'_{1/2}$	$J'_{3/2}$	$J'_{5/2}$
1.1656	2.4605	3.6328
4.6042	6.0293	7.3670

Half-integral Bessel functions can be shown to reduce to simpler forms:

$$J_{n+1/2}(x)=\sqrt{\frac{2}{\pi}}x^{n+1/2}\left(-\frac{1}{x}\right)^n\frac{\mathrm{d}^n}{\mathrm{d}x^n}\left(\frac{\sin x}{x}\right), \tag{5.4}$$

$$Y_{n+1/2}(x)=(-1)^{n+1}\sqrt{\frac{2}{\pi}}x^{n+1/2}\left(\frac{1}{x}\right)^n\frac{\mathrm{d}^n}{\mathrm{d}x^n}\left(\frac{\cos x}{x}\right). \tag{5.5}$$

In particular, we can show from Equation 5.4 that

$$J_{1/2}(x)=\sqrt{\frac{2}{\pi x}}\sin x, \tag{5.6}$$

$$J_{3/2}(x)=\sqrt{\frac{2}{\pi x}}\left(\left(\frac{\sin x}{x}\right)-\cos x\right). \tag{5.7}$$

From Equations 5.6 and 5.7, we can obtain the zeros of half-integral Bessel functions and their derivatives. These are listed in Tables 5.1 and 5.2.

5.2 Solutions of Scalar Helmholtz Equation

The scalar Helmholtz equations

$$\nabla^2F+k^2F=0 \tag{5.8}$$

and

$$\frac{1}{r^2}\frac{\partial}{\partial r}\left(r^2\frac{\partial F}{\partial r}\right)+\frac{1}{r^2\sin\theta}\frac{\partial}{\partial\theta}\left(\sin\theta\frac{\partial F}{\partial\theta}\right)+\frac{1}{r^2\sin^2\theta}\frac{\partial^2F}{\partial\phi^2}+k^2F=0$$

can be solved in spherical coordinates by the usual technique of separation of variables. Let

$$F = (r,\theta,\phi) = f_1(r) f_2(\theta) f_3(\phi). \tag{5.9}$$

Substituting Equations 5.9 into 5.8 and manipulating the equation, we obtain the ordinary differential equations for f_1, f_2, and f_3 involving two separation constants m and n:

$$\frac{d}{dr}\left(\frac{r^2 df_1}{dr}\right) + \left[k^2 r^2 - n(n+1)\right] f_1 = 0, \tag{5.10}$$

$$\frac{1}{\sin\theta}\frac{d}{d\theta}\left(\frac{\sin\theta\, df_2}{d\theta}\right) + \left[n(n+1) - \frac{m^2}{\sin^2\theta}\right] f_2 = 0, \tag{5.11}$$

$$\frac{d^2 f_3}{d\phi^2} + m^2 f_3 = 0. \tag{5.12}$$

By substituting

$$r = \frac{u}{k} \quad \text{and} \quad f_1 = \frac{f}{\sqrt{r}} = \sqrt{\frac{k}{u}} f \tag{5.13}$$

in Equation 5.10, we obtain Equation 5.2, with f thus given by a linear combination of half-integral Bessel functions given by Equation 5.3. The solution for f_1 is given by

$$f_1(r) = \frac{f(u)}{\sqrt{r}} = \frac{f(kr)}{\sqrt{r}}. \tag{5.14}$$

Defining the spherical Bessel function b_n in terms of the half-integral Bessel functions $B_{n+1/2}$, we obtain

$$b_n(kr) = \sqrt{\frac{\pi}{2kr}} B_{n+1/2}(kr), \tag{5.15}$$

where
b_n stands for any of the spherical Bessel functions j_n, y_n, $h_n^{(1)}$, and $h_n^{(2)}$
$B_{n+1/2}$ stands for the half–integral Bessel functions, $J_{n+1/2}$, $Y_{n+1/2}$, $H_{n+1/2}^{(1)}$, and $H_{n+1/2}^{(2)}$

Equation 5.11 is called the associate Legendre equation whose solutions are linear combinations of associate Legendre polynomials of the first kind (P_n^m) and the second kind (Q_n^m) of degree n and order m. These polynomials are given by

$$P_n^m(x) = (x^2 - 1)^{m/2} \frac{d^m}{dx^m} P_n(x), \tag{5.16}$$

$$P_n(x) = \frac{1}{2^n n!}\frac{\mathrm{d}^n}{\mathrm{d}x^n}(x^2-1)^n, \tag{5.17}$$

$$Q_n^m(x) = (x^2-1)^{m/2}\frac{\mathrm{d}^m}{\mathrm{d}x^m}Q_n(x), \tag{5.18}$$

$$Q_n(x) = \frac{1}{2}P_n(x)\ln\frac{1+x}{1-x} - \sum_{l=1}^{n}\frac{1}{l}P_{l-1}(x)P_{n-1}(x). \tag{5.19}$$

Associated Legendre polynomials of zero order ($m = 0$) are called Legendre polynomials. In Equations 5.16 through 5.19, $x = \cos\theta$. Note that the order m is an integer between 0 and n. Plots and expressions for a few Legendre polynomials are given in Figure 5.1 and Table 5.3. Note that on the polar axis, that is, $\theta = 0$ or π, $Q_n \to \infty$. So, for the problem that includes a positive or negative polar axis in the field region, the function P_n^m alone is the suitable solution. The solution for Equation 5.12 is a linear combination of $\cos m\phi$ and $\sin m\phi$. Putting together the solutions for f_1, f_2, and f_3, the solution for Equation 5.8 may be written as

$$F(r,\theta,\phi) = \begin{Bmatrix} j_n(kr) \\ y_n(kr) \\ h_n^{(1)}(kr) \\ h_n^{(1)}(kr) \end{Bmatrix} \begin{Bmatrix} P_n^m(\cos\theta) \\ Q_n^m(\cos\theta) \end{Bmatrix} \begin{Bmatrix} \cos m\phi \\ \sin m\phi \end{Bmatrix}. \tag{5.20}$$

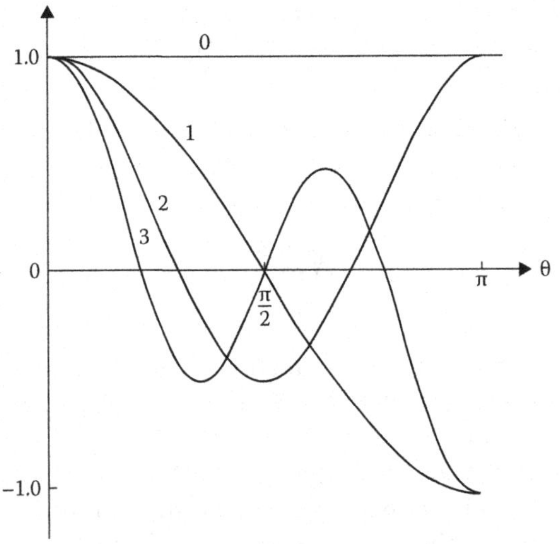

FIGURE 5.1
Polynomials $P_n(\cos\theta)$. The degree n is indicated on the graph.

TABLE 5.3

Legendre Functions

$P_0(\cos\theta) = 1$
$P_1(\cos\theta) = \cos\theta$
$P_2(\cos\theta) = \frac{1}{2}(3\cos^2\theta - 1)$
$P_3(\cos\theta) = \frac{1}{2}(5\cos^2\theta - 3\cos\theta)$
$\int_0^a P_n(\cos\theta) P_q(\cos\theta)\sin\theta d\theta = 0, \quad n \neq q$
$\int_0^a [P_n(\cos\theta)]^2 \sin\theta d\theta = \frac{2}{2n+1}$
$Q_0(\cos\theta) = \frac{1}{2}\ln\frac{1+\cos\theta}{1-\cos\theta}$
$Q_1(\cos\theta) = \frac{\cos\theta}{2}\ln\frac{1+\cos\theta}{1-\cos\theta} - 1$
$Q_2(\cos\theta) = \frac{3\cos^2\theta - 1}{4}\ln\frac{1+\cos\theta}{1-\cos\theta} - \frac{3\cos\theta}{2}$
$P_1^1(\cos\theta) = (1-\cos^2\theta)^{1/2}$
$P_2^1(\cos\theta) = 3\cos\theta s(1-\cos^2\theta)^{1/2}$
$P_2^2(\cos\theta) = 3(1-\cos^2\theta)$
$\int_0^a P_n^m(\cos\theta) P_q^m(\cos\theta)\sin\theta d\theta = 0, \quad n \neq q$
$\int_0^a [P_n^m(\cos\theta)]^2 \sin\theta d\theta = \frac{2}{2n+1}\frac{(n+m)!}{(n-m)!}$

5.3 Vector Helmholtz Equation

The electric and magnetic fields in a sourceless region satisfy the vector Helmholtz equations

$$\nabla \times \nabla \times \tilde{E} - k^2\tilde{E} = \nabla^2\tilde{E} + k^2\tilde{E} = 0, \tag{5.21}$$

$$\nabla \times \nabla \times \tilde{H} - k^2\tilde{H} = \nabla^2\tilde{H} + k^2\tilde{H} = 0. \tag{5.22}$$

Each Cartesian component of $\tilde{E}$ and $\tilde{H}$ satisfies the scalar Helmholtz equation (5.8), but the spherical components do not satisfy Equation 5.8.

It is convenient once more to think of simple solutions of these equations in terms of TE^r, TM^r, and TEM^r modes in the context of spherical geometry.

5.4 TM^r Modes

These modes may be constructed by considering a vector field

$$\tilde{M} = \nabla \times \left[\hat{r} r \tilde{F}_e \right], \tag{5.23}$$

where $\tilde{F}_e$ satisfies Equation 5.8. It can be shown that $\tilde{M}$ satisfies the vector Helmholtz equation. Identifying $\tilde{M}$ with $\tilde{H}$ and noting from Equation 5.23 that

$$H_r = 0, \tag{5.24}$$

$$\tilde{H}_\theta = \frac{1}{\sin\theta} \frac{\partial F_e}{\partial \phi}, \tag{5.25}$$

$$\tilde{H}_\phi = -\frac{\partial F_e}{\partial \theta}, \tag{5.26}$$

we recognize all the above as the magnetic field components of TM^r modes in terms of the electric Debye potential $\tilde{F}_e$. The corresponding electric field components can be obtained from Maxwell's equation

$$\nabla \times \tilde{H} = j\omega\varepsilon\tilde{E} \tag{5.27}$$

and are given as

$$\tilde{E}_r = -\frac{j}{\omega\varepsilon} \left[\frac{\partial^2}{\partial r^2} r\tilde{F}_e + k^2 r\tilde{F}_e \right], \tag{5.28}$$

$$\tilde{E}_\theta = -\frac{j}{\omega\varepsilon} \frac{1}{r} \frac{\partial^2}{\partial r \partial \theta} r\tilde{F}_e, \tag{5.29}$$

$$\tilde{E}_\phi = -\frac{j}{\omega\varepsilon} \frac{1}{r\sin\theta} \frac{\partial^2}{\partial r \partial \phi} r\tilde{F}_e. \tag{5.30}$$

5.5 TE^r Modes

These modes may be constructed by considering the vector function

$$\tilde{N} = \nabla \times \left[\hat{r} r \tilde{F}_m \right], \tag{5.31}$$

where $\tilde{F}_m$ satisfies Equation 5.8. It can be shown that $\tilde{N}$ satisfies the vector Helmholtz equation. Identifying $\tilde{N}$ with the $\tilde{E}$ field and noting from Equation 5.31 that

$$\tilde{E}_r = 0, \tag{5.32}$$

$$\tilde{E}_\theta = \frac{1}{\sin\theta}\frac{\partial F_m}{\partial\phi}, \tag{5.33}$$

$$\tilde{E}_\phi = -\frac{\partial F_m}{\partial\theta}, \tag{5.34}$$

we recognize the above as the electric field expressions of TE^r modes. The corresponding magnetic field components can be obtained from Maxwell's equation

$$\nabla\times\tilde{E} = -j\omega\mu\tilde{H} \tag{5.35}$$

and are given by

$$\tilde{H}_r = \frac{j}{\omega\mu}\left[\frac{\partial^2}{\partial r^2}r\tilde{F}_m + k^2 r\tilde{F}_m\right], \tag{5.36}$$

$$\tilde{H}_\theta = \frac{j}{\omega\mu}\frac{1}{r}\frac{\partial^2}{\partial r\partial\theta}r\tilde{F}_m, \tag{5.37}$$

$$\tilde{H}_\phi = \frac{j}{\omega\mu}\frac{1}{r\sin\theta}\frac{\partial^2}{\partial r\partial\phi}r\tilde{F}_m. \tag{5.38}$$

5.6 Spherical Cavity

We illustrate the construction of solutions in spherical coordinates by determining the resonant frequencies of a spherical cavity with PEC boundary conditions at $r = a$ (Figure 5.2).

For *TM* modes $\tilde{F}_e$ is given by Equation 5.20.

Since $\theta = 0$ or π are part of the field region, we choose P_n^m for θ variation. Since we need an oscillatory function that is finite at the origin, we choose $j_n(kr)$ for n variation.

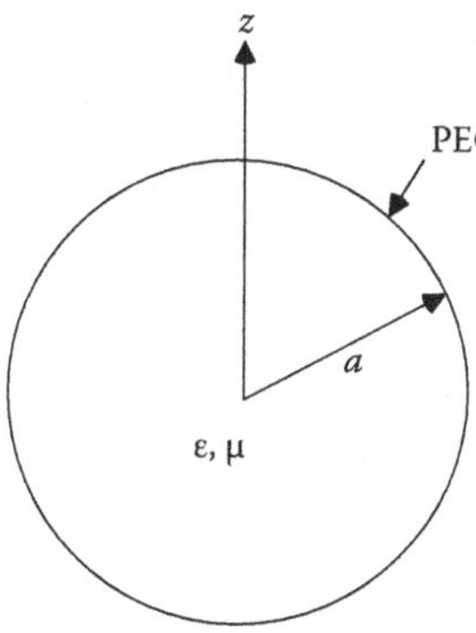

FIGURE 5.2
Spherical cavity with PEC boundary.

The r variation of rF_e is given by

$$rj_n(kr) = r\sqrt{\frac{\pi}{2kr}}J_{n+1/2}(kr) = \sqrt{r}\sqrt{\frac{\pi}{2k}}J_{n+1/2}(kr). \tag{5.39}$$

To suit such applications as above, Schelkunoff defined another set of spherical Bessel functions:

$$\hat{B}_n(x) = \sqrt{\frac{\pi x}{2}}B_{n+1/2}(x). \tag{5.40}$$

The roots of $\hat{J}_n(\zeta)$ are given in Table 5.4.

We may as well now list in Table 5.5 the roots of $\hat{J}'_n(y) = 0$.

The PEC boundary conditions are $\tilde{E}_\theta = 0$ and $\tilde{E}_\phi = 0$ at $r = a$, which leads to the boundary condition

$$\frac{\partial}{\partial r}\left(r\tilde{F}_e\right)\bigg|_{r=a} = 0, \quad \text{that is, } \hat{J}'_n(ka) = 0, \tag{5.41}$$

$$(k)^{TM^r}_{nmp} = \frac{\zeta'_{np}}{a}, \tag{5.42}$$

$$(f_r)^{TM^r}_{nmp} = \frac{1}{2\pi\sqrt{\mu\varepsilon}}\frac{\zeta'_{np}}{a}, \quad n,p = 1,2,\ldots,\infty. \tag{5.43}$$

Note that n cannot be zero since

$$P_0^0 = 1. \tag{5.44}$$

TABLE 5.4

Zeros of $\hat{J}_n$

ζ_{np}	$n = 1$	$n = 2$	$n = 3$
$p = 1$	4.493	5.763	6.988
$p = 2$	7.725	9.095	10.417
$p = 3$	10.904	12.323	13.698

TABLE 5.5

Zeros of $\hat{J}'_n$

ζ'_{np}	$n = 1$	$n = 2$	$n = 3$
$p = 1$	2.744	3.87	4.973
$p = 2$	6.117	7.443	8.722
$p = 3$	9.317	10.713	12.064

And from Equations 5.29 and 5.30, $\tilde{E}_\theta$ and $\tilde{E}_\phi$ are zero not only on the boundary but also everywhere inside the cavity. Thus, $n = 0$ gives a trivial solution. The lowest resonant frequency of *TM* modes is given by

$$(f_r)_{101}^{TM^r} = \frac{1}{2\pi\sqrt{\mu\varepsilon}}\frac{\zeta_{11}'}{a} = \frac{2.744}{2\pi\sqrt{\mu\varepsilon}a}. \tag{5.45}$$

Let us next consider the TE^r modes. The boundary conditions in this case, from Equations 5.33 and 5.34, are

$$\left.\frac{\partial F_m}{\partial\theta}\right|_{r=a} = 0, \tag{5.46}$$

$$\left.\frac{\partial F_m}{\partial\phi}\right|_{r=a} = 0, \tag{5.47}$$

which translate to

$$\hat{J}_n(ka) = 0, \tag{5.48}$$

$$(k)_{nmp}^{TE^r} = \zeta_{np}, \tag{5.49}$$

$$(f_r)_{nmp}^{TE^r} = \frac{1}{2\pi\sqrt{\mu\varepsilon}}\frac{\zeta_{np}}{a}. \tag{5.50}$$

The lowest value of the above occurs when $n = 1$ and $p = 1$,

$$(f_r)_{101}^{TE^r} = \frac{1}{2\pi\sqrt{\mu\varepsilon}}\frac{4.493}{a}. \tag{5.51}$$

The lowest resonant frequency of all the modes is thus given by Equation 5.45. The modes in a spherical cavity are highly degenerate, that is, for the same resonant frequency we can have many different field distributions.

6

*Laplace Equation: Static and Low-Frequency Approximations**

In an ideal transmission line, the voltage phasor $\tilde{V}$ satisfies the ordinary differential equation (2.54):

$$\frac{\partial^2 \tilde{V}}{\partial z^2} + \beta^2 \tilde{V} = 0,$$

where

$$\beta^2 = \omega^2 L'C' = \frac{\omega^2}{v^2}.$$

The time-harmonic equation in three dimensions (Helmholtz equation) is given by

$$\nabla^2 \tilde{\mathbf{E}} + k^2 \tilde{\mathbf{E}} = 0,$$

where

$$k^2 = \frac{\omega^2}{v^2} = \omega^2 \mu\varepsilon$$

and ∇^2 is the Laplacian operator.

If the frequency is zero ($f = 0$) or low such that $\beta^2 = k^2 \approx 0$, the Helmholtz equation can be approximated by the Laplace equation.

Static or low-frequency problems satisfy the Laplace equation

$$\nabla^2 \Phi = 0, \tag{6.1}$$

where Φ is called the potential. In the first undergraduate course in electromagnetics, the one-dimensional solution of the Laplace equation in various coordinate systems is discussed and illustrated through simple examples. In the following section, we list these solutions for completeness.

* For chapter appendices, see 6A in Appendices section.

6.1 One-Dimensional Solutions

Cartesian coordinates:

$$\Phi(x) = A_1 x + A_2. \tag{6.2}$$

Cylindrical coordinates:

$$\Phi(\rho) = A_1 \ln \rho + A_2, \tag{6.3}$$

$$\Phi(\phi) = A_1 \phi + A_2. \tag{6.4}$$

Spherical coordinates:

$$\Phi(r) = \frac{A_1}{r} + A_2, \tag{6.5}$$

$$\Phi(\theta) = A_1 \ln\big(\tan(\theta/2)\big) + A_2. \tag{6.6}$$

6.2 Two-Dimensional Solutions

6.2.1 Cartesian Coordinates

In Chapter 3, we solved the waveguide problem by solving the Helmholtz equation

$$\frac{\partial^2 F}{\partial x^2} + \frac{\partial^2 F}{\partial y^2} + k_c^2 F = 0 \tag{6.7}$$

by assuming

$$F = f_1(x) f_2(y), \tag{6.8}$$

where $f_1(x)$ and $f_2(y)$ are given in the form of a template, given, respectively, by Equations 3.15 and 3.16, where $k_c^2 = k_x^2 + k_y^2$. In the case of the Laplace equation, $k_c^2 = 0$, giving $k_x^2 = -k_y^2$.

The implication is that if one of the first four elements in Equation 3.15 is chosen for the x-variation, then one of the last four elements in Equation 3.16 has to be chosen for the y-variation, and $|k_x| = |K_y|$. The choice of $f_1(x)$ and $f_2(y)$ can be reversed, if the boundary conditions require such a choice. The point is illustrated through an example.

Example 6.1: Figure 6.1 shows the geometry for the example.

Let the boundary conditions be

B.C.1 $\Phi = 0, \quad y = 0,$

B.C.2 $\Phi = 0, \quad y = b,$

B.C.3 $\Phi = 0, \quad x = 0,$

B.C.4 $\Phi = f(y) = 100 \sin \dfrac{3\pi y}{b}, \quad x = a.$

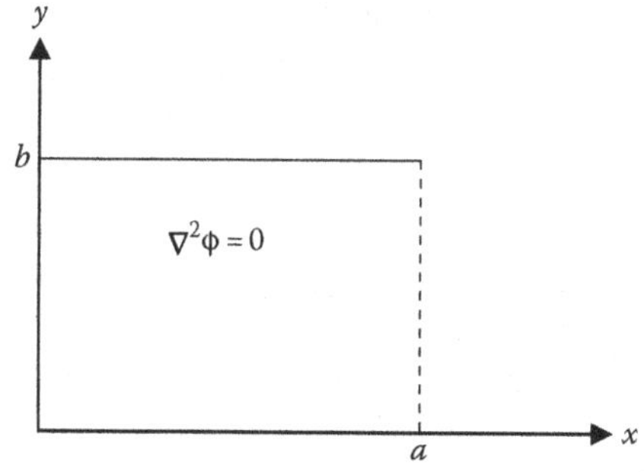

FIGURE 6.1
Geometry for Example 6.1.

Solution

From B.C.1 and B.C.2, the y-variation is

$$f_1(y) \sim \sin\frac{n\pi y}{b}, \quad n = 1, 2, \ldots, \infty.$$

From B.C.3, the x-variation has to be

$$f_1(x) \sim \sinh\frac{n\pi x}{b}, \quad n = 1, 2, \ldots, \infty.$$

B.C.4 is satisfied if $n = 3$. Thus,

$$\Phi = A_3 \sinh\frac{3\pi x}{b} \sin\frac{3\pi y}{b}.$$

From B.C.4,

$$100 \sin\frac{3\pi y}{b} = A_3 \sinh\frac{3\pi a}{b} \sin\frac{3\pi y}{b},$$

$$\therefore A_3 = \frac{100}{\sinh(3\pi a/b)},$$

$$\Phi = 100\frac{\sinh(3\pi x/b)}{\sinh(3\pi a/b)} \sin\frac{3\pi y}{b}. \tag{6.9}$$

A little thought would show that we could have written the solution (Equation 6.9) by inspection. The factor $\sinh(3\pi a/b)$ in the denominator neutralizes the value of $\sinh(3\pi x/b)$, when $x = a$ is substituted in the process of implementing B.C.4. The constant 100 is the Fourier coefficient of the third harmonic. B.C.4 requires only the third harmonic of the Fourier series. All other Fourier coefficients are zero if we considered the expansion of $f(y) = 100\sin(3\pi y/b)$ in Fourier series. Suppose $f(y)$ is a more general function than given in B.C.4. Obviously, the Fourier coefficients of other harmonic can be nonzero and

$$\Phi = f(y) = \sum_{n=1}^{\infty} B_n \sin\frac{3\pi y}{b}. \tag{6.10}$$

Let us illustrate such Fourier series solution through Example 6.2.

Example 6.2: Same as Example 6.1, except that

$$\text{B.C.4}\ \Phi = f(y) = 100.$$

Solution

The boundary at $x = a$ is a conductor of constant voltage of 100 V.

We have to determine the Fourier coefficient B_n when $f(y)$ is defined to be a constant in the interval $0 < y < b$ (Figure 6.2).

We will make two interesting points about the Fourier expansion being sought in this problem. The Fourier series is for a periodic function. However, the function $f(y)$ is only defined over half of the period of the fundamental $\sin(\pi y/b)$. The full period of the fundamental is $2b$. Since the desired Fourier series is to have only sine terms, the function has to be an odd function. The function can be continued for the full period, from $(-b)$ to $(+b)$, as shown in Figure 6.3.

Note that the solution is valid only in the field region $0 < x < a$ and $0 < y < b$; the continuation of $f(y)$ in the region beyond $0 < y < b$, is immaterial to the validity of the solution in the field region; however, the condition is important in determining the Fourier coefficients [1].

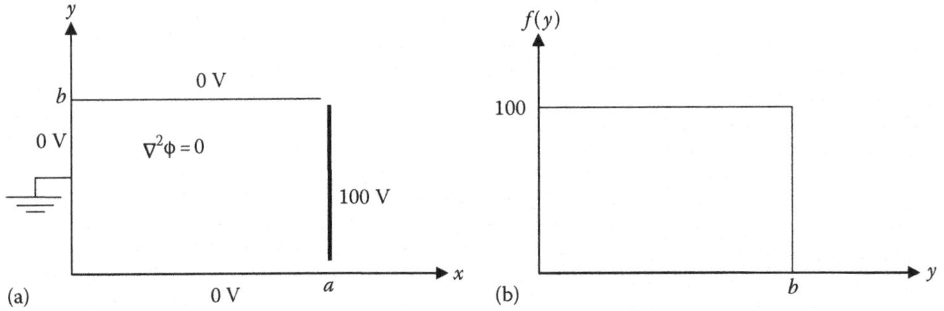

FIGURE 6.2
(a) Geometry for Example 6.2 and (b) function $f(y)$ in B.C.4.

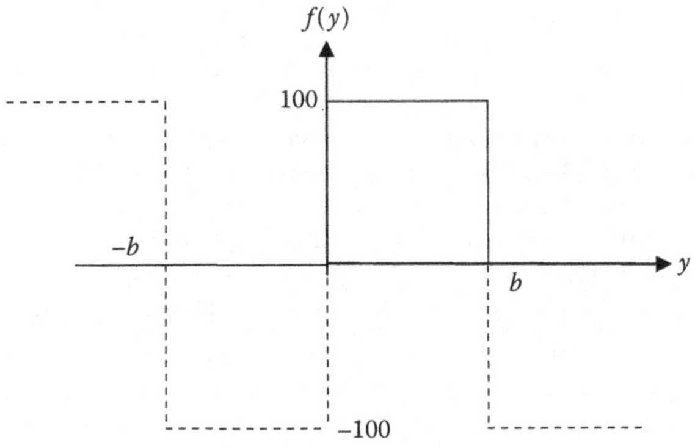

FIGURE 6.3
Periodic continuation of $f(y)$ of Example 6.2 for having only sine terms in the Fourier series expansion.

For the problem on hand,

$$B_n = \frac{2}{\text{period}} \int_{\text{period}} f(y) \sin \frac{n\pi y}{b} dy,$$

$$B_n = \frac{2}{2b} \int_{-b}^{b} f(y) \sin \frac{n\pi y}{b} dy. \tag{6.11}$$

Since $f(y)$ is an odd function of y and $\sin(n\pi y/b)$ is an odd function of y, the product is an even function and evaluating for the given $f(y)$:

$$B_n = \frac{2}{b} \int_{0}^{b} f(y) \sin \frac{n\pi y}{b} dy,$$

$$B_n = \frac{200}{b} \frac{b}{n\pi} \cos \frac{n\pi y}{b} \bigg|_0^b,$$

$$B_n = \begin{cases} \dfrac{400}{n\pi}, & n \text{ odd}, \\ 0, & n \text{ even}. \end{cases} \tag{6.12}$$

Thus,

$$\Phi(x, y) = \sum_{n=1,\text{odd}}^{\infty} \frac{400}{n\pi} \frac{\sinh(n\pi x/b)}{\sinh(n\pi a/b)} \sin\left(\frac{n\pi y}{b}\right), \quad \begin{matrix} 0 < x < a, \\ 0 < y < b. \end{matrix} \tag{6.13}$$

Example 6.2 brings out several interesting points.

The boundary conditions 1, 2, and 3 dictated the type of the restricted Fourier series expansion, in the example sine series only, and the continuation of the periodic function, beyond the original domain, can be constructed to achieve the restricted Fourier series. In Appendix 6A, we examine the possible periodic continuation of a function $x(t)$, defined over a finite range $0 < t < t_1$, so that the Fourier series will have

A. odd harmonic only;
B. sine terms only (Example 6.2);
C. cosine and odd harmonic only;
D. sine and odd harmonic only.

In subsequent sections, we will be discussing the expansion of an arbitrary function defined over a finite interval in Bessel series and other orthonormal functions.

Example 6.3: Determine the potential in the field region described by the geometry given in Figure 6.4.

Solution

Let us write the solution for the problem by inspection, based on the experience we gained in solving Examples 6.1 and 6.2:

$$\Phi(x, y) = \sum_{m=1,\text{odd}}^{\infty} \frac{400}{m\pi} \frac{\sinh\left(m\pi(b-y)/a\right)}{\sinh(m\pi b/a)} \sin \frac{m\pi x}{a}. \tag{6.14}$$

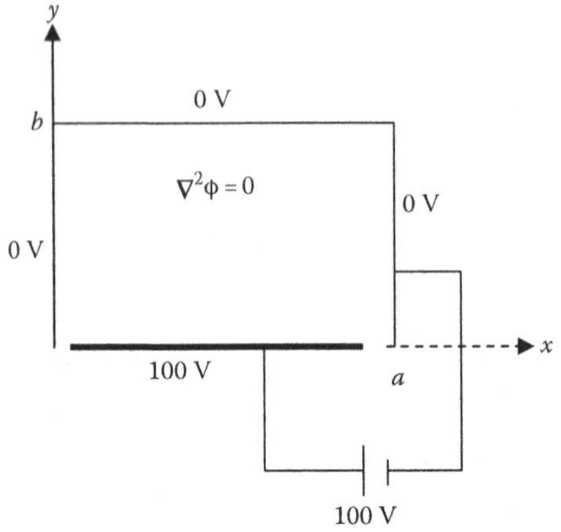

FIGURE 6.4
Geometry for Example 6.3.

The steps that allowed us to write this solution by inspection are as follows.

Step 1: The requirement of multiple zeros on the x-axis including a zero at $x = 0$ leads to the choice of

$$f_1(x) = \sin\frac{m\pi x}{a}. \tag{6.15}$$

Step 2: Since $\Phi = 0$ when $y = b$, we choose

$$f_2(y) = \sinh\frac{m\pi}{a}(b-y). \tag{6.16}$$

Note that $\sinh(m\pi/a)(b-y)$ is a linear combination of $\sinh(m\pi y/a)$ and $\cosh(m\pi y/a)$, which are in the template of functions for $f_2(y)$.

Step 3: The term $\sinh(m\pi b/a)$ in the denominator is the neutralizing (normalizing) factor in satisfying the boundary condition $\Phi = 100$, when $y = 0$, so that $B_m = 400/m\pi$ is the Fourier coefficient.

Example 6.4: Determine the potential in the field region described in the geometry of Figure 6.5.

Solution

$$\Phi_1(x,y) = \sum_{n=1,\text{odd}}^{\infty} \frac{400}{n\pi} \frac{\sinh(n\pi x/b)}{\sinh(n\pi a/b)} \sin\frac{n\pi y}{b}, \tag{6.17}$$

$$\Phi_2(x,y) = \sum_{m=1,\text{odd}}^{\infty} \frac{200}{m\pi} \frac{\sinh(m\pi y/a)}{\sinh(m\pi b/a)} \sin\frac{m\pi x}{a}, \tag{6.18}$$

$$\Phi = \Phi_1 + \Phi_2. \tag{6.19}$$

Example 6.4 is solved by using superposition (Figure 6.6).

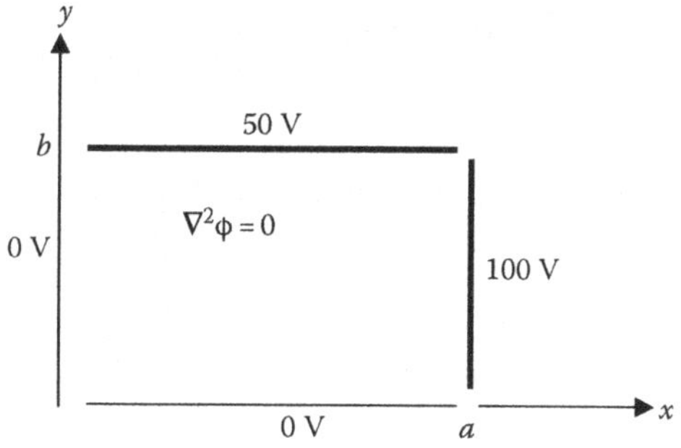

FIGURE 6.5
Geometry for Example 6.4.

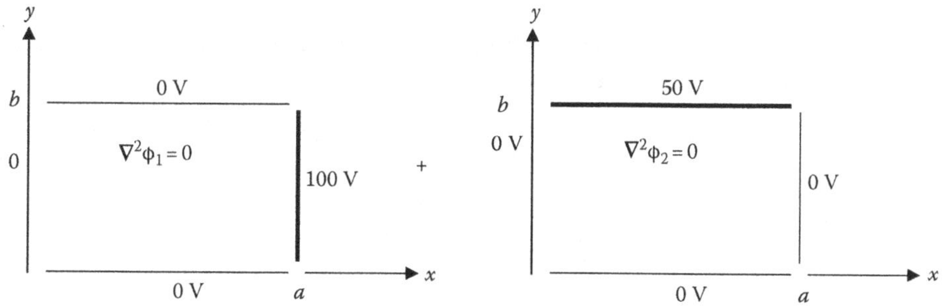

FIGURE 6.6
The use of superposition for Example 6.4.

In Chapter 3, we simply studied the various modes that can exist in a waveguide with a view to get the cutoff frequencies of various modes. We did not study the excitation of the various modes by a source in a waveguide. The amplitudes of the various modes can be obtained by expanding the source function in the various modes. The modes often are the orthonormal functions permitting such expansions. Reference 2 makes use of such expansions.

6.2.2 Circular Cylindrical Coordinates

Case 1:

$$\Phi = \Phi(\rho, \phi). \tag{6.20}$$

Let us first consider that the potential is a function of the ρ and ϕ coordinates. Then

$$\nabla^2\Phi(\rho, \phi) = \frac{1}{\rho}\frac{\partial}{\partial\rho}\left(\rho\frac{\partial\Phi}{\partial\rho}\right) + \frac{1}{\rho^2}\frac{\partial^2\Phi}{\partial\phi^2} = 0. \tag{6.21}$$

Let

$$\Phi(\rho, \phi) = f(\rho)g(\phi), \tag{6.22}$$

$$\frac{1}{\rho}\frac{\partial}{\partial\rho}\left(\rho\frac{\partial[f(\rho)g(\phi)]}{\partial\rho}\right)+\frac{f(\rho)}{\rho^2}\frac{\partial^2 g(\phi)}{\partial\phi^2}=0,$$

$$\frac{g(\phi)}{\rho}\frac{\partial}{\partial\rho}\left(\rho\frac{df(\rho)}{d\rho}\right)+\frac{f(\rho)}{\rho^2}\frac{\partial^2 g(\phi)}{\partial\phi^2}=0.$$

Multiplying the above equation by $\rho^2/(f(\rho)g(\phi))$, we obtain

$$\underbrace{\frac{\rho}{f(\rho)}\frac{\partial}{\partial\rho}\left(\rho\frac{df(\rho)}{d\rho}\right)}_{\text{term }1=n^2}+\underbrace{\frac{1}{g(\phi)}\frac{\partial^2 g(\phi)}{\partial\phi^2}}_{\text{term }2=-n^2}=0. \tag{6.23}$$

Term 1 is entirely a function of ρ and term 2 is entirely a function of ϕ, and hence each must be a constant and the sum of the constants must be zero. Denoting the separation constant by n^2, Equation 6.23 can be written as two ordinary differential equations:

$$\rho^2\frac{d^2 f(\rho)}{d\rho^2}+\rho\frac{df(\rho)}{d\rho}-n^2 f(\rho)=0, \tag{6.24}$$

$$\frac{d^2 g(\phi)}{d\phi^2}+n^2 g(\phi)=0. \tag{6.25}$$

The solution for $f(\rho)$ can be written as a linear combination of ρ^n and ρ^{-n} and thus

$$f(\rho)=\begin{Bmatrix}\rho^n\\ \rho^{-n}\end{Bmatrix},\qquad g(\phi)=\begin{Bmatrix}\cos n\phi\\ \sin n\phi\end{Bmatrix}. \tag{6.26}$$

Example 6.5: Commutator problem

Figure 6.7 gives the geometry for the commutator problem. The field region is $0 < \rho < a$ and the upper part of the long cylinder is at the potential V_0 and the bottom part is at the potential $-V_0$.

Solution

From the physics of the problem, we know that the potential is finite at $\rho = 0$. The second term in the template for $f(\rho)$, ρ^{-n}, blows up as ρ tends to zero. Hence, we choose ρ^n as the ρ-variation. The boundary condition $\Phi(a, \phi) = F(\phi)$, sketched in Figure 6.8, is an odd function of ϕ.

So, by inspection, we write

$$\Phi(\rho,\phi)=\sum_{n=1}^{\infty}B_n\left(\frac{\rho}{a}\right)^n\sin n\phi \tag{6.27}$$

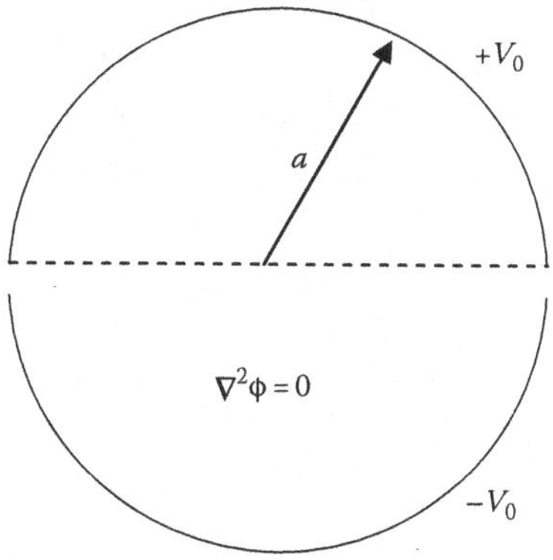

FIGURE 6.7
Geometry for Example 6.5.

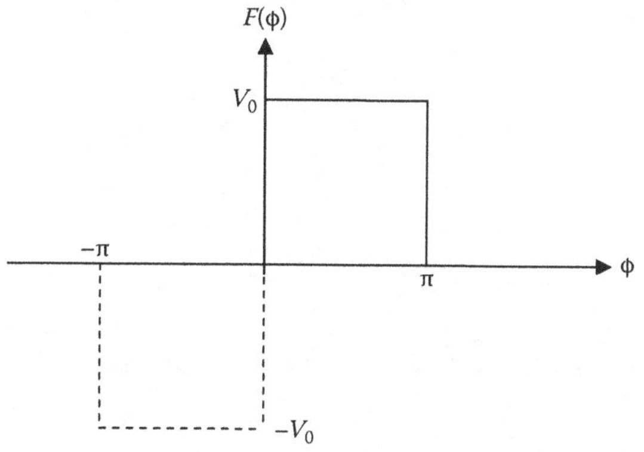

FIGURE 6.8
Sketch of the function $F(\phi)$ for Example 6.5.

and

$$B_n = \frac{2}{\pi}\int_0^{\pi} F(\phi)\sin n\phi \, d\phi$$
$$= \begin{cases} \dfrac{4V_0}{n\pi}, & n \text{ odd}, \\ 0, & n \text{ even}. \end{cases} \tag{6.28}$$

Case 2:

$$\Phi = \Phi(\rho, z). \tag{6.29}$$

Using the separation of variable technique, two ordinary differential equations can be obtained, in terms of the separation constant k^2:

$$\frac{d^2 f_1}{d\rho^2} + \frac{1}{\rho}\frac{df_1}{d\rho} + k^2 f_1 = 0, \tag{6.30}$$

$$\frac{d^2 f_2}{dz^2} - k^2 f_2 = 0. \tag{6.31}$$

Equation 6.30 is the same as that of Equation 3.59 with $n = 0$. Therefore, the solution can be written as

$$f_1 = \begin{Bmatrix} J_0(k\rho) \\ Y_0(k\rho) \end{Bmatrix}, \quad f_2 = \begin{Bmatrix} \cosh kz \\ \sinh kz \\ e^{\mp kz} \end{Bmatrix} \tag{6.32}$$

or

$$f_1 = \begin{Bmatrix} I_0(K\rho) \\ K_0(K\rho) \end{Bmatrix}, \quad f_2 = \begin{Bmatrix} \cos Kz \\ \sin Kz \end{Bmatrix}, \tag{6.33}$$

where

$$K^2 = -k^2. \tag{6.34}$$

Example 6.6: In the field region $0 < \rho < a$ and $0 < z < l$, $0 < \phi < 2\pi$, the Laplace equation is satisfied.

The curved surface, $\rho = a$, $0 < z < l$, is at a constant potential V_0 as shown in Figure 6.9. The ends of the can are grounded.

Determine $\Phi(\rho, \phi, z)$.

Solution

Note the symmetry with respect to the ϕ-coordinate and hence

$$\Phi(\rho, \phi, z) = \Phi(\rho, z). \tag{6.35}$$

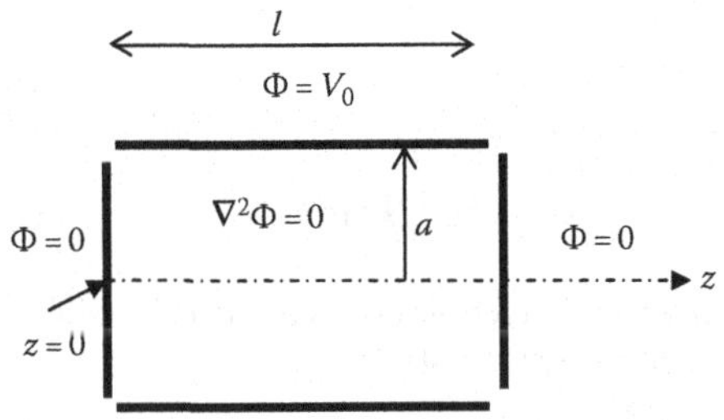

FIGURE 6.9
Geometry for Example 6.6.

Since we need multiple zeros on the z-axis and the potential is zero for $z = 0$, we choose

$$f_2 = \sin Kz, \tag{6.36}$$

$$K = \frac{n\pi}{l}. \tag{6.37}$$

From Equation 6.33, noting K_0 blows up when $\rho \to 0$, we choose I_0 for ρ variation and write by inspection

$$\Phi(\rho, z) = \sum_{n=1,\text{odd}}^{\infty} \frac{4V_0}{n\pi} \frac{I_0(n\pi\rho/l)}{I_0(n\pi a/l)} \sin\frac{n\pi z}{l}. \tag{6.38}$$

Example 6.7: The field region is the same as that of Example 6.6. However, the boundary conditions are different and are shown in Figure 6.10:

$$\Phi(a, z) = 0, \tag{6.39}$$

$$\Phi(\rho, 0) = 0, \tag{6.40}$$

$$\Phi(\rho, l) = F(\rho) = V_0. \tag{6.41}$$

The boundary condition Φ $(a,z) = 0$ requires an oscillatory function for ρ-variation. Hence, we choose, the template of Equation 6.32. Moreover, f_1 has to be finite on the z-axis ($\rho = 0$). By inspection, we write

$$\Phi(\rho, z) = \sum_{m=1}^{\infty} \frac{A_m J_0(p_m(\rho/a))}{\sinh\left(\frac{p_m l}{a}\right)} \sinh\frac{p_m z}{a}, \tag{6.42}$$

where $p_m = x_{om}$ is the mth root of J_0. A few of these values are given in Table 2.4.

The boundary condition Φ $(\rho, l) = F(\rho) = V_0$ can be satisfied if

$$\Phi(\rho, l) = F(\rho) = V_0 = \sum_{m=1}^{\infty} A_m J_0\left(\frac{x_{om}\rho}{a}\right). \tag{6.43}$$

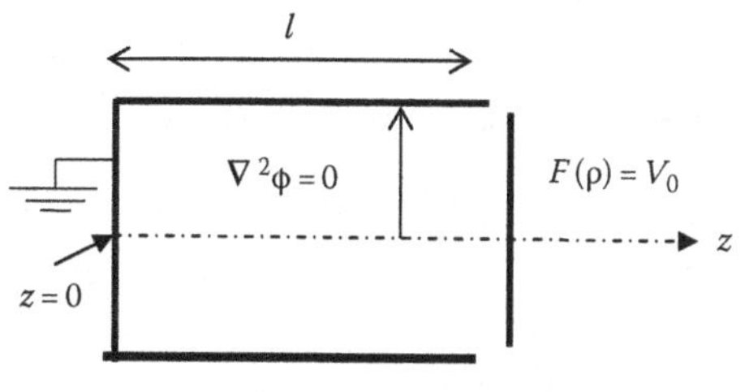

FIGURE 6.10
Geometry for Example 6.7.

Equation 6.43 is an expansion of an arbitrary function in a Bessel series. Such an expansion, analogous to the Fourier series in trigonometric functions, is possible since these Bessel functions also possess orthogonal properties:

$$\int_0^a \rho J_0\left(\frac{p_m\rho}{a}\right) J_0\left(\frac{p_q\rho}{a}\right) \mathrm{d}\rho = 0, \quad m \neq q, \tag{6.44}$$

$$\int_0^a \rho \left[J_0\left(\frac{p_m r}{a}\right) J_0\left(\frac{p_q r}{a}\right)\right] \mathrm{d}\rho, \quad m = q$$

$$= \frac{a^2}{2} J_1^2(p_q). \tag{6.45}$$

Multiplying Equation 6.43 by $\rho J_0(p_q\rho/a)$ and integrating from 0 to a, we obtain

$$\int_0^a F(\rho)\rho J_0\left(\frac{p_q\rho}{a}\right) \mathrm{d}\rho$$

$$= \int_a \rho \sum_{m=1}^{\infty} A_m J_0\left(p_m \frac{\rho}{a}\right) J_0\left(p_q \frac{\rho}{a}\right) \mathrm{d}\rho$$

$$= A_q \frac{a^2}{2} J_1^2(p_q).$$

Thus,

$$A_q = \frac{2}{a^2 J_1^2(p_q)} \int_0^a F(\rho)\rho J_0\left(\frac{p_q\rho}{a}\right) \mathrm{d}\rho \tag{6.46}$$

When $F(\rho) = V_0$

$$A_q = \frac{2V_0}{a^2 J_1^2(p_q)} \int_0^a \rho J_0\left(\frac{p_q\rho}{a}\right) \mathrm{d}\rho. \tag{6.47}$$

Using the integral

$$\int x J_0(x) \mathrm{d}x = x J_1(x), \tag{6.48}$$

we obtain

$$\int_0^a \rho J_0\left(\frac{p_q\rho}{a}\right) \mathrm{d}\rho = \frac{a^2}{p_q} J_1(p_q). \tag{6.49}$$

Substituting Equation 6.49 into Equation 6.47, we obtain

$$A_q = \frac{2V_0}{p_q J_1(p_q)}, \tag{6.50}$$

$$\Phi(\rho, z) = \sum_{m=1}^{\infty} \frac{((2V_0/p_m)/J_1(p_m)) J_0(p_m(\rho/a))}{\sinh(p_m l/a)} \sinh \frac{p_m z}{a}. \tag{6.51}$$

6.3 Three-Dimensional Solution

6.3.1 Cartesian Coordinates

The three-dimensional solution of the Laplace equation in a closed box may be obtained on the same line as the two-dimensional solution discussed in Section 6.2.1. One can also view the solution from the viewpoint of the rectangular cavity problem discussed in Section 4.1. The central point is that in the case of the Laplace equation,

$$k^2 = k_x^2 + k_y^2 + k_z^2 = 0, \tag{6.52}$$

which in turn means that if $f_1(x)$ and $f_2(y)$ are trigonometric functions, then $f_3(z)$ necessarily has to be the hyperbolic type. Example 6.8 illustrates the technique.

Example 6.8

In a rectangular box of dimensions a, b, and h, shown in Figure 4.1, the faces $x = 0, a; y = 0, b; z = 0$ are all at zero potential. For the face $z = h$, the potential is V_0. Determine the potential Φ inside the field region $0 < x < a$, $0 < y < b$, and $0 < z < h$.

Assume that the Laplace equation is satisfied in the field region.

Solution

The solution can be written by inspection as

$$\Phi(x, y, z) = \sum_{m=1}^{\infty}\sum_{n=1}^{\infty} B_{mn} \frac{\sinh\left[(m\pi/a)^2 + (n\pi/b)^2\right]^{1/2} z}{\sinh\left[(m\pi/a)^2 + (n\pi/b)^2\right]^{1/2} h} \sin\frac{m\pi x}{a} \sin\frac{n\pi y}{b}. \tag{6.53}$$

B_{mn} is the Fourier coefficient of a double Fourier series expansion of the function

$$\Phi(x, y, h) = F(x, y) = V_0. \tag{6.54}$$

It can easily be evaluated using the orthogonality property and is given by

$$B_{mn} = \begin{cases} \dfrac{16V_0}{mn\pi^2}, & m \text{ and } n \text{ are odd,} \\ 0 & \text{otherwise.} \end{cases} \tag{6.55}$$

6.3.2 Cylindrical Coordinates

Instead of trigonometric and hyperbolic functions, the template contains the Bessel functions and the modified Bessel functions for the ρ-variation.

6.3.3 Spherical Coordinates

By substituting $k = 0$ in Equation 5.10, the equation for f_1 can be obtained. Its solution gives

$$\mathbf{f}_1 = \begin{cases} \mathbf{r}^{\mathrm{n}}, \\ \mathbf{r}^{-(\mathrm{n}+1)}. \end{cases} \tag{6.56}$$

The equations for f_2 and f_3 are unchanged and are given, respectively, by Equations 5.11 and 5.13, and thus the three-dimensional solution of the Laplace equation in spherical coordinates is given by

$$\Phi(r,\theta,\phi)=\begin{Bmatrix} r^n \\ r^{-(n+1)} \end{Bmatrix}\begin{Bmatrix} P_n^m(\cos\theta) \\ Q_n^m(\cos\theta) \end{Bmatrix}\begin{Bmatrix} \cos m\phi \\ \sin m\phi \end{Bmatrix}. \tag{6.57}$$

The properties of the Legendre polynomials are given in Chapter 5.

Example 6.9

In a spherical volume of radius a, the Laplace equation is satisfied. The top-half of the surface at $r = a$, $0 < \theta < \pi/2$, is at the voltage V_0 and the bottom-half of the surface at $r = a$, $\pi/2 < \theta < \pi$, is at $-V_0$, as shown in Figure 6.11. Determine the potential inside and outside the sphere.

Solution

The boundary condition can be stated as

$$\Phi(a,\theta)=F(\theta), \tag{6.58}$$

which is sketched in Figure 6.11b.

From the symmetry, it is obvious that the potential does not depend on the ϕ-coordinate. Thus, it is a two-dimensional problem. By choosing $f_3 = \cos m\phi$ and $m = 0$, the ϕ dependence is removed:

$$\Phi(r,\theta)=\begin{Bmatrix} r^n \\ r^{-(n+1)} \end{Bmatrix}\begin{Bmatrix} P_n^0(\cos\theta) \\ Q_n^0(\cos\theta) \end{Bmatrix}. \tag{6.59}$$

Note that the associated Legendre polynomials of zero degree ($m = 0$) P_n^0 or Q_n^0 are called Legendre polynomials P_n or Q_n, respectively. Figure 5.1 presents the graphs of these functions. Table 5.3 consists of the polynomial expressions, in terms of $\cos\theta$, of these functions.

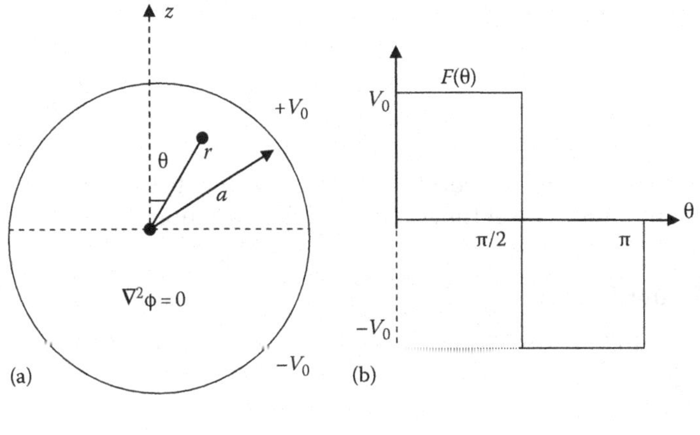

FIGURE 6.11
(a) Geometry and (b) sketch of the boundary condition for Example 6.9.

For Example 6.9, the solution can be obtained separately for inside and outside the sphere.

Case 1: $r < a$.

$$\Phi(r,\theta) = \sum_{m=0}^{\infty} A_n \left(\frac{r}{a}\right)^n P_n. \tag{6.60}$$

From the boundary condition,

$$\Phi(a,\theta) = F(\theta) = \sum_{m=0}^{\infty} A_n P_n. \tag{6.61}$$

The coefficient A_n can be determined based on the orthogonality property listed in Table 5.3. Multiplying both sides by $P_m \sin\theta$ and integrating, we have

$$\int_0^{\pi} F(\theta) P_m \sin\theta d\theta = \int_0^{\pi} \sum_{n=0}^{\infty} A_m P_m P_n \sin\theta d\theta = A_m \frac{2}{2m+1}.$$

Thus,

$$A_m = \frac{2m+1}{2} \int_0^{\pi} F(\theta) P_m \sin\theta d\theta. \tag{6.62}$$

Substituting for $f(\theta)$, we obtain

$$A_m = \frac{2m+1}{2} \left[\int_0^{\pi/2} P_m \sin\theta d\theta - \int_{\pi/2}^{\pi} P_m \sin\theta d\theta \right].$$

Evaluating,

$$A_0 = 0,$$

$$A_1 = \frac{3}{4} V_0.$$

The solution is completed by evaluating other A_m coefficients.

Case 2: $r > a$.

$$\Phi(r,\theta) = \sum_{n=0}^{\infty} A_n \left(\frac{r}{a}\right)^{-(n+1)} P_n. \tag{6.63}$$

Example 6.10

This example [1] allows us to determine the field's off-axis, knowing the fields on the axis of a circular loop. Consider the direct current in a circular loop of radius a, in the x–y-plane with the center of the loop at the origin. It is well known that the magnetic field at a point on the z-axis is given by

$$\mathbf{H} = \hat{z}H_z,$$

$$H_z = \frac{a^2 I}{2\left(a^2 + z^2\right)^{3/2}} = \frac{I}{2a}\left(1 + \frac{z^2}{a^2}\right)^{-3/2}.$$

From the binomial expansion,

$$\left(1+u\right)^{-3/2} = 1 - \frac{3u}{2} + \frac{15u^2}{8} - \frac{105u^3}{48} + \cdots, \quad |u| < 1.$$

Thus, for $z < a$,

$$H_z\Big|_{\text{axis}} = \frac{I}{2a}\left[1 - \frac{3u}{2} + \frac{15u^2}{8} - \frac{105u^3}{48} + \cdots\right]. \tag{6.64}$$

In a region where there are no currents, we have

$$\nabla \times \mathbf{H} = 0, \tag{6.65}$$

$$\mathbf{H} = -\nabla \Phi_m, \tag{6.66}$$

where Φ_m is called the scalar magnetic potential and since

$$\nabla \cdot \mathbf{H} = 0, \tag{6.67}$$

$$\nabla^2 \Phi_m = 0. \tag{6.68}$$

First consider $r < a$:

$$\Phi(r, \theta) = \sum_{m=0}^{\infty} b_n \left(\frac{r}{a}\right)^n P_n, \tag{6.69}$$

$$\Phi(r, \theta)\Big|_{\theta=0} = \sum_{m=0}^{\infty} b_n \left(\frac{z}{a}\right)^n. \tag{6.70}$$

Note $P_n\,|_{\cos 0} = 1$.
Comparing Equation 6.70 with Equation 6.64, we obtain

$$b_0 = \left(\frac{I}{2a}\right)(1),$$

$$b_1 = 0,$$

$$b_2 = -\frac{3}{2}\left(\frac{I}{2a}\right),$$

$$b_3 = 0,$$

$$b_4 = \frac{15}{8}\left(\frac{I}{2a}\right).$$

Thus,

$$H_z(r,\theta) = \frac{I}{2a}\left[1 - \frac{3}{2}\left(\frac{r}{a}\right)^2 P_2 + \frac{15}{8}\left(\frac{r}{a}\right)^4 P_4 + \cdots\right]. \tag{6.71}$$

There are several homework problems illustrating the use of the solution in spherical coordinates for practical static problems.

References

1. Ramo, S., Whinnery, J. R., and Van Duzer, T. V., *Fields and Waves in Communication Electronics*, 3rd edn., Wiley, New York, 1994.
2. Ji, C., Approximate analytical techniques in the study of quadruple-ridged waveguide (QRW) and its modifications, Doctoral thesis, University of Massachusetts Lowell, Lowell, MA, 2007.

7

Miscellaneous Topics on Waves*

Most of the problems we have dealt with till now are monochromatic waves unbounded in space. Practical signals involve pulse waves or beam waves. The simple solutions obtained for monochromatic waves can be used to construct solutions for more difficult problems. In this connection, we first explore the concept of group velocity.

7.1 Group Velocity v_g

Group velocity is the velocity with which a group of waves travel. Let us consider that at $z = 0$, we have two sinusoidal oscillations:

$$E(0,t) = \sin(\omega_0 - \Delta\omega)t + \sin(\omega_0 + \Delta\omega)t. \tag{7.1}$$

The two oscillations differ in frequency by $2\Delta\omega$.

Equation 7.1 may be also written as

$$E(0,t) = 2\sin\omega_0 t\cos\Delta\omega t. \tag{7.2}$$

A sketch of Equation 7.2 is shown in Figure 7.1.

It shows a carrier of frequency ω_0 being modulated by a signal of frequency $\Delta\omega$. The oscillation propagates as a wave that can be written as

$$E(Z,t) = \sin\left[(\omega_0 - \Delta\omega)t - (\beta - \Delta\beta)z)\right] + \sin\left[(\omega_0 + \Delta\omega)t - (\beta + \Delta\beta)z)\right], \tag{7.3}$$

$$E(z,t) = 2\sin(\omega_0 t - \beta_0 z)\cos(\Delta\omega t - \Delta\beta z). \tag{7.4}$$

From Equation 7.4, it is clear that while the carrier propagates with the phase velocity ω_0/β_0, the group travels with the group velocity:

$$v_g = \frac{d\omega}{d\beta}, \tag{7.5}$$

$$v_p = \frac{\omega}{\beta}. \tag{7.6}$$

* For chapter appendices, see 7A in Appendices section.

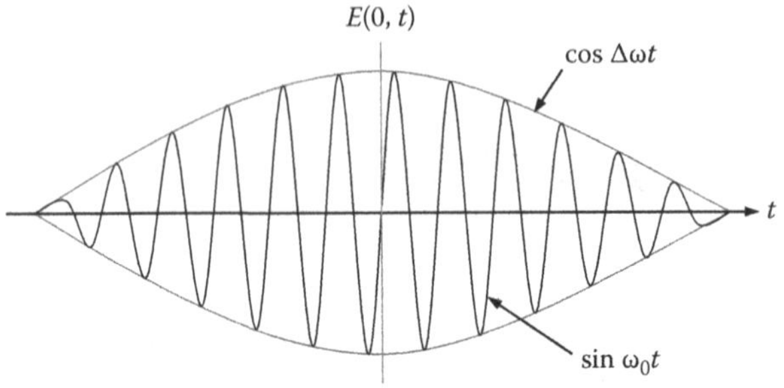

FIGURE 7.1
Two oscillations with a small beat frequency.

The importance of the ω–β diagram now becomes clear. If we plot ω versus β, then the phase velocity is given by the ratio of the vertical coordinate to the horizontal coordinate, whereas the slope of the tangent gives the group velocity. If the ω–β diagram is not a straight line, then the velocity with which the members of the group travel will be different and hence the signal gets distorted. Group velocity, in many cases, is also equal to the velocity of energy transport or simply energy velocity v_E defined by

$$v_E = \frac{S}{w}, \tag{7.7}$$

where S is the power density and w the total energy density. The energy velocity represents the velocity with which the wave energy is transported.

From Equations 7.5 and 7.6, it can be shown that

$$v_g = \frac{v_p}{1 - (\omega/v_p)(dv_p/d\omega)}. \tag{7.8}$$

Several examples of calculation of group velocity are given as problems.

7.2 Green's Function

Green's function is the response of a system to an impulse input. In circuits, it is usually denoted by $h(t)$, whereas in field theory it is denoted by G. The following one-dimensional problem will be solved to illustrate the essential aspects of constructing a Green's function. G is the solution of the following differential equation:

$$\frac{d^2G}{dx^2} = -\delta(x - x'), \quad 0 < x < L, \tag{7.9}$$

subject to the boundary conditions

$$G = 0, \quad x = 0, \tag{7.10}$$

$$G = 0, \quad x = L. \tag{7.11}$$

We expect G to depend on x as well as x', that is, $G = G(x, x')$. Let us first solve the problem mathematically and then we will interpret the solution. Let

$$G = G_1, \quad 0 < x < x', \tag{7.12}$$

$$G = G_2, \quad x' < x < L. \tag{7.13}$$

Since

$$\delta(x - x') = 0 \quad \text{if } x \neq x', \tag{7.14}$$

$$G_1 = A_1 x + B_1, \tag{7.15}$$

$$G_2 = A_2 x + B_2. \tag{7.16}$$

To determine the four unknowns A_1, B_1, A_2, and B_2, we need four equations. Two are derived from the boundary conditions

$$G_1 = 0, \quad x = 0, \tag{7.17}$$

$$G_2 = 0, \quad x = L. \tag{7.18}$$

We obtain two more equations from "source conditions," at $x = x'$:

$$G_1 = G_2, \quad x = x', \tag{7.19}$$

$$\left.\left(\frac{\partial G_2}{\partial x} - \frac{\partial G_1}{\partial x}\right)\right|_{x=x'} = -1. \tag{7.20}$$

Equation 7.19 comes from the physics of the problem and Equation 7.20 follows from the differential equation (7.9): integrating Equation 7.9 with respect to x over the interval $0 < x < L$ gives

$$\int_0^L \frac{\mathrm{d}^2 G}{\mathrm{d}x^2}\mathrm{d}x = -\int_0^L \delta(x - x')\mathrm{d}x = -1. \tag{7.21}$$

The last part of Equation 7.21 follows from the definition of the impulse function. The LHS of Equation 7.21 may be written as

$$\int_0^L \frac{\mathrm{d}^2 G}{\mathrm{d}x^2}\mathrm{d}x = \lim_{\varepsilon \to 0} \int_{x'-\varepsilon}^{x'+\varepsilon} \frac{\mathrm{d}^2 G}{\mathrm{d}x^2}\mathrm{d}x = \left.\left(\frac{\partial G_2}{\partial x} - \frac{\partial G_1}{\partial x}\right)\right|_{x=x'}. \tag{7.22}$$

Note that $\mathrm{d}^2G/\mathrm{d}x^2 = 0$, except for a small interval 2ε around $x = x'$. From Equations 7.10, 7.11, 7.19, and 7.20, we can determine A_1, B_1, A_2, and B_2. Thus, we get

$$G_1 = x\frac{(L - x')}{L}, \quad 0 < x < x', \tag{7.23}$$

$$G_2 = x'\frac{(L - x)}{L}, \quad x' < x < L. \tag{7.24}$$

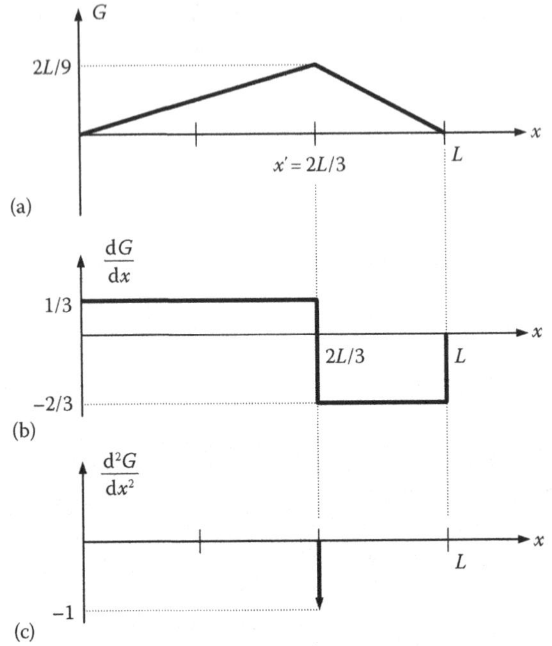

FIGURE 7.2
Sketches of Green's function and its derivatives: (a) G, (b) dG/dx, and (c) d^2G/dx^2.

Figure 7.2 shows the plot of G, dG/dx, and d^2G/dx^2 versus x, for the case of $x' = 2L/3$.

Equation 7.9 is the normalized form of the differential equation of a string under tension clamped at the ends. G is the downward displacement of the string when a point load is applied at $x = 2L/3$. The physical picture is shown in Figure 7.3. The advantage of developing Green's function for a problem is that the response to an arbitrary excitation may be written as a superposition integral. If the input is $-f(x)$, for the problem, that is,

$$
\begin{aligned}
&\frac{d^2 y}{dx} = -f(x), \quad 0 < x < L, \\
&y = 0, \quad x = 0, \\
&y = 0, \quad x = L,
\end{aligned}
\tag{7.25}
$$

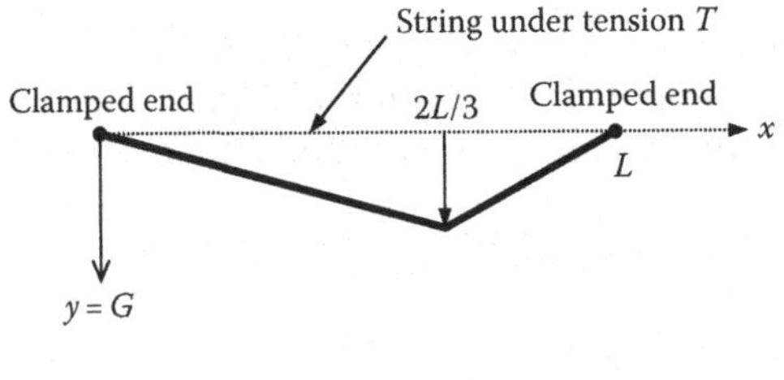

FIGURE 7.3
Clamped string under tension.

then

$$y(x)=\int_{0}^{L}f(x')G(x,x')\mathrm{d}x'. \tag{7.26}$$

We note that Green's function has the following properties:

1. It is symmetric, that is, $G(x, x') = G(x', x)$.
2. It is continuous.
3. The Derivative of G has a Unit Discontinuity at $x = x'$. Note that the first derivative of G is one order less than the highest-order derivative in the differential equation for G.

The general procedure for constructing Green's function may be stated as follows:

1. Determining the solution of the differential equation with the RHS equal to zero. The solution will involve undetermined constants.
2. Determining the constants using the boundary conditions and the source conditions.

Using the above, it can be shown that Green's function G of the one-dimensional Helmholtz equation

$$\frac{\mathrm{d}^2G}{\mathrm{d}x^2}+k^2G=-\delta(x-x'), \tag{7.27}$$

subject to the boundary conditions,

$$G=0, \quad x=0, \tag{7.28}$$

$$G=0, \quad x=L, \tag{7.29}$$

is given by

$$G=\begin{cases}\dfrac{1}{k}\dfrac{\sin kx\sin k(L-x')}{\sin kL}, & 0<x<x',\\[2ex] \dfrac{1}{k}\dfrac{\sin kx'\sin k(L-x)}{\sin kL}, & x'<x<L.\end{cases} \tag{7.30}$$

Appendix 7A deals with the advanced aspects of the Green's function.

7.3 Network Formulation

Classical network theory deals with the interconnection of simple individual components. In the context of high-frequency electrical engineering, the individual components are wave types including transmission lines, waveguides, and cavity resonators. After briefly

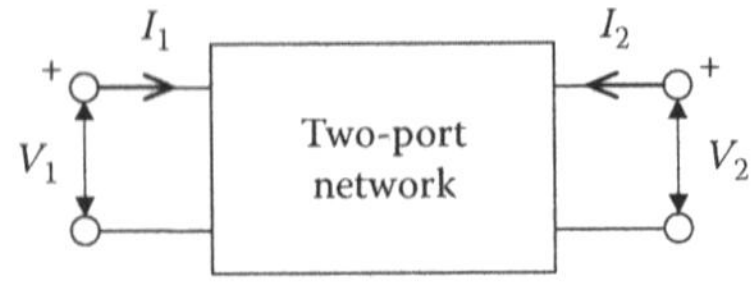

FIGURE 7.4
Two-port network.

reviewing the network parameter description of two-port networks, the S-parameter description will be explained. Figure 7.4 shows a two-port network. The input and output variables V_1, V_2, I_1, and I_2 are related by a 2 × 2 matrix whose elements are the parameters of the network. For example,

$$\begin{bmatrix} V_1 \\ V_2 \end{bmatrix} = \begin{bmatrix} Z_{11} & Z_{12} \\ Z_{21} & Z_{22} \end{bmatrix} \begin{bmatrix} I_1 \\ I_2 \end{bmatrix} \tag{7.31}$$

describes open-circuit impedance parameters, called Z parameters.

Since $Z_{11} = V_1/I_1|_{I2=0}$, it is the input impedance when the output is open-circuited. In the elementary circuit theory, one might have learnt about the description of a network through other parameters: short-circuit Y parameters, hybrid H parameters, and inverse hybrid G parameters. Two more sets, $ABCD$ parameters and scattering S parameters, are particularly useful in high-frequency work. Any one set of parameters can be transformed into another set. Books on circuit theory give tables of conversion formulas. In this section, we will concentrate on $ABCD$ parameters and S parameters.

7.3.1 *ABCD* Parameters

The matrix definition of $ABCD$ parameters is given by

$$\begin{bmatrix} V_1 \\ I_1 \end{bmatrix} = \begin{bmatrix} A & B \\ C & D \end{bmatrix} \begin{bmatrix} V_2 \\ -I_2 \end{bmatrix}. \tag{7.32}$$

If two networks are cascaded as shown in Figure 7.5a, then the equivalent $ABCD$ network is as shown in Figure 7.5b, where

$$\begin{bmatrix} A & B \\ C & D \end{bmatrix} = \begin{bmatrix} A_1 & B_1 \\ C_1 & D_1 \end{bmatrix} \begin{bmatrix} A_2 & B_2 \\ C_2 & D_2 \end{bmatrix}. \tag{7.33}$$

The advantage of using $ABCD$ parameters is evident from Equation 7.33. The overall $ABCD$ parameters of a number of two ports connected in cascade is the product of the

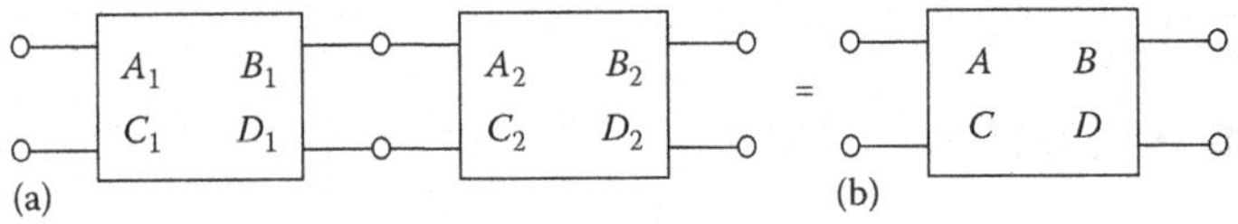

FIGURE 7.5
Networks in cascade. (a) Two networks in cascade and (b) equivalent network.

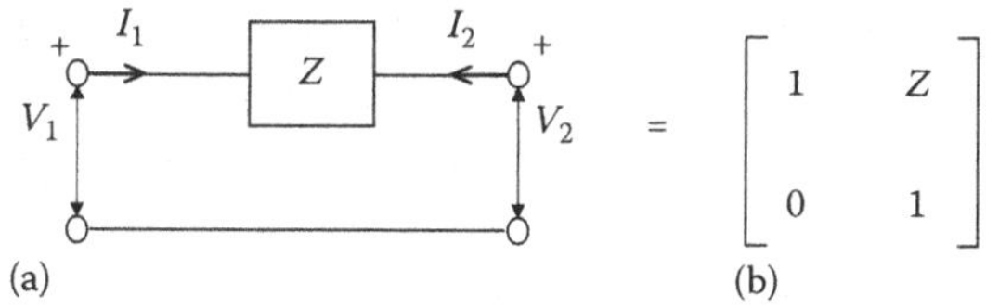

FIGURE 7.6
ABCD parameters of series element. (a) Series element and (b) *ABCD* parameters.

matrices describing the individual two-port network. As a simple example of finding *ABCD* parameters of a two-port network, consider a series impedance Z as shown in Figure 7.6.

Since $V_1 = V_2 + ZI_1$ and $I_1 = -I_2$,

$$\begin{bmatrix} V_1 \\ I_1 \end{bmatrix} = \begin{bmatrix} 1 & Z \\ 0 & 1 \end{bmatrix} \begin{bmatrix} V_2 \\ -I_2 \end{bmatrix}. \tag{7.34}$$

Let us consider next the lossless transmission line (Figure 7.7).

Taking the reference to be at $z = 0$:

$$\begin{bmatrix} V_2 \\ -I_2 \end{bmatrix} = \begin{bmatrix} 1 & 1 \\ Z_0^{-1} & -Z_0^{-1} \end{bmatrix} \begin{bmatrix} V_0^+ \\ V_0^- \end{bmatrix}. \tag{7.35}$$

Also

$$\begin{bmatrix} V_1 \\ I_1 \end{bmatrix} = \begin{bmatrix} e^{j\beta l} & e^{-j\beta l} \\ e^{j\beta l}/Z_0 & -e^{-j\beta l}/Z_0 \end{bmatrix} \begin{bmatrix} V_0^+ \\ V_0^- \end{bmatrix}. \tag{7.36}$$

From Equations 7.35 and 7.36, we obtain

$$\begin{bmatrix} V_1 \\ I_1 \end{bmatrix} = \begin{bmatrix} \cos\beta l & jZ_0 \sin\beta l \\ \dfrac{j}{Z_0}\sin\beta l & \cos\beta l \end{bmatrix} \begin{bmatrix} V_2 \\ -I_2 \end{bmatrix}. \tag{7.37}$$

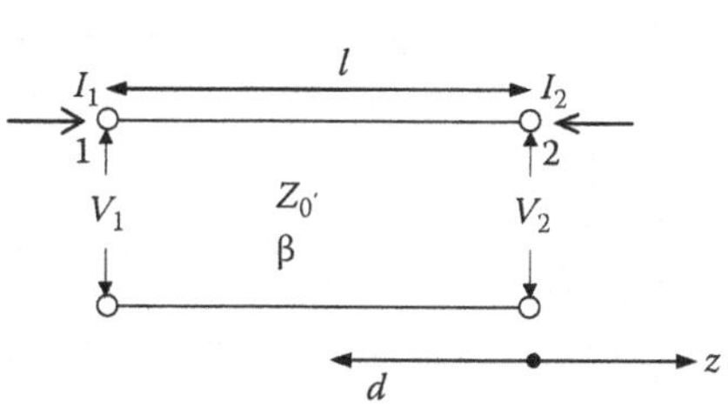

FIGURE 7.7
Lossless transmission line.

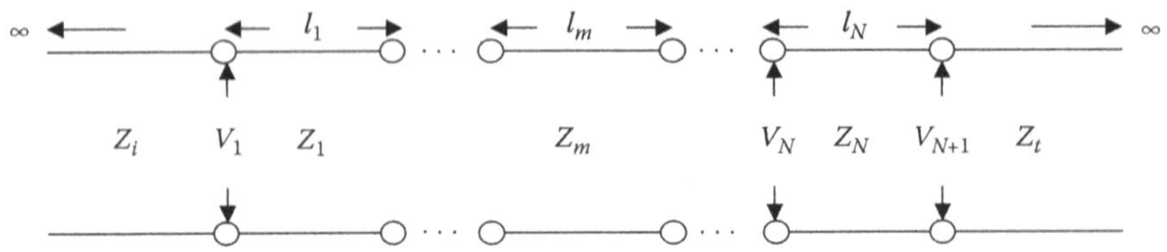

FIGURE 7.8
N transmission lines in tandem.

For a line with losses,

$$\begin{bmatrix} V_1 \\ I_1 \end{bmatrix} = \begin{bmatrix} \cosh \gamma l & Z_0 \sinh \gamma l \\ \dfrac{\sinh \gamma l}{Z_0} & \cosh \gamma l \end{bmatrix} \begin{bmatrix} V_2 \\ -I_2 \end{bmatrix}, \tag{7.38}$$

where $\gamma = \alpha + j\beta$ is the propagation constant.

Let us next consider N transmission lines in tandem, fed by the input transmission line i and feeding the output transmission line t as shown in Figure 7.8. Then

$$\begin{bmatrix} V_1 \\ I_1 \end{bmatrix} = \begin{bmatrix} A & B \\ C & D \end{bmatrix} \begin{bmatrix} V_{N+1} \\ -I_{N+1} \end{bmatrix}, \tag{7.39}$$

where

$$\begin{bmatrix} A & B \\ C & D \end{bmatrix} = \begin{bmatrix} A_1 & B_1 \\ C_1 & D_1 \end{bmatrix} \begin{bmatrix} A_2 & B_2 \\ C_2 & D_2 \end{bmatrix} \cdots \begin{bmatrix} A_N & B_N \\ C_N & D_N \end{bmatrix}. \tag{7.40}$$

If V_0^+ is the incident voltage, Γ is the reflection coefficient, and T is the transmission coefficient, then

$$V_1 = V_i = V_0^+ \left[1 + \Gamma\right], \tag{7.41}$$

$$I_1 = I_i = \frac{V_0^+}{Z_i}\left[1 - \Gamma\right], \tag{7.42}$$

$$V_{N+1} = V_t = TV_0^+, \tag{7.43}$$

$$-I_{N+1} = -I_t = \frac{TV_0^+}{Z_t}, \tag{7.44}$$

$$\begin{bmatrix} V_0^+ (1 + \Gamma) \\ \dfrac{V_0^+}{Z_i}(1 - \Gamma) \end{bmatrix} = \begin{bmatrix} A & B \\ C & D \end{bmatrix} \begin{bmatrix} TV_0^+ \\ \dfrac{T}{Z_t} V_0^+ \end{bmatrix}. \tag{7.45}$$

Arranging in a matrix form,

$$\begin{bmatrix} -1 & A+\dfrac{B}{Z_t} \\ \dfrac{1}{Z_i} & C+\dfrac{D}{Z_t} \end{bmatrix} \begin{bmatrix} \Gamma \\ T \end{bmatrix} = \begin{bmatrix} 1 \\ Z_i \end{bmatrix}. \tag{7.46}$$

Solving for Γ and T yields

$$\Gamma = \frac{A + B/Z_t - Z_i(C + D/Z_t)}{A + B/Z_t + Z_i(C + D/Z_t)}, \tag{7.47}$$

$$T = \frac{2}{A + B/Z_t + Z_i(C + D/Z_t)}. \tag{7.48}$$

7.3.2 *S* Parameters

At low frequencies only, say less than 0.5 GHz, voltage and current can be measured and *ABCD* parameters can be calculated. Also if the system under investigation contains active devices such as diodes or transistors, placing an open- or short-circuit calibration piece on one port, needed to measure the *ABCD* parameters of the device, may result in instability. The high-frequency method involves measuring incident and reflected wave quantities rather than total voltage and current. The corresponding parameters are called scattering parameters. Let us first define scattering parameters for a one-port network (Figure 7.9).

Let the incident voltage be V_{I1} and reflected voltage V_{R1}. Define the normalized voltages of the incident and scattered waves as

$$a_1 = \frac{V_{I1}}{\sqrt{Z_0}} \tag{7.49}$$

and

$$b_1 = \frac{V_{R1}}{\sqrt{Z_0}}. \tag{7.50}$$

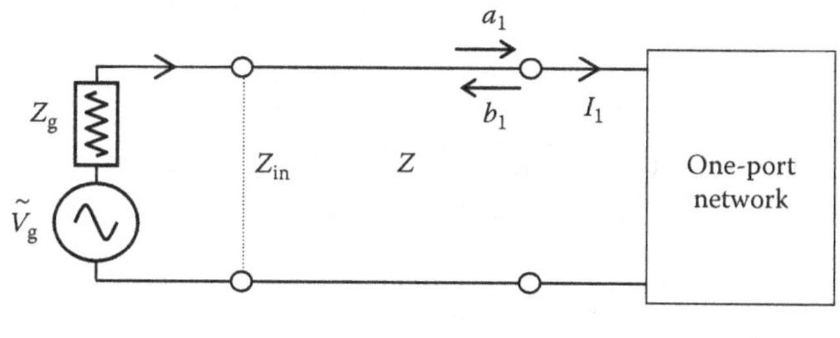

FIGURE 7.9
One-port network.

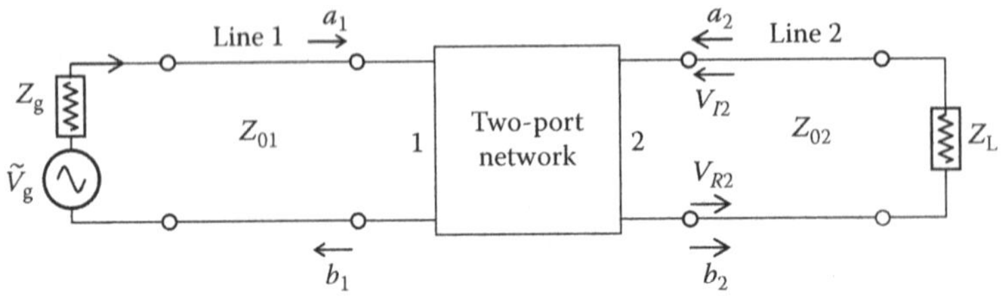

FIGURE 7.10
Two-port network.

The incident power is $|a_1|^2$ and the scattered power is $|b_1|^2$. At high frequencies, power measurements can be made with relative ease. Also b_1 can be set to zero by matching. In the one-port case, we define

$$S_{11} = \frac{b_1}{a_1} = \frac{V_{R1}}{V_{I1}} = \Gamma_0. \tag{7.51}$$

In the case of two ports (Figure 7.10), we define

$$a_2 = \frac{V_{I2}}{\sqrt{Z_{02}}}, \tag{7.52}$$

$$b_2 = \frac{V_{R2}}{\sqrt{Z_{02}}}, \tag{7.53}$$

and relate the b's with a's with the help of the scattering matrix:

$$\begin{bmatrix} b_1 \\ b_2 \end{bmatrix} = \begin{bmatrix} S_{11} & S_{12} \\ S_{21} & S_{22} \end{bmatrix} \begin{bmatrix} a_1 \\ a_2 \end{bmatrix}. \tag{7.54}$$

Since

$$S_{11} = \left.\frac{b_1}{a_1}\right|_{a_2=0}, \tag{7.55}$$

the measurement is made by terminating the output side of the two-port network with a matched load ($a_2 = 0$). S_{11} is the input reflection coefficient. Since

$$S_{21} = \left.\frac{b_2}{a_1}\right|_{a_2=0}, \tag{7.56}$$

TABLE 7.1

Transformation of *S–ABCD–S* Parameters

S	$\rightarrow$	$ABCD$
$\begin{bmatrix} S_{11} & S_{12} \\ S_{21} & S_{22} \end{bmatrix}$	$\rightarrow$	$\frac{1}{A+B+C+D}\begin{bmatrix} A+B-C-D & 2(AD-BC) \\ 2 & -A+B-C+D \end{bmatrix}$
S	$\leftarrow$	$ABCD$
$\frac{1}{2S_{21}}\begin{bmatrix} (1+S_{11})(1-S_{22})+S_{12}S_{21} & (1+S_{11})(1+S_{22})-S_{12}S_{21} \\ (1-S_{11})(1-S_{22})-S_{12}S_{21} & (1-S_{11})(1+S_{22})+S_{12}S_{21} \end{bmatrix}$	$\leftarrow$	$\begin{bmatrix} A & B \\ C & D \end{bmatrix}$

it becomes the forward transmission coefficient for the network. It measures the attenuation for passive circuits and measures the gain for an active network, say an amplifier. When the input side of the network is matched ($a_1 = 0$), we measure S_{22} and S_{12}

$$S_{22} = \left.\frac{b_2}{a_2}\right|_{a_1=0}, \tag{7.57}$$

$$S_{12} = \left.\frac{b_1}{a_2}\right|_{a_1=0}. \tag{7.58}$$

Here, S_{22} is the output reflection coefficient and S_{12} is the reverse transmission coefficient. Table 7.1 shows the relationship between S and $ABCD$ parameters.

It is impractical to connect measuring equipment to the two-port device under test directly. Measurements are usually carried out by introducing sections of transmission line between the device under test and the measuring equipment. These lines introduce known phase shifts, which have to be taken into account in obtaining the S parameters of the device from the measured S' parameters:

$$S' = T^{-1}ST, \tag{7.59}$$

where

$$T = \begin{bmatrix} \exp(-j\gamma l_1) & 0 \\ 0 & \exp(-j\gamma l_2) \end{bmatrix}. \tag{7.60}$$

In the above, γ is the propagation constant of the transmission line.

7.4 Stop Bands of a Periodic Media

Equation 9.127 derived from the first principles is the dispersion relation between ω and β of a periodic medium. The results are illustrated in Figures 9.12 and 9.13. Another way of obtaining the dispersion relation, using *ABCD* parameters, is discussed here. As an illustration, we use two transmission lines in cascade as the two layers in a unit cell shown in Figure 7.11.

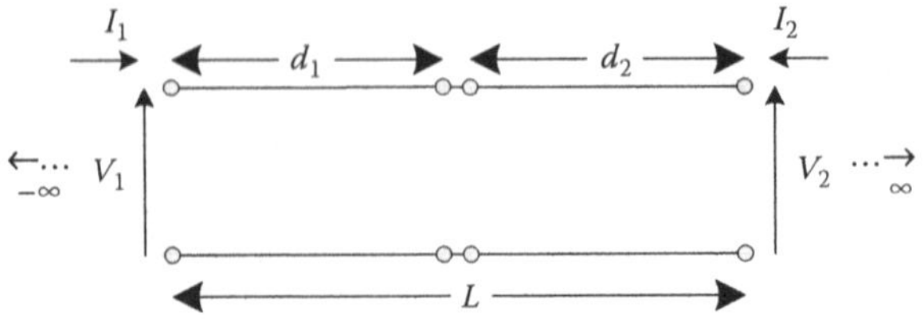

FIGURE 7.11
Unit cell of a periodic medium of infinite extent.

Let the *ABCD* parameters of the first transmission line be denoted by

$$\begin{bmatrix} A_1 & B_1 \\ C_1 & D_1 \end{bmatrix}$$

and the second transmission line by

$$\begin{bmatrix} A_2 & B_2 \\ C_2 & D_2 \end{bmatrix}.$$

The overall *ABCD* parameters are given by Equation 7.33 relating the input voltage and current with the output voltage and current as given by Equation 7.32. However, since the medium is periodic, from Equations 9.104 and 9.105, we get

$$V_2 = e^{-j\beta L} V_1, \tag{7.61}$$

$$-I_2 = +e^{-j\beta L} I_1, \tag{7.62}$$

or

$$\begin{bmatrix} V_1 \\ I_1 \end{bmatrix} = e^{+j\beta L} \begin{bmatrix} V_2 \\ -I_2 \end{bmatrix} = \lambda \begin{bmatrix} V_2 \\ -I_2 \end{bmatrix}, \tag{7.63}$$

where

$$\lambda = e^{j\beta L} \tag{7.64}$$

is the eigenvalue as will be shown shortly. From Equation 7.32,

$$\begin{bmatrix} V_1 \\ I_1 \end{bmatrix} = \begin{bmatrix} A & B \\ C & D \end{bmatrix} \begin{bmatrix} V_2 \\ -I_2 \end{bmatrix}. \tag{7.65}$$

From Equations 7.63 and 7.65,

$$\begin{bmatrix} A & B \\ C & D \end{bmatrix} \begin{bmatrix} V_2 \\ -I_2 \end{bmatrix} = \lambda \begin{bmatrix} V_2 \\ -I_2 \end{bmatrix}. \tag{7.66}$$

The eigenvalue $\lambda = e^{j\beta L}$ is now determined by the eigenvalue equation

$$\begin{vmatrix} A-\lambda & B \\ C & D-\lambda \end{vmatrix} = 0. \tag{7.67}$$

The solution for λ is given by

$$(A-\lambda)(D-\lambda) - BC = 0, \tag{7.68}$$

$$AD - \lambda(A+D) + \lambda^2 - BC = 0. \tag{7.69}$$

For a reciprocal network, $AD - BC = 1$. Hence,

$$\lambda^2 - \lambda(A+D) + 1 = 0, \tag{7.70}$$

$$\lambda = \frac{A+D}{2} \pm j\sqrt{1 - \left(\frac{A+D}{2}\right)^2}, \tag{7.71}$$

$$\beta L = -j\ln\left[\frac{A+D}{2} \pm j\sqrt{1 - \left(\frac{A+D}{2}\right)^2}\right] = \cos^{-1}\frac{A+D}{2}, \tag{7.72}$$

or

$$\cos\beta L = \frac{A+D}{2}. \tag{7.73}$$

A typical sketch of the ω–β diagram is shown in Figure 7.12.

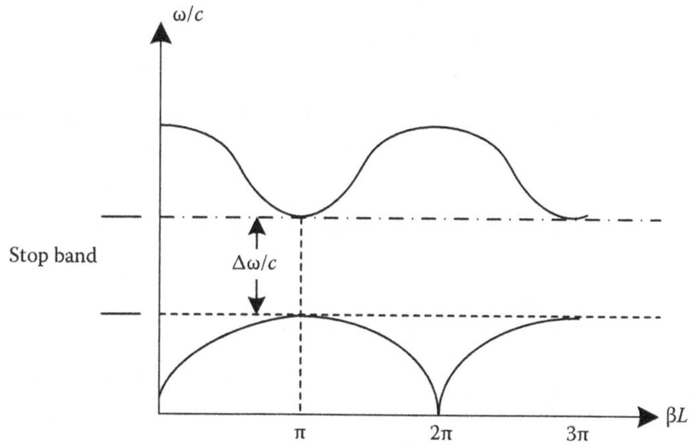

FIGURE 7.12
First stop band.

Since $|\cos \beta L| \leq 1$, the edges of pass and stop bands may be located by solving

$$\left|\frac{A+D}{2}\right| = 1. \tag{7.74}$$

The stop and pass bands may be located by solving Equation 7.74 for ω. The first stop band shown in the figure is marked as Δω and is obtained by solving Equation 7.74 for $\beta L = \pi$. The first nonzero value shown in Figure 7.12 can arise due to the natural cutoff frequency of the unbounded medium as was the case shown in Figure 9.12.

7.5 Radiation

In Section 1.3.1, we considered radiation in free space, when the source is a Hertzian dipole excited by an impulse current source. In this section, we will discuss radiation due to a time-harmonic current source. The starting point is the evaluation of Equation 1.50:

$$\tilde{\mathbf{A}}(r) = \frac{\mu}{4\pi} \iiint\limits_{\text{Source}} \mathrm{J}(r') \frac{e^{-jk|\mathbf{r}-\mathbf{r}'|}}{|\mathbf{r}-\mathbf{r}'|} dv' \tag{1.50}$$

assuming that the current distribution of the source is known.

We can make some approximations based on distances [1]. If the source dimensions are of the order of d and $d \ll \lambda$, then the field region can be classified as

The near-field (static) zone: $d \ll r \ll \lambda$,
The intermediate (induction) zone: $d \ll r \sim \lambda$,
The far (radiation) zone: $d \ll \lambda \ll r$.

In the far zone, the term $|\boldsymbol{r} - \boldsymbol{r}'|$ in the exponential of Equation 1.50 is replaced by

$$|\boldsymbol{r}-\boldsymbol{r}'| \approx r - \hat{r}\cdot\boldsymbol{r}' \tag{7.75}$$

and the term $|\boldsymbol{r} - \boldsymbol{r}'|$ in the denominator is replaced by r. This approximation can be stated as

$$\lim_{kr\to\infty} \tilde{\mathrm{A}}(r) = \frac{\mu}{4\pi}\frac{e^{-jkr}}{r} \iiint\limits_{\text{Source}} \tilde{\mathrm{J}}(\mathrm{r}') e^{+jk\hat{r}\cdot r'} dv'. \tag{7.76}$$

The term outside the integral shows the spherical nature of the vector potential, and the integral term shows angular dependence.

Expanding the exponent in Equation 7.76,

$$\tilde{A}(r) = \frac{\mu}{4\pi}\frac{-e^{-jkr}}{r} \sum_n \frac{(\mathrm{jk})^n}{n!} \iiint\limits_{\text{Source}} \tilde{\mathrm{J}}(\mathrm{r}')(\hat{r}\cdot\boldsymbol{r}')^n dv'. \tag{7.77}$$

If the source dimensions are of the order of d, and from our assumption that $kd \ll 1$, then the dominant term is the first nonvanishing term in Equation 7.77.

If that term is $n = 0$ term, then

$$\tilde{\mathbf{A}}(r) = \frac{\mu}{4\pi}\frac{e^{-jkr}}{r}\iiint_{v'} \tilde{\mathbf{J}}(\mathbf{r}')dv'. \tag{7.78}$$

For the filamentary current, $\mathbf{J}\ dv'$ can be replaced by $\boldsymbol{I}\ dl'$:

$$\tilde{\mathbf{A}}(r) = \frac{\mu}{4\pi}\frac{e^{-jkr}}{r}\int_{\text{Source}} \tilde{I}dl'. \tag{7.79}$$

If the filament is along the z-axis, then

$$\tilde{\mathbf{A}}(r) = \hat{z}\frac{\mu}{4\pi}\frac{e^{-jkr}}{r}\int_{\text{Source}} \tilde{I}dz'. \tag{7.80}$$

For a volume source, the integrand in Equation 7.78 can be manipulated:

$$\begin{aligned}\iiint_{v'} \tilde{\mathbf{J}}(\mathbf{r}')dv' &= -\iiint_{v'} \mathbf{r}'\left(\nabla'\cdot\tilde{\mathbf{J}}\right)dv' \\ &= j\omega\iiint_{v'} \mathbf{r}'\tilde{\rho}_v\left(r'\right)dv'.\end{aligned} \tag{7.81}$$

In order to obtain the last term in Equation 7.81, we use the continuity equation

$$\nabla\cdot\mathbf{J} + \frac{\partial\rho_v}{\partial t} = 0. \tag{7.82}$$

The middle term in Equation 7.81 is obtained through integration by parts, or by evaluating one of the components:

$$\begin{aligned}\iiint_{v'} \tilde{\mathbf{J}}_x dv' &= \iiint_{v'} \left(\tilde{\mathbf{J}}\cdot\hat{x}\right)dv' = \iiint_{v'} \left(\tilde{\mathbf{J}}\cdot\nabla'x'\right)dv' \\ &= \iiint_{v'} \nabla'\cdot\left(x'\tilde{\mathbf{J}}\right)dv - \iiint_{v'} x'\cdot\left(\nabla'\cdot\tilde{\mathbf{J}}\right)dv', \\ \iiint_{v'} \tilde{\mathbf{J}}_x dv'dv' &= \oint\!\!\oint_{s'} x'\cdot\tilde{\mathbf{J}}ds' - \iiint_{v'} x'\left(\nabla'\cdot\tilde{\mathbf{J}}\right)dv'.\end{aligned} \tag{7.83}$$

The first term on the RHS of Equation 7.83 is zero since the current density is tangent to the surface s'. Thus, we have

$$\begin{aligned}\iiint_{v'} \tilde{\mathbf{J}}_x dv' &= -\iiint_{v'} x'(\nabla'\cdot\mathbf{J})dv', \\ &= \iiint_{v'} \mathbf{r}'(\nabla'\cdot\mathbf{J})dv'.\end{aligned} \tag{7.84}$$

From Equations 7.78 and 7.81,

$$\tilde{\mathbf{A}}(\boldsymbol{r}) = \frac{\mathrm{j}\mu\omega}{4\pi}\frac{\mathrm{e}^{-jkr}}{r}\tilde{p}_e, \tag{7.85}$$

where $\tilde{p}_e$ is given by

$$\tilde{p}_\mathrm{e} = \iiint_{v'} \boldsymbol{r}'\rho_v(\boldsymbol{r}')\mathrm{d}v', \tag{7.86}$$

which is called the electric dipole moment.

7.5.1 Hertzian Dipole

A Hertzian dipole is a very short filament of length d, and if one considers the current to be constant, then Equation 7.79 becomes

$$\tilde{\mathbf{A}}(\boldsymbol{r}) = \hat{z}\frac{\mu(\tilde{\mathrm{I}}d)}{4\pi}\frac{\mathrm{e}^{-jkr}}{r}. \tag{7.87}$$

If $\tilde{I}$ is considered a constant, then the continuity equation requires point charges Q and $-Q$ at the ends of the filament. The dipole moment $\tilde{p}_\mathrm{e}$ of this pair of charges, when the filament is along the z-axis, is given by

$$\tilde{p}_\mathrm{e} = \hat{z}Qd. \tag{7.88}$$

Equation 7.86 gives the same result when evaluated for this very small length filament along the z-axis.

Evaluating $\tilde{\mathbf{B}}$ and $\tilde{\mathbf{H}}$ from the vector potential $\tilde{\mathbf{A}}$, in the far zone, we obtain

$$\tilde{\mathbf{H}} = \frac{u_\mathrm{p}k^2}{4\pi}(\tilde{r}\times\tilde{p}_\mathrm{e})\frac{\mathrm{e}^{-jkr}}{r} \tag{7.89}$$

$$\tilde{\mathbf{E}} = \eta\tilde{\mathbf{H}}\times\hat{r}, \tag{7.90}$$

where $u_\mathrm{p} = c = 3\times10^8$ m/s and $\eta = \eta_\mathrm{o} = 120\pi\ \Omega$ for a free-space medium. The equivalence of $\tilde{I}$ and $\tilde{Q}$ are clearly seen as

$$\begin{aligned} Q &= \int \mathbf{I}\mathrm{d}t, \\ \tilde{I} &= \mathrm{j}\omega\tilde{Q}. \end{aligned} \tag{7.91}$$

In Problem P1.5, we studied the Hertzian dipole (point dipole) making the approximations of a Hertzian dipole right at the beginning of the analysis.

The result for the case of a Hertzian dipole at the origin along the z-axis, that is, $\tilde{p}_\mathrm{e} = \tilde{I}_0 d/\mathrm{j}\omega$, is given as

$$\tilde{\mathbf{H}} = \frac{ck^2}{4\pi}\left(\tilde{r}\times\frac{\tilde{\mathrm{I}}_0 d}{\mathrm{j}\omega}\hat{z}\right)\frac{\mathrm{e}^{-jkr}}{r}. \tag{7.92}$$

Since $\hat{r}\times\hat{z}=(\hat{\rho}\rho+\hat{z}z)/r\times\hat{z}=-\rho\hat{\phi}/r=-\hat{\phi}\sin\theta$,

$$\tilde{\mathbf{H}}=\phi j\tilde{I}_0 d\frac{k^2}{4\pi}\frac{e^{-jkr}}{kr}\sin\theta, \tag{7.93}$$

$$\tilde{\mathbf{E}}=\eta_o\tilde{\mathbf{H}}\times\hat{r}. \tag{7.94}$$

The time-averaged power density is

$$\begin{aligned}\mathbf{S}_{av}&=\frac{1}{2}\mathrm{Re}\left[\tilde{\mathbf{E}}\times\tilde{\mathbf{H}}^*\right],\\ \mathbf{S}_{av}&=\hat{r}\eta_0\frac{k^2\tilde{I}_0^2d^2}{32\pi^2r^2}\sin^2\theta.\end{aligned} \tag{7.95}$$

The power coming out of a sphere of radius r is

$$P_{\mathrm{rad}}=\iint_s \mathbf{S}_{\mathrm{av}}\cdot ds, \tag{7.96}$$

When equated to $1/2\tilde{I}_0^2R_{\mathrm{rad}}$ (radiation resistance) gives

$$R_{\mathrm{rad}}=80\pi^2\left(\frac{d}{\lambda}\right)^2\ (\Omega). \tag{7.97}$$

The AC resistance of the wire of radius a and conductivity σ_c is given by

$$R_{\mathrm{AC}}=\frac{d}{\sigma_c 2\pi a\delta}, \tag{7.98}$$

where δ is the skin depth. For example [2], if $f = 75$ MHz, σ_c(copper) = 5.8×10^7, $a = 0.4$ mm, and $d = 4$ cm, then $R_{\mathrm{AC}} = 0.036\ \Omega$ and $R_{\mathrm{rad}} = 0.08\ \Omega$. Obviously the radiation efficiency is low, in the sense that the power loss as heat is about 50% of the power radiated.

7.5.2 Half-Wave Dipole

We can show that there will be a dramatic improvement in efficiency by increasing the length of the wire. To illustrate this aspect, consider a center-fed wire antenna of length $\lambda/2$. The current $\tilde{I}(z)$ given by

$$\tilde{I}(z)=\tilde{I}_0\cos kz,\quad -\frac{\lambda}{4}\le z\le\frac{\lambda}{4}, \tag{7.99}$$

is assumed to be a standing wave, and such an assumption supported by experiments simplifies the problem. A rigorous approach requires formulation as a boundary value problem [1] (Figure 7.13).

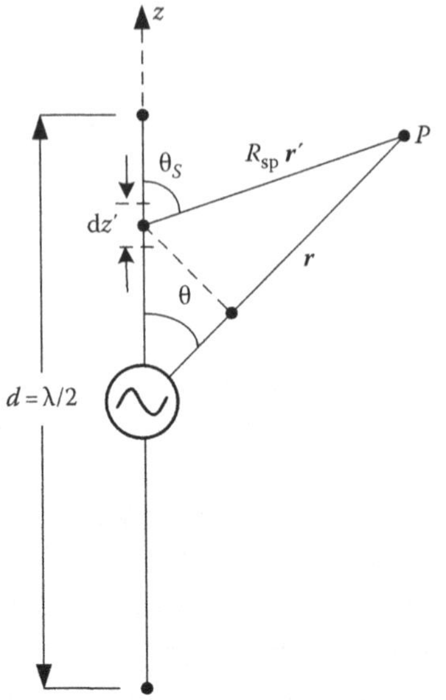

FIGURE 7.13
Half-wave center-fed dipole antenna.

A simple approach is to consider a differential length dz′ of this wire as a Hertzian point dipole. The radiated electric field in the far zone due to this differential source can be written, from Equations 7.93 and 7.94, as

$$d\tilde{\mathbf{E}}_\theta = \frac{jk_0\eta_0}{4\pi}\tilde{I}(z')dz'\frac{e^{-jkR_{sp}}}{R_{sp}}\sin\theta_s, \tag{7.100}$$

$$d\tilde{\mathbf{H}}_\phi = \frac{d\tilde{\mathbf{E}}_\theta}{\eta_0}. \tag{7.101}$$

In the far zone, we can make the following approximations

$$\theta_s \approx \theta, \tag{7.102}$$

$$R_{sp} = r - z'\cos\theta \tag{7.103}$$

in the exponent.

A further approximation for R_{sp} in the denominator can be made:

$$R_{sp} = r \tag{7.104}$$

giving

$$\tilde{E}_\theta = \frac{jk\eta_0}{4\pi}\frac{e^{-jkr}}{r}\tilde{I}_0\sin\theta\int_{z'=-\lambda/4}^{\lambda/4}\cos kz' e^{jkz'\cos\theta}dz' \tag{7.105}$$

$$\tilde{E}_\theta = j60\tilde{I}_0\left[\frac{\cos\left(\dfrac{\pi}{2}\cos\theta\right)}{\sin\theta}\right]\frac{e^{-jkr}}{r} \quad (7.106)$$

$$\tilde{H}_\phi = \frac{\mathbf{E}_\theta}{\eta_o}, \quad (7.107)$$

The approximations made above are equivalent to those made in Equation 7.76. Also note that in Equation 7.93, the expression for $\tilde{\mathbf{H}}$ was given first and $\tilde{\mathbf{E}}$ is expressed in terms of $\tilde{\mathbf{H}}$. In Equation 7.100, $d\tilde{\mathbf{E}}_\theta$ is given first and $d\tilde{\mathbf{H}}_\phi$ is then expressed in terms of $d\tilde{\mathbf{E}}_\theta$ [2].

The time-averaged power density in the far zone is given by

$$\mathbf{S}_{av} = \hat{r}S_o F(\theta, \phi), \quad (7.108a)$$

$$S_o = \frac{15\tilde{I}_o^2}{\pi r^2}, \quad (7.108b)$$

$$F(\theta, \phi) = \left[\frac{\cos\left(\dfrac{\pi}{2}\cos\theta\right)}{\sin\theta}\right]^2, \quad (7.108c)$$

where $F(\theta, \phi)$ is called normalized radiation intensity. Qualitatively, the pattern (Figure 7.14a) has the same features as that of the Hertzian dipole.

However, the radiated power evaluated by Equation 7.96, using Equation 7.108a, is given by

$$P_{rad} = 36.6\tilde{I}_0^2, \quad (7.109)$$

which in turn gives

$$R_{rad} = \frac{2P_{rad}}{\tilde{I}_0^2} \approx 73\ \Omega. \quad (7.110)$$

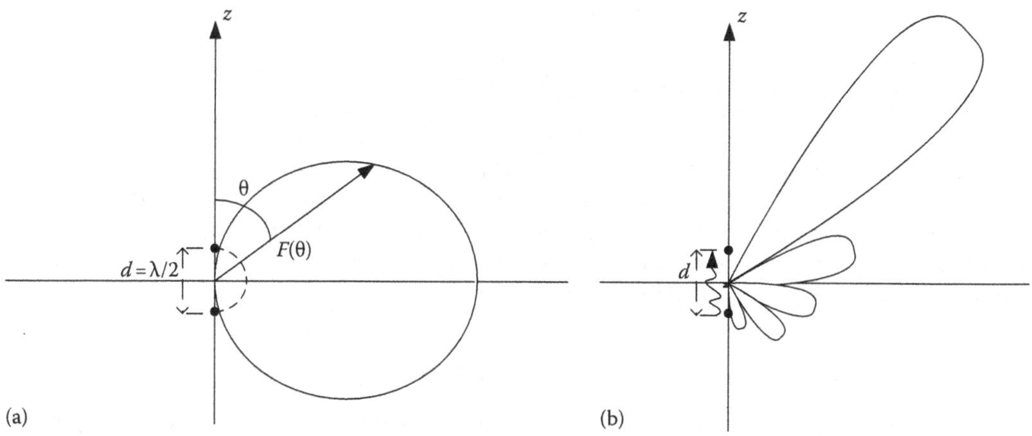

FIGURE 7.14
Radiation pattern $F(\theta)$ for (a) a standing wave source $d = \lambda/2$ and (b) a traveling wave source.

In this case, R_{loss} of the example given for Hertzian dipole (except for changing $d = \lambda/2$) gives [2]

$$R_{\text{loss}} = 1.8\ \Omega.$$

The efficiency of this antenna is quite good.

It can be noted that a monopole antenna with $d = \lambda/4$ over a ground plane has the same radiated fields as that of the half-wave dipole, for $z > 0$. The radiated power will be half of the half-wave dipole and the radiation resistance $R_{\text{rad}} \approx 36.5\ \Omega$.

7.5.3 Dipoles of Arbitrary Length

For an arbitrary length d relative to λ, for a center-fed antenna, the current distribution can be assumed as

$$\tilde{I}(z) = I_0 \sin\left[k\left(\frac{d}{z} - z\right)\right], \quad \text{for } 0 \le z \le \frac{d}{2}, \tag{7.111a}$$

$$= I_0 \sin\left[k\left(\frac{d}{2} + z\right)\right], \quad \text{for } -\frac{d}{2} \le z \le 0. \tag{7.111b}$$

And using the same technique as for the half-wave dipole, one can obtain

$$F(\theta, \phi) = \left[\frac{\cos\left((\pi d/\lambda)\cos\theta\right) - \cos(\pi d/\lambda)}{\sin\theta}\right]^2. \tag{7.112}$$

A plot of the radiation pattern for $d = \lambda/2$, $d = \lambda$, and $d = 3\lambda/2$ can be found in [2].

7.5.4 Shaping the Radiation Pattern

Ulaby [2] gives an elementary discussion of aperture antennas, antenna arrays, and other topics of practical interest in shaping the radiation pattern.

7.5.5 Antenna Problem as a Boundary Value Problem

A rigorous formulation of the antenna problem is a boundary value problem. The assumption that the current distribution is known simplifies the mathematics and allows us to compute the radiated fields. At a more basic level, what is known is the field distribution across a gap in the antenna; say, one can know the electric field in the gap. The additional boundary conditions are that the current vanishes at the end of the wires. Such a formulation is used in computational electromagnetics.

7.5.6 Traveling Wave Antenna and Cerenkov Radiation

The thin-wire antenna problem discussed earlier is based on an assumed standing wave current distribution. It will be interesting to study the effect of a traveling wave current distribution.

Let $\tilde{I}(z)$ be given by

$$\tilde{I}(z) = I_0 e^{-jpkz}. \tag{7.113}$$

The radiation vector **N** defined as [3] the integral,

$$\tilde{\mathbf{N}} = \iiint_{\text{Source}} \tilde{\mathbf{J}}(\mathbf{r}') e^{jk\hat{r}\cdot r'} dv'. \tag{7.114}$$

For this case, it becomes

$$\tilde{\mathbf{N}} = \hat{z} I_0 \int_{-d/2}^{d/2} e^{jkz'\cos\theta} e^{-jpkz'} dz'. \tag{7.115}$$

And the vector potential in the far zone, from Equation 7.76, is given by

$$\tilde{\mathbf{A}}(r) = \frac{\mu_0}{4\pi} \frac{e^{-jkr}}{r} \tilde{\mathbf{N}}. \tag{7.116}$$

Evaluating $\tilde{E}_\theta$, $\tilde{H}_\phi$, and S_{av} as before, the normalized radiation intensity for the traveling wave current excitation can be shown [3] to be

$$S_{av} = \hat{r} \frac{\eta I_0^2}{8\pi^2 r^2} F(\theta, \phi), \tag{7.117a}$$

$$F(\theta, \phi) = \left\{ \sin\theta \frac{\sin\left[(\pi d/\lambda)(p - \cos\theta)\right]}{p - \cos\theta} \right\}^2. \tag{7.117b}$$

The normalized radiation intensity $F(\theta, \phi)$ for the traveling wave source given by Equation 7.117b differs from the radiation intensity for the standing wave source given by Equation 7.112 in one important aspect. The former is asymmetrical with respect to the equatorial plane $\theta = \pi/2$, whereas the latter is symmetrical. The traveling wave source creates a major lobe in the radiation pattern in the forward direction. Figure 7.14 shows the sketch of the radiation pattern for two cases.

The maximum radiation is in the neighborhood of $\theta = 0$ and the minimum radiation is in the neighborhood of $\theta = \pi$.

For the traveling wave case, the maximum radiation appears as a cone in the forward direction of the wave traveling, and the half-angle of the cone decreases as p increases or as d/λ increases.

Papas [3] points out the similarity of the conical beam radiation to that of Cerenkov radiation from fast electrons.

7.5.7 Small Circular Loop Antenna

We will assume that the circumference of the loop is small compared to the wavelength. At a field point far away from the loop, the magnetic field due to steady current is similar to the electric field due to a static electric dipole.

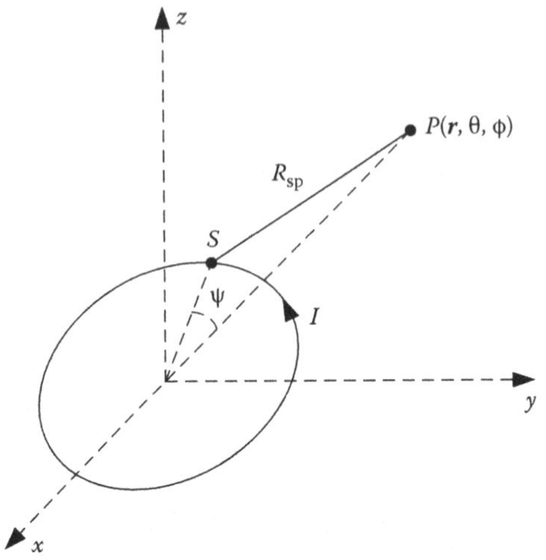

FIGURE 7.15
Magnetic dipole antenna.

See Section 8.9. Hence this antenna is called the magnetic dipole antenna. In applying Equation 7.114 to this case, we replace $\tilde{\mathbf{J}}dv'$ by $\hat{\phi}\tilde{I}a\cos\phi'd\phi'$ [4]. This replacement is easily explained if we consider the coordinates of the field point as $P(r, \theta, 0)$ without loss of generality in the light of ϕ symmetry of the source.

The ϕ component of $\tilde{\mathbf{A}}$ is the same as the y component of $\tilde{\mathbf{A}}$ at this point. The y component of $\tilde{\mathbf{A}}$ can be computed from knowing the y component of $\hat{\mathbf{J}}$.

Thus, $\tilde{\mathbf{J}}dv' = \hat{y}\tilde{J}_y\, dx'\, dz'\, dy' = \hat{y}\tilde{I}\, dy'$. Since $y' = a \sin \phi'$, $dy' = a \cos \phi' \, d\phi'$. Moreover, $k\hat{r}\mathbf{r}'$ in (7.114) and (7.76) is given by $a \cos \psi$, where ψ is the angle shown in Figure 7.15 and is given by

$$\cos\psi = \cos\theta\cos\theta' + \sin\theta\sin\theta'\cos(\phi - \phi'). \tag{7.118}$$

When the source is in the x–y-plane, $\theta' = \pi/2$.

For the computation of $\mathbf{A}_y$, $\phi = 0$ and we obtain for the case under consideration

$$\cos\psi = \sin\theta\cos\phi'. \tag{7.119}$$

Thus, we obtain

$$N_\phi = \tilde{I}\int_0^{2\pi} e^{jka\sin\theta\cos\phi'} a\cos\phi' d\phi'. \tag{7.120}$$

Assuming that ka is small, we can approximate (7.120) as

$$\begin{aligned} N_\phi &\approx \tilde{I}\int_0^{2\pi} (1 + jka\sin\theta\cos\phi')a\cos\phi' d\phi' \\ &= jk\pi a^2\tilde{I}\sin\theta, \\ N_\phi &\approx jkp_m\sin\theta, \end{aligned} \tag{7.121}$$

where $p_m = \pi a^2\tilde{I}$ is the magnetic dipole moment.

Computing the fields in the far zone, the time-averaged power density S_{av}, the total power radiated P, and equating to $(1/2)I^2R_{rad}$, one can obtain [4,5]

$$R_{rad} = 20\pi^2 (ka)^4 \ (\Omega). \tag{7.122}$$

7.5.8 Other Practical Radiating Systems

Ramo et al. [4] have listed the steps involved in systemization of calculations for obtaining the radiated power, originally given by Schelkunoff [6], and applied them to many radiating systems.

7.6 Scattering

Let us assume that a time-harmonic electromagnetic wave is propagating in an unbounded medium whose electric and magnetic fields are $\tilde{\mathbf{E}}^i$ and $\tilde{\mathbf{H}}^i$, respectively. Let the fields be $\tilde{\mathbf{E}}^t$ and $\tilde{\mathbf{H}}^t$, when a structure is present in the medium. The interaction of the incident wave (superscript i) with the structure could produce charges and currents in the structure and on the surface of the structure. These sources in turn produce scattering fields $\tilde{\mathbf{E}}^s$ and $\tilde{\mathbf{H}}^s$. The total fields, in the presence of the structure, are given by

$$\tilde{\mathbf{E}}^t = \tilde{\mathbf{E}}^i + \tilde{\mathbf{E}}^s, \tag{7.123a}$$

$$\tilde{\mathbf{H}}^t = \tilde{\mathbf{H}}^i + \tilde{\mathbf{H}}^s. \tag{7.123b}$$

We can interpret that the scattering fields are those produced by the induced sources in the structure due to the interaction of the incident fields with the structure. The relationship between the incident fields and the induced sources in the structure can be found by using, say, the boundary conditions; as a second step, we can compute the scattered fields. Balanis [7] has several examples of such calculations. In this section, we will only do one example of such calculations, that is, the case of the two-dimensional plane wave TM (Transverse Magnetic) scattering by a circular conducting long cylinder [7].

First, we will list the relation between plane waves and cylindrical wave functions.

7.6.1 Cylindrical Wave Transformations [7]

$$e^{-j\beta x} = e^{-j\beta\rho\cos\phi} = \sum_{n=-\infty}^{\infty} (j)^{-n} J_n(\beta\rho) e^{jn\phi}, \tag{7.124a}$$

$$e^{+j\beta x} = e^{j\beta\rho\cos\phi} = \sum_{n=-\infty}^{\infty} (j)^{n} J_n(\beta\rho) e^{jn\phi}, \tag{7.124b}$$

$$H_0^{(1,2)}(\beta|\rho-\rho'|) = \sum_{n=-\infty}^{\infty} J_n(\beta\rho) H_n^{(1,2)}(\beta\rho') e^{jn(\phi-\phi')}, \quad \rho \le \rho', \tag{7.125a}$$

$$H_o^{(1,2)}\left(\beta\left|\rho-\rho'\right|\right)=\sum_{n=-\infty}^{\infty}J_n\left(\beta\rho'\right)H_n^{(1,2)}\left(\beta\rho\right)e^{jn(\phi-\phi')},\quad \rho\geq\rho', \tag{7.125b}$$

$$J_o\left(\beta\left|\rho-\rho'\right|\right)=\sum_{n=-\infty}^{\infty}J_n\left(\beta\rho'\right)J_n\left(\beta\rho\right)e^{jn(\phi-\phi')}, \tag{7.126}$$

$$Y_o\left(\beta\left|\rho-\rho'\right|\right)=\sum_{n=-\infty}^{\infty}Y_n\left(\beta\rho'\right)J_n\left(\beta\rho\right)e^{jn(\phi-\phi')},\quad \rho\leq\rho', \tag{7.127a}$$

$$Y_o\left(\beta\left|\rho-\rho'\right|\right)=\sum_{n=-\infty}^{\infty}Y_n\left(\beta\rho\right)J_n\left(\beta\rho'\right)e^{jn(\phi-\phi')},\quad \rho\geq\rho'. \tag{7.127b}$$

7.6.2 Calculation of Current Induced on the Cylinder [7]

Let the electric field of the incident wave be

$$\tilde{\mathbf{E}}^i=\hat{z}\mathbf{E}_o e^{-j\beta x}=\hat{z}\mathbf{E}_o e^{-j\beta\rho\cos\phi}. \tag{7.128}$$

From Equation 7.124a, we have

$$\tilde{\mathbf{E}}^i=\hat{z}\mathbf{E}_o\sum_{n=-\infty}^{\infty}(j)^{-n}J_n\left(\beta\rho\right)e^{jn\phi}=\hat{z}\mathbf{E}_o\sum_{n=0}^{\infty}(-j)^{n}\varepsilon_n J_n\left(\beta\rho\right)\cos n\phi, \tag{7.129a}$$

where

$$\varepsilon_n=\begin{cases}1, & n=0,\\ 2, & n\neq 0.\end{cases} \tag{7.129b}$$

The scattered field $\tilde{E}^s$ can be written as

$$\tilde{E}^s=\hat{z}E_z^s=\hat{z}E_o\sum_{n=-\infty}^{\infty}c_n H_n^{(2)}\left(\beta\rho\right). \tag{7.130}$$

In writing Equation 7.130, we note that the scattered field has to be an outgoing wave as $\beta\rho\to\infty$, and hence we choose $H_n^{(2)}$ as discussed in Section 2.15.

The PEC boundary condition at $\rho=a$ gives

$$E_z^t\Big|_{\rho=a}=0=E_z^i+E_z^s\Big|_{\rho=a}, \tag{7.131}$$

$$E_o\sum_{n=-\infty}^{\infty}\left[c_n H_n^{(2)}\left(\beta a\right)+(j)^{-n}J_n\left(\beta a\right)e^{jn\phi}\right]=0,$$
$$c_n=-(j)^{-n}\frac{J_n\left(\beta a\right)}{H_n^{(2)}\left(\beta a\right)}e^{jn\phi}. \tag{7.132}$$

The scattered electric field E_z^s is given by

$$E_z^s = -E_o \sum_{n=0}^{\infty} (-j)^n \varepsilon_n \frac{J_n(\beta a)}{\mathbf{H}_n^{(2)}(\beta a)} H_n^{(2)}(\beta\rho)\cos n\phi. \tag{7.133}$$

The z component of the electric field at any $\rho \geq a$ is given by

$$E_z^t = E_o \sum_{n=-\infty}^{\infty} (j)^{-n} \left[J_n(\beta a) - \frac{J_n(\beta a)}{H_n^{(2)}(\beta a)} H_n^{(2)}(\beta\rho) \right] e^{jn\phi}. \tag{7.134}$$

From Maxwell's equation,

$$\tilde{\mathbf{H}} = -\frac{1}{j\omega\mu} \nabla \times \tilde{\mathbf{E}}. \tag{7.135}$$

One can obtain $\tilde{\mathbf{H}}$ (*note* $E_\rho = E_\phi = 0$). The $\tilde{H}_z$ component will be zero. The other tangential component $\tilde{H}_\phi$ can be found. The surface current **K** on the surface of the cylinder is given by

$$\mathbf{K} = \hat{n} \times \tilde{\mathbf{H}}^t \Big|_{\rho=a} = \hat{z} \mathrm{H}_\phi^t \Big|_{\rho=a}, \tag{7.136}$$

which comes out to be [7]

$$\mathbf{K} = \hat{z} \frac{2E_0}{\pi a \omega\mu} \sum_{n=-\infty}^{\infty} (j)^{-n} \frac{e^{jn\phi}}{H_n^{(2)}(\beta a)}. \tag{7.137}$$

For ($a \ll \lambda$) very thin wires, the first term is dominant:

$$\mathbf{K} \approx \hat{z} \frac{2E_o}{\pi a \omega\mu} \frac{1}{H_n^{(2)}(\beta a)}, \quad a \ll \lambda. \tag{7.138}$$

We can approximate further by using the asymptotic value of $H_n^{(2)}$ for small βa [7]:

$$\tilde{\mathbf{K}} = \hat{z} j \frac{E_o}{a\omega\mu} \frac{1}{\ln[1.781\beta a/2]}, \quad a \ll \lambda. \tag{7.139}$$

7.6.3 Scattering Width

Scattering by the target is quantified by a parameter called *echo area, radar cross section* (RCS) σ. The RCS is formally defined as the area intercepting the amount of power that, when scattered isotropically, produces at the receiver a density scattered by the actual target [7]. For a three-dimensional target,

$$\sigma_{3D} = \lim_{r\to\infty} \left[4\pi r^2 \frac{S^s}{S^i} \right]. \tag{7.140}$$

For a two-dimensional target, the parameter of interest is called scattering width, σ_{2D}:

$$\sigma_{2D} = \lim_{\rho\to\infty}\left[2\pi\rho\frac{S^s}{S^i}\right] \tag{7.141a}$$

$$= \lim_{\rho\to\infty}\left[2\pi\rho\frac{|E^s|^2}{|E^i|^2}\right]. \tag{7.141b}$$

Since $\rho \to \infty$, we can make far-zone approximations and replace

$$H_n^{(2)}(\beta\rho) \approx \sqrt{\frac{2j}{\pi\beta\rho}}\, j^n e^{-j\beta\rho}.$$

Evaluating Equation 7.141b [7],

$$\sigma_{2D} = \frac{2\lambda}{\pi}\left|\sum_{n=0}^{\infty}\varepsilon_n \frac{J_n(\beta a)}{H_n^{(2)}(\beta a)}\cos n\phi\right|^2, \tag{7.142a}$$

where

$$\begin{cases}\varepsilon_n = 1, & n = 0,\\ \varepsilon_n = 2, & n \neq 0.\end{cases} \tag{7.142b}$$

Note the ϕ dependence of the scattering width. ϕ is the azimuthal angle of the observation point with reference to the direction of the propagation vector of the incident plane wave. Such a directional-dependent scattering width is called *bistatic* width.

For $\phi = 180°$, it is called monostatic width or backscatter width:

$$\sigma_{2D} = \frac{2\lambda}{\pi}\sum_{n=0}^{\infty}\varepsilon_n\left|\frac{J_n(\beta a)}{H_n^{(2)}(\beta a)}\right|^2. \tag{7.143}$$

For the low-frequency limit, βa is small, the $n = 0$ term is dominant, and

$$\sigma_{2D} = \frac{\pi\lambda}{2}\left|\frac{1}{\ln(0.89\beta a)}\right|^2, \quad a \ll \lambda. \tag{7.144}$$

For higher frequencies, the convergence of the series in Equation 7.143 is slow. For $\beta a = 3$, six terms give satisfactory results. For $\beta a = 100$, over 100 terms are needed [8].

The moment method is successfully used to study the scattering problems of PEC and dielectric objects [9].

7.7 Diffraction

Figure 7.16 shows a PEC plate with a circular hole of radius a. If a plane electromagnetic (EM) wave normally incident on it from the left is considered as a photon particle, then the photons in the circular area of radius a will go through the hole and the rest of them will be totally reflected. A circular beam of radius a will emerge into the right-half of the space. The wave front remains to be a plane as shown in Figure 7.16a. This is "the geometrical optics" approximation of the ray theory. The wave nature of the photon, however, will cause "diffraction" and cause the divergence of the beam that emerges from the hole. A diffracted ray follows a path which cannot be interpreted through reflection or refraction. Figure 7.16b shows a spherical wave front. The beam divergence can be quantified by calculating the angle θ_0 [4,10] shown in Figure 7.16b. The theory of diffraction is explained through "physical optics," making use of Maxwell equations. Various approximations are possible based on the size of (a/λ), where λ is the wavelength.

A rigorous theory of diffraction is mathematically intense. In this section, we give the theory for the calculation of θ_0 approximately. In the process, we introduce the concept of magnetic current, $\mathbf{J}_m$, electric vector potential **F**, and the magnetic radiation vector **L**.

7.7.1 Magnetic Current and Electric Vector Potential

Since magnetic monopoles do not exist in nature, the concept of magnetic charge density (Wb/m^3) and the magnetic current density (V/m^2) is artificial. However, their introduction into Maxwell equations brings out certain symmetry to the curl equations and helps us to formulate the duality theorem [7]. As we see in the next section, it also helps us to solve practical boundary value problems. The modified Maxwell equations are

$$\nabla \cdot \mathbf{D} = \rho_{ev}, \tag{7.145}$$

$$\nabla \cdot \mathbf{B} = \rho_{mv}, \tag{7.146}$$

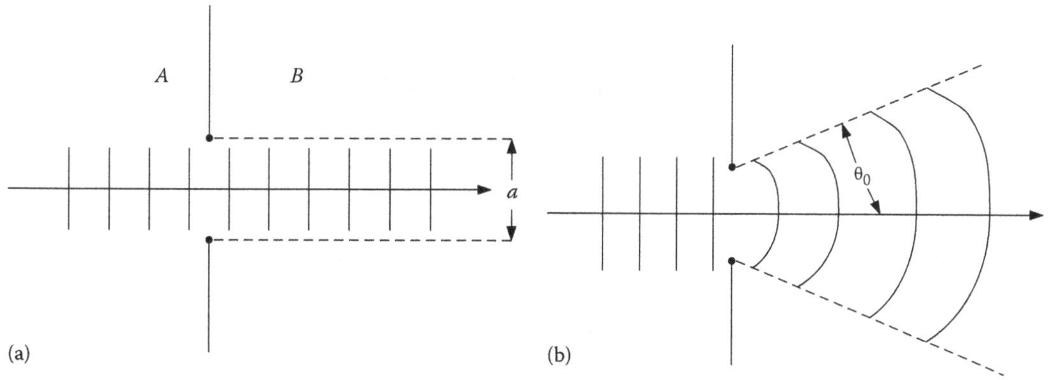

FIGURE 7.16
Propagation of the beam through a circular hole of radius a. (a) Geometrical optics. (b) Physical optics: the beam diverges.

$$-\nabla \times \mathbf{E} = +\mathbf{J}_m + \frac{\partial \mathbf{B}}{\partial t}, \tag{7.147}$$

$$\nabla \times \mathbf{H} = +\mathbf{J}_e + \frac{\partial \mathbf{D}}{\partial t}. \tag{7.148}$$

The boundary conditions Equations 1.14 through 1.17 will be modified as

$$\hat{n}_{12} \times (\mathbf{D}_2 - \mathbf{D}_1) = \rho_{es}, \tag{7.149}$$

$$\hat{n}_{12} \cdot (\mathbf{B}_2 - \mathbf{B}_1) = \rho_{ms}, \tag{7.150}$$

$$-\hat{n}_{12} \times (\mathbf{E}_2 - \mathbf{E}_1) = \mathbf{M}, \tag{7.151}$$

$$\hat{n}_{12} \times (\mathbf{H}_2 - \mathbf{H}_1) = \mathbf{K}, \tag{7.152}$$

where **M** (V/m) is the surface magnetic current density. The concepts **M** and **K** will be useful in replacing the tangential electric and magnetic fields in an aperture, as the one shown in Figure 7.16 and **E** and **H** are the fields on the *A* side of the opening. Placing the fictitious currents in the opening reduces the fields to zero in the region *A* and give rise, in region *B*, the fields that would have been there due to the original sources in region *A*.

Equation 1.50, when applied to a surface current source, gives us, for the time-harmonic case, the magnetic vector potential

$$\tilde{\mathbf{A}}(r) = \frac{\mu}{4\pi} \iint_S \tilde{\mathbf{K}}(\mathbf{r}') \frac{e^{-jk|r-r'|}}{|r-\mathbf{r}'|} ds'. \tag{7.153}$$

In a source-free region,

$$\nabla \cdot \tilde{\mathbf{D}} = 0. \tag{7.154}$$

And we can define an electric vector potential related to

$$\tilde{\mathbf{D}}_F = -\nabla \times \tilde{\mathbf{F}}, \tag{7.155}$$

$$\mathbf{E}_F = -\frac{1}{\varepsilon} \nabla \times \tilde{\mathbf{F}}. \tag{7.156}$$

The electric vector potential $\tilde{\mathbf{F}}$ can be related to its magnetic current source

$$\tilde{\mathbf{F}}(r) = \frac{\varepsilon}{4\pi} \iint_S \tilde{\mathbf{M}}(r') \frac{e^{-jk|r-r'|}}{|r-r'|} ds'. \tag{7.157}$$

The fields $\tilde{\mathbf{E}}$ and $\tilde{\mathbf{H}}$ can be computed from $\tilde{\mathbf{A}}$ and $\tilde{\mathbf{F}}$:

$$\tilde{\mathbf{E}} = -j\omega\tilde{\mathbf{A}} - \frac{j\omega}{k^2}\nabla\left(\nabla\cdot\tilde{\mathbf{A}}\right) - \frac{1}{\varepsilon}\nabla\times\tilde{\mathbf{F}}, \tag{7.158}$$

$$\tilde{\mathbf{H}} = -j\omega\tilde{\mathbf{F}} - \frac{j\omega}{k^2}\nabla\left(\nabla\cdot\tilde{\mathbf{F}}\right) + \frac{1}{\mu}\nabla\times\tilde{\mathbf{A}}. \tag{7.159}$$

The middle term on the RHS of Equation 7.158 comes from the Lorentz condition:

$$\nabla\cdot\mathbf{A} + \mu\varepsilon\frac{\partial\phi}{\partial t} = 0, \tag{7.160}$$

$$\begin{aligned} &\nabla\cdot\tilde{\mathbf{A}} + \mu\varepsilon j\omega\tilde{\phi} = 0, \\ &\tilde{\phi} = -\frac{1}{j\omega\mu\varepsilon}\nabla\cdot\tilde{\mathbf{A}}, \end{aligned} \tag{7.161}$$

$$\begin{aligned} &\mathbf{E} = -\nabla\phi - \frac{\partial A}{\partial t}, \\ &\tilde{\mathbf{E}} = -\nabla\tilde{\phi} - j\omega\tilde{\mathbf{A}}, \end{aligned} \tag{7.162}$$

$$\begin{aligned} &\tilde{\mathbf{E}} = \frac{1}{j\omega\mu\varepsilon}\nabla\left(\nabla\cdot\tilde{\mathbf{A}}\right) - j\omega\tilde{\mathbf{A}}, \\ &\tilde{\mathbf{E}} = -j\frac{\omega}{k^2}\nabla\left(\nabla\cdot\tilde{\mathbf{A}}\right) - j\omega\tilde{\mathbf{A}}. \end{aligned} \tag{7.163}$$

The last term in Equation 7.158 comes from the contribution of the electric vector potential.

Equation 7.159 can be obtained on similar lines.

In Equations 7.153 and 7.157, we can substitute $\left(\hat{n}\times\tilde{\mathbf{H}}\right)$ for $\tilde{\mathbf{K}}$ and $\left(-\hat{n}\times\tilde{\mathbf{E}}\right)$ for $\tilde{\mathbf{M}}$. Thus, the potentials are given in terms of the corresponding tangential components of the fields.

We can define a magnetic radiation vector **L**, analogous to **N** given by Equation 7.114 by replacing the electric current density by the magnetic current density:

$$\tilde{\mathbf{L}} = \iiint_{\text{Source}} \tilde{\mathbf{J}}_m(\mathbf{r}')e^{jk\hat{r}\cdot r'}dv'. \tag{7.164}$$

And the electric vector potential $\tilde{\mathbf{F}}$ in the far zone is given by

$$\tilde{\mathbf{F}} = \varepsilon\frac{e^{-jkr}}{4\pi r}\tilde{\mathbf{L}}. \tag{7.165}$$

7.7.2 Far-Zone Fields and Radiation Intensity [4]

The electric and magnetic fields that do not decrease faster than $1/r$ are now obtained from Equations 7.158 and 7.159 as

$$\tilde{E}_\theta = -j\frac{e^{-jkr}}{2\lambda r}\left(\eta N_\theta + L_\phi\right), \tag{7.166}$$

$$\tilde{H}_\phi = \frac{\tilde{E}_\theta}{\eta}, \tag{7.167}$$

$$\tilde{E}_\phi = j\frac{e^{-jkr}}{2\lambda r}\left(-\eta N_\phi + L_\theta\right), \tag{7.168}$$

$$\tilde{H}_\theta = \frac{-\tilde{E}_\phi}{\eta}. \tag{7.169}$$

Note that the factor $\omega\mu/4\pi$ that appears in some of the electric formulas can also be written as

$$\frac{\omega\mu}{4\pi} = \frac{f\mu}{2} = \frac{f}{2}\sqrt{\mu}\sqrt{\varepsilon}\frac{\sqrt{\mu}}{\sqrt{\varepsilon}} = \frac{f}{2c}\eta = \frac{\eta}{2\lambda}.$$

The time-averaged power density $\boldsymbol{S}_{av}$ can be calculated from which the total power radiated per unit solid angle $dP/d\Omega$ called radiation intensity can be calculated:

$$S_{av}r^2 = \frac{dP}{d\Omega} = \frac{\eta}{8\lambda^2}\left[\left|N_\theta + \left(\frac{L_\phi}{\eta}\right)\right|^2 + \left|N_\phi - \left(\frac{L_\theta}{\eta}\right)\right|^2\right]. \tag{7.170}$$

Note that the vector $\hat{r}\cdot\boldsymbol{r}'$ in Equation 7.164 is given by

$$\hat{r}\cdot\boldsymbol{r}' = \boldsymbol{r}'\cos\psi. \tag{7.171}$$

When ψ, the angle between $\boldsymbol{r}'$ and $\boldsymbol{r}$, is expressed in terms of the spherical coordinates, we have

$$\cos\psi = \cos\theta\cos\theta' + \sin\theta\sin\theta'\cos(\phi - \phi'). \tag{7.172}$$

7.7.3 Elemental Plane Wave Source and Radiation Intensity [4]

Let us consider a differential surface element ds on a uniform plane wave, having E_x and H_y and propagating in the z-direction. The equivalent current sheets have

$$K_x = -H_y = -\frac{E_x}{\eta}, \quad M_y = -E_x. \tag{7.173}$$

From Equation 7.114, for a differential surface source, noting that $\psi = 90°$,

$$N_x = -\frac{E_x ds}{\eta}. \tag{7.174}$$

And from Equation 7.164, for a differential surface source,

$$L_y = -E_x ds, \tag{7.175}$$

$$N_\theta = N_x\cos\phi\cos\theta, \tag{7.176}$$

$$N_\phi = -N_x \sin\phi, \tag{7.177}$$

$$L_\theta = L_y \sin\phi \cos\theta, \tag{7.178}$$

$$L_\phi = L_y \cos\phi. \tag{7.179}$$

From Equation 7.170,

$$\frac{dP}{d\Omega} = \frac{E_x^2 (ds)^2}{2\eta\lambda^2} \cos^4 \frac{\theta}{2}. \tag{7.180}$$

Note that the plane wave differential surface source is equivalent to crossed electric and magnetic point dipoles.

7.7.4 Diffraction by the Circular Hole

At the source point $S(r', \pi/2, \phi')$, the electric field of the uniform plane wave is given by

$$E_x = E_0 e^{jk\hat{r}\cdot\bar{r}'} = E_0 e^{jkr'\cos\psi}, \tag{7.181}$$

and since $\theta' = \pi/2$ from Equation 7.118, we have

$$\cos\psi = \sin\theta \cos(\phi - \phi'). \tag{7.182}$$

We can compute $dP/d\Omega$ due to the fields in the circular area by integrating Equation 7.180 over the circular area:

$$\frac{dP}{d\Omega} = \frac{\mathbf{E}_0^2}{2\eta\lambda^2} \cos^4 \frac{\theta}{2} \left| \int_0^{2\pi}\int_0^{a} e^{jkr'\sin\theta\cos(\phi-\phi')} r'\, dr'\, d\phi' \right|^2. \tag{7.183}$$

It can be shown that

$$\int_0^{2\pi} e^{jq\cos\phi}\, d\phi = 2\pi J_o(q). \tag{7.184}$$

Thus,

$$\frac{dP}{d\Omega} = \frac{E_0^2}{2\eta\lambda^2} \cos^4 \frac{\theta}{2} \left| 2\pi \int_0^{a} J_o(kr'\sin\theta) r'\, dr' \right|^2. \tag{7.185}$$

From the integral

$$\int v^\nu J_{\nu-1}(v)\, dv = \nu^\nu J_\nu(\nu), \tag{7.186}$$

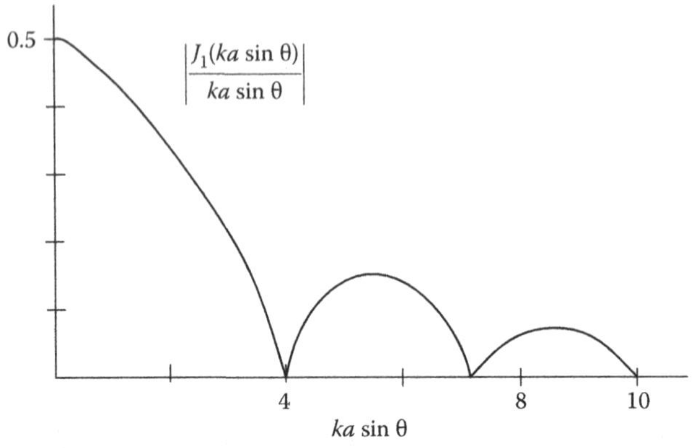

FIGURE 7.17
Approximate form of radiation pattern of a circular aperture for small θ.

Equation 7.185 is evaluated:

$$\frac{dP}{d\Omega} = \frac{2E_0^2\pi^2a^4}{\eta\lambda^2}\cos^4\frac{\theta}{2}\left[\frac{J_1(ka\sin\theta)}{ka\sin\theta}\right]^2. \tag{7.187}$$

For small θ, it is the second term in θ that establishes the pattern. A plot of bracketed term is shown in Figure 7.17.

The main lobe has $\theta = \theta_o$, given by the first zero in the graph of Figure 7.17:

$$\begin{aligned} ka\sin\theta_0 &= 3.832, \\ \frac{2\pi}{\lambda}a\theta_0 &= 3.832. \\ \theta_0 &= 0.61\frac{\lambda}{a}. \end{aligned} \tag{7.188}$$

The above result is valid for $ka \gg 1$ and the beam divergence is reduced by increasing a/λ and becomes zero in the limit.

The total power transmitted by the hole can be obtained from Equation 7.183:

$$P = \int_0^{2\pi}\int_0^{\pi/2} \frac{dP}{d\Omega}\sin\theta d\theta d\phi. \tag{7.189}$$

Note that the upper limit for θ integration is $\theta = \pi/2$ and not π. The incident power P_i is given by

$$P_i = \frac{E_0^2}{2\eta}\pi a^2. \tag{7.190}$$

And the power transmission coefficient is

$$\tau = \frac{P}{P_i}. \tag{7.191}$$

In the two extreme cases of $ka \gg 1$ and $ka \ll 1$, it can be shown [1] that

$$\tau = 1, \quad ka \gg 1, \tag{7.192a}$$

$$\tau = \frac{1}{3}(ka)^2, \quad ka \ll 1. \tag{7.192b}$$

Equation 7.192b has some inconsistent approximations, but shows that the transmission is small for small holes. We illustrate the order of the power transmission coefficient for a screen in the front door window of the microwave oven consumer product which has a metal screen with small holes.

At the operating frequency of 2.45 GHz ($\lambda = 12$ cm), $a = 0.5$ mm, from Equation 7.192b,

$$\tau = \frac{1}{3}\left(2\pi\frac{0.5\times10^{-3}}{0.12}\right)^2 \approx 2\times10^{-6}.$$

At microwave frequencies, the screen in the window is practically a conductor preventing the microwave radiation from escaping from the inside of the microwave oven. This is the principle of shielding using the Faraday's case [11].

For visual frequencies, say $\lambda = 0.5 \times 10^{-6}$ m, $ka \approx 6 \times 10^3$, and from Equation 7.192a, $\tau = 1$. A person can easily see through the microwave oven window into the food chamber.

Schmitt [11] has illustrative graphs of simulated diffraction patterns of the radiation intensity for apertures of various widths with an incident radiation of 100 MHz ($\lambda = 3$ m), patterns measured 10 m from the aperture.

A complimentary screen to the circular aperture is a circular disk. Babinet's principle of complementary screens [1] can be used to study the interactions with the complimentary screen in terms of the solution of the screen.

Diffraction radiation in the far zone is called Fraunhoffer diffraction. Fresnel diffraction is the radiation close to the aperture.

The rigorous treatment of scattering and diffraction are mathematically intensive and Jackson [1] and Balanis [7] have such a treatment embedded in these textbooks on electromagnetic theory.

References

1. Jackson, J. D., *Classical Electrodynamics*, 3rd edn., Wiley, New York, 1999.
2. Ulaby, F. T., *Applied Electromagnetics*, Prentice Hall, Englewood Cliffs, NJ, 2001.
3. Papas, C. H., *Theory of Electromagnetic Wave Propagation*, McGraw-Hill, New York, 1965.
4. Ramo, S., Whinnery, J. R., and Van Duzer, T., *Fields and Waves in Communication Electromagnetics*, Wiley, New York, 1965.
5. Lorrain, P., Corson, D. P., and Lorrain, F., *Electromagnetic Fields and Waves*, 3rd edn., W.H. Freeman, New York, 1988.
6. Schelkunoff, S. A., A general radiation formula, *Proc. I.R.E.*, 27, 660–666, 1939.

7. Balanis, C. A., *Advanced Engineering Electromagnetics*, Wiley, New York, 1989.
8. Van Bladel, J., *Electromagnetic Fields*, McGraw-Hill, New York, 1964.
9. Harrington, R. F., *Field Computation by Moment Methods*, Macmillan, New York, 1968.
10. Schwarz, S. E., *Electromagnetics for Engineers*, Sanders College Publishing, Philadelphia, PA, 1990.
11. Schmitt, R., *Electromagnetics Explained*, Newnes, Amsterdam, the Netherlands, 2002.

Part II

Electromagnetic Equations of Complex Media

8

Electromagnetic Modeling of Complex Materials*

The models we used till now assumed that the electromagnetic parameters σ, μ, and ε are scalar constants. In this section, we examine the circumstances under which the electromagnetic parameters are not scalar constants. In particular, we study the frequency dependence of the dielectric constant. In Section 8.1, we first study the interaction of an external electric field with a dielectric material by considering a volume of electric dipoles in free space. In Section 8.2, we study the frequency dependence of a dielectric material by considering a mechanical equivalent of an atom in a dielectric.

8.1 Volume of Electric Dipoles

Let us briefly revise the concept of a dipole. An elementary electric dipole is constructed by a negative and a positive charges of equal value separated by a small distance d (Figure 8.1).

The dipole moment $\boldsymbol{p}$ of this system is given by

$$\boldsymbol{p} = q\boldsymbol{d} = \hat{z}qd. \tag{8.1}$$

The potential fields Φ_p due to such a dipole, when $r \gg d$, is given by

$$\Phi_P = \frac{\boldsymbol{p}\cdot\hat{r}}{4\pi\varepsilon_0 r^2} = -\frac{1}{4\pi\varepsilon_0}\boldsymbol{p}\cdot\nabla\left[\frac{1}{r}\right], \quad r \gg d, \tag{8.2}$$

and the electric field is given by

$$\mathbf{E}_P = \frac{p}{4\pi\varepsilon_0 r^3}\left[\hat{r}\,2\cos\theta + \hat{\theta}\sin\theta\right]. \tag{8.3}$$

For a more general distribution of volume charges in a localized volume, where the total charge is zero, the dipole moment (Figure 8.2) is given by

$$p = \iiint\limits_{V'} \mathrm{d}V' r' \rho_V(r'). \tag{8.4}$$

* For chapter appendices, see 8A and 8B in Appendices section.

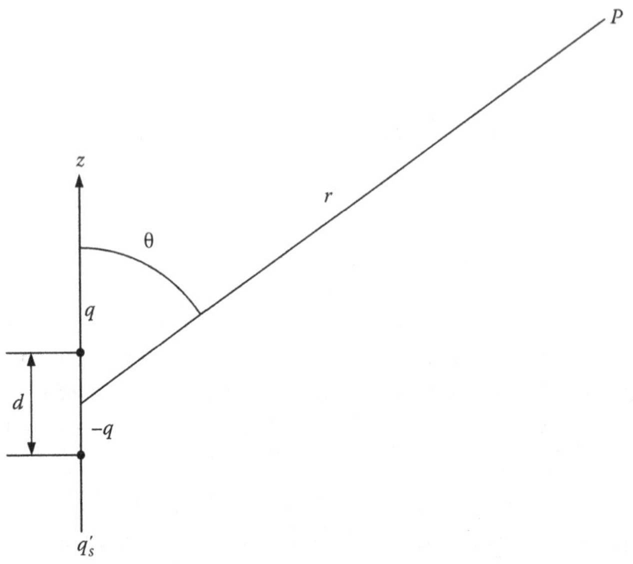

FIGURE 8.1
Electric dipole.

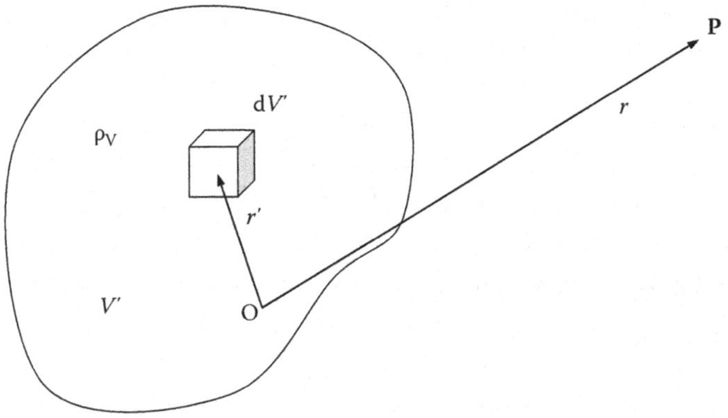

FIGURE 8.2
Dipole moment for a general distribution of volume charges.

The potential due to a volume of dipoles in V' of dipole moment **P** per unit volume, also called polarization vector (C/m^2) (see Figure 8.3), is given by

$$\Phi_P = -\frac{1}{4\pi\varepsilon_0}\iiint_{V'} dV'\mathbf{P}\cdot\nabla\left[\frac{1}{R_{SP}}\right]. \tag{8.5}$$

Equation 8.5 may be transformed into

$$\Phi_P = \frac{1}{4\pi\varepsilon_0}\left[\oiint_s \frac{P\cdot ds'}{R_{SP}} - \iiint_{V'} dV'\frac{\nabla'\cdot P}{R_{SP}}\right]. \tag{8.6}$$

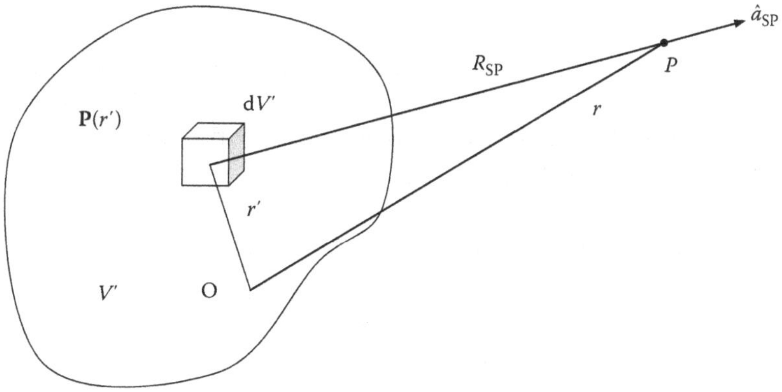

FIGURE 8.3
Volume of dipoles of polarization P.

By comparing Equation 8.6 with similar expressions for potential due to monopole distributions, the following equivalence may be drawn: a volume of dipoles of dipole moment density **P** is equivalent to a monopole volume charge of density ρ_{Vb} and a surface charge density ρ_{Sb}:

$$\rho_{Vb} = -\nabla \cdot \mathbf{P}, \tag{8.7}$$

$$\rho_{Sb} = \mathbf{P} \cdot \hat{n} \tag{8.8}$$

(see Figure 8.4).

Gauss's law in free space with volume of dipoles can be written as

$$\nabla \cdot (\varepsilon_0 \mathbf{E}) = \rho_V + \rho_{Vb}, \tag{8.9}$$

where ρ_V is the volume charge density due to free charges and ρ_{Vb} is an equivalent volume charge density due to the volume of dipoles.

By substituting Equation 8.7 into Equation 8.9, we obtain

$$\nabla \cdot (\varepsilon_0 \mathbf{E} + \mathbf{P}) = \rho_V. \tag{8.10}$$

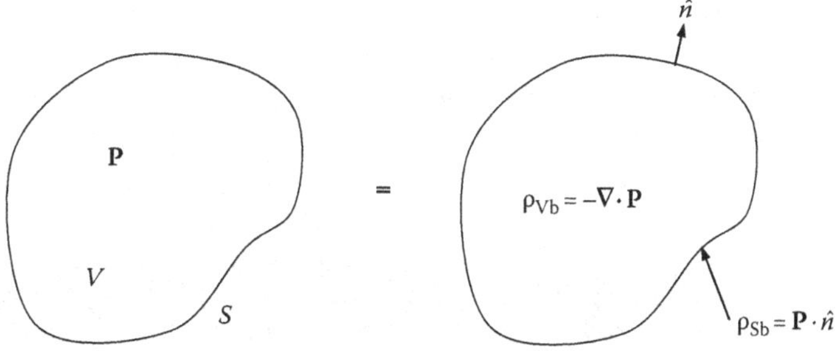

FIGURE 8.4
Equivalence of volume of electric dipoles with monopole distributions.

Let us denote

$$\varepsilon_0\mathbf{E} + \mathbf{P} = \mathbf{D}. \tag{8.11}$$

For linear materials, **P** is proportional to **E** and let us write this linear dependence in terms of electric susceptibility χ_e:

$$\mathbf{P} = \varepsilon_0\chi_e\mathbf{E}. \tag{8.12}$$

Substituting Equation 8.12 into Equation 8.11, we obtain

$$\mathbf{D} = \varepsilon_0(1+\chi_e)\mathbf{E}, \tag{8.13}$$

$$\mathbf{D} = \varepsilon_0\varepsilon_r\mathbf{E}, \tag{8.14}$$

where

$$\varepsilon_r = 1 + \chi_e \tag{8.15}$$

and is called relative permittivity.

8.2 Frequency-Dependent Dielectric Constant

Under the influence of an electric field, the positive and negative charges inside each atom are displaced from their equilibrium position. Since the mass of the nucleus M is much larger than that of an electron m, $M \gg m$, the positive charge is assumed to be stationary. The electron moves and exhibits a friction coefficient ν. The mechanical equivalence of the system is shown [1] in Figure 8.5.

When a time-harmonic electric field of angular frequency ω, $E = E_0\,e^{j\omega t}$, is applied to the atom, the force equation is given by

$$m\frac{d^2x}{dt^2} + m\nu\frac{dx}{dt} + kx = -qE_0e^{j\omega t}, \tag{8.16}$$

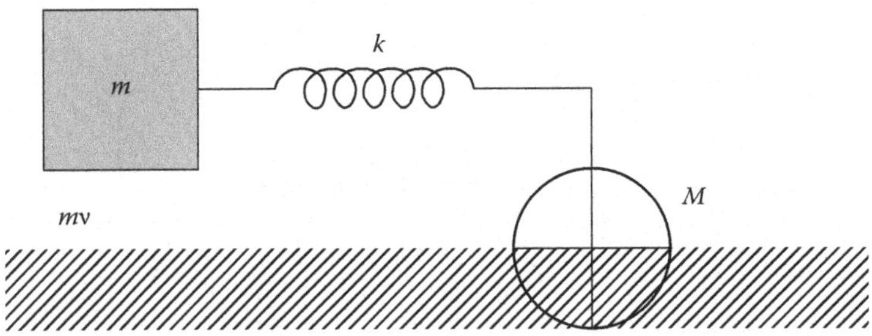

FIGURE 8.5
Mechanical equivalence of an atom in a dielectric.

where x is the position of the electron relative to the atom. To find the particular solution, let $x = x_0\,e^{j\omega t}$, then

$$x_0 = -\frac{qE_0}{m\left(\omega_0^2 - \omega^2 + j\nu\omega\right)}, \tag{8.17}$$

where

$$\omega_0 = \sqrt{\frac{k}{m}}, \tag{8.18}$$

and q is the absolute value of the charge and m is the mass of the electron. The displacement of the charge centers gives rise to a dipole moment $p = -qx$. If there are N dipoles created per unit volume, then the polarization P, which is the dipole moment per unit volume, is given by

$$P = Np = \frac{Nq^2E_0\,e^{j\omega t}}{m\left(\omega_0^2 - \omega^2 + j\nu\omega\right)} = \frac{Nq^2E}{m\left(\omega_0^2 - \omega^2 + j\nu\omega\right)}. \tag{8.19}$$

The electric susceptibility χ_e defined by $P = \varepsilon_0\chi_e E$ can be computed as

$$\chi_e = \frac{P}{\varepsilon_0 E} = \frac{Nq^2/m\varepsilon_0}{\omega_0^2 - \omega^2 + j\nu\omega}. \tag{8.20}$$

The complex dielectric constant $\varepsilon_r = \varepsilon_p = 1 + \chi_e = n^2$. Because of the presence of ν, ε_r and n are complex. The parameter ν is called the collision frequency and has the unit of radians/second. Denoting the numerator of the right-most part of Equation 8.20 by ω_p^2, the dielectric (constant) function is

$$\varepsilon_p = n^2 = 1 + \frac{\omega_p^2}{\omega_0^2 - \omega^2 + j\nu\omega}, \tag{8.21}$$

where

$$\omega_p = \sqrt{\frac{Nq^2}{m\varepsilon_0}}. \tag{8.22}$$

In simple metals, there are free electrons that are not bound ($k = 0$) but are free to move, that is, $\omega_0 = 0$, and

$$\varepsilon_p = 1 - \frac{\omega_p^2}{\omega(\omega - j\nu)}\left(\omega_0 = 0, \text{ simple metals}\right). \tag{8.23}$$

On the other hand, for some dielectrics, there are multiple resonances and for such materials, the most general form of the dielectric function is given by

$$\varepsilon_p(\omega) = \varepsilon_\infty + \sum_{R=1}^{N} \frac{\left(\varepsilon_s - \varepsilon_\infty\right)\omega_R^2}{\omega_R^2 + 2j\omega\delta_R - \omega^2}. \tag{8.24}$$

In the above, ε_∞ is the dielectric constant at $\omega = \infty$, ε_s the relative permittivity at DC, ω_R a resonant frequency, and δ_R a damping constant. A sketch of the real part of a dielectric constant of a hypothetical material is given in Figure 8.10. A rapid change in the real part of the dielectric constant around a resonant frequency is indicated in the figure. Around this resonant frequency, the imaginary part of the dielectric function would have a significant value and the electromagnetic wave is strongly absorbed.

8.3 Modeling of Metals

Equation 8.23 gives the dielectric function of simple metals such as sodium. Since the electron density N is of the order of $10^{23}/\text{cm}^3$, the plasma frequency $\omega_p \approx 2 \times 10^{16}$ rad/s. The collision frequency ν is of the order of 10^{13}. Since $\nu/\omega_p \approx 10^{-3}$, the metal is modeled as a low-loss plasma. Modeling of such metals may be further subdivided into three frequency regions.

8.3.1 Case 1: $\omega < \nu$ and $\nu^2 \ll \omega_p^2$ (Low-Frequency Region)

For this case, $\varepsilon_p = 1 - j\omega_p^2/\omega\nu$, and the imaginary part of ε_p dominates. It is more appropriate to describe this medium in terms of the conductivity parameter rather than the dielectric function. The relation between the two is easily obtained by noting

$$\sigma\tilde{E} = j\omega\varepsilon_0\varepsilon_p\tilde{E}, \tag{8.25}$$

$$\sigma = j\omega\varepsilon_0\varepsilon_p \approx j\omega\varepsilon_0\left(\frac{-jNq^2}{m\varepsilon_0}\right)\frac{1}{\omega\nu}, \tag{8.26}$$

$$\sigma = \frac{Nq^2}{m\nu}. \tag{8.27}$$

Since this expression is obtained at low frequencies, this is called DC conductivity which is given by

$$\sigma_{DC} = \frac{Nq^2}{m\nu}, \tag{8.28}$$

and is independent of the frequency. This case is treated as a simple conducting medium. We have discussed this model in detail in Section 2.2. The dominant features of wave propagation in this medium are given by Equations 2.12 and 2.13 and repeated here for convenience:

$$\alpha = \beta = \sqrt{\pi f \mu\sigma} = \frac{1}{\delta}, \tag{8.29}$$

$$\eta = \frac{\sqrt{2}}{\sigma\delta}\angle 45^\circ. \tag{8.30}$$

8.3.2 Case 2: $\nu < \omega < \omega_p$ (Intermediate-Frequency Region)

It can be shown that, in this region, called cutoff region, the complex refractive index $n = n_R - jn_I$ can be approximated as given below [2]:

$$n_R = \frac{\nu\omega_p}{2\omega^2}\left[1 - \frac{5\nu^2}{8\omega^2} + \frac{\omega^2}{2\omega_p^2}\right], \tag{8.31}$$

$$n_I = \frac{\omega_p}{\omega}\left[1 - \frac{3\nu^2}{8\omega^2} - \frac{\omega^2}{2\omega_p^2}\right], \tag{8.32}$$

$$\delta = \frac{c}{\omega_p}\left[1 + \frac{3\nu^2}{8\omega^2} + \frac{\omega^2}{2\omega_p^2}\right]. \tag{8.33}$$

8.3.3 Case 3: $\omega > \omega_p$ (High-Frequency Region)

In this region, the plasma becomes a low-loss dielectric. In the limit $\nu^2 \ll \left(\omega^2 - \omega_p^2\right)$ and $\nu^2 \ll \omega^2\left(\omega^2 - \omega_p^2\right)/\omega_p^4$, we can obtain the following approximations:

$$n_R = \left(1 - \frac{\omega_p^2}{\omega^2}\right)^{1/2}, \tag{8.34}$$

$$n_I = \frac{\nu\omega_p^2}{2\omega^3}\left[1 - \frac{\omega_p^2}{\omega^2}\right]^{-1/2}, \tag{8.35}$$

$$\delta = \frac{c}{\omega_p}\left(\frac{2\omega^2}{\nu\omega_p}\right)\left(1 - \frac{\omega_p^2}{\omega^2}\right)^{1/2}. \tag{8.36}$$

Figure 8.6 shows the variation of n_R, n_I, α, and β for a laboratory plasma of electron density $N = 10^{13}/\text{cm}^3$ corresponding to a plasma frequency $\omega_p \approx 3 \times 10^{10}$. In metals, the electron density is much higher but the features of conducting, cutoff, and dielectric regions are all present.

8.4 Plasma Medium

In Equation 8.22, we defined a new quantity ω_p, called the angular plasma frequency, which is proportional to the square root of the electron density. In this section, we will explore the characteristics of the medium, called plasma medium. Plasma state is a fourth state of matter, the other three being solid, liquid, and gaseous states. On heating a solid material, we can successfully obtain liquid and gaseous states. If we heat the material further, we can obtain an ionized gas. The electrons are stripped away from the neutral atoms and in this state of matter we have positive ions and free electrons. Though the electrons are no longer

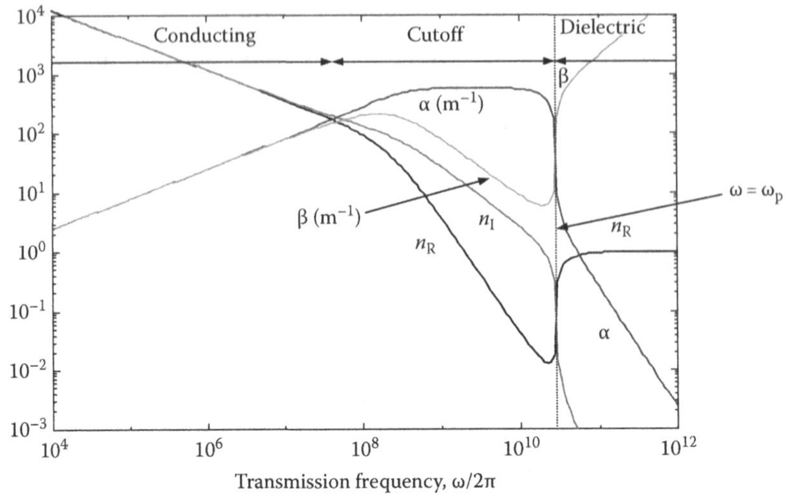

FIGURE 8.6
Plasma medium: conducting, cutoff, and dielectric regions.

bound in the atom, they are still subject to some restoring forces due to Coulomb forces between positive charges and negative charges. The effect of these long-range Coulomb forces, which vary as $1/r^2$, may be understood by considering the following one-dimensional problem. Consider a plasma medium that has positive ions and electrons with overall charge neutrality to begin with. As shown in Figure 8.7, let us assume that we disturbed the equilibrium situation by displacing a layer of electrons by a small distance z. At one end, there will be an excess of electric charge, $-Nqz$ per unit area and at the other end, there will be an excess of positive ion charge $+Nqz$ per unit area. The electric field created by this imbalance from charge neutrality at the ends may be calculated from the boundary condition $D_n = \rho_s = Nqz$. Hence,

$$E = \frac{Nqz}{\varepsilon_0}. \tag{8.37}$$

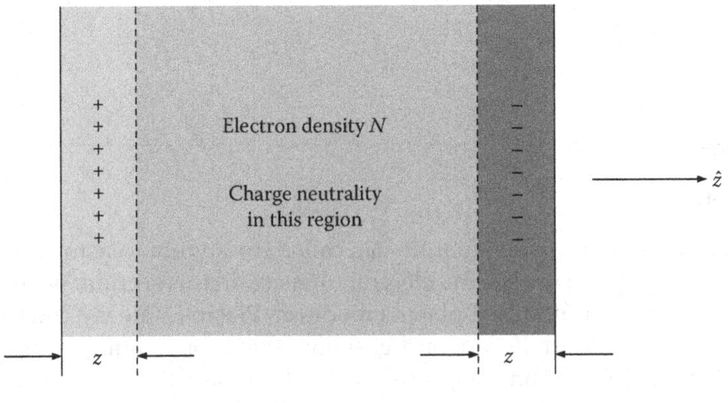

FIGURE 8.7
Oscillation of an electronic charge layer in a plasma medium.

The force experienced by an electron in the presence of this electric field is given by

$$\mathbf{F} = -q\mathbf{E} = -\hat{z}\frac{Nq^2 z}{\varepsilon_0}. \tag{8.38}$$

Equating this to the inertial force, we can write the force balance equation as follows:

$$m\frac{d^2 z}{dt^2} = -\frac{Nq^2}{\varepsilon_0}z, \tag{8.39}$$

$$\frac{d^2 z}{dt^2} + \frac{Nq^2}{m\varepsilon_0}z = 0. \tag{8.40}$$

Equation 8.40 is an equation of oscillatory motion with the solution

$$z = z_0\, e^{j\omega_p t}, \tag{8.41}$$

where ω_p^2 is given by $Nq^2/m\varepsilon_0$. Thus, ω_p gives the natural frequency of oscillation of the electronic charge layer.

8.5 Polarizability of Dielectrics

If there are N molecules for unit volume, then the polarization may be written as

$$\mathbf{P} = \varepsilon_0 \chi_e \mathbf{E} = N\alpha_T \mathbf{E}_{loc} = N\alpha_T g\mathbf{E}, \tag{8.42}$$

where α_T is the molecular polarizability and g the ratio between local field $\mathbf{E}_{loc}$ acting on the molecule and the applied field $\mathbf{E}$. These two fields differ because of the presence of surrounding molecules. Since $\varepsilon_r = 1 + \chi_e$, we obtain from Equation 8.42,

$$\varepsilon_r = 1 + \frac{N\alpha_T g}{\varepsilon_0}. \tag{8.43}$$

It can be shown that

$$g = \frac{2+\varepsilon_r}{3}, \tag{8.44}$$

if the surrounding molecules act in a spherically symmetric fashion. Figure 8.8 shows the applied and local fields for the spherically symmetrical case.

The polarization $\mathbf{P} = \hat{z}P$ creates equivalent charges on the walls. The equivalent charge dq on the walls is given by $\mathbf{P} \cdot d\mathbf{s}$, where $d\mathbf{s}$ is the vector differential surface element on the sphere given by $\hat{r}r_0^2 \sin\theta d\theta d\phi$. Therefore, the electric field $d\mathbf{E}_p$ at the center of the sphere is given by

$$d\mathbf{E}_p = \frac{\hat{z}P \cdot \hat{r}r_0^2 \sin\theta d\theta d\phi}{4\pi\varepsilon_0 r_0^2}(-\hat{r}) = \frac{P\cos\theta \sin\theta d\theta d\phi}{4\pi\varepsilon_0}(-\hat{r}), \tag{8.45}$$

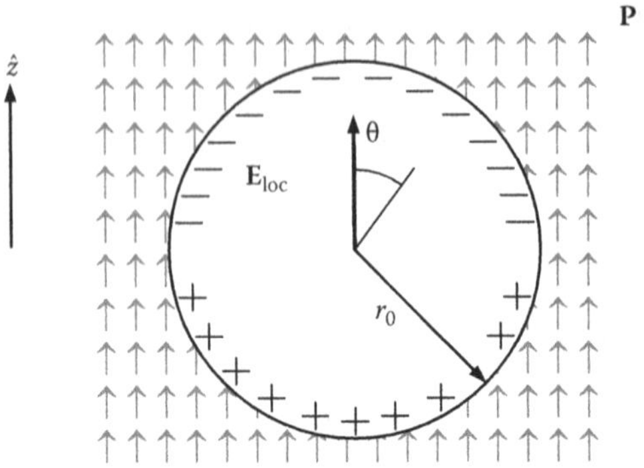

FIGURE 8.8
Molecular polarizability and local field.

$$\mathbf{E}_p = -\frac{P}{4\pi\varepsilon_0}\int_0^{2\pi} d\phi \int_0^{\pi} \hat{r}\cos\theta\sin\theta d\theta = -\frac{P}{2\varepsilon_0}\int_0^{\pi} \hat{r}\cos\theta\sin\theta d\theta. \tag{8.46}$$

Since

$$\hat{r} = (\hat{r}\cdot\hat{\rho})\hat{\rho} + (\hat{r}\cdot\hat{\phi})\hat{\phi} + (\hat{r}\cdot\hat{z})\hat{z} = \sin\theta\hat{\rho} + 0 + \cos\theta\hat{z}, \tag{8.47}$$

$$\int_0^{\pi} \cos\theta\sin^2\theta d\theta = \frac{1}{3}\sin^3\theta\Big|_0^{\pi} = 0, \tag{8.48}$$

$$\int_0^{\pi} \cos^2\theta\sin\theta d\theta = -\frac{1}{3}\cos^3\theta\Big|_0^{\pi} = -\frac{2}{3}, \tag{8.49}$$

$$\mathbf{E}_p = \hat{z}\frac{P}{3\varepsilon_0}, \tag{8.50}$$

and

$$g = \frac{\mathbf{E}_{loc}}{\mathbf{E}} = \frac{\mathbf{E}_p + \mathbf{E}}{\mathbf{E}} = 1 + \frac{\mathbf{E}_p}{\mathbf{E}} = 1 + \frac{P/3\varepsilon_0}{P/\varepsilon_0\chi_e} = 1 + \frac{\chi_e}{3} = \frac{3+\chi_e}{3} = \frac{3+\varepsilon_r-1}{3} = \frac{2+\varepsilon_r}{3}. \tag{8.51}$$

From Equations 8.43 and 8.44, we obtain

$$\alpha_T = \frac{3\varepsilon_0}{N}\frac{\varepsilon_r - 1}{\varepsilon_r + 2}, \tag{8.52}$$

$$\varepsilon_r = \frac{1 + 2N\alpha_T/3\varepsilon_0}{1 - N\alpha_T/3\varepsilon_0}. \tag{8.53}$$

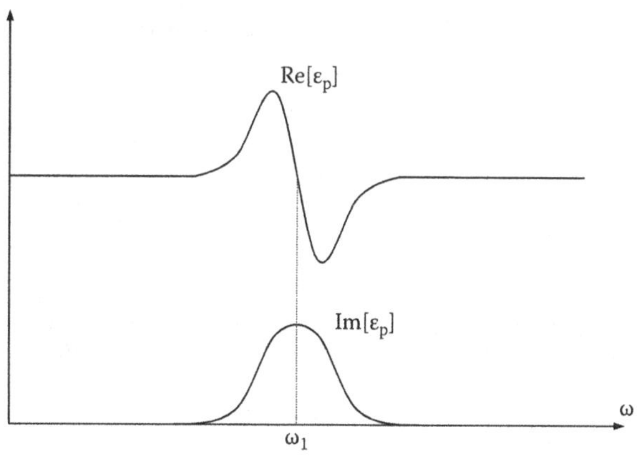

FIGURE 8.9
Dispersion and complex permittivity.

This is called the *Clausius–Mossotti* relation. We have noted from the above equations that the local field is different from the applied electric field. However, in writing the force Equation 8.16, the RHS is electric force on the charge due to the applied electric force. If we use, instead, the force due to local electric field, then Equation 8.20 has to be modified: in Equation 8.20, replace ω_0 by ω_1, where

$$\omega_1^2 = \omega_0^2 - \frac{Nq^2}{3\varepsilon_0 m}. \tag{8.54}$$

Figure 8.9 shows the general shape of ε_p versus ω_1, where ε_p is given by Equation 8.21 with ω_1 replacing ω_0.

The complex permittivity

$$\varepsilon_0\varepsilon_p = \varepsilon'(\omega) - j\varepsilon''(\omega) \tag{8.55}$$

has analytical properties given by the Kramer–Kronig relations:

$$\varepsilon'(\omega) = \varepsilon_0 + \frac{2}{\pi}\int_0^\infty \frac{\omega'\varepsilon''(\omega')}{\omega'^2 - \omega^2}\,d\omega', \tag{8.56}$$

$$\varepsilon''(\omega) = -\frac{2\omega}{\pi}\int_0^\infty \frac{\varepsilon'(\omega') - \varepsilon_0}{\omega'^2 - \omega^2}\,d\omega'. \tag{8.57}$$

The polarization α_T has contributions from several effects and may be written as

$$\alpha_T = \alpha_e + \alpha_i + \alpha_d, \tag{8.58}$$

where the subscripts e, i, and d stand for electronic, ionic, and permanent dipole contributions to the polarizability, respectively. The electronic and ionic components may be generalized in the form

$$\alpha_j = \frac{F_j}{\left(\omega_j^2 - \omega^2\right) + j\omega v_j}, \tag{8.59}$$

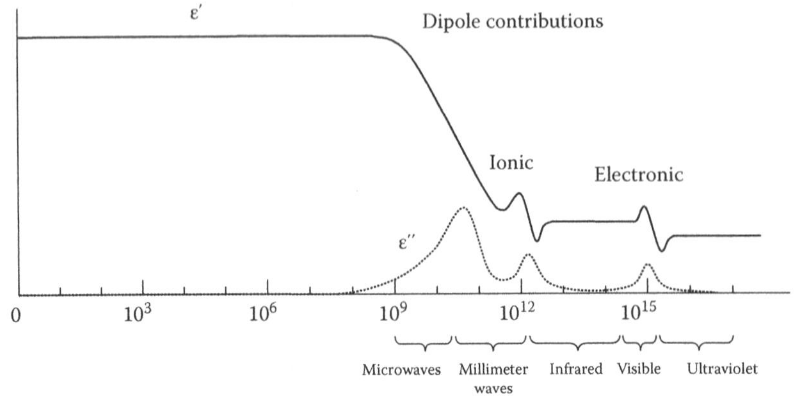

FIGURE 8.10
Frequency response of a hypothetical dielectric.

where F_j is the measure of the strength of the jth resonance. The permanent dipole contribution is different in the sense that the force opposing the complete alignment of the dipoles in the direction of the applied electric field is of thermal nature:

$$\alpha_d = \frac{p^2}{3k_B T(1+j\omega\tau)}, \tag{8.60}$$

where T is the temperature, k_B the Boltzmann constant, p the permanent dipole moment, and τ the relaxation time. Figure 8.10 [4] is a sketch of frequency response of a hypothetical dielectric.

In addition to the three polarization mechanisms discussed above, interfacial polarization or space–charge polarization [3] gives rise to a complex dielectric constant:

$$\varepsilon_r = \frac{-j\sigma_0}{\omega\varepsilon_0} + \sum_{m=1}^{M}\left(a_m + \frac{b_m - a_m}{1+j\omega\tau_m}\right). \tag{8.61}$$

This polarization is due to the large-scale field distortions caused by the piling of space charges in the volume or of the surface charges at the interfaces between different small portions of materials with different characteristics. Geophysical media are examples [4] of such materials.

8.6 Mixing Formula

Mixing formula gives the effective dielectric constant of dielectric inclusions of a material of dielectric constant ε_{r2} embedded in a host dielectric material of dielectric constant ε_{r1}. Let the inclusions be spheres of radius a. Let the radius a be small when compared to the wavelength. This assumption permits us to ignore the effects of scattering. Let the spheres be sparsely distributed in the host material so that the fractional volume f, which is a fraction of the volume occupied by the spheres, is lower than a few percent, $f \ll 1$. This assumption permits us to ignore the consideration of a correlation between spheres. The situation described here is similar to the situation described in Section 8.5, where the dielectric constant of the material consisting of many dipoles is given by

$$\varepsilon_r = \frac{1+2N\alpha/3\varepsilon_0}{1-N\alpha/3\varepsilon_0}, \tag{8.62}$$

where N is the number of dipoles per unit volume and α the polarizability of the dipole. In our case, the effective dielectric constant ε_e is given by

$$\frac{\varepsilon_e}{\varepsilon_{r1}} = \frac{1+2N\alpha/3\varepsilon_{r1}}{1-N\alpha/3\varepsilon_{r1}}, \tag{8.63}$$

where N is the number of spheres per unit volume. The polarizability α of the sphere is given by [3]

$$\alpha = \frac{3\varepsilon_{r1}\left(\varepsilon_{r2}-\varepsilon_{r1}\right)V}{\varepsilon_{r2}+2\varepsilon_{r1}}. \tag{8.64}$$

Equation 8.64 is obtained after making several simplifying assumptions, some of which are stated above.

Since

$$f = NV, \tag{8.65}$$

where V is the volume of the sphere, we obtain the effective dielectric constant as

$$\varepsilon_e = \varepsilon_{r1}\frac{1+2fy}{1-fy}. \tag{8.66}$$

In the above equation, we have

$$y = \frac{\varepsilon_{r2}-\varepsilon_{r1}}{\varepsilon_{r2}+2\varepsilon_{r1}}. \tag{8.67}$$

This is called the *Maxwell–Garnet* mixing formula. If the inclusion has a nonspherical shape, then the expression for α for that shape should be used. For more information on mixing formulas, see [3,5].

8.7 Good Conductors and Semiconductors

The transport of charge gives rise to the flow of current, and the constitutive relation of a simple conductor is given by $\mathbf{J} = \sigma\mathbf{E}$. Based on the magnitude of σ, we classify conductors into

semiconductors ($\sigma \sim 10^{-2}$ S/m) and
conductors (metals and alloys) ($\sigma \sim 10^{-2} - 10^{8}$ S/m).

In general, the charge carriers in a conductor are free electrons, whereas in a semiconductor the charge carriers are electrons as well as holes. We have seen that $\sigma_{DC} = Nq^2/m\nu$.

We can use this expression for the conductivity σ provided $\omega < \nu$ and $\nu^2 \ll \omega_p^2$. These conditions are satisfied in metals for frequencies right up to optical frequencies. The conductivity of metals may also be written in terms of its electron mobility μ_e, where

$$\sigma = Nq\mu_e\left(\text{S/m}\right). \tag{8.68}$$

The power dissipation per unit volume is

$$w = \sigma\left|\mathbf{E}\right|^2\left(\text{W/m}^3\right). \tag{8.69}$$

Let us discuss the collision frequency ν a little more in the context of conductors. The mean free path Λ of an electron can be related to the average drift velocity of the electron:

$$\Lambda = \left|v\right|\tau. \tag{8.70}$$

Here τ is the average collision time and is related to the collision frequency ν by

$$\nu = \frac{1}{\tau}. \tag{8.71}$$

In the context of conductors, τ is also called relaxation time. Since the conductivity depends on the average collision time, the electrical resistivity $\rho = 1/\sigma$ increases with temperature. It may be expressed in terms of its value ρ_{RT} at room temperature T_R as

$$\rho = \rho_{RT}\left[1 + \alpha_R\left(T - T_R\right)\right]. \tag{8.72}$$

For pure metals, α_R, the temperature coefficient of resistivity, is about 0.004/°C. For a table of values for α_R, see [5]. As the absolute temperature T approaches zero, the pure metal has zero resistivity and becomes typically a superconductor. The temperature dependence of conductivity σ in pure metals as dictated by the mobility of electrons may be written as

$$\sigma = \begin{cases} \dfrac{A}{T} & \text{in the high-temperature regime,} \\ \dfrac{B}{T^5} & \text{in the low-temperature regime.} \end{cases} \tag{8.73}$$

Intrinsic semiconductors belong to group IV elements that include silicon and germanium. The conductivity of intrinsic semiconductors may be expressed as

$$\sigma = N_n\mu_n q + N_p\mu_p q, \tag{8.74}$$

where N_n and N_p are the intrinsic electron and hole densities, respectively. Theoretical and experimental studies show that σ can be expressed in terms of forbidden gap energy E_g [5]:

$$\ln\sigma = \frac{A - E_g}{1.7\times10^{-4}\,T}. \tag{8.75}$$

Germanium (Ge) has a gap energy of 0.67 eV, while silicon (Si) has an E_g of 1.12 eV. Ge has a higher mobility than Si, hence Ge devices can be operated at higher frequencies. However, Ge devices are more sensitive to temperature changes. Extrinsic semiconductors are made by adding impurities (dopants) to the intrinsic semiconductors. The dopants are from group III or group V elements. The electrical properties of extrinsic semiconductors are strongly influenced by the dopants. N-type semiconductors have dopants, from group V (e.g., phosphorous) and have a set of easily activated electrons in the conduction band. Thus, donor impurities add to the free-electron population in the conduction band, facilitating an increased conductivity of the material. The P-type semiconductors are constituted by the addition of impurity atoms from group III elements (e.g., aluminum) and the impurity atoms are called acceptors. The conductivity of N-type is given by

$$\sigma_{\text{N-type}} = \mu_e N_D q, \tag{8.76}$$

where N_D is the donor concentration. The conductivity of P-type is given by

$$\sigma_{\text{P-type}} = \mu_e N_A q, \tag{8.77}$$

where N_A is the acceptor concentration.

Effective masses of electrons and holes in semiconductors are different from free electron mass due to various interactive force fields experienced by the charge carriers. As an example, for silicon, $m_e^*/m_0 \approx 0.190 - 0.260$ and $m_h^*/m_0 = 0.5$, where m_e^* is the effective mass of the electron, m_h^* the effective mass of the hole, and m_0 the mass of the free electron. For more details, see [5]. When a semiconductor is subjected to crossed electric (**E**) and magnetic (**H**) fields, a small potential field ($\mathbf{E}_H$) orthogonal to **H** and **E** is induced. This phenomenon is called the Hall effect and is discussed in Chapter 11.

8.8 Perfect Conductors and Superconductors

The resistivity of many metals and alloys drops abruptly when they are cooled to a sufficiently low-temperature T_c (Figure 8.11a). The superconductivity state [5–7] is destroyed and normal conductivity state is restored if the current density J_c or the magnetic field H_c exceeds a threshold value. An empirical law relating H_c with the temperature T is given by [5]

$$H_c(T) = H_{c0}\left[1 - \left(\frac{T}{T_c}\right)^2\right], \tag{8.78}$$

where H_{c0} is the critical magnetic field at $T = 0$. The boundary between superconductivity state and normal conducting state is shown in Figure 8.11b.

The empirical relation between B_{max} and T_{max} is given by

$$B_{max} = 0.02 T_{max}. \tag{8.79}$$

Provided that the current density, the magnetic field, and the temperature are less than the threshold values, the resistivity drops to zero and the conductor may be treated as a

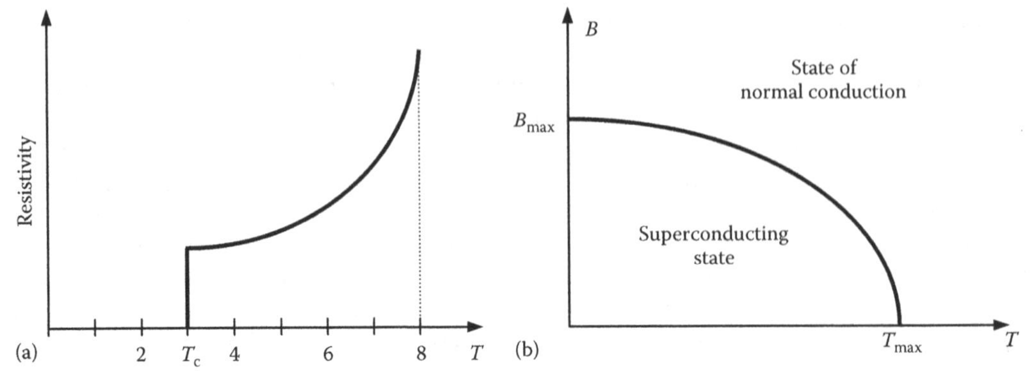

FIGURE 8.11
Superconductor: (a) resistivity versus temperature and (b) limits of B_{max} and T_{max} for superconducting state.

perfect conductor. As we noted in the previous section, a DC magnified field H_{DC} can exist in a conductor even though $E_{DC} = 0$. However, in a perfect superconductor, it will be shown that $H_{DC} = 0$. Thus, a perfect superconductor is not only a perfect conductor, but also a perfect diamagnet. The magnetic field in a specimen is expelled as the temperature is brought below the critical temperature T_c. This phenomenon is called the *Meissner* effect. We investigate the electromagnetic phenomena in a superconductor based on a macroscopic theory given by London.

Instead of the usual Ohm's law for normal conductors, that is,

$$\mathbf{J}_n = \sigma_n \mathbf{E}\left(\frac{\mathrm{A}}{\mathrm{m}^2}\right), \tag{8.80}$$

London postulated that in a superconductor, the current density $\mathbf{J}_s$ is given by

$$\mathbf{J}_s = C\mathbf{A}, \tag{8.81}$$

where C is a constant characteristic of a given material and $\mathbf{A}$ the usual vector potential. Taking the curl of Equation 8.81, we have

$$\nabla \times \mathbf{J}_s = C\nabla \times \mathbf{A} = C\mathbf{B}. \tag{8.82}$$

From Ampere's law of direct current, we have

$$\nabla \times \mathbf{B} = \mu \mathbf{J}_s, \tag{8.83}$$

$$\nabla \times \nabla \times \mathbf{B} = \mu \nabla \times \mathbf{J}_s = \mu C\mathbf{B}, \tag{8.84}$$

$$\nabla(\nabla \cdot \mathbf{B}) - \nabla^2 \mathbf{B} = \mu C\mathbf{B}. \tag{8.85}$$

Since $\nabla \cdot \mathbf{B} = 0$,

$$\nabla^2 \mathbf{B} = -\mu C\mathbf{B}. \tag{8.86}$$

The solution of Equation 8.86 for a semi-infinite superconductor occupying the space $z > 0$ is given by

$$B(z) = B(0)e^{-z/\lambda_L}, \tag{8.87}$$

where

$$\lambda_L^2 = -\frac{1}{\mu C}, \tag{8.88}$$

and B is in x–y-plane. We will show later that

$$C = -\frac{n_s q^2}{m}, \tag{8.89}$$

where n_s is the density of superconducting electrons, and q and m are the charge and mass of an electron, respectively. Thus,

$$\lambda_L = \left(\frac{m}{\mu n_s q^2}\right)^{1/2}, \tag{8.90}$$

which is a real quantity. Equation 8.87 shows that λ_L is the length characterizing the decay of the flux density from its value at the surface of the superconductor. It is called the London penetration depth. For lead, λ_L = 146 Å. In a good normal conductor, the time-harmonic magnetic field penetrates to a depth δ:

$$\delta = \frac{1}{\sqrt{\pi f \mu\sigma}}, \tag{8.91}$$

where f is the frequency. For direct current, $\delta = \infty$, signifying uniform distribution of DC magnetic field. In a superconductor, even for DC, λ_L is very small; this explains the *Meissner* effect. Let us next give a phenomenological basis for Equation 8.81. The current density $\mathbf{J}_s$ in the superconductor is given by

$$\mathbf{J}_s = n_s^* q_s \mathbf{v}_s, \tag{8.92}$$

where n_s^* is the density of carriers of charge q_s and $\mathbf{v}_s$ is their drift velocity. In the presence of a magnetic field $\mathbf{B}_s = \nabla \times \mathbf{A}_s$, the total momentum $\mathbf{P}_s$ is

$$\mathbf{P}_s = m_s \mathbf{v}_s + q_s \mathbf{A}_s, \tag{8.93}$$

where m_s is the mass of the charge carriers. The last term is the additional momentum due to the presence of the magnetic field. From Equations 8.92 and 8.93, we have

$$\mathbf{J}_s = \frac{n_s^* q_s}{m_s}\mathbf{P}_s - \frac{n_s^* q_s^2}{m_s}\mathbf{A}_s. \tag{8.94}$$

For super electrons, the total momentum is zero, provided

$$C = -\frac{n_s^* q_s^2}{m_s}. \tag{8.95}$$

Equation 8.95 is obtained by making use of Equation 8.81. To calculate λ_L, we need to find the values to be used for m_s, n_s^*, and q_s. In a superconducting state, the electrons are loosely associated as pairs [6]. The electrons in a given pair have momentum $\hbar k$ and $-\hbar k$, so that the net pair momentum is zero. The electron pair behaves like a single particle, giving rise to

$$m_s = 2m, \tag{8.96}$$

$$q_s = 2q, \tag{8.97}$$

$$n_s^* = \frac{n_s}{2}, \tag{8.98}$$

where m, q, and n_s are the mass, absolute value of the charge, and the number density of the electrons in a superconducting state, respectively. For a superconductor, the charge carriers are paired electrons. Substituting Equations 8.96 through 8.98 into Equation 8.95, we see that Equation 8.95 is the same as Equation 8.89. From Equation 8.90, we see that λ_L should always be finite. Perfect superconductor, which is a perfect diamagnetic (*Meissner* effect of flux exclusion), is an idealization of a practical situation where $\lambda_L \ll l$, where l is the relevant dimension of the sample.

A two-fluid model of superconductors at high frequencies is developed by considering that a fraction of the electrons are in superconductive state and they form pairs. In normal conductors, free electrons are assumed to move under the influence of the electric field and experience collisions as described earlier. The superconducting electron pairs, however, are immune from collisions. In the two-fluid model, we write separate momentum equations for each of the fluids, the normal fluid denoted by the subscript n and the superconducting fluid by the subscript s:

$$m_s \frac{d\mathbf{v}_s}{dt} = -q_s\mathbf{E} \quad \text{or} \quad m\frac{d\mathbf{v}_s}{dt} = -q\mathbf{E}, \tag{8.99}$$

$$m\frac{d\mathbf{v}_n}{dt} + m\nu\mathbf{v}_n = -q\mathbf{E}, \tag{8.100}$$

$$\mathbf{J} = \mathbf{J}_s + \mathbf{J}_n, \tag{8.101}$$

$$\mathbf{J}_s = -n_s^* q_s \mathbf{v}_s = -n_s q \mathbf{v}_s, \tag{8.102}$$

$$\mathbf{J}_n = -n_n q \mathbf{v}_n, \tag{8.103}$$

$$n = n_s + n_n. \tag{8.104}$$

For a time-harmonic field, we have from Equations 8.99, 8.100, 8.102, and 8.103 that

$$J_s = -j\frac{q^2 n_s}{m\omega}\mathbf{E}, \tag{8.105}$$

$$\mathbf{J}_n = -j\frac{q^2 n_n}{m\omega[1 - j/\omega\tau]}\mathbf{E}, \tag{8.106}$$

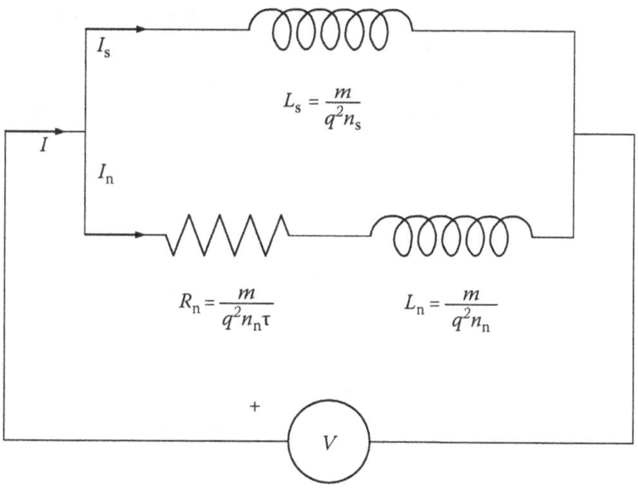

FIGURE 8.12
Circuit analogy for a superconductor based on the two-fluid model.

where $\tau = 1/\nu$ is the momentum relaxation time. An equivalent circuit for the above may easily be constructed as shown in Figure 8.12, where the analogy is

$$V \leftrightarrow \mathbf{E}, \tag{8.107}$$

$$I_s \leftrightarrow \mathbf{J}_s, \tag{8.108}$$

$$I_n \leftrightarrow \mathbf{J}_n. \tag{8.109}$$

In the above circuit, $V = j\omega L_s I_s$ and from the analogy

$$L_s = \frac{m}{q^2 n_s}. \tag{8.110}$$

From the analogy of the second branch,

$$I_n = \frac{1}{R_n + j\omega L_n} V \tag{8.111}$$

and

$$\frac{1}{R_n + j\omega L_n} = \frac{-jq^2 n_n}{m\omega[1 - j/\omega\tau]} = \frac{\omega\tau q^2 n_n}{jm\omega[\omega\tau - j]} = \frac{1}{[m/(q^2 n_n \tau)][j\omega\tau + 1]} = \frac{1}{(jm\omega/q^2 n_n) + (m/q^2 n_n \tau)}. \tag{8.112}$$

By comparison,

$$R_n = \frac{m}{q^2 n_n \tau}, \tag{8.113}$$

$$L_n = \frac{m}{q^2 n_n}. \tag{8.114}$$

Note that for $\omega = 0$, $\omega L_n = 0$, and R_n is short-circuited by ωL_s. Thus, for $T < T_c$, although there are normal as well as superconducting electrons, the resistivity is identically zero. If ω is finite, then R_n is no longer shorted.

Maxwell's equations for the superconductor, based on the two-fluid theory, are written as

$$\nabla \times \mathbf{E} = -\frac{\partial \mathbf{B}}{\partial t}, \tag{8.115}$$

$$\nabla \times \mathbf{H} = \mathbf{J} + \frac{\partial \mathbf{D}}{\partial t}, \tag{8.116}$$

$$\nabla \cdot \mathbf{B} = 0, \tag{8.117}$$

$$\nabla \cdot \mathbf{D} = 0, \tag{8.118}$$

$$\mathbf{D} = \varepsilon_0 \mathbf{E}, \tag{8.119}$$

$$\mathbf{B} = \mu_0 \mathbf{H}, \tag{8.120}$$

$$\mathbf{J} = \mathbf{J}_s + \mathbf{J}_n, \tag{8.121}$$

$$\mathbf{J}_n = \sigma_n \mathbf{E}, \tag{8.122}$$

$$\nabla \times \mathbf{J}_s = C\mathbf{B} = -\frac{1}{\lambda_L^2}\mathbf{H}, \quad \lambda_L = \left(\frac{m}{\mu_0 n_s q^2}\right)^{1/2}. \tag{8.123}$$

The wave equation for $\mathbf{H}$ may be written as

$$\nabla^2 \mathbf{H} - \frac{1}{\lambda_L^2}\mathbf{H} - \mu_0 \sigma_n \frac{\partial \mathbf{H}}{\partial t} - \frac{1}{c^2}\frac{\partial^2 \mathbf{H}}{\partial t^2} = 0, \tag{8.124}$$

Note that $\lambda_L = \infty$, if $n_s = 0$. The last term on the LHS is negligible for a good conductor. This term arises because of the displacement current in the conductor. Let us obtain a one-dimensional time-harmonic solution, assuming

$$\mathbf{H} = \hat{y} H_y(z) e^{j\omega t}. \tag{8.125}$$

The magnetic field satisfies the equation

$$\frac{\partial^2 H_y}{\partial z^2} - \frac{1}{\lambda_L^2} H_y - \frac{j2}{\delta^2} H_y = 0, \tag{8.126}$$

where $\delta = 1/\sqrt{\pi f \mu_0 \sigma_n}$ is the skin depth.

Assuming further

$$H_y(z) = H_0 e^{-\gamma z}, \tag{8.127}$$

$$\gamma^2 = \frac{1}{\lambda_L^2} + \frac{2j}{\delta^2} = \frac{1}{\lambda_L^2}\left[1 + 2j\left(\frac{\lambda_L}{\delta}\right)^2\right], \tag{8.128}$$

$$\gamma = \frac{1}{\lambda_L}\left[1+2j\left(\frac{\lambda_L}{\delta}\right)^2\right]^{1/2}, \tag{8.129}$$

$$\gamma = \frac{1}{\lambda_L}\left[\left(\frac{\theta+1}{2}\right)^{1/2} + j\left(\frac{\theta-1}{2}\right)^{1/2}\right], \tag{8.130}$$

where

$$\theta = \left[1+4\left(\frac{\lambda_L}{\delta}\right)^4\right]^{1/2}. \tag{8.131}$$

For the direct current case, $\delta = \infty$, $\theta = 1$, and $\gamma = 1/\lambda_L$. λ_L is the penetration depth of H_{DC} in a superconductor. This parameter is the quantitative measure of the *Meissner* effect. If it is a normal conductor and $\omega \neq 0$, then δ is finite, λ_L is large, and

$$\theta \approx 2\left(\frac{\lambda_L}{\delta}\right)^2, \tag{8.132}$$

$$\left(\frac{\theta+1}{2}\right)^{1/2} \approx \left(\frac{\theta-1}{2}\right)^{1/2} \approx \left(\frac{\theta}{2}\right)^{1/2} = \frac{\lambda_L}{\delta}, \tag{8.133}$$

thus

$$\gamma = \frac{1}{\delta}\left[1+j\right], \quad n_s = 0, \quad \omega \neq 0 \tag{8.134}$$

as expected.

The surface resistance may be calculated, following the steps in Section 2.5. Alternatively, for a unit width and unit length along y and x axes, respectively,

$$Z_s = \frac{E_0}{\int_0^\infty J_x dz}, \tag{8.135}$$

where

$$J_x = -\frac{\partial H_y}{\partial z}. \tag{8.136}$$

From Equations 8.135 and 8.136,

$$Z_s = \frac{E_0}{\int_0^\infty \gamma H_0 e^{-\gamma z} dz} = \frac{E_0}{H_0}, \tag{8.137}$$

Since, E_0 is the electric field on the surface at $z = 0$,

$$E_x(z) = E_0 e^{-\gamma z}, \tag{8.138}$$

and from Maxwell's equation 8.115,

$$-\frac{\partial E_x}{\partial z} = -\mu_0 j\omega H_y. \tag{8.139}$$

Thus, we obtain the relation between E_0 and H_0:

$$E_0 = \frac{j\omega\mu_0 H_0}{\gamma}. \tag{8.140}$$

The surface impedance is

$$Z_s = R_s + jX_s = \omega\mu_0\lambda_L\left(\frac{\theta-1}{2\theta^2}\right)^{1/2} + j\omega\mu_0\lambda_L\left(\frac{\theta+1}{2\theta^2}\right)^{1/2}. \tag{8.141}$$

8.9 Magnetic Materials

Magnetic materials [4–6] get magnetized. The quantitative measure of magnetization is given by the magnetization vector **M** (A/m³), which is the net magnetic dipole moment p_m per unit volume. It is analogous to the polarization **P** (C/m²), which is the net electric dipole moment per unit volume.

One can think of an analogous situation with reference to magnetic charges and define a magnetic dipole moment

$$\mathbf{p}_m = q_m \mathbf{d}, \tag{8.142}$$

where the magnetic charges $+q_m$ and $-q_m$ are separated by a small distance d. It is worth noting at this time that we cannot isolate magnetic charges, they can appear only as a pair, and the monopole magnetic charge density is zero. When we view currents as the sources of magnetic fields, we can draw the following analogy of a magnetic dipole and a loop of current (see Figure 8.13). At a greater distance, they have the same vector potential:

$$\mathbf{A}_p = \frac{\mu_0}{4\pi r^2} p_m \times \hat{r} = -\frac{\mu_0}{4\pi} p_m \times \nabla\left(\frac{1}{r}\right), \tag{8.143}$$

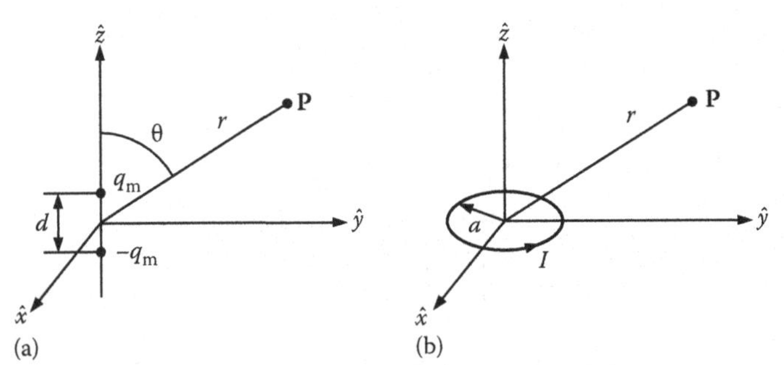

FIGURE 8.13
Equivalence of (a) magnetic dipole with a (b) small circular loop of current.

where

$$p_{\mathrm{m}} = \hat{z}q_{\mathrm{m}}d = \hat{z}I\pi a^2. \tag{8.144}$$

For a more general loop of current,

$$p_{\mathrm{m}} = \frac{1}{2}I\oint \boldsymbol{r}' \times \mathrm{d}l'. \tag{8.145}$$

If the loop is planar and the origin is in the plane (Figure 8.14),

$$\frac{1}{2}r' \times \mathrm{d}l' = \mathrm{d}s' \tag{8.146}$$

and

$$\mathrm{d}p_{\mathrm{m}} = I\,\mathrm{d}s'. \tag{8.147}$$

If the source is a volume current **J**, then

$$p_{\mathrm{m}} = \frac{1}{2}\iiint\limits_{V'} r' \times \mathbf{J}\,\mathrm{d}V'. \tag{8.148}$$

The vector magnetic potential due to a volume of magnetic dipoles (Figure 8.15) of dipole density **M** (A/m) is given by

$$\mathbf{A}_{\mathrm{p}} = -\frac{\mu_0}{4\pi}\iiint\limits_{V'} M \times \nabla\left(\frac{1}{R_{\mathrm{SP}}}\right)\mathrm{d}V'. \tag{8.149}$$

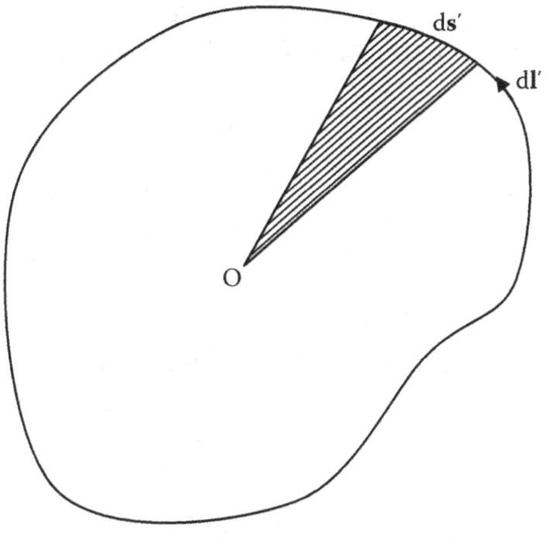

FIGURE 8.14
Planar loop.

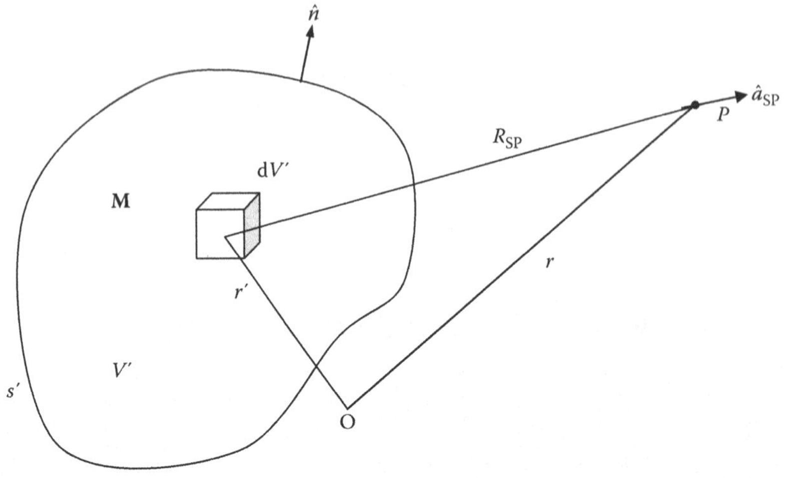

FIGURE 8.15
Volume of magnetic dipoles.

Equation 8.149 may be manipulated to yield

$$\mathbf{A}_{\mathrm{p}} = \frac{\mu_0}{4\pi}\left[\oiint_{s'} \frac{\hat{n}\times M(r')}{R_{\mathrm{SP}}}\mathrm{d}s' + \iiint_{V'} \frac{\nabla\times M(r')}{R_{\mathrm{SP}}}\mathrm{d}V'\right]. \tag{8.150}$$

By comparing Equation 8.150 with the expressions for **A** due to volume current of density $\mathbf{J}_b$ and surface current of density $\mathbf{K}_b$, the following equivalence may be given (Figure 8.16):

$$\mathbf{J}_b = \nabla\times\mathbf{M}\left(\mathrm{A/m^2}\right), \tag{8.151}$$

$$\mathbf{K}_b = \hat{n}\times\mathbf{M}\left(\mathrm{A/m}\right). \tag{8.152}$$

A topic of interest connected to a magnetic dipole is the torque **T** experienced by the dipole when it is brought into an external **B** field. It is given by

$$\mathbf{T} = p_{\mathrm{m}}\times\mathbf{B}. \tag{8.153}$$

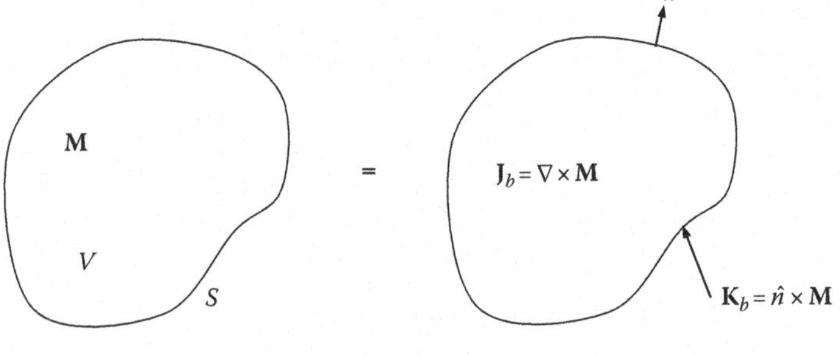

FIGURE 8.16
Equivalence of volume of magnetic dipoles with current distributions.

A current loop experiences a torque in the presence of an external **B** field. The loop will rotate and when the normal to the plane of the loop is parallel to **B**, an equilibrium is achieved. An electron in an orbit constitutes current. If an electron of charge $-q$ executes an orbital motion of angular velocity ω, the corresponding current is given by

$$I = -q\left(\frac{\omega}{2\pi}\right), \tag{8.154}$$

where ω depends on the environment to which the electron is subjected. Consider the simplest case of an electron circulating around a proton of charge $+q$ in the presence of a magnetic field H_0 perpendicular to the circular orbit (Figure 8.17).

From Lorentz force equation,

$$\mathbf{F} = -\hat{\rho}\frac{q^2}{4\pi\varepsilon_0 r^2} - q\left(\hat{\phi}\mathrm{v} \times \mu_0 H_0 \hat{z}\right) = -\hat{\rho}\frac{q^2}{4\pi\varepsilon_0 r^2} - \hat{\rho} q \mathrm{v} \mu_0 H_0. \tag{8.155}$$

Assuming that the massive proton is stationary, we have

$$\frac{q^2}{4\pi\varepsilon_0 r^2} + q\omega r \mu_0 H_0 = m\omega^2 r, \tag{8.156}$$

where we have used the relation $v = \omega r$. The term $m\omega^2 r$ on the RHS of the above equation is the counterbalancing centrifugal force due to the circular motion of the electron. Solving for ω,

$$\omega = \omega_L \pm \left(\omega_L^2 + \omega_0^2\right)^{1/2}, \tag{8.157}$$

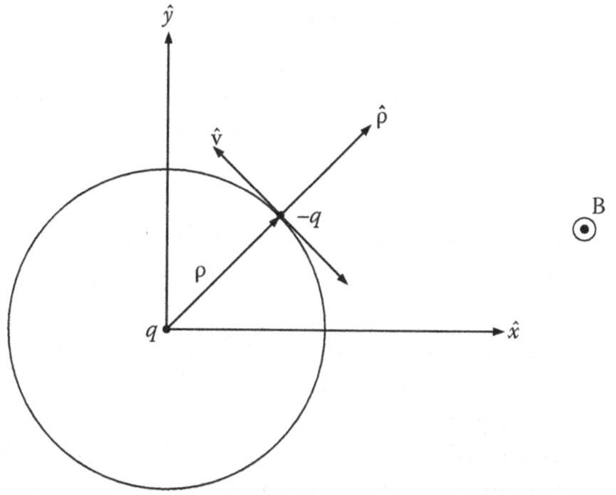

FIGURE 8.17
Electron circulating around a proton in the presence of a magnetic field perpendicular to the circular orbit.

where

$$\omega_L = \frac{q\mu_0 H_0}{2m}, \tag{8.158}$$

$$\omega_0 = \left(\frac{q^2}{4\pi m \varepsilon_0 r^3} \right)^{1/2}. \tag{8.159}$$

In the above, ω_0 is the frequency of circulation in the absence of H_0. If $\omega_L \gg \omega_0$, then corresponding to the case of large H_0 or large r,

$$\omega = 2\omega_L = \frac{q\mu_0 H_0}{m} = \omega_b, \quad \omega_L \gg \omega_0, \tag{8.160}$$

where ω_L is called the Larmor frequency and ω_b is called the cyclotron frequency. On the other hand, if H_0 is small or r is small ($\omega_L \ll \omega_0$), then the case for electrons orbiting in an atom is

$$\omega \approx \omega_0 + \omega_L. \tag{8.161}$$

Diamagnetic materials have μ_r slightly less than 1. It occurs in materials where the spin magnetization, explained later, is zero. The magnetization is only due to orbital motion. The magnetization due to the electrons spinning on their axis is zero due to even number of spins. The orbital magnetic dipole moment is given by

$$\left(p_m\right)_{orb} = -q\frac{\omega}{2\pi}\pi\rho^2\hat{z}. \tag{8.162}$$

The net dipole moment per unit volume **M** that arises from the orbital motion of the electrons is opposite to the original **H** field and when related to the field through magnetic susceptibility χ_m,

$$\mathbf{M} = \chi_m \mathbf{H}, \tag{8.163}$$

where χ_m will be a negative value. This results in a relative permeability μ_r:

$$\mu_r = 1 + \chi_m, \tag{8.164}$$

which is slightly less than 1. The above explanation is oversimplified but serves the purpose of showing that magnetization due to the orbital motion only leads to diamagnetization.

A simple explanation for paramagnetism in which μ_r is slightly larger than 1 is in terms of magnetization due to electron spin. The electrons of an atom with an even number of electrons usually exist in pairs, with the members of pairs having opposite spin directions, thereby canceling each other's spin magnetic moment. If the number of electrons is odd, individual atoms have net magnetic moment. If an external magnetic field is applied, then the individual atoms will align with it subject to the randomizing effect of thermal motion. The magnetic moments partially line up with the external fields giving rise to positive susceptibility. However, in paramagnetic material, the coupling between the magnetic moments of various atoms is relatively weak.

Ferrimagnetic materials have magnetized domains. A magnetized domain of a material (of 10^{-10} m^3 size) has atoms that are aligned parallel to each other (typically of the order of 10^{10} atoms in a domain). The domains are randomly oriented but in the presence of external fields they align. Examples of ferromagnetic materials are iron, nickel, and cobalt. The relative permeability is of the order of 3000. They exhibit nonlinearity, residual magnetism, and hysteresis. Soft ferromagnetic materials have narrow hysteresis loops and hence they can be more easily magnetized and demagnetized. Ferromagnetic materials are, in general, metallic in nature and their low resistivity gives rise to high eddy current losses at microwave frequencies. These losses are proportional to the square of the frequency and also proportional to the cube of the sample thickness. Laminating usually reduces these losses. At microwave frequencies, the losses are still large for vacuum-deposited thin ferromagnetic films of ~100–10,000 Å.

Ferromagnetic materials, known as ferrites and garnets, have high resistivity of up to $10^7\ \Omega$ m when compared to $10^{-7}\ \Omega$ m for iron. This material is a mixture of metallic oxides of high resistivity. They have a lower magnetization and are not used in applications requiring high flux density as in power transformers. They are very useful at microwave frequencies. The high-resistivity property arises because, unlike ferromagnetic materials, the constituents of ferrite materials, both metallic and oxygen ions, have no free 3s or 4s electrons available for conduction. Ferrites in the presence of a static magnetic field behave as an anisotropic magnetic material. The wave propagation in such a material is discussed in Appendix 11B.

8.10 Chiral Medium [8,9]

Figure 8.18 shows an approximation to a single turn left-handed and right-handed helix. The approximation consists of a circular conducting loop in a plane (x–y-plane) of radius a and a conductor of length $2l$ along the normal to the plane (z-axis). The distance between the turns (pitch) of the helix is $2l$.

Assume a time-harmonic incident electric excitation field E_0 along the z-axis. The voltage induced is given by

$$\text{emf} = \int \mathbf{E} \cdot d\boldsymbol{l} = 2lE_0. \tag{8.165}$$

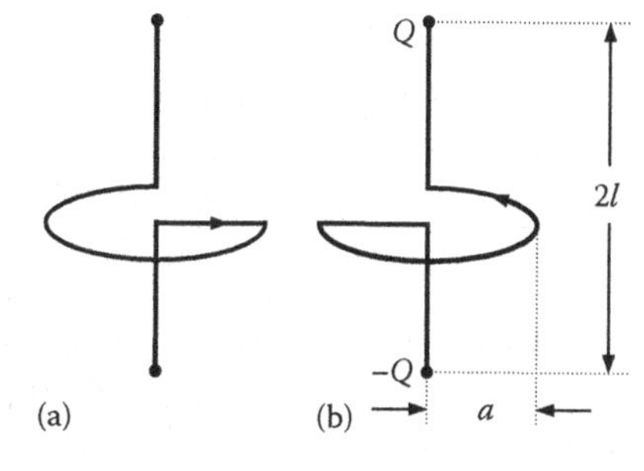

FIGURE 8.18
Approximate model for a helix: (a) left-handed and (b) right-handed.

The current is

$$I = \frac{\text{emf}}{Z_H} = \frac{2lE_0}{Z_H}, \tag{8.166}$$

where Z_H is the impedance of the helix. The electric dipole moment due to electric excitation is

$$p_{ee} = \hat{z}2lQ = \hat{z}2l\frac{I}{j\omega} = \hat{z}\frac{(2l)(2l)E_0}{j\omega Z_H}. \tag{8.167}$$

Note that for a time-harmonic field,

$$Q = \int I \mathrm{d}t = \frac{I}{j\omega}. \tag{8.168}$$

The magnetic dipole moment, due to electric excitation, in this case is

$$p_{me} = \hat{z}\pi a^2 I = \hat{z}\pi a^2 \frac{2lE_0}{Z_H}. \tag{8.169}$$

Next consider the case of time-harmonic magnetic excitation along the z-axis, $\hat{z}H_0$. From Faraday's law, we have

$$\text{emf} = \int \mathbf{E}\cdot \mathrm{d}l = \iint_s -\frac{\partial \mathbf{B}}{\partial t}\cdot \mathrm{d}s = \hat{z}\left[-\pi a^2 \mu j\omega H_0\right], \tag{8.170}$$

and the induced current I in the loop is given by

$$I = \frac{\text{emf}}{Z_H} = \frac{-\pi a^2 \mu j\omega H_0}{Z_H}. \tag{8.171}$$

The magnetic dipole moment due to magnetic excitation is

$$p_{mm} = \hat{z}\pi a^2 I = \hat{z}\frac{-j\omega\left(\pi a^2\right)^2 \mu H_0}{Z_H}. \tag{8.172}$$

The electric dipole moment due to magnetic excitation is

$$p_{em} = \hat{z}Q(2l) = \hat{z}\frac{I}{j\omega}(2l) = -\hat{z}\frac{\pi a^2 \mu 2lH_0}{Z_H}. \tag{8.173}$$

The point of the above calculation for the helix is to show that an electric excitation produces not only electric dipole moment, but also magnetic dipole moment. Since polarization is electric dipole moment per unit volume and magnetization is magnetic dipole moment per unit volume, we have shown that a material which has helices embedded in it experiences not only polarization, but also magnetization in the presence of an electric excitation. Similar remark can be made regarding magnetic excitation. Such a material is bi-isotropic in the sense that the constitutive relations are given by

$$\boldsymbol{D} = \varepsilon \boldsymbol{E} - \mathrm{j}\xi_c \boldsymbol{B}, \tag{8.174}$$

$$\boldsymbol{H} = -\mathrm{j}\xi_c \boldsymbol{E} + \frac{\boldsymbol{B}}{\mu}, \tag{8.175}$$

where ε and μ are, respectively, the scalar permittivity and permeability of the media and ξ_c is the chirality admittance. The word chiral in Greek means handedness and chiral materials include elements having a handedness property. An artificial chiral medium may be constructed by embedding metal helices of one handedness in a host medium (Figure 8.19). Axes of the helices are randomly oriented to create an isotropic chiral medium.

Bahr and Clansing [9] give the following details of an artificial chiral material.

Helices

n (helix density) = $20 \times 10^6/\mathrm{m}^3$,

a (radius of the helix) = 0.625 mm,

b (wire diameter) = 0.1524 mm,

$2l$ (pitch of the helix) = 0.667 mm,

N (number of turns in a helix) = 3.

Nonchiral host material

Silicon rubber: constitutive parameters at 7 GHz,

$\mu_h = \mu_0\,(1 - \mathrm{j}0)$,

$\varepsilon_h = \varepsilon_0\,(2.95 - \mathrm{j}0.07)$.

In the above μ_h and ε_h are, respectively, the permeability and permittivity of the host medium at the operating frequency. The constitutive relations for a chiral medium are given some times in the form of

$$\boldsymbol{D} = \varepsilon_c \boldsymbol{E} - \mathrm{j}\kappa\sqrt{\mu_h \varepsilon_h}\,\boldsymbol{H}, \tag{8.176}$$

$$\boldsymbol{B} = \mu_c \boldsymbol{H} + \sqrt{\mu_h \varepsilon_h}\,\boldsymbol{E}. \tag{8.177}$$

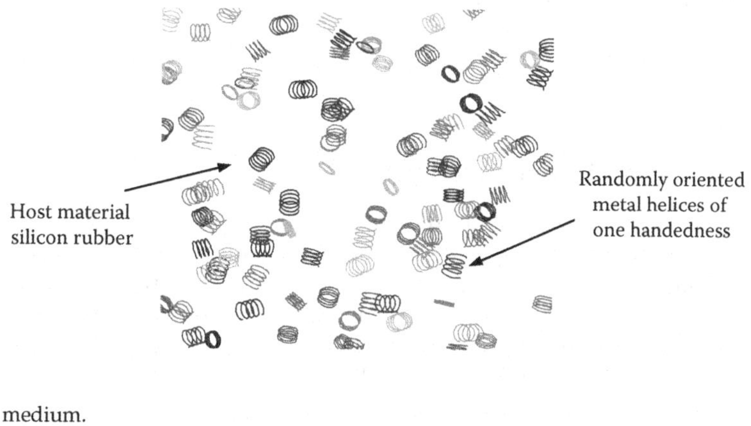

FIGURE 8.19
Artificial chiral medium.

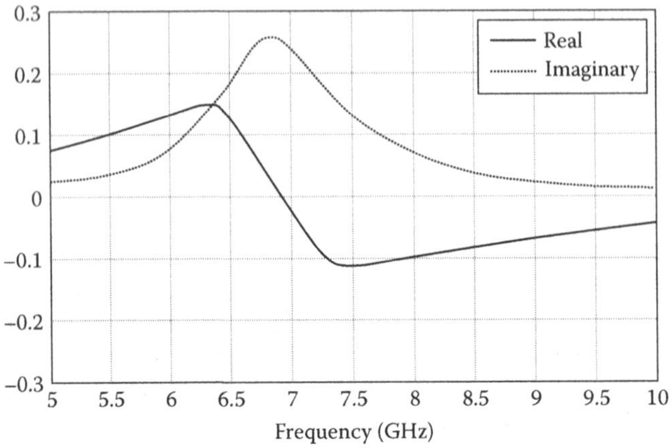

FIGURE 8.20
Chirality parameter κ versus frequency.

In this notation, favored by Lindell–Sihvola [8], ε_c and μ_c are the effective permittivity and permeability of the chiral material, respectively, and κ is a scalar chiral parameter. The above-mentioned artificial chiral material has the following parameter values at $f = 5$ GHz:

$$\varepsilon = 1.3\varepsilon_0, \tag{8.178}$$

$$\mu = 1.05\mu_0, \tag{8.179}$$

$$\kappa = 0.07. \tag{8.180}$$

Bahr and Clansing [9] have given the following results for κ as a function of frequency (Figure 8.20). The chirality parameter is negligible much below 5 GHz and also much above 10 GHz. This is understandable since the artificial medium has a length scale of cm.

Example of a material that exhibits chiral properties at optical frequencies is sugar. Such a property arises since these molecules have an inherently helical character and the length scale is much shorter than the artificial medium mentioned above. A discussion of wave propagation in a chiral medium is given in Appendix 8A. Kalluri and Rao [10] discuss chiral filters. DeMartinis [11] discusses artificial chiral materials with conical inclusions.

8.11 Plasmonics and Metamaterials

Plasmonics, a broad area of research in the subject of "Electromagnetics and Plasmas," is having a resurgence of interest because of new applications [12,13]. Metamaterials are engineered artificial materials "with properties and functionalities not found in nature" [14], helping to remove or decrease some limitations of the devices constructed from purely natural materials. Of particular interest are those that have negative permittivity and negative permeability in some frequency bands. In such a frequency band, the material is called left-handed. This material is different from the chiral left-handed material discussed in the previous section. The qualifier 'left-handed' is used here in a different sense and is explained in Appendix 8B.

Much is claimed and expected from these materials. They are under intense investigations by many research groups. The Appendix 8B based on transmission line analogies of electromagnetic metamaterials [15] is a simple and short account to establish the connection with the transmission line theory.

References

1. Balanis, C. A., *Advanced Engineering Electromagnetics*, Wiley, New York, 1989.
2. Heald, M. A. and Wharton, C. B., *Plasma Diagnostics with Microwaves*, Wiley, New York, 1965.
3. Ishimaru, A., *Electromagnetic Wave Propagation, Radiation and Scattering*, Prentice Hall, Englewood Cliffs, NJ, 1991.
4. Ramo, S., Whinnery, J. R., and Van Duzer, T., *Fields and Waves in Communication Electronics*, Wiley, New York, 1994.
5. Neelakanta, P. S., *Handbook of Electromagnetic Materials*, CRC Press, Boca Raton, FL, 1995.
6. Soohoo, R. F., *Microwave Magnetics*, Harper & Row, New York, 1985.
7. Kadin, A. M., *Introduction to Superconducting Circuits*, Wiley, New York, 1999.
8. Lindell, I. V., Sihvola, A. H., Tretyakov, S. A., and Viitanen, A. J., *Electromagnetic Waves in Chiral and Bi-Isotopic Media*, Artech House, Boston, MA, 1994.
9. Bahr, A. J. and Clansing, K. R., An approximate model for artificial chiral material, *IEEE Trans. Antennas Propag.*, AP-42, 1592, 1994.
10. Kalluri, D. K. and Rao, T. C. K., Filter characteristics of periodic chiral layers, *Pure Appl. Opt.*, 3, 231, 1994.
11. DeMartinis, G., Chiral media using conical coil wire inclusions, Doctoral thesis, University of Massachusetts Lowell, Lowell, MA, 2008.
12. Maier, S. A., *Plasmonics: Fundamentals and Applications*, Springer, New York, 2007.
13. Veksler, D. B., Plasma wave electronic devices for THZ detection, Doctoral thesis, Rensselaer Polytechnic Institute, Troy, NY, 2007.
14. Noginov, M. A. and Podolskiy, V. A., *Tutorials in Metamaterials*, CRC Press, Taylor & Francis Group, Boca Raton, FL, 2012.
15. Caloz, C. and Itoh, T., *Electromagnetic Metamaterials*, Wiley, Hoboken, NJ, 2006.

9

*Waves in Isotropic Cold Plasma: Dispersive Medium**

In Section 8.4, we briefly discussed the plasma medium and characterized it by the associated parameter called the plasma frequency ω_p, which is related to the electron density N through Equation 8.22. In Chapter 10, we will discuss quantitatively the various technical terms used in describing the plasma state of matter.

In this chapter, we consider the wave propagation in an isotropic cold (thermal effects neglected) plasma medium. We start with the equation for the force experienced by the electron in the plasma medium due to the wave electric and magnetic fields, given by

$$m\frac{\mathrm{d}\boldsymbol{v}}{\mathrm{d}t} = -q\left[\boldsymbol{E} + \boldsymbol{v} \times \boldsymbol{B}\right]. \tag{9.1a}$$

It can be shown that the force due to the wave magnetic field is negligible compared to the force due to the wave electric field.

The plasma current density $\boldsymbol{J}$ is related to its density N by

$$\boldsymbol{J} = -qN\boldsymbol{v} \tag{9.1b}$$

After neglecting the $\boldsymbol{v} \times \boldsymbol{B}$ term in (9.1a), and using (8.22) and (9.1b), one can obtain the equation relating $\boldsymbol{J}$ and $\boldsymbol{E}$:

$$\frac{\mathrm{d}\boldsymbol{J}}{\mathrm{d}t} = \varepsilon_0 \omega_p^2 \boldsymbol{E} \tag{9.1c}$$

Let us assume that the electron density varies only in space (inhomogeneous isotropic plasma):

$$\omega_p^2 = \omega_p^2(\mathbf{r}). \tag{9.1d}$$

9.1 Basic Equations

One can then construct basic solutions by assuming that the field variables vary harmonically in time:

$$\boldsymbol{F}(\mathbf{r}, t) = \mathbf{F}(\mathbf{r})\exp(\mathrm{j}\omega t). \tag{9.2}$$

* For chapter appendices, see 9A in Appendices section.

In the above, $\mathbf{F}(\mathbf{r})$ stands for any of $\mathbf{E}$, $\mathbf{H}$, or $\mathbf{J}$ phasors. The basic field equations are then given by

$$\nabla \times \mathbf{E} = -j\omega\mu_0\mathbf{H}, \tag{9.3}$$

$$\nabla \times \mathbf{H} = j\omega\varepsilon_0\mathbf{E} + \mathbf{J}, \tag{9.4}$$

$$j\omega\mathbf{J} = \varepsilon_0\omega_p^2(\mathbf{r})\mathbf{E}. \tag{9.5}$$

Combining Equations 9.4 and 9.5, we can write

$$\nabla \times \mathbf{H} = j\omega\varepsilon_0\varepsilon_p(\mathbf{r}, \omega)\mathbf{E}, \tag{9.6}$$

where ε_p is the dielectric constant of the isotropic plasma and is given by

$$\varepsilon_p(\mathbf{r}, \omega) = 1 - \frac{\omega_p^2(\mathbf{r})}{\omega^2}. \tag{9.7}$$

The vector wave equations are

$$\nabla^2\mathbf{E} + \frac{\omega^2}{c^2}\varepsilon_p(\mathbf{r})\mathbf{E} = \nabla(\nabla \cdot \mathbf{E}), \tag{9.8}$$

$$\begin{aligned}\nabla^2\mathbf{H} + \frac{\omega^2}{c^2}\varepsilon_p(\mathbf{r})\mathbf{H} &= -\frac{\varepsilon_0}{j\omega}\nabla\omega_p^2(\mathbf{r}) \times \mathbf{E} \\ &= -j\omega\varepsilon_0\nabla\varepsilon_p(\mathbf{r}) \times \mathbf{E}.\end{aligned} \tag{9.9}$$

The scalar one-dimensional equations take the form

$$\frac{\partial E}{\partial z} = -j\omega\mu_0 H, \tag{9.10}$$

$$-\frac{\partial H}{\partial z} = j\omega\varepsilon_0 E + J, \tag{9.11}$$

$$j\omega J = \varepsilon_0\omega_p^2(z)E. \tag{9.12}$$

Combining Equations 9.11 and 9.12, or from Equations 9.6, we have

$$-\frac{\partial H}{\partial z} = j\omega\varepsilon_0\varepsilon_p(z, \omega)E, \tag{9.13}$$

where

$$\varepsilon_p(z, \omega) = 1 - \frac{\omega_p^2(z)}{\omega^2}. \tag{9.14}$$

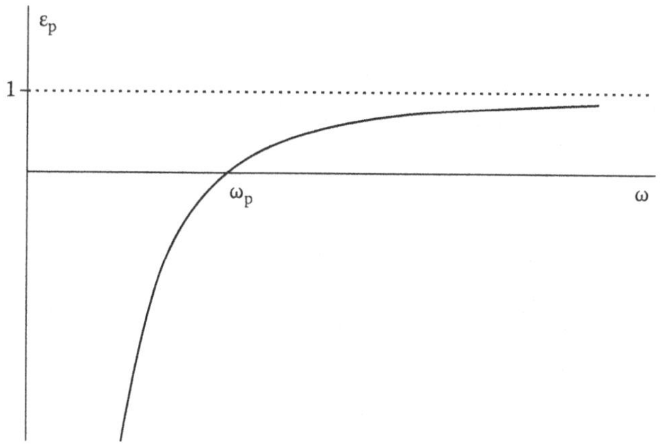

FIGURE 9.1
Dielectric constant versus frequency for a plasma medium.

Figure 9.1 shows the variation of ε_p with ω. ε_p is negative for $\omega < \omega_p$, 0 for $\omega = \omega_p$, and positive but less than 1 for $\omega > \omega_p$. The scalar one-dimensional wave equations, in this case, reduce to Equations 9.15 and 9.16:

$$\frac{d^2E}{dz^2} + \frac{\omega^2}{c^2}\varepsilon_p(z, \omega)E = 0, \tag{9.15}$$

$$\frac{d^2H}{dz^2} + \frac{\omega^2}{c^2}\varepsilon_p(z, \omega)H = \frac{1}{\varepsilon_p(z, \omega)}\frac{d\varepsilon_p(z, \omega)}{dz}\frac{dH}{dz}. \tag{9.16}$$

Equation 9.15 can easily be solved if we further assume that the dielectric is homogeneous (ε_p is not a function of z). Such a solution is given by

$$E = E^+(z) + E^-(z), \tag{9.17}$$

$$E = E_0^+ \exp(-jkz) + E_0^- \exp(+jkz), \tag{9.18}$$

where

$$k = \frac{\omega}{c}\sqrt{\varepsilon_p} = k_0\sqrt{\varepsilon_p} = k_0 n. \tag{9.19}$$

Here, n is the refractive index and k_0 is the free-space wave number. The first term on the RHS of Equation 9.17 represents a traveling wave in the positive z-direction (positive-going wave) of angular frequency ω and wave number k and the second term a similar but negative-going wave. The z-direction is the direction of phase propagation, and the velocity of phase propagation v_p (phase velocity) of either of the waves is given by

$$v_p = \frac{\omega}{k}. \tag{9.20}$$

From Equation 9.10,

$$H^{+}(z)=\frac{1}{(-j\omega\mu_0)}(-jk)E_0^{+}\exp(-jkz)=\frac{1}{\eta}E^{+}(z), \tag{9.21}$$

$$H^{-}(z)=-\frac{1}{\eta}E^{-}(z), \tag{9.22}$$

where

$$\eta=\sqrt{\frac{\mu_0}{\varepsilon_0\varepsilon_p}}=\frac{\eta_0}{n}=\frac{120\pi}{n}(\Omega). \tag{9.23}$$

Here η is the intrinsic impedance of the medium. The waves described above which are one-dimensional solutions of the wave equation in an unbounded isotropic homogeneous medium are called *uniform plane waves* in the sense that the phase and the amplitude of the waves are constant in a plane. Such one-dimensional solutions in a coordinate-free description are called transverse electric and magnetic waves (TEM) and for an arbitrarily directed wave their properties can be summarized as follows:

$$\hat{\mathbf{E}}\times\hat{\mathbf{H}}=\hat{k}, \tag{9.24}$$

$$\hat{\mathbf{E}}\cdot\hat{k}=0, \quad \hat{\mathbf{H}}\cdot\hat{k}=0, \tag{9.25}$$

$$E=\eta H. \tag{9.26}$$

In the above, $\hat{k}$ is a unit vector in the direction of phase propagation. In other words, the properties are:

1. Unit electric field vector, unit magnetic field vector, and the unit vector in the direction of (phase) propagation form a mutually orthogonal system.
2. There is no component of the electric or the magnetic field vector in the direction of propagation.
3. The ratio of the electric field amplitude to the magnetic field amplitude is given by the intrinsic impedance of the medium.

Figure 9.2 shows the variation of the dielectric constant ε_p with frequency for a typical real dielectric. For such a medium, in a broad frequency band, ε_p can be treated as not varying with ω and denoted by the dielectric constant ε_r. We will next consider the step profile approximation for a dielectric profile. In each of the media 1 and 2, the dielectric can be treated as homogeneous. The solution for this problem is well known but is included here to provide a comparison with the solutions in Chapters 10 through 12.

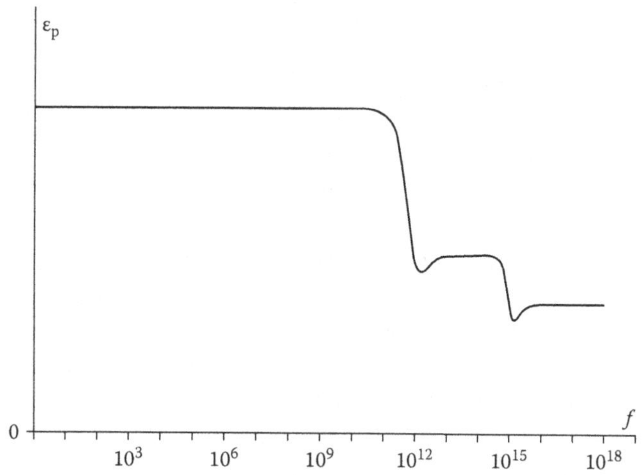

FIGURE 9.2
Sketch of the dielectric constant of a typical material.

9.2 Dielectric–Dielectric Spatial Boundary

The geometry of the problem is shown in Figure 9.3. Let the incident wave in medium 1, also called the source wave, have the fields

$$\mathbf{E}_i(x, y, z, t) = \hat{x}E_0 \exp\left[j(\omega_i t - k_i z)\right], \tag{9.27}$$

$$\mathbf{H}_i(x, y, z, t) = \hat{y}\frac{E_0}{\eta_1} \exp\left[j(\omega_i t - k_i z)\right], \tag{9.28}$$

where ω_i is the frequency of the incident wave and

$$k_i = \omega_i\sqrt{\mu_0\varepsilon_1} = \frac{\omega_i}{v_{p1}}. \tag{9.29}$$

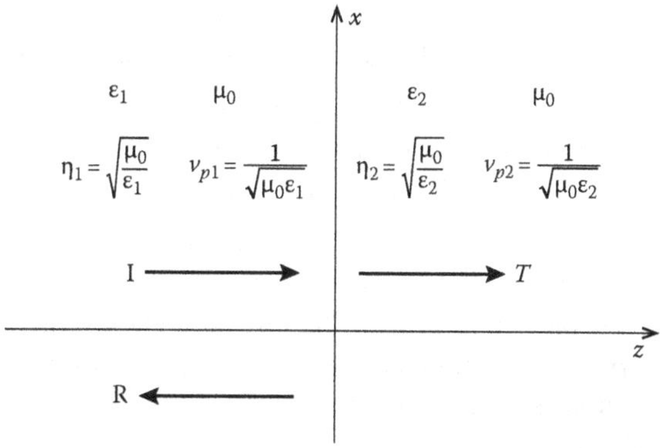

FIGURE 9.3
Dielectric–dielectric spatial boundary problem.

The fields of the reflected wave are given by

$$\mathbf{E}_r(x, y, z, t) = \hat{x}E_r \exp\left[j(\omega_r t + k_r z)\right], \tag{9.30}$$

$$\mathbf{H}_r(x, y, z, t) = -\hat{y}\frac{E_r}{\eta_1}\exp\left[j(\omega_r t + k_r z)\right], \tag{9.31}$$

where ω_r is the frequency of the reflected wave and

$$k_r = \omega_r\sqrt{\mu_0\varepsilon_1}. \tag{9.32}$$

The fields of the transmitted wave are given by

$$\mathbf{E}_t(x, y, z, t) = \hat{x}E_t \exp\left[j(\omega_t t - k_t z)\right], \tag{9.33}$$

$$\mathbf{H}_t(x, y, z, t) = \hat{y}\frac{E_t}{\eta_2}\exp\left[j(\omega_t t - k_t z)\right], \tag{9.34}$$

where ω_r is the frequency of the transmitted wave and

$$k_t = \omega_t\sqrt{\mu_0\varepsilon_2} = \frac{\omega_t}{v_{p2}}. \tag{9.35}$$

The boundary condition of the continuity of the tangential component of the electric field at the interface $z = 0$ can be stated as

$$\mathbf{E}(x, y, 0^-, t) \times \hat{z} = \mathbf{E}(x, y, 0^+, t) \times \hat{z}, \tag{9.36}$$

$$\left[\mathbf{E}_i(x, y, 0^-, t) + \mathbf{E}_r(x, y, 0^-, t)\right] \times \hat{z} = \left[\mathbf{E}_t(x, y, 0^+, t)\right] \times \hat{z}. \tag{9.37}$$

The above must be true for all x, y, and t. Thus, we have

$$E_0 \exp\left[j\omega_i t\right] + E_r \exp\left[j\omega_r t\right] = E_t \exp\left[j\omega_t t\right]. \tag{9.38}$$

Since Equation 9.38 must be satisfied for all t, the coefficients of t in the exponents of Equation 9.38 must match:

$$\omega_i = \omega_r = \omega_t = \omega. \tag{9.39}$$

The above result can be stated as follows: the frequency ω is conserved across a spatial discontinuity in the properties of an electromagnetic medium. As the wave crosses from one medium to the other in space, the wave number k changes as dictated by the change in the phase velocity (see Figure 9.4).

From Equations 9.38 and 9.39, we obtain

$$E_0 + E_r = E_t. \tag{9.40}$$

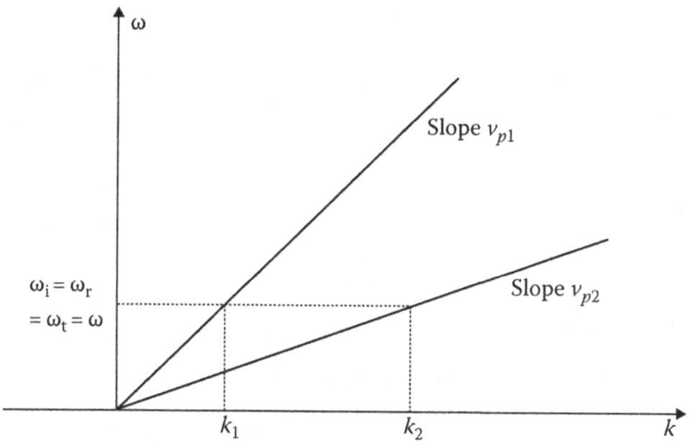

FIGURE 9.4
Conservation of frequency across a spatial boundary.

The second independent boundary condition of continuity of the tangential magnetic field component is written as

$$\left[\mathbf{H}_i\left(x, y, 0^-, t\right) + \mathbf{H}_r\left(x, y, 0^-, t\right)\right] \times \hat{z} = \left[\mathbf{H}_t\left(x, y, 0^+, t\right)\right] \times \hat{z}. \tag{9.41}$$

The reflection coefficient $R_A = E_r/E_0$ and the transmission coefficient $T_A = E_t/E_0$ are determined using Equations 9.40 and 9.41. From Equation 9.41, we have

$$\frac{E_0 - E_r}{\eta_1} = \frac{E_t}{\eta_2}. \tag{9.42}$$

The results are as follows:

$$R_A = \frac{\eta_2 - \eta_1}{\eta_2 + \eta_1} = \frac{n_1 - n_2}{n_1 + n_2}, \tag{9.43}$$

$$T_A = \frac{2\eta_2}{\eta_2 + \eta_1} = \frac{2n_1}{n_1 + n_2}. \tag{9.44}$$

The significance of the subscript A in the above is to distinguish from the reflection and transmission coefficients due to a temporal discontinuity in the properties of the medium.

We next show that the time-averaged power density of the source wave is equal to the sum of the time-averaged power density of the reflected and the transmitted waves:

$$\left|\frac{1}{2}\mathrm{Re}\left[\mathbf{E}_i \times \mathbf{H}_i^* \cdot \hat{z}\right]\right| = \left|\frac{1}{2}\mathrm{Re}\left[\mathbf{E}_r \times \mathbf{H}_r^* \cdot \hat{z}\right]\right| + \left|\frac{1}{2}\mathrm{Re}\left[\mathbf{E}_t \times \mathrm{H}_t^* \cdot \hat{z}\right]\right|. \tag{9.45}$$

The LHS is

$$\frac{1}{2}E_0 H_0 = \frac{1}{2}\frac{E_0^2}{\eta_1}, \tag{9.46}$$

whereas the RHS is

$$\frac{1}{2}E_0^2\left[\frac{|R_A|^2}{\eta_1}+\frac{|T_A|^2}{\eta_2}\right]=\frac{1}{2}\frac{E_0^2}{\eta_1}\left[\frac{(\eta_2-\eta_1)^2+4\eta_1\eta_2}{(\eta_2+\eta_1)^2}\right]=\frac{1}{2}\frac{E_0^2}{\eta_1}. \tag{9.47}$$

9.3 Reflection by a Plasma Half-Space

Let an incident wave of frequency ω traveling in the free space ($z < 0$) be incident normally on the plasma half-space ($z > 0$) of plasma frequency ω_p. The intrinsic impedance of the plasma medium is given by

$$\eta_p=\frac{\eta_0}{n_p}=\frac{120\pi\omega}{\sqrt{\omega^2-\omega_p^2}}. \tag{9.48}$$

From Equation 9.43, the reflection coefficient R_A is given by

$$R_A=\frac{\eta_p-\eta_0}{\eta_p+\eta_0}=\frac{1-n_p}{1+n_p}=\frac{\Omega-\sqrt{\Omega^2-1}}{\Omega+\sqrt{\Omega^2-1}}, \tag{9.49}$$

where

$$\Omega=\frac{\omega}{\omega_p} \tag{9.50}$$

is the source frequency normalized with respect to the plasma frequency. The power reflection coefficient $\rho = |R_A|^2$ and the power transmission coefficient τ ($=1-\rho$) versus Ω are shown in Figure 9.5. In the frequency band $0 < \omega < \omega_p$, the characteristic impedance η_p

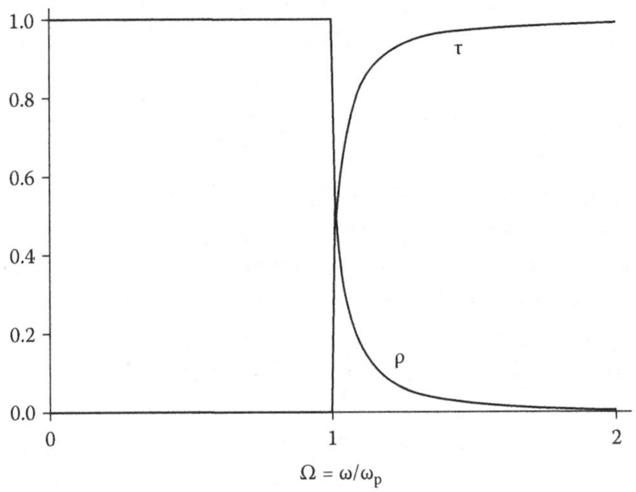

FIGURE 9.5
Sketch of the power reflection coefficient (ρ) and the power transmission coefficient (τ) of a plasma half-space.

is imaginary. The electric field $\mathbf{E}_P$ and the magnetic field $\mathbf{H}_P$ in the plasma are in *time quadrature*; the real part of $\left(\mathbf{E}_P \times \mathbf{H}_P^*\right)$ is zero. The wave in plasma is an *evanescent wave* [1] and carries no real power. The source wave is totally reflected and $\rho = 1$. In the frequency band $\omega_p < \omega < \infty$, the plasma behaves as a dielectric and the source wave is partially transmitted and partially reflected. The time-averaged power density of the incident wave is equal to the sum of the time-averaged power densities of the reflected and transmitted waves.

Metals at optical frequencies are modeled as plasmas with plasma frequency ω_p of the order of 10^{16}. Refining the plasma model by including the collision frequency will lead to three frequency domains of conducting, cutoff, and dielectric phenomena [2]. A more comprehensive account of modeling metals as plasmas is given in References [3,4]. The associated phenomena of attenuated total reflection, surface plasmons, and other interesting topics are based on modeling metals as plasmas. A brief account of surface wave is given in Section 9.8.

9.4 Reflection by a Plasma Slab [5]

In this section, we will consider oblique incidence to add to the variety to the problem formulation. The geometry of the problem is shown in Figure 9.6. Let x–z be the plane of incidence and $\hat{k}$ be the unit vector along the direction of propagation. Let the magnetic field H^I be along y and the electric field E^I lie entirely in the plane of incidence. Such a wave is described in the literature by various names: (1) p wave, (2) TM wave, and (3) parallel-polarized wave. Since the boundaries are at $z = 0$ and d, it is necessary to use x–y–z-coordinates in formulating the problem and express $\hat{k}$ in terms of x- and z-coordinates.

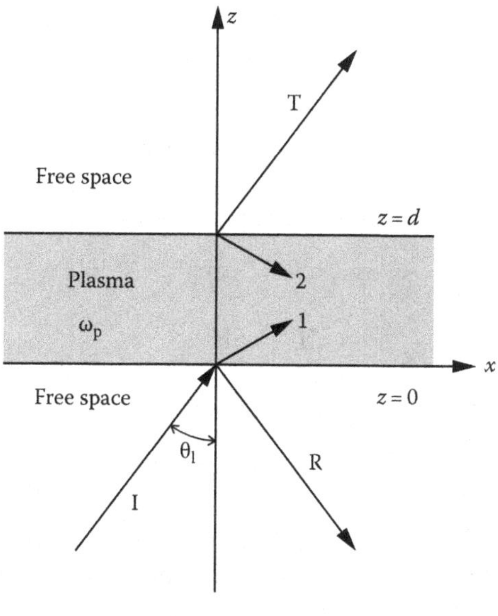

FIGURE 9.6
Plasma slab problem.

The problem, though one-dimensional, appears to be two-dimensional. Equations 9.51 through 9.56 describe the incident wave:

$$\mathbf{H}^{\mathrm{I}} = \hat{y} H_y^{\mathrm{I}} \exp\left[-jk_0 \hat{k}\cdot\mathbf{r}\right], \tag{9.51}$$

$$\hat{k} = \hat{x}S + \hat{z}C, \tag{9.52}$$

$$S = \sin\theta_{\mathrm{I}}, \quad C = \cos\theta_{\mathrm{I}}, \tag{9.53}$$

$$\mathbf{E}^{\mathrm{I}} = \left(\hat{y}\times\hat{k}\right)E^{\mathrm{I}} \exp\left[-jk_0\hat{k}\cdot\mathbf{r}\right] = \left(\hat{x}E_x^{\mathrm{I}} + \hat{z}E_z^{\mathrm{I}}\right)\Psi^{\mathrm{I}}, \tag{9.54}$$

$$E_x^{\mathrm{I}} = CE^{\mathrm{I}}, \quad E_z^{\mathrm{I}} = -SE^{\mathrm{I}}, \quad E^{\mathrm{I}} = \eta_0 H_y^{\mathrm{I}}, \tag{9.55}$$

$$\Psi^{\mathrm{I}} = \exp\left[-jk_0\left(Sx + Cz\right)\right]. \tag{9.56}$$

The reflected wave is written as

$$\mathbf{E}^{\mathrm{R}} = \left(\hat{x}E_x^{\mathrm{R}} + \hat{z}E_z^{\mathrm{R}}\right)\Psi^{\mathrm{R}}, \tag{9.57}$$

$$\mathbf{H}^{\mathrm{R}} = \hat{y}H_y^{\mathrm{R}}\Psi^{\mathrm{R}}, \tag{9.58}$$

$$\Psi^{\mathrm{R}} = \exp\left[-jk_0\left(Sx - Cz\right)\right]. \tag{9.59}$$

The boundary conditions at $z = 0$ require that the x dependence in the region $z > 0$ remains the same as in the region $z < 0$. Since the region $z > d$ is also a free space, the exponential factor for the transmitted wave will be the same as the incident wave and the fields of the transmitted wave are

$$\mathbf{E}^{\mathrm{T}} = \left(\hat{x}E_x^{\mathrm{T}} + \hat{z}E_z^{\mathrm{T}}\right)\Psi^{\mathrm{T}}, \tag{9.60}$$

$$\mathbf{H}^{\mathrm{T}} = \hat{y}H_y^{\mathrm{T}}\Psi^{\mathrm{T}}, \tag{9.61}$$

$$\Psi^{\mathrm{T}} = \exp\left[-jk_0\left(Sx + Cz\right)\right]. \tag{9.62}$$

All the field amplitudes in the free space can be expressed in terms of E_x^{I}, E_x^{R}, or E_x^{T}:

$$CE_z^{\mathrm{I,T}} = -SE_x^{\mathrm{I,T}}, \tag{9.63}$$

$$CE_z^{\mathrm{R}} = SE_x^{\mathrm{R}}, \tag{9.64}$$

$$C\eta_0 H_y^{\mathrm{I,T}} = E_x^{\mathrm{I,T}}, \tag{9.65}$$

$$C\eta_0 H_y^{\mathrm{R}} = -E_x^{\mathrm{R}}. \tag{9.66}$$

We consider next the waves in the homogeneous plasma region $0 < z < d$. The x and the z dependence for waves in the plasma comes from the exponential factor Ψ^{P}:

$$\Psi^{\mathrm{P}} = \exp\left[-jk_0\left(Sx + qz\right)\right], \tag{9.67}$$

where q has to be determined. The wave number in the plasma is given as $k_p = k_0\sqrt{\varepsilon_p}$. Thus, we have

$$q^2 + S^2 = \varepsilon_p = 1 - \frac{\omega_p^2}{\omega^2}, \tag{9.68}$$

$$q_{1,2} = \pm\sqrt{C^2 - \frac{\omega_p^2}{\omega^2}} = \pm\sqrt{\frac{C^2\Omega^2 - 1}{\Omega^2}}. \tag{9.69}$$

The two values for q indicate the excitation of two waves in the plasma, the first is the positive wave and the second is a negative wave. For $\Omega < 1/C$, q is imaginary and the two waves in the plasma are evanescent. The next section deals with this frequency band.

The fields in the plasma can be written as

$$\mathbf{E}^P = \sum_{m=1}^{2}\left(\hat{x}E_{xm}^P + \hat{z}E_{zm}^P\right)\Psi_m^P, \tag{9.70}$$

$$\mathbf{H}^P = \sum_{m=1}^{2}\hat{y}H_{ym}^P\Psi_m^P, \tag{9.71}$$

$$\Psi_m^P = \exp\left[-jk_0\left(Sx + q_m z\right)\right]. \tag{9.72}$$

All the field amplitudes in the plasma can be expressed in terms of E_{x1}^P and E_{x2}^P:

$$\eta_0 H_{ym}^P = \frac{\varepsilon_p}{q_m}E_{xm}^P = \eta_{ym}E_{xm}^P, \tag{9.73}$$

$$E_{zm}^P = -\frac{S}{q_m}E_{xm}^P. \tag{9.74}$$

The above relations are obtained from Equations 9.3 and 9.6 by noting $\partial/\partial x = -jk_0S$ and $\partial/\partial z = -jk_0q$; they can also be written from inspection. Assuming that E_x^I is known, the unknowns reduce to four, that is, E_{x1}^P, E_{x2}^P, E_x^R, and E_x^T. They can be determined from the four boundary conditions of continuity of the tangential components E_x and H_y at $z = 0$ and $z = d$. In matrix form,

$$\begin{bmatrix} 1 & 1 & -1 & 0 \\ C\eta_{y1} & C\eta_{y2} & 1 & 0 \\ \lambda_1 & \lambda_2 & 0 & -1 \\ \lambda_1 C\eta_{y1} & \lambda_2 C\eta_{y2} & 0 & -1 \end{bmatrix}\begin{bmatrix} E_{x1}^P \\ E_{x2}^P \\ E_x^R \\ E_x^T \end{bmatrix} = \begin{bmatrix} 1 \\ 1 \\ 0 \\ 0 \end{bmatrix}E_x^I, \tag{9.75}$$

where

$$\begin{aligned} \lambda_1 &= \exp\left[jk_0\left(C - q_1\right)d\right], \\ \lambda_2 &= \exp\left[jk_0\left(C - q_2\right)d\right]. \end{aligned} \tag{9.76}$$

Solving Equation 9.75,

$$E_{x1}^{\mathrm{P}} = 2\lambda_2\left(1 - C\eta_{y2}\right)E_x^{\mathrm{I}}/\Delta, \tag{9.77}$$

$$E_{x2}^{\mathrm{P}} = -2\lambda_1\left(1 - C\eta_{y1}\right)E_x^{\mathrm{I}}/\Delta, \tag{9.78}$$

$$E_x^{\mathrm{R}} = \left(1 - C\eta_{y1}\right)\left(1 - C\eta_{y2}\right)\left(\lambda_2 - \lambda_1\right)E_x^{\mathrm{I}}/\Delta, \tag{9.79}$$

$$E_x^{\mathrm{T}} = 2\lambda_1\lambda_2 C\left(\eta_{y1} - \eta_{y2}\right)E_x^{\mathrm{I}}/\Delta, \tag{9.80}$$

where

$$\Delta = \lambda_2\left(1 + C\eta_{y1}\right)\left(1 - C\eta_{y2}\right) - \lambda_1\left(1 - C\eta_{y1}\right)\left(1 + C\eta_{y2}\right). \tag{9.81}$$

The power reflection coefficient $\rho = \left|E_x^{\mathrm{R}}/E_x^{\mathrm{I}}\right|^2$ is given by

$$\rho = \frac{1}{1 + \left(2C\varepsilon_{\mathrm{p}}q_1 / \left(C^2\varepsilon_{\mathrm{p}}^2 - q_1^2\right)\mathrm{cosec}\left(2\pi\Omega q_1 d_{\mathrm{p}}\right)\right)^2}, \tag{9.82}$$

where

$$d_{\mathrm{p}} = \frac{d}{\lambda_{\mathrm{p}}}, \tag{9.83}$$

$$\lambda_{\mathrm{p}} = \frac{2\pi c}{\omega_{\mathrm{p}}}. \tag{9.84}$$

In the above, λ_{p} is the free space wavelength corresponding to the plasma frequency and is used to normalize the slab width. Substituting for q_1, in the range of real q_1,

$$\rho = \frac{1}{1 + B\,\mathrm{cosec}^2 A}, \quad \frac{1}{C} < \Omega < \infty, \tag{9.85}$$

and in the range of imaginary q_1,

$$\rho = \frac{1}{1 - B\,\mathrm{cosech}^2|A|}, \quad 0 < \Omega < \frac{1}{C}, \tag{9.86}$$

where A and B are given by

$$A = 2\pi\Omega q_1 d_{\mathrm{p}} = 2\pi\sqrt{C^2\Omega^2 - 1}\,d_{\mathrm{p}}, \tag{9.87}$$

$$B = \frac{4C^2\varepsilon_{\mathrm{p}}^2 q_1^2}{\left(C^2\varepsilon_{\mathrm{p}}^2 - q_1^2\right)^2} = \frac{4C^2\Omega^2\left(\Omega^2 - 1\right)^2\left(C^2\Omega^2 - 1\right)}{\left(2C^2\Omega^2 - \Omega^2 - C^2\right)^2}. \tag{9.88}$$

The power transmission coefficient $\tau = (1 - \rho)$ and is given by

$$\tau = \frac{|B|}{|B| + \sin^2 |A|}, \quad \frac{1}{C} < \Omega < \infty, \tag{9.89}$$

$$\tau = \frac{|B|}{|B| + \sinh^2 |A|}, \quad 0 < \Omega < \frac{1}{C}. \tag{9.90}$$

From Equation 9.85, $\rho = 0$ when $\sin A = 0$ or

$$A = n\pi, \quad n = 0, 1, 2, \ldots. \tag{9.91}$$

There is one more value of Ω for which $\rho = 0$. From Equation 9.88, when $C\varepsilon_p = q_1$, $B = \infty$, and from Equation 9.89, $\tau = 1$ and $\rho = 0$. This point corresponds to a frequency Ω_B given by

$$\Omega_B^2 = \frac{C^2}{2C^2 - 1}. \tag{9.92}$$

It is easily shown that Ω_B exists only for the *p* wave and is greater than $1/C$. In fact, this point corresponds to the *Brewster angle* [4]. Figure 9.7 shows Ω_B versus $\cos \theta_B$, where θ_B is the Brewster angle.

Thus, in the frequency band $\Omega > 1/C$, the variation of ρ with the slab width is oscillatory. Stratton [6] discussed these oscillations for a dielectric slab and associated them with the interference of the internally reflected waves in the slab. In the case of the plasma slab, the variation of ρ with the source frequency is also oscillatory, since in this range the plasma behaves like a dispersive dielectric. The maxima of ρ are less than or equal to 1 and occur for Ω satisfying the transcendental equation

$$\tan A = fA, \tag{9.93}$$

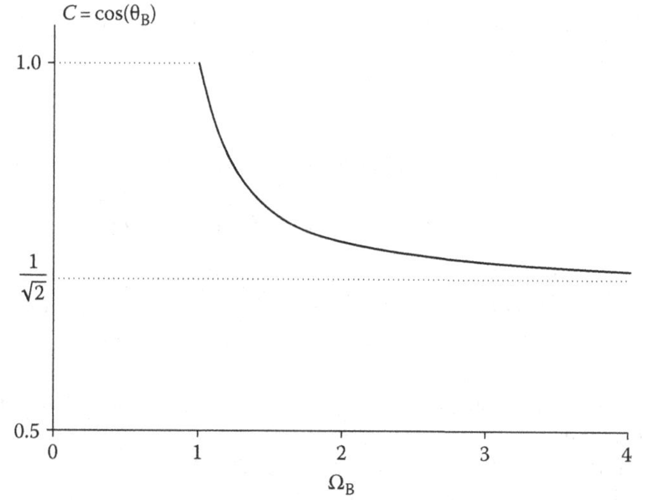

FIGURE 9.7
Brewster angle for a plasma medium.

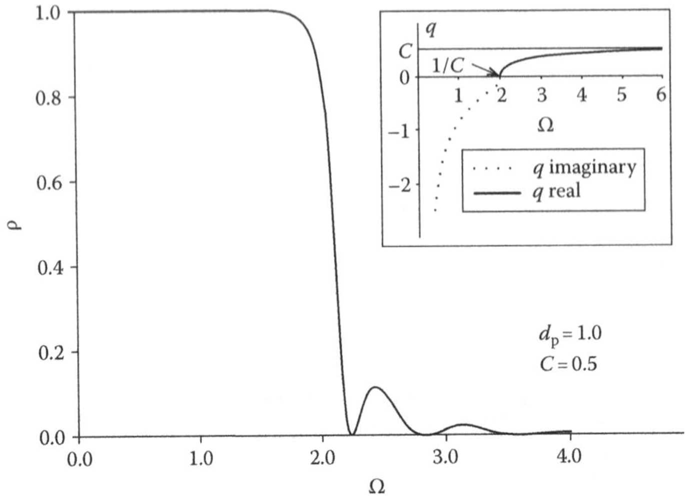

FIGURE 9.8
ρ versus Ω for isotropic plasma slab (parallel polarization).

where

$$f = \frac{C^2\Omega^2\left(\Omega^2 - 1\right)\left(C^2 + \Omega^2 - 2C^2\Omega^2\right)}{\left(2C^2\Omega^4 - 2C^2\Omega^2 + C^2 - \Omega^2\right)\left(1 + \Omega^2 - 2C^2\Omega^2\right)}. \tag{9.94}$$

Figure 9.8 shows ρ versus Ω for a normalized slab width $d_p = 1.0$ and $C = 0.5$ ($\theta_I = 60°$). The inset shows the variation of q with Ω. The oscillations can be clearly seen in the real range of q. Heald and Wharton [2] show similar curves for normal incidence on a lossy plasma slab using parameters normalized with reference to the source wave quantities.

In Figure 9.8, $\rho \approx 1$ in the imaginary range of q showing total reflection of the source wave. However, it will be shown that there can be a considerable *tunneling of power* for sufficiently thin plasma slabs.

9.5 Tunneling of Power through a Plasma Slab [5], [7]

For the frequency band $\Omega < 1/C$, the characteristic roots are imaginary and the waves excited in the plasma are evanescent. The incident wave is completely reflected by the semi-infinite plasma. However, in the case of a plasma slab, some power gets transmitted through it even in this frequency band. It can be shown that this tunneling effect is due to the interaction of the electric field of the positive wave with the magnetic field of the negative wave and vice versa.

The above statement is supported by examining the power flow through the Poynting vector calculation. From Equations 9.77 through 9.81, one can obtain

$$E_x^{\mathrm{T}} = 2C\varepsilon_p q_1 \exp\left[jk_0 Cd\right] E_x^{\mathrm{I}}/\Delta, \tag{9.95}$$

$$E_{x1}^{\mathrm{P}} = \left(q_1 + C\varepsilon_p\right) q_1 \exp\left[\mathrm{j}k_0 q_1 d\right] E_x^{\mathrm{I}}/\Delta, \tag{9.96}$$

$$E_{x2}^{\mathrm{P}} = -\left(-q_1 + C\varepsilon_{\mathrm{p}}\right) q_1 \exp\left[-\mathrm{j}k_0 q_1 d\right] E_x^{\mathrm{I}}/\Delta, \tag{9.97}$$

where

$$\Delta = 2q_1 C\varepsilon_{\mathrm{p}} \cos\left(k_0 q_1 d\right) + \mathrm{j}\left(q_1^2 + C^2\varepsilon_{\mathrm{p}}^2\right)\sin\left(k_0 q_1 d\right). \tag{9.98}$$

The magnetic field components H_{y1}^{P}, H_{y2}^{P}, and H_y^{T} can be obtained from Equations 9.65 and 9.73.

The power crossing any plane in the plasma parallel to the interface can be found by calculating the *z*-component of the Poynting vector associated with the plasma waves. This is given by

$$S_z^{\mathrm{P}} = \frac{1}{2}\mathrm{Re}\left[\left(E_{x1}^{\mathrm{P}}\Psi_1^{\mathrm{P}} + E_{x2}^{\mathrm{P}}\Psi_2^{\mathrm{P}}\right)\left(H_{y1}^{\mathrm{P}*}\Psi_1^{\mathrm{P}*} + H_{y2}^{\mathrm{P}*}\Psi_2^{\mathrm{P}*}\right)\right], \tag{9.99}$$

which on expansion gives

$$\begin{aligned} S_z^{\mathrm{P}} = {} & \frac{1}{2}\mathrm{Re}\left[E_{x1}^{\mathrm{P}}\Psi_1^{\mathrm{P}}H_{y1}^{\mathrm{P}*}\Psi_1^{\mathrm{P}*}\right] + \frac{1}{2}\mathrm{Re}\left[E_{x2}^{\mathrm{P}}\Psi_2^{\mathrm{P}}H_{y1}^{\mathrm{P}*}\Psi_1^{\mathrm{P}*}\right] \\ & + \frac{1}{2}\mathrm{Re}\left[E_{x1}^{\mathrm{P}}\Psi_1^{\mathrm{P}}H_{y2}^{\mathrm{P}*}\Psi_2^{\mathrm{P}*}\right] + \frac{1}{2}\mathrm{Re}\left[E_{x2}^{\mathrm{P}}\Psi_2^{\mathrm{P}}H_{y2}^{\mathrm{P}*}\Psi_2^{\mathrm{P}*}\right]. \end{aligned} \tag{9.100}$$

where Ψ_1^{P} and Ψ_2^{P} are given by Equation 9.72. The contribution from each of the four terms on the RHS of Equation 9.100 is now discussed for the two cases of q_1 real $\Omega > 1/C$ and q_1 imaginary $\Omega < 1/C$. It is to be noted that the second and the third terms give the contributions due to the cross-interaction of the fields in the positive and negative waves.

Case 1: q_1 *real*

It is easy to see that the second and the third terms being equal but opposite in sign, there is no contribution to the net power flow from the cross-interaction. The sum of the first and fourth terms is equal to $(1/2)\mathrm{Re}\left[\left(E_x^{\mathrm{T}}\Psi^{\mathrm{T}}\right)\left(H_y^{\mathrm{T}*}\Psi^{\mathrm{T}*}\right)\right]$, where Ψ^{T} is given by Equation 9.62. Thus, the power crossing any plane parallel to the slab interface gives exactly the power that emerges into the free space at the other boundary of the slab.

Case 2: q_1 *imaginary*

It can be shown that each of the first and the fourth terms is zero. This is because the corresponding electric and magnetic fields are in time quadrature. Furthermore, the second term is equal to the third term and each is equal to $(1/4)\mathrm{Re}\left[\left(E_x^{\mathrm{T}}\Psi^{\mathrm{T}}\right)\left(H_y^{\mathrm{T}*}\Psi^{\mathrm{T}*}\right)\right]$. Thus, the power flow in the tunneling frequency band ($\Omega < 1/C$) comes entirely from the cross-interaction.

At $\Omega = 1$, $\varepsilon_{\mathrm{p}} = 0$, and from Equation 9.88, $B = 0$ but $A \neq 0$. From Equation 9.89, $\tau = 0$. This point exists only for the *p* wave.

At $\Omega = 1/C$, A = 0, and B = 0. Evaluating the limit, we get the power transmission coefficient τ_{c} at this point:

$$\tau_{\mathrm{c}} = \frac{1}{1 + S^4\pi^2 d_{\mathrm{p}}^2}, \quad \Omega = \frac{1}{C}. \tag{9.101}$$

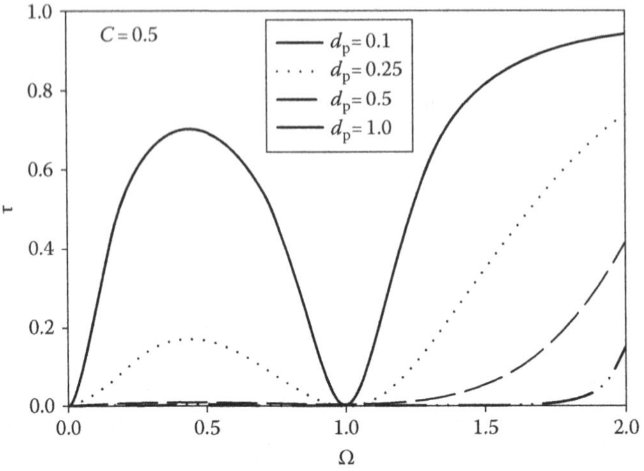

FIGURE 9.9
Transmitted power for an isotropic plasma slab (parallel polarization) in the tunneling range ($\Omega < 1/C$) for various d_p. The right side of the bottom two curves can be seen more clearly than the left side and they are for $d_p = 0.5$ and $d_p = 1.0$.

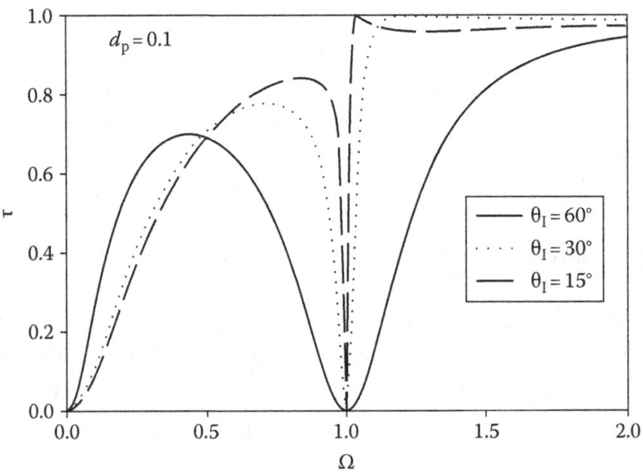

FIGURE 9.10
Transmitted power for an isotropic plasma slab (parallel polarization) in the tunneling range ($\Omega < 1/C$) for various angles of incidence. The broken line curve is for $\theta_I = 15°$.

Numerical results of τ versus Ω are presented in Figures 9.9 and 9.10 for the tunneling frequency band. In Figure 9.9, the angle of incidence is taken to be 60° and the curves are given for various d_p. Between $\Omega = 0$ and 1, there is a value of $\Omega = \Omega_{max}$ where the transmission is maximum. The maximum value of the transmitted power decreases as d_p increases.

In Figure 9.10, τ versus Ω is shown with the angle of incidence as the parameter. It is seen that the point of maximum transmission moves to the right as the angle of incidence decreases and merges with $\Omega = 1$ for $\theta_I = 0°$ (normal incidence). The maximum value of the transmitted power decreases as θ_I increases. This point is given by the solution of the transcendental equation

$$\tanh|A| = f|A|, \tag{9.102}$$

where f and A are given in Equations 9.94 and 9.87, respectively. It is suggested that by measuring Ω_{max} experimentally, one can determine either the plasma frequency or the slab width, knowing the angle of incidence.

9.6 Inhomogeneous Slab Problem

Let us next consider the case where ε_p is a function of z in the region $0 < z < d$. To simplify and focus on the effect of the inhomogeneity of the properties of the medium, let us revert to the normal incidence. The differential equation for the electric field is given by Equation 9.15. This equation cannot be solved exactly for a general $\varepsilon_p(z)$ profile.

However, there are a small number of profiles for which exact solutions can be obtained in terms of *special functions*. As an example, the problem of the *linear profile* can be solved in terms of *Airy functions*. An account of such solutions for a dielectric profile $\varepsilon_r(z)$ are given in [8,9] and for a plasma profile $\varepsilon_p(z, \omega)$ in [2,10]. Abramowitz and Stegun [11] give detailed information on special functions.

A vast amount of literature is available on the approximate solution for an inhomogeneous media problem. The following techniques are emphasized in this book. The actual profile is considered as a perturbation of a profile for which the exact solution is known. The effect of the perturbation is calculated through the use of Green's function. The problem of a *fast profile* with finite range—length can be so handled by using a step profile as the reference. At the other end of the approximation scale, the *slow profile* problem can be handled by adiabatic and WKBJ techniques. See [1,8] for dielectric profile examples and [2,10] for plasma profile examples. Numerical approximation techniques can also be used which are extremely useful in obtaining specific numbers for a given problem and in validating the theoretical results obtained from the physical approximations.

9.7 Periodic Layers of Plasma

Next, we consider a particular case of an inhomogeneous plasma medium which is periodic. Let the dielectric function

$$\varepsilon_p(z,\omega) = \varepsilon_p(z + mL,\omega), \tag{9.103}$$

where m is an integer and L is the spatial period. If the domain of m is all integers, $-\infty < m < \infty$, then we have an unbounded periodic media. The solution of Equation 9.15 for the infinite periodic structure problem [1,9,12] will be such that the electric field $E(z)$ differs from the electric field at $E(z + L)$ by a constant:

$$E(z + L) = CE(z). \tag{9.104}$$

The complex constant C be written as

$$C = \exp(-j\beta L), \tag{9.105}$$

where β is the complex propagation constant of the periodic media. Thus, one can write

$$E(z) = E_\beta(z)\exp(-j\beta z). \tag{9.106}$$

From Equation 9.106,

$$E(z+L) = E_\beta(z+L)\exp(-j\beta(z+L)). \tag{9.107}$$

Also from Equations 9.104 and 9.106,

$$E(z+L) = E(z)\exp(-j\beta L) = E_\beta(z)\exp(-j\beta z)\exp(-j\beta L) = E_\beta(z)\exp(-j\beta(z+L)). \tag{9.108}$$

From Equations 9.107 and 9.108, it follows that $E_\beta(z)$ is periodic:

$$E_\beta(z+L) = E_\beta(z). \tag{9.109}$$

Equation 9.106, where $E_\beta(z)$ is periodic, is called the Bloch wave condition. Such a periodic function can be expanded in a Fourier series:

$$E_\beta(z) = \sum_{m=-\infty}^{\infty} A_m \exp(-j2m\pi z/L), \tag{9.110}$$

and $E(z)$ can be written as

$$E(z) = \sum_{m=-\infty}^{\infty} A_m \exp\left[-j(\beta + 2m\pi/L)z\right] = \sum_{m=-\infty}^{\infty} A_m \exp(-j\beta_m z), \tag{9.111}$$

where

$$\beta_m = \beta + \frac{2m\pi}{L}. \tag{9.112}$$

Taking into account both positive-going and negative-going waves, $E(z)$ can be expressed as [1]

$$E(z) = \sum_{m=-\infty}^{\infty} A_m \exp(-j\beta_m z) + \sum_{m=-\infty}^{\infty} B_m \exp(+j\beta_m z). \tag{9.113}$$

Let us next consider the wave propagation in a periodic layered media with plasma layers alternating with the free space. The geometry of the problem is shown in Figure 9.11. Adopting the notation of [12] a unit cell consists of free space from $-l < z < l$ and a plasma

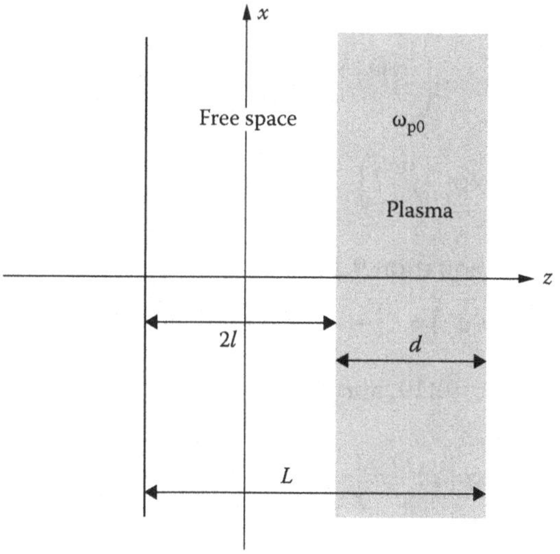

FIGURE 9.11
Unit cell of unbounded periodic media consisting of layer of plasma.

layer of plasma frequency ω_{p0} from $l < z < l + d$. The thickness of the unit cell is $L = 2l + d$. The electric field $E(z)$ in the two layers of the unit cell can be written as

$$E(z) = A\exp\left(-j\frac{\omega}{c}z\right) + B\exp\left(j\frac{\omega}{c}z\right), \quad -l \le z \le l, \tag{9.114}$$

$$E(z) = C\exp\left[-jn\frac{\omega}{c}(z-l)\right] + D\exp\left[jn\frac{\omega}{c}(z-l)\right], \quad l \le z \le l+d, \tag{9.115}$$

where n is the refractive index of the plasma medium:

$$n = \sqrt{\varepsilon_p} = \sqrt{1-\omega_{p0}^2/\omega^2}. \tag{9.116}$$

The continuity of tangential electric and magnetic fields at the boundary translates into the continuity of E and $\partial E/\partial z$ at the interfaces:

$$E(l^-) = E(l^+), \tag{9.117}$$

$$\frac{\partial E}{\partial z}(l^-) = \frac{\partial E}{\partial z}(l^+), \tag{9.118}$$

$$E(l+d^-) = E(l+d^+), \tag{9.119}$$

$$\frac{\partial E}{\partial z}(l+d^-) = \frac{\partial E}{\partial z}(l+d^+). \tag{9.120}$$

From Equations 9.117 and 9.118,

$$A\exp\left(-j\frac{\omega}{c}l\right)+B\exp\left(j\frac{\omega}{c}l\right)=C+D, \tag{9.121}$$

$$A\exp\left(-j\frac{\omega}{c}l\right)-B\exp\left(j\frac{\omega}{c}l\right)=n(C-D). \tag{9.122}$$

Since $l + d^+ = -l^+ + L$, from Equation 9.104,

$$E\left(l+d^+\right)=E\left(-l^+ +L\right)=\exp(-j\beta L)E\left(-l^+\right). \tag{9.123}$$

From Equations 9.114, 9.115, 9.119, and 9.123

$$C\exp\left(-jn\frac{\omega}{c}d\right)+D\exp\left(jn\frac{\omega}{c}d\right)=\exp(-j\beta L)\left[A\exp\left(-j\frac{\omega}{c}l\right)+B\exp\left(j\frac{\omega}{c}l\right)\right]. \tag{9.124}$$

From Equations 9.114, 9.115, 9.120, and 9.123

$$\begin{aligned}&-n\frac{\omega}{c}C\exp\left(-jn\frac{\omega}{c}d\right)+n\frac{\omega}{c}D\exp\left(jn\frac{\omega}{c}d\right)\\&=\exp(-j\beta L)\left[-j\frac{\omega}{c}A\exp\left(-\frac{j\omega l}{c}\right)+j\frac{\omega}{c}B\exp\left(j\frac{\omega}{c}l\right)\right].\end{aligned} \tag{9.125}$$

Equations 9.121, 9.122, 9.124, and 9.125 can be arranged as a matrix as

$$\begin{bmatrix} e^{-j\omega l/c} & e^{+j\omega l/c} & -1 & -1 \\ e^{-j\omega l/c} & -e^{j\omega l/c} & -n & n \\ e^{-j\omega l/c} & e^{+j\omega l/c} & -e^{j(\beta L-n\omega d/c)} & -e^{j(\beta L+n\omega d/c)} \\ e^{-j\omega l/c} & -e^{+j\omega l/c} & -ne^{j(\beta L-n\omega d/c)} & ne^{j(\beta L+n\omega d/c)} \end{bmatrix}\begin{bmatrix} A \\ B \\ C \\ D \end{bmatrix}=0. \tag{9.126}$$

A nonzero solution for the fields can be obtained by equating the determinant of the square matrix in Equation 9.126 to zero. This leads to the dispersion relation

$$\cos\beta L=\cos\frac{n\omega d}{c}\cos\frac{2\omega l}{c}-\frac{1}{2}\left[n+\frac{1}{n}\right]\sin\frac{n\omega d}{c}\sin\frac{2\omega l}{c}. \tag{9.127}$$

Section 7.4 gives an alternative way of obtaining the dispersion relation based on transmission line analogy and the ABCD parameters.

The above equation can be studied by using reference values β_r and $\omega_r = \beta_r c = 2\pi c/\lambda_r$. Thus, Equation 9.127, in the normalized form, can be written as

$$\cos\left(\frac{\beta}{\beta_r}\frac{2\pi L}{\lambda_r}\right)=\cos\left(\frac{n\omega}{\omega_r}\frac{2\pi d}{\lambda_r}\right)\cos\left(\frac{2\omega}{\omega_r}\frac{2\pi l}{\lambda_r}\right)-\frac{1}{2}\left(n+\frac{1}{n}\right)\sin\left(\frac{n\omega}{\omega_r}\frac{2\pi d}{\lambda_r}\right)\sin\left(\frac{2\omega}{\omega_r}\frac{2\pi l}{\lambda_r}\right). \tag{9.128}$$

Figure 9.12 shows the graph ω/ω_r versus β/β_r for the following values of the parameters: $L = 0.6\lambda_r$, $d = 0.2\lambda_r$, and $\omega_{p0} = 1.2\omega_r$. We note from Figure 9.12 that the wave is evanescent in the

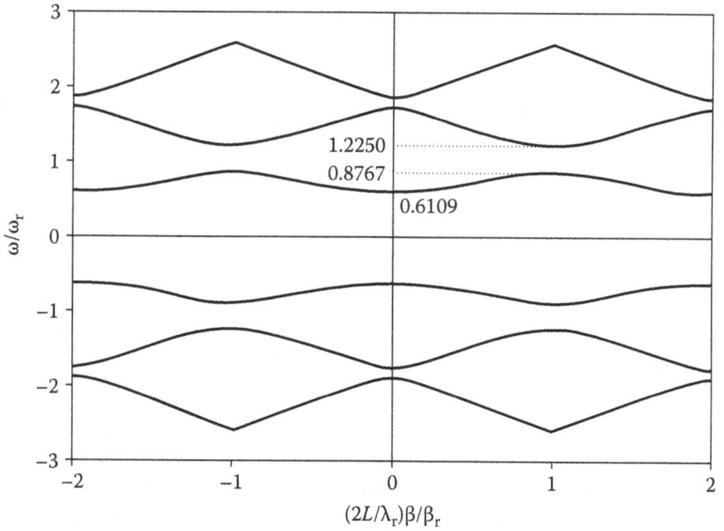

FIGURE 9.12
Dispersion relation of a periodic plasma medium for $d = 0.2\lambda_r$ and $L = 0.6\lambda_r$.

frequency band $0 < \omega/\omega_r < 0.611$. This stop band is due to the plasma medium in the layers. If the layers are dielectric, this stop band will not be present. We have again a stop band in the frequency domain $0.877 < \omega/\omega_r < 1.225$. This stop band is due to the periodicity of the medium. Periodic dielectric layers do exhibit such a forbidden band. This principle is used in optics to construct *Bragg reflectors* with extremely large reflectance [9]. Figure 9.13 shows the ω–β diagram for dielectric layers with the refractive index $n = 1.8$. The first stop band in this case is given by $0.537 < \omega/\omega_r < 0.779$.

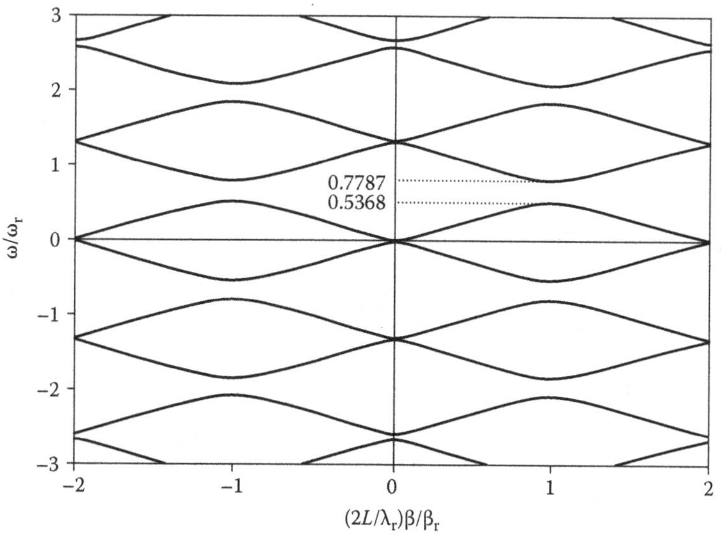

FIGURE 9.13
Dispersion relation of periodic dielectric layers for $d = 0.2\lambda_r$ and $L = 0.6\lambda_r$. The refractive index of the dielectric layer is 1.8.

9.8 Surface Waves

Let us backtrack a little bit and consider the oblique incidence of a p wave on a plasma half-space. In this case, only the outgoing wave will be excited in the plasma half-space and Equation 9.75 becomes

$$\begin{bmatrix} 1 & -1 \\ C\eta_{y1} & 1 \end{bmatrix} \begin{bmatrix} E_{x1}^{P} \\ E_x^{R} \end{bmatrix} = \begin{bmatrix} 1 \\ 1 \end{bmatrix} E_x^{I}. \tag{9.129}$$

Solution of Equation 9.129 gives the fields of the reflected wave and the wave in the plasma half-space in terms of the fields of the incident wave. If E_x^I is zero (no incident wave), we expect $E_{x1}^{P} = E_x^{R} = 0$. This is true in general with one exception. The exception occurs when the determinant of the square matrix on the LHS of Equation 9.129 is zero. In such a case, E_{x1}^{P} and E_x^{R} can be nonzero even if E_x^{I} is zero, indicating the possibility of the existence of fields in free space even if there is no incident field. For the exceptional solution, the reflection coefficient is infinity, that is, $E_x^{R} \neq 0$, but $E_x^{I} = 0$. The solution is an eigenvalue solution and gives the dispersion relation for surface plasmons. From Equations 9.73 and 9.129, the dispersion relation is obtained as

$$1 + C\eta_{y1} = 1 + C\frac{\varepsilon_p}{q_1} = 0. \tag{9.130}$$

Noting that

$$k_0 q_1 = \sqrt{k_0^2 \varepsilon_p - k_x^2} \tag{9.131}$$

and

$$k_0^2 C^2 = k_0^2 - k_x^2, \tag{9.132}$$

where

$$k_0 = \frac{\omega}{c}, \tag{9.133}$$

$$k_x = k_0 S, \tag{9.134}$$

the dispersion relation can be written as

$$k_x^2 = \frac{\omega^2}{c^2}\frac{\varepsilon_p}{1+\varepsilon_p}. \tag{9.135}$$

The exponential factors for $z < 0$ and $z > 0$ can be written as

$$\psi^{P1} = \exp\left[-jk_x x - jk_0 q_1 z\right], \quad z > 0, \tag{9.136}$$

$$\psi^{R} = \exp\left[-jk_x x + jk_0 C z\right], \quad z < 0. \tag{9.137}$$

It can be shown that k_x will be real and both

$$\alpha_1 = jk_0 q_1 \tag{9.138}$$

and

$$\alpha_2 = jk_0C = j\sqrt{k_0^2 - k_x^2} \tag{9.139}$$

will be real and positive if

$$\varepsilon_p < -1. \tag{9.140}$$

When Equation 9.140 is satisfied

$$\psi^{P1} = \exp(-\alpha_1 z)\exp(-jk_x x), \quad z > 0, \tag{9.141}$$

$$\psi^{R} = \exp(\alpha_2 z)\exp(-jk_x x), \quad z < 0, \tag{9.142}$$

where α_1 and α_2 are positive real quantities. Equations 9.141 and 9.142 show that the waves, while propagating along the surface, attenuate in the direction normal to the surface. For this reason, the wave is called a surface wave. In a plasma when $\omega < \omega_p/\sqrt{2}$, ε_p will be less than −1. Thus, an interface between free space and a plasma medium can support a surface wave.

Defining the refractive index of the surface mode as n, that is,

$$n = \frac{c}{\omega}k_x, \tag{9.143}$$

Equation 9.135 can be written as

$$n^2 = \frac{\Omega^2 - 1}{2\left(\Omega^2 - 1/2\right)}, \quad \Omega < \frac{1}{\sqrt{2}}. \tag{9.144}$$

For $0 < \omega < \omega_p/\sqrt{2}$, n is real and the surface wave propagates. In this interval of ω, the dielectric constant $\varepsilon_p < -1$. The surface wave mode is referred to, in the literature, as Fano mode [13] and also as a nonradiative surface plasmon. In the interval $1 < \Omega < \infty$, n^2 is positive and less than 0.5. However, the α-values obtained from Equations 9.138 and 9.139 are imaginary. The wave is not bound to the interface and the mode, called Brewster mode, in this case is radiative and referred to as a radiative surface plasmon. Figure 9.14 sketches $\varepsilon_p(\omega)$ and $n^2(\omega)$. Another way of presenting the information is through the Ω–K diagram, where Ω and K are normalized frequency and wave number, respectively, that is,

$$\Omega = \frac{\omega}{\omega_p}, \tag{9.145}$$

$$K = \frac{ck_x}{\omega_p} = n\Omega. \tag{9.146}$$

From Equations 9.143 and 9.144, we obtain the relation

$$K^2 = n^2\Omega^2 = \frac{\Omega^2\left(\Omega^2 - 1\right)}{2\Omega^2 - 1}. \tag{9.147}$$

Figure 9.15 shows the Ω–K diagram.

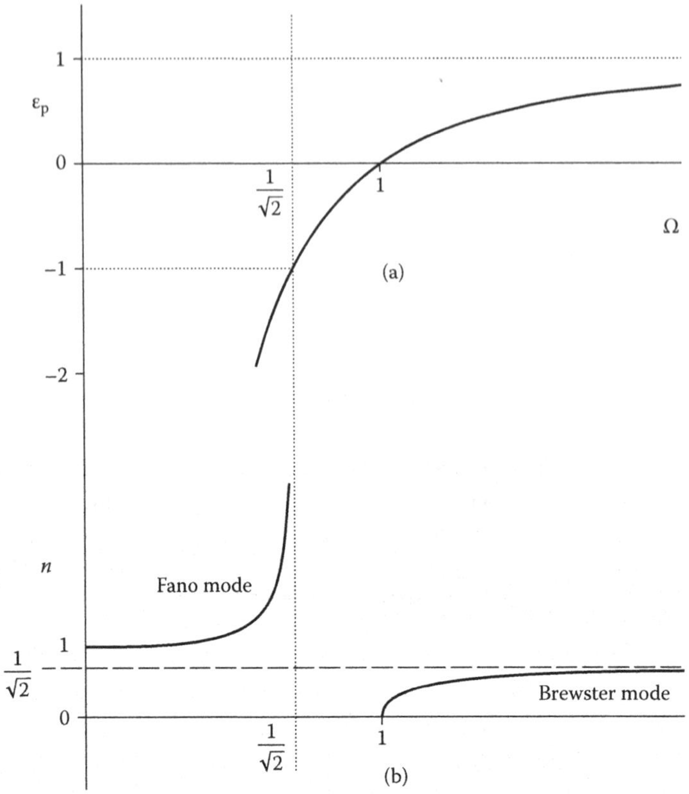

FIGURE 9.14
Refractive index for surface plasmon modes.

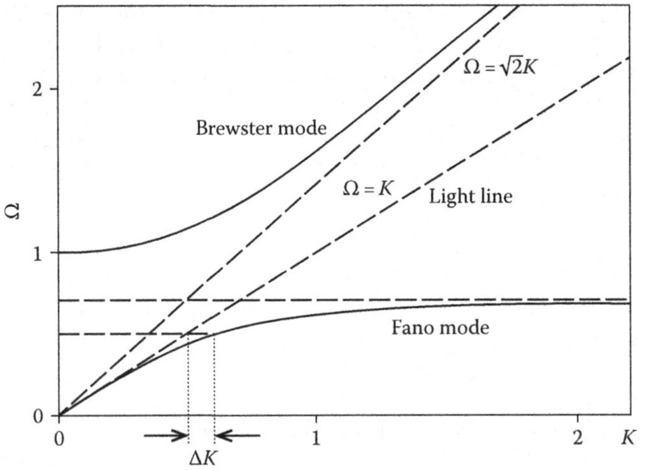

FIGURE 9.15
Ω–K diagram for surface plasmon modes.

The Fano mode is a surface wave called the surface plasmon and exists for $\omega < \omega_p/\sqrt{2}$. The surface mode does not have a low-frequency cutoff. It is guided by an open-boundary structure. It has a phase velocity less than that of light. The eigenvalue spectrum is continuous. This is in contrast to the eigenvalue spectrum of a conducting-boundary waveguide that has an infinite number of discrete modes of propagation at a given frequency.

The application of photons (electromagnetic light waves) to excite the surface plasmons meets with the difficulty that the dispersion relation of the Fano mode lies to the right of the light line $\Omega = K$ (see Figure 9.15). At a given photon energy $\hbar\omega$, the photon momentum $\hbar k_x$ has to be increased by $\hbar\Delta k_x = \hbar(\omega_p/c)\Delta K$ in order to transform the photons into surface plasmons. The two techniques commonly used to excite surface plasmons are (a) grating coupler (b) attenuated total reflection (ATR) method. A complete account of surface plasmons can be found in [13,14].

Surface waves can also exist at the interface of a dielectric with a plasma layer or a plasma cylinder. The theory of surface waves on a gas-discharge plasma is given in [15]. Kalluri [16] used this theory to explore the possibility of a backscatter from a plasma plume see Appendix 9A for details. Moissan et al. [17] achieved plasma generation through surface plasmons in a device called *Surfatron*. It is a highly efficient device for launching surface plasmons that produce plasma at a microwave frequency.

9.9 Transient Response of a Plasma Half-Space

Reflection of transient pulses from a sharply bounded plasma half-space can be obtained by finding the response of the plasma to a time-harmonic wave. For example, if $R(j\omega)$ is the reflection coefficient, then the response to an impulsive plane wave $e_r(t)$ can be obtained [18].

Let $e_0(t) = \delta(t)$ be the electric field of the impulsive plane wave at $z = 0$. Its Laplace transform is

$$E_0(s) = \mathcal{L}\left[e_0(t)\right] = \mathcal{L}\left[\delta(t)\right] = 1. \tag{9.148}$$

If the reflected field is designated at the point of incidence on the interface $e_r(t)$, then its Laplace transform $E_r(s)$ is given by

$$E_r(s) = R(s)E_0(s) = R(s), \tag{9.149}$$

where $R(s)$ is obtained by replacing $(j\omega)$ by s in $R(j\omega)$: the Laplace inverse of $R(s)$ gives $e_r(t)$:

$$e_r(t) = \mathcal{L}^{-1}R(s) = \frac{1}{2\pi j}\int_{\sigma_1 - j\infty}^{\sigma_1 + j\infty} R(s)e^{st}\,ds. \tag{9.150}$$

A direct analytical evaluation of Equation 9.150 is modified to suit the numerical evaluation [19]

$$e_r(t) = \frac{2}{\pi}\int_0^{\infty} \mathrm{Re}\left[\mathrm{R}(j\omega)\right]\cos\omega t\,d\omega, \tag{9.151}$$

where is assumed $e_r(t) = 0$ for $t < 0$ and Re stands for real.

9.9.1 Isotropic Plasma Half-Space s Wave

Consider oblique incidence of an *s* wave. It can be shown [18] that

$$R_s(j\omega) = \frac{j\omega C - j^2\omega^2 C^2 + \omega_p^2}{j\omega C + j^2\omega^2 C^2 + \omega_p^2}, \tag{9.152}$$

$$R_s(s) = \frac{\left[s - \left(s^2 + \omega_{p1}^2\right)^{1/2}\right]^2}{\omega_{p1}^2}, \tag{9.153a}$$

where

$$\omega_{p1} = \frac{\omega_p}{C}. \tag{9.153b}$$

$$C = \cos\theta, \tag{9.153c}$$

and θ is the angle of incidence. From a standard Laplace transform table [11],

$$\mathcal{L}^{-1}\left[\left(s^2 + b^2\right)^{1/2} - s\right]^{\nu} = \frac{\upsilon b^{\nu}}{t} J_{\nu}(bt), \quad \nu > 0. \tag{9.154}$$

Thus, one obtains the Laplace inverse of Equation 9.153, leading to

$$\omega_p^{-1} e_r(t) = -\frac{2}{\omega_p t} J_2\left(\frac{\omega_p t}{C}\right). \tag{9.155}$$

Figure 9.16 shows the impulse response for various angles of incidence. The effect of dispersion is clearly visible. There is an appreciable effect of the angle of incidence on the

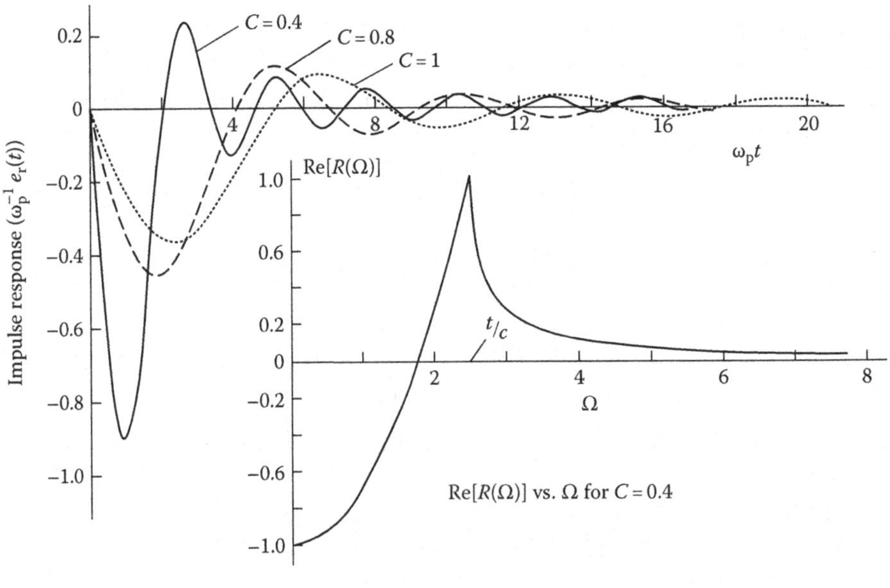

FIGURE 9.16
Impulse response of isotropic plasma half-space to an *s*-wave incidence.

amplitude and the period of oscillation of the response. It is seen that as the angle of incidence decreases (or C increases), the amplitude of the response (especially the early part) decreases, whereas the period of the oscillation increases. The variation of $\mathrm{Re}[R(\Omega)]$ with $\Omega = \omega/\omega_p$ is shown as an inset for $C = 0.4$.

9.9.2 Impulse Response of Several Other Cases Including Plasma Slab

References [18–26] and the references in these citations give access to the additional literature on this topic.

9.10 Solitons [27]

This topic deals with a desirable aspect of the EMW transformation by the balancing act of two complexities each of which by itself produces an undesirable effect. The balancing of the wave-breaking feature of a nonlinear medium complexity with the flattening feature of the dispersion leads to a soliton solution. Hirose and Longren [27] give a transmission line analogy to explain the balancing of the two dominant effects of the complexities of dispersion and nonlinearity. Ricketts and Ham [28] discuss the theory, design, and application of electrical solitons.

9.11 Perfect Dispersive Medium

As we have seen in Figure 9.16, a dispersive medium distorts a pulse by modulating the instantaneous frequency contents of the pulse. Gupta and Caloz [29] have explained the concept of a perfect dispersive medium with a perfectly flat transmission coefficient magnitude over the entire spectrum leading to an improved real-time signal processing technology. It is an engineered medium of "stacked loss-gain metasurfaces" [29].

Several such new ideas are being pursued under the broad banner of dispersion-engineered space-time modulated systems.

References

1. Ishimaru, A., *Electromagnetic Wave Propagation, Radiation, and Scattering*, Prentice Hall, Englewood Cliffs, NJ, 1991.
2. Heald, M. A. and Wharton, C. B., *Plasma Diagnostics with Microwaves*, Wiley, New York, 1965.
3. Forstmann, F. and Gerhardts, R. R., *Metal Optics Near the Plasma Frequency*, Springer, New York, 1986.
4. Boardman, A. D., *Electromagnetic Surface Modes*, Wiley, New York, 1982.
5. Kalluri, D. K. and Prasad, R. C., Thin film reflection properties of isotropic and uniaxial plasma slabs, *Appl. Sci. Res. (Netherlands)*, 27, 415, 1973.

6. Stratton, J. A., *Electromagnetic Theory*, McGraw-Hill, New York, 1941.
7. Kalluri, D. K. and Prasad, R. C., Transmission of power by evanescent waves through a plasma slab, *1980 IEEE International Conference on Plasma Science*, Madison, WI, p. 11, May 19–21, 1980.
8. Lekner, J., *Theory of Reflection*, Kluwer, Boston, MA, 1987.
9. Yeh, P., *Optical Waves in Layered Media*, Wiley, New York, 1988.
10. Budden, K. G., *Radio Waves in the Ionosphere*, Cambridge University Press, Cambridge, U.K., 1961.
11. Abramowitz, M. and Stegun, I. A., *Handbook of Mathematical Functions*, Dover Publications, New York, 1965.
12. Kuo, S. P. and Faith, J., Interaction of an electromagnetic wave with a rapidly created spatially periodic plasma, *Phys. Rev. E*, 56, 1, 1997.
13. Boardman, A. D., Hydrodynamic theory of plasmon–polaritons on plane surfaces, in: Boardman, A. D. (Ed.), *Electromagnetic Surface Modes*, Wiley, New York, 1982, Chap. 1, pp. 1–76.
14. Raether, H., *Surface Plasmons*, Springer, New York, 1988.
15. Shivarova, A. and Zhelyazkov, I., Surface waves in gas-discharge plasmas, in: Boardman, A. D. (Ed.), *Electromagnetic Surface Modes*, Wiley, New York, 1982, Chap. 12.
16. Kalluri, D. K., Backscattering from a plasma plume due to excitation of surface waves, Final Report Summer Faculty Research Program, Air Force Office of Scientific Research, Arlington Virginia, 1994.
17. Moissan, C., Beandry, C., and Leprince, P., *Phys. Lett.*, 50A, 125, 1974.
18. Prasad, R. C., Transient and frequency response of a bounded plasma, Doctoral thesis, Birla Institute of Technology, Ranchi, India, 1976.
19. Ley, B. J., *Computer Aided Analysis and Design for Electrical Engineering*, Holt, Rinehart and Winston, New York, 1970.
20. Kalluri, D. and Prasad, R. C., Transient response of a cold lossless plasma half-space in the presence of transverse static magnetic field, *J. Appl. Phys.*, 49(6), 3593–3594, 1978.
21. Kalluri, D. and Prasad, R. C., Reflection of an impulsive plane wave by a plasma half-space moving perpendicular to the plane of incidence, *J. Appl. Phys.*, 49(5), 2696–2699, 1978.
22. Prasad, R. C. and Kalluri, D., Reflection of an obliquely incident impulsive plane wave from isotropic and uniaxial cold lossy plasma half-spaces, *IEEE Trans. Plasma Sci.*, PS-6(4), 543–549, 1978.
23. Kalluri, D. and Prasad, R. C., On impulse response of plasma half-space moving normal to the interface-I: Formulation of the solution, *1979 IEEE International Conference on Plasma Science*, Montreal, Quebec, Canada, Conference Record-Abstracts, June 4–6, 1979, p. 155.
24. Kalluri, D. and Prasad, R. C., On impulse response of plasma half-space moving normal to the interface-II: Results, *1979 IEEE International Conference on Plasma Science*, Montreal, Quebec, Canada, Conference Record-Abstracts, June 4–6, 1979, pp. 154–155.
25. Kalluri, D. and Prasad, R. C., Transient response of a cold lossless plasma half-space in the presence of transverse static magnetic field, *J. Appl. Phys.*, 50(4), 2959–2961, 1979.
26. Prasad, R. C. and Kalluri, D., Impulse response of a lossy magnetoplasma half-space moving along the magnetic field, *1980 IEEE International Conference on Plasma Science*, Madison, WI, Conference Record-Abstracts, May 19–21, 1980, pp. 11–12.
27. Hirose, A. and Longren, K. E., *Introduction to Wave Phenomena*, Wiley, New York, 1985.
28. Ricketts, D. S. and Ham, D., *Electrical Solitons*, CRC Press, Boca Raton, FL, 2011.
29. Gupta, S., and Caloz, C., Perfect dispersive medium for real-time signal processing, *IEEE Trans. Antennas Propag.*, 64(12), 5299–5308, December 2016.

10

*Spatial Dispersion and Warm Plasma**

In Chapter 9, we discussed wave propagation in cold isotropic plasma. The effects of dispersion, to be more precise, temporal dispersion, are emphasized. In literature, cold plasma is sometimes referred to as "temperate plasma" [1,2]. The permittivity of warm plasma, in addition to being the function of the frequency ω, is also a function of the wave number k. For this reason, the warm plasma is said to be spatially as well as temporally dispersive [1–3]. Metal at optical frequencies can be modeled as warm plasma [4], the Fermi velocity playing the role of the thermal velocity of the warm plasma. While the model based on the cold plasma is said to represent local optics, the warm plasma model represents nonlocal optics. Section 10.7 gives rudimentary explanation of the technical terms used in describing the plasma state and the references give a deeper exposure.

10.1 Waves in a Compressible Gas

A compressible gas satisfies the equation of state

$$P = k_B N T, \tag{10.1}$$

where P is the pressure, N the number density, T the temperature, and k_B Boltzmann's constant (1.38×10^{-23} J/K).

Let

$$P = P_0 + p, \tag{10.2}$$

$$N = N_0 + n. \tag{10.3}$$

The values with subscript zero are the average values and the lower-case ones are a.c. values.

For an isothermal process (T constant), from Equations 10.1 through 10.3,

$$P_0 = k_B N_0 T,$$

$$p = k_B n T.$$

* For chapter appendices, see 10A through 10C in Appendices section.

For acoustic waves, the adiabatic process where no heat transfer is taking place is more appropriate; P and N satisfy the relation

$$\frac{P}{N^{\gamma}} = \frac{P_0}{N_0^{\gamma}}, \tag{10.4}$$

where γ is the ratio of specific heats.

The value of γ depends on the type of gas (monatomic, diatomic, or polyatomic). For a one-dimensional plasma problem, $\gamma = 3$ is appropriate.

From Equations 10.2 through 10.4, we have

$$p = \gamma k_B n T. \tag{10.5}$$

The hydrodynamic equation (Navier–Stokes equation) is

$$m\frac{dv}{dt} = -\frac{1}{N}\nabla p, \tag{10.6}$$

where m is the mass of the particle.

The continuity equation is

$$\nabla \cdot [Nv] + \frac{\partial N}{\partial t} = 0. \tag{10.7}$$

The RHS of Equation 10.6 is nonzero in a compressible gas.

For small-signal a.c. values, after linearizing, Equations 10.6 and 10.7 are approximated as

$$m\frac{dv}{dt} = -\frac{1}{N_0}\nabla p, \tag{10.8}$$

$$N_0 \nabla \cdot v = -\frac{\partial n}{\partial t}. \tag{10.9}$$

Taking the divergence of Equation 10.8, we obtain

$$\frac{\partial (\nabla \cdot v)}{\partial t} = -\frac{\nabla^2 p}{N_0 m}. \tag{10.10}$$

From Equations 10.9 and 10.5, we obtain

$$\nabla \cdot v = -\frac{1}{N_0 \gamma k_B T}\frac{\partial p}{\partial t}. \tag{10.11}$$

Substituting Equation 10.11 into Equation 10.10, we obtain the wave equation for the pressure:

$$\nabla^2 p - \frac{1}{a^2}\frac{\partial^2 p}{\partial t^2} = 0, \tag{10.12}$$

where the acoustic velocity a is given as

$$a = \left(\frac{\gamma k_B T}{m}\right)^{1/2}. \tag{10.13}$$

10.2 Waves in Warm Plasma

Let us consider the case where the gas has charged particles. In Chapter 9, we considered free electron gas, called cold plasma, and the force equation was based on the Lorentz force:

$$m\frac{\mathrm{d}v}{\mathrm{d}t} = -q\mathbf{E}. \tag{10.14}$$

In the cold plasma case, we ignored the term on the RHS of Equation 10.6 assuming that it is negligible compared to the Lorentz force. The assumption is valid as long as the temperature of the gas is low and the gas is considered as incompressible. In the warm plasma case, we consider the contributions to the force due to pressure gradients and we modify Equation 10.6 for the warm plasma case as

$$m\frac{\mathrm{d}v}{\mathrm{d}t} = -q\mathbf{E} - \frac{1}{N}\nabla p. \tag{10.15}$$

Taking the divergence of Equation 10.15 (and approximating $\mathrm{d}/\mathrm{d}t \to \partial/\partial t$),

$$\frac{\partial(\nabla\cdot v)}{\partial t} = -\frac{q}{m}(\nabla\cdot\mathbf{E}) - \frac{1}{Nm}\nabla^2 p. \tag{10.16}$$

Using Equation 10.11 in Equation 10.16, we obtain the wave equation for the pressure:

$$\nabla^2 p - \frac{1}{a^2}\frac{\partial^2 p}{\partial t^2} + N_0 q\nabla\cdot\mathbf{E} = 0. \tag{10.17}$$

The variable $\mathbf{E}$ in Equation 10.17 can be eliminated by using the Maxwell equation

$$\nabla\times\mathbf{H} = \varepsilon_0\frac{\partial\mathbf{E}}{\partial t} - N_0 q v, \tag{10.18}$$

$$\begin{aligned}
0 = \nabla\cdot(\nabla\times\mathbf{H}) &= \varepsilon_0\frac{\partial(\nabla\cdot\mathbf{E})}{\partial t} - N_0 q(\nabla\cdot v),\\
0 &= \varepsilon_0\frac{\partial(\nabla\cdot\mathbf{E})}{\partial t} + \frac{q}{\gamma k_\mathrm{B}T}\frac{\partial p}{\partial t},\\
\nabla\cdot\mathbf{E} &= -\frac{q}{\gamma k_\mathrm{B}T\varepsilon_0}p,\\
N_0 q\nabla\cdot\mathbf{E} &= -\frac{N_0 q^2}{\gamma k_\mathrm{B}T\varepsilon_0}p = -\frac{N_0 q^2}{m\varepsilon_0}\frac{m}{\gamma k_\mathrm{B}T}p,\\
N_0 q\nabla\cdot\mathbf{E} &= -\frac{\omega_\mathrm{p}^2}{a^2}p.
\end{aligned} \tag{10.19}$$

Thus, the wave equation for the pressure variable p for the warm plasma case is given by

$$\nabla^2 p - \frac{1}{a^2}\frac{\partial^2 p}{\partial t^2} - \frac{\omega_p^2}{a^2} p = 0. \tag{10.20}$$

To obtain the wave equation for the electric field, we start with the Maxwell equation

$$\bar{\nabla} \times \mathbf{E} = -\mu_0 \frac{\partial \mathbf{H}}{\partial t} \tag{10.21}$$

and take the curl on both sides

$$\nabla \times \nabla \times \mathbf{E} = -\mu_0 \frac{\partial(\nabla \times \mathbf{H})}{\partial t}$$

and from Equation 10.18, we obtain

$$\nabla \times \nabla \times \mathbf{E} = -\mu_0 \frac{\partial}{\partial t}\left(\varepsilon_0 \frac{\partial \mathbf{E}}{\partial t} - N_0 q v\right),$$

and from Equation 10.15

$$-\nabla \times \nabla \times \mathbf{E} - \mu_0 \varepsilon_0 \frac{\partial^2 \mathbf{E}}{\partial t^2} + N_0 \mu_0 q\left(-\frac{q}{m}\mathbf{E} - \frac{1}{N_0 m}\nabla p\right) = 0,$$

$$-\nabla \times \nabla \times \mathbf{E} - \frac{1}{c^2}\frac{\partial^2 \mathbf{E}}{\partial t^2} - \mu_0 \varepsilon_0 \omega_p^2 \mathbf{E} - \frac{\mu_0 q}{m}\nabla p = 0.$$

Using Equation 10.19 to eliminate p from the above,

$$-\nabla \times \nabla \times \mathbf{E} - \frac{1}{c^2}\frac{\partial^2 \mathbf{E}}{\partial t^2} - \frac{\omega_p^2}{c^2}\mathbf{E} + \frac{\mu_0 q}{m}\bar{\nabla}\left(\frac{N_0 q a^2}{\omega_p^2}\nabla \cdot \mathbf{E}\right) = 0.$$

By simplifying, we obtain the wave equation for **E** as

$$-\nabla \times \nabla \times \mathbf{E} - \frac{1}{c^2}\frac{\partial^2 \mathbf{E}}{\partial t^2} - \frac{\omega_p^2}{c^2}\mathbf{E} + \mu_0 \varepsilon_0 a^2 \nabla(\nabla \cdot \mathbf{E}) = 0. \tag{10.22}$$

We should be able to recover the wave equation for **E** for the cold plasma case by substituting $a = 0$ in Equation 10.22:

$$-\nabla \times \nabla \times \mathbf{E} - \frac{1}{c^2}\frac{\partial^2 \mathbf{E}}{\partial t^2} - \frac{\omega_p^2}{c^2}\mathbf{E} = 0,$$

$$-\nabla(\nabla \cdot \mathbf{E}) + \nabla^2 \mathbf{E} - \frac{1}{c^2}\frac{\partial^2 \mathbf{E}}{\partial t^2} - \frac{\omega_p^2}{c^2}\mathbf{E} = 0.$$

For the cold homogeneous plasma case (incompressible gas), $p = 0$ and $\nabla \cdot \mathbf{E} = 0$, and hence we obtain

$$\nabla^2 \bar{E} - \frac{1}{c^2}\frac{\partial^2 \bar{E}}{\partial t^2} - \frac{\omega_p^2}{c^2}\bar{E} = 0. \tag{10.23}$$

Let us solve Equation 10.22 for a harmonic wave with a phase factor $e^{j\omega t}\,e^{-j\mathbf{k}\cdot\mathbf{r}}$. Noting that ∇ can be replaced by $(-j\mathbf{k})$, for such a phase factor, Equation 10.22 becomes

$$\begin{aligned} -(-j\mathbf{k})\times(-j\mathbf{k})\times\mathbf{E}+\frac{\omega^2}{c^2}\mathbf{E}-\frac{\omega_p^2}{c^2}\mathbf{E}+\frac{a^2}{c^2}(-j\mathbf{k})(-j\mathbf{k}\cdot\mathbf{E})&=0,\\ \mathbf{k}\times\mathbf{k}\times\mathbf{E}+\frac{\omega^2}{c^2}\mathbf{E}-\frac{\omega_p^2}{c^2}\mathbf{E}-\frac{a^2}{c^2}\mathbf{k}(\mathbf{k}\cdot\mathbf{E})&=0. \end{aligned} \tag{10.24}$$

Since $\mathbf{k}\times\mathbf{k}\times\mathbf{E}=\mathbf{k}(\mathbf{k}\cdot\mathbf{E})-k^2\mathbf{E}$, we can write Equation 10.24 as

$$\mathbf{k}(\mathbf{k}\cdot\mathbf{E})\left(1-\frac{a^2}{c^2}\right)-\left[k^2-k_0^2\left(1-\frac{\omega_p^2}{\omega^2}\right)\right]\mathbf{E}=0. \tag{10.25}$$

Equation 10.24 can be written separately for $E_\parallel$ and $E_\perp$, where $E_\parallel$ is a component parallel to $\mathbf{k}$ and $E_\perp$ is normal to $\mathbf{k}$.

For the parallel component $\mathbf{k}\cdot\mathbf{E}=kE_\parallel$, we have

$$\begin{aligned} k^2E_\parallel\left(1-\frac{a^2}{c^2}\right)-\left[k^2-k_0^2\left(1-\frac{\omega_p^2}{\omega^2}\right)\right]E_\parallel&=0,\\ k^2\frac{a^2}{c^2}+k_0^2\left(1-\frac{\omega_p^2}{\omega^2}\right)&=0,\\ k^2=k_0^2\frac{\left(1-\left(\omega_p^2/\omega^2\right)\right)}{a^2/c^2}&=\frac{\omega^2-\omega_p^2}{a^2}. \end{aligned} \tag{10.26}$$

Let

$$\frac{a}{c}=\delta, \tag{10.27}$$

$$\frac{\omega_p^2}{\omega^2}=X, \tag{10.28}$$

Then

$$k^2=k_0^2\frac{1-X}{\delta^2}. \tag{10.29}$$

From Equation 10.21, we obtain

$$\mathbf{H}=-\frac{1}{\mu_0 j\omega}\mathbf{k}\times\mathbf{E}, \tag{10.30}$$

and since $\mathbf{k}$ and $\mathbf{E}$ are parallel for this case

$$\mathbf{H}=0. \tag{10.31}$$

The parallel mode under discussion has an electric field component in the direction of propagation and the zero magnetic field. It is also called an electron plasma wave. It is an

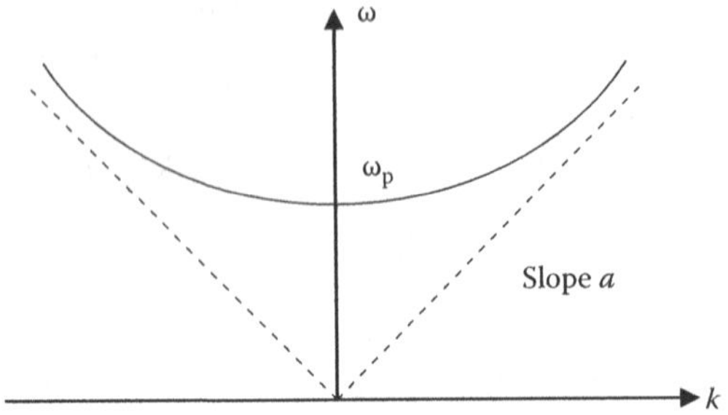

FIGURE 10.1
ω–*k* diagram for the electron plasma wave.

acoustic wave, like a sound wave, modified by the pressure of a charged compressible fluid. Its phase velocity is given by

$$V_{\mathrm{p}} = \frac{\omega}{k} = \frac{a}{\sqrt{1-X}} = c\frac{\delta}{\sqrt{1-X}}. \tag{10.32}$$

The ω–*k* diagram for the electron plasma wave is shown in Figure 10.1.

Note that the phase velocity a becomes negligible for the cold plasma approximation and the electron plasma wave becomes a plasma oscillation at the plasma frequency ω_{p}. A simple way of interpreting the electron plasma wave is to say that the plasma oscillation becomes a longitudinal electron plasma wave when the plasma is considered as warm.

The electron plasma wave propagates for only $\omega > \omega_{\mathrm{p}}$. For $\omega < \omega_{\mathrm{p}}$, this wave is evanescent. Now let us examine the wave propagation when **E** is perpendicular to **k**.

From Equation 10.24,

$$\left[k^2 - k_0^2\left(1 - \frac{\omega_{\mathrm{p}}^2}{\omega^2}\right)\right]E_{\perp} = 0, \tag{10.33}$$

$$k^2 = k_0^2\left(1 - \frac{\omega_{\mathrm{p}}^2}{\omega^2}\right). \tag{10.34}$$

Equation 10.34 is the same as the dispersion relation for the cold isotropic plasma, discussed in Chapter 9. One can obtain $H_{\perp}$ from Equation 10.30, which shows that

$$E_{\perp} = \eta_{\mathrm{p}} H_{\perp}, \tag{10.35}$$

where

$$\eta_{\mathrm{p}} = \frac{\eta_0}{\sqrt{1-X}}. \tag{10.36}$$

While the electron plasma wave and TEM wave can exist independently in an unbounded warm plasma, the amplitudes of the two will be determined in a bounded warm plasma by the boundary conditions. Appendix 10A discusses such a problem. Appendices 10B and 10C deal with such problems for the cases of warm magnetoplasma media.

10.3 Constitutive Relation for a Lossy Warm Plasma

For a lossy warm plasma, Equation 10.15 will be modified as

$$m\frac{dv}{dt}+m\nu v=-q\mathbf{E}-\frac{1}{N}\nabla p. \tag{10.37}$$

Since

$$\mathbf{J}=-N_0 q v, \tag{10.38}$$

Equation 10.37 can be written as (multiply Equation 10.37 by $-N_0q/m$)

$$\frac{d\mathbf{J}}{dt}+\nu\mathbf{J}=\varepsilon_0\omega_p^2\mathbf{E}+\frac{q}{m}\nabla p,$$

$$\nabla p=\gamma k_B T\nabla n.$$

Since

$$\varepsilon_0\nabla\cdot\mathbf{E}=\rho_v=-qn,$$

$$\nabla n=-\frac{\varepsilon_0}{q}\nabla(\nabla\cdot\mathbf{E}),$$

$$\nabla p=-\gamma k_B T\frac{\varepsilon_0}{q}\nabla(\nabla\cdot\mathbf{E}),$$

$$\frac{q}{m}\nabla p=-\frac{\gamma k_B T\varepsilon_0}{m}\nabla(\nabla\cdot\mathbf{E})=-\varepsilon_0 a^2\nabla(\nabla\cdot\mathbf{E}).$$

Thus, the constitutive relation is given by

$$\frac{d\mathbf{J}}{dt}+\nu\mathbf{J}=\varepsilon_0\omega_p^2\mathbf{E}-\varepsilon_0 a^2\nabla(\nabla\cdot\mathbf{E}). \tag{10.39}$$

One can model the warm plasma as a dielectric by using Equation 10.39 and the Maxwell equation

$$\nabla\times\mathbf{H}=\mathbf{J}+\varepsilon_0\frac{\partial\mathbf{E}}{\partial t}. \tag{10.40}$$

For harmonic variations in space and time $e^{j(\omega t-\mathbf{k}\cdot\mathbf{r})}$, Equation 10.39 becomes

$$j(\omega+v)\mathbf{J}=\varepsilon_0\omega_p^2\mathbf{E}-\varepsilon_0 a^2(-j\mathbf{k})(-j\mathbf{k}\cdot\mathbf{E}).$$

Since

$$\mathbf{J}=J_{\parallel}+J_{\perp}\quad(\text{with reference to }\mathbf{k}),$$

$$\mathbf{E} = E_{\parallel} + E_{\perp},$$
$$(j\omega + \nu) J_{\parallel} = \varepsilon_0 \omega_p^2 E_{\parallel} + \varepsilon_0 a^2 k_{\parallel}^2 E_{\parallel}, \tag{10.41}$$

$$(j\omega + \nu) J_{\perp} = \varepsilon_0 \omega_p^2 E_{\perp} + 0. \tag{10.42}$$

$$(\bar{\nabla} \times \bar{H})_{\perp} = J_{\perp} + j\omega\varepsilon_0 E_{\perp} = \left(\frac{\varepsilon_0 \omega_p^2}{j\omega + \nu} + j\omega\varepsilon_0 \right) E_{\perp},$$
$$(\bar{\nabla} \times \bar{H})_{\perp} = j\omega\varepsilon_{p\perp} E_{\perp}, \tag{10.43}$$

where

$$\varepsilon_{p\perp} = 1 - \frac{\omega_p^2}{\omega(\omega - j\nu)}. \tag{10.44}$$

Let us derive next an expression for $\varepsilon_{p\parallel}$.
From Equation 10.41,

$$J_{\parallel} = \frac{\varepsilon_0 \omega_p^2}{j\omega + \nu} E_{\parallel} + \frac{\varepsilon_0 a^2 k_{\parallel}^2}{j\omega + \nu} E_{\parallel},$$
$$(\bar{\nabla} \times \bar{H})_{\parallel} = J_{\parallel} + j\omega\varepsilon_0 E_{\parallel}$$
$$= \frac{\varepsilon_0 \omega_p^2}{j\omega + \nu} E_{\parallel} + \frac{\varepsilon_0 a^2 k_{\parallel}^2}{j\omega + \nu} E_{\parallel} + j\omega\varepsilon_0 E_{\parallel}, \tag{10.45}$$
$$(\bar{\nabla} \times \bar{H})_{\parallel} = j\omega\varepsilon_0 \varepsilon_{p\parallel} E_{\parallel},$$

where

$$\varepsilon_{p\parallel} = 1 - \frac{\omega_p^2}{\omega(\omega - j\nu)} - \frac{a^2 k_{\parallel}^2}{\omega(\omega - j\nu)}. \tag{10.46}$$

If we neglect collisions, then

$$\varepsilon_{p\parallel} = 1 - \frac{\omega_p^2 + a^2 k_{\parallel}^2}{\omega^2}. \tag{10.47}$$

Substituting Equation 10.26 for $k_{\parallel}$, we obtain

$$\varepsilon_{p\parallel} = 0. \tag{10.48}$$

This will be true even if collisions are taken into account.
Equation 10.47 implies that $E_{\parallel}$ can be nonzero even if $D_{\parallel} = 0$.
The dispersion relation Equation 10.26 and the ω k diagram in Figure 10.1 can be obtained by imposing Equation 10.47.

10.4 Dielectric Model of Warm Loss-Free Plasma

Maxwell's equations and the constitutive relations for warm plasma, viewed as a dielectric medium, are given below:

$$\nabla \times \mathbf{E} = -\frac{\partial \mathbf{B}}{\partial t}, \tag{10.49}$$

$$\nabla \times \mathbf{H} = -\frac{\partial \mathbf{D}}{\partial t}, \tag{10.50}$$

$$\nabla \cdot \mathbf{D} = 0, \tag{10.51}$$

$$\nabla \cdot \mathbf{B} = 0, \tag{10.52}$$

$$\mathbf{B} = \mu_0 \mathbf{H}, \tag{10.53}$$

$$\mathbf{D} = \eta_0 \bar{\varepsilon}_{\mathbf{p}} \cdot \mathbf{E}, \tag{10.54}$$

$$\bar{\varepsilon}_{\mathbf{p}} = \begin{bmatrix} \varepsilon_{p\perp} & 0 \\ 0 & \varepsilon_{p\parallel} \end{bmatrix}, \tag{10.55}$$

where

$$\varepsilon_{p\perp} = 1 - \frac{\omega_p^2}{\omega^2}, \tag{10.56}$$

$$\varepsilon_{p\parallel} = 1 - \frac{\omega_p^2}{\omega^2 - a^2 k_\parallel^2}. \tag{10.57}$$

Note that Equations 10.47 and 10.57 are the same when $\varepsilon_{p\parallel} = 0$.

10.5 Conductor Model of Warm Lossy Plasma

Maxwell's equations in the time domain and the constitutive relations for a warm lossy plasma, viewed as a conducting medium, are given below:

$$\nabla \times \mathbf{E} = -\mu_0 \frac{\partial \mathbf{H}}{\partial t}, \tag{10.58}$$

$$\nabla \times \mathbf{H} = \mathbf{J} + \varepsilon_0 \frac{\partial \mathbf{E}}{\partial t}, \tag{10.59}$$

$$\nabla \cdot \mathbf{H} = 0, \tag{10.60}$$

$$\frac{d\mathbf{J}}{dt} + \nu \mathbf{J} = \varepsilon_0 \omega_p^2 \mathbf{E} - \varepsilon_0 a^2 \nabla (\nabla \cdot \mathbf{E}). \tag{10.61}$$

Note that

$$\varepsilon_0 \nabla \cdot \mathbf{E} = \rho, \tag{10.62}$$

$$\rho = -qn, \tag{10.63}$$

$$\nabla p = \gamma k T \nabla n. \tag{10.64}$$

In Equations 10.62 and 10.63, ρ is the volume charge density.
For time-harmonic fields, Equation 10.61 can be written as

$$\mathbf{J} = \bar{\sigma} \cdot \mathbf{E}, \tag{10.65}$$

$$\bar{\sigma} = \begin{bmatrix} \sigma_\perp & 0 \\ 0 & \sigma_\parallel \end{bmatrix}, \tag{10.66}$$

$$\sigma_\perp = -j\varepsilon_0 \frac{\omega_p^2}{\omega - j\nu}, \tag{10.67}$$

$$\sigma_\parallel = -j\omega\varepsilon_0 \frac{\omega_p^2}{\omega(\omega - j\nu) - ak_\parallel^2}. \tag{10.68}$$

By noting

$$\bar{\varepsilon}_p = \bar{I} + \frac{\bar{\sigma}}{j\omega\varepsilon_0}, \tag{10.69}$$

we can obtain the conductivity tensor $\bar{\sigma}$ from $\bar{\varepsilon}_p$ and vice versa.

10.6 Spatial Dispersion and Nonlocal Metal Optics

The longitudinal wave in warm plasma has the dispersion relation

$$\varepsilon_{p\parallel} = 0 = 1 - \frac{\omega_p^2}{\omega(\omega - j\nu) - a^2k^2} \tag{10.70}$$

and the ω–k diagram is sketched in Figure 10.1. In the sense that

$$\varepsilon_{p\parallel} = \varepsilon_{p\parallel}(\omega, k), \tag{10.71}$$

the warm plasma is spatially as well as temporally dispersive.

A simple metal, such as sodium, is modeled as a lossy plasma in Section 8.3. Typical values for sodium are $\omega_p = 8.2 \times 10^{15}$ rad/s and $\nu/\omega_p = 3 \times 10^{-3}$. The plasma frequency is in the optical range. Due to the high electron density in metals, it can be shown that a parameter called Fermi velocity v_f is also important in modeling metal as plasma [5]. A typical value of v_f for sodium is $v_f = 9.8 \times 10^5$ m/s. The Fermi velocity v_f plays the role of the acoustic velocity in warm plasma. Thus, the modeling of metal for studying the interaction of thin metal films with optical waves with frequencies near the plasma frequency require a warm plasma model, which takes into account the spatial dispersion due to the excitation of the electron plasma wave in the metal, when the source wave is a p wave. Spatial dispersion will have important consequences and such a study is labeled as the study of nonlocal optics in contrast to the local optics (Fresnel Optics). A comprehensive account of the nonlocal optical studies using the warm plasma model is given by Forstmann and Gerhardts [4].

10.7 Technical Definition of Plasma State

We have earlier given a vague definition of plasma medium as an ionized medium with positive ions and electrons with an "overall" charge neutrality. We also used the terms cold plasma and warm plasma. While reading books on plasma science, we also come across terms such as collective behavior, temperate plasma, hot plasma, nonneutral plasma, Debye length and plasma sheath, and so on. This section is used to explain the various terms quantitatively in the context of the theory discussed in Sections 8.3 and 8.4, Chapter 9, and this chapter. The one-dimensional electron–proton plasma with one degree of freedom is assumed.

10.7.1 Temperate Plasma

In cold plasma, we neglect the effect of the temperature. The thermal velocity V_{th} is related to temperature T by

$$\frac{1}{2} m V_{th}^2 = \frac{1}{2} k_B T. \tag{10.72}$$

The RHS of Equation 10.72 is one-half of an electron volt (eV) and substituting for k_B, the Boltzmann constant, we obtain the equivalence

$$1\,\text{eV} = 11000^\circ\, K. \tag{10.73}$$

In linearizing the Maxwell equations and the force equation, we assume that the electron velocity v is small compared to V_{th}. The cold plasma approximation neglects the spatial dispersion, which will be a valid approximation if V_{th} is less than the phase velocity $V_{ph} = \omega/k$ of the wave. So the cold plasma approximation is valid, if [1]

$$v \ll V_{th} \ll V_{ph}. \tag{10.74}$$

Thus, $v \ll V_{th}$ is implied by calling cold plasma as temperate plasma.

A plasma is called a warm plasma if the temperature is such that V_{th} is approximately equals V_{ph} of the electron plasma wave.

10.7.2 Debye Length, Collective Behavior, and Overall Charge Neutrality

A plasma medium shields out any excessive fields in the medium. A characteristic distance, called "Debye length" λ_D, is the length-scale parameter, which ensures the shielding of the excessive electric fields in plasma [6]. The Debye length depends on the temperature and the electron density and is given by

$$\lambda_D = \left(\frac{\varepsilon_0 k_B T}{Nq^2}\right)^{1/2} = \frac{a}{\sqrt{\gamma}\omega_p}, \tag{10.75}$$

$$\lambda_D = 69\left(\frac{T}{N}\right)^{1/2}. \tag{10.76}$$

It can be shown [6] that any excess potential Φ in the plasma will decay as

$$\Phi = \Phi_0 e^{-(|z|/\lambda_D)}. \tag{10.77}$$

For $|z| > 5\lambda_D$, $\Phi = \Phi_0 e^{-5} \approx 0$, and hence the excess potential is screened out in about five times λ_D.

If the plasma system dimension L is much larger than λ_D, the bulk of the plasma has a charge neutrality, that is, the net charge in a macroscopic volume is zero:

$$N_e \approx N_i \approx N. \tag{10.78}$$

In Equation 10.78, N_i is the ion density and N is often referred to as the plasma density. The qualifying word "overall" before the word neutrality is used to indicate that the plasma is not so neutral to blot out all interesting electromagnetic forces. This shielding becomes possible, because the electric Coulomb force is inversely proportional to the square of the distance. This force is considered long range compared to the other short-range forces. If we assume that the electron is a point, then the charges in a volume $(4\pi/3)\lambda_D^3$ collectively shield the effect of this electron. It is this collective behavior, in contrast to the collision between single particles, that control the motion of the charged particles. The collective behavior can occur only when the particle density is such that there are enough particles in a Debye volume.

10.7.3 Unneutralized Plasma

The phenomenon of Debye shielding in a modified form can occur in an electron stream in klystron, and a proton beam in a cyclotron. These are not strictly plasmas but because they use the concepts of collective behavior, and shielding and use the mathematical tools of plasma physics, they are referred to as unneutralized plasmas [7].

References

1. Allis, W. P., Buchsbaum, S. J., and Bens, A., *Waves in Anisotropic Plasmas*, MIT Press, Cambridge, MA, 1963.
2. Heald, M. A. and Wharton, C. B., *Plasma Diagnostics with Microwaves*, Wiley, New York, 1965.
3. Akira, I., *Electromagnetic Wave Propagation, Radiation, and Scattering*, Prentice Hall, Englewood Cliffs, NJ, 1991.

4. Forstmann, F. and Gerhardts, R. R., *Metal Optics Near the Plasma Frequency*, Springer, New York, 1986.
5. Kittel, C., *Introduction to Solid State Physics*, Wiley, New York, 2005.
6. Chen, F. F., *Introduction to Plasma Physics and Controlled Fusion*, Vol. 1: Plasma Physics, Springer, New York, 1983.
7. Davidson, R. C., *Physics of Nonneutral Plasmas*, Addison-Wesley, New York, 1990.

11

*Wave in Anisotropic Media and Magnetoplasma**

11.1 Introduction

Plasma medium in the presence of a static magnetic field behaves like an anisotropic dielectric. Therefore, the rules of electromagnetic wave propagation are the same as those of the light waves in crystals. Of course, in addition, we have to take into account the dispersive nature of the plasma medium.

The well-established *Magnetoionic Theory* [1–3] is concerned with the study of wave propagation of an arbitrarily polarized plane wave in cold anisotropic plasma, where the direction of phase propagation of the wave is at an arbitrary angle to the direction of the static magnetic field. As the wave travels in such a medium, the polarization state continuously changes. However, there are specific normal modes of propagation in which the state of polarization is unaltered. Waves with the left (L wave) or the right (R wave) circular polarization are the normal modes in the case of phase propagation along the static magnetic field. Such propagation is labeled as *longitudinal propagation.* Ordinary wave (O wave) and the extraordinary wave (X wave) are the normal modes for *transverse propagation*. The properties of these waves are explored in Sections 11.3 through 11.5.

11.2 Basic Field Equations for a Cold Anisotropic Plasma Medium

In the presence of a static magnetic field $\mathbf{B}_0$, the force equation needs modification due to the additional magnetic force term:

$$m\frac{\mathrm{d}\mathbf{v}}{\mathrm{d}t} = -q\left[\mathbf{E} + \mathbf{v}\times\mathbf{B}_0\right]. \tag{11.1}$$

The corresponding modification of the constitutive relation in terms of **J** is given by

$$\frac{\mathrm{d}\mathbf{J}}{\mathrm{d}t} = \varepsilon_0\omega_p^2(\mathbf{r},t)\mathbf{E} - \mathbf{J}\times\omega_b, \tag{11.2}$$

* For chapter appendices, see 11A and 11B in Appendices section.

where

$$\omega_b = \frac{q\mathbf{B}_0}{m} = \omega_b \hat{\mathbf{B}}_0. \tag{11.3}$$

In the above, $\hat{\mathbf{B}}_0$ is a unit vector in the direction of $\mathbf{B}_0$ and ω_b is the absolute value of the electron gyrofrequency.

Equations 1.1 and 1.2 are repeated here for convenience:

$$\nabla \times \mathbf{E} = -\mu_0 \frac{\partial \mathbf{H}}{\partial t}, \tag{11.4}$$

$$\nabla \times \mathbf{H} = \varepsilon_0 \frac{\partial \mathbf{E}}{\partial t} + \mathbf{J}, \tag{11.5}$$

and Equation 11.2 is the basic equation that will be used in discussing the electromagnetic wave transformation by magnetized cold plasma.

Taking the curl of Equation 11.4 and eliminating **H**, the wave equation for **E** can be derived:

$$\nabla^2 \mathbf{E} - \nabla(\nabla \cdot \mathbf{E}) - \frac{1}{c^2}\frac{\partial^2 \mathbf{E}}{\partial t^2} - \frac{1}{c^2}\omega_p^2(\mathbf{r},t)\mathbf{E} + \mu_0 \mathbf{J} \times \omega_b = 0. \tag{11.6}$$

Similar efforts will lead to a wave equation for the magnetic field:

$$\nabla^2 \dot{\mathbf{H}} - \frac{1}{c^2}\frac{\partial^2 \dot{\mathbf{H}}}{\partial t^2} - \frac{1}{c^2}\omega_p^2(\mathbf{r},t)\dot{\mathbf{H}} + \varepsilon_0 \nabla \omega_p^2(\mathbf{r},t) \times \mathbf{E} + \nabla \times (\mathbf{J} \times \mathbf{E}) = 0, \tag{11.7}$$

where

$$\dot{\mathbf{H}} = \frac{\partial \mathbf{H}}{\partial t}. \tag{11.8}$$

If ω_p^2 and ω_b vary only with t, then Equation 11.7 becomes

$$\nabla^2 \dot{\mathbf{H}} - \frac{1}{c^2}\frac{\partial^2 \dot{\mathbf{H}}}{\partial t^2} - \frac{1}{c^2}\omega_p^2(t)\dot{\mathbf{H}} + \varepsilon_0 \omega_b(t)(\nabla \cdot \dot{\mathbf{E}}) = 0. \tag{11.9}$$

Equations 11.6 and 11.7 involve more than one field variable. It is possible to convert them into higher-order equations in one variable. In any case, it is difficult to obtain meaningful analytical solutions to these higher-order vector partial differential equations. The equations in this section are useful in developing numerical methods.

We will consider next, as in the isotropic case, simple solutions to particular cases where we highlight one parameter or one aspect at a time. These solutions will serve as building blocks for the more involved problems.

11.3 One-Dimensional Equations: Longitudinal Propagation and L and R Waves

Let us consider the particular case where (1) the variables are functions of one spatial coordinate only, say the z-coordinate, (2) the electric field is circularly polarized, (3) the static magnetic field is z-directed, and (4) the variables are denoted by

$$\mathbf{E} = (\hat{x} \mp \mathrm{j}\hat{y}) E(z,t), \tag{11.10}$$

$$\mathbf{H} = (\pm \mathrm{j}\hat{x} + \hat{y}) H(z,t), \tag{11.11}$$

$$\mathbf{J} = (\hat{x} \mp \mathrm{j}\hat{y}) J(z,t), \tag{11.12}$$

$$\omega_p^2 = \omega_p^2(z,t), \tag{11.13}$$

$$\omega_b = \hat{z}\omega_b(z,t). \tag{11.14}$$

The basic equations for E, H, and J take the following simple form:

$$\frac{\partial E}{\partial z} = -\mu_0 \frac{\partial H}{\partial t}, \tag{11.15}$$

$$-\frac{\partial H}{\partial z} = \varepsilon_0 \frac{\partial E}{\partial t} + J, \tag{11.16}$$

$$\frac{\mathrm{d}J}{\mathrm{d}t} = \varepsilon_0 \omega_p^2(z,t) E \pm \mathrm{j}\omega_b(z,t) J, \tag{11.17}$$

$$\frac{\partial^2 E}{\partial z^2} - \frac{1}{c^2}\frac{\partial^2 E}{\partial t^2} - \frac{1}{c^2}\omega_p^2(z,t)E \mp \frac{\mathrm{j}}{c^2}\omega_b(z,t)\frac{\partial E}{\partial t} \mp \mathrm{j}\mu_0\omega_b(z,t)\frac{\partial H}{\partial z} = 0, \tag{11.18}$$

$$\begin{aligned}&\frac{\partial^2 \dot{H}}{\partial z^2} - \frac{1}{c^2}\frac{\partial^2 \dot{H}}{\partial t^2} - \frac{1}{c^2}\omega_p^2(z,t)\dot{H} + \varepsilon_0 \frac{\partial}{\partial z}\omega_b(z,t) E \mp \mathrm{j}\omega_b(z,t)\frac{\partial^2 H}{\partial t^2} \\ &\pm \frac{1}{c^2}\mathrm{j}\omega_b(z,t)\frac{\partial^2 H}{\partial t^2} \mp \mathrm{j}\frac{\partial \omega_b(z,t)}{\partial z}\frac{\partial H}{\partial z} \mp \mathrm{j}\varepsilon_0 \frac{\partial \omega_b(z,t)}{\partial z}\frac{\partial E}{\partial t} = 0.\end{aligned} \tag{11.19}$$

If ω_p and ω_b are functions of time only, then Equation 11.19 reduces to

$$\frac{\partial^2 \dot{H}}{\partial z^2} - \frac{1}{c^2}\frac{\partial^2 \dot{H}}{\partial t^2} - \frac{1}{c^2}\omega_p^2(t)\dot{H} \mp \mathrm{j}\omega_b(t)\frac{\partial^2 H}{\partial z^2} \pm \frac{1}{c^2}\mathrm{j}\omega_b(t)\frac{\partial^2 H}{\partial t^2} = 0. \tag{11.20}$$

Let us next look for the plane wave solutions in a homogeneous, time-invariant unbounded magnetoplasma medium, that is,

$$f(z,t) = \exp[\mathrm{j}(\omega t - kz)], \tag{11.21}$$

$$\omega_p^2(z,t) = \omega_p^2, \tag{11.22}$$

$$\omega_b(z,t) = \omega_b, \tag{11.23}$$

where f stands for any of the field variables E, H, or J. From Equations 11.16 and 11.17, it is shown that

$$jkH = j\omega\varepsilon_0\varepsilon_{pR,L}E, \tag{11.24}$$

where

$$\varepsilon_{pR,L} = 1 - \frac{\omega_p^2}{\omega(\omega \mp \omega_b)}. \tag{11.25}$$

This shows clearly that the magnetized plasma, in this case, can be modeled as a dielectric with the dielectric constant given by Equation 11.25. The dispersion relation is obtained from

$$k^2 = \frac{\omega^2}{c^2}\varepsilon_{pR,L}, \tag{11.26}$$

which when expanded gives

$$\omega^3 \mp \omega_b\omega^2 - \left(k^2c^2 + \omega_p^2\right)\omega \pm k^2c^2\omega_b = 0. \tag{11.27}$$

The expression for the dielectric constant can be written in an alternate fashion in terms of ω_{c1} and ω_{c2} which are called cutoff frequencies: the dispersion relation can also be recast in terms of the cutoff frequencies as

$$\varepsilon_{pR,L} = \frac{(\omega \pm \omega_{c1})(\omega \mp \omega_{c2})}{\omega(\omega \mp \omega_b)}, \tag{11.28}$$

$$\omega_{c1,c2} = \mp\frac{\omega_b}{2} + \sqrt{\left(\frac{\omega_b}{2}\right)^2 + \omega_p^2}, \tag{11.29}$$

$$k_{R,L}^2c^2 = \omega^2\varepsilon_{pR,L} = \frac{\omega(\omega \pm \omega_{c1})(\omega \mp \omega_{c2})}{\omega \mp \omega_b}. \tag{11.30}$$

In the above, the top sign is for the right circular polarization (R wave) and the bottom sign is for the left circular polarization. Figure 11.1 gives ε_p versus ω and Figure 11.2 the ω–k diagram, respectively, for the R wave. Figures 11.3 and 11.4 do the same for the L wave. R and L waves propagate in the direction of the static magnetic field, without any change in the polarization state of the wave and are called the characteristic waves of longitudinal propagation. For these waves, the medium behaves like an isotropic plasma except that the dielectric constant ε_p is influenced by the strength of the static magnetic field. Particular

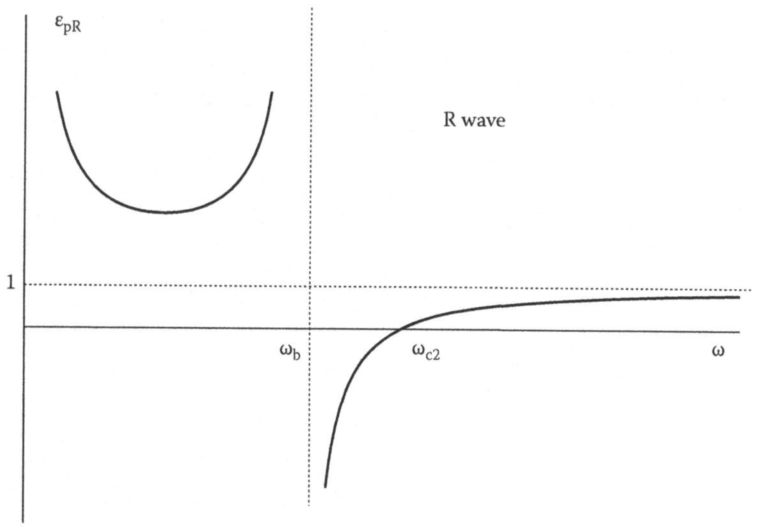

FIGURE 11.1
Dielectric constant for an R wave.

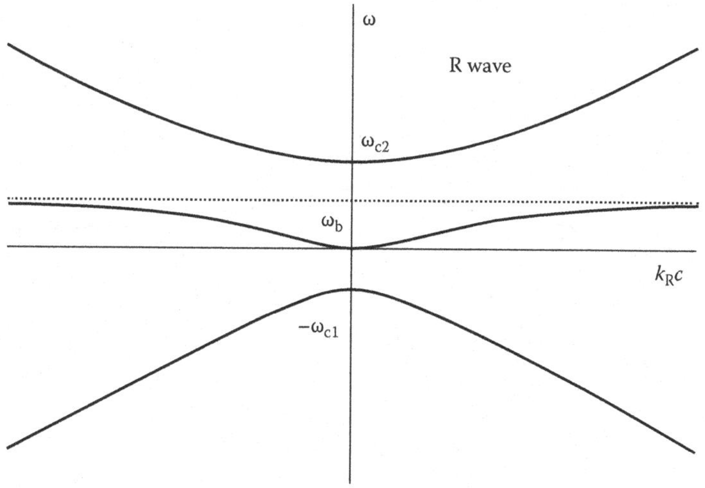

FIGURE 11.2
ω–k diagram for an R wave.

attention is drawn to Figure 11.1 which shows that the dielectric constant $\varepsilon_p > 1$ for the R wave in the frequency band $0 < \omega < \omega_b$. This mode of propagation is called *whistler mode* in the literature on ionospheric physics and helicon mode in the literature on solid-state plasma. Note that an isotropic plasma does not support a whistler wave.

The longitudinal propagation of a linearly polarized wave is accompanied by the *Faraday rotation* of the plane of polarization. This phenomenon is easily explained in terms of the propagation of the R and L characteristic waves. A linearly polarized wave is the superposition of R and L waves each of which propagates without change of its polarization state but each with a different phase velocity. (Note from Equation 11.25 that ε_p for a given ω, ω_p, and ω_b is different for R and L waves due to the sign difference in the denominator.)

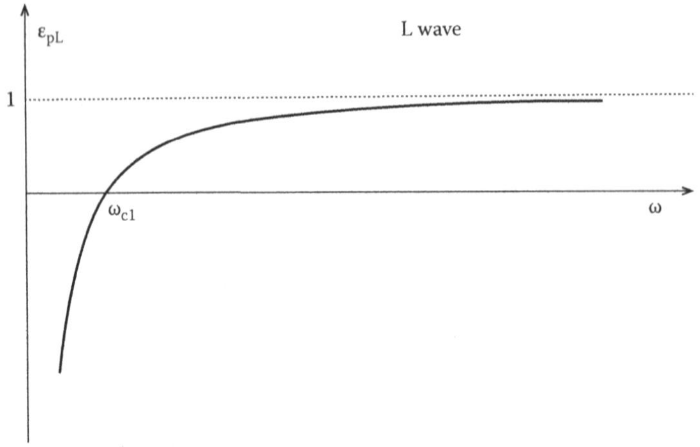

FIGURE 11.3
Dielectric constant for an L wave.

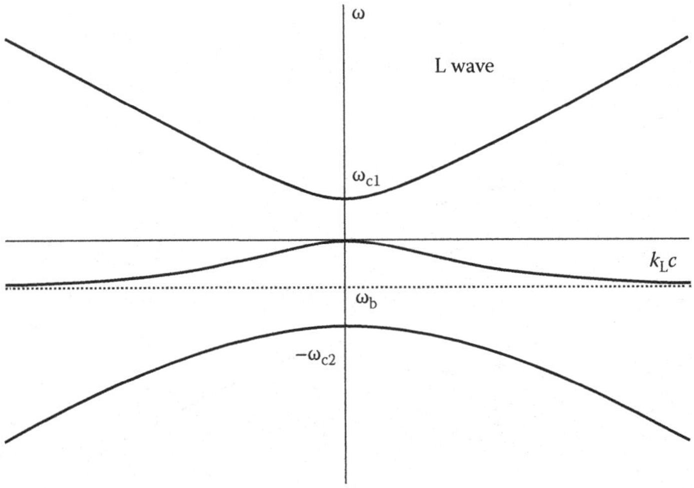

FIGURE 11.4
ω–k diagram for an L wave.

The result of combining the two waves, after traveling a distance d, is a linearly polarized wave again but with the plane of polarization rotated by an angle ψ given by

$$\Psi = \frac{1}{2}\left(k_{\mathrm{L}} - k_{\mathrm{R}}\right)d. \tag{11.31}$$

Appendix 11A discusses Faraday rotation in more detail.

The high value of ε_{p} in the lower-frequency part of the whistler mode can be demonstrated by the following calculation given in [2]; for $f_{\mathrm{p}} = 10^{14}$ Hz, $f_{\mathrm{b}} = 10^{10}$ Hz, and $f = 10$ Hz, $\varepsilon_{\mathrm{p}} = 9 \times 10^{16}$. The wavelength of the signal in the plasma is only 10 cm. See [4–6] for literature on the associated phenomena.

The resonance of the R wave at $\omega = \omega_b$ is of special significance. Around this frequency, it can be shown that even a low-loss plasma strongly absorbs the energy of a source electromagnetic wave and heats the plasma. This effect is the basis of radio frequency heating of *Fusion Plasmas* [7]. It is also used to experimentally determine the effective mass of an electron in a crystal [8].

For the L wave, Figure 11.3 does not show any resonance effect. In fact, the L wave has resonance at the ion gyrofrequency. In the Lorentz plasma model, the ion motion is neglected. A simple modification, by including the ion equation of motion, will extend the EMW transformation theory to a low-frequency source wave [2,9–11].

11.4 One-Dimensional Equations: Transverse Propagation—O Wave

Let us next consider that the electric field is linearly polarized in the y-direction, that is,

$$\mathbf{E} = \hat{y}E(z,t), \tag{11.32}$$

$$\mathbf{H} = -\hat{x}H(z,t), \tag{11.33}$$

$$\mathbf{J} = \hat{y}J(z,t), \tag{11.34}$$

$$\omega_p^2 = \omega_p^2(z,t), \tag{11.35}$$

$$\omega_b = \hat{y}\omega_b(z,t). \tag{11.36}$$

The last term on the RHS of Equation 11.2 is zero since the current density and the static magnetic field are in the same direction. The equations then are no different from those of the isotropic case and the static magnetic field has no effect. The electrons move in the direction of the electric field and give rise to a current density in the plasma in the same direction as that of the static magnetic field. In such a case, the electrons do not experience any magnetic force and their orbit is not bent and they continue to move in the direction of the electric field. The one-dimensional solution for a plane wave in such a medium is called an *ordinary wave* or O wave. Its characteristics are the same as those discussed in the previous chapters. In this case, unlike the case considered in Section 11.2, the direction of phase propagation is perpendicular to the direction of the static magnetic field and hence this case comes under the label of transverse propagation.

11.5 One-Dimensional Solution: Transverse Propagation—X Wave

The more difficult case of the transverse propagation is when the electric field is normal to the static magnetic field. The trajectories of the electrons that start moving in the direction of the electric field get altered and bent due to the magnetic force. Such a motion gives rise to an additional component of the current density in the direction of phase propagation and to obtain a self-consistent solution we have to assume that a component of the electric field exists also in the direction of phase propagation. Let

$$\mathbf{E} = \hat{x}E_x(z,t) + \hat{z}E_z(z,t), \tag{11.37}$$

$$\mathbf{H} = \hat{y}H(z,t), \tag{11.38}$$

$$\mathbf{J} = \hat{x}J_x(z,t) + \hat{z}J_z(z,t), \tag{11.39}$$

$$\omega_p^2 = \omega_p^2(z,t), \tag{11.40}$$

$$\omega_b = \hat{y}\omega_b(z,t). \tag{11.41}$$

The basic equations for E, H, and J take the following form:

$$\frac{\partial E_x}{\partial z} = -\mu_0 \frac{\partial H}{\partial t}, \tag{11.42}$$

$$-\frac{\partial H}{\partial z} = \varepsilon_0 \frac{\partial E_x}{\partial t} + J_x, \tag{11.43}$$

$$\varepsilon_0 \frac{\partial E_z}{\partial t} = -J_z, \tag{11.44}$$

$$\frac{dJ_x}{dt} = \varepsilon_0 \omega_p^2(z,t) E_x + \omega_b(z,t) J_z, \tag{11.45}$$

$$\frac{dJ_z}{dt} = \varepsilon_0 \omega_p^2(z,t) E_z - \omega_b(z,t) J_x. \tag{11.46}$$

Let us look again for the plane wave solution in a homogeneous, time-invariant, unbounded magnetoplasma medium by applying Equation 11.21 to set Equations 11.42 through 11.46:

$$-jkE_x = -\mu_0 j\omega H, \tag{11.47}$$

$$jkH = j\omega\varepsilon_0 E_x + J_x, \tag{11.48}$$

$$\varepsilon_0 j\omega E_z = -J_z, \tag{11.49}$$

$$j\omega J_x = \varepsilon_0 \omega_p^2 E_x + \omega_b J_z, \tag{11.50}$$

$$j\omega J_z = \varepsilon_0 \omega_p^2 E_z - \omega_b J_x. \tag{11.51}$$

From Equations 11.49 through 11.51, we get the following relation between J_x and E_x:

$$J_x = \varepsilon_0 \frac{\omega_p^2}{\left[1 - \dfrac{\omega_b^2}{\left(\omega^2 - \omega_p^2\right)}\right]} \frac{E_x}{j\omega}. \tag{11.52}$$

Substituting Equation 11.52 into Equation 11.48,

$$jkH = j\omega\varepsilon_0\varepsilon_{pX}E_x, \tag{11.53}$$

$$\varepsilon_{pX} = 1 - \frac{\omega_p^2/\omega^2}{\left[1 - \frac{\omega_b^2}{(\omega^2 - \omega_p^2)}\right]}. \tag{11.54}$$

An alternate expression for ε_{pX} can be written in terms of ω_{c1}, ω_{c2}, and ω_{uh}:

$$\varepsilon_{pX} = \frac{(\omega^2 - \omega_{c1}^2)(\omega^2 - \omega_{c2}^2)}{\omega^2(\omega^2 - \omega_{uh}^2)}. \tag{11.55}$$

The cutoff frequencies ω_{c1} and ω_{c2} are defined earlier and ω_{uh} is the upper hybrid frequency:

$$\omega_{uh}^2 = \omega_p^2 + \omega_b^2. \tag{11.56}$$

The dispersion relation $k^2 = (\omega^2/c^2)\varepsilon_{pX}$, when expanded, gives

$$\omega^4 - (k^2c^2 + \omega_b^2 + 2\omega_p^2)\omega^2 + \left[\omega_p^4 + k^2c^2(\omega_b^2 + \omega_p^2)\right] = 0. \tag{11.57}$$

In obtaining Equation 11.57, the expression for ε_{pX} given by Equation 11.54 is used. For the purpose of sketching the ω–k diagram, it is more convenient to use Equation 11.54 for ε_{pX} and write the dispersion relation as

$$k^2c^2 = \frac{(\omega^2 - \omega_{c1}^2)(\omega^2 - \omega_{c2}^2)}{\omega^2 - \omega_{uh}^2}. \tag{11.58}$$

Figures 11.5 and 11.6 sketch ε_{pX} versus ω, and ω versus kc, respectively. Substituting Equations 11.49 and 11.52 into Equation 11.51, we can find the ratio of E_z to E_x:

$$\frac{E_z}{E_x} = -j\omega\frac{\omega_b}{\omega^2 - \omega_p^2}(\varepsilon_{pX} - 1). \tag{11.59}$$

This shows that the polarization in the x–z-plane, whether linear, circular, or elliptic, depends on the source frequency and the plasma parameters. Moreover, using Equations 11.52 and 11.49, the relation between **J** and **E** can be written as

$$\mathbf{J} = \bar{\sigma}\cdot\mathbf{E}, \tag{11.60}$$

$$\begin{bmatrix} J_x \\ J_z \end{bmatrix} = j\omega\varepsilon_0\begin{bmatrix} \varepsilon_{pX} - 1 & 0 \\ 0 & -1 \end{bmatrix}\begin{bmatrix} E_x \\ E_z \end{bmatrix}. \tag{11.61}$$

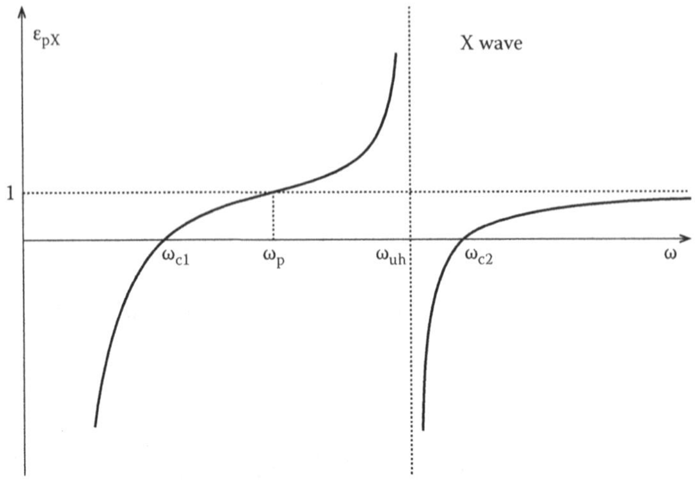

FIGURE 11.5
Dielectric constant for an X wave.

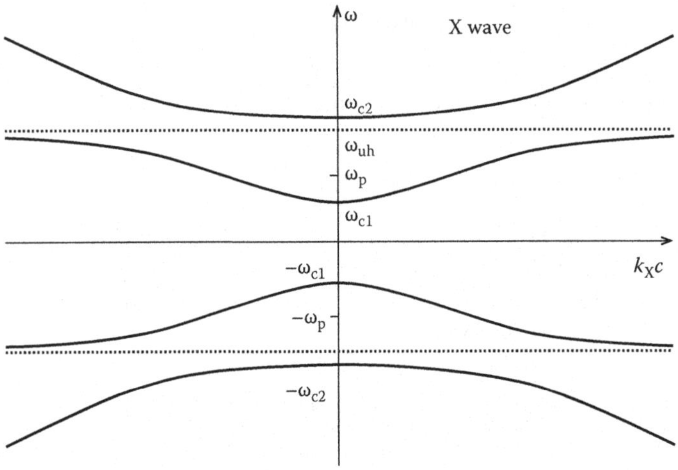

FIGURE 11.6
ω–k diagram for an X wave.

where $\bar{\bar{\sigma}}$ is the conductivity tensor for the case under consideration. The dielectric modeling of the plasma can be deduced from

$$\mathbf{D} = \varepsilon_0 \bar{\bar{\mathbf{K}}} \cdot \mathbf{E}, \quad (11.62)$$

where the dielectric tensor $\bar{\bar{\mathbf{K}}}$ is related to $\bar{\bar{\sigma}}$ by

$$\bar{\bar{\mathbf{K}}} = \bar{\bar{\mathbf{I}}} + \frac{\bar{\bar{\sigma}}}{j\omega\varepsilon_0}, \quad (11.63)$$

which gives, for this case,

$$\begin{bmatrix} D_x \\ D_z \end{bmatrix} = \begin{bmatrix} \varepsilon_{pX} & 0 \\ 0 & 0 \end{bmatrix} \begin{bmatrix} E_x \\ E_z \end{bmatrix}. \quad (11.64)$$

This shows that $D_z = 0$, even when $E_z \neq 0$. For the X wave, **D**, **H**, and the direction of propagation are mutually orthogonal. While **E** and **H** are orthogonal, **E** and $\hat{\mathbf{k}}$ (the unit vector in the direction of propagation) are not orthogonal. We can summarize by stating

$$\hat{\mathbf{E}} \times \hat{\mathbf{H}} \neq \hat{\mathbf{k}}, \quad \hat{\mathbf{D}} \times \hat{\mathbf{H}} = \hat{\mathbf{k}}, \tag{11.65}$$

$$E_x = \eta_{pX} H_y = \frac{\eta_0}{\sqrt{\varepsilon_{pX}}} = \frac{\eta_0}{n_{pX}}, \tag{11.66}$$

$$\mathbf{E} \cdot \hat{\mathbf{k}} \neq 0, \quad \mathbf{D} \cdot \hat{\mathbf{k}} = 0. \tag{11.67}$$

In the above, n_{pX} is the refractive index. Figure 11.7 is a geometrical sketch of the directions of various components of the X wave. The direction of the power flow is given by the Poynting vector $\mathbf{S} = \mathbf{E} \times \mathbf{H}$ which, in this case, is not in the direction of phase propagation. The result that $D_z = 0$ for the plane wave comes from the general equation

$$\nabla \cdot \mathbf{D} = 0, \tag{11.68}$$

in a sourceless medium. In an anisotropic medium, it does not necessarily follow that the divergence of the electric field is also zero because of the tensorial nature of the constitutive relation. For example, from the expression

$$D_z = \varepsilon_0 \left[K_{zx} E_x + K_{zy} E_y + K_{zz} E_z \right], \tag{11.69}$$

D_z can be zero without E_z being zero.

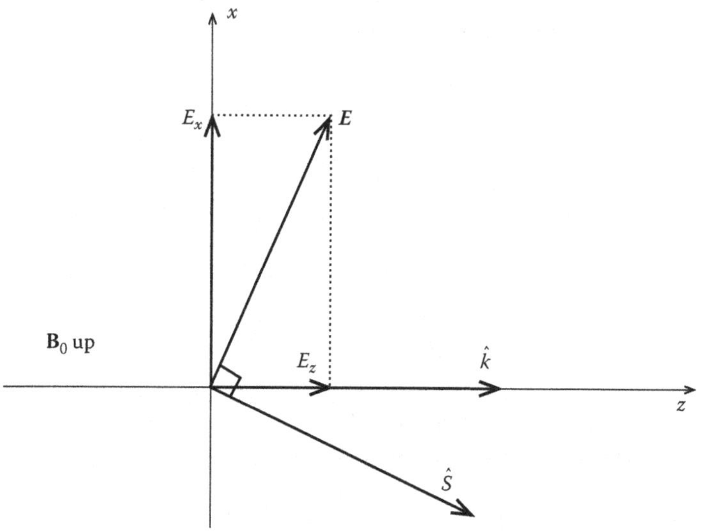

FIGURE 11.7
Geometrical sketch of the directions of various components of the X wave.

11.6 Dielectric Tensor of a Lossy Magnetoplasma Medium

The constitutive relation for a lossy plasma with collision frequency ν (rad/s) is obtained by modifying Equation 11.2 further:

$$\frac{d\mathbf{J}}{dt} + \nu\mathbf{J} = \varepsilon_0\omega_p^2(\mathbf{r},t)\mathbf{E} - \mathbf{J}\times\mathbf{w}_b. \tag{11.70}$$

Assuming time-harmonic variation exp(jωt) and taking the z-axis as the direction of the static magnetic field,

$$\omega_b = \hat{z}\omega_b, \tag{11.71}$$

we can write Equation 11.70 as

$$(\nu + j\omega)\begin{bmatrix} J_x \\ J_y \end{bmatrix} = \varepsilon_0\omega_p^2\begin{bmatrix} E_x \\ E_y \end{bmatrix} - j\omega_b\begin{bmatrix} J_y \\ -J_x \end{bmatrix}, \tag{11.72}$$

and

$$J_z = \frac{\varepsilon_0\omega_p^2}{\nu + j\omega}E_z. \tag{11.73}$$

Equation 11.72 can be written as

$$\begin{bmatrix} \nu + j\omega & \omega_b \\ -\omega_b & \nu + j\omega \end{bmatrix}\begin{bmatrix} J_x \\ J_y \end{bmatrix} = \varepsilon_0\omega_p^2\begin{bmatrix} E_x \\ E_y \end{bmatrix}, \tag{11.74}$$

$$\begin{bmatrix} J_x \\ J_y \end{bmatrix} = \varepsilon_0\omega_p^2\begin{bmatrix} \nu + j\omega & \omega_b \\ \omega_b & \nu + j\omega \end{bmatrix}^{-1}\begin{bmatrix} E_x \\ E_y \end{bmatrix}. \tag{11.75}$$

Combining Equation 11.75 with (11.73), we obtain

$$\begin{bmatrix} J_x \\ J_y \\ J_z \end{bmatrix} = \bar{\sigma}\cdot\mathbf{E}, \tag{11.76}$$

where

$$\bar{\sigma} = \begin{bmatrix} \sigma_\perp & -\sigma_H & 0 \\ \sigma_H & \sigma_\perp & 0 \\ 0 & 0 & \sigma_\parallel \end{bmatrix}. \tag{11.77}$$

In the above, $\sigma_\perp$, σ_H, and $\sigma_\parallel$ are called perpendicular conductivity, Hall conductivity, and parallel conductivity, respectively. From Equation 11.63, the dielectric tensor $\bar{\mathbf{K}}$ can now be obtained:

$$\bar{\mathbf{K}} = \begin{bmatrix} \varepsilon_{r\perp} & -j\varepsilon_{rH} & 0 \\ j\varepsilon_{rH} & \varepsilon_{r\perp} & 0 \\ 0 & 0 & \varepsilon_{r\parallel} \end{bmatrix}, \tag{11.78}$$

where

$$\varepsilon_{r\perp} = 1 - \frac{(\omega_p/\omega)^2(1 - j\nu/\omega)}{\left(1 - j(\nu/\omega)\right)^2 - (\omega_b/\omega)^2}, \tag{11.79}$$

$$\varepsilon_{rH} = \frac{(\omega_p/\omega)^2(\omega_b/\omega)}{(1 - j\nu/\omega)^2 - (\omega_b/\omega)^2}, \tag{11.80}$$

$$\varepsilon_{r\parallel} = 1 - \frac{(\omega_p/\omega)^2}{1 - j\nu/\omega}. \tag{11.81}$$

11.7 Periodic Layers of Magnetoplasma

Layered semiconductor–dielectric periodic structures in dc magnetic field can be modeled as periodic layers of magnetoplasma. Electromagnetic wave propagating in such structures can be investigated by using the dielectric tensor derived in Section 11.6. The solution is obtained by extending the theory of Section 9.7 to the case where the plasma layer is magnetized and hence has the anisotropy complexity [12]. Brazis and Safonova [13–15] investigated the resonance and absorption bands of such structures.

11.8 Surface Magnetoplasmons

The theory of Section 9.8 can be extended to the propagation of surface waves at a semiconductor–dielectric interface in the presence of a static magnetic field. The semiconductor in a dc magnetic can be modeled as an anisotropic dielectric tensor. Wallis [16] discussed at length the properties of surface magnetoplasmons on semiconductors.

11.9 Surface Magnetoplasmons in Periodic Media

Wallis et al. [17,18] discuss the propagation of surface magnetoplasmons in truncated superlattices. This model is a combination of the models used in Sections 11.7 and 11.8.

11.10 Permeability Tensor

Ferrite material (see Section 8.9) in the presence of a static magnetic field behaves like an anisotropic magnetic media. The wave propagation in such a medium has properties similar to the magnetoplasma. The permeability is a tensor. Appendix 11B discusses the permeability tensor.

11.11 Reflection by a Warm Magnetoplasma Slab

We discussed the slab problem in the case of a cold plasma slab in Section 9.4 and a warm plasma slab in Appendix 10A. The more general problem is that of a warm magnetoplasma slab. The essence of this problem can be understood by considering the normal incidence involving the coupling of an electromagnetic wave with an electron plasma wave, as shown in Appendix 11C.

References

1. Heald, M. A. and Wharton, C. B., *Plasma Diagnostics with Microwaves*, Wiley, New York, 1965.
2. Booker, H. G., *Cold Plasma Waves*, Kluwer Academic Publishers, Hingham, MA, 1984.
3. Swanson, D. G., *Plasma Waves*, Academic Press, New York, 1989.
4. Steele, M. C., *Wave Interactions in Solid State Plasmas*, McGraw-Hill, New York, 1969.
5. Bowers, R., Legendy, C., and Rose, F., Oscillatory galvanometric effect in metallic sodium, *Phys. Rev. Lett.*, 7, 339, 1961.
6. Aigrain, P. R., *Proceedings of the International Conference on Semiconductor Physics*, Czechoslovak Academy Sciences, Prague, Czech Republic, p. 224, 1961.
7. Miyamoto, K., *Plasma Physics for Nuclear Fusion*, The MIT Press, Cambridge, MA, 1976.
8. Solymar, L. and Walsh, D., *Lectures on the Electrical Properties of Materials*, 5th edn., Oxford University Press, Oxford, U.K., 1993.
9. Madala, S. R. V. and Kalluri, D. K., Longitudinal propagation of low frequency waves in a switched magnetoplasma medium, *Radio Sci.*, 28, 121, 1993.
10. Dimitrijevic, M. M. and Stanic, B. V., EMW transformation in suddenly created two-component magnetized plasma, *IEEE Trans. Plasma Sci.*, 23, 422, 1995.
11. Madala, S. R. V., Frequency shifting of low frequency electromagnetic waves using magnetoplasmas, Doctoral thesis, University of Massachusetts Lowell, Lowell, MA, 1993.
12. Kalluri, D. K., *Electromagnetics of Time Varying Complex Media: Frequency and Polarization Transformer*, CRC Press/Taylor & Francis Group, Boca Raton, FL, 2010.
13. Brazis, R. S. and Safonova, L. S., Resonance and absorption band in the classical magnetoactive semiconductor–insulator superlattice, *Int. J. Infrared Millim. Waves*, 8, 449, 1987.
14. Brazis, R. S. and Safonova, L. S., Electromagnetic waves in layered semiconductor–dielectric periodic structures in DC magnetic fields, *Proc. SPIE*, 1029, 74, 1988.
15. Brazis, R. S. and Safonova, L. S., In-plane propagation of millimeter waves in periodic magnetoactive semiconductor structures, *Int. J. Infrared Millim. Waves*, 18, 1575, 1997.
16. Wallis, R. F., Surface magnetoplasmons on semiconductors, in: Boardman, A. D. (Ed.), *Electromagnetic Surface Modes*, Wiley, New York, 1982, Chap. 2.
17. Wallis, R. F., Szenics, R., Quihw, J. J., and Giuliani, G. F., Theory of surface magnetoplasmon polaritons in truncated superlattices, *Phys. Rev. B*, 36, 1218, 1987.
18. Wallis, R. F. and Quinn, J. J., Surface magnetoplasmon polaritons in truncated superlattices, *Phys. Rev. B*, 38, 4205, 1988.

12

Optical Waves in Anisotropic Crystals

An optical crystal is an anisotopic dielectric and the dielectric tensor of such a crystal in a coordinate system that coincides with the principal axes of the crystal will have zero off-diagonal elements and is given by

$$\bar{\varepsilon} = \begin{bmatrix} \varepsilon_1 & 0 & 0 \\ 0 & \varepsilon_2 & 0 \\ 0 & 0 & \varepsilon_3 \end{bmatrix}. \tag{12.1}$$

The principal refractive indices of the crystal are given by

$$n_j = \left(\frac{\varepsilon_j}{\varepsilon_0}\right)^{1/2}, \quad j = 1,2,3. \tag{12.2}$$

The crystals are classified as biaxial, uniaxial, or isotropic depending on the number of equalities involving the principal refractive indices:

$$n_1 \neq n_2 \neq n_3 \quad \text{(biaxial crystal)}, \tag{12.3}$$

$$n_1 = n_2 = n_\mathrm{o}, \quad n_3 = n_\mathrm{e} \quad \text{(uniaxial crystal)}, \tag{12.4}$$

$$n_1 = n_2 = n_3 = n_\mathrm{o} \quad \text{(isotropic crystal)}. \tag{12.5}$$

Properties of wave propagation in isotropic crystals are discussed in Chapter 2. In the following section, we discuss wave propagation in crystals in the order of increasing complexity of analysis [1–3].

12.1 Wave Propagation in a Biaxial Crystal along the Principal Axes

Let the principal axes be aligned along the Cartesian coordinate system. A uniform plane wave with a harmonic variation in space and time given by

$$f(\mathbf{r},t) = \exp\left[\mathrm{j}(\omega t - k \cdot \mathbf{r})\right] \tag{12.6}$$

satisfies following equations:

$$k \times \mathbf{E} = \omega \mathbf{B}, \tag{12.7}$$

$$k \times \mathbf{H} = -\omega \mathbf{D}, \tag{12.8}$$

$$k \cdot \mathbf{B} = 0, \tag{12.9}$$

$$k \cdot \mathbf{D} = 0, \tag{12.10}$$

$$\mathbf{B} = \mu_0 \mathbf{H}, \tag{12.11}$$

$$D_x = \varepsilon_0 n_x^2 E_x, \quad D_y = \varepsilon_0 n_y^2 E_y, \quad D_z = \varepsilon_0 n_z^2 E_z. \tag{12.12}$$

Equations 12.7 through 12.12 are obtained from Maxwell's equations and the constitutive relation for the crystal. From Equations 12.9 and 12.10, we obtain

$$B_k = D_k = 0 \tag{12.13}$$

and from Equations 12.7 and 12.8, we obtain

$$\boldsymbol{k} \times \boldsymbol{k} \times \mathbf{E} = \omega(\boldsymbol{k} \times \mathbf{B}) = -\mu_0 \omega^2 \mathbf{D}. \tag{12.14}$$

For the particular case of

$$\boldsymbol{k} = \hat{z}k, \tag{12.15}$$

$$\mathbf{E} = \hat{x}E, \tag{12.16}$$

$$\mathbf{H} = \hat{y}H, \tag{12.17}$$

we obtain

$$\left(-k^2 + \mu_0 \varepsilon_0 \omega^2 n_x^2\right) E = 0. \tag{12.18}$$

Therefore, the dispersion relation in this case is given by

$$k^2 = n_x^2 \omega^2 \mu_0 \varepsilon_0. \tag{12.19}$$

The results obtained in this particular case may be stated in words as follows: the propagation of a wave linearly polarized in the direction of a principal axis and propagating along the direction of another principal axis is similar to that of wave propagation in an isotropic case except that the effective refractive index is given by the refractive index in the direction of polarization.

We next consider an example of change of wave polarization by a biaxial crystal. Let the electric field **E** of a wave propagating in the z-direction in this crystal, at $z = 0$, be given by

$$\mathbf{E}(0) = E_0 \left[\hat{x} \cos 45^\circ + \hat{y} \sin 45^\circ\right], \quad z = 0. \tag{12.20}$$

This wave is linearly polarized at 45° (with the x-axis) in the $z = 0$ plane. If we consider this wave as a superposition of two waves, one with linear polarization in the x-direction and another with linear polarization in the y-direction, then we can write the electric field of this wave at $z = d$ as

$$\mathbf{E}(d) = \hat{x}E_0 \cos 45^\circ \exp\left[j(\omega t - n_x k_0 d)\right] + \hat{y}E_0 \sin 45^\circ \exp\left[j(\omega t - n_y k_0 d)\right]. \quad (12.21)$$

Note that the phase of the wave changes with d since $n_x k_0 d \neq n_y k_0 d$. Hence, the wave polarization will not remain linear.

12.2 Propagation in an Arbitrary Direction

The notation and the formulations in Sections 12.2 through 12.4 closely follow that of reference [3]. Let the unit vector in the direction of propagation be designated by $\hat{e}_3$ (see Figure 12.1):

$$\boldsymbol{k} = \hat{e}_3 k. \quad (12.22)$$

Let $\hat{e}_1$ be a unit vector in the x–y-plane. The direction of $\hat{e}_1$ is obtained by rotating the projection of $\hat{e}_3$ in the x–y-plane by 90° as shown in the figure. It is obvious that

$$\hat{e}_3 \cdot \hat{e}_1 = 0 \quad (12.23)$$

and

$$\hat{e}_1 = (\hat{e}_1 \cdot \hat{x})\hat{x} + (\hat{e}_1 \cdot \hat{y})\hat{y} = \cos(90 - \phi)\hat{x} + \cos(180 - \phi)\hat{y} = \hat{x}\sin\phi - \hat{y}\cos\phi. \quad (12.24)$$

Let

$$\hat{e}_2 = \hat{e}_3 \times \hat{e}_1 = \hat{x}\cos\theta\cos\phi + \hat{y}\cos\theta\sin\phi - \hat{z}\sin\theta. \quad (12.25)$$

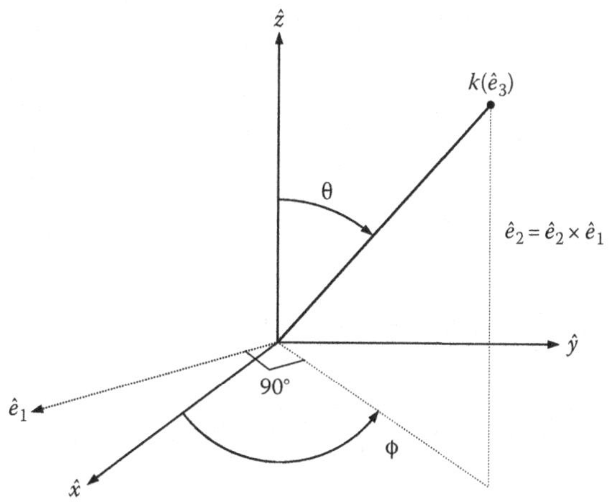

FIGURE 12.1
Orthogonal coordinate system with one coordinate along the wave normal.

The unit vectors $\hat{e}_1$, $\hat{e}_2$, and $\hat{e}_3$ are mutually orthogonal. A vector **A** may be resolved along x–y–z coordinates as well as along the directions of $\hat{e}_1$, $\hat{e}_2$, and $\hat{e}_3$. In the latter case, let us designate the vector by $\mathbf{A}_k$ with components A_1, A_2, and A_3. The transformation that relates the components of **A** with components of $\mathbf{A}_k$ may be written as

$$\mathbf{A}_k = \bar{\mathbf{T}} \cdot \mathbf{A}, \tag{12.26}$$

$$\mathbf{A} = \bar{\mathbf{T}}^{-1} \cdot \mathbf{A}_k. \tag{12.27}$$

The transformation is easily determined by noting

$$A_1 = \hat{e}_1 \cdot \mathbf{A}_k = \hat{e}_1 \cdot \hat{x}A_x + \hat{e}_1 \cdot \hat{y}A_y + \hat{e}_1 \cdot \hat{z}A_z = A_x \sin\phi - A_y \cos\phi. \tag{12.28}$$

Similar considerations will lead to the determination of the transformation matrix $\bar{\mathbf{T}}$:

$$\bar{\mathbf{T}} = \begin{bmatrix} \sin\phi & -\cos\phi & 0 \\ \cos\theta\cos\phi & \cos\theta\sin\phi & -\sin\theta \\ \sin\theta\cos\phi & \sin\theta\sin\phi & \cos\theta \end{bmatrix}. \tag{12.29}$$

By calculating $\bar{\mathbf{T}}^{-1}$, we can show that

$$\bar{\mathbf{T}}^{-1} = \left(\bar{\mathbf{T}}\right)^{\mathrm{T}}, \tag{12.30}$$

where the superscript T stands for the transpose operation. Defining an impermittivity tensor $\bar{\eta}$ (here $\bar{\eta}$ is not characteristic impedance),

$$\eta = \varepsilon_{\mathrm{p}}^{-1}, \tag{12.31}$$

where

$$\bar{\varepsilon} = \varepsilon_0 \bar{\varepsilon}_{\mathrm{p}}, \tag{12.32}$$

we obtain

$$\mathbf{E} = \frac{1}{\varepsilon_0} \bar{\eta} \cdot \mathbf{D}, \tag{12.33}$$

$$\bar{\mathbf{T}}^{-1} \cdot \mathbf{E}_k = \frac{1}{\varepsilon_0} \bar{\eta} \cdot \bar{\mathbf{T}}^{-1} \cdot \mathbf{D}_k, \tag{12.34}$$

$$\mathbf{E}_k = \frac{1}{\varepsilon_0} \left[\bar{\mathbf{T}} \cdot \bar{\eta} \cdot \bar{\mathbf{T}}^{-1}\right] \cdot \bar{\mathbf{D}}_k, \tag{12.35}$$

$$\bar{\mathbf{E}}_k - \frac{1}{\varepsilon_0} \bar{\eta}_k \cdot \bar{\mathbf{D}}_k. \tag{12.36}$$

If the elements of $\overline{\eta}_k$ are designated by η_{ij}, then Equation 12.33 may be written as

$$\begin{bmatrix} E_1 \\ E_2 \\ E_3 \end{bmatrix} = \frac{1}{\varepsilon_0} \begin{bmatrix} \eta_{11} & \eta_{12} & \eta_{13} \\ \eta_{21} & \eta_{22} & \eta_{23} \\ \eta_{31} & \eta_{32} & \eta_{33} \end{bmatrix} \begin{bmatrix} D_1 \\ D_2 \\ 0 \end{bmatrix}. \tag{12.37}$$

From Equations 12.7, 12.8, and 12.37, we obtain

$$\omega B_2 = kE_1 = \frac{k}{\varepsilon_0}\left[\eta_{11}D_1 + \eta_{12}D_2\right], \tag{12.38}$$

$$\omega B_1 = -kE_2 = -\frac{k}{\varepsilon_0}\left[\eta_{21}D_1 + \eta_{22}D_2\right], \tag{12.39}$$

$$\omega D_2 = -kH_1 = -\frac{kB_1}{\mu_0}, \tag{12.40}$$

$$\omega D_1 = kH_2 = \frac{kB_2}{\mu_0}. \tag{12.41}$$

By eliminating all other variables, we can obtain the dispersion relation

$$\begin{bmatrix} \eta_{11} - \dfrac{u^2}{c^2} & \eta_{12} \\ \eta_{21} & \eta_{22} - \dfrac{u^2}{c^2} \end{bmatrix} \begin{bmatrix} D_1 \\ D_2 \end{bmatrix} = 0, \tag{12.42}$$

where $u = \omega/k$ is the phase velocity.

12.3 Propagation in an Arbitrary Direction: Uniaxial Crystal

As an example of further calculations, we restrict ourselves to a uniaxial crystal. For this case,

$$\overline{\eta}_k = \begin{bmatrix} \eta & 0 & 0 \\ 0 & \eta\cos^2\theta + \eta_z \sin^2\theta & (\eta - \eta_z)\sin\theta\cos\theta \\ 0 & (\eta - \eta_z)\sin\theta\cos\theta & \eta\sin\theta + \eta_z\cos\theta \end{bmatrix}, \tag{12.43}$$

where

$$\eta = \frac{1}{n_o^2}, \tag{12.44}$$

$$\eta_z = \frac{1}{n_e^2}. \tag{12.45}$$

Thus,

$$\eta_{11} = \eta, \tag{12.46}$$

$$\eta_{12} = \eta_{21} = 0, \tag{12.47}$$

$$\eta_{22} = \eta\cos^2\theta + \eta_z \sin^2\theta. \tag{12.48}$$

From Equations 12.42 and 12.46 through 12.48, we obtain the following results.
Case 1:

$$\left(\eta - \frac{u^2}{c^2}\right) = 0 \quad (D_1 \neq 0, D_2 = 0). \tag{12.49}$$

Case 2:

$$\eta_{22} = \eta\cos^2\theta + \eta_z \sin^2\theta = \frac{u^2}{c^2} \quad (D_1 = 0,\ D_2 \neq 0). \tag{12.50}$$

In the first case, the phase velocity is

$$u = \pm\frac{c}{n_o} \quad (\text{ordinary wave}) \tag{12.51}$$

and in the second case it is

$$\frac{u}{c} = \frac{1}{n} = \sqrt{\frac{\cos^2\theta}{n_o^2} + \frac{\sin^2\theta}{n_e^2}} \quad (\text{extraordinary wave}). \tag{12.52}$$

In the latter case, the phase velocity depends on θ.

12.4 *k*-Surface

Let *k* be the scalar wave number and is given by

$$k_z = k\cos\theta, \tag{12.53}$$

$$k_s = k\sin\theta, \tag{12.54}$$

where k_s is the transverse wave number.

For Case 1,

$$u = \frac{c}{n_o} = \frac{\omega}{k}, \tag{12.55}$$

$$k_z^2 + k_s^2 = k^2 = \frac{\omega^2 n_o^2}{c^2}. \tag{12.56}$$

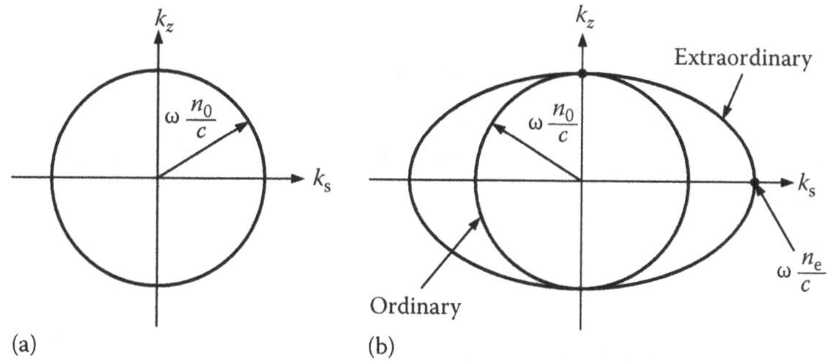

FIGURE 12.2
k-Surface. (a) Ordinary and (b) ordinary and extraordinary.

Figure 12.2a shows this circle in $k_s - k_z$ plain.
For Case 2,

$$\frac{u^2}{c^2} = \frac{\omega^2}{k^2c^2} = \frac{\cos^2\theta}{n_o^2} + \frac{\sin^2\theta}{n_e^2}, \tag{12.57}$$

which may be written as

$$\frac{k_s^2}{n_e^2\left(\omega^2/c^2\right)} + \frac{k_z^2}{n_o^2\left(\omega^2/c^2\right)} = 1. \tag{12.58}$$

Thus, the curve in $k_s - k_z$ plane is an ellipse as shown in Figure 12.2.

The phase velocity is given by ω/k and for the ordinary wave it is the same for all θ. For the extraordinary wave, ω/k depends on θ, since the k curve is an ellipse. The group velocity for this case can be defined as a vector:

$$\mathbf{u}_g = \nabla_k\omega = \hat{k}_s\frac{\partial\omega}{\partial k_s} + \hat{z}\frac{\partial\omega}{\partial k_z}. \tag{12.59}$$

In three-dimensional k-space, we obtain a sphere for the ordinary wave and ellipsoid for the extraordinary wave.

For a uniaxial crystal, the refractive index is the same along the two principal directions (n_o in x- and y-directions in our notation) and different along the axis (n_e along the z-axis) called the optical axis. Note that the refractive index surfaces touch along the k_z-axis.

12.5 Group Velocity as a Function of Polar Angle

The angle between the optical axis and the wave normal (direction of phase propagation) is the polar angle θ. From Equation 12.58, we obtain

$$\omega^2 = c^2\left[\frac{k_z^2}{n_o^2} + \frac{k_s^2}{n_e^2}\right],$$

$$\omega = c\left[\frac{k_z^2}{n_o^2} + \frac{k_s^2}{n_e^2}\right]^{1/2},$$

where c is the velocity of light in free space, $k_z = k\cos\theta$, and $k_s = k\sin\theta$:

$$\frac{\partial\omega}{\partial k_z} = \frac{c}{2}\left[\frac{k_z^2}{n_o^2} + \frac{k_s^2}{n_e^2}\right]^{-1/2}\left[\frac{2k_z}{n_o^2}\right], \tag{12.60}$$

$$\frac{\partial\omega}{\partial k_s} = \frac{c}{2}\left[\frac{k_z^2}{n_o^2} + \frac{k_s^2}{n_e^2}\right]^{-1/2}\left[\frac{2k_s}{n_e^2}\right]. \tag{12.61}$$

Noting that

$$\left[\frac{k_z^2}{n_o^2} + \frac{k_s^2}{n_e^2}\right]^{-1/2} = \left(\frac{\omega^2}{c^2}\right)^{-1/2} = \frac{c}{\omega},$$

$$\frac{\partial\omega}{\partial k_z} = \frac{c}{2}\frac{2k_z}{n_o^2}\frac{c}{\omega} = \frac{k_z c^2}{\omega n_o^2} = \frac{c^2 k\cos\theta}{\omega n_o^2}.$$

Similarly,

$$\frac{\partial\omega}{\partial k_s} = \frac{c^2 k\sin\theta}{\omega n_e^2},$$

$$\mathbf{u}_g = c^2\frac{k}{\omega}\left[\frac{\sin\theta}{n_e^2}\hat{a}_s + \frac{\cos\theta}{n_o^2}\hat{a}_z\right]. \tag{12.62}$$

The magnitude of the group velocity $|\mathbf{u}_g|$ is given by

$$|\mathbf{u}_g| = c^2\frac{k}{\omega}\left[\frac{\sin^2\theta}{n_e^4} + \frac{\cos^2\theta}{n_o^4}\right]^{1/2}. \tag{12.63}$$

Since $\omega/k = u$,

$$|\mathbf{u}_g|u = c^2\left[\frac{\sin^2\theta}{n_e^4} + \frac{\cos^2\theta}{n_o^4}\right]^{1/2}. \tag{12.64}$$

If the angle of $\mathbf{u}_g$ with the polar axis is θ_g, then

$$\tan\theta_g = \frac{u_{gs}}{u_{gz}} = \frac{\sin\theta/n_e^2}{\cos\theta/n_o^2} = \tan\theta\frac{n_o^2}{n_e^2}.$$

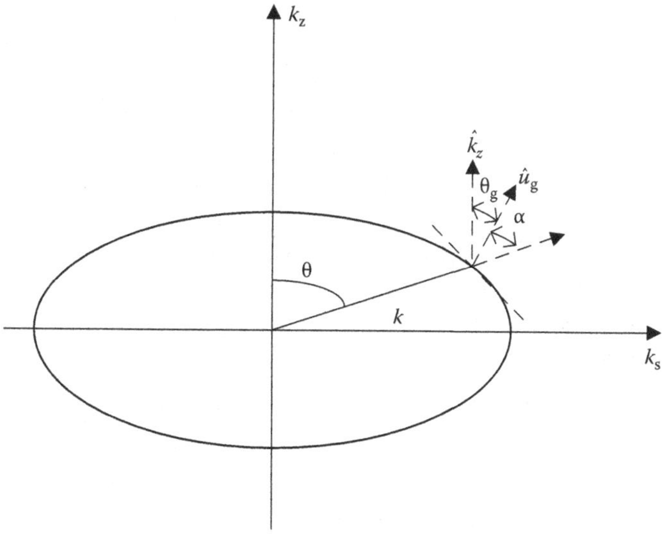

FIGURE 12.3
Various angles for the extraordinary wave.

If $\alpha = \theta - \theta_g$, then

$$\tan\alpha = \tan(\theta - \theta_g) = \frac{\tan\theta - \tan\theta_g}{1 + \tan\theta\tan\theta_g} = \frac{\left(1 - n_o^2/n_e^2\right)\tan\theta}{1 + \left(n_o^2\tan^2\theta\right)/n_e^2}. \tag{12.65}$$

Note that if $\theta = 0$, then $\tan\alpha = 0$, and $\alpha = 0$.

If $\theta = \pi/2$, then $\tan\alpha = \infty/\infty^2 = 1/\infty = 0$, thus $\alpha = 0$ (Figure 12.3).

To find the value of θ so that α is maximum, let $\tan\theta = x$, and we have to find x that maximizes $x/\left(1 + \left(n_o^2x^2\right)/n_e^2\right)$. Differentiating with respect to x and equating to zero (use $v\mathrm{d}u - u\mathrm{d}v = 0$), we obtain

$$\left[1 + \frac{n_o^2}{n_e^2}x^2\right] - x\left[\frac{n_o^2}{n_e^2}2x\right] = 0,$$

$$1 - \frac{n_o^2}{n_e^2}x^2 = 0,$$

$$x = \frac{n_e}{n_o}. \tag{12.66}$$

α is maximum when $\tan\theta = n_e/n_o$:

$$\tan(\alpha_{max}) = \frac{n_e/n_o\left(1 - n_o^2/n_e^2\right)}{1 + \left(n_o^2/n_e^2\right)\left(n_e^2/n_o^2\right)} = \frac{n_e/n_o - n_o/n_e}{2} = \frac{n_e^2 - n_o^2}{2n_on_e} = \frac{(n_e + n_o)(n_e - n_o)}{2n_on_e},$$

$$\tan(\alpha_{max}) = \frac{(n_e + n_o)(n_e - n_o)}{2n_on_e}. \tag{12.67}$$

For example, for quartz,

$$\tan(\alpha_{max}) = \frac{(1.553+1.544)(1.553-1.544)}{2(1.553)(1.544)} = 5.812\times10^{-3},$$

$$\alpha_{max} = 0.333^{\circ}.$$

A further approximation can be made for the case $n_o \approx n_e$.
For $n_o \approx n_e$, $\tan\theta \approx 1$ for α to be maximum.
Hence, $\theta \approx 45°$ for α to be maximum.
From Equation 12.67, we obtain

$$\tan(\alpha_{max}) \approx \frac{2n_e(n_e-n_o)}{2n_o n_e} = \frac{n_e-n_o}{n_o} = \frac{\Delta n}{n} = \frac{0.009}{1.544} = 5.829\times10^{-3}.$$

$\alpha_{max} = 0.334°$ nearest to the previous answer.
$\tan(\alpha_{max})$ is proportional to $|n_o - n_e|$.

12.6 Reflection by an Anisotropic Half-Space

When the transmission medium is anisotropic, we have seen that there are two transmitted waves (Figure 12.4). For oblique incidence, each transmitted wave has its own transmission angle. Since the boundary condition is still that k_x is conserved for all waves, Snell's law may be written as

$$k_x = k_o \sin\theta_i = k_1 \sin\theta_{T1} = k_2 \sin\theta_{T2}, \tag{12.68}$$

where k_1 and k_2 are the wave numbers of the respective transmitted waves. We have noted that for the extraordinary wave, the k-surface is not a sphere. In such a case, the value of k

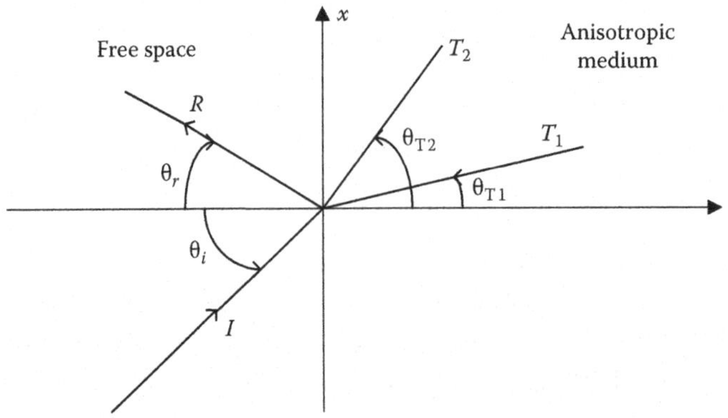

FIGURE 12.4
Reflection and transmission by an anisotropic medium.

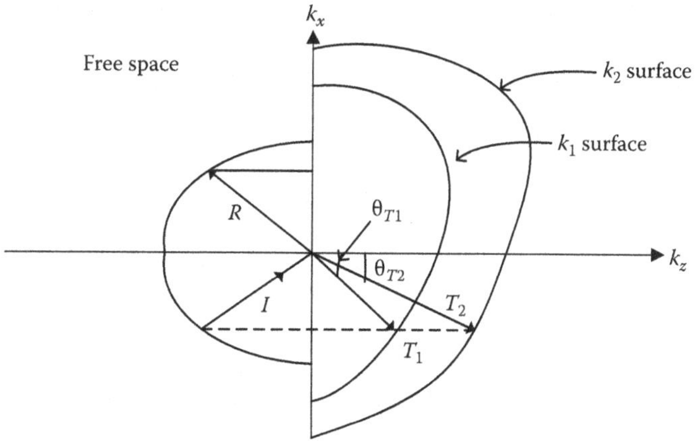

FIGURE 12.5
Computation of the angle of transmission.

depends on the angle of refraction. A graphical interpretation of the technique of obtaining the solution is shown in Figure 12.5.

References

1. Saleh, B. E. A. and Teich, M. C., *Fundamentals of Photonics*, Wiley, New York, 1991.
2. Yariv, A. and Yeh, P., *Optical Waves in Crystals*, Wiley, New York, 1984.
3. Kong, J. U., *Electromagnetic Wave Theory*, Wiley, New York, 1990.

13

Time-Domain Solutions

13.1 Introduction

Section 1.3 briefly dealt with the solution of Maxwell's equations in time domain. A simple example of obtaining the radiated fields when a point electric dipole is excited by an impulse current was discussed. Most of the rest of the chapters till now dealt with time-harmonic fields using the concept of phasors. We assumed that the transient response when needed can be obtained from the frequency response using a superposition integral. Using the Laplace transform technique and replacing ($j\omega$) by the Laplace transform variable s in the frequency response, we obtained the impulse response of a plasma half-space when excited by an impulse s wave in Section 9.9.

In this chapter, we will explore in a systematic way the direct solution of the Maxwell's equations in the time domain. Let us begin with the solution of the one-dimensional scalar wave equation for the scalar potential $\psi(z,t)$ in free space given by (1B.38) and repeated here as (13.1):

$$\frac{\partial^2 \psi(z,t)}{\partial z^2} - \frac{1}{c^2}\frac{\partial^2 \psi(z,t)}{\partial t^2} = 0. \tag{13.1}$$

With zero initial conditions, we showed that the scalar potential is given by (1B.43) and repeated here as (13.2):

$$\psi(z,t) = f_1\left(t - \frac{z}{c}\right) + f_2\left(t + \frac{z}{c}\right). \tag{13.2}$$

The functions $f_1(t)$ and $f_2(t)$ will be determined by the two boundary conditions needed to solve the second-order partial differential equation (PDE) (13.1). We illustrated through Figure 1B.4 that the first term on the right-hand side (RHS) of (13.2) is a forward-going impulse wave traveling with a velocity c in the positive z-direction, if $f_1(t)$ is an impulse $\delta(t)$. The voltage of an ideal transmission line satisfies a wave equation similar to (13.1) and is given by (2.52). In the next section, we will exploit the interpretation of the solution (13.2) to study the propagation of pulses on a bounded transmission line excited at the left end of the line and loaded at the right end.

13.2 Transients on Bounded Ideal Transmission Lines

Figure 13.1 shows an ideal transmission line, excited by a source at the source end $z = 0$, which is switched on at $t = 0$. The source is modeled by an ideal voltage source whose internal impedance is R_g. The $z = l$ is the load end of the transmission line and R_L is the load resistance.

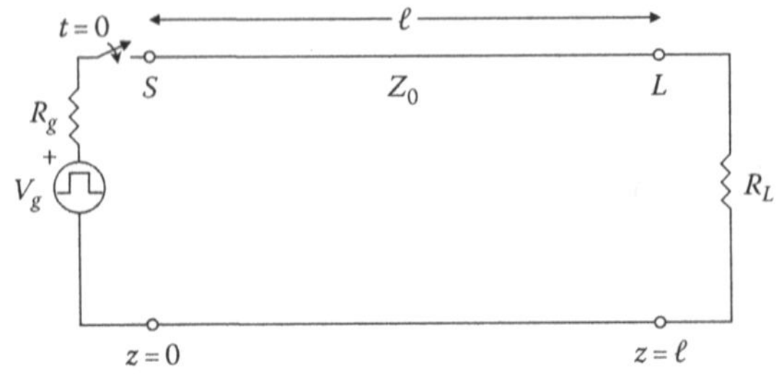

FIGURE 13.1
Geometry to study the transients on a transmission line.

The wave equation for voltage $V(z, t)$ of the ideal transmission line is given by (2.52). Figure 13.1 is the modified version of Figure 2.5, modified to study the propagation of transient waves on ideal transmission lines as opposed to the steady-state analysis of Section 2.8. The solution can be written as

$$V(z,t)=V^{+}\left(t-\frac{z}{v}\right)+V^{-}\left(t+\frac{z}{v}\right), \tag{13.3}$$

where v is given by (2.53). It can be shown from (2.51) that the current $I(z, t)$, with $G' = 0$, is given by

$$I(z,t)=I^{+}\left(t-\frac{z}{v}\right)+I^{-}\left(t+\frac{z}{v}\right)=\frac{1}{Z_0}\left[V^{+}\left(t-\frac{z}{v}\right)-V^{-}\left(t+\frac{z}{v}\right)\right], \tag{13.4}$$

where, with $R'=0$,

$$Z_0=\frac{1}{C'v}=\sqrt{\frac{L'}{C'}}. \tag{13.5}$$

Note the negative sign in front of the last term in (13.4), giving the correct sign for the power VI in the negative-going wave. To summarize,

$$V^{+}(z,t)=Z_0I^{+}(z,t) \tag{13.6}$$

$$V^{-}(z,t)=-Z_0I^{-}(z,t). \tag{13.7}$$

13.2.1 Step Response for Resistive Terminations

Several examples follow to illustrate as well as establish the well-known techniques of bookkeeping of the effect of reflections of the waves at the source end and the load end. However, it is instructive to have a qualitative discussion of the broad principles that govern the transient response of bounded transmission line problem as given in Figure 13.1. For the purpose of this qualitative discussion, let us assume that the source is a battery of voltage V_0 and the transmission line is uncharged with both voltage and current zero for $t = 0^-$. When the switch is closed at $t = 0$, the source end S will be immediately, at $t = 0^+$, excited with a step function, that is, $V(0, 0^+)= Au(t)$. The amplitude A

has to be determined taking into account the effect of R_g. Since the excited positive-going wave $V^+(z, t)$ travels with a finite velocity and the line is uncharged, it only sees the impedance Z_0, and the current $I^+(z, t)$ will be $V_g/(R_g + Z_0)$ and $A = V_g Z_0/(R_g + Z_0)$. Thus, a positive-going wave $Au(t - z/v)$ is launched on the line. When this wave reaches the load end in a period $T = l/v$, it encounters the lumped element load with a specified voltage–current relationship, which may not be satisfied by the voltage–current relationship of the positive-going wave. We can consider the positive-going wave as the incident wave at the load end. This imbalance is resolved by superposition of a suitable negative-going wave $V^-(z, t)$, which we can call as reflected wave at the load. The ratio of the voltage of the reflected wave to that of the incident wave at the load end can be defined as Γ_L^V. Similar ratio for currents, defined as Γ_L^I, can easily be shown to be the negative of Γ_L^V. At $t = 2T$, the wave reaches the source end S. Now the negative-going wave becomes the incident wave at the source end, and to satisfy the voltage–current relationship at the source end, another positive-going wave will be generated. The parameters Γ_S^V and Γ_S^I will relate the voltage and current of the newly generated positive-going wave. This process of bouncing of waves at both the ends describes qualitatively the successive generation and traveling of the positive-going and negative-going waves on the bounded transmission line. The expressions for the reflection coefficients described above can be obtained from the boundary conditions as given below.

At the load end L,

$$R_L I_L(t) = V_L(t) \tag{13.8}$$

$$\frac{R_L}{Z_0}\left[V^+(l,t) - V^-(l,t)\right] = \left[V^+(l,t) + V^-(l,t)\right]. \tag{13.9}$$

From (13.8) and (13.9), the expressions for the reflection coefficients can be obtained:

$$\Gamma_L^V = \frac{V^-(l,t)}{V^+(l,t)} = \frac{R_L - Z_0}{R_L + Z_0} \tag{13.10}$$

$$\Gamma_L^I = \frac{I^-(l,t)}{I^+(l,t)} = \frac{V^-(l,t)/(-Z_0)}{V^+(l,t)/(Z_0)} = -\Gamma_L^V. \tag{13.11}$$

At the source end S,

$$V_s(t) = V_g(t) - R_g I_s(t) = V^+(0,t) + V^-(0,t) \tag{13.12}$$

$$I_s(t) = \left[I^+(0,t) + I^-(0,t)\right] = \frac{1}{Z_0}\left[V^+(0,t) - V^-(0,t)\right]. \tag{13.13}$$

Solving for $V^+(0,t)$,

$$V^+(0,t) = \frac{Z_0}{R_g + Z_0} V_g(t) + \frac{R_g - Z_0}{R_g + Z_0} V^-(0,t). \tag{13.14}$$

The first term on the RHS of (13.14) can be identified with the amplitude A of the unit step mentioned above and the coefficient of the second term on the RHS of (13.14) with Γ_s^V:

$$\Gamma_s^V = \frac{R_g - Z_0}{R_g + Z_0}. \tag{13.15}$$

These definitions of the reflection coefficients in the time domain are in line with our understanding of the definitions of the reflection coefficients in Section 2.8. The source end definitions are in accordance with Thevenin's theorem.

A bounce diagram, shown in Figure 13.2, is a useful technique to keep track of effects of the successive reflections from the load and source ends. The horizontal axis is the z-axis and the vertical axis is the t-axis. At $t = 0^+$, a positive-going step wave of voltage function V_1^+ of amplitude A is launched, and it travels on the line with a velocity v as shown by the straight line with a slope v. It reaches the load end L in a time period $T = l/v$. It is this incident wave that encounters the impedance mismatch and gives rise to a reflected wave whose voltage is denoted by $V_1^- = \Gamma_L^V V_1^+$. At $t = 2T$, this wave reaches the source end and will be reflected at the source end giving rise to $V_2^+ = \Gamma_L^V \Gamma_S^V V_1^+$. At $t = 3T$, a second reflection takes place at the load end giving rise to $V_2^- = \Gamma_L^V \left(\Gamma_L^V\right)^2 V_1^+$.

The voltages of the successive reflected waves at the source and load ends are as shown in Figure 13.2. It is now easy to show [1] that

$$V(z,\infty) = V_1^+ \frac{1+\Gamma_L^V}{1-\Gamma_L^V \Gamma_S^V}. \tag{13.16}$$

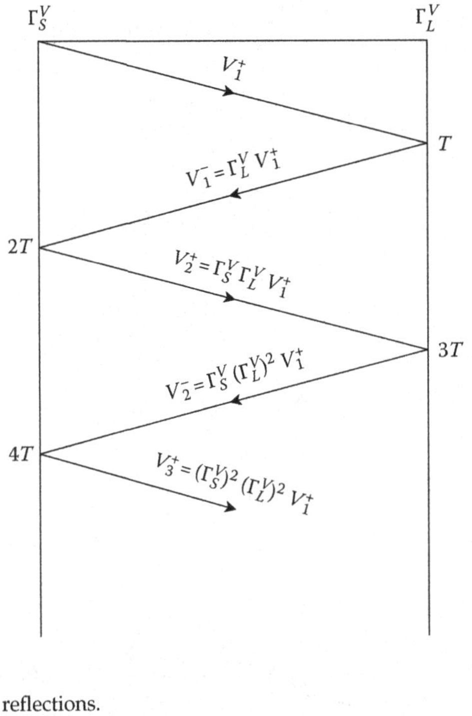

FIGURE 13.2
Bounce diagram of successive reflections.

From (13.10), (13.15), and the expression for $V_1^+ = A = \dfrac{Z_0}{R_g + Z_0} V_g(t)$, we obtain

$$V(z,\infty) = V_g \frac{R_L}{R_g + R_L}. \tag{13.17}$$

Equation 13.17 confirms that in steady state, the voltage on the transmission line at any z is the same as the voltage drop across the load resistance, as expected when the source is a DC voltage V_g and the transmission line acts merely as connecting wire.

A bounce diagram for the current can be drawn on similar lines as above using the current reflection coefficients instead of the voltage reflection coefficients. From (13.11), the current reflection coefficient at the load end L is the negative of the voltage reflection coefficient. Similar statement is also true for the source end.

An alternative to the bounce diagram is a graphical method called Bergeron diagram [2–4], to represent the effects of successive reflections. Bergeron diagram is not that useful for linear loads and sources but is the best way to treat nonlinear loads and sources. However, its explanation is facilitated by looking at linear loads and sources. Figure 13.3 shows the relevant graphs for the linear case.

The horizontal axis is a quantity proportional to the current and labeled as $Z_0 I$. The vertical axis is the voltage. The two solid lines marked as S (source end) line and L (load end) line are the linear equations relating the V–I relationships at S and L points on the transmission lines. The operating point (intersection point of these two straight lines) represents the voltage, and current on the line in the limit t approaches infinity, when the interconnects (transmission line) act merely as connecting wires. However, the operating point is reached by successive reflections at the source and load ends.

The point S_1 applicable at $t = 0$ is the intersection of the S-solid line with the dotted 45° line starting from the origin. The equations are

$$V_S = V_g - \frac{R_g}{Z_0}(Z_0 I_S) \quad S\text{—solid line} \tag{13.18}$$

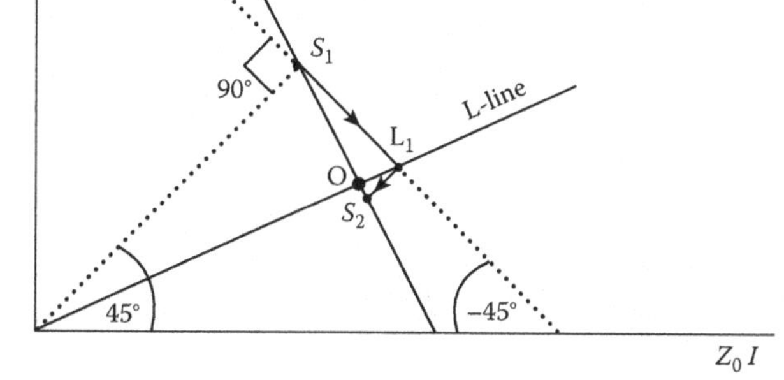

FIGURE 13.3
Bergeron diagram for a linear load.

$$V^+ = Z_0 I^+ \quad 45^\circ \text{ line.} \tag{13.19}$$

The point L_1 applicable at $t = T$ is located as the intersection of the two lines with the following two equations:

$$V_L = \frac{R_L}{Z_0}(Z_0 I_S) \quad L\text{—solid line} \tag{13.20}$$

$$V^- = -Z_0 I^- \quad -45^\circ \text{ line.} \tag{13.21}$$

The broken straight line S_1L_1 in Figure 13.3 is drawn from S_1 and continuing at -45° till we reach the intersection point L_1 on the solid L-line. We next draw L_2S_2 at 45° starting from L_1 and ending at S_2 on the S-solid line. Point S_2 represents the voltage–current relationship at $t = 2T$. We continue this process of locating L_2, S_3, L_3, etc., till we reach the operating point within a prescribed error or reach the limitation of the accuracy of the graphical method.

Some numerical examples are given below. In these examples, $T = l/v$.

Example 13.1

In Figure 13.1, assume $R_g = 75$ ohms (Ω), $R_L = 100$ Ω, and the source is a step voltage of 100 V. Let the length of the line l be 10 cm and the velocity of the traveling waves on the line as $2/3c$, where c is the velocity of light. Let Z_0 be 50 Ω. Let the initial conditions on the transmission line be zero. Sketch $V(0.25l, t)$.

Solution

We calculated $T = 0.5$ ns, $\Gamma_S^V = 1/5, \Gamma_L^V = 1/3$. We can now fill in the bounce diagram with numbers using the formulas developed above.

$V_1^+ = 40, V_1^- = 40/3, V_2^+ = 8/3, V_2^- = 8/9, V_3^+ = 8/45, \ldots V(z,\infty) = 400/7 = 57.07$ V. From the bounce diagram, we can sketch $V(0.25l, t)$ versus t as shown in Figure 13.4 and it reaches 56 at $t = 9T/4$, that is $t = 1.125$ ns. If we continue to graph further, the voltage will reach the asymptotic value of 57.07 V.

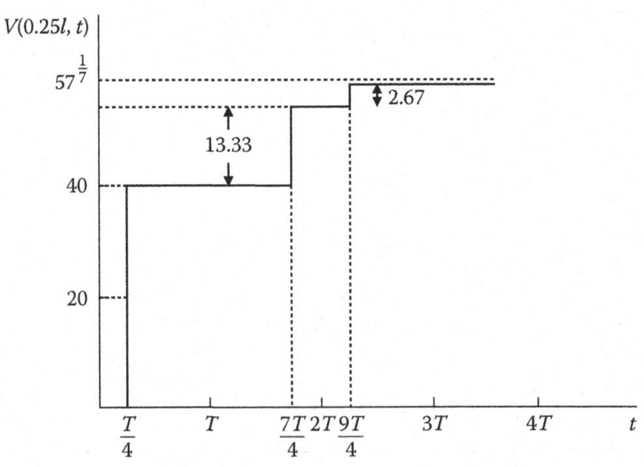

FIGURE 13.4
Sketch of the voltage $V(0.25l, t)$ for the data of Example 13.1.

Example 13.2

Assume the same data as for Example 13.1 except $R_L = Z_0 = 50\ \Omega$. Load end is matched to the line. Sketch $V(0.25l, t)$ and $I(0.25l, t)$.

Solution

$$V_1^+ = 40\ \text{V},\quad \Gamma_L^V = 0,\quad V_1^- = 0,$$

$$I_1^+ = 0.8\ \text{A},\quad \Gamma_L^I = 0.$$

$$V(0.25l,\infty) = 40\ \text{V},\quad I(0.25l,\infty) = 0.8\text{A}.$$

The sketches $V(0.25l, t)$ and $I(0.25l, t)$ are shown in Figure 13.5.

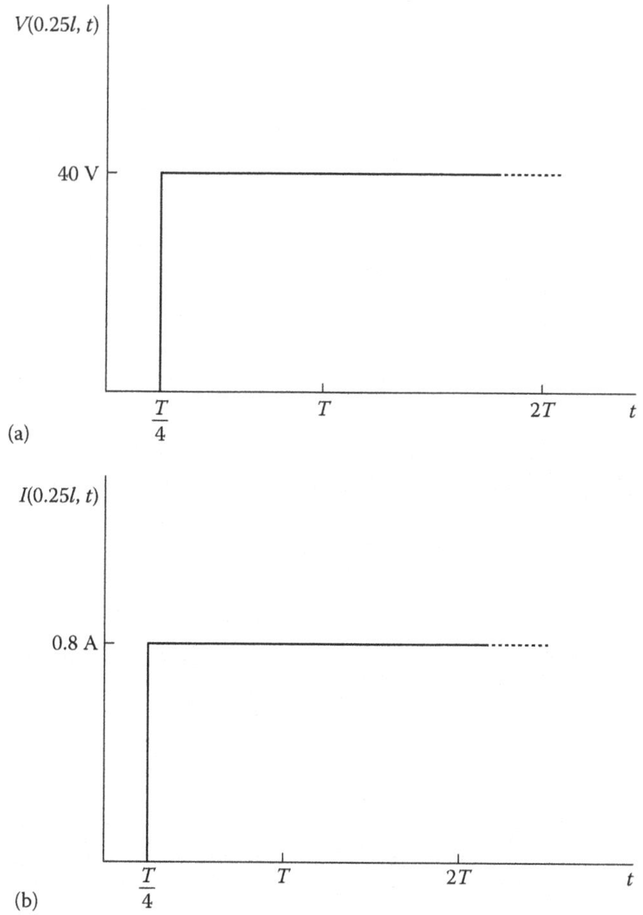

FIGURE 13.5
Sketches of (a) voltage and (b) current at $z = l/4$ for the data of Example 13.2.

Example 13.3

Assume the same data as for Example 13.1 except $R_g = Z_0 = 50\ \Omega$. The source end is matched to the line. Sketch $V(0.25l, t)$ and $I(0.25l, t)$.

Solution

$$V_1^+ = 50,\quad \Gamma_L^V = 1/3,\quad V_1^- = 50/3,\quad \Gamma_S^V = 0,\quad V_2^+ = 0,$$

$$I_1^+ = 1\ \text{A},\quad \Gamma_L^I = -1/3,\quad I_1^- = 1/3\ \text{A},\quad \Gamma_S^I = 0,\quad I_2^+ = 0$$

$$V(z,\infty) = 200/3\ \text{V},\quad I(z,\infty) = 2/3\ \text{A}.$$

The sketches $V(0.25l, t)$ and $I(0.25l, t)$ are shown in Figure 13.6.

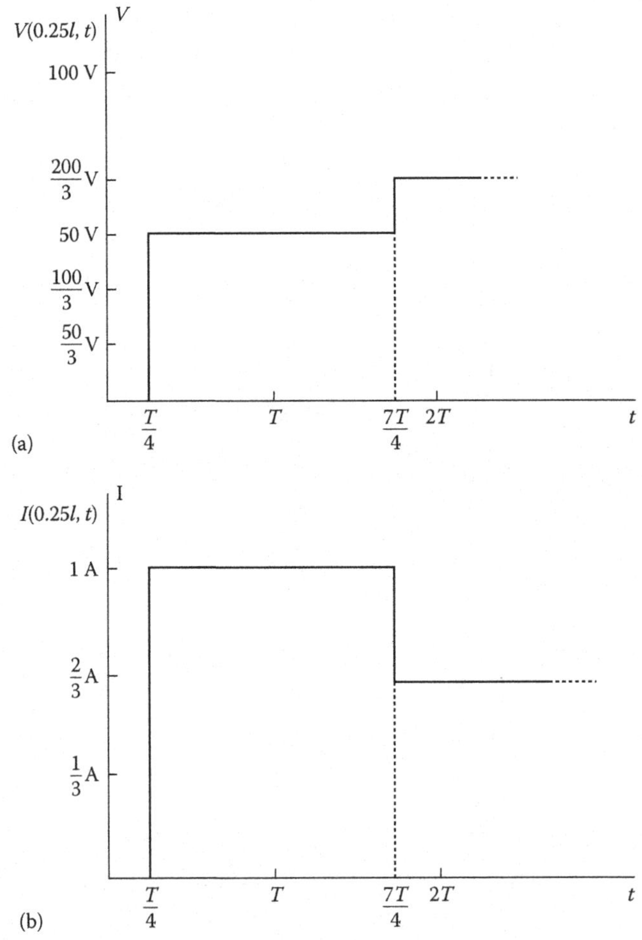

FIGURE 13.6
Sketches of (a) voltage and (b) current at $z = l/4$ for the data of Example 13.3.

Example 13.4

Assume the same data as for Example 13.1 except $R_S = Z_0 = R_L = 50\ \Omega$. The source end as well as the load end is matched to the line. Sketch $V(0.25l, t)$ and $I(0.25l,t)$.

Solution

$$V_1^+ = 50\ \text{V},\quad \Gamma_L^V = 0,\quad V_1^- = 0$$

$$I_1^+ = 1\ \text{A},\quad \Gamma_L^I = 0,\quad I_1^- = 0$$

$$V(z,\infty) = 50\ \text{V},\quad I(z,\infty) = 1\ \text{A}.$$

The sketches $V(0.25l, t)$ and $I(0.25l, t)$ are shown in Figure 13.7.

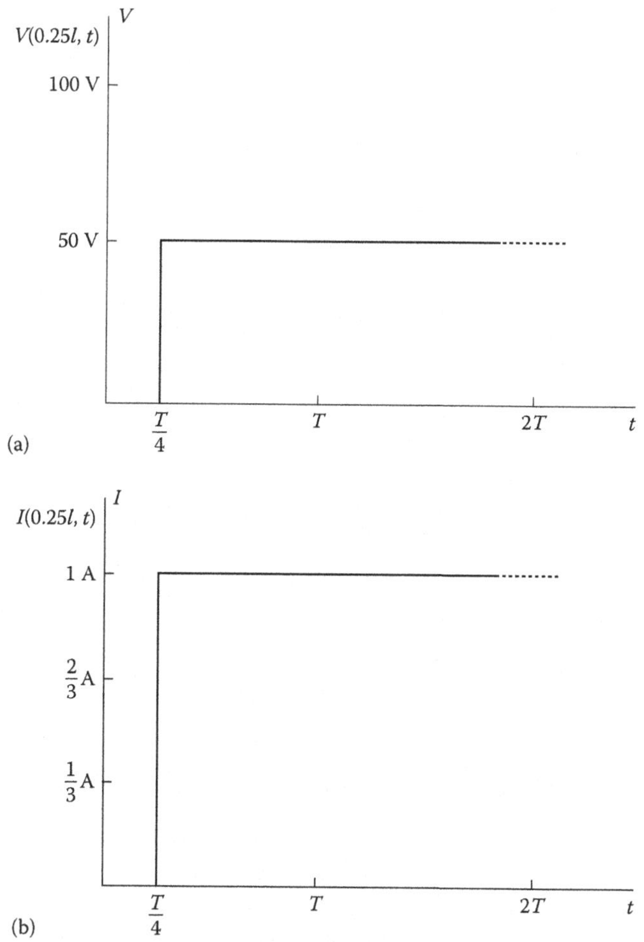

FIGURE 13.7
Sketches of (a) voltage and (b) current at $z = l/4$ for the data of Example 13.4.

Example 13.5

Assume the same data as for Example 13.3 except that $R_L = \infty$. Sketch $V(0, t)$ and $I(0, t)$.

Solution

$$V_1^+ = 50\ \text{V},\quad \Gamma_L^V = 1,\quad V_1^- = 50\ \text{V},\quad \Gamma_S^V = 0,\quad V_2^+ = 0$$

$$I_1^+ = 1\ \text{A},\quad \Gamma_L^1 = -1,\quad I_1^- = -1\ \text{A},\quad \Gamma_S^I = 0,\quad I_2^+ = 0$$

$$V(z,\infty) = 100\ \text{V},\quad I(z,\infty) = 0.$$

The sketches $V(0, t)$ and $I(0, t)$ are shown in Figure 13.8.

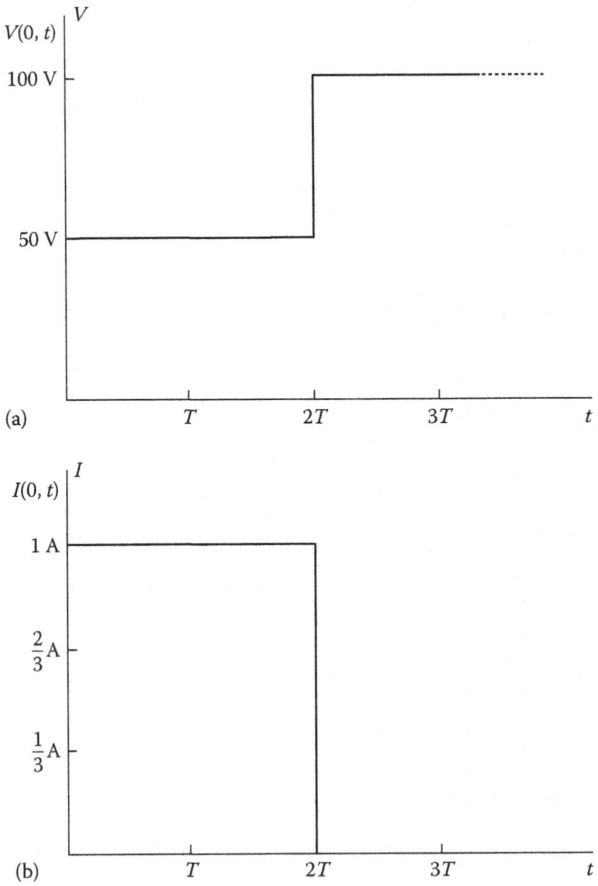

FIGURE 13.8
Sketches of (a) voltage and (b) current at $z = 0$ for the data of Example 13.5. At the load end, there is an open circuit.

Example 13.6

Assume the same data as for Example 13.3 except that $R_L = 0$. Sketch $V(0, t)$ and $I(0, t)$.

Solution

$$V_1^+ = 50 \text{ V}, \quad \Gamma_L^V = -1, \quad V_1^- = -50 \text{ V}, \quad \Gamma_S^V = 0, \quad V_2^+ = 0$$

$$I_1^+ = 1 \text{ A}, \quad \Gamma_L^I = 1, \quad I_1^- = 1 \text{ A}, \quad \Gamma_S^I = 0, \quad I_2^+ = 0$$

$$V(z,\infty) = 0, \quad I(z,\infty) = 2.$$

The sketches $V(0, t)$ and $I(0, t)$ are shown in Figure 13.9.

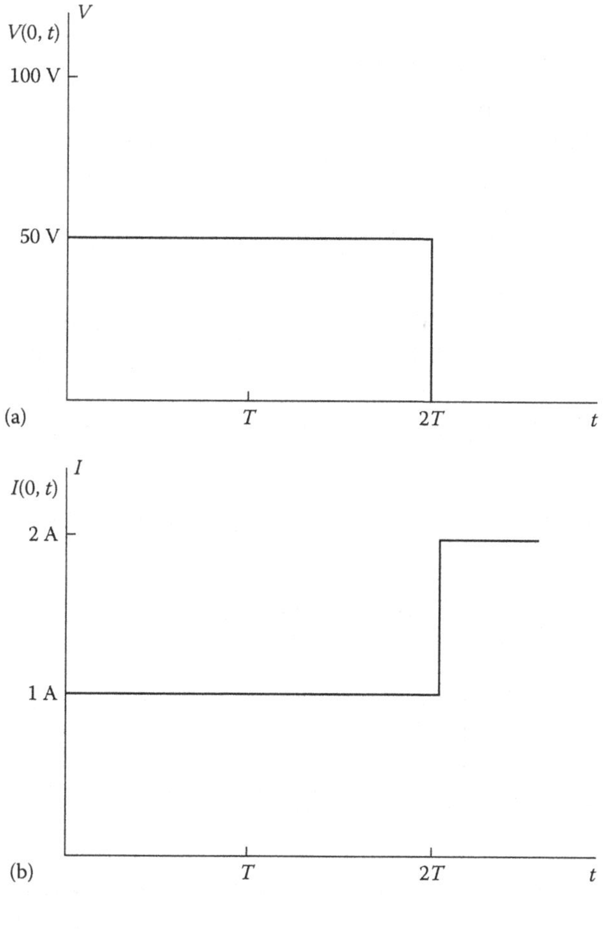

FIGURE 13.9
Sketches of (a) voltage and (b) current at source end for the data of Example 13.6. The source end is matched to the line but the load end is shorted.

Example 13.7

Assume the data of Example 13.4 of a transmission line matched at the source as well as the load end. However, in this example, we will consider the effect of a fault, represented by a parallel fault resistance R_f of unknown value, on the line at an unknown distance d from the source end. The sketch of the voltage $V(0, t)$ is as given in Figure 13.10. Determine d and R_f.

Solution

From Figure 13.10, we note that the change in the voltage at the source end occurs at $t = T/3$, which is the delay time for the signal to travel from the source to the fault and the reflected wave generated at the fault to travel from the fault to the source end. Hence,

$$2d/v = T/3 = (l/v)/3$$

$$d = l/6.$$

At $z = d, \Gamma_f^V = -(50 - 20)/50 = -3/5 = (Z_d - Z_0)/(Z_d + Z_0)$, solving which we get $Z_d = (1/4)Z_0$.
Since Z_d is the impedance of the parallel combination of R_f and Z_0,
$(R_f Z_0)/(R_f + Z_0) = Z_0/4$, solving which we get $R_f = Z_0/3 = 50/3\ \Omega$.

13.2.2 Response to a Rectangular Pulse

A rectangular pulse can be considered as the superposition of a unit step function with the negative of a delayed unit step function. Let us assume that the source voltage in Figure 13.1 is given by

$$V_g(t) = u(t) - u(t - t_s). \tag{13.22}$$

One can draw two bounce diagrams for each of the sources $u(t)$ and $u(t - ts)$ and obtain the voltage at any point on the line as a function of time by appropriately combining the bounce diagrams. If the line is matched at the source end as well as the load end, the rectangular pulse of half the amplitude will be delivered to the load. There will be no further reflections, since the reflection coefficients at both the ends are zero.

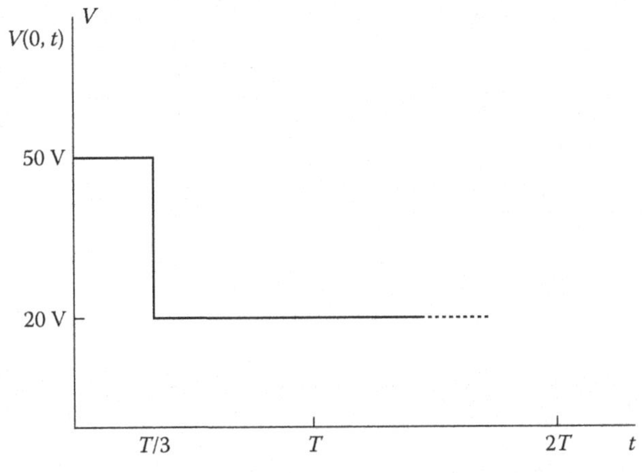

FIGURE 13.10
Voltage at the sending end for the data of Example 13.7. The line is matched to the load as well as the source, but a fault occurs at an unknown location.

13.2.3 Response to a Pulse with a Rise Time and Fall Time

If the line is matched at both the ends appropriately, the pulse will be delivered faithfully to the load. However, if in addition to the load resistance of Z_0, the load terminals might give rise to parasitic capacitance in parallel with Z_0. In this case, the pulse will be distorted. One can study the effect of the capacitance on the voltage at the source end as well as the load end. We will give next an example simplifying the problem by assuming that the source is a step-like function with a ramp transition [4] defined in the example. We will use the Laplace transform technique.

13.2.4 Source with Rise Time: Response to Reactive Load Terminations

To illustrate the effect of the rise time and parasitic capacitance at the load terminal, we will consider the following example.

Example 13.8

Assume the data of Example 13.4 of a transmission line matched at the source as well as the load end. However, we will make two changes: (a) The source is a step-like function with a ramp transition shown in Figure 13.11, defined as follows:

$$V_g(t)=0, \quad t<0$$
$$V_g(t)=t/a, \quad 0<t<a \tag{13.23}$$
$$V_g(t)=1, \quad t>a.$$

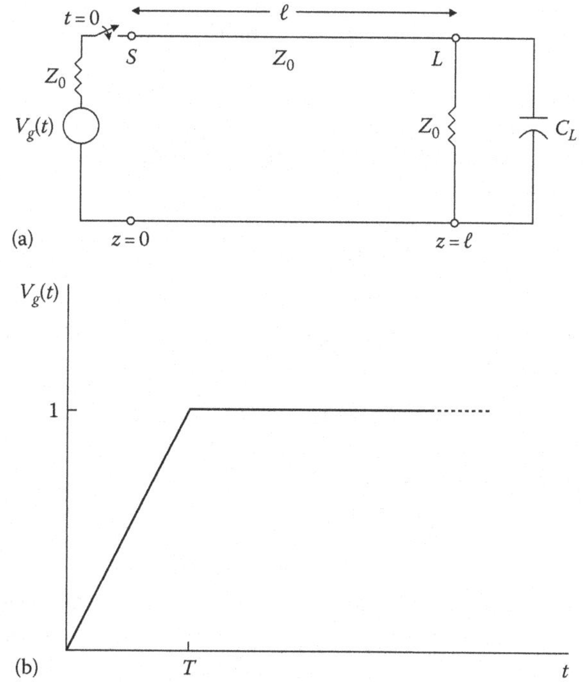

FIGURE 13.11
Effect of parasitic capacitance at the load end of a matched transmission line. Excited by a step-like source with a ramp transition (a) sketch of the transmission line and (b) sketch of the exciting pulse. See Example 13.8.

The load is a parallel combination of Z_0 and capacitance C_L.

Solve analytically using the Laplace transform technique and also sketch the voltage waveform at the load end L and source end S.

Solution

The load $Z_L(s)$, which is a parallel combination of Z_0 and $1/C_L s$, can be written as

$$Z_L(s)=\frac{Z_0}{Z_0 C_L s+1}. \tag{13.24}$$

We can now calculate the voltage reflection coefficient and the voltage transmission coefficient T_L^V at the load end:

$$\Gamma_L^V=\frac{Z_L(s)-Z_0}{Z_L(s)+Z_0} \tag{13.25}$$

$$T_L^V(s)=1+\Gamma_L^V=\frac{1/t_L}{s+1/t_L}, \tag{13.26a}$$

where

$$t_L=\frac{Z_0 C_L}{2}. \tag{13.26b}$$

The voltage at the source end S is of $V_g(t)$ given in (13.23) and can be expressed as the difference between two ramp functions:

$$V_g(t)=\frac{t}{a}u(t)-\frac{t-a}{a}u(t-a). \tag{13.27}$$

The Laplace transform of the first term on the RHS of (13.27) is obtained by making use of the Laplace transform of ramp function and the Laplace transform of the second term by using the shifting theorem. We thus obtain the Laplace transform of the voltage at the source end:

$$V_S(s)=\frac{1}{2as^2}\left(1-e^{-as}\right). \tag{13.28}$$

The bounce diagram in the Laplace transform domain is shown in Figure 13.12a. The load voltage in s domain is given by

$$V_L(s)=V_S(s)e^{-Ts}\left[1+\Gamma_L^V(s)\right]. \tag{13.29}$$

Substituting (13.28) and (13.26a) in (13.29),

$$V_L(s)=\frac{1}{2as^2}\left(1-e^{-as}\right)e^{-Ts}\frac{1/t_L}{s+1/t_L}. \tag{13.30}$$

The Laplace-transformed term (13.31), which appears as part of the RHS of (13.30), can be easily inverted by residue technique or partial fraction expansion by simplifying as given below:

$$\frac{1}{s^2}\frac{1/t_L}{s+1/t_L}=\frac{1}{s^2}-\frac{t_L}{s}+\frac{t_L}{s+1/t_L}. \tag{13.31}$$

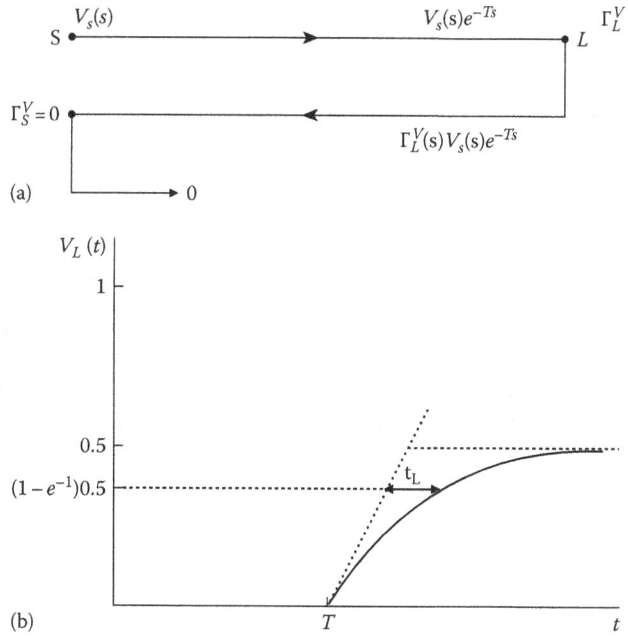

FIGURE 13.12
(a) Bounce diagram for Example 13.8, (b) sketch of load voltage for Example 13.8.

Substituting (13.31) in (13.30) and inverting

$$V_L(t) = \frac{1}{2a}\left\{(t-T) - t_L\left[1 - e^{-(t-T)/t_L}\right]\right\}u(t-T)$$
$$-\frac{1}{2a}\left\{(t-T-a) - t_L\left[1 - e^{-(t-T-a)/t_L}\right]\right\}u(t-T-a). \quad (13.32)$$

The load voltage as a function of time is sketched in Figure 13.12b. If the capacitance is not present, due to the matching at both the ends, the load end voltage $V_L(t)$ will be half the amplitude of $V_g(t)$. The front arrives at $t = T$ as shown with broken line curve in Figure 13.12b.

Note that the broken line curve retains the shape of a step-like function with a ramp transition. The solid curve shows the effect of the parasitic capacitance C_L. The ramp transition is bent into an exponential curve adding t_L given by (13.26b) to the delay when the voltage is $0.5 \times (1 - e^{-1})$.

13.2.5 Response to Nonlinear Terminations

As the clock speed of the computers increases into 10's of GHz, the interconnect lengths of even centimeters will necessitate treating the interconnects as transmission lines (Chapter 4 of Reference 4). It is thus of interest to find a way to determine the evolution of the voltage or current pulse when the load is nonlinear, for example, a diode [3]. It is possible that both the source and the load have nonlinear V–I characteristics, for example, logic gates [4], where the interconnect has to be treated as a transmission line.

Example 13.9

Let the load in Figure 13.11 be a nonlinear resistor and the generator voltage be a step voltage as shown in Figure 13.13a.

Draw a Bergeron diagram to determine the evolution of the source end voltage V_S and the load end voltage V_L.

Solution

The diagram is shown in Figure 13.13b and is similar to Figure 13.3 except that the load line (L-line) is a curved line instead of a straight line.

The solid L-line is the V–I characteristic of the nonlinear resistor, and the solid line labeled as S-line is the V–I characteristic of the source terminal similar to the S-line in Figure 13.3. The dashed curves have slopes of plus or minus 45 degrees. The points S_1, S_2, etc. are the successive intersection points with the S-line and the 45-degree line. The points L_1, L_2, etc. are the intersection points of the L-line with the minus 45-degree line. The steady-state point (t approaches infinity) coincides with the intersection point of the S-line with the L-line, as expected.

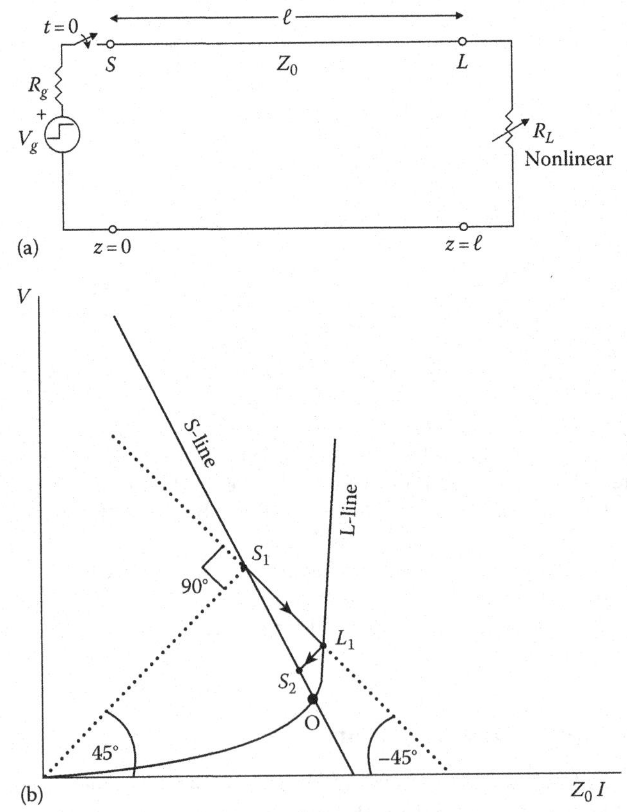

FIGURE 13.13
(a) Transmission line with a nonlinear resistive load. (b) Bergeron diagram for a nonlinear resistive load.

13.2.6 Practical Applications of the Theory [1,2]

The simple theory presented above can be used to explain the principles of practical measuring and diagnostic devices to detect faults in inaccessible transmission lines like underground cables. The device is called time-domain reflectometer. Its principle is illustrated through Example 13.7. Some more examples are given as problems at the end of the chapter. More details can be obtained from [1,2].

13.3 Transients on Lossy Transmission Lines

Equations 2.29 and 2.51 give the time-domain first-order coupled differential equations for the voltage $V(z, t)$ and the current $I(z, t)$ of a transmission line. These equations in s domain are given by

$$-\frac{\partial V(z,s)}{\partial z} = (R' + sL')I(z,s) - L'I_0(z) \tag{13.33}$$

$$-\frac{\partial I(z,s)}{\partial z} = (G' + sL')I(z,s) - C'V_0(z), \tag{13.34}$$

where $I_0(z)$ is the initial current distribution and $V_0(z)$ is the initial voltage distribution over the transmission line. Capital letters V and I are used for the time-domain functions as well as the s domain functions and where necessary, to avoid confusion, the argument of t or s is given in the parenthesis.

Assuming the line parameters are constant, we can obtain the second-order differential equations for the voltage:

$$\begin{gathered}\frac{d^2V(z,s)}{dz^2} - k^2(s)V(z,s) = L'\frac{d}{dz}I_0(z) - C'(R' + sL')V_0(z) \\ k^2(s) = (R' + sL')(G' + sC').\end{gathered} \tag{13.35}$$

The solution of this second-order ordinary differential equation can now be obtained [5], the homogeneous part being

$$V_h(z,s) = V_0^+(s)e^{-k(s)z} + V_0^-(s)e^{+k(s)z} \tag{13.36}$$

and the particular integral being

$$V_p(z, s) = \frac{1}{2k(s)}\left[e^{k(s)z}\int e^{-k(s)z}M_v(z)dz - e^{-k(s)z}\int e^{k(s)z}M_v(z)dz\right], \tag{13.37}$$

where

$$k(s) = \frac{1}{v_0}\sqrt{\left[[(s+\mu]^2 - \nu^2\right]} \tag{13.38}$$

$$\mu = \frac{1}{2}\left[\frac{R'}{L'} + \frac{G'}{C'}\right] \tag{13.39}$$

$$\nu = \frac{1}{2}\left[\frac{R'}{L'} - \frac{G'}{C'}\right] \tag{13.40}$$

$$v_0 = \sqrt{\frac{1}{L'C'}} \tag{13.41}$$

and

$$M_v(z) = L'\frac{d}{dz}I_0(z) - C'(R' + sL')V_0(z). \tag{13.42}$$

In (13.38), (13.39), and (13.40), μ is not the permeability and ν is not the collision frequency. In this section, these symbols are used with a different meaning to be consistent with [6], whose approach is used to solve the problem by using the Laplace transform technique.

13.3.1 Solution Using the Laplace Transform Technique

Let us assume that the line does not have any initial energy.

To obtain the main results of such a solution, let us assume that the line has zero initial energy, that is,

$$V_0(z) = 0 \tag{13.43}$$

$$I_0(z) = 0. \tag{13.44}$$

Then the solution for $V(z, s)$ is given by (13.36). The solution for $I(z, s)$ is obtained by substituting (13.36) in (13.33) and is given by

$$I(z,s) = I^+(z,s) + I^-(z,s) \tag{13.45}$$

$$I^+(z,s) = \frac{V^+(z,s)}{Z_c(s)} \tag{13.46}$$

$$I^-(z,s) = -\frac{V^-(z,s)}{Z_c(s)}, \tag{13.47}$$

where

$$Z_c(s) = \sqrt{\frac{R' + sL'}{G' + sC'}} \tag{13.48}$$

$$Z_c(s) = Z_0\sqrt{\frac{s + (\mu + \nu)}{s + (\mu - \nu)}} \tag{13.49}$$

$$Z_0 = \frac{L'}{C'}. \tag{13.50}$$

13.3.1.1 Loss-Free Line

If the line is loss free ($R' = 0, G' = 0$), we recover the equations we used for the ideal transmission line, since

$$\mu = 0, \quad \nu = 0, \tag{13.51}$$

$$k(s) = \frac{s}{v_0}, \tag{13.52}$$

$$V(z,s) = V_0^+(s)e^{-sz/v_0} + V_0^-(s)e^{+sz/v_0}. \tag{13.53}$$

If $V_0^+(s)$ is the Laplace transform of $V_0^+(t)$ and $V_0^-(s)$ is the Laplace transform of $V_0^-(t)$, the time-domain solution can be obtained as

$$V(z,t) = V_0^+\left(t - \frac{z}{v_0}\right) + V_0^-\left(t + \frac{z}{v_0}\right). \tag{13.54}$$

The first term on the RHS of (13.54) is the positive-going wave and the second term is the negative-going wave.

13.3.1.2 Distortionless Line

The second simple case occurs when

$$\frac{R'}{L'} = \frac{G'}{C'} = \mu \quad \text{and} \quad \nu = 0, \tag{13.55}$$

which was the condition for distortionless transmission line. In such a case

$$k(s) = \frac{s + \mu}{v_0}. \tag{13.56}$$

The positive-going wave has the Laplace transform

$$V_0^+(z,s) = V_0^+(s)e^{-(s+\mu)z/v_0}$$

$$V_0^+(z,s) = V_0^+(s)e^{-\mu z/v_0}e^{-sz/v_0} \tag{13.57}$$

$$V_0^+(z,t) = e^{-\mu z/v_0}V_0^+\left(t - \frac{z}{v_0}\right). \tag{13.58}$$

As expected, the wave attenuates but travels without distortion.

13.3.1.3 Lossy Line

When nu is not equal to zero, the signal gets distorted as it travels. Hence, ν is called the distortion constant. To study this quantitatively, we will take the Laplace inverse of $\exp[-k(s)z]$, that is, we take $V_0^+(s) = 1$ and $V_0^+(t) = \delta(t)$.

The Laplace inverse of exp[−k(s)z], where k(s) is given by (13.38), can be written as the sum of *term*1 and *term*2 [6]:

$$term1 = e^{-\mu z/v_0}\delta\left(t - \frac{z}{v_0}\right) \tag{13.59}$$

$$term2 = \nu^2 a(z,t)\, u\left(t - \frac{z}{v_0}\right), \tag{13.60}$$

where

$$a(z,t) = \left(\frac{z}{v_0}\right) e^{-\mu t} \frac{I_1\left[\nu\sqrt{t^2 - (z/v_0)^2}\right]}{\nu\sqrt{t^2 - (z/v_0)^2}}. \tag{13.61}$$

In (13.60) u is the unit step function, and in (13.61) I is the modified Bessel function of the first kind. The term1 is the original impulse traveling along z with attenuation but no distortion and thus is the same solution as that of a distortionless line. The second term is the "wake" behind the distortionless wave. The wake is due to the dispersion and is sketched in Figure 13.14. The wake is zero for $t < z/v_0$. At $t = z/v_0$, it reaches a peak value:

$$\text{Peak value} = \frac{\nu^2}{2}\frac{z}{v_0} e^{-\mu z/v_0}. \tag{13.62}$$

For $t > z/v_0$, it decays with an asymptotic value given by Equation 4.44 of Reference [6] (Figure 13.14).

The solution for the current can be obtained on similar lines, and Reference [6] discusses this and other aspects of the solution of the lossy transmission line problem using the Laplace transform technique.

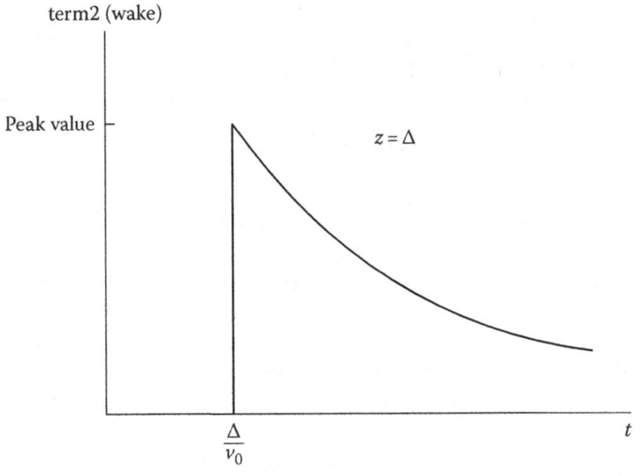

FIGURE 13.14
Sketch of the term2 given by (13.60), which is the wake behind the undistorted wave. The peak value is given by (13.62).

13.4 Direct Solution in Time Domain: Klein–Gordon Equation

From Equations 2.49 and 2.51, when the transmission line parameters are constant, one can obtain easily the second-order PDE for the voltage $V(z,t)$.

$$\frac{\partial^2 V}{\partial t^2} - v_0^2 \frac{\partial^2 V}{\partial z^2} + \left(\frac{R'}{L'} + \frac{G'}{C'}\right)\frac{\partial V}{\partial t} + \left(\frac{R'}{L'}\frac{G'}{C'}\right)V = 0. \tag{13.63}$$

The solution in the time domain can be obtained by the following substitution:

$$V(z,t) = e^{-\mu t}U(z,t) \tag{13.64}$$

in the time-domain equation for $V(z,t)$. The function $U(z,t)$ satisfies

$$\frac{\partial^2 U}{\partial t^2} - v_0^2 \frac{\partial^2 U}{\partial z^2} = \nu^2 U. \tag{13.65}$$

Equation 13.33 is like the well-known *Klein–Gordon* equation of mathematical physics and is used to model a number of physical problems. Two simple examples are discussed below.

13.4.1 Examples of Klein–Gordon Equation

A simple example is the model for the motion of a string of mass density ρ per meter length, under tension T, in an elastic medium like a thin rubber sheet of spring constant K. The PDE for the displacement ψ is given by [7]

$$\rho \frac{\partial^2 \psi}{\partial t^2} = T \frac{\partial^2 \psi}{\partial z^2} - K\psi. \tag{13.66}$$

This equation can be written as

$$\frac{\partial^2 \psi}{\partial z^2} - \frac{1}{v_s^2}\frac{\partial^2 \psi}{\partial t^2} = \mu_s^2 \psi, \tag{13.67a}$$

where

$$v_s^2 = \frac{T}{\rho} \tag{13.67b}$$

$$\mu_s^2 = \frac{K}{T}. \tag{13.67c}$$

The dispersion equation for harmonic variations in space and time is given by

$$\omega^2 = v_s^2 k^2 + \mu_s^2 v_s^2. \tag{13.68}$$

Equation 13.68 is similar to the dispersion relation for a plasma medium or a waveguide with the feature of a cutoff frequency ω_c. In this case,

$$\omega_c = \mu_s v_s. \tag{13.69}$$

If the string is pinned down at $z = 0$ and $z = L$ (Dirichlet BC), the solution can be easily found as

$$\psi = \psi_0 \sin\frac{n\pi z}{L} e^{j\omega_n t}, \tag{13.70}$$

where

$$\omega_n^2 = v_s^2\left[(n\pi/L)^2 + \mu_s^2\right]. \tag{13.71}$$

The allowed frequencies are larger because of the quantity μ_s^2, which is proportional to the spring constant of the elastic medium [7].

A second example is the modeling of propagation of a uniform plane wave in an isotropic plasma medium of plasma frequency ω_p. Equations 9.10 through 9.12 give the relevant equations for E, H, and J in the plasma medium. Eliminating H and J, we obtain

$$\frac{\partial^2 E}{\partial z^2} - \frac{1}{c^2}\frac{\partial^2 E}{\partial t^2} = \frac{\omega_p^2}{c^2} E. \tag{13.72}$$

The dispersion relation and the cutoff frequency in this case are given by

$$\omega^2 = k^2 c^2 + \omega_p^2 \tag{13.73}$$

$$\omega_c = \omega_p. \tag{13.74}$$

Equation 13.72 can be transformed into the standard form of the KG equation

$$\frac{\partial^2 f}{\partial \eta^2} - \frac{\partial^2 f}{\partial \tau^2} = f \tag{13.75}$$

by the following substitutions given in (13.76a), (13.76b), and (13.76c):

$$\eta = \frac{\omega_p z}{c} \tag{13.76a}$$

$$\tau = \omega_p t \tag{13.76b}$$

$$E = E_0 f. \tag{13.76c}$$

The nonsinusoidal time-domain exact solution of the KG equation (13.75) is given by Shvartsburg [8]:

$$f = \sum_q E_q f_q(\tau,\eta) \tag{13.77}$$

$$f_q(\tau,\eta) = \frac{1}{2}\left[g_{q-1}(\tau,\eta) - g_{q+1}(\tau,\eta)\right] \tag{13.78}$$

$$g_q(\tau,\eta)=\left(\frac{\tau-\eta}{\tau+\eta}\right)^{q/2} J_q\left(\sqrt{\tau^2-\eta^2}\right), \tag{13.79}$$

where J is the Bessel function of the first kind.

By changing η to $j\eta$ and τ to $j\tau$, Equation 13.75 can be transformed into

$$\frac{\partial^2 U}{\partial \eta^2}-\frac{\partial^2 U}{\partial \tau^2}=-U. \tag{13.80}$$

Its solution is given by (13.77) through (13.79) except that (13.79) changes to (13.81):

$$g_q(\tau,\eta)=\left(\frac{\tau-\eta}{\tau+\eta}\right)^{q/2} I_q\left(\sqrt{\tau^2-\eta^2}\right), \quad \tau\geq\eta\geq 0. \tag{13.81}$$

13.5 Nonlinear Transmission Line Equations and KdV Equation

The circuit representation of a differential length of a nonlinear transmission line with dispersion that will be discussed in this section is shown in Figure 13.15.

Let us take a specific example where

$$C'_{NL}=C'-C'_N V. \tag{13.82}$$

It can be shown that $V(z,t)$ satisfies the nonlinear PDE [9]

$$\frac{\partial^2 V}{\partial z^2}-L'C'\frac{\partial^2 V}{\partial t^2}+L'C'_d\frac{\partial^4 V}{\partial z^2 \partial t^2}+L'C'_N\frac{\partial^2\left(V^2\right)}{\partial t^2}=0. \tag{13.83}$$

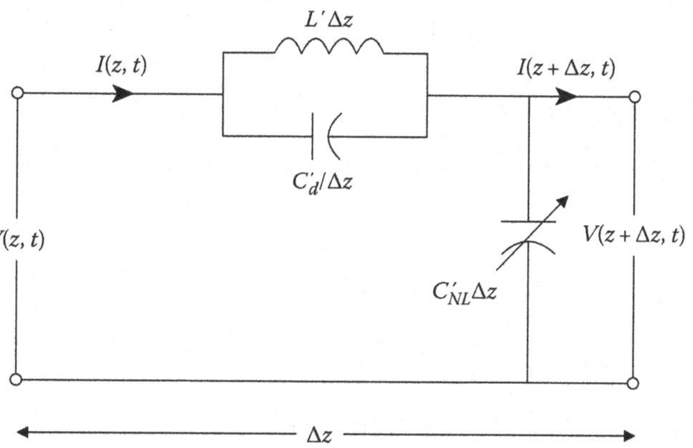

FIGURE 13.15
Circuit representation of a differential length of a nonlinear transmission line.

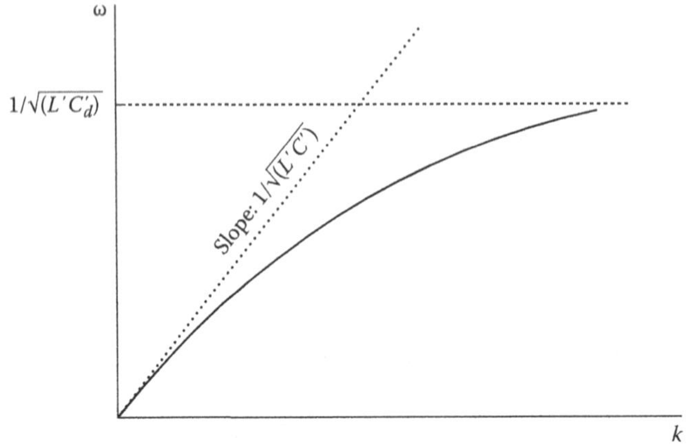

FIGURE 13.16
Sketch of the dispersion relation given by Equation 13.84.

The first two terms on the LHS of (13.83) describe the wave propagation in an ideal transmission line. The third term arises because of the presence of the capacitor C_d' parallel to L'. Its effect can be studied by making $C_N' = 0$ in (13.83). In such a case, the dispersion equation becomes

$$\frac{\omega}{k} = \frac{1}{\sqrt{L'C_d'k^2 + L'C'}}. \tag{13.84}$$

The sketch of the dispersion relation (13.84) is given in Figure 13.16.

The third term causes the dispersion, while the fourth term provides the nonlinearity. In Section 9.10, it is mentioned that the balancing of the wave-steepening feature of nonlinearity can compensate the wave-flattening feature of the dispersion to preserve the signal shape as the wave propagates. This "soliton" aspect can be quantitatively studied by this model. In "TODA" model of the NLTL [10], the dispersion arises due to the use of a finite number of sections to represent the transmission line [10], where $\Delta z = h$, the length of each section. Such discretization provides numerical dispersion [11]. The study of the required parameter relationships between L', C', C_d', and C_N' to obtain the needed balance between dispersion and nonlinearity to preserve the wave shape is facilitated by understanding the standard form of Korteweg-de-Vries (KdV) equation, whose solution is a "soliton."

13.5.1 Korteweg-de-Vries (KdV) Equation and Its Solution

The standard form of KdV equation and its solution are given below:

$$\frac{\partial u}{\partial t} + 6u\frac{\partial u}{\partial z} + \frac{\partial^3 u}{\partial z^3} = 0 \tag{13.85}$$

$$U(z,t) = A\sec h^2\left[\sqrt{\frac{A}{2}}(z - 2At)\right]. \tag{13.86}$$

By direct substitution of (13.86) in the third term on the LHS of (13.85), it can be shown [10] that

$$U_{zzz} = -2AU_z - 6UU_z, \tag{13.87}$$

where we follow the standard notation of the addition of a subscript to the symbol to denote its partial derivative with respect to the subscripted independent variable, that is,

$$U_z = \frac{\partial U}{\partial z}. \tag{13.88}$$

Equation 13.85 can now be written as

$$U_t + 6UU_z - 2AU_z - 6UU_z = 0. \tag{13.89}$$

The nonlinear term $6UU_z$, the second term in (13.89), cancels with the fourth term, which comes from the dispersion term U_{zzz} given in (13.87). We are now left with a linear equation

$$U_t - 2AU_z = 0, \tag{13.90}$$

whose solution is given by (13.86). This can be shown by substitution of (13.86) in (13.90). A sketch of (13.86) is given in Figure 13.17. Its velocity is $2A$, and its amplitude is A.

The solution (13.86) can also be obtained by solving (13.85) in the wave frame [9,10]

$$\varsigma = z - v_s t, \tag{13.91}$$

where v_s is the velocity of the "solitary wave"; the velocity could be a function of the amplitude of the wave. Equation 13.85 transforms into

$$-v_s u_\varsigma + 6uu_\varsigma + u_{\varsigma\varsigma\varsigma} = 0, \tag{13.92}$$

where the subscript ζ is used as usual to denote the derivative with respect to ζ. Equation 13.92 is now an ordinary differential equation. Integrating it with respect to ζ, we get

$$-v_s u + 3u^2 + u_{\varsigma\varsigma} = c_1. \tag{13.93}$$

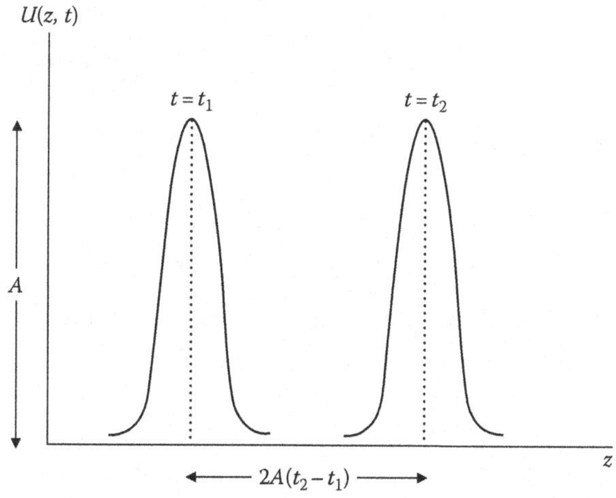

FIGURE 13.17
Sketch of a soliton given by Equation 13.86.

The pulse solution that satisfies zero values for u and its first- and second-order derivatives as the magnitude of ζ tends to infinity make $c_1 = 0$.

Multiplying (13.93) by u_ς, and recognizing the term $u_{\varsigma\varsigma}u_\varsigma$ as $\frac{1}{2}\frac{d}{d\varsigma}(u_\varsigma)^2$, we get

$$-v_s u u_\varsigma + 3u^2 u_\varsigma + \frac{1}{2}\frac{d}{d\varsigma}(u_\varsigma)^2 = 0. \tag{13.94}$$

Integration of (13.94) with respect to ζ yields (13.95), where the integration constant c_2 is set to zero for the same reason that c_1 is set to zero.

$$-v_s \frac{u^2}{2} + u^3 + \frac{1}{2}(u_\varsigma)^2 = 0. \tag{13.95}$$

Solving (13.95) for u_ζ, we get (13.96)

$$\frac{du}{d\varsigma} = \left(v_s u^2 - 2u^3\right)^{1/2}. \tag{13.96}$$

The solution of this first-order differential equation gives the relation of ζ with u. By transforming back to the lab frame in z, t, we get the solution given in (13.86).

13.5.2 KdV Approximation of NLTL Equation

Equation 13.83 can be approximated by KdV equation for small amplitudes of the wave. Reductive perturbation method [9,10,12,13] is used to make this approximation. It begins by transforming from lab coordinates (z, t) to those of (S, T) in a reference frame that moves with a velocity v_0:

$$v_0 = \frac{1}{\sqrt{L'C'}} \tag{13.97}$$

$$S = \varepsilon^{1/2}(z - v_0 t) \tag{13.98a}$$

$$T = \varepsilon^{3/2} t. \tag{13.98b}$$

In the above, ε is called the "bookkeeping" parameter. The relations between the derivatives are

$$\frac{\partial}{\partial z} = \varepsilon^{1/2}\frac{\partial}{\partial S} \tag{13.99a}$$

$$\frac{\partial}{\partial t} = -\varepsilon^{1/2} v_0 \frac{\partial}{\partial S} + \varepsilon^{3/2} v_0 \frac{\partial}{\partial T}. \tag{13.99b}$$

The voltage V will be expressed in a perturbation series in powers of ε:

$$V = \varepsilon u^{(1)} + \varepsilon^2 u^{(2)} + \ldots \tag{13.100}$$

Substituting (13.97) through (13.100) in (13.83) and collecting terms of similar power in ε, one can obtain a series of equations. The lowest-order equation for $u^{(1)}$ is given by

$$v_0 \frac{\partial u^{(1)}}{\partial T} + \frac{C_N'}{C'} u^{(1)} \frac{\partial u^{(1)}}{\partial S} + \frac{C_d'}{2C'}\frac{\partial^3 u^{(1)}}{\partial S^3} = 0. \tag{13.101}$$

By comparing (13.85), the standard form of KdV equation, with (13.101), the solution of (13.85) can be adapted to obtain the solution of (13.101). On transforming back to the lab coordinates z and t, we obtain [9]

$$U(z,t) = U_0 \sec h^2 \left[\sqrt{\frac{C_N' U_0}{6C_d'}} \left(z - \left(1 + \frac{C_N' U_0}{3C'}\right) \frac{t}{\sqrt{L'C'}} \right) \right]. \quad (13.102)$$

The velocity v_S of the wave solution in (13.102), called solitary wave, is given by [9]

$$v_S = \left(1 + \frac{C_N' U_0}{3C'}\right) \frac{1}{\sqrt{L'C'}}. \quad (13.103)$$

The velocity increases with amplitude of the wave.

It can be noted that KdV equation (13.101) is a small-amplitude approximation of (13.83), since we approximated V by the first term $u(1)$ in (13.100). For a comprehensive study of the theory, design, and applications of solitons, Reference [10] can be consulted. A few pointers from [10] that connect well with the material of this section are given below.

The "soliton" solution of the standard KdV equation (13.86) has a single parameter A, the amplitude of the soliton. The velocity of this pulse is amplitude dependent and is given by

$$v_S = 2A. \quad (13.104)$$

A taller pulse travels faster than a shorter pulse. If A_1 and A_2 are the parameters of two solitons, and $A_2 > A_1$, the second (taller) pulse can collide with and overtake the first (shorter) pulse. While the two pulses interact nonlinearly during the collision, they can come out unaltered after the collision. This characteristic distinguishes a soliton [14] from other solitary wave solutions with hyperbolic secant profiles. Two such examples of differential equations are given in [10].

13.6 Charged Particle Dynamics [15,16]

13.6.1 Introduction

Electromagnetic force on a charged particle of charge q in electric and magnetic fields is given by (1.6) and is repeated here as (13.105)

$$\boldsymbol{F} = q(\boldsymbol{E} + \mathrm{v} \times \boldsymbol{B}). \quad (13.105)$$

One can derive from (13.105) the concepts of and equations relating force density $\boldsymbol{f}$, Maxwell's stress tensor $\bar{\boldsymbol{T}}$, and electromagnetic momentum $\boldsymbol{g}$ in a volume V bounded by a surface s due to charge density ρ and current density $\boldsymbol{J}$.

This aspect is discussed in Appendix 13A rather than here to maintain the narrative of the charged *particle* dynamics. The Newton's second law (see Appendix 13B, for discussion of the validity of Newton's third law to forces on charged particles) gives

$$\boldsymbol{F} = m\boldsymbol{a}, \quad (13.106)$$

where

$$a = \frac{dv}{dt} = \frac{d^2 r}{dt^2}. \tag{13.107}$$

In (13.107) r is the position vector of the point where the charged particle is located. Thus, we will write (13.108) as the starting equation to study the charged particle dynamics.

$$m\frac{dv}{dt} = m\frac{d^2 r}{dt^2} = q\left(E + \mathrm{v} \times B\right). \tag{13.108}$$

We used (13.108) with appropriate approximations to model dielectrics, conductors, and plasmas in Chapters 8, 9, and 11. In Section 13.8, we will use an appropriate approximation of (13.108) to model magnetohydrodynamics (MHD), which is a study of moving liquid conductors. However, in this section, we study the pertinent aspects of the dynamics of a charged particle. First we study in the next section the kinematics, that is, the expressions for the velocity v and acceleration a given the trajectory, thus separating this aspect from the aspect of specific forces that produced this trajectory.

13.6.2 Kinematics

Figure 13.18 gives a sketch of a general curve and a typical point P on the curve with reference to the origin O of a coordinate system.

In Cartesian coordinate system, the position vector r, the directed line segment OP, shown in the figure is given by

$$r(t) = \hat{x}x(t) + \hat{y}y(t) + \hat{z}z(t). \tag{13.109}$$

The time derivatives d/dt (sometimes denoted by a dot on the top) of the unit vectors x, y, and z are zero since they are constant unit vectors. Thus,

$$v(t) = \frac{d}{dt}r(t) = \dot{r}(t) = \hat{x}\dot{x}(t) + \hat{y}\dot{y}(t) + \hat{z}\dot{z}(t) \tag{13.110}$$

$$a(t) = \dot{v}(t) = \frac{d^2}{dt^2}r(t) = \ddot{r}(t) = \hat{x}\ddot{x}(t) + \hat{y}\ddot{y}(t) + \hat{z}\ddot{z}(t). \tag{13.111}$$

In cylindrical coordinates, the trajectory will be given in terms of $\rho(t)$, $\phi(t)$, $z(t)$ and the unit vectors $\hat{\rho}$, $\hat{\phi}$, $\hat{z}$. Note that the unit vectors in this case are not constant and depend on the azimuthal coordinate ϕ. See Appendix 1A. The position vector is given by

$$r = \hat{\rho}(\phi)\rho + \hat{z}z, \tag{13.112}$$

where

$$\hat{\rho}(\phi) = \hat{x}\cos\phi(t) + \hat{y}\sin\phi(t) \tag{13.113}$$

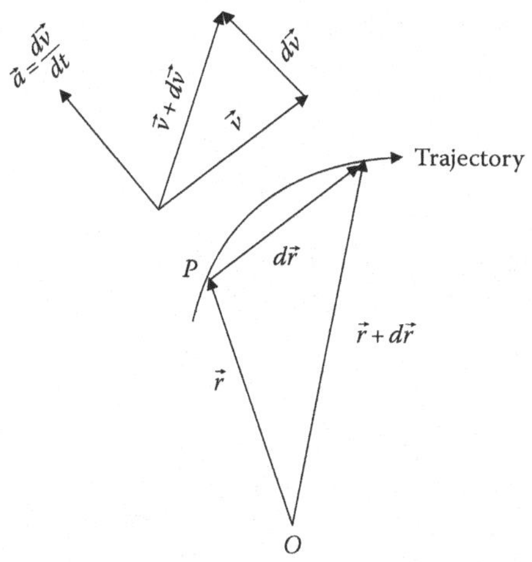

FIGURE 13.18
Sketch showing the position vector $\boldsymbol{r}$, velocity $\boldsymbol{v}$, and acceleration $\boldsymbol{a}$ of a particle moving along a trajectory.

$$\hat{\phi}(\phi) = -\hat{x}\sin\phi(t) + \hat{y}\cos\phi(t). \tag{13.114}$$

From (13.113),

$$\begin{aligned}\frac{d}{dt}\hat{\rho}(\phi) &= \hat{x}\frac{d}{dt}\cos\phi(t) + \hat{y}\frac{d}{dt}\sin\phi(t)\\ &= \hat{x}\frac{d}{d\phi}\cos\phi\frac{d\phi}{dt} + \hat{y}\frac{d}{d\phi}\sin\phi\frac{d\phi}{dt}\\ &= -\hat{x}\sin\phi\frac{d\phi}{dt} + \hat{y}\cos\phi\frac{d\phi}{dt}.\end{aligned}$$

From (13.114),

$$\frac{d}{dt}\hat{\rho}(\phi) = \hat{\phi}\frac{d\phi}{dt}. \tag{13.115}$$

Proceeding on similar lines, we can obtain Equations 13.116 through 13.120:

$$\frac{d}{dt}\hat{\phi}(\phi) = -\hat{\rho}\frac{d\phi}{dt} \tag{13.116}$$

$$\frac{d^2}{dt^2}\hat{\rho}(\phi) = -\hat{\rho}\left[\dot{\phi}\right]^2 + \hat{\phi}\frac{d^2\phi}{dt^2} \tag{13.117}$$

$$\frac{d^2}{dt^2}\hat{\phi}(\phi) = -\hat{\rho}\frac{d^2\phi}{dt^2} - \hat{\phi}\left[\dot{\phi}\right]^2 \tag{13.118}$$

$$\boldsymbol{v}(t) = \frac{d^2}{dt^2}\boldsymbol{r}(t) = \dot{\boldsymbol{r}}(t) = \hat{\rho}\dot{\rho} + \hat{\phi}\rho\dot{\phi} + \hat{z}\dot{z} \tag{13.119}$$

$$\boldsymbol{a}(t) = \dot{\boldsymbol{v}}(t) = \frac{d^2}{dt^2}\boldsymbol{r}(t) = \ddot{\boldsymbol{r}}(t) = \hat{\rho}\left[\ddot{\rho} - \rho\left(\dot{\phi}\right)^2\right] + \hat{\phi}\left[\rho\ddot{\phi} + 2\dot{\rho}\dot{\phi}\right] + \hat{z}\ddot{z}. \tag{13.120}$$

The second term in the multiplier of the unit vector $\hat{\rho}$ in (13.120) gives the so-called centripetal acceleration $\boldsymbol{a}_{cp}$:

$$\boldsymbol{a}_{cp} = \hat{\rho}\left[-\rho\left(\dot{\phi}\right)^2\right], \tag{13.121}$$

which, because of the negative sign, is directed toward the axis of the cylinder.

The second term in the multiplier of the unit vector $\hat{\phi}$ is called Coriolis acceleration $\boldsymbol{a}_{CL}$:

$$\boldsymbol{a}_{CL} = \hat{\phi}\left[2\dot{\rho}\dot{\phi}\right]. \tag{13.122}$$

For a circular motion in the x–y plane ($\rho=\rho_0$, $z = 0$), the expressions given in (13.112), (13.119), and (13.120) for the position vector, velocity, and acceleration, respectively, can be further simplified and are given by (13.123) through (13.125):

$$\boldsymbol{r} = \hat{\rho}\left(\phi\right)\rho_0 \tag{13.123}$$

$$\boldsymbol{v}\left(t\right) = \hat{\phi}\rho_0\omega \tag{13.124}$$

$$\boldsymbol{a}\left(t\right) = \hat{\rho}\left[-\rho_0\left(\omega\right)^2\right] + \hat{\phi}\rho_0\alpha, \tag{13.125}$$

where $\omega = \dot{\phi}$ is the instantaneous angular velocity and $\alpha = \ddot{\phi}$ is the instantaneous angular acceleration.

13.6.3 Conservation of Particle Energy due to Stationary Electric and Magnetic Fields

Let us start with the assumption that the electric and magnetic fields are not varying in time. By taking the dot product of (13.108) with $\boldsymbol{v}$, and using $\boldsymbol{E} = -\nabla\Phi$, we obtain the following energy conservation law:

$$\frac{d}{dt}\left(\frac{1}{2}mv^2 + q\Phi\right) = 0. \tag{13.126}$$

The first term in the parenthesis is the kinetic energy and the second term is the potential energy of the particle. The total energy of the particle is conserved due to its motion, in electric and magnetic fields not varying in time.

13.6.4 Constant Electric and Magnetic Fields

Let us consider the case of constant electric and magnetic fields. Without loss of generality, we can assume the magnetic field is in the z-direction. Then we can write the vector differential equation (13.108) as three scalar differential equations:

$$\dot{v}_x = \frac{q}{m}\left(E_y + Bv_y\right) \tag{13.127}$$

$$\dot{v}_y = \frac{q}{m}\left(E_y - Bv_x\right) \tag{13.128}$$

$$\dot{v}_z = \frac{q}{m}E_z. \tag{13.129}$$

Equations 13.127 and 13.128 are coupled and by differentiating (13.127) and substituting for $\dot{v}_y$, we obtain

$$\ddot{v}_x + \omega_b^2 v_x = \omega_b^2 \frac{E_y}{B}. \tag{13.130}$$

Equation 13.130 is a second-order differential equation and its solution is easily obtained:

$$v_x = A\sin\left(\omega_b t - \delta\right) + \frac{E_y}{B}. \tag{13.131}$$

The first term on the RHS is the homogeneous part of the solution with two undetermined constants A and δ. The second term is the particular integral. After obtaining $\dot{v}_x$ from (13.131) and substituting it in (13.127), one can obtain v_y:

$$v_y = A\cos\left(\omega_b t - \delta\right) - \frac{E_x}{B}. \tag{13.132}$$

The particular integrals in (13.131) and (13.132) can be written in a more general vectorial form

$$\boldsymbol{v}_{DE} = \frac{\boldsymbol{E}\times\boldsymbol{B}}{B^2}, \tag{13.133}$$

where v_{DE} is called the drift velocity due to the electric field, which is independent of the mass and charge of the particle. Both the electrons and positive ions drift along together. The homogeneous part of the solutions given in (13.131) and (13.132) are oscillations at the gyrofrequency ω_b. The drift velocity is zero if $\boldsymbol{E}$ and $\boldsymbol{B}$ are in the same direction or if $\boldsymbol{E}$ is zero. Let us consider in more detail the motion of the particle for the second case of $\boldsymbol{E} = 0$.

13.6.4.1 Special Case of E = 0

Since the potential energy is zero, from (13.126), the kinetic energy of the particle is a constant of motion in the x–y plane (plane perpendicular to the B field). The magnitude of the velocity v is a constant of motion.

Let us take a concrete case by specifying the following initial conditions $v_x(0) = 0$ and $v_y(0) = v_0$. Since $E = 0$, the harmonic solution for v_x and v_y will be chosen from sin or cos type from the template like (3.15) that satisfies the given initial conditions:

$$v_x\left(t\right) = v_0 \sin\omega_b t \tag{13.134}$$

$$v_y\left(t\right) = v_0 \cos\omega_b t. \tag{13.135}$$

The expressions for x and y coordinates of the particle can be obtained by integrating (13.134) and (13.135) and will be again trigonometric type multiplied by of v_0/ω_b. If the initial conditions are $x(0) = -\rho_0$ and $y(0) = 0$, the trajectory of the charged particle in the x–y plane is given by

$$x(t) = -\rho_0 \cos \omega_b t \tag{13.136}$$

$$y(t) = \rho_0 \sin \omega_b t, \tag{13.137}$$

where

$$\rho_0 = \frac{v_0}{\omega_b}. \tag{13.138}$$

This trajectory is a circle of radius ρ_0 centered at the origin in the transverse plane as sketched in Figure 13.19.

For an ion of positive charge q and mass M, the particle rotates clockwise as shown by the solid line. For an electron of negative charge $(-q)$ and mass m, ω_b is negative and larger in magnitude than for the case of ion, and the electron rotates counterclockwise as sketched in the figure with a dotted line. The radius of the circle ρ_0 is called radius of gyration or cyclotron radius of the particle. This circular motion is already described from the kinematic viewpoint in Equations 13.123 and 13.124 when we identify the angular velocity ω with the particle gyrofrequency ω_b.

Equation 13.129 for the case of $\boldsymbol{E} = 0$ has a simple solution, and for the assumed initial conditions of $v_z(0) = v_1$ and $z(0) = 0$,

$$v_z(t) = v_1 \tag{13.139}$$

$$z(t) = v_1 t. \tag{13.140}$$

Thus, the particle trajectory is helical in the three-dimensional space.

Let us state in words the results we obtained so far in this section. The combination of the parallel component of E and the B field produces only uniform motion of the charged

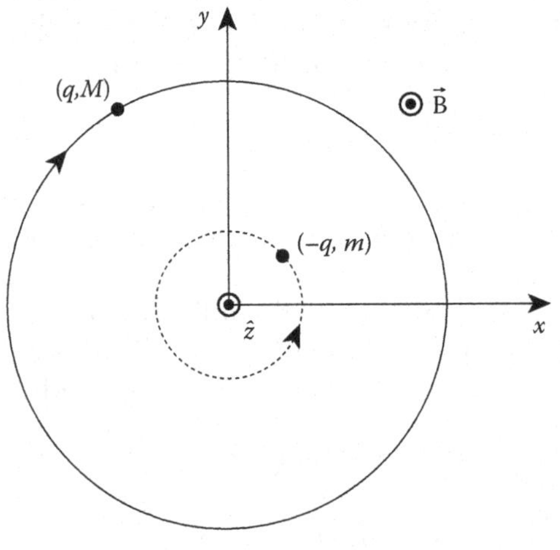

FIGURE 13.19
Sketch of circular trajectories of charged particles in the plane perpendicular to the static magnetic field. Note that the radius and the sense of rotation for an electron is different from those for a positive ion like a proton.

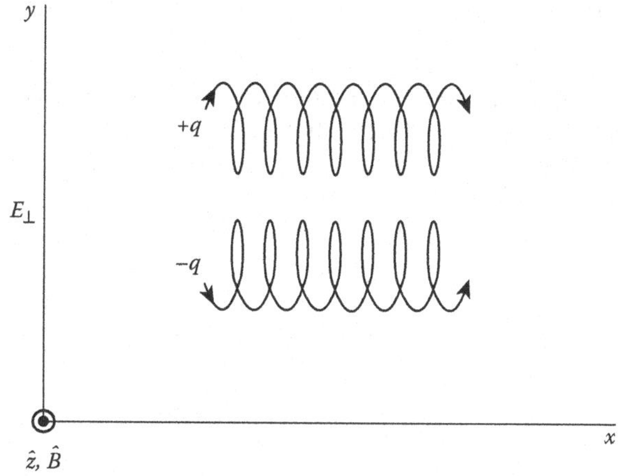

FIGURE 13.20
Sketches of the trajectories of positive and negative charges. The B field is in the z-direction, and the perpendicular component of the electric field is in the y-direction.

particle. The B field has no effect. The combination of the perpendicular component of E and the B field produces a moving circular orbit whose center, called the guiding center, moves with a drift velocity given by (13.133). Note that the drift takes place in the transverse plan normal to the perpendicular component of the E field. Magnitude of the drift velocity does not depend on the charge or mass of the particle. It depends on the ratio of the magnitude of the transverse component of the electric field to that of the B field. See Figure 13.20.

13.6.5 Constant Gravitational Field and Magnetic Field

In this section, we replace the electric field with the gravitational field and consider the solution of the equation

$$m\frac{d\boldsymbol{v}}{dt} = m\frac{d^2\boldsymbol{r}}{dt^2} = m\boldsymbol{g} + q(\boldsymbol{v}\times\boldsymbol{b}), \tag{13.141}$$

where $\boldsymbol{g}$ is the acceleration due to gravity. Comparing (13.108) with (13.141), we note that the drift velocity in this case will be given by

$$\boldsymbol{v}_{Dg} = \frac{m}{q}\frac{\boldsymbol{g}\times\boldsymbol{B}}{B^2}. \tag{13.142}$$

In this case the drift velocity does depend on the mass and the algebraic value of the charge. The electrons and ions have drift in opposite directions. We can generalize the formula for the drift velocity when we replace the electric force $\boldsymbol{E}$ in (13.133) by a general external force $\boldsymbol{F}$:

$$\boldsymbol{v}_{DF} = \frac{1}{q}\frac{\boldsymbol{F}\times\boldsymbol{B}}{B^2}. \tag{13.143}$$

13.6.6 Drift Velocity in Nonuniform *B* Field

In this section, we will discuss the motion of a charged particle in a slightly nonuniform but static *B* field. As a specific example, let us calculate the drift velocity, the velocity of the guiding center of the trajectory, in a *z*-directed *B* field with a small gradient in the *x*-direction. If the imposed *B* field is in free space, it has to satisfy the following two equations:

$$\nabla \cdot \boldsymbol{B} = 0 \tag{13.144}$$

$$\nabla \times \boldsymbol{B} = 0. \tag{13.145}$$

Tannenbaum [15] considers the following expression for the **B** field, which satisfies (13.144) and (13.145):

$$\boldsymbol{B} = B_0\left[\hat{x}\alpha z + \hat{z}(1+\alpha x)\right]. \tag{13.146}$$

If the parameter α is zero, we have the uniform *B* case considered in the previous section. If αz and αx are small compared to one, we have slightly nonuniform and nearly uniform static magnetic field. Since

$$\nabla B_z = \hat{x}\alpha B_0, \quad (113) \tag{13.147}$$

the second term in the *z*-component of the *B* field in (13.146) will be called the "gradient" term. The first term on the RHS of (13.146), which is the *x* component of **B**, will be called the "curvature" term. A sketch of the *B* field line near the origin that goes through the origin of the *x*–*z* plane is shown in Figure 13.21. Its equation is given by [15]

$$\left(x - \frac{1}{\alpha}\right)^2 + z^2 = \frac{1}{\alpha^2}. \tag{13.148}$$

The solution for the velocities is obtained using a perturbation technique for the case $\alpha x \ll 1$ and $\alpha z \ll 1$. The end result for the drift velocity is given by

$$v_D = \hat{y}\frac{\alpha}{\omega_b}\left(\frac{1}{2}v_0^2 + v_1^2\right). \tag{13.149}$$

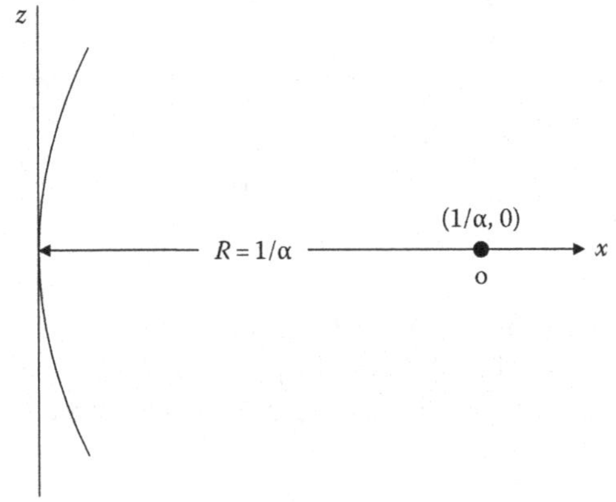

FIGURE 13.21
Curved *B* field line passing through the origin. See Equation 13.148.

The term ω_b in (13.149) shows that the drift velocity for a positive charge is in opposite direction to that of the negative charge.

The generalization of (13.149) is obtained by noting that the factor $\hat{y}\alpha$ in (13.149) can be written as

$$\hat{y}\alpha = \frac{\boldsymbol{B}_0 \times \nabla B_z}{B_0^2} = \frac{\boldsymbol{R} \times \boldsymbol{B}_0}{R^2 B_0}, \tag{13.150}$$

where the radius of curvature

$$\boldsymbol{R} = \frac{-\hat{x}}{\alpha}, \tag{13.151}$$

as shown in Figure 13.21. Equation 13.149 can now be written as [15]

$$\boldsymbol{v}_D = \frac{v_0^2}{2\omega_b} \frac{\boldsymbol{B}_0 \times \nabla B_z}{B_0^2} + \frac{v_1^2}{\omega_b} \frac{\boldsymbol{R} \times \boldsymbol{B}_0}{R^2 B_0}. \tag{13.152}$$

The first term on the RHS of (13.152) is called *Grad-B* drift, when *Grad-B* is perpendicular to B. The second term is called curvature drift.

13.6.7 Time-Varying Fields and Adiabatic Invariants

We showed earlier that the charged particle trajectory in a uniform B field is a circle of radius ρ_0, relabeled here as ρ_L for Larmor radius. The circular motion is periodic, with a period

$$t_B = \frac{1}{f_B} = \frac{2\pi}{\omega_B} = \frac{2\pi m}{|q| B}. \tag{13.153}$$

The velocity of the particle will be denoted by $v_\perp$ and its kinetic energy by $w_\perp$:

$$w_\perp = \frac{1}{2} m v_\perp^2, \tag{13.154}$$

where

$$w_\perp = \omega_B \rho_L. \tag{13.155}$$

If B is uniform, $w_\perp$ is a constant of motion.

If B is slowly varying, the trajectory is not a closed circle; nevertheless one can still approximate it as a closed path. In such an approximation, we can use the concept of action integral defined in (13.183) and use the property of constant of motion identifying the generalized momentum p with the transverse component of the particle momentum and the generalized differential coordinate dq with the differential length element of the closed path. The slowness can be defined by

$$\Delta B = \frac{\partial B}{\partial t} t_B \propto \omega B t_B \propto \frac{t_B}{t_s} B = \frac{\omega}{\omega_b} B, \tag{13.156}$$

where ΔB is the change in B in one gyration. In the above equation, t_s is the characteristic time scale of the B field variation. The slow change is called "adiabatic" and the conserved variables are called "adiabatic invariants" denoted by the symbol J, not to be confused with the volume current density. These are stated in various ways [17]:

$$J_1 \propto B\rho_L^2 \tag{13.157}$$

$$J_2 \propto \frac{p_\perp^2}{B} \tag{13.158}$$

$$J_3 \propto \mu_m. \tag{13.159}$$

In the above, the symbol $p_\perp$ is the linear momentum of the charged particle in circular motion and μ_m is the magnetic dipole moment (abbreviated as magnetic moment) defined as p_{m} in (8.144). For the relativistic velocities (Chapter 14), the RHS of (13.159) will have a multiplier of the Lorentz factor γ.

We shall next consider the evaluation of (8.114) for the magnetic moment for a particle of positive charge q with a circular trajectory, shown in Figure 13.19, in a constant B field. Since its trajectory is clockwise, the direction of the vector area **S** of the circle is in opposite direction of the **B** field. The current I is the charge q multiplied by f_B.

$$\mu_m IS = q\left(\frac{\omega_b}{2\pi}\right)\pi\rho_L^2\left(\hat{B}\right) = -\frac{1}{2}\frac{mv_\perp^2}{B}\hat{B} \tag{13.160}$$

$$\mu_m = \frac{w_\perp}{B}. \tag{13.161}$$

If we use for the generalized coordinate p the angular momentum $mv_\perp\rho_L$ and for q the coordinate ϕ in (13.183), we obtain

$$J_3 = \oint p\,dq = \oint mv_\perp\rho_L\,d\phi = \frac{2\pi mv_\perp^2}{\omega_b} = 4\pi\frac{m}{q}\mu_m. \tag{13.162}$$

Thus, μ_m is a constant of motion. However, for relativistic velocities it is $\gamma\mu_m$ that is a constant of motion. The magnetic flux Φ_m through a Larmor orbit is

$$\Phi_m = \pi\rho_L^2 B = \frac{2\pi m}{q^2}\mu_m. \tag{13.163}$$

It is a constant of motion as long as μ_m is a constant of motion. Equation 13.157 is thus proved. Consequently, as the B field changes, the Larmor radius of the particle orbit changes suitably to preserve the magnetic flux through the area of the orbit.

It should be noted that the B field here is an imposed one and there is no imposed electric field. However, the changing magnetic field, according to Faraday's law (1.1), gives rise to an induced electric field, denoted by $\mathbf{E}_\perp$. Consequently the equation of motion

$$\frac{d}{dt}\left(\frac{1}{2}mv_\perp^2\right) = qv_\perp \cdot \mathbf{E}_\perp. \tag{13.164}$$

It can be shown [18] that the change of the kinetic energy per gyration, $\Delta w_\perp$, is given by

$$\Delta w_\perp = \Delta B\frac{w_\perp}{B} + O\left(\varepsilon^2\right)w_\perp. \tag{13.165}$$

13.6.8 Lagrange and Hamiltonian Formulations of Equations of Motion

The equations of motion can be obtained by more advanced formulations than Newtonian mechanics, which are essentially based on $\boldsymbol{F}=m\boldsymbol{a}$. Lagrange formulation is based on D'Alembert's principle and the principle of virtual work [19]. The relevant equation in generalized coordinates is

$$\frac{d}{dt}\left(\frac{\partial T}{\partial \dot{q}_j}\right)-\left(\frac{\partial T}{\partial q_j}\right)=Q_j \tag{13.166}$$

$$T=\sum \frac{1}{2}m_i v_i^2. \tag{13.167}$$

In the above equation, q_j is the generalized coordinate, which need not have the dimension of length. Q_j is the component of the generalized force. It can be shown that (13.166) reduces to Newton's second law $\boldsymbol{F}=m\boldsymbol{a}=d\boldsymbol{p}/dt$; if q is the position coordinate, Q is the corresponding component of the force $\boldsymbol{F}$, and $\boldsymbol{p}=m\boldsymbol{v}$ is the momentum.

If Q_j can be written in terms of a generalized potential U, that is,

$$Q_j=\frac{d}{dt}\left(\frac{\partial U}{\partial \dot{q}_j}\right)-\left(\frac{\partial U}{\partial q_j}\right), \tag{13.168}$$

then (13.166) can be written as

$$\frac{d}{dt}\left(\frac{\partial L}{\partial \dot{q}_j}\right)-\left(\frac{\partial L}{\partial q_j}\right)=0. \tag{13.169}$$

Here the Lagrangian L is given by

$$L=T-U. \quad (136) \tag{13.170}$$

We shall next obtain the Lagrangian for describing the motion of a charged particle in an electromagnetic field. The electric field $\boldsymbol{E}$ in terms of the electric scalar potential Φ and the vector magnetic potential $\boldsymbol{A}$ is given by (1.8) and repeated here for convenience:

$$\boldsymbol{E}=-\nabla\phi-\frac{\partial \boldsymbol{A}}{\partial t}. \tag{13.171}$$

Substituting (13.171) and (1.7) in the force equation (1.6), we get

$$\boldsymbol{F}=q\left(-\nabla\Phi-\frac{\partial \boldsymbol{A}}{\partial t}+\boldsymbol{v}\times\nabla\times\boldsymbol{A}\right). \tag{13.172}$$

Equation 13.172 can be written in a more convenient form. Its x-component

$$F_x=-\frac{\partial U}{\partial x}+\frac{d}{dt}\left(\frac{\partial U}{\partial v_x}\right), \tag{13.173}$$

where

$$U = q\Phi - q\boldsymbol{A}\cdot\boldsymbol{v}. \tag{13.174}$$

From (13.168), (13.173), and (13.174), it is obvious that U is a generalized potential. If we consider the case of motion of a charged particle in an electrostatic field, then

$$U = q\Phi, \quad (\text{electrostatic case}), \tag{13.175}$$

which is the potential energy, and the Lagrangian is equal to the kinetic energy minus the potential energy. The Lagrangian for the electromagnetic case is given by

$$L = \frac{1}{2}mv^2 + q\boldsymbol{A}\cdot\boldsymbol{v} - q\Phi. \tag{13.176}$$

In the relativistic case, discussed in Chapter 14, since mass varies with velocity, L_{rel} is given by [18]

$$L_{rel} = -m_0\left(1 - \frac{v^2}{c^2}\right) + q\boldsymbol{A}\cdot\boldsymbol{v} - q\Phi, \tag{13.177}$$

where m_0 is the rest mass.

13.6.8.1 Hamiltonian Formulation

In the Hamiltonian formulation of equations of motion, the independent generalized coordinates are (q,p,t) instead of $(q,\dot{q},t)$, where p's are called generalized momenta, defined by

$$p_i = \frac{\partial L(q_j,\dot{q}_j,t)}{\partial \dot{q}_i}. \tag{13.178}$$

It can be shown [19], using Legendre transformation, that the following first-order differential equations describe the motion:

$$\frac{dq_i}{dt} = \frac{\partial \mathcal{H}}{\partial p_i} \tag{13.179}$$

$$\frac{dp_i}{dt} = -\frac{\partial \mathcal{H}}{\partial q_i}, \tag{13.180}$$

where the Hamiltonian H is related to the Lagrangian L by

$$\mathcal{H}(p,q,t) = \sum_i \dot{q}_i p_i - L(q,\dot{q},t) \tag{13.181}$$

$$\frac{\partial \mathcal{H}}{\partial t} = -\frac{\partial L}{\partial t}. \tag{13.182}$$

Equations 13.179 and 13.180 are called canonical equations of Hamilton. The canonical variables p and q are independent variables. For each coordinate q, which is periodic, the action integral J is

$$J = \oint p\,dq \tag{13.183}$$

and is invariant.

We can further show that if L, and hence $\mathcal{H}$, is not an explicit function of t,

$$\frac{d\mathcal{H}}{dt} = \frac{\partial \mathcal{H}}{\partial t}. \tag{13.184}$$

For a charged particle,

$$\mathcal{H} = \frac{1}{2m}\left|(\boldsymbol{p} - q\boldsymbol{A})\right|^2 + q\Phi. \tag{13.185}$$

In the above equation, $\boldsymbol{p}$ is the mechanical momentum.

For an electrostatic field, $\boldsymbol{A}$ is zero, and the Hamiltonian is kinetic energy (K.E.) plus potential energy (P.E.):

$$\mathcal{H} = \frac{p^2}{2m} + q\Phi = \frac{1}{2}mv^2 + q\Phi \tag{13.186}$$

$$\mathcal{H} = K.E. + P.E. \quad \text{(electrostatic field)}. \tag{13.187}$$

For the same case, the Lagrangian L

$$L = K.E. - P.E. \quad \text{(electrostatic case)}. \tag{13.188}$$

For the relativistic case,

$$\mathcal{H}_{rel} = \left[m_0^2c^4 + c^2\left|(\boldsymbol{p} - q\boldsymbol{A})\right|^2\right]^{1/2} + q\Phi. \tag{13.189}$$

An example [20] of using the generalized coordinates and Hamiltonian formulation is given below.

13.6.8.2 Photon Ray Theory

Photon ray theory is based on the geometrical optics concepts of propagation of electromagnetic wave packets in a medium. We will start with an x-polarized uniform plane wave propagating in a uniform medium in z-direction. Its electric field is given by

$$E_x(z,t) = E_0(z,t)e^{j(\omega t - kz)}. \tag{13.190}$$

If the propagation is taking place in a slowly varying medium, we can modify (13.190) by

$$E_x(z,t) = E_0(z,t)e^{j\psi(z,t)}. \tag{13.191}$$

Where ψ is the wave phase and E_0 is a slowly varying amplitude. We can now define instantaneous values of frequency ω and wave number k by

$$k = -\frac{\partial \psi}{\partial z} \tag{13.192a}$$

$$\omega = \frac{\partial \psi}{\partial t}. \tag{13.192b}$$

The local dispersion relation

$$\omega = \omega(k,z,t) \tag{13.193}$$

can be established when the properties of the medium (the refractive index n for a dielectric and the plasma frequency ω_p for a plasma medium) are known.

$$\frac{\partial k}{\partial t} = -\frac{\partial}{\partial t}\frac{\partial \psi}{\partial z} = -\frac{d}{dz}\frac{\partial \psi}{\partial t} = -\frac{d\omega}{dz} = -\left[\frac{\partial \omega}{\partial z} + \frac{\partial \omega}{\partial k}\frac{\partial k}{\partial z}\right] \tag{13.194}$$

$$\frac{\partial k}{\partial t} = -\left[\frac{\partial \omega}{\partial z} + v_g \frac{\partial k}{\partial z}\right], \tag{13.195}$$

where we define v_g as the group velocity:

$$v_g = \frac{\partial \omega}{\partial k}. \tag{13.196}$$

The last term on the RHS of (13.194) arises since ω is also a function of k. Equation 13.193 can also be written as

$$\frac{\partial k}{\partial t} + v_g \frac{\partial k}{\partial z} = -\frac{\partial \omega}{\partial z}. \tag{13.197}$$

If v_g is also the group velocity of the propagating wave packet, the LHS of (13.197) can be written as the total derivative dk/dt. Thus, we get the Equation 13.198 akin to the second of the Hamiltonian pair (13.180):

$$\frac{dk}{dt} = -\frac{\partial \omega}{\partial z}. \tag{13.198}$$

The group velocity v_g can also be written in terms of the position of the centroid of the wave packet, that is, dr/dt. In our example of the one-dimensional wave, it is given by dz/dt. Thus, we get the first of the pair of the Hamiltonian canonical equations, like (13.179), by equating v_g obtained in two ways:

$$\frac{dz}{dt} = \frac{\partial \omega}{\partial k}. \tag{13.199}$$

In this example, the generalized coordinates are z and k, and the Hamiltonian is $\omega(k,z,t)$. The energy E of a photon is $hf = h\omega/2\pi = \hbar\omega$ and the momentum of the photon is $\hbar k$.

From (13.181) and the Hamiltonian analogy, the Lagrangian L for a photon can be written as

$$L(z,v_g,t) = v_g k - \omega(z,k,t). \tag{13.200}$$

The Lagrange equation for the photon is given by

$$\frac{d}{dt}\left(\frac{\partial L}{\partial v_g}\right) - \left(\frac{\partial L}{\partial z}\right) = 0. \tag{13.201}$$

Note from (13.200)

$$\frac{\partial L}{\partial v_g} = k \tag{13.202a}$$

$$\frac{\partial L}{\partial z} = -\frac{\partial \omega}{\partial z}. \tag{13.202b}$$

Thus, (13.201) is the same as (13.198), repeated here as (13.203):

$$\frac{dk}{dt} = -\frac{\partial \omega}{\partial z}. \tag{13.203}$$

If k is like momentum and ω is like energy, Equation 13.203 is like Newton's law of motion. Also note that (13.203) is akin to the second of the Hamiltonian canonical pair for the photon given by (13.180).

Photon in vacuum has zero mass and zero charge. However, one can consider it having an effective mass m_{eff} and equivalent charge q_{ph} in a medium [20]. For example, Mendonca [20] arrives at the following values for these in an isotropic cold electron plasma of plasma frequency ω_p :

$$m_{eff} = \frac{\omega_p \hbar}{c^2} \tag{13.204}$$

$$q_{ph} = -\hbar \frac{e k_p^2}{m_{eff}\omega_0} \tag{13.205}$$

$$k_p = \frac{\omega_p}{v_g}, \tag{13.206}$$

where m_{eff} and e are the mass and absolute value of the charge, respectively, of an electron. ω_0 is the central frequency and v_g is the group velocity of the wave packet. These concepts help explain [20] the "ponderomotive force," photon-beam plasma instabilities, etc. Ponderomotive force is due to the radiation pressure of the photon gas exerted on the electrons of the plasma.

Equations 13.204 and 13.205 give the effective mass and the equivalent charge of a "dressed" particle in contrast to the characterization of a photon in vacuum as a bare particle [20] of zero mass and zero charge.

13.6.8.3 Space and Time Refraction Explained through Photon Theory [20]

We can use the above mentioned photon theory based on the Hamiltonian to explain the constancy of the frequency ω at a spatial discontinuity and the constancy of the wave number k at a temporal discontinuity of a medium. A temporal discontinuity consequently causes a change in the frequency. This aspect will be pursued more extensively in Section 13.9 based on Maxwell's equations and full wave theory.

We generalize the direction of wave propagation by writing Equations 13.191 through 13.192b in a generalized form:

$$E_x(z,t) = E_0(z,t)e^{j\psi(r,t)} \tag{13.207}$$

$$k = -\nabla\psi \quad (32a) \tag{13.208a}$$

$$\omega = \frac{\partial\psi}{\partial t}. \tag{13.208b}$$

The Hamiltonian pair is given by

$$\frac{d\mathbf{k}}{dt} = -\nabla\omega \tag{13.209}$$

$$\frac{d\mathbf{r}}{dt} = \nabla_k\omega. \tag{13.210}$$

As long as ω is not an *explicit* function of time, we also have

$$\frac{d\omega}{dt} = \frac{\partial\omega}{\partial t}. \quad (35) \tag{13.211}$$

Consider a dielectric medium [20] with

$$\omega = \frac{kc}{n}. \tag{13.212}$$

From (13.210),

$$\frac{d\mathbf{r}}{dt} = \nabla_k\omega = \frac{\partial}{\partial k}\frac{kc}{n}\hat{k} = \frac{c}{n}\hat{k} = \frac{ck}{n}\frac{\mathbf{k}}{k^2} = \frac{\omega\mathbf{k}}{k^2}$$

$$\frac{d\mathbf{r}}{dt} = \frac{\omega\mathbf{k}}{k^2}. \tag{13.213}$$

From (13.209),

$$\begin{aligned}\frac{d\mathbf{k}}{dt} &= -\nabla\omega = \frac{\partial}{\partial r}\frac{kc}{n}\hat{r} = \frac{kc}{n^2}\frac{\partial n}{\partial r}\hat{r} = \frac{\omega}{n}\frac{\partial n}{\partial r}\hat{r} = \omega\nabla\ln(n)\\ \frac{d\mathbf{k}}{dt} &= \omega\nabla\ln(n).\end{aligned} \tag{13.214}$$

From (13.211),

$$\begin{aligned}\frac{d\omega}{dt}&=\frac{\partial\omega}{\partial t}=\frac{\partial}{\partial t}\left(\frac{kc}{n}\right)=-\frac{kc}{n^2}\frac{\partial n}{\partial t}=-\frac{\omega}{n}\frac{\partial n}{\partial t}=-\omega\frac{\partial}{\partial t}\left[\ln(n)\right]\\ \frac{d\omega}{dt}&=-\omega\frac{\partial}{\partial t}\left[\ln(n)\right].\end{aligned}\tag{13.215}$$

Let us now consider two separate cases of n being a function of only one independent variable.

Case 1 Let

$$n(\boldsymbol{r},t)=n(z),\tag{13.216}$$

where $n(z)$ is a step function shown in Figure 13.22.

From (13.215),

$$\frac{d\omega}{dt}=0\tag{13.217a}$$

$$\omega=\text{constant}.\tag{13.217b}$$

Thus, we get the well-known result that the frequency of the incident, reflected, and the transmitted waves is the same, as stated in (2.90). From (13.214) and (13.216), since n is a function of z, the x-component of $k=k_x$ satisfies

$$\frac{dk_x}{dt}=0\tag{13.218a}$$

$$k_x=\text{constant}.\tag{13.218b}$$

The x-component of $\boldsymbol{k}$ is the same for the incident, reflected, and the transmitted waves, from which we get the well-known Snell's law, normally stated as in (2.91) and shown in Figure 13.22.

Case 2 Let

$$n(\boldsymbol{r},t)=n(t)\tag{13.219}$$

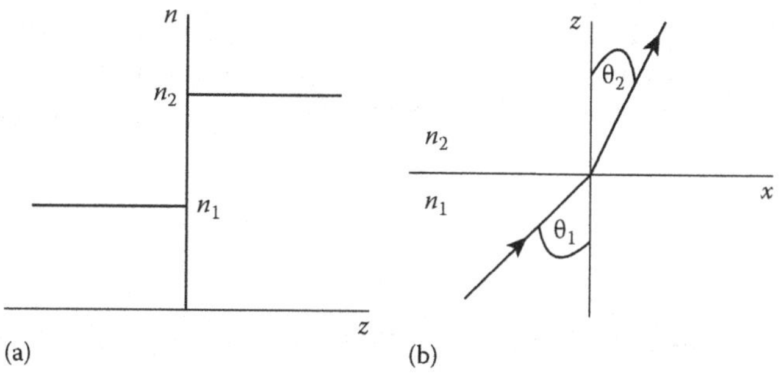

FIGURE 13.22
Refraction at a spatial discontinuity: (a) spatial step profile of the refractive index and (b) refraction at the spatial boundary.

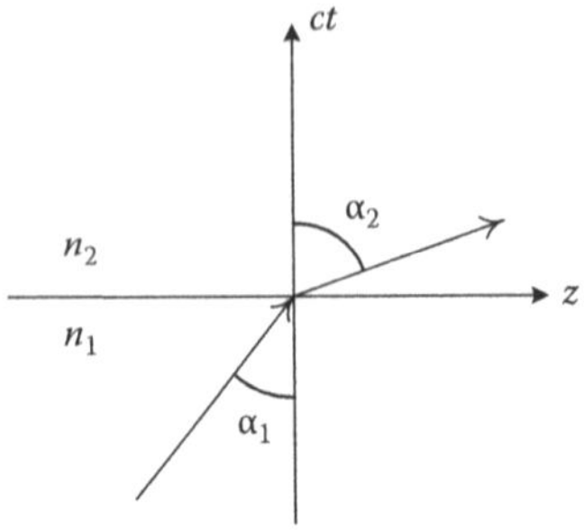

FIGURE 13.23
Refraction of a photon at a temporal discontinuity.

where $n(t)$ is a step function and the wave is propagating in the z-direction, that is, $\mathbf{k} = \hat{z}k$.

Since, in this case, n is not a function of z, from (13.214)

$$\frac{d\mathbf{k}}{dt} = 0 \tag{13.220a}$$

$$k = \text{constant.} \tag{13.220b}$$

Thus, at a temporal discontinuity, it is the wave number that is conserved. Figure 13.23 shows the refraction in time (vertical axis ct), and the angles α_1 and α_2 are akin to the θ_1 and θ_2 in Figure 13.22b [21].

The relationships between α, ω, n, and the group velocity v_g of the wave packet can be stated as

$$\frac{dz}{dt} = \frac{c}{n(t)} = v_g(t) = \tan\alpha, \tag{13.221}$$

and for a time step profile,

$$n_1\omega_1 = n_2\omega_2 \tag{13.222}$$

$$\frac{\tan\alpha_2}{\tan\alpha_1} = \frac{\omega_2}{\omega_1}. \tag{13.223}$$

Mendonca [20] calls it time refraction and discusses further the space–time refraction from the viewpoint of photon ray theory. These and other problems can be discussed using "full wave" theory [21]. A brief account [22, Appendix 10C] of using Maxwell's equations in the time domain to calculate the amplitudes and frequencies of the waves generated by "switching the medium," particularly the plasma medium, is given in Section 13.9.

13.7 Nuclear Electromagnetic Pulse and Time-Varying Conducting Medium

A 400 km high-altitude nuclear test conducted in July 1962 on the mid-Pacific Ocean, named Star Fish, produced an electromagnetic pulse over a large area. This pulse of about 1 μs rise time (t_r) is called HEMP-E1 (High Altitude Electromagnetic Pulse, early-time), which knocked out street lighting as far away as Hawaii and destroyed some electronic equipment [23]. Though Fermi [24] anticipated such an electromagnetic pulse as early as 1945, the strength and the vast area over which the HEMP had a destructive effect surprised many. A proper scientific explanation for HEMP was offered by Longmire [25] in 1966. A detailed discussion was given in a report prepared for Oakridge National Laboratory in 2010 [26]. It can be noted that the intermediate-time E2 HEMP ($t_r = 1$ μs to 1 s) and late-time E3 HEMP ($t_r = 1$ s to 1000s) are not discussed in this section.

Nuclear bombs emit a small percentage of their energy in gamma rays, which are electromagnetic waves of frequency 10^{20}–10^{21} Hz. These gamma rays traveling with the speed of light, interacting with air atoms at about 50 km height through "Compton scattering" [25,26], produce energetic (about 1 MeV) electrons moving in the forward direction. These directed electrons constitute Compton current J_c. One way of explaining EMP is to consider that the Compton current radiates the EMP. The 1-MeV electrons have relativistic velocities with $v/c = \beta = 0.94$.

The Compton electrons turn [25] (Figure 13.24) in the geomagnetic field (say about 0.56 gauss) with a gyroradius $\rho_L = 85$ m. However, these Compton electrons collide with the electrons in air atoms as they enter the thicker atmosphere at about 50 km height and lose their energy by the time they reach down to, say, 30 km. The energy lost by the 1-MeV electron in stopping produces about 30,000 secondary electrons distributed along its path.

The secondary electrons have randomly distributed velocities and by themselves do not produce a current. However, in the presence of an electric field, they drift, and this effect is best modeled by Ohm's law $J = \sigma E$. The conductivity varies with time with an initial zero value. The current J_c produced by the Compton electrons is in opposition to the conduction current due to the secondary electrons produced by the Compton electrons being stopped by the increasingly dense atmosphere as the gammas travel downwards.

One can use a planar geometry model (Figure 13.25) to approximately calculate and describe the shape of the HEMP-E1.

Let

$$\mathbf{E} = \hat{x} E_x(T) \tag{13.224}$$

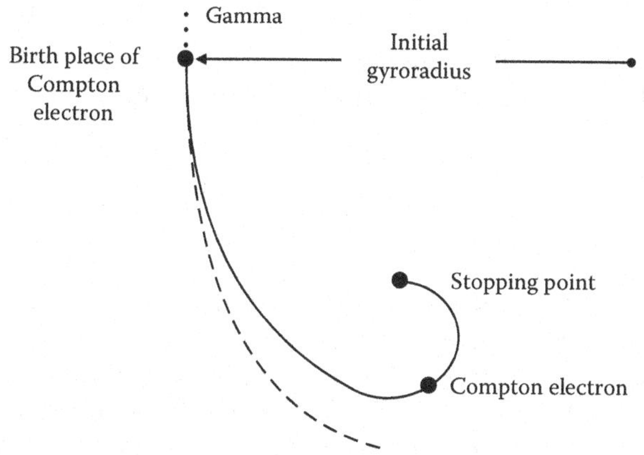

FIGURE 13.24
Birth and turning of the Compton electron in a geomagnetic field.

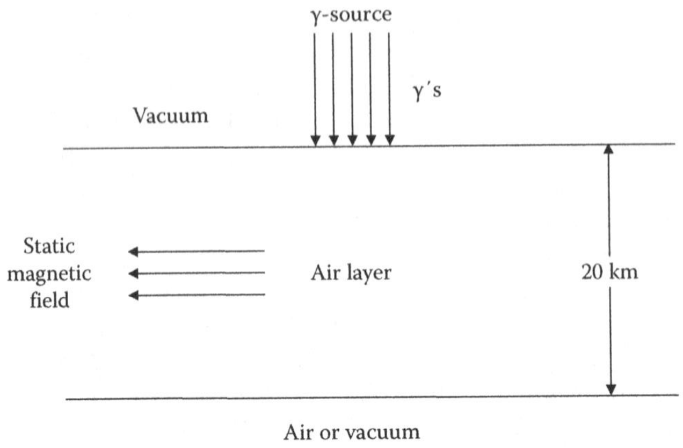

FIGURE 13.25
Planar geometry to simulate E1 HEMP.

$$T = t - \frac{z}{c}. \tag{13.225}$$

Thus, E_x is the electric field an observer at z would see if the person "triggered the horizontal sweep on the arrival of the gamma pulse" [25]. However, the field $E_x(T,z)$ will be the electric field observed by different observers (distinguished by their z coordinates) on their gamma-triggered scopes.

The outgoing wave equation for the simplified planar model is given by Longmire [25] from the conservation of the energy principle:

$$\frac{\partial E_x}{\partial z} + \frac{\eta_0 \sigma}{2} E_x = -\frac{\eta_0}{2} J_{cx}. \tag{13.226}$$

If there is no Compton current ($J_{cx} = 0$), then E_x attenuates with distance according to

$$E_x(T,z) = E_{x0}(T) e^{-\frac{\eta_0 \sigma}{2} z}. \tag{13.227}$$

On the other hand, the very early-time approximation will be based on the conductivity being negligible, and the electric field is entirely due to the Compton current and is given by

$$E_x(T,z) = -\frac{\eta_0}{2} \int_0^z J_{cx}(T,z') dz'. \tag{13.228}$$

Based on these approximations, one can deduce that the shape of the pulse will be as shown by the solid line in Figure 13.26. A qualitative explanation for the shape is as follows.

The electromagnetic pulse is due to the two competing effects: the Compton current generating the electric field as per (13.228), shown as a broken line curve in Figure 13.26, and the secondary electrons making conductivity that tries to quench the electric field.

This leads to an effect called saturation [26], shown in Figure 13.26 as a dotted curve, which will not permit the electric field to become too high. The peak value for the electric

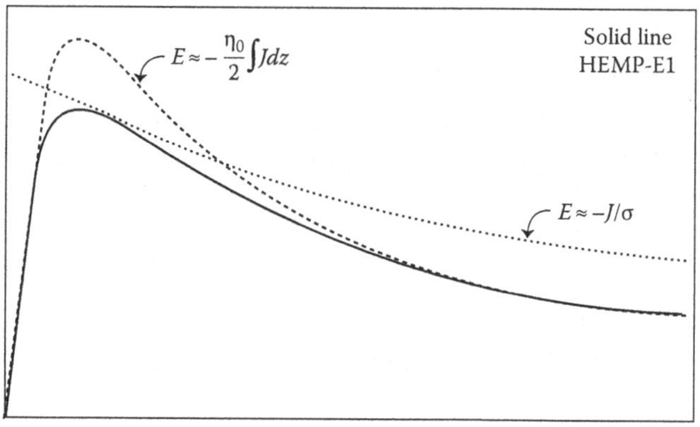

FIGURE 13.26
The broken line curve is the electric field due to the Compton current and the dotted line curve is due to saturation. The solid line curve is the HEMP-E1.

field will increase with the fast buildup of E1 by Compton current and also will increase with the slow buildup of the conductivity by the secondary electrons. The peak value is less dependent on the yield of the nuclear blast than on the relative rates of E1 buildup due to Compton electrons, and the buildup of the conductivity due to the secondary electrons. The later buildup depends on the air chemistry as discussed in [26].

A mathematical expression [26] for a generic waveform of HEMP-E1 given by International Electrotechnical Commission (IEC) is given in (13.229):

$$E_1(t) = 0, \quad t < 0 \tag{13.229a}$$

$$E_1(t) = E_{01}k_1\left(e^{-b_1 t} - e^{-a_1 t}\right), \quad t > 0 \tag{13.229b}$$

$$E_{01} = 50 \text{ kV m}^{-1} \quad k_1 = 1.3 \quad a_1 = 6\times10^8 \text{s}^{-1} \quad b_1 = 4\times10^7 \text{ s}^{-1}. \tag{13.229c}$$

The peak value is about 73,500 times the amplitude of a typical FM signal. A vast amount of information is available on the topic of the HEMP, and two additional references [27,28] are chosen to lead to some of them.

An electrical circuit analogy to the EMP production is offered in Figure 13.27.

A problem at the end of the chapter pursues this analogy in explaining the competing processes in determining the peak value of the HEMP-E1 and the time of its occurrence.

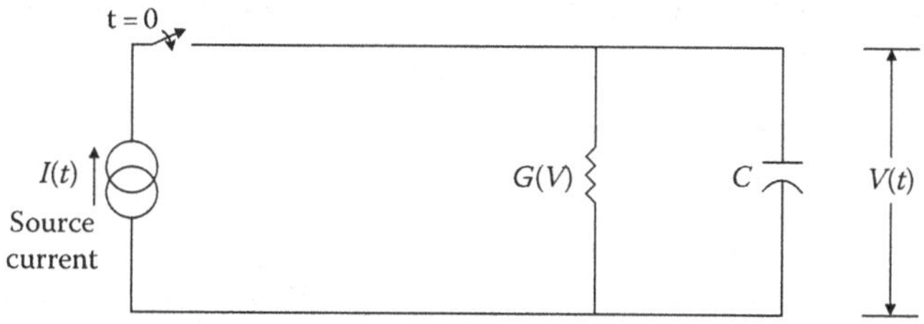

FIGURE 13.27
Circuit analogy to show the competing effects of the Compton electron current flowing through the capacitor and the saturation effect due to the current flow in the voltage-dependent conductance element in parallel. The source current decays with time.

13.8 Magnetohydrodynamics (MHD)

MHD is the science of the interaction of the electric and magnetic fields with a moving conducting fluid. Electric fields arise due to the motion of the conducting fluid in the presence of a magnetic field. The electric field drives a current, which in turn modifies the B field. The mechanical motion of the system is described in terms of the hydrodynamic variables ρ (mass density), v (velocity), and p (pressure), and the electromagnetic variables E, B, and J. The relevant equations are [15,17]

$$\frac{\partial \rho}{\partial t}+\nabla\cdot(\rho \boldsymbol{v})=0 \tag{13.230}$$

$$\rho\frac{d\boldsymbol{v}}{dt}=-\nabla p+\boldsymbol{J}\times\boldsymbol{B}+\boldsymbol{F}_v+\rho g \tag{13.231}$$

$$\nabla\times\boldsymbol{E}=-\frac{\partial \boldsymbol{B}}{\partial t} \tag{13.232}$$

$$\nabla\times\boldsymbol{B}=\mu_0\boldsymbol{J}. \tag{13.233}$$

In Equation 13.231, $\boldsymbol{F}_v$ is the viscous force. In Equation 13.233 we did not include the term due to the displacement current, since it is much smaller than the term due to the conduction current for most of the MHD applications.

A simplified constitutive equation (Ohm's law) for this case is given by

$$J=\sigma(\boldsymbol{E}+\boldsymbol{v}\times\boldsymbol{B}). \tag{13.234}$$

Equation 13.234 can be obtained by noting that in the rest frame moving with nonrelativistic velocity v, $\boldsymbol{J}'=\sigma\boldsymbol{E}'$ and $\boldsymbol{J}=\boldsymbol{J}'+\rho_e\boldsymbol{v}$. But ρ_e, the net charge density, is zero, since in this one component model, both ions and electrons move with the same velocity and there is no charge accumulation. Thus, $\boldsymbol{J}=\boldsymbol{J}'$. However,

$$\boldsymbol{E}'=(\boldsymbol{E}+\boldsymbol{v}\times\boldsymbol{B}). \tag{13.235}$$

For some applications, instead of (13.234), one may have to use a more generalized Ohm's law [15].

13.8.1 Evolution of the *B* Field

In MHD applications, the study of the evolution of the B field by itself is important.

From (13.232) and (13.234),

$$\frac{\partial \boldsymbol{B}}{\partial t}=-\nabla\times\boldsymbol{E}=-\nabla\times\left[\frac{\boldsymbol{J}}{\sigma}-\boldsymbol{v}\times\boldsymbol{B}\right] \tag{13.236}$$

$$\frac{\partial \boldsymbol{B}}{\partial t}=-\nabla\times\left[\frac{\nabla\times\boldsymbol{B}}{\sigma\mu_0}\right]+\nabla\times(\boldsymbol{v}\times\boldsymbol{B}) \tag{13.237}$$

$$\frac{\partial \boldsymbol{B}}{\partial t}=-\frac{1}{\sigma\mu_0}\left[\nabla(\nabla\cdot\boldsymbol{B})-\nabla^2\boldsymbol{B}\right]+\nabla\times(\boldsymbol{v}\times\boldsymbol{B}). \tag{13.238}$$

Since $\nabla \cdot \boldsymbol{B}$ is zero, the equation for $\boldsymbol{B}$ can be written as

$$\frac{\partial \boldsymbol{B}}{\partial t} = \nu_m \nabla^2 \boldsymbol{B} + \nabla \times (\boldsymbol{v} \times \boldsymbol{B}), \tag{13.239}$$

where ν_m is called magnetic viscosity given by

$$\nu_m = \frac{1}{\sigma \mu_0}. \tag{13.240}$$

The first term on the RHS of (13.239) is called the diffusion term, and the second term is called the flow term.

The behavior of the solution of (13.239) largely depends on the conductivity σ. Let us study the effect of each term on the RHS of (13.239) on the solution for $\boldsymbol{B}$.

Case 1 ($v = 0$) fluid at rest

For the case of the fluid at rest, $\boldsymbol{B}$ satisfies the following equation:

$$\frac{\partial \boldsymbol{B}}{\partial t} = \nu_m \nabla^2 \boldsymbol{B}. \tag{13.241}$$

Equation 13.241 is a classical diffusion equation showing that an initial configuration of $\boldsymbol{B}$ vector will decay away in a diffusion time τ_d. The expression for the τ_d can be obtained as follows. If L is a characteristic length, then (13.241) may be approximated as

$$\frac{\partial B}{\partial t} \approx \frac{\nu_m}{L^2} B \tag{13.242}$$

and its solution as

$$B \approx B_0 e^{-\frac{\nu_m}{L^2} t} = B_0 e^{-\frac{t}{\tau_d}}, \tag{13.243}$$

where

$$\tau_d = \frac{L^2}{\nu_m} = \sigma \mu_0 L^2. \tag{13.244}$$

Based on (13.244), it is estimated [17] that it takes 10^4 years for the magnetic field to decay in the molten core of the earth and 10^{10} years in the sun.

Case 2 σ large

The other limit is that sigma is large, and thus, ν_m is small. One can quantify how small it needs to be by defining the magnetic Reynolds number R_m:

$$R_m = \frac{vL}{\nu_m} = \frac{v\tau_d}{L}. \tag{13.245}$$

The Case 2 belongs to the high Reynolds number. In this case, the first term in (13.239) is neglected and the equation for $\boldsymbol{B}$ vector is given by

$$\frac{\partial \boldsymbol{B}}{\partial t} = \nabla \times (\boldsymbol{v} \times \boldsymbol{B}). \tag{13.246}$$

Integrating (13.246) over an open surface S

$$\iint_S \frac{\partial \boldsymbol{B}}{\partial t}\cdot d\boldsymbol{s} = \iint_S \nabla\times(\boldsymbol{v}\times\boldsymbol{B})\cdot d\boldsymbol{s} = \oint_C (\boldsymbol{v}\times\boldsymbol{B})\cdot d\boldsymbol{l} = -\oint_C (\boldsymbol{B})\cdot(\boldsymbol{v}\times d\boldsymbol{l}). \tag{13.247}$$

Thus, we have

$$\iint_S \frac{\partial \boldsymbol{B}}{\partial t}\cdot d\boldsymbol{s} + \oint_C (\boldsymbol{B})\cdot(\boldsymbol{v}\times d\boldsymbol{l}) = 0. \tag{13.248}$$

The LHS of (13.248) can be combined into one term [15] giving us

$$\frac{d}{dt}\int_S \boldsymbol{B}\cdot d\boldsymbol{s} = 0. \tag{13.249}$$

Equation 13.249 states that the magnetic flux passing through the moving surface stays constant. This is known as the "frozen field condition."

The proof that (13.249) can be obtained from (13.248) is given as a problem since such a proof helps to understand the *emf* calculation by transformer *emf* and motional *emf* stated in (1.61) of Section 1.6.

13.9 Time-Varying Electromagnetic Medium

In Section 13.6.8, under photon ray theory, we considered space and time refraction separately and showed that a temporal discontinuity in the properties of a medium causes a change in the frequency. In this section, we briefly discuss the theory and applications of this frequency-shifting property of a time-varying medium from the viewpoint of full wave theory and Maxwell's equations. A more extensive discussion, particularly the aspect of amplitude calculations of the newly created frequency-shifted waves, is given in Kalluri [21].

13.9.1 Frequency Change due to a Temporal Discontinuity in the Medium Properties

Let us consider normal incidence on a spatial discontinuity in the dielectric properties of a medium of a plane wave propagating in the z-direction. The spatial step profile of the permittivity ε is shown at the top of Figure 13.28a.

The permittivity suddenly changes from ε_1 to ε_2 at $z=0$. Let us also assume that the permittivity profile is time invariant. The phase factors of the incident, reflected, and transmitted waves are expressed as $\psi_A = e^{j(\omega_A t - k_A z)}$, where the subscript $A=I$ for the incident wave, $A=R$ for the reflected wave, and $A=T$ for the transmitted wave. The boundary condition of the continuity of the tangential component of the electric field at $z=0$ *for all t* requires

$$\omega_I = \omega_R = \omega_T = \omega_A. \tag{13.250}$$

This result can be stated as follows: The frequency ω is conserved across a spatial discontinuity in the properties of the electromagnetic medium. As the wave crosses from one

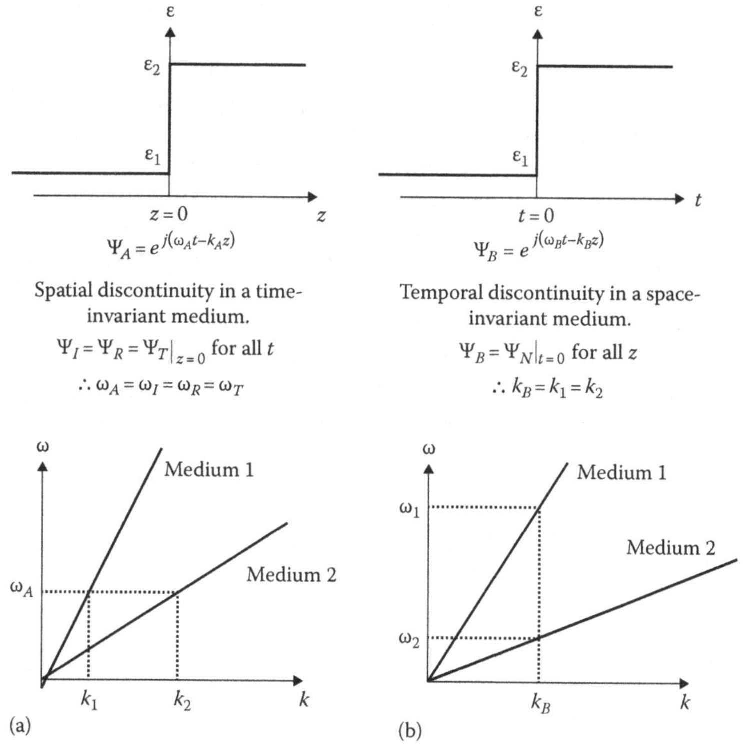

FIGURE 13.28
Comparison of the effects of (a) spatial and (b) temporal discontinuities.

medium to the other in space, the wave number k changes as dictated by the change in the phase velocity, not considering absorption here. The bottom part of Figure 13.28a illustrates this aspect graphically. The slopes of the two straight lines in the ω–k diagram are the phase velocities in the two mediums. Conservation of ω is implemented by drawing a horizontal line, which intersects the two straight lines. The k values of the intersection points give the wave numbers in the two mediums.

A dual problem can be created by considering a temporal discontinuity in the properties of the medium. Let an unbounded medium (in space) undergo a sudden change in its permittivity at $t = 0$. The continuity of the electric field at $t = 0$ now requires that the phase factors of the wave existing before the discontinuity occurs, called a source wave, must match with phase factors, Ψ_N, of the newly created waves in the altered or switched medium, when $t = 0$ is substituted in the phase factors. This must be true for all values of z. Thus, it leads to the requirement that k is conserved across a temporal discontinuity in a spatially invariant medium. Conservation of k is implemented by drawing a vertical line in the ω–k diagram as shown in the bottom part of Figure 13.28b. The ω values of the intersection points give the frequencies of the newly created waves [29–32].

13.9.2 Effect of Switching an Unbounded Isotropic Plasma Medium

Wave propagation in isotropic plasmas is discussed in Chapter 9. Given below is a summary to act as a self-supporting background of this topic in reference to the time-varying media topic under discussion here.

Any plasma is a mixture of charged particles and neutral particles. The mixture is characterized by two independent parameters for each of the particle species. These are the particle density N and the temperature T. There is a vast amount of literature on plasmas. A few references of direct interest to the reader of this chapter are provided in [15,33,34]. The models are adequate in exploring some of the applications where the medium can be considered to have time-invariant electromagnetic parameters.

There are some applications in which the thermal effects are unimportant. Such a plasma is called *cold plasma*. A *Lorentz plasma* [33] is a further simplification of the medium. It is assumed that the electrons interact with each other in a Lorentz plasma only through collective space charge forces and that the heavy positive ions and neutral particles are at rest. The positive ions serve as a background that ensures the overall charge neutrality of the mixture. In this section, the Lorentz plasma is the dominant model used to explore the major effects of a nonperiodically time-varying electron density profile $N(t)$.

The constitutive relations for this simple model viewed as a dielectric medium are given by

$$\mathbf{D} = \varepsilon_o \varepsilon_p \mathbf{E}, \tag{13.251}$$

where

$$\varepsilon_p = 1 - \frac{\omega_p^2}{\omega^2}, \tag{13.252}$$

and

$$\omega_p^2 = \frac{q^2 N}{m \varepsilon_o}. \tag{13.253}$$

In these equations, q and m are the absolute values of the charge and mass of the electron, respectively, and ω_p^2 is the square of the plasma frequency proportional to the electron density.

A sketch of ε_p versus ω is given in Figure 9.1. The relative permittivity ε_p is real-positive valued only if the signal frequency is larger than the plasma frequency. Hence, ω_p is a cutoff frequency for the isotropic plasma. Above cutoff, a Lorentz plasma behaves as a dispersive dielectric with the relative permittivity lying between 0 and 1.

The relative permittivity of the medium can be changed by changing the electron density N, which in turn can be accomplished by changing the ionization level. The sudden change in the permittivity shown in Figure 13.28b is an idealization of a rapid ionization. Quantitatively, the sudden-change approximation can be used if the period of the source wave is much larger than the rise time of the temporal profile of the electron density (subcycle time–varying medium). A step change in the profile is referred to in the literature as sudden creation [21,30] or flash ionization [21].

Experimental realization of a small rise time is not easy. A large region of space has to be ionized uniformly at a given time. Joshi et al. [35], Kuo [36], Kuo and Ren [37], as well as Rader et al. [38] developed ingenious experimental techniques to achieve these conditions and demonstrated the principle of frequency shifting using isotropic plasma (see Part IV of Reference 21). One of the earliest pieces of experimental evidence of frequency shifting quoted in the literature is a seminal paper by Yablonovitch [39]. Savage et al. [40] used the ionization front to upshift the frequencies. Ionization front is a moving boundary between unionized medium and the plasma [41]. Such a front can be created by a source of ionizing radiation pulse, say, a strong laser pulse. As the pulse travels in a neutral gas, it converts it into plasma, thus creating a moving boundary between the plasma and the unionized medium. However, the ionization-front problem is somewhat different from the moving-plasma problem. In the front problem, the boundary alone is moving and the plasma is not moving with the boundary.

The constitutive relation (13.251), based on the dielectric model of a plasma, does not explicitly involve the current density **J** in the plasma. The constitutive relations that involve the plasma current density **J** are given by

$$\mathbf{D} = \varepsilon_o \mathbf{E}, \tag{13.254}$$

$$\mathbf{J} = -qN\mathbf{v}. \tag{13.255}$$

The velocity **v** of the electrons is given by the force equation

$$m\frac{d\mathbf{v}}{dt} = -q\mathbf{E}. \tag{13.256}$$

In (13.256), the magnetic force due to the wave's magnetic field **H** is neglected since it is much smaller [33] than the force due to the wave's electric field. The magnetic force term $(-q\mathbf{v}\times\mathbf{H})$ is nonlinear. Stanic [42] studied this problem as a weakly nonlinear system.

Since ion motion is neglected, (13.255) does not contain ion current. Such an approximation is called radio approximation [34]. It is used in the study of radio wave propagation in the ionosphere. Low-frequency wave propagation studies take into account the ion motion [34].

Substituting (13.254) through (13.256) in the Ampere–Maxwell equation

$$\nabla\times\mathbf{H} = \mathbf{J} + \frac{\partial\mathbf{D}}{\partial t}, \tag{13.257}$$

we obtain

$$\nabla\times\mathbf{H} = j\omega\varepsilon_o\varepsilon_p(\omega)\mathbf{E}, \tag{13.258}$$

where $\varepsilon_p(\omega)$ is given by (13.252) and an $\exp(j\omega t)$ time dependence has been assumed. For an arbitrary temporal profile of the electron density $N(t)$, (13.255) is not valid [21,43,44]. The electron density $N(t)$ increases because of the new electrons born at different times. The newly born electrons start with zero velocity and are subsequently accelerated by the fields. Thus, all the electrons do not have the same velocity at a given time during the creation of the plasma. Therefore,

$$\mathbf{J}(t) \neq -qN(t)\mathbf{v}(t), \tag{13.259}$$

but

$$\Delta\mathbf{J}(t) = -q\Delta N_i\mathbf{v}_i(t), \tag{13.260}$$

instead. Here, ΔN_i is the electron density added at t_i, and $\mathbf{v}_i(t)$ is the velocity at time t of these ΔN_i electrons born at time t_i. Thus, $\mathbf{J}(t)$ is given by the integral of (13.260) and not by

(13.259). The integral of (13.260), when differentiated with respect to t, gives the constitutive relation between **J** and **E** as follows:

$$\frac{d\mathbf{J}}{dt} = \varepsilon_o \omega_p^2(\mathbf{r},t)\mathbf{E}(\mathbf{r},t). \tag{13.261}$$

Equations 13.257 and 13.261 and the Faraday equation

$$\nabla \times \mathbf{E} = -\mu_o \frac{\partial \mathbf{H}}{\partial t} \tag{13.262}$$

are needed to describe the electromagnetics of isotropic plasmas.

Propagation of an electromagnetic wave traveling in the z-direction with $\mathbf{E} = \hat{\mathbf{x}}E$ and $\mathbf{H} = \hat{\mathbf{y}}H$ can be studied by assuming that the components of the field variables have harmonic space variation, that is,

$$F(z,t) = f(t)e^{-jkz}. \tag{13.263}$$

Substituting (13.263) in (13.257), (13.261), and (13.262), we obtain the wave equations

$$\frac{d^2E}{dt^2} + \left[k^2c^2 + \omega_p^2(t)\right]E = 0, \tag{13.264}$$

and

$$\frac{d^3H}{dt^3} + \left[k^2c^2 + \omega_p^2(t)\right]\frac{dH}{dt} = 0, \tag{13.265}$$

for E and H.

13.9.2.1 Sudden Creation of an Unbounded Plasma Medium

The geometry of the problem is shown in Figure 13.29. A plane wave of frequency ω_o is propagating in free space in the z-direction. Suddenly at $t = 0$, an unbounded plasma medium of plasma frequency ω_p is created. Thus arises a temporal discontinuity in the properties of the medium. The solution of (13.265), when ω_p is a constant, can be obtained as

$$H(t) = \sum_{m=1}^{3} H_m \exp(j\omega_m t), \tag{13.266}$$

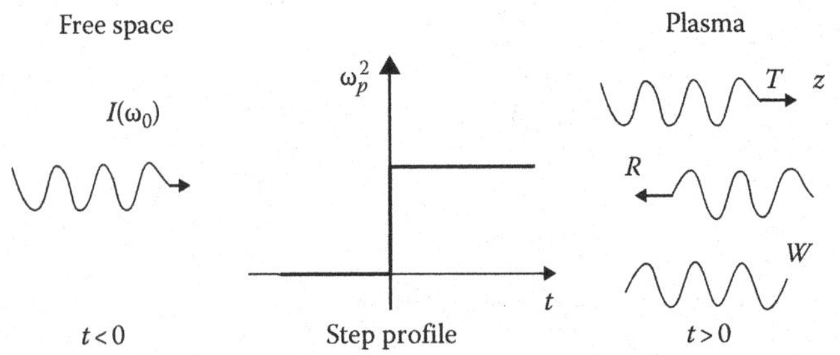

FIGURE 13.29
Suddenly created unbounded plasma medium.

where

$$\omega_m\left[\omega_m^2-\left(k^2c^2+\omega_p^2\right)\right]=0. \tag{13.267}$$

The ω–k diagram [2] for the problem under discussion is obtained by graphing (13.267). Figure 13.30 shows the ω–k diagram, where the top and the bottom branches are due to the factor in the square brackets equated to zero, and the horizontal line is due to the factor $\omega = 0$. The line k = constant is a vertical line that intersects the ω–k diagram at the three points marked as 1, 2, and 3. The third mode is the *wiggler* mode [21,30,45]. Its real-valued fields are

$$\mathbf{E}_3(x,y,z,t)=0, \tag{13.268}$$

$$\mathbf{H}_3(x,y,z,t)=\hat{\mathbf{y}}H_o\frac{\omega_p^2}{\omega_o^2+\omega_p^2}\cos(kz), \tag{13.269}$$

$$\mathbf{J}_3(x,y,z,t)=\hat{\mathbf{x}}H_ok\frac{\omega_p^2}{\omega_o^2+\omega_p^2}\sin(kz). \tag{13.270}$$

It is of zero frequency but varies in space. Such wiggler fields are used in a free electron laser (FEL) to generate coherent radiation [46]. Its electric field is zero but has a magnetic field due to the plasma current $\mathbf{J}_3$. In the presence of a static magnetic field in the z-direction, the third mode becomes a traveling wave with a downshifted frequency. This aspect is discussed below in Section 13.9.4, under *Frequency-shifting characteristics of various R waves.*

The modes 1 and 2 have frequencies given by

$$\omega_{1,2}=\pm\sqrt{\left(\omega_p^2+k^2c^2\right)}, \tag{13.271}$$

where ω_2 has a negative value. Since the harmonic variations in space and time are expressed in the phase factor exp($\omega t-kz$), a negative value for ω gives rise to a wave propagating in the negative z-direction. It is a backward-propagating wave or, for

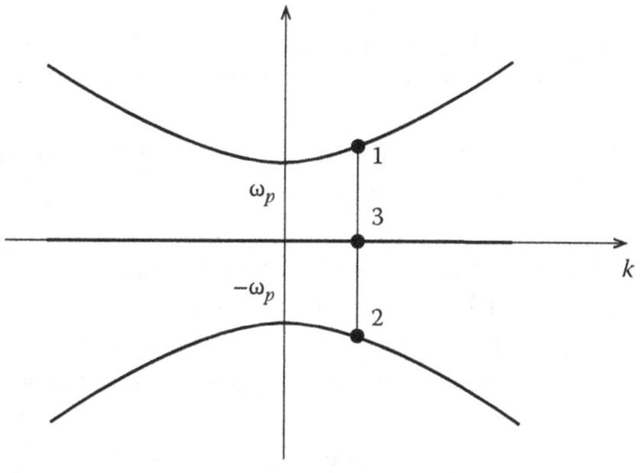

FIGURE 13.30
ω–k diagram and wiggler magnetic field.

convenience, can be referred to as a wave reflected by the discontinuity. The modes 1 and 2 have higher frequencies than the source wave. These are upshifted waves.

13.9.3 Sudden Creation of a Plasma Slab

The interaction of an electromagnetic wave with a plasma slab is experimentally more realizable than with an unbounded plasma medium. When an incident wave enters a pre-existing plasma slab, the wave must experience a spatial discontinuity. If the plasma frequency is lower than the incident wave frequency, then the incident wave is partially reflected and transmitted. When the plasma frequency is higher than that of the incident wave, the wave is totally reflected because the relative permittivity of the plasma is less than zero.

However, if the plasma slab is sufficiently thin, the wave can be transmitted by tunneling effect [21]. For this time-invariant plasma, the reflected and transmitted waves have the same frequency as the source wave frequency, and they are called as *A* waves. The wave inside the plasma has a different wave number but the same frequency due to the requirement of the boundary conditions.

When a source wave is propagating in free space and suddenly a plasma slab is created, the wave inside the slab region experiences a temporal discontinuity in the properties of the medium. Hence, the switching action generates new waves whose frequencies are upshifted, and then the waves propagate out of the slab. They are called as *B* waves. The phenomenon is illustrated in Figure 13.31a. In Figure 13.31a, the source wave of frequency ω_o is propagating in free space. At $t = 0$, a slab of plasma frequency ω_p is created. The *A* waves in Figure 13.31b have the same frequency as that of the source wave. The *B* waves are newly created waves due to the sudden switching of the plasma slab and have upshifted frequencies $\omega_1 = \sqrt{\left(\omega_o^2 + \omega_p^2\right)} = -\omega_2$. The negative value for the frequency of the second *B* wave shows that it is a backward-propagating wave. These waves, however, have the same wave number as that of the source wave as long as they remain in the slab region. As the *B* waves come out of the slab, they encounter a spatial discontinuity, and therefore the wave number changes accordingly. The *B* waves are only created at the time of switching and, hence, exist for a finite time. In Figure 13.31, the waves are sketched in the time domain to show frequency changes. Thus, the arrow head symbol in the sketches show the *t* variable. See Figure 13.31b for a more accurate description. Note that the vertical axis is *ct*.

13.9.4 Time-Varying Magnetoplasma Medium

Wave propagation in an anisotropic magnetoplasma is discussed in Chapter 11. Given below is a summary to act as a self-supporting background of this topic in reference to the time-varying media topic under discussion here.

A plasma medium in the presence of a static magnetic field behaves like an anisotropic dielectric [21]. Therefore, the theory of electromagnetic wave propagation in this medium is similar to the theory of light waves in crystals. Of course, in addition, account has to be taken of the highly dispersive nature of the plasma medium.

A cold magnetoplasma is described by two parameters: the electron density *N* and the quasistatic magnetic field. The first parameter is usually given in terms of the plasma frequency ω_p. The strength and the direction of the quasistatic magnetic field have significant effect on the dielectric properties of the plasma. The parameter that is proportional to the static magnetic field is the electron gyrofrequency ω_b defined below. The cutoff frequency of the magnetoplasma is influenced by ω_p as well as ω_b. An additional important aspect of the dielectric properties of the magnetoplasma medium is the existence of a resonant frequency.

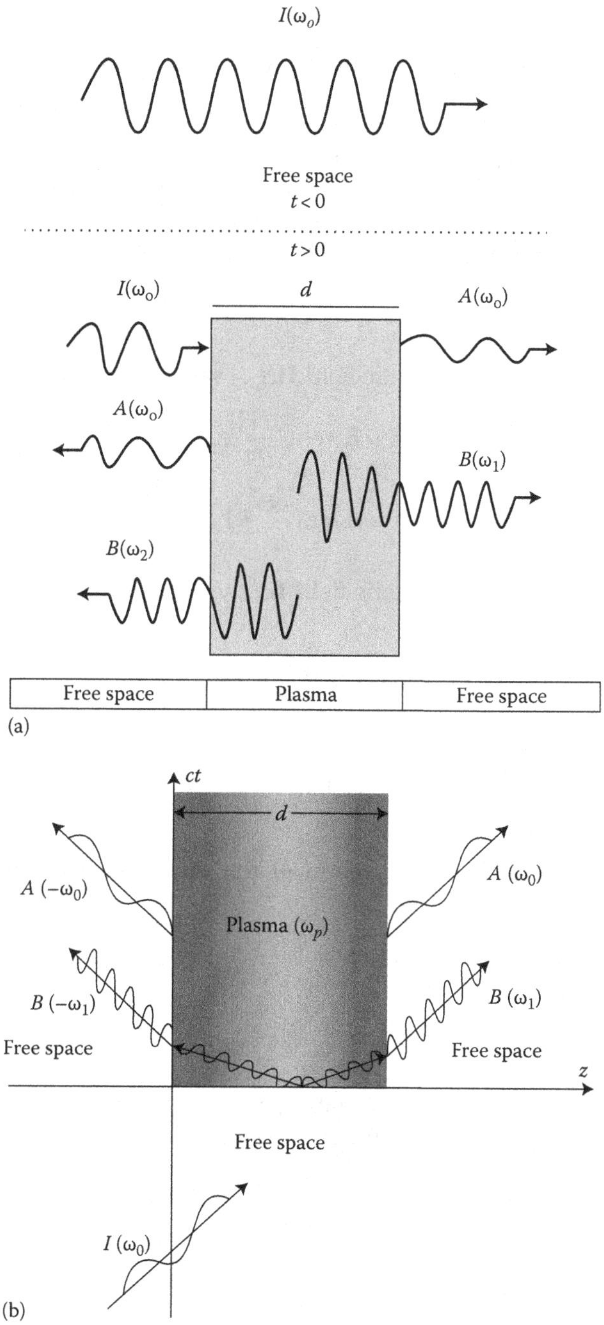

FIGURE 13.31
Effect of switching an isotropic plasma slab. *A* waves have the same frequency as the incident wave frequency (ω_o), and *B* waves have upshifted frequency $\omega_1 = \sqrt{\left(\omega_o^2 + \omega_p^2\right)} = -\omega_2$. (a) The waves are sketched in the time domain to show the frequency changes. (b) Shows a more accurate description.

At the resonant frequency, the relative permittivity ε_p goes to infinity. As an example, for longitudinal propagation defined under *characteristic waves*, resonance occurs when the signal frequency ω is equal to the electron gyrofrequency ω_b. For the frequency band, $0<\omega<\omega_b$, $\varepsilon_p>1$ and can have very high values for certain combinations of ω, ω_p, and ω_b; for instance, when $f_p = \frac{\omega_p}{2\pi} = 10^{14}$ Hz, $f_b = \frac{\omega_b}{2\pi} = 10^{10}$ Hz, and $f=10$ cm, $\varepsilon=9\times10^{16}$ [34]. A big change in ε_p can thus be obtained by collapsing the electron density, thus converting the magnetoplasma medium into free space. A big change in ε_p can also be obtained by collapsing the background quasistatic magnetic field, thus converting the magnetoplasma medium into an isotropic plasma medium. These aspects are discussed in the remaining parts of this section.

13.9.4.1 Basic Field Equations

The electric field $\mathbf{E}(\mathbf{r},t)$ and the magnetic field $\mathbf{H}(\mathbf{r},t)$ satisfy the Maxwell curl equations:

$$\nabla\times\mathbf{E} = -\mu_0\frac{\partial\mathbf{H}}{\partial t}, \tag{13.272}$$

$$\nabla\times\mathbf{H} = \varepsilon_0\frac{\partial\mathbf{E}}{\partial t} + \mathbf{J}. \tag{13.273}$$

In the presence of a quasistatic magnetic field $\mathbf{B}_0$, the constitutive relation for the current density is given by

$$\frac{d\mathbf{J}}{dt} = \varepsilon_0\omega_p^2(\mathbf{r},t)\mathbf{E} - \mathbf{J}\times\boldsymbol{\omega}_b(\mathbf{r},t), \tag{13.274}$$

where

$$\boldsymbol{\omega}_b = \frac{q\mathbf{B}_0}{m} = \omega_b\hat{\mathbf{B}}_0. \tag{13.275}$$

Therein, $\hat{\mathbf{B}}_0$ is a unit vector in the direction of the quasistatic magnetic field, ω_b is the gyrofrequency, and

$$\omega_p^2(\mathbf{r},t) = \frac{q^2N(\mathbf{r},t)}{m\varepsilon_o}. \tag{13.276}$$

13.9.4.2 Characteristic Waves

Next, the solution for a plane wave propagating in the z-direction in a homogeneous, time-invariant unbounded magnetoplasma medium can be obtained by assuming

$$f(z,t) = \exp\left[j(\omega t - kz)\right], \tag{13.277}$$

$$\omega_p^2(z,t) = \omega_p^2, \tag{13.278}$$

$$\omega_b(z,t) = \omega_b, \tag{13.279}$$

where f stands for the components of the field variables **E**, **H**, or **J**.

The well-established *magnetoionic theory* [33,34] and [15] is concerned with the study of plane wave propagation of an arbitrarily polarized plane wave in a cold, anisotropic plasma, where the direction of phase propagation of the plane wave is at an arbitrary angle to the direction of the static magnetic field. As the plane wave travels in such a medium, the polarization state continuously changes. However, there are specific normal modes of propagation in which the state of polarization is unaltered. Plane waves with left (L wave) or right (R wave) circular polarization are the normal modes in the case of wave propagation along the quasistatic magnetic field. Such propagation is labeled as *longitudinal propagation*. The ordinary wave (O wave) and the extraordinary wave (X wave) are the normal modes for *transverse propagation*, where the direction of propagation is perpendicular to the static magnetic field. In this chapter, propagation of the R wave in a time-varying plasma is discussed. An analysis of the propagation of other characteristic waves can be found elsewhere [21].

13.9.4.3 R-Wave Propagation

The relative permittivity for R-wave propagation is [21,33]

$$\varepsilon_{pR} = 1 - \frac{\omega_p^2}{\omega(\omega - \omega_b)} = \frac{(\omega + \omega_{c1})(\omega - \omega_{c2})}{\omega(\omega - \omega_b)}, \tag{13.280}$$

where ω_{c1} and ω_{c2} are the cutoff frequencies given by

$$\omega_{c1,c2} = \mp\frac{\omega_b}{2} + \sqrt{\left(\frac{\omega_b}{2}\right)^2 + \omega_p^2}. \tag{13.281}$$

Equation 13.280 is obtained by eliminating J from (13.273) with the help of (13.274) and recasting it in the form of (13.258).

The dispersion relation is obtained from

$$k_R^2c^2 = \omega^2\varepsilon_{pR} = \frac{\omega(\omega + \omega_{c1})(\omega - \omega_{c2})}{(\omega - \omega_b)}, \tag{13.282}$$

where k_R is the wave number for the R wave and c is the speed of light in free space. When expanded, this equation becomes

$$\omega^3 - \omega_b\omega^2 - \left(k_R^2c^2 + \omega_p^2\right)\omega + k_R^2c^2\omega_b = 0. \tag{13.283}$$

Figure 13.32, which is the same as Figure 11.1, shows a graph of ε_p versus ω, while Figure 13.33 gives the ω–k diagram for R-wave propagation. The R wave is a characteristic

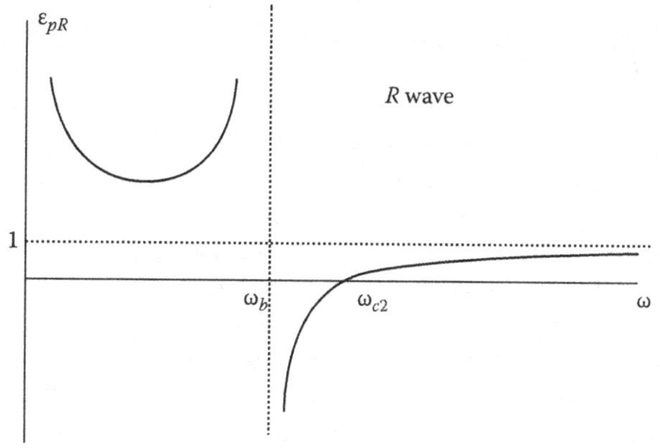

FIGURE 13.32
Relative permittivity for a R-wave propagation.

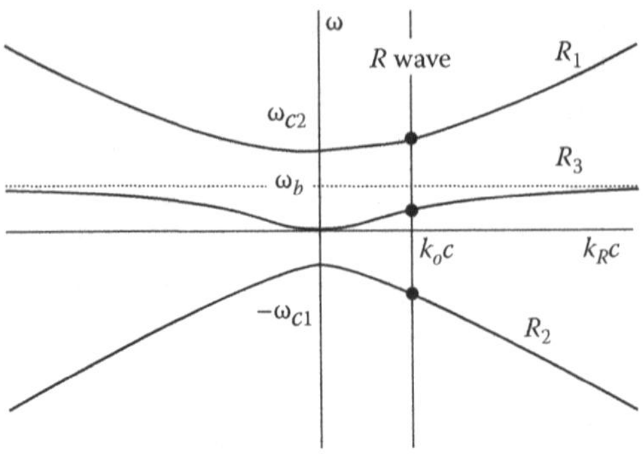

FIGURE 13.33
ω–k diagram for a R-wave propagation.

wave of longitudinal propagation. For this wave, the medium behaves like an isotropic plasma except that ε_p is influenced by the strength of the quasistatic magnetic field. Particular attention is drawn to the specific feature, visible in Figure 13.32, showing $\varepsilon_p > 1$ for the R wave in the frequency band $0 < \omega < \omega_b$. This mode of propagation is called *whistler mode* in the literature on ionospheric physics and *helicon mode* in the literature on solid-state plasmas. Section 10C7.6 and 10C7.7 of Reference [22] deal with the transformation of the whistler wave by a transient magnetoplasma medium and the consequences of such a transformation. An isotropic plasma medium does not support a whistler wave.

13.9.4.4 Sudden Creation

In this section, the problem of sudden creation of the plasma in the presence of a static magnetic field in z-direction is analyzed. The geometry of the problem is given in Figure 13.34. The source wave is assumed to be an R wave. The sudden creation is equivalent to creating a temporal discontinuity in the dielectric properties of the medium. In such a case, the wave number k_0 is conserved across the temporal discontinuity. For a given k_0 of the source wave, we draw a vertical line that intersects the branches in the ω–k diagram in Figure 13.33 at three points. The frequencies of these waves are different from the source frequency. The medium switching, in this case, creates three R waves labeled as R_1, R_2, and R_3. Whereas R_1 and R_3 are transmitted waves, R_2 is a reflected wave (Figure 13.34).

A physical interpretation of the waves can be given in the following way: The electric and magnetic fields of the incident wave and the quasistatic magnetic field accelerate the electrons in the newly created magnetoplasma, which in turn radiate new waves. The frequencies of the new waves and their fields can be obtained by adding contributions from the many electrons whose positions and motions are correlated by the collective effects supported by the magnetoplasma medium. Such a detailed calculation of the radiated fields seems to be quite involved. A simple, but less accurate, description of the plasma effect is obtained by modeling the magnetoplasma medium as a dielectric medium whose refractive index is computed through magnetoionic theory [34]. The frequencies of the new waves are constrained by the requirements that the wave number k_0 is conserved over the temporal discontinuity and the refractive index n is the one that is applicable to the type of wave propagation in the magnetoplasma. This gives a conservation law [47] $k_0c = \omega_0 = n(\omega)$ from which ω can be determined. Solution of the associated electromagnetic initial value problem gives the electric and magnetic fields of the new waves.

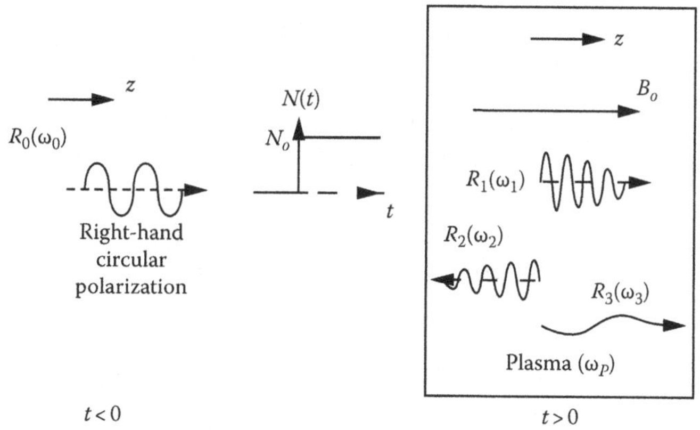

FIGURE 13.34
Effect of switching an unbounded magnetoplasma medium. Sketches of the *B* waves generated in the plasma are given for *R* incidence. The waves are sketched in the time domain to show the frequency changes.

13.9.4.5 Frequency-Shifting Characteristics of Various R Waves

The shift ratio and the efficiency of the frequency-shifting operation can be controlled by the parameters ω_p and ω_b. The results are presented by normalizing all frequency variables with respect to the source wave frequency ω_0. This normalization is achieved by taking $\omega_0 = 1$ in numerical work.

For *R* waves, the curves of $\omega - \omega_p$ and $\omega - \omega_b$ are sketched in Figure 13.35a and b, respectively. In Figure 13.36, results are presented for the R_1 wave: The values on the vertical axis give a frequency-shift ratio since the frequency variables are normalized with respect to ω_0. This is an upshifted wave and the shift ratio increases with ω_p as well as ω_b. From Figure 13.36 it appears that by a suitable choice of ω_p and ω_b, one can obtain any desired large frequency shift. However, the wave generated can have weak fields associated

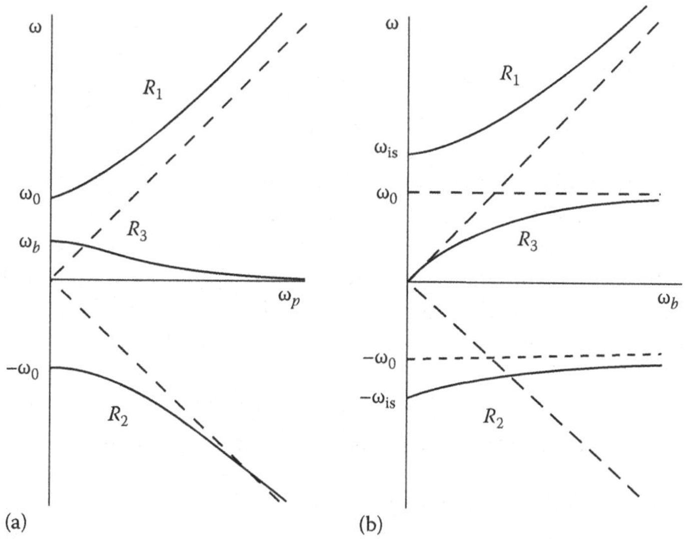

FIGURE 13.35
Frequency shifting of *R* waves. Sketches of (a) ω versus ω_p, (b) ω versus ω_b; $\omega_{is} = \left(\omega_o^2 + \omega_p^2\right)^{1/2}$.

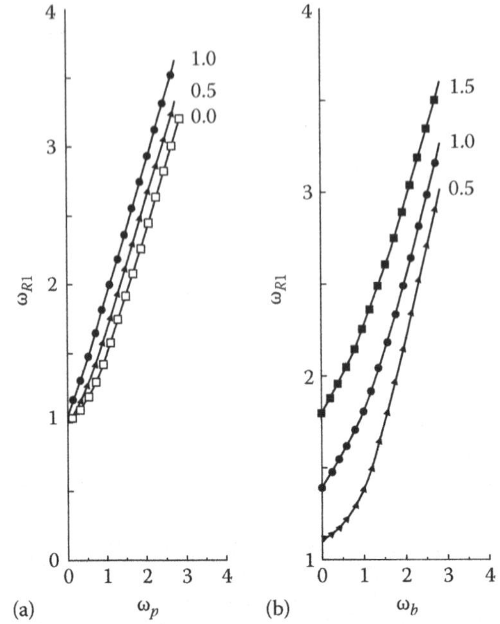

FIGURE 13.36
Frequency shifting of the R_1 wave. The frequency variables are normalized with respect to the source wave frequency by taking $\omega_0 = 1$. Shown is the frequency-shift ratio versus (a) ω_p and (b) ω_b. The numbers on the curves are (a) ω_b and (b) ω_p.

TABLE 13.1

R_1 Wave-Shift Ratio and Power Density for Two Sets of (ω_p, ω_b)

ω_o	ω_p	ω_b	E_1/E_o	H_1/H_o	S_1/S_o	ω_{R1}/ω_o
1	0.5	0.5	0.83	0.69	0.57	1.20
1	2.0	2.0	0.39	0.18	0.07	3.33

with it and the power density S_1 can be low. This point is illustrated in Table 13.1 by considering two sets of values for the parameters (ω_p, ω_b). For the set (0.5, 0.5), the shift ratio is 1.2, but the power density ratio S_1/S_0 is 0.57; whereas for the set (2.0, 2.0), the shift ratio is 3.33 but the power density ratio is only 0.07. Similar remarks apply to other waves.

The R_2 wave is a reflected wave. This is an upshifted wave and the shift ratio increases with ω_b but decreases with ω_b [21]. The R_3 wave in Figure 13.37 is a transmitted wave that is downshifted. The shift ratio decreases with ω_p and increases with ω_b. When $\omega_b = 0$, ω_{R3} becomes zero. The electric field E_3 becomes zero, and the magnetic field degenerates to the wiggler magnetic field [48]. This result is in conformity with the result for the isotropic case discussed in Section 13.9.2.1.

13.9.5 Modeling of Building Up Plasma versus Collapsing Plasma

13.9.5.1 Building Up Magnetoplasma

For the case of building-up plasma, the total current density, $\mathbf{J}(t)$, cannot simply be taken as $-qN(t)\mathbf{v}(t)$ since the electron velocity, $\mathbf{v}(t)$, depends on the time of electron birth. The newly born electrons start with zero velocity and are subsequently accelerated by the fields.

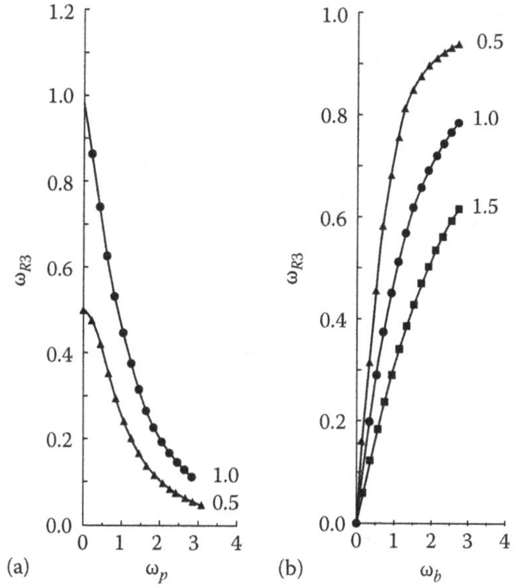

FIGURE 13.37
Frequency shifting of the R_3 wave. $\omega_0 = 1$. Shown is the frequency-shift ratio versus (a) ω_p and (b) ω_b. The numbers on the curves are (a) ω_b and (b) ω_p.

Thus, all the electrons do not have the same velocity at a given time during the creation or building-up of the plasma. When the electron density is increasing due to electrons born with zero velocity at different times, the change in current density at t due to the electrons born at t_i can be computed as follows [21]:

$$\Delta\mathbf{J}(t) = -q\Delta N_i \mathbf{v}_i(t). \tag{13.284}$$

where ΔN_i is the electron density added at t_i and $\mathbf{v}_i(t)$ is the velocity at t of these ΔN_i electrons born at t_i. To obtain $\mathbf{J}(t)$, Equation 13.284 must be integrated to account for the current density contributions from electrons born at all times t_i [Appendix A of Reference 21]. This results in the constitutive equation for the plasma creation case:

$$\frac{d\mathbf{J}}{dt} + \nu\mathbf{J} = \varepsilon_0\omega_p^2(\mathbf{r},t)\mathbf{E} - \mathbf{J}\times\boldsymbol{\omega}_b(\mathbf{r},t) + \frac{q}{m}\nabla p. \tag{13.285}$$

Here the electron cyclotron frequency (gyrofrequency) $\boldsymbol{\omega}_b$ is given by

$$\boldsymbol{\omega}_b = \frac{q\mathbf{B}_0}{m} = \omega_b\hat{\mathbf{B}}_0, \tag{13.286}$$

where $\hat{\mathbf{B}}_0$ is a unit vector in the direction of the static magnetic field and q and m are the absolute values of the charge and mass of the electron, respectively. The square of the plasma frequency, ω_p^2, is proportional to the electron density, $N(\mathbf{r}, t)$, and given by

$$\omega_p^2(\mathbf{r},t) = \frac{q^2 N(\mathbf{r},t)}{m\varepsilon_0}. \tag{13.287}$$

The gradient term in (13.285) is to account for the warm plasma case, discussed in Chapter 10.

13.9.5.2 Collapsing Magnetoplasma

In the case of collapsing magnetoplasma due to the collapse of the ionization, the building-up plasma formulation of the previous section is not appropriate [21,49,50] because as the plasma decays the number of free electrons decreases, and therefore, ultimately, the current vector density J should disappear. However, using the above model, the collapsing plasma makes $dJ/dt = 0$ and does not make $J = 0$ ([21, p. 148] and [49,50]). An alternative model is needed to implement this condition.

One can consider the following model for the collapse of the plasma: the decrease in the electron density takes place by the process of sudden removal of $\Delta N(t)$ electrons; the velocities of all the remaining electrons are unaffected by this capture and have the same instantaneous velocity $\mathbf{v}(t)$. The model has to ensure the continuity of $\mathbf{v}(t)$. The modification of (13.284) for the collapsing plasma is obtained as follows. Since the electrons are assumed to all have the same velocity $\mathbf{v}(t)$, the current density is given by

$$\mathbf{J}(t) = -qN(\mathbf{r},t)\mathbf{v}(t). \tag{13.288}$$

Taking the derivative of (13.288) with respect to time results in

$$\frac{d\mathbf{J}}{dt} = -q\frac{\partial N(\mathbf{r},t)}{\partial t}\mathbf{v}(t) - qN(\mathbf{r},t)\frac{\partial \mathbf{v}(t)}{\partial t}. \tag{13.289}$$

Substituting (13.288) and (13.289) into the momentum equation (13.285), and using (13.286) and (13.287) results in the constitutive relation for the collapsing plasma case

$$\frac{d\mathbf{J}}{dt} + \left[\nu - \frac{\frac{\partial}{\partial t}\left[\omega_p^2(t)\right]}{\omega_p^2(t)}\right]\mathbf{J} = \varepsilon_0\omega_p^2(\mathbf{r},t)\mathbf{E} - \mathbf{J}\times\omega_b(\mathbf{r},t) + \frac{q}{m}\nabla p. \tag{13.290}$$

Even if it is assumed that the plasma is loss free ($\nu = 0$), the term involving the time derivative of $\omega\omega_p^2(t)$ ensures the current density $\mathbf{J}$ ultimately disappears when the plasma collapses completely. The parameter ν in (13.290) represents the energy dissipated as heat, whereas the $\left[\omega_p^2(t)\right]'/\omega_p^2(t)$ term represents the energy loss due to the $\Delta N(t)$ electrons lost in the plasma decay process. Neglecting this term can lead to errors as shown in [50]. If ν_1 is defined as

$$\begin{aligned} \nu_1 &= \nu \quad \text{building-up plasma} \\ \nu_1 &= \nu - \frac{\frac{\partial}{\partial t}\left[\omega_p^2(t)\right]}{\omega_p^2(t)} \quad \text{collapsing plasma.} \end{aligned} \tag{13.291}$$

the constitutive equation for either case can be written as

$$\frac{d\mathbf{J}}{dt} + \nu_1\mathbf{J} = \varepsilon_0\omega_p^2(\mathbf{r},t)\mathbf{E} - \mathbf{J}\times\omega_b(\mathbf{r},t) + \frac{q}{m}\nabla p. \tag{13.292}$$

The pressure p can also be related to the electric field $\mathbf{E}$ by the following equation [22]:

$$\nabla p = -\gamma k_B T\frac{\varepsilon_0}{q}\nabla(\nabla\cdot\mathbf{E}), \tag{13.293}$$

and equation (13.292) can be written as

$$\frac{d\mathbf{J}}{dt} + \nu_1 \mathbf{J} = \varepsilon_0 \omega_p^2(\mathbf{r},t)\mathbf{E} - \mathbf{J} \times \omega_b(\mathbf{r},t) - \varepsilon_0 a^2 \nabla(\nabla \cdot \mathbf{E}), \tag{13.294}$$

where

γ is the ratio of specific heats
k_B is the Boltzmann constant [1.38066×10^{-23} J/K]
T is the temperature [K]
ε_0 is the permittivity of free space [$8.85418782 \times 10^{-12}$ F/m]
a is the acoustic velocity [m/s]

The acoustic velocity (a) is given by

$$a = \left(\frac{\gamma k_B T}{m}\right)^{\frac{1}{2}}. \tag{13.295}$$

Several interesting results are presented using this model [51].

13.9.6 Applications

Many applications for the frequency-shifting research are developing. An obvious application is for frequency transformers. The source wave can be generated in a frequency band using standard equipment, and the switched plasma device converts the source wave into a new wave in a frequency band not easily accessible by other methods. The frequency-shifting mechanism can be applied for plasma cloaking of satellites and aircrafts and for producing short-chirped pulses as ultra-wideband signals [55]. Application to photonics has been dealt in detail by Nerukh et al. [56]. One particular application of great significance is mentioned next.

13.9.6.1 Application: Frequency Transformer 10–1000 GHz

An application of the phenomena of frequency upshifting (by a ratio of 100) of a source wave of 10–1000 GHz by a collapsing magnetoplasma in a cavity is discussed in Appendix 13C. The material developed in Sections 13.9.4 and 13.9.5 supplies the background material that will help in understanding the details omitted in Appendix 13C.

As explained before, the interaction of an electromagnetic wave with a time-varying medium is governed by the property of conservation of the wave number. This property can be utilized to construct a frequency transformer. It is shown that switching off the ionization source and creating a decaying magnetoplasma medium in a cavity will upshift the source frequency of 10–1000 GHz for the appropriate choice of initial magnetoplasma parameters. The electric field of the output wave is comparable to that of the source wave. Moreover, the switching angle can alter the polarization of the output wave.

One of the examples given in Appendix 13C has the following parameters: $f_0 = 10$ GHz, $f_{p0} = 30$ GHz, and $f_b = 10.009$ GHz, and a plate separation in the cavity of $d = 0.15$ mm. When the ionization source is switched off, the output mode has a frequency of 1000 GHz. The electric field of the output wave is comparable to the input wave.

The results presented as FDTD solution of this ideal problem has one puzzling aspect: If the frequency of the wave changes from 10 to 1000 GHz, it will go through the cyclotron resonance at $f = f_b = 10.009$ GHz, should there not be a strong absorption, when the plasma is assumed lossy. The simulation results do not show any such strong absorption even for $\nu/\omega_p = 0.1$. A more detailed picture of the variation of the amplitudes and frequencies of the three modes with the decaying plasma parameter $\omega_p(t)$ will clarify the results. The interesting physics is explained in Figure 13C.12, which shows the variation of the frequency (a) and the amplitude (b) of the electric field of the three modes labeled as R_1, R_2, and R_3 (for each of the initial components of the standing wave). As ω_p decreases and becomes zero, ω_1 becomes $n_0\omega_0$, ω_2 becomes $-n_0\omega_0$, and ω_3 becomes ω_b.

The puzzle is now solved. At the instant of switching off the plasma, that is at $t = 0+$, the three modes are excited; the first mode has frequency higher than $n_0\omega_0$, but the amplitude of the electric field is zero. The amplitude builds up as ω_p and decreases reaching a final value E_f. Mode 2 is a negative-going wave but otherwise has similar characteristics as mode 1 and builds up as the plasma collapses. Mode 3 is a positive-going wave with an initial frequency ω_0 and amplitude E_0 of the source R wave. However, as plasma decays, ω_3 increases from ω_0 to ω_b and E_3 amplitude decreases from E_0 to 0. Even if we assume no losses in the plasma medium, this mode dies. In the presence of collisions, it is absorbed at a faster rate. The absorption in this case serves a useful purpose in removing the unwanted mode 3.

The presence of a cavity introduces the switching angle ϕ_0 as a controlling parameter, controlling the relative amplitudes and phases of the x and y components of the electric fields of the output wave. The polarization of the output wave can be altered by changing the switching angle. Figure 13C.11 shows that the output wave is linearly x-polarized for an initial switching angle of $\phi_0 = 90°$. However, if the switching angle is zero degrees, the output wave will be y-polarized. Thus the case under study for the chosen parameters (see Figure 13C.11) gives a frequency and polarization transformer, transforming 10 GHz input wave to 1000 GHz output wave with a change of polarization from right circularly polarized to x-polarized.

13.9.7 Subcycle Time-Varying Medium

In Sections 13.9.2.1 and 13.9.4.4, we modeled the fast temporal change in the parameters as sudden switching to simplify the problem. In case the rise or decay time is not zero, we can calculate the scattering coefficients by using perturbation and variational techniques to obtain the corrections. Their validity is based on the rise or decay time being small compared to the period of the source wave. The time-varying medium under these restrictions is recognized in [52] as a "subcycle time-varying medium." A fundamental difference is the requirement of causality in the case of the time-varying medium [21]. A related technique, "comparison identities," is used to calculate the amplitudes [53,54,55,56].

13.9.8 Periodically Time-Varying Parameter, Mathieu Equation, and Parametric Resonance

Equation 13.264, when kc is the signal frequency ω_0 in free space for $t < 0$, can be written as

$$\frac{d^2E}{dt^2} + \omega^2(t)E = 0 \tag{13.296a}$$

$$\omega^2(t) = \left[\omega_0^2 + \omega_p^2(t)\right]. \tag{13.296b}$$

It is a linear equation but in general not exactly solvable for a general temporal profile $\omega_p^2(t)$, in which case it is called Hill's equation [57]. If the profile is periodic, given by

$$\omega_p^2(t) = \omega_{p0}^2 \cos \omega_x t, \tag{13.297a}$$

Equation 13.296b becomes

$$\omega^2(t) = \omega_0^2 [1 + X \cos \omega_x t], \tag{13.297b}$$

where

$$X = \frac{\omega_{p0}^2}{\omega_0^2}. \tag{13.297c}$$

Equation 13.296a, when (13.297b) is inserted for $\omega^2(t)$ is called Mathieu equation [58]. For $X \ll 1$, small amplitude oscillation at the frequency ω_0 grows in amplitude exponentially in time if ω_x is close to $2\omega_0/n$, where n is a positive integer. This phenomenon is called parametric resonance [16,57–63], since it occurs due to the periodic variation of a parameter. The amplitude is most strong when $n = 1$. Parametric resonance is known to occur in various physical processes. A parametric pendulum, where the pivot of a simple frictionless pendulum is subjected to an alternating vertical motion as shown in Figure 13.38, is discussed in [58].

An *L-R-C* circuit with a periodically time-varying capacitor (Figure 13.39)

$$C(t) = C_0 [1 + a \cos \omega_a t] \tag{13.298}$$

is discussed in [59].

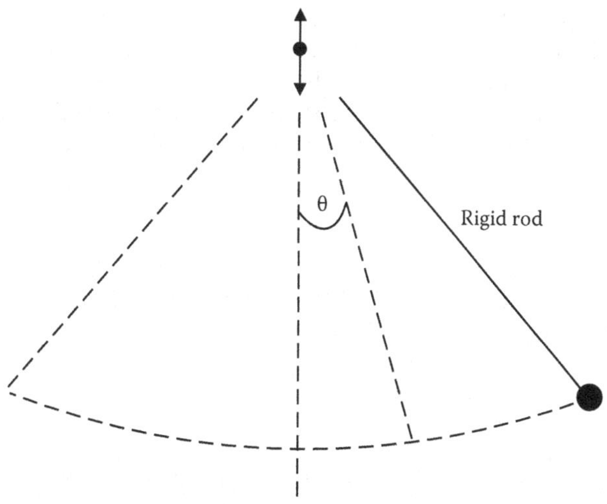

FIGURE 13.38
Geometry for parametric pendulum.

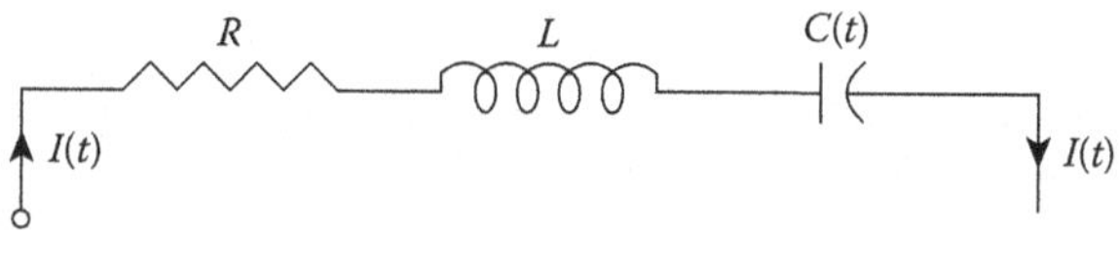

FIGURE 13.39
A series *L-R-C* circuit with a periodically time-varying capacitor.

The differences between forced resonance phenomena and the parametric resonance need to be carefully noted. The equation for the forced resonance is of the canonical form

$$\frac{d^2E}{dt^2} + \omega_0^2 E = E_0 \cos \omega_d t. \tag{13.299}$$

In the forced resonance case, the source frequency ω_d is tuned to the natural frequency of the resonance ω_0. In the parametric resonance case, it is the tuning of the frequency of the time-varying parameter $\omega_x = 2\omega_0/n$ to the frequency of the source ω_0 that brings resonance. In the forced resonance case, the buildup of the amplitude of oscillation diverges linearly with time, whereas in the case of the parametric resonance, the buildup of the amplitude diverges exponentially with time [59].

The interesting features of the solution of Mathieu's equation is shown in [58] through the example of a parametric pendulum (Figure 13.38).

The Mathieu equation solved for the pendulum angle θ is

$$\frac{d^2\theta}{dt^2} + \omega_0{}^2\left[1 + h\cos(2\omega t)\right]\theta = 0. \tag{13.300}$$

The forcing term in this case is of amplitude h and period $T_{exc} = 2\pi/2\omega$, and the natural period T_{nat} of the pendulum is $2\pi/2\omega_0$. Because of the periodic parameter in (13.300), Floquet's theorem, discussed in Section 9.7, applies and

$$\theta(t + T) = e^{\mu T}\theta(t). \tag{13.301}$$

The above can be interpreted as the following: Theta gets rescaled by $e^{\mu T}$ each period. The solution of (13.300), Mathieu equation, will be the product of exponential growth/decay term and a periodic function of time. Note that $\theta = \dot{\theta} = 0$ is a rest state of (13.300), but it is an unstable state if μ is positive. The instability is strongest when the excitation frequency ω in (13.300) is twice the natural frequency ω_0 of the pendulum. One can investigate this instability by taking

$$\omega = 2\omega_0 + \varepsilon, \quad \varepsilon \ll 1. \tag{13.302}$$

Assuming the excitation amplitude h as small, the solution of (13.300) with ω as given in (13.302) can be written as

$$\theta(t) = a(t)\cos\left(\omega_0 + \frac{\varepsilon}{2}\right) + b(t)\sin\left(\omega_0 + \frac{\varepsilon}{2}\right). \tag{13.303}$$

Substituting (13.303) in (13.300), retaining terms that are first order in ε and neglecting second-order derivatives of a and b [58], we can show that

$$\mu^2 = \frac{1}{4}\left[\left(\frac{1}{2}h\omega_0\right)^2 - \varepsilon^2\right]. \tag{13.304}$$

Parametric resonance occurs when μ is real and positive, which gives the bounds for the relation between ε and h:

$$-\frac{1}{2}h\omega_0 < \varepsilon < \frac{1}{2}h\omega_0. \tag{13.305}$$

The inequality relation (13.305) means that the parametric resonance takes place not only when ε is zero but also slightly away from the resonance within the region specified by the two straight lines in the $\varepsilon - h$ phase diagram with a slope of $\pm h\omega_0/2$.

The limit to the amplitude of the parametric oscillation arises due to the damping γ- term in the modified Mathieu equation given below:

$$\frac{d^2\theta}{dt^2} + 2\gamma\frac{d\theta}{dt} + \omega_0^2\left[1 + h\cos\left(\left(2\omega_0 + \varepsilon\right)t\right)\right]\theta = 0. \tag{13.306}$$

In this case, the inequality relation (13.305) gets modified to

$$-\left[\left(\frac{1}{2}h\omega_0\right)^2 - 4\gamma^2\right]^{1/2} < \varepsilon < \left[\left(\frac{1}{2}h\omega_0\right)^2 - 4\gamma^2\right]^{1/2}. \tag{13.307}$$

Also, the parametric resonance is possible only if $h > h_c$, where

$$h_c = \frac{4\gamma}{\omega_0}. \tag{13.308}$$

The above is a brief account of an insightful explanation of the parametric pendulum given in [58]. It is available for free download on the Internet, from which one can obtain more details. Reference [59], also available for free downloading from the Internet, has simulations to show the difference between forced resonance and parametric resonance through an L-C-R-example. Reference [57] has a good account of various physical phenomena of self-oscillation, which sometimes is wrongly attributed to either forced resonance or parametric resonance. Swanson [63] and Chen [16] discuss parametric instabilities in plasmas.

13.10 Statistical Mechanics and Boltzmann Equation

Boltzmann equation [15,16,33,64–66] is a PDE for a particle distribution function f used to describe particle trajectories in "phase space." The phase space is a six-dimensional space that consists of both physical space (also called configuration space) denoted by d^3x and velocity space denoted by d^3v. In the terminology of probability theory, f will be called a density function.

The expression [64]

$$f\left(\boldsymbol{r},\boldsymbol{v},t\right)d^3x d^3v \tag{13.309}$$

will denote the number of particles located in the phase space between $(\boldsymbol{r},\boldsymbol{v})$ and $(\boldsymbol{r} + d\boldsymbol{r}, \boldsymbol{v} + d\boldsymbol{v})$. The variable $N(\boldsymbol{r}, t)$, the number of particles per unit volume of physical space, can be written as

$$N\left(\boldsymbol{r},t\right) = \int f\left(\boldsymbol{r},\boldsymbol{v},t\right)d^3v. \tag{13.310}$$

The limits of integration when not specified are assumed to be over all possible values of velocities in the velocity volume space.

The RHS of (13.310) assumes a simpler form for an isotropic distribution function, that is,

$$f(\mathbf{r},v,t) = f(\mathbf{r},v^2,t). \tag{13.311}$$

In such a case we can write (13.310) as

$$N(\mathbf{r},t) = 4\pi \int_{v=0}^{\infty} f(\mathbf{r},v^2,t)v^2 dv. \tag{13.312}$$

The above is obtained by writing the differential volume element d^3v in velocity space as

$$d^3v = v^2 dv \sin\theta d\theta d\phi. \tag{13.313}$$

Combining the terms that depend on the scalar velocity variable, we can define a new distribution function g:

$$g(\mathbf{r},v^2,t) = 4\pi v^2 f(\mathbf{r},v^2,t). \quad (1c4) \tag{13.314}$$

Note that the function g is zero at $v = 0$ and attains its maximum elsewhere [16].

If f is normalized so that $\hat{f}$

$$\int_{-\infty}^{\infty} \hat{f}(\mathbf{r},v,t) d^3v = 1, \tag{13.315}$$

then

$$f(\mathbf{r},v,t) = N(\mathbf{r},t)\hat{f}(\mathbf{r},v,t). \tag{13.316}$$

13.10.1 Maxwell Distribution $\hat{f}_M$ and Kinetic Definition of Temperature T

An often used simple example of $\hat{f}$ is called the Maxwellian distribution $\hat{f}_M$:

$$\hat{f}_M = \left(\frac{m}{2\pi k_B T}\right)^{3/2} \exp\left(-\frac{mv^2}{2k_B T}\right), \tag{13.317}$$

which satisfies (13.315). It can be shown that the root mean square (RMS) velocity, using Maxwellian distribution $\hat{f}_M$, is given by

$$v_{rms} = \left(< v^2 >\right)^{1/2} = \left(\frac{3k_B T}{m}\right)^{1/2}. \tag{13.318}$$

See (13.323) for a quantitative definition of the average $<\cdot>$ given in (13.318). Thus, $(3/2)k_B T$ is the average kinetic energy of the particles. In physics, it is customary to give the energy of the particles in electron volts (eV) rather than in joules:

$$1 \text{ eV} = \frac{e}{k_B} = \frac{1.6\times 10^{-19}}{1.38\times 10^{-23}} = 11,600 \text{ K}. \tag{13.319}$$

13.10.2 Boltzmann Equation

The Boltzmann equation states that the rate of change of f along the trajectory of the particles is due to collisions among particles including processes like ionization and recombination of particles and is written as

$$\frac{df}{dt} = \left(\frac{\partial f}{\partial t}\right)_{col}. \tag{13.320}$$

A graphical interpretation of the Boltzmann equation is given in [16]. One example of a distribution function is the Maxwellian distribution of an ideal gas, discussed earlier, in defining the thermodynamic variable, temperature T, in degrees Kelvin (K) and its relationship with an electron volt (eV) of energy. On the other hand, an ideal beam of particles has an impulse distribution function $\delta\,(v)$.

The total rate of change of f with time (df/dt) along the trajectory can be expressed as

$$\frac{df}{dt} = \frac{\partial f}{\partial t} + \boldsymbol{v}\cdot\nabla_r f + \boldsymbol{a}\cdot\nabla_v f, \tag{13.321}$$

where ∇_r is the gradient in physical space and ∇_v is the gradient in the velocity space. If the acceleration vector $\boldsymbol{a}$ is due to the electromagnetic forces only,

$$\boldsymbol{a} = \frac{\boldsymbol{F}}{m} = \frac{q(\boldsymbol{E} + \boldsymbol{v}\times\boldsymbol{B})}{m}. \tag{13.322}$$

Based on the above equations, one can develop a complete and more accurate kinetic theory of electromagnetic wave propagation in plasmas. However, major difficulties are encountered in evaluating the RHS of (13.320) for various collision processes. Moreover, the appropriate distribution function for a given problem is not always known or available, hence the popularity of the so-called fluid theory, which we used extensively in Chapters 8 through 11 and Section 13.8. A few experimental results like Landau damping are not explained by the fluid theory but require the use of the kinetic theory with appropriate approximations, however retaining the essential physics. The fluid theory used earlier assumed that all the particles have the same velocity whereas a distribution function gives the probabilities of particles possessing various velocities. Another approach of obtaining the plasma equations is to use the kinetic theory to derive the "transport equations" for the average velocities of each of the particle species and consider the plasma as a mixture of the various particle species [15]. The notation and the definition of average of a quantity Q for alpha type of particles is given by

$$< Q >_\alpha = \frac{1}{N_\alpha}\int f_\alpha Q d^3 v. \tag{13.323}$$

If Q is not a function of the velocity variable, that is, $Q = Q(\boldsymbol{r}, t)$, then

$$< Q >_\alpha = Q. \tag{13.324}$$

We will apply later on this definition of the average, when $Q = v_x$, the x-component of the velocity variable for an electron:

$$< v_x > = \frac{1}{N}\int f v_x d^3 v. \tag{13.325}$$

13.10.3 Boltzmann–Vlasov Equation

The simplest approximation of the Boltzmann equation is to ignore the collision term on the RHS of (13.320) all together:

$$\frac{\partial f}{\partial t}+\boldsymbol{v}\cdot\nabla_r f+\frac{q}{m}(\boldsymbol{E}+\boldsymbol{v}\times\boldsymbol{B})\cdot\nabla_v f=0. \tag{13.326}$$

Equation (13.326) is called Boltzmann–Vlasov equation.

13.10.4 Krook Model for Collisions

The next level of approximation is to assume that the in the absence of external forces, the distribution returns to the equilibrium distribution in a relaxation time τ. The RHS of (13.320) is assumed to be [15]

$$\left(\frac{\partial f}{\partial t}\right)_{col}=-\nu(v)(f-f_0), \tag{13.327}$$

where the collision frequency

$$\nu(v)=\frac{1}{\tau}, \tag{13.328}$$

and f_0 is the equilibrium distribution.

The solution of even the Vlasov equation is not easy. One can use approximations by expanding the distribution function in a series and retain order terms [66]:

$$f(\boldsymbol{r},\boldsymbol{v},t)=f_0(\boldsymbol{r},v^2)+\alpha_1 f_1(\boldsymbol{r},\boldsymbol{v},t)+\alpha_2 f_2(\boldsymbol{r},\boldsymbol{v},t)+\ldots, \tag{13.329}$$

where α's measure small departures from the first term on the RHS. This first term is the equilibrium isotropic distribution and is independent of time and can be expressed as

$$f_0(\boldsymbol{r},v^2)=N(\boldsymbol{r})\hat{f}_M. \tag{13.330}$$

The convergence of this expansion is slow when the perturbation is anisotropic. Allis [66] has successfully used the expansion of f in orthogonal Legendre polynomials (P's in Table 5.3) in velocity space, valid when v has a preferred direction in velocity space, say x's direction in velocity space, that is, $v_x=v\cos\theta$:

$$f(\boldsymbol{r},v,t)=\sum_i f_i(\boldsymbol{r},v,t)P_i(\cos\theta), \tag{13.331}$$

$$f(\boldsymbol{r},v,t)=f_0(v^2)+f_1(r,v^2,t)\cos\theta+f_2(r,v^2,t)\frac{3\cos^2\theta-1}{2}+\cdots, \tag{13.332}$$

where f_0 is an equilibrium distribution like a Maxwellian distribution. Using the orthogonal property and from (13.310), we can write

$$\int f(\boldsymbol{r},\boldsymbol{v},t)d^3v=4\pi\int_0^\infty f_0(\boldsymbol{r},v^2)v^2dv=\int_0^\infty g_0(\boldsymbol{r},v^2)dv=N(\boldsymbol{r}). \tag{13.333}$$

13.10.5 Isotropic Dielectric Constant of Plasma

As a simple example of the use of the kinetic theory of plasma, let us calculate the relative permittivity of the isotropic plasma medium [33,64,66] using f as the electron distribution function. We will start with the Boltzmann equation with the simplified collision term on the RHS.

$$\frac{\partial f}{\partial t}-\frac{q}{m}E_x\frac{\partial f}{\partial v_x}=-\nu\left(f-f_0\right). \tag{13.334}$$

As mentioned before (Chapter 9), we neglect the force on the electron due to the wave magnetic field compared to the force due to the wave electric field E_x in the x-polarized plane wave in the plasma. Keeping only the first two terms in (13.332) and noting $\cos\theta=v_x/v$, f can be approximated, for linear theory, as

$$f\left(v,t\right)=f_0\left(v\right)+\frac{v_x}{v}f_1\left(v\right)\exp\left(j\omega t\right). \tag{13.335}$$

From (13.334) and (13.335),

$$f_1=\frac{\left(q/m\right)E_0}{\nu\left(v\right)+j\omega}\frac{\partial f_0}{\partial v}, \tag{13.336}$$

where E_0 is the amplitude of the x-component of the wave electric field. Since

$$J_x=-Nqv_x \tag{13.337a}$$

$$\sigma=\frac{J_x}{E_x}, \tag{13.337b}$$

one can write an expression for the conductivity σ of the plasma [33] as

$$\sigma=-\frac{4\pi}{3}\frac{Nq^2}{m}\int_0^{\infty}\frac{1}{\nu\left(v\right)+j\omega}\frac{df_0\left(v\right)}{dv}v^3dv. \tag{13.338}$$

The relative permittivity ε_p is given by [33]

$$\varepsilon_p=1-\frac{j\sigma}{\varepsilon_0\omega}=1+\frac{\omega_p^2}{\omega}\frac{4\pi}{3}\int_0^{\infty}\frac{1}{\omega-j\nu\left(v\right)}\frac{df_0\left(v\right)}{dv}v^3dv, \tag{13.339}$$

where the electron plasma frequency ω_p is given by (8.22). Please note carefully the notations, the Greek symbol ν for collision frequency and the roman letter v for the velocity in these equations.

We can check on this equation by considering the special case of the collision frequency ν independent of the velocity variable v. It can be shown that (13.339) reduces to (8.23) repeated here for convenience:

$$\varepsilon_p=1-\frac{\omega_p^2}{\omega\left(\omega-j\nu\right)}. \tag{13.340}$$

Kinetic theory formulation allows us to take into account the more realistic situation of ν being a function of the velocity variable. Because of the simplicity of the expression in (13.340), even for the case when ν is a function of the velocity, ε_p is written as

$$\varepsilon_p = 1 - \frac{\omega_p^2}{\omega\left(\omega - j\nu_{eff}\right)}, \tag{13.341}$$

where ν_{eff} is called the effective collision frequency. From (13.339) and (13.341), we relate ν_{eff} with the integral in (13.339) in the following way:

$$\frac{1}{\omega - j\nu_{eff}(\omega)} = -\frac{4\pi}{3}\int_0^{\infty} \frac{1}{\omega - j\nu(v)} \frac{df_0(v)}{dv} v^3 dv. \tag{13.342}$$

It can be shown that for $\omega \gg \nu$ [33],

$$\nu_{eff} = -\frac{4\pi}{3}\int_0^{\infty} \nu(v) \frac{df_0(v)}{dv} v^3 dv. \tag{13.343}$$

Furthermore, for a Maxwellian form for $f_0(v)$ and the collision frequency $\nu(v) = C_0 v^l$,

$$\nu_{eff} \cong C_0 \left(\frac{2k_B T}{m}\right)^{1/2} \frac{\Gamma\left(\frac{l+5}{2}\right)}{\Gamma\left(\frac{5}{2}\right)} \tag{13.344}$$

where Γ is the gamma function [67].

From experiments, it is determined that for a weakly ionized dry air, $l = 2$ [64,68].

13.10.6 Plasma Dispersion Function and Landau Damping

Certain experimental results involving plasma medium cannot be explained by the fluid theory. Prominent among them is Landau damping, which is the damping of waves in a plasma medium even when we assume the collision frequency ν is zero [64,66]. Another is cyclotron damping and Bernstein waves. In this connection, we come across a tabulated special function [67] called plasma dispersion function defined in [65]:

$$Z(\varsigma) = \frac{1}{\sqrt{\pi}}\int_{-\infty}^{\infty} \frac{\exp\left(-\xi^2\right)}{\xi - \varsigma} d\xi, \quad Im(\varsigma) > 0. \tag{13.345}$$

Its properties and its relation to other special functions are discussed in [65].

Section 13.9 deals with wave propagation in temporally modulated electromagnetic medium. A more general modulation, namely, space-time modulation of the medium, has many more novel applications and is of current research interest. To give a flavor of this research, some of the works on space-time modulation are listed in References [69–76].

References

1. Ulabi, F. T., *Fundamentals of Applied Electromagnetics*, Prentice Hall, Upper Saddle River, NJ, 2001.
2. Inan, U. S. and Inan, A. S., *Engineering Electromagnetics*, Addison-Wesley, Menlo Park, CA, 1999.
3. Magnusson, P. C., Alexander, G. C., Tripathi, V. K., and Weisshaar, A., *Transmission Lines and Wave Propagation*, 4th edn., CRC Press, Boca Raton, FL, 2001.
4. Rosenstark, S., *Transmission Lines in Computer Engineering*, McGraw-Hill, New York, 1994.
5. Weber, E., *Linear Transient Analysis*, Volume II, New York, 1956.
6. Miono, G. and Maffucci, A., *Transmission Lines and Lumped Circuits*, Academic Press, San Diego, CA, 2001.
7. Morse, P. M. and Feshbach, H., *Methods of Theoretical Physics, Part II*, McGraw-Hill, New York, 1953.
8. Shvartsburg, A. B., *Impulse Time-domain Electromagnetics of Continuous Media*, Birkhauser, Boston, MA, 1999.
9. Hirose, A. and Lonngren, K. E., *Introduction to Wave Phenomena*, Wiley, New York, 1985.
10. Ricketts, D. S. and Ham, D., *Electrical Solitons*, CRC Press, Boca Raton, FL, 2011.
11. Afshari, E., Bhat, H. S., Hajimiri, A., and Marsden, J. E., Extremely wideband signal shaping using one- and two-dimensional nonuniform nonlinear transmission lines, *J. Appl. Phys.*, 99, 054901, 2006.
12. Remoissenet, M., *Waves Called Solitons: Concepts and Experiments*, Springer, New York, 1999.
13. Taniuti, T. and Wei, C. C., Reductive perturbation method in nonlinear wave propagation, *J. Phys. Soc. Jpn.*, 21, 209–212, 1968.
14. Zabusky, N. J. and Kruskal, M. D., Interaction of solitons in a collisionless plasma and the recurrence of initial states, *Phys. Rev. Lett.*, 15(6), 240–243, 1965.
15. Tannenbaum, B. S., *Plasma Physics*, McGraw-Hill, New York, 1967.
16. Chen, F. F., *Introduction to Plasma Physics and Controlled Fusion*, Volume 1: Plasma Physics, Plenum Press, New York, 1988.
17. Jackson, J. D., *Classical Electrodynamics*, Wiley, New York, 1962.
18. Miyamoto, K., *Plasma Physics for Nuclear Fusion Revised Edition*, MIT Press, Cambridge, MA, 1976.
19. Goldstein, R., *Classical Mechanics*, Addison-Wesley, Reading, MA, 1959.
20. Mendonca, J. T., *Theory of Photon Acceleration*, IOP Publishing, Bristol, U.K., 2001.
21. Kalluri, D. K., *Electromagnetics of Time Varying Complex Media*, 2nd edition, CRC Press, Taylor & Francis Group, Boca Raton, FL, 2010.
22. Kalluri, D. K., *Electromagnetics of Waves, Material, and Computation with MATLAB*, CRC Press, Taylor & Francis Group, Boca Raton, FL, 2012. Accessed on January 15, 2017.
23. Longmire, C. L., EMP on Honolulu from the Starfish event, Theoretical note 353, March 1985, www.ece.unm.edu/summa/notes.
24. Longmire, C. L., On the electromagnetic pulse produced by nuclear explosions, *IEEE Trans. Antennas Propag.*, Ap26(1), 3–13, 1978.
25. Longmire C. L., Justification and verification of high-altitude EMP theory Part I, Report Contract No: LLNL-9323905, Lawrence Livermore National Laboratory, Livermore, CA, pp. 1–63, June 1986.
26. Savage, E., Gilbert, J., and Radasky, W., The early time (E1)high-altitude electromagnetic pulse (HEMP) and its impact on the U.S. power grid, Metatech Corporation Subcontract 6400009137. Report prepared for Oakridge National Laboratory, Goleta, CA, January 2010.
27. Baum, C. E., From the electromagnetic pulse to high-power electromagnetic, *Proc. IEEE*, 80(6), 789–817, June 1992.
28. Giri, D. V. and Prather, W. D., High-altitude electromagnetic pulse (HEMP) evolution of rise time and technology standards exclusively for E1 environment, *IEEE Trans. Electromagn. Compat.*, 55, 484–491, March 2013.

29. Auld, B. A., Collins, J. H., and Zapp, H. R., Signal processing in a nonperiodically time-varying magnetoelastic medium, *Proc. IEEE*, 56, 258–272, 1968.
30. Jiang, C. L., Wave propagation and dipole radiation in a suddenly created plasma, *IEEE Trans. Antennas. Propag.*, 23, 83–90, 1975.
31. Felsen, B. L. and Whitman, G. M., Wave propagation in time-varying media, *IEEE Trans. Antennas Propag.*, 18, 242–253, 1970.
32. Fante, R. L., Transmission of electromagnetic waves into time-varying media, *IEEE Trans. Antennas Propag.*, 19, 417–424, 1971.
33. Heald, M. A. and Wharton, C. B., *Plasma Diagnostics with Microwaves*, Wiley, New York, 1965.
34. Booker, H. G., *Cold Plasma Waves*, Kluwer, Higham, MA, 1984.
35. Joshi, C. J., Clayton, C. E., Marsh, K., Hopkins, D. B., Sessler, A., and Whittum, D., Demonstration of the frequency upshifting of microwave radiation by rapid plasma creation, *IEEE Trans. Plasma Sci.*, 18, 814–818, 1990.
36. Kuo, S. P., Frequency up-conversion of microwave pulse in a rapidly growing plasma, *Phys. Rev. Lett.*, 65, 1000–1003, 1990.
37. Kuo, S. P. and Ren, A., Experimental study of wave propagation through a rapidly created plasma, *IEEE Trans. Plasma Sci.*, 21, 53–56, 1993.
38. Rader, M., Dyer, F., Matas, A., and Alexeff, I., Plasma-induced frequency shifts in microwave beams, *IEEE International Conference on Plasma Science*, Oakland, CA, Conference Record Abstracts, p. 171, 1990.
39. Yablonovitch, E., Spectral broadening in the light transmitted through a rapidly growing plasma, *Phys. Rev. Lett.*, 31, 877–879, 1973.
40. Savage, R. L., Jr., Joshi, C. J., and Mori, W. B., Frequency up-conversion of electromagnetic radiation upon transmission into an ionization front, *Phys. Rev. Lett.*, 68, 946–949, 1992.
41. Lampe, M. and Walker, J. H., Interaction of electromagnetic waves with a moving ionization front, *Phys. Fluids*, 21, 42–54, 1978.
42. Stanic, B., Drljaca, P., and Boskoic, B., Electron plasma waves generation in suddenly created isotropic plasma, *IEEE Trans. Plasma Sci.*, 26, 1514–1519, 1998.
43. Stepanov, N. S., Dielectric constant of unsteady plasma, *Sov. Radiophys. Quan. Electron.*, 19, 683–689, 1976.
44. Banos, A., Jr., Mori, W. B., and Dawson, J. M., Computation of the electric and magnetic fields induced in a plasma created by ionization lasting a finite interval of time, *IEEE Trans. Plasma Sci.*, 21, 57–69, 1993.
45. Wilks, S. C., Dawson, J. M., and Mori, W. B., Frequency up-conversion of electromagnetic radiation with use of an overdense plasma, *Phys. Rev. Lett.*, 61, 337–340, 1988.
46. Granastein, V. L. and Alexeff, I., *High-Power Microwave Sources*, Artech House, Boston, MA, 1987.
47. Kalluri, D. K., Effect of switching a magnetoplasma medium on a traveling wave: Conservation law for frequencies of newly created waves, *IEEE International Conference on Plasma Science*, Oakland, CA, Conference Record Abstracts, p. 129, 1990.
48. Kalluri, D. K., Effect of switching a magnetoplasma medium on a travelling wave: longitudinal propagation, *IEEE Trans. Antennas Propag.*, 37, 1638–1642, 1989.
49. Lee, J. H., Kalluri, D. K., and Nigg, G. C., FDTD simulation of electromagnetic wave transformation in a dynamic magnetized plasma, *Int. J. Infrared Millimeter Waves*, 21(8), 1223–1253, 2000.
50. Lade, R. K., Lee, J. H., and Kalluri, D. K., Frequency transformer: Appropriate and different models for a building-up and collapsing magnetoplasma medium, *J. Infrared Millimeter Terahertz Waves*, 31(7), 960–972, 2010.
51. Lade, R. K., Electromagnetic wave interactions in a time varying medium: Appropriate and different models for a building-up and collapsing magnetoplasma medium, PhD thesis, University of Massachusetts Lowell, Lowell, MA, 2012.
52. Salem, M. A. and Caloz, C., Sub-cycle time-varying electromagnetic systems, *Radio Science Conference (URSI AT RASC)*, Gran Caneria, Spain, 2015.

53. Kalluri, D. K. and Chen, J., Comparison identities for wave propagation in a time varying plasma medium, *IEEE Trans. Antennas Propag.*, 57(9), 2698–2705, 2009.
54. Jinming, C., Wave propagation of transverse modes in a time varying magnetoplasma medium: Amplitude calculations using comparison identities, PhD thesis, University of Massachusetts Lowell, Lowell, MA, 2009.
55. Mori, W. and Editor, B., Special Issue on generation of coherent radiation using plasmas, *IEEE Trans. Plasma Sci.*, 21, 1–208, 1993.
56. Nerukh, A. G., Scherbatko, L. V., and Marciniak, M., *Electromagnetics of Modulated Media with Applications to Photonics*, IEEE/LEOS Poland chapter, National Institute of Telecommunications, Dept. of Transmission and Fiber Technology, Warsaw, Poland, 2001.
57. Jenkins, A., Self oscillation, https://arxiv.org/pdf/1109.6640.pdf. Accessed on January 15, 2017.
58. Rothman, D., Parametric Oscillator, Lecture 6 of the MIT course 12.006J Fall 2006, 2006. https://dspace.mit.edu/bitstream/handle/1721.1/84612/12-006j-fall-2006/contents/lecture-notes/lecnotes6.pdf.
59. Franklin, D. and Amador, H. E., Simulation of a parametric resonance circuit, www.aias.us, www.atomicprecision.com, www.upitec.org, March 30, 2012.
60. Morrison, T. M., PhD thesis, Three Problems in Nonlinear Dynamics with 2:1 Parametric Excitation, Cornell University, May 2006, http://www.math.cornell.edu/~rand/randdocs/Doctoral_theses/Tina%20Morrison.pdf.
61. Mandelshtam, L. I. and Papuleski, N. D., On the parametric excitation of electric oscillations, *Zhurnal Technicheskoy Fiziki*, 4(1), 5–29 (1934). Translated from Russian by T. Watt in February 1968. http://www.implosionamp.com/ParametricExcitationNASA.pdf. Accessed on January 15, 2017.
62. Butikov, E. I., Parametric Resonance in a Linear Oscillator, Lecture Notes, St. Petersburg State University, St. Petersburg, Russia, 1996. http://butikov.faculty.ifmo.ru/Applets/Par_resonance.pdf.
63. Swanson, D. G., *Plasma Waves*, Academic Press, San Diego, CA, pp. 348–362, 1989.
64. Wait, J. R., *Electromagnetics and Plasmas*, Holt, Rinehart and Winston, New York, 1968.
65. Swanson, D. G., *Plasma Waves*, Academic Press, Harcourt Brace Jovanovich, Boston, MA, 1989.
66. Allis, P. W., Buchsbaum, S. J., and Bers, A., *Waves in Anisotropic Plasmas*, MIT Press, Cambridge, MA, 1963.
67. Abramowitz, M. and Stegun, I. A. (Eds.), *Handbook of Mathematical Functions*, Dover, New York, 1972.
68. Phelps, A. V., Propagation constants for electromagnetic waves in a weakly ionized dry air, *J. Appl. Phys.*, 31, 1723, 1960.
69. Cassedy, E. S. and Oliner, A. A., Dispersion relations in time-space periodic media: Part I – stable interactions, *Proc. IEEE*, 51(10), 1342–1359, October 1963.
70. Cassedy, E. S., Dispersion relations in time-space periodic media: Part II—Unstable interactions, *Proc. IEEE*, 55(7), 1154–1168, July 1967.
71. Salem, M. A. and Caloz, C., Space-time cross-mapping and applications to wave scattering, arXiv: 1504.02012v1 [physics.optics], 1–9, April 7, 2015.
72. Chamanara, N., Taravati, S., Deck-Leger, Z.-L., and Caloz, C., Optical isolation based on space-time engineered asymmetric photonic band gaps, arXiv:1612.08398v2 [physics.optics], 1–14, March 29, 2017.
73. Chamanara, N. and Caloz, C., Electromagnetic wave amplification based on dispersion engineered intraband photonic transitions, private communication.
74. Hadad, Y., Sounas, D. L., and Alu, A., Space-time gradient metasurfaces, archive.org/pdf/1506.00690.pdf, date of access June, pp. 1–18.
75. Biacahana, F., Amann, A., Uskov, V., and O'Reilly, E. P., Dynamics of light propagation in spatiotemporal dielectric structures, *Phys. Rev. E*, 046607, 1–12, 2007.
76. Deck-Leger, Z. L. and Caloz, C., Wave scattering by superluminal space-time modulated slabs and symmetries with the subluminal case, private communication.

14

Electromagnetics of Moving Media: Uniform Motion

14.1 Introduction

The key work in the study of electromagnetics of moving media is the Lorentz transformation (LT) law. Under these transformation laws, Maxwell's equation has the same form in two reference frames moving with respect to each other with a uniform velocity v_0. The velocity of light in free space remains the same value $c = 3 \times 10^8$ m/s irrespective of the velocity of the observer measuring the velocity of light. The postulate of special velocity that the physical laws are form invariant among uniformly moving observers is consistent with LTs. A vast amount of literature is available, out of which the author has selected [1–30] as a sample. In this chapter, we study the electromagnetic wave interactions with a moving media. The moving frame of reference in which the medium is at rest is assumed to be inertial [1]; it does not accelerate and it does not rotate.

14.2 Snell's Law

Let us consider the oblique incidence of a plane s-wave discussed in Section 2.13 for the case of a stationary medium. Figure 14.1 shows the same geometry, except that the medium is now moving with a velocity v_0 along the positive z-axis as shown in Figure 14.1. Before we discuss Snell's law for this case, let us revisit the question of the stationary medium and show that the frequency of the reflected wave will be the same as the frequency of the incident wave. The above-mentioned property as well as Snell's law $\theta_i = \theta_r$ follows from the satisfaction of the boundary condition at $z = 0$, for all x and t.

Let

$$\mathbf{E}_s^I = \hat{y}E_0 \cos\left(\omega_i t - x\frac{\omega_i}{c}\sin\theta_i - z\frac{\omega_i}{c}\cos\theta_i\right), \tag{14.1}$$

$$\mathbf{E}_s^R = \hat{y}E_r \cos\left(\omega_r t - x\frac{\omega_r}{c}\sin\theta_r + z\frac{\omega_r}{c}\cos\theta_r\right), \tag{14.2}$$

where

c is the velocity of light in free space
ω_i is the frequency of the incident wave
ω_r is the frequency of the reflected wave

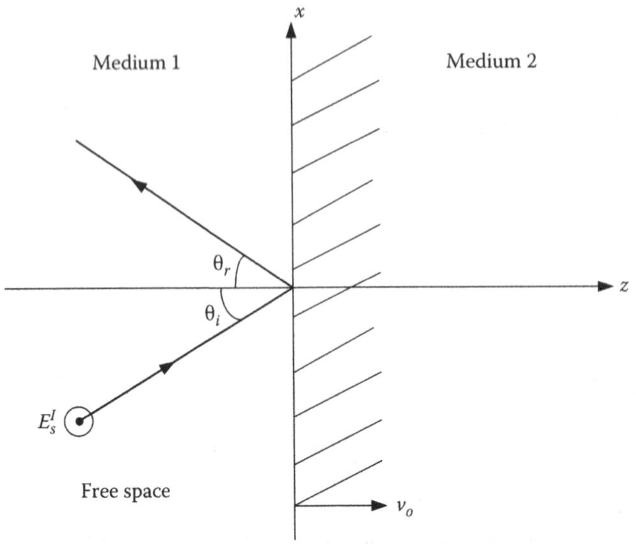

FIGURE 14.1
Reflection by a moving mirror.

The boundary condition at PEC wall $z = 0$ is that $E_{\tan} = 0$ for all x and t:

$$E_{\tan}\big|_{z=0^-} = 0 \quad \text{for all } x \text{ and } t. \tag{14.3}$$

Since the y component is the tangential component, the left-hand side (LHS) of (14.3) is obtained by adding (14.1) and (14.2) and substituting $z = 0$:

$$E_0 \cos\left(\omega_i t - x\frac{\omega_i}{c}\sin\theta_i\right) + E_r \cos\left(\omega_r t - x\frac{\omega_r}{c}\sin\theta_r\right) = 0. \tag{14.4}$$

Since (14.4) is not an equation for x and t but an identity in these two variables

$$\omega_i = \omega_r = \omega, \tag{14.5a}$$

$$\frac{\omega_i}{c}\sin\theta_i = \frac{\omega_r}{c}\sin\theta_r, \tag{14.5b}$$

$$\theta_i = \theta_r. \tag{14.5c}$$

Let us now study the case where $v_0 \neq 0$ but $v_0 \ll c$. We are considering "nonrelativistic" velocities. Thus, we consider the so-called Galilean transformations (GTs). Assuming that the boundary is $z = 0$, at $t = 0$, at any other time

$$z = v_0 t. \tag{14.6}$$

In this case, the boundary condition $E_{\tan} = 0$ is implemented at $z = v_0 t$. Substituting this value in (14.1) and (14.2),

$$E_0 \cos\left(\omega_i t - x\frac{\omega_i}{c}\sin\theta_i - v_0 t\frac{\omega_i}{c}\cos\theta_i\right) + E_r \cos\left(\omega_r t - x\frac{\omega_r}{c}\sin\theta_r + v_0 t\frac{\omega_r}{c}\cos\theta_r\right) = 0 \tag{14.7}$$

Since (14.7) is an identity in x as well as t, we get

$$\frac{\omega_i}{c}\sin\theta_i = \frac{\omega_r}{c}\sin\theta_r, \tag{14.8a}$$

$$\omega_i - v_0\frac{\omega_i}{c}\cos\theta_i = \omega_r + v_0\frac{\omega_r}{c}\cos\theta_r. \tag{14.8b}$$

Denoting

$$\beta = \frac{v_0}{c}, \tag{14.9}$$

Equation 14.8 can be written as

$$\frac{\sin\theta_r}{\sin\theta_i} = \frac{\omega_i}{\omega_r} = \frac{1+\beta\cos\theta_r}{1-\beta\cos\theta_i}. \tag{14.10}$$

Of course, $\beta = 0$ gives the result of (14.5a) and (14.5c) for the stationary boundary. When $\beta \neq 0$, the frequency of the reflected wave is different from the frequency of the incident wave. This frequency difference is the basis of "Doppler radar," which is used to measure the speed of the moving conducting object, for example, the speed at which you are driving the car. When the coefficient of t and x are the same in the two terms on the LHS of (14.7), the cosine term can be cancelled and we get

$$E_0 + E_r = 0.$$

$$\Gamma_0 = \frac{E_r}{E_0} = -1. \tag{14.11}$$

We still have total reflection from the conductor (PEC) even if it is moving. In this section, we showed that Snell's law is obtained by implementing the boundary conditions at the interface for all points on the boundary (any x in the example) and for all times (any t in the example).

14.3 Galilean Transformation

Let Σ be the laboratory frame of reference and Σ' be the frame of reference in which the moving medium is stationary as shown in Figure 14.2.

In the usual formulation, where t is "absolute,"

$$t' = t \tag{14.12}$$

$$x' = x \tag{14.13}$$

$$y' = y \tag{14.14}$$

$$z' = z - v_0 t. \tag{14.15}$$

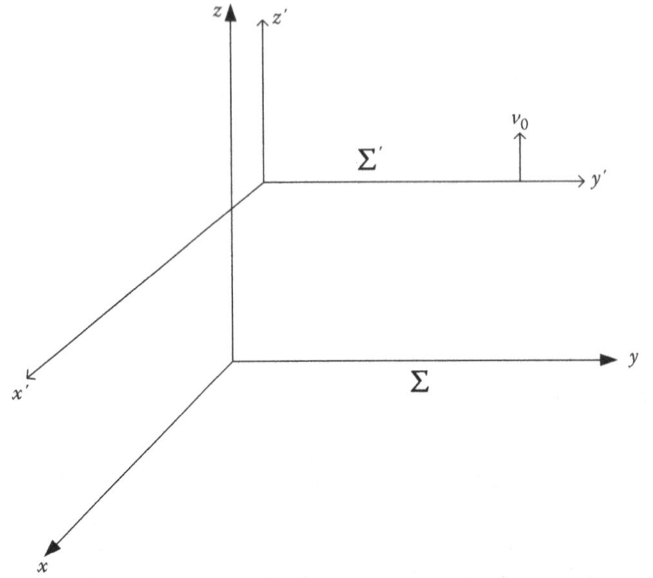

FIGURE 14.2
Two frames of reference $\sum$ and $\sum'$ is moving with a uniform velocity v_0 in the z direction with reference to $\sum$. At $t = 0$, the two frames coincide ($z = z' = 0$).

Let us now consider Faraday's law in differential form for a moving medium.

Faraday's law in integral form is usually stated in the form

$$\oint_C \mathbf{E}'\cdot d\mathbf{l} = emf = -\frac{d\Phi_m}{dt} = -\frac{d}{dt}\iint_S \mathbf{B}\cdot d\mathbf{s} \tag{14.16}$$

or

$$\int_C \mathbf{E}'\cdot d\mathbf{l} = \iint_S \left(-\frac{\partial \mathbf{B}}{\partial t}\right)\cdot d\mathbf{s} + \oint_C (v\times\mathbf{B})\cdot d\mathbf{l}, \tag{14.17}$$

where C is a closed curve bounding the open surface S. Figure 14.3 describes the geometry to help identify the various terms in (14.17).

The last term on the right-hand side (RHS) of (14.16) gives the total electromotive force (emf) induced in the closed circuit C, which could be changing with time, also the point function **B** could as well be changing with time. Equation 14.17 separates the contributions due to time-changing magnetic flux density field [called transformer *emf*, the first term on the RHS of (14.17)] and due to time-varying circuit C [motional *emf*, the second term on the RHS of (14.17)]. The last term on the RHS of (14.17) can be changed to $-\oint_C \mathbf{B}\cdot(v\times d\mathbf{l})$. The term $v\times d\mathbf{l}$ is the area swept by the element in 1 s. The last term is thus equal to $-d\Phi_m/dt$ due to the motion.

Applying Stoke's theorem, the closed line integrals in (14.17) can be converted to surface integrals:

$$\iint_S \nabla\times\mathbf{E}'\cdot d\mathbf{s} = \iint_S -\frac{\partial \mathbf{B}}{\partial t}\cdot d\mathbf{s} + \iint_S \nabla\times(v\times\mathbf{B})\cdot d\mathbf{s}. \tag{14.18}$$

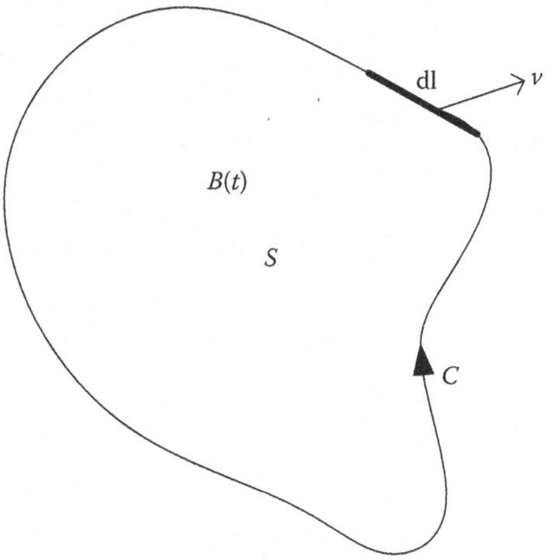

FIGURE 14.3
Geometry for Faraday's law statement of (14.17).

Since (14.18) is true for any S, provided it is the same S on all the integrals in (14.18), the integrands must be the same on both sides of (14.18).

$$\nabla \times \mathbf{E}' = -\frac{\partial \mathbf{B}}{\partial t} + \nabla \times (v \times \mathbf{B}). \tag{14.19}$$

Equation 14.18 is the same as (1.61) except that we used in (14.18) the symbol $\mathbf{E}'$ instead of $\mathbf{E}$ to emphasize that the electric field in the definition of the emf for a moving circuit has to be the effective electric field. The expression for $\mathbf{E}'$ can be obtained from (1.6) from its definition as the force for unit charge and is given by

$$\mathbf{E}' = \mathbf{E} + (v \times \mathbf{B}). \tag{14.20}$$

Substitution of (14.20) in (14.19) results in (1.1), the differential form of Faraday's law in the laboratory frame.

We can also show [10] that (14.20) and (14.22) are the consequence of imposing Galilean invariance of the form of Faraday's law in differential form, that is,

$$\nabla' \times \mathbf{E}' = -\frac{\partial \mathbf{B}'}{\partial t'}. \tag{14.21}$$

$$\mathbf{B}' = \mathbf{B}. \tag{14.22}$$

Similarly, we can show [10] that the form invariance of Ampere's law (1.2) under GT gives us the following additional relationships among the fields in inertial and laboratory frames:

$$\mathbf{D}' = \mathbf{D} \tag{14.23}$$

$$\mathbf{H}' = \mathbf{H} - v_0 \times \mathbf{D}. \tag{14.24}$$

There is some inconsistency among Equations 14.20 and 14.22 through 14.24. For example, let us apply these equations to free space and show the inconsistency. From (14.20),

$$\frac{\mathbf{D}'}{\varepsilon_0} = \mathbf{E}' = \mathbf{E} + v_0 \times \mathbf{B}$$

$$\mathbf{D}' = \mathbf{D} + v_0 \times \varepsilon_0\mu_0\mathbf{H}$$

$$\mathbf{D}' = \mathbf{D} + \frac{v_0 \times \mathbf{H}}{c^2}. \tag{14.25}$$

Equation 14.25 is in variance with (14.23).

Similarly, from (14.23),

$$\frac{\mathbf{B}'}{\mu_0} = \mathbf{H}' = \mathbf{H} - v_0 \times \mathbf{D}$$

$$\mathbf{B}' = \mathbf{B} - v_0 \times \mu_0\varepsilon_0\mathbf{E}$$

$$\mathbf{B}' = \mathbf{B} - \frac{1}{c^2} v_0 \times \mathbf{E}. \tag{14.26}$$

Equation 14.26 is in variance with (14.22). Note that they become the same if c is taken as infinity. We have to reject the GT of Maxwell's equations for moving media even for the nonrelativistic velocities of the medium. We show next that LTs given in the next section overcome these difficulties. For nonrelativistic velocities we can then formulate first-order Lorentz transformations (FOLT) as consistent approximations [4] discussed in Section 14.18.

GT does not alter the Newton's force equation since

$$\mathbf{F} = \frac{\mathrm{d}}{\mathrm{d}t}(mv). \tag{14.27}$$

$$\mathbf{F}' = \frac{\mathrm{d}}{\mathrm{d}t'}(m'v') \tag{14.28}$$

since

$$t = t' \tag{14.29}$$

$$m = m' \tag{14.30}$$

$$\frac{\mathrm{d}v}{\mathrm{d}t} = \frac{\mathrm{d}(v - v_0)}{\mathrm{d}t} = \frac{\mathrm{d}v'}{\mathrm{d}t'}$$

$$\mathbf{F} = \mathbf{F}'. \tag{14.31}$$

However, under LTs, to be discussed in the next section, we show that $t \neq t'$, $v \neq v' + v_0$, $m \neq m'$, $F \neq F'$, and that GT is valid only for nonrelativistic velocities. For arbitrary value of the parameter $\beta = v_0/c$, one needs to consider relativistic mechanics based on LTs. LTs are connected with the theory of special relativity and a four-dimensional space.

14.4 Lorentz Transformation

Since $t = t'$, in the GT, this transformation leaves the distance between two physical points unchanged in all coordinate systems.

The velocity of light as measured by a moving observer on z axis (at rest in $\sum'$) will be $(c - v_0)$ for the geometry shown in Figure 14.2. However, experiments show that the velocity of light as measured by a stationary observer or a moving observer will remain the same. This first postulate of the theory of special relativity is well confirmed. LTs in four-dimensional space x, y, z, ict will preserve the first postulate.

Let us, for convenience, use the nomenclature x_μ, $\mu = 1, 2, 3, 4$ to designate the coordinates in the four-dimensional space:

$$x_1 = x \tag{14.32}$$

$$x_2 = y \tag{14.33}$$

$$x_3 = z \tag{14.34}$$

$$x_4 = ict. \tag{14.35}$$

Note that we write the time component with the subscript 4 and designate it as imaginary. This should be carefully distinguished from [4] the imaginary notation in "quantum theory and wave theory" in some books. An alternative definition of x_4 without i in its definition is discussed in Section 14.19.

The distance between two points in the four-dimensional space will remain the same provided

$$R^2 = \sum_{\mu=1}^{4} x_\mu^2 = \sum_{\mu=1}^{4} x_\mu'^2 \tag{14.36}$$

For the geometry shown in Figure 14.2,

$$x' = x \tag{14.37}$$

$$y' = y \tag{14.38}$$

$$z' = a_1 z + a_2 t \tag{14.39}$$

$$t' = b_1 z + b_2 t. \tag{14.40}$$

Equation 14.36 will be satisfied if

$$(a_1 z + a_2 t)^2 - c^2 (b_1 z + b_2 t)^2 = z^2 - c^2 t^2 \tag{14.41}$$

Moreover, in the limit $v_0 \to 0$

$$\underset{v_0 \to 0}{\text{Limit}} \begin{bmatrix} a_1 \\ a_2 \\ b_1 \\ b_2 \end{bmatrix} = \begin{bmatrix} 1 \\ 0 \\ 0 \\ 1 \end{bmatrix} \tag{14.42}$$

The origin of $\sum'$ moves with a velocity v_0 with respect to the origin $\sum$. Thus, we have the position of the origin of $\sum'$ as $z = vt$. Substituting $z' = 0$ in (14.39),

$$0 = a_1 z + a_2 t \tag{14.43a}$$

$$0 = a_1 vt + a_2 t \tag{14.43b}$$

$$a_2 = -va_1. \tag{14.44}$$

Solving the relevant equations with signs chosen to agree with (14.42), we obtain

$$a_1 = b_1 = \frac{1}{\sqrt{1-\frac{v_0^2}{c^2}}} \tag{14.45a}$$

$$b_2 = -\frac{v_0}{c^2} a_1. \tag{14.45b}$$

Thus, we have

$$z' = \gamma(z - \beta ct) \tag{14.46}$$

$$t' = \gamma\left(t - \frac{\beta}{c} z\right), \tag{14.47}$$

where

$$\beta = \frac{v_0}{c} \tag{14.48}$$

$$\gamma = \frac{1}{\sqrt{1-\beta^2}}. \tag{14.49}$$

Generalization of the above, where the relative velocity between $\sum$ and $\sum'$ is v_0 instead of $\hat{z}v_0$, is straightforward and is given by

$$x'_{\parallel} = \gamma(x_{\parallel} - v_0 t), \quad x'_{\perp} = x_{\perp} \tag{14.50}$$

$$t' = \gamma\left(t - \frac{v_0 \cdot x}{c^2}\right) \tag{14.51}$$

Equation 14.50 can be written as

$$\mathbf{x}' = \mathbf{x} + (\gamma - 1)\frac{(x \cdot v_0)v_0}{v_0^2} - \gamma v_0 t. \tag{14.52}$$

Equations 14.46 and 14.47 give rise to the length contraction and time dilation

$$L_0 = \gamma L \tag{14.53a}$$

$$\tau = \gamma \tau_0 \tag{14.53b}$$

In (14.52) [3] $L = z_2 - z_1$, where z_2 and z_1 are the instantaneous coordinates of the end points of the rod, observed at the same time t; $L_0 = z_2' - z_1'$, where z_2' and z_1' are the coordinates at the end points in $\sum'$.

Equation 14.53b [3] is interpreted as time dilation. A clock moving relative to an observer is found to run more slowly than one at rest relative to the observer. Considering unstable elementary particles as a clock, one can explain the time dilation concept by examining the lifetime of the particles. In (14.53), τ_0 is the lifetime of the particle at rest in the system $\sum'$. The particle is moving with a uniform velocity v_0 relative to the system $\sum$. When viewed from $\sum$, the moving particle lives longer than a particle at rest in $\sum$. The "clock" in motion is observed to run more slowly than an identical one at rest. Reference [2] has a number of graphs and examples to illustrate length contraction and time dilation.

14.5 Lorentz Scalars, Vectors, and Tensors

Equation 14.36 is a constraint involved in the rotation of coordinates in the four-dimensional space. R is an invariant under LTs. LTs are rotations in a four-dimensional Euclidean space. This orthogonal transformation can be written as

$$x_\mu' = \sum a_{\mu\nu} x_\nu, \quad \mu = 1,2,3,4. \tag{14.54a}$$

For the geometry of Figure 14.2, the transformation matrix is given by

$$\left(a_{\mu\nu}\right) = \begin{vmatrix} 1 & 0 & 0 & 0 \\ 0 & 1 & 0 & 0 \\ 0 & 0 & \nu & i\gamma\beta \\ 0 & 0 & -i\gamma\beta & \gamma \end{vmatrix}. \tag{14.54b}$$

Figure 14.4a shows the geometry of rotation of the coordinates in the $x_3 - x_4$ plane by an angle ψ that achieves the LT.

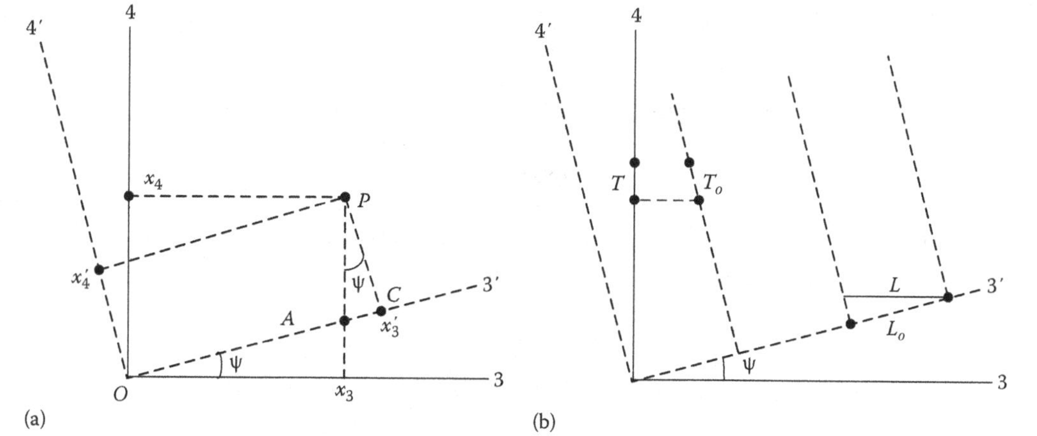

FIGURE 14.4
Lorentz transformation: (a) rotation angle for the orthogonal transformation and (b) time dilation and the length contraction of the transformation.

An inverse transformation is obtained by reversing the sign of β, which gives $a_{\nu\mu}$. Thus we have

$$x_\mu = \sum_{\nu=1}^{4} x'_\nu a_{\nu\mu}. \tag{14.54c}$$

We also have

$$\left[a_{\mu\nu}(-\beta)\right] = \left[a_{\mu\nu}(\beta)\right]^{-1} = \left[a_{\mu\nu}(\beta)^{T}\right]. \tag{14.55}$$

Equations 14.54a and 14.54b show that LT is an orthogonal transformation.
From the geometry (Figure 14.4), $OC = OA + AC$, giving

$$x'_3 = (\cos\psi)x_3 + (\sin\psi)x_4. \tag{14.56a}$$

Similarly,

$$x'_4 = (-\sin\psi)x_3 + (\cos\psi)x_4. \tag{14.56b}$$

Comparing (14.56) with (14.54c), we get

$$\tan\psi = i\beta. \tag{14.57}$$

Note that the angle ψ is complex and its cosine, $\cos\psi = \gamma \geq 1$. Thus, LT as a rotation, by an angle ψ, is a concept rather than a real angular rotation. Figure 14.4b shows L, L_0, T, T_0. The distance L_0 in the frame $\sum'$ is observed as L in the frame $\sum$ at constant time in $\sum$. Though L appears larger than L_0 in the figure, because of the complex nature of ψ,

$$L_0 = L\cos\psi \tag{14.58a}$$

$$L = \frac{L_0}{\gamma} < L_0. \tag{14.58b}$$

The time interval T_0 in the frame $\sum'$ is seen as T in the frame $\sum$, where

$$T = T_0\cos\psi = \gamma T_0 > T_0. \tag{14.59}$$

A quantity that does not change under LT is called a Lorentz scalar, say Φ. Its derivative (can be called four-gradient) with respect to x'_μ is a four-vector.

$$\frac{\partial\Phi}{\partial x'_\mu} = \sum_{\nu=1}^{4} \frac{\partial\Phi}{\partial x_\nu}\frac{\partial x_\nu}{\partial x'_\nu}$$

$$\frac{\partial\Phi}{\partial x'_\mu} = \sum_{\nu=1}^{4} a_{\mu\nu}\frac{\partial\Phi}{\partial x_\nu}. \tag{14.60}$$

The derivative transforms like a four-vector.

The four-divergence of a four-vector is Lorentz invariant:

$$\square' \cdot \mathbf{A}' = \sum_{\nu=1}^{4} \frac{\partial A'_\nu}{\partial x'_\nu} = \sum_{\mu=1}^{4} \frac{\partial A_\mu}{\partial x_\mu}. \tag{14.61}$$

If $A_\mu = \partial \Phi / \partial x_\mu$, (14.61) shows that the four-dimensional Laplacian operator is a Lorentz invariant operator.

$$\square'^2 \Phi = \sum_{\nu=1}^{4} \frac{\partial^2 \Phi}{\partial x'^2_\nu} = \sum_{\mu=1}^{4} \frac{\partial^2 \Phi}{\partial x^2_\mu} = \square^2 \Phi, \tag{14.62}$$

where $\square'$ is the vector operator in four dimensions in (14.61) and $\square^2 = \square \cdot \square$ in (14.62).

The scalar product of two four-vectors is a scalar invariant

$$(\mathbf{A}' \cdot \mathbf{B}') = (\mathbf{A} \cdot \mathbf{B}). \tag{14.63}$$

Lorentz scalar can be considered as a tensor of zero rank and four-vector as a tensor of rank one.

A second-rank tensor is a set of 16 quantities that transforms as

$$T'_{\mu\theta} = \sum_{\lambda=1}^{4} \sum_{\sigma=1}^{4} a_{\mu\lambda} a_{\theta\sigma} T_{\lambda\sigma}. \tag{14.64}$$

An example of such a tensor is the field strength four-tensor $F_{\mu\nu}$, which can be used to combine two of the Maxwell equations, as shown in the next section.

14.6 Electromagnetic Equations in Four-Dimensional Space

In the four-dimensional space,

$$x_\mu = (x_1, x_2, x_3, x_4) = (x, y, z, ict) = (\mathbf{r}, ict). \tag{14.65}$$

One can write the electromagnetic equations in terms of suitably chosen scalars, four-vectors, and four-tensors of second rank. The quantity in four-dimensional space will be denoted by an italic symbol with a subscript μ. Also a shorthand notation, that a repeated Greek letter subscript implies summation over that subscript as it varies from 1 to 4, will be used.

Let us start with the continuity equation (1.5)

$$\nabla \cdot \mathbf{J} + \frac{\partial \rho_v}{\partial t} = 0. \tag{1.5}$$

Let

$$\mathbf{J}_\mu = (\mathbf{J}, ic\rho_v)$$

$$(J_1, J_2, J_3, J_4) = (J_x, J_y, J_z, J_{ict}) \tag{14.66}$$

Equation 1.5 is

$$\frac{\partial J_x}{\partial x}+\frac{\partial J_y}{\partial y}+\frac{\partial J_z}{\partial z}+\frac{\partial J(ic\rho_v)}{\partial(ict)}=0$$

$$\frac{\partial J_1}{\partial x_1}+\frac{\partial J_2}{\partial x_2}+\frac{\partial J_3}{\partial x_3}+\frac{\partial J_4}{\partial x_4}=0$$

$$\sum_{\mu=1}^{4}\frac{\partial J_\mu}{\partial x_\mu}=0=\frac{\partial J_\mu}{\partial x_\mu}. \tag{14.67}$$

Since μ is a Greek symbol that is repeated, we can use the shorthand notation given as the last term in (14.67).

Moreover, if we definite the four-vector differential operator □

$$\square=\left(\nabla,\frac{\partial}{\partial(ict)}\right), \tag{14.68}$$

the continuity equation can be written, in terms of four divergence, as

$$\square\cdot\boldsymbol{J}_\mu=0, \tag{14.69}$$

where the bold italic $\boldsymbol{J}_\mu$ is a four-vector current density.

Let us next consider the Lorentz condition, when the medium is free space, from (1.21):

$$\nabla\cdot\mathbf{A}+\frac{1}{c^2}\frac{\partial\Phi}{\partial t}=0. \tag{14.70}$$

Let us consider the four-vector potential $\mathbf{A}_\mu$:

$$\mathbf{A}_\mu=\left(\mathbf{A},i\frac{\Phi}{c}\right). \tag{14.71}$$

Equation 14.71 can be written as

$$\square\cdot\mathbf{A}_\mu=0, \tag{14.72}$$

The wave equation in free space

$$\nabla^2\mathbf{A}-\frac{1}{c^2}\frac{\partial^2\mathbf{A}}{\partial t^2}=-\mu_0\mathbf{J}, \tag{14.73a}$$

$$\nabla^2\Phi-\frac{1}{c^2}\frac{\partial^2\Phi}{\partial t^2}=-\frac{\rho_v}{\varepsilon_0}, \tag{14.73b}$$

can be written as

$$\square^2 \mathbf{A}_\mu = -\mu_0 \mathbf{J}_\mu. \tag{14.74}$$

Note that (14.73b) is obtained from

$$\left[\nabla^2 + \left(\frac{\partial}{\partial(\mathrm{i}ct)}\right)^2\right] A_4 = -\mu_0 J_4,$$

$$\nabla^2 \Phi - \frac{1}{c^2}\frac{\partial^2 \Phi}{\partial t^2} = -\frac{\mu_0 \mathrm{i} c^2}{\mathrm{i}} \rho_v = -\frac{\rho_v}{\varepsilon_0}.$$

If the source is a point charge Q moving with a relativistic velocity v along a specified trajectory, the scalar and vector potentials, called Lienard–Wiechert potentials, can be calculated and discussed in Appendix 14G. We touched on this topic in Chapter 13 also.

The field strength tensor $F_{\mu\nu}$ can be constructed so that the equations

$$\mathbf{B} = \nabla \times \mathbf{A} \tag{14.75}$$

$$\mathbf{E} = -\nabla \Phi - \frac{\partial \mathbf{A}}{\partial t} \tag{14.76}$$

are satisfied. For example, from (14.75) and (14.76) we can write

$$B_1 = B_x = \frac{\partial A_z}{\partial y} - \frac{\partial A_y}{\partial z} = \frac{\partial A_3}{\partial x_2} - \frac{\partial A_2}{\partial x_3}. \tag{14.77}$$

If (14.75) and (14.76) can be written (in the form of a four-dimensional curl operation) as

$$F_{\mu\nu} = \frac{\partial A_\nu}{\partial x_\mu} - \frac{\partial A_\mu}{\partial x_\nu}, \tag{14.78}$$

then (14.77) is (14.78) with $\mu = 3$, $\nu = 2$, and $F_{\mu\nu} = F_{32} = -B_1 = -B_x$.

One can verify that (14.75) and (14.76) become (14.78) if $F_{\mu\nu}$, a 4×4 matrix, has the following elements expressed in terms of the components of **E** and **B**.

$$(F_{\mu\nu}) = \begin{vmatrix} 0 & B_z & -B_y & -\mathrm{i}\dfrac{E_x}{c} \\ -B_z & 0 & B_x & -\mathrm{i}\dfrac{E_y}{c} \\ B_y & -B_x & 0 & -\mathrm{i}\dfrac{E_z}{c} \\ \mathrm{i}\dfrac{E_x}{c} & \mathrm{i}\dfrac{E_y}{c} & \mathrm{i}\dfrac{E_z}{c} & 0 \end{vmatrix}. \tag{14.79}$$

Though we only verified the element $F_{32} = -B_1$, one can check the correctness of the other elements.

With this field strength tensor, we can show that the two Maxwell equations in free space

$$\nabla \cdot \mathbf{E} = \frac{\rho_v}{\varepsilon_0}, \tag{14.80}$$

$$\nabla \times \mathbf{B} = \mu_0 \mathbf{J} + \mu_0 \varepsilon_0 \frac{\partial \mathbf{E}}{\partial t}, \tag{14.81}$$

take the form, in Lorentz space,

$$\frac{\partial F_{\mu\nu}}{\partial x_\nu} = J_\mu. \tag{14.82}$$

The two Maxwell equations, valid in any medium,

$$\nabla \cdot \mathbf{B} = 0 \tag{14.83}$$

$$\nabla \times \mathbf{E} = -\frac{\partial \mathbf{B}}{\partial t} \tag{14.84}$$

can be expressed in Lorentz space

$$\frac{\partial F_{\mu\nu}}{\partial x_\lambda} + \frac{\partial F_{\lambda\mu}}{\partial x_\nu} + \frac{\partial F_{\nu\lambda}}{\partial x_\mu} = 0. \tag{14.85}$$

For this reason $F_{\mu\nu}$ is called field strength tensor. From (14.79), we note it is an antisymmetric tensor. The other two Maxwell equations, valid in any medium

$$\nabla \times \mathbf{H} = \mathbf{J} + \frac{\partial \mathbf{D}}{\partial t} \tag{14.86a}$$

$$\nabla \cdot \mathbf{D} = \rho_v \tag{14.86b}$$

can be combined as [10]

$$\frac{\partial G_{\mu\nu}}{\partial x_\nu} = J_\mu, \tag{14.87}$$

where

$$G_{\mu\nu} = \begin{vmatrix} 0 & H_z & -H_y & -icD_x \\ -H_z & 0 & H_x & -icD_y \\ H_y & -H_x & 0 & -icD_z \\ icD_x & icD_y & icD_z & 0 \end{vmatrix}. \tag{14.88}$$

Since $G_{\mu\nu}$ is a second-rank tensor, it transforms as

$$G'_{\mu\nu} = \sum_{\lambda=1}^{4} \sum_{\sigma=1}^{4} a_{\mu\lambda} a_{\nu\sigma} G_{\lambda\sigma}. \tag{14.89}$$

It is an antisymmetric tensor and can be called as excitation tensor.

14.7 Lorentz Transformation of the Electromagnetic Fields

The field strength $F_{\mu\nu}$ is a tensor; its transformation is given by (14.62).

$$F'_{\mu\nu} = a_{\mu\lambda}\,\alpha_{\nu\sigma}\,F_{\lambda\sigma} \tag{14.90}$$

$$E'_1 = \gamma\left(E_1 - \beta c B_2\right), \quad B'_1 = \gamma\left(B_1 + \left(\frac{\beta}{c}\right)E_2\right) \tag{14.91a}$$

$$E'_2 = \gamma\left(E_2 + \beta c B_1\right), \quad B'_2 = \gamma\left(B_2 - \left(\frac{\beta}{c}\right)E_1\right) \tag{14.91b}$$

$$E'_3 = E_3, \quad B'_3 = B_3. \tag{14.91c}$$

The inverse transformation is obtained by interchanging primed and unprimed quantities and replacing β by −(β). For a general velocity $\boldsymbol{v}_o$, one can generalize (14.90)–(14.91) by

$$\mathbf{E}'_{\parallel} = \mathbf{E}_{\parallel}, \quad \mathrm{B}'_{\parallel} = \mathrm{B}_{\parallel} \tag{14.92a}$$

$$\mathbf{E}'_{\perp} = \gamma\left(\mathbf{E}_{\perp} + v_0 \times \mathbf{B}\right), \quad \mathbf{B}'_{\perp} = \gamma\left(\mathbf{B}_{\perp} - \frac{v_0}{c^2} \times \mathbf{E}\right) \tag{14.92b}$$

14.8 Frequency Transformation and Phase Invariance

A four-vector wave number $\mathbf{k}_\mu$ can be constructed

$$\mathbf{k}_\mu = \left(\mathbf{k}, \mathrm{i}\frac{\omega}{c}\right) \tag{14.93a}$$

so that

$$\mathbf{k}_\mu \cdot \mathbf{x}_\mu = \mathbf{k}\cdot\mathbf{r} - \omega t. \tag{14.93b}$$

Since (14.93b) is a Lorenz scalar, it is Lorentz invariant giving the concept of phase invariance. Thus, we obtain (14.94)

$$\mathbf{k}\cdot\mathbf{r} - \omega t = \mathbf{k}'\cdot\mathbf{r}' - \omega' \mathrm{t}'. \tag{14.94}$$

Equation 14.94 leads to the results given by (14.95):

$$\omega' = \gamma\omega\left(1 - \beta\cos\theta\right) \tag{14.95a}$$

$$\tan\theta' = \frac{\sin\theta}{\gamma\left(\cos\theta - \beta\right)} = \frac{\tan\theta}{\gamma\left(1 - \beta\sec\theta\right)}, \tag{14.95b}$$

where θ and θ′ are the angles of $\mathbf{k}$ and $\mathbf{k}'$ relative to the direction of $\boldsymbol{v}_o$. Equation 14.95a is the relativistic formula for the frequency shift, called relativistic Doppler effect.

Equation 14.95b is the relativistic formula for aberration [4]. Equations 14.95 can be more easily seen by considering the motion of the frame to be on the z-direction, and the wave is propagating in the x–z plane.

$$k_x' = k_x \tag{14.96a}$$

$$k' \sin\theta' = k \sin\theta \tag{14.96b}$$

$$\gamma\left(k_z - \frac{\beta}{c}\omega\right) \tag{14.96c}$$

$$k' \cos\theta' = \gamma\left(k \cos\theta - \frac{\beta}{c}\omega\right)$$

$$k' \cos\theta' = \gamma\left(k \cos\theta - \beta k\right)$$

$$k' \cos\theta' = \gamma k\left(\cos\theta - \beta\right). \tag{14.96d}$$

Dividing (14.96b) by (14.96d), we obtain (14.95b).

From (14.94), we obtain

$$\omega' = \gamma\left(\omega - \beta c k_z\right)$$

$$\omega' = \gamma\left(\omega - \beta c k \cos\theta\right),$$

which is the same as (14.95a).

14.9 Reflection from a Moving Medium

Let us solve the problem of determining the power reflection coefficient α of a moving medium. We will use the LT technique to solve the problem. The incident wave will be transformed to the primed frame in which the medium 2 is at rest, and this reflection problem in the rest frame with a stationary boundary can be solved using the techniques developed in Chapter 2. After obtaining the primed reflection coefficient R', we will relate it to the unprimed R.

14.9.1 Incident s-Wave

Let the s-wave in free space have the electric field

$$\mathbf{E}_s^{\mathrm{I}} = \hat{y}\mathbf{E}_o \psi_{\mathrm{I}}$$

$$\psi_{\mathrm{I}} = e^{j(\omega t - k_x x - k_z z)}$$

$$k_x = k_o \sin\theta = k_o S$$

$$k_z = k_o \cos\theta = k_o C$$

$$k_o = \frac{\omega}{c}.$$

The $\sum'$ frame is attached to the moving conductor which is moving with a velocity

$$v_o = \hat{z}v_o$$

with reference to the laboratory frame $\sum$. The geometry is as shown in Figure 14.1.

We will transform the incident wave to $\sum'$ frame in which the moving conductor is at rest:

$$\left(\mathbf{E}_s^I\right)' = \hat{y}\mathbf{E}_o'\psi_I'$$

$$\psi_I' = e^{j\left(\omega' t' - k_x' x' - k_z' z'\right)}$$

$$k_x' = k_o' \sin\theta' = k_o' S'$$

$$k_z' = k_o' \cos\theta' = k_o' C'$$

$$k_o' = \frac{\omega'}{c}.$$

From the LT (14.96),

$$k_x' = k_x \tag{14.97a}$$

$$\gamma\left(k_z - \frac{v_o}{c^2}\omega\right) \tag{14.97b}$$

$$\omega' = \gamma(\omega - v_o k_z. \tag{14.97c}$$

This leads to

$$\omega' = p_z\omega \tag{14.98a}$$

$$k_x' = k_o' S' = k_o S \tag{14.98b}$$

$$k_z' = k_o' C' = \gamma k_o (C - \beta), \tag{14.98c}$$

where

$$S' = \frac{S}{p_z}, \quad C' = \frac{C - \beta}{1 - C\beta}, \tag{14.98d}$$

where

$$p_z = \gamma(1 - C\beta). \tag{14.98e}$$

The wave vector and the frequency of the reflected wave can be determined through (14.97).

From the phase invariance principle, the phase of the reflected wave as seen in $\sum'$ and $\sum$ are the same:

$$\omega' t' - k_x' x' + k_z z' = \omega_r t - k_{xr} x + k_{zr} z \tag{14.99}$$

From (14.99), (14.46), (14.47), and (14.50),

$$\omega_r = \gamma\left(\omega' - \beta c k_z'\right) \tag{14.100}$$

$$k_{zr} = \gamma\left(k_z' - \beta c \omega'\right) \tag{14.101}$$

$$k_{xr} = k_o' S' = k_o S. \tag{14.102}$$

Substituting (14.97) in (14.100)–(14.102), we obtain the following:

$$\omega_r = \gamma^2 \omega\left(1 + \beta^2 - 2\beta C\right) \tag{14.103a}$$

$$k_z^r = \gamma^2 k_o\left(\left(1 + \beta^2\right)C - 2\beta\right) \tag{14.103b}$$

$$\cos\theta_r = \frac{C}{\beta'} \tag{14.104a}$$

$$\sin\theta_r = \frac{S}{p_z}, \tag{14.104b}$$

where

$$\beta' = \frac{1 + \beta^2 - 2\beta C}{1 + \beta^2 - 2\beta/C}. \tag{14.104c}$$

As an aside, let us list the results for the case of the medium moving with velocity v_o along x-axis, $\mathbf{v} = \hat{x} v_o$, and we obtain the following (the student is advised to derive these equations):

$$k_x' = \gamma\left(k_x - \frac{v_o}{c^2}\omega\right) \tag{14.105a}$$

$$k_z' = k_z \tag{14.105b}$$

$$\omega' = \gamma\left(\omega - v_o k_x\right) \tag{14.105c}$$

$$\omega' = p_x \omega \tag{14.105d}$$

$$S' = \frac{S - \beta}{1 - S\beta} \tag{14.105e}$$

$$C' = \frac{C}{p_x}, \tag{14.105f}$$

where

$$p_x = \gamma(1 - S\beta) \tag{14.105g}$$

$$\omega_r = \omega \tag{14.105h}$$

$$k_x^r = k_x = k_o S \tag{14.105i}$$

$$k_z^r = k_z = k_o C. \tag{14.105j}$$

14.9.2 Field Transformations

In Σ' the medium is stationary, and $R_s' = \left(E_y^r\right)' / \left(E_y'^I\right)$ can be calculated using the theories developed for the stationary media. The electromagnetic properties of the medium 2 in Σ' determine R_s'. For example, if the medium 2 is a perfect electric conductor $R_s' = -1$.

The relation between R_s' and R_s, where

$$R_s = \frac{E_y^r}{E_y^I}, \tag{14.106}$$

can be obtained from LT (14.92) for the fields. For the case of $v_o = \hat{z}v_o$,

$$E_y'^I = \gamma(1 - \beta C) E_y^I \tag{14.107a}$$

$$E_y'^R = \gamma\left(1 + \frac{\beta C}{\beta'}\right) E_y^R. \tag{14.107b}$$

Dividing (14.107b) by (14.107a), we obtain

$$R_s' = \left(\frac{1 + \beta C/\beta'}{1 - \beta C}\right) R_s \tag{14.108a}$$

$$R_s = \left(\frac{1 - \beta C}{1 + \beta C/\beta'}\right) R_s'. \tag{14.108b}$$

It can be noted that the factor in the brackets () in (14.108) comes from the transformation of fields in the incident medium, depending on the velocity of the moving second medium and the angle of incidence. It is not influenced by the properties of medium 2.

14.9.3 Power Reflection Coefficient of a Moving Mirror for s-Wave Incidence

If the medium 2 is a PEC, $R_s' = -1$, and the power reflection coefficient α is given by

$$\alpha = |R_s|^2 \frac{\cos\theta_r}{\cos\theta_i} = \frac{1}{\beta'} |R_s|^2 \tag{14.109a}$$

$$\alpha = \frac{1}{\beta'}\left[\frac{1 - \beta C}{1 + \dfrac{\beta C}{\beta'}}\right]^2. \tag{14.109b}$$

Simplifying (14.109b), the power reflection coefficient for the case of s-wave incidence on a PEC moving perpendicular to the interface is given by

$$\alpha = \frac{\left(1+\beta^2-2\beta C\right)\left(1+\beta^2-\dfrac{2\beta}{C}\right)}{\left(1-\beta^2\right)^2}. \tag{14.110}$$

It is worth remarking that the PEC medium is a perfect reflector and α is the highest power reflection coefficient that can be achieved by any second medium. If the second medium is not a perfect reflector, for example, if the second medium is a dielectric or a plasma, certain amount of power can be transmitted into the second medium and $\left(R_s'\right)^2$ will not be one. However, the power reflection coefficient ρ can be expressed for such a case as α times $\left(R_s'\right)^2$

Figure 14.5 shows the plot of α vs. β. The parameter $C = 0.5$, that is, the angle of incidence $\theta = 60°$. The angle of reflection θ' as seen in $\sum'$ and the angle of reflection θ_r as seen in $\sum$ are also shown.

Several interesting points can be noted from Figure 14.5. α is zero at $\beta = \dfrac{1-S}{C}$, since the angle of reflection θ_{r}, as seen in $\sum$, is 90° at this value of β. In the range $\dfrac{1-S}{C} < \beta < C, \theta_r > 90°$ and α is thus negative.

This means that in this range of β, the reflected wave travels towards the medium 2 as seen from $\sum$. For an interpretation of Figure 14.5, for the range $C<\beta<1$, see [12], which is given as Appendix 14A.

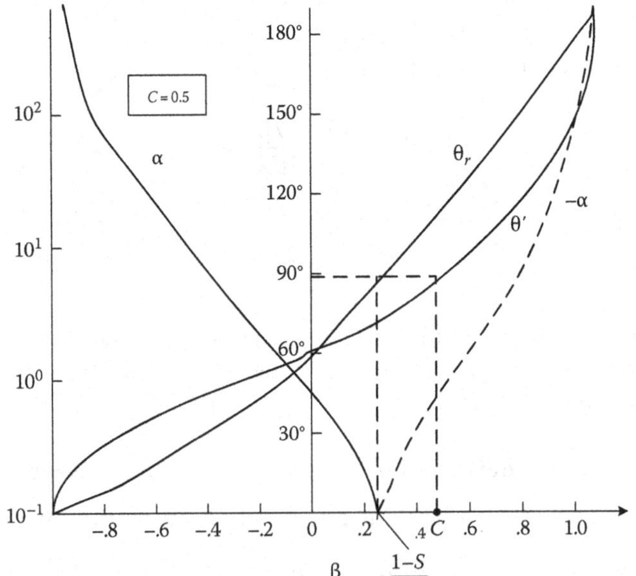

FIGURE 14.5
The power reflection coefficient α, the angle of reflection θ' as seen in $\sum'$, and the angle of reflection θ_r as seen in $\sum$ as a function of β are shown. The angle of incidence $\theta = 60°$, given $C = \cos\theta = 0.5$. α scale is logarithmic. Note that $\alpha > 1$ for $\beta < 0$. Also note that $\theta_r = 90°$ for $\beta = \dfrac{1-S}{C}$, where $S = \sin\theta$.

Another surprising result is that α is greater than one for the range of $-1<\beta<0$. This phenomenon can be explained by noting that mechanical power is supplied to the system to maintain the uniform motion of the conductor against the radiation pressure. A detailed calculation of the electric power P_{el}, the rate of the decrease of the stored energy in the fields P_s, and the mechanical power supplied to the medium to overcome the radiation pressure force P_m for the conductor as well as the dielectric half-space are given in Appendix 14B [23].

This appendix contains many more references on the topic.

14.10 Constitutive Relations for a Moving Dielectric

Minkowsky theory is developed based on

$$\mathbf{D}' = \varepsilon'\mathbf{E}' \tag{14.111}$$

$$\mathbf{B}' = \mu'\mathbf{H}' \tag{14.112}$$

for a simple loss-free medium, where ε' and μ' are the parameters measured in $\sum'$. Using field transformations, one can obtain the constitutive relations of the Minkowsky theory.

$$\mathbf{D} + \frac{1}{c^2} v_o \times \mathbf{H} = \varepsilon'(\mathbf{E} + v_o \times \mathbf{B}) \tag{14.113}$$

$$\mathbf{B} - \frac{1}{c^2} v_o \times \mathbf{E} = \mu'(\mathbf{H} - v_o \times \mathbf{D}). \tag{14.114}$$

Equations 14.113 and 14.114 are obtained using (14.92) and (14.115):

$$\mathbf{D}' = \gamma\left(\mathbf{D} + \frac{1}{c^2} v_o \times \mathbf{H}\right) + (1-\gamma)\frac{\mathbf{D}\cdot v_o}{v_o^2} v_o \tag{14.115a}$$

$$\mathbf{H}' = \gamma\left(\mathbf{H} + \frac{1}{c^2} v_o \times \mathbf{D}\right) + (1-\gamma)\frac{\mathbf{H}\cdot v_o}{v_o^2} v_o. \tag{14.115b}$$

Equations 14.115 are obtained from (14.85) [10].

The constitutive relations in $\sum$ can now be written as [6]

$$\mathbf{D} = \varepsilon'\alpha \cdot \mathbf{E} + \Omega \times \mathbf{H} \tag{14.116}$$

$$\mathbf{B} = \mu'\alpha \cdot \mathbf{H} - \Omega \times \mathbf{E}, \tag{14.117}$$

where

$$\Omega = \frac{(n^2-1)}{(1-n^2\beta^2)c}\hat{z} \tag{14.118}$$

$$n = \left[\frac{\mu'\varepsilon'}{\mu_o\varepsilon_o}\right]^{1/2} \tag{14.119}$$

$$\alpha = \begin{bmatrix} a & 0 & 0 \\ 0 & a & 0 \\ 0 & 0 & 1 \end{bmatrix} \tag{14.120a}$$

$$a = \frac{1-\beta^2}{1-n^2\beta^2}. \tag{14.120b}$$

In the above, it is assumed that $v_o = \hat{z}v_o$.

14.11 Relativistic Particle Dynamics

A brief review of the relativistic particle dynamics is given below. A more complete account can be found in [3]. The charge of an electron q is invariant. However, the mass of an electron m is given by

$$m = \gamma m_o, \tag{14.121}$$

where m_o is the rest mass.

The relativistic momentum of a mass m moving with velocity v_o

$$\mathbf{p} = mv_o = \gamma m_o v_o. \tag{14.122}$$

In a collision it is this momentum that is conserved and not the classical momentum mv_o. The relativistic force

$$\mathbf{F} = \frac{dp}{dt} = \frac{d}{dt}(mv_o). \tag{14.123}$$

The four momentum of a particle of energy ε

$$p_\mu = \left(\mathbf{p}, i\frac{\varepsilon}{c}\right) \tag{14.124}$$

transforms as a four-vector, giving

Given

$$p_1 = p_1', \quad p_2 = p_2' \tag{14.125a}$$

$$p_3 = \gamma\left(p_3' + \beta\frac{\varepsilon'}{c}\right) \tag{14.125b}$$

$$\mathscr{E} = \gamma\left(\varepsilon' + \beta c p_3'\right). \tag{14.126}$$

The inverse transformation can be obtained by changing $\beta \rightarrow -\beta$ and interchanging the primed and unprimed quantities.

In the above, $\sum'$ is moving in the direction of $z = x_3$. The length of the four-vector p_u is a scalar invariant:

$$p_u \cdot p_u = p_u' \cdot p_u' = p_1^2 + p_2^2 + p_3^2 - \frac{\mathscr{E}^2}{c^2}$$

$$= p_1'^2 + p_2'^2 + p_3'^2 - \frac{\mathscr{E}'^2}{c^2}. \tag{14.127}$$

In the rest frame of the particle, $\mathbf{p}' = 0$, and we get the equation

$$p_1^2 + p_2^2 + p_3^2 - \frac{\varepsilon^2}{c^2} = \frac{-(\varepsilon')^2}{c^2} = \frac{-(m_o c^2)^2}{c^2} = -m_o^2 c^2. \tag{14.128}$$

If the particle is moving along z-axis

$$v_o = \hat{z} v_o,$$

then

$$p_1 = p_2 = 0, \quad p_3 = p_z = p$$

$$p^2 - \frac{\varepsilon^2}{c^2} = -m_o^2 c^2 \tag{14.129}$$

$$m^2 c^4 = \varepsilon^2 = p^2 c^2 + m_o^2 c^4. \tag{14.130}$$

Equation 14.130 can be interpreted as a relationship between three sides of a right-angled triangle. The hypotenuse is the relativistic energy and of the two sides making the right angle, one of them is the rest energy $m_o c^2$ [1].

The total energy ε is the sum of the rest energy $m_o c^2$ and the kinetic energy T given by

$$T = (\gamma - 1) m_o c^2. \tag{14.131}$$

To summarize, a free particle with rest mass m_o moving with a velocity v_o in a reference frame Σ has a momentum $\mathbf{p}$ and energy ε given by

$$\mathbf{p} = \gamma m_o v_o \tag{14.132}$$

$$\varepsilon = \gamma m_o c^2. \tag{14.133}$$

A photon in free space has an energy ε:

$$\varepsilon = hf = \hbar 2\pi f = \hbar\omega, \tag{14.134}$$

where h is Plank's constant and f is the frequency in Hz, $\hbar = h/2\pi$, and ω is the frequency in radians/s.

Its velocity is c. Its momentum is given by

$$p = \frac{\varepsilon}{c} = \frac{hf}{c} = \frac{h}{\lambda} = \frac{\hbar 2\pi}{\lambda} = \hbar k. \tag{14.135}$$

14.12 Transformation of Plasma Parameters

The plasma frequency in the respective frames are given by

$$\omega_p'^2 = \frac{(q')^2 N'}{\varepsilon_o m'} \tag{14.136}$$

$$\omega_p^2 = \frac{(q)^2 N}{\varepsilon_o m}, \tag{14.137}$$

where $(-q', m', N')$ and $(-q, m, N)$ are the electron charge, mass, and volume electron number density in their respective frames.

Since

$$q' = q \tag{14.138}$$

$$m' = m/\gamma \tag{14.139}$$

$$V' = \gamma V \tag{14.140}$$

$$N' = N/\gamma, \tag{14.141}$$

we obtain

$$\omega_p' = \omega_p \tag{14.142}$$

Thus,

$$\varepsilon_r' = 1 - \frac{\omega_p'^2}{(\omega')^2} = 1 - \frac{\omega_p^2}{(\omega')^2}. \tag{14.143}$$

14.13 Reflection by a Moving Plasma Slab

In Section 14.9, we discussed the key steps in solving the moving media problem. The first step is to obtain the reflection coefficient R' in the rest frame. The second step is to relate the reflection coefficient R' to R by considering the transformation of the fields and reflection angle in the laboratory frame and to determine the factor α, which was identified as the power reflection coefficient of a moving mirror. The power reflection coefficient ρ for any moving, transmitting medium can then be expressed as

$$\rho = \alpha |R'|^2. \tag{14.144}$$

In Section 14.9 we showed, for s-wave and the mirror movement perpendicular to the interface,

$$R_s = \left(\frac{1-\beta C}{1+\dfrac{\beta C}{\beta'}} \right) R_s' \tag{14.145}$$

and the power reflection coefficient for the moving mirror

$$\alpha_s = \frac{\left(1+\beta^2-2\beta C\right)\left(1+\beta^2-\dfrac{2\beta}{C}\right)}{\left(1-\beta^2\right)^2}. \tag{14.146}$$

It can be shown that for p-wave

$$R_p = \left(\frac{C-\beta}{C+\beta\beta'} \right) R_p', \tag{14.147a}$$

where

$$R_p = \frac{E_x^R}{E_x^I} \tag{14.147b}$$

$$R_p' = \frac{\left(E_x^R\right)'}{\left(E_x^I\right)'}. \tag{14.147c}$$

Substituting $R_p' = -1$ for the p-wave incidence on a mirror in (14.147), the power reflection coefficient when the mirror is moving perpendicular to the interface is given by

$$\alpha_p = \left|R_p\right|^2 \frac{\cos\theta_i}{\cos\theta_r} = \beta' \left[\frac{(C-\beta)}{(C+\beta\beta')} \right]^2. \tag{14.148}$$

On simplification, (14.148) will be the same as (14.146).

Appendix 14A discusses the isotropic moving plasma slab problem by computing R_p'. Section 9.4 also deals with the stationary plasma slab problem. Appendix 14C [14] discusses the uniaxial plasma slab problem with

$$\varepsilon_r = \left(\hat{x}\hat{x} + \hat{y}\hat{y} + \varepsilon_r'\hat{z}\hat{z}\right), \tag{14.149}$$

where ε_r' is given by (14.143).

14.14 Brewster Angle and Critical Angle for Moving Plasma Medium

Appendix 14D [18] discusses the Brewster angle for a plasma medium moving at a relativistic speed. Appendix 14E [22] discusses the total reflection of electromagnetic waves from moving plasmas.

14.15 Bounded Plasmas Moving Perpendicular to the Plane of Incidence

Appendix 14F [20] discusses the reflection properties of a plasma moving perpendicular to the plane of incidence. This reference is included in this book as Appendix 14F.

14.16 Waveguide Modes of Moving Plasmas

References 11,16,17,19,21,28, and 29 discuss the waveguide modes of moving plasmas.

14.17 Impulse Response of a Moving Plasma Medium

References 24–26 and 30 discuss the reflection of an impulse plane wave by moving plasma half-spaces.

14.18 First-Order Lorentz Transformation

One is tempted to use the GT for nonrelativistic velocities. However, we have shown in Section 14.3 that such an attempt can lead to inconsistent approximations. Since Maxwell's equations are covariant with LTs, the best way to solve a moving media electromagnetic problem is to solve the problem using LTs, and for nonrelativistic velocities, drop the terms of the order of β^2 and higher. The resulting transformations are called [4] FOLT. In FOLT, γ is approximated by one. Listed below are the FOLT approximations of some of the equations given earlier using LT.

LT		FOLT	
$z' = \gamma(z - \beta ct)$	(14.46)	$z' = (z - \beta ct)$	(14.46F)
$t' = \gamma\left(t - \frac{\beta}{c} z\right)$	(14.47)	$t' = \left(t - \frac{\beta}{c} z\right)$	(14.47F)
$\mathbf{E}'_\perp = \gamma(\mathbf{E}_\perp + v_0 \times \mathbf{B}),\ \mathbf{B}'_\perp = \gamma\left(\mathbf{B}_\perp - \frac{v_0}{c^2} \times \mathbf{E}\right)$	(14.92b)	$\mathbf{E}'_\perp = (\mathbf{E}_\perp + v_0 \times \mathbf{B}),\ \mathbf{B}'_\perp = \left(\mathbf{B}_\perp - \frac{v_0}{c^2} \times \mathbf{E}\right)$	(14.92bF)

It can be noted that one can obtain GT as an approximation of LT if the limit c approaches infinity rather than ignoring terms of β^2 and higher for nonrelativistic velocities involved in applying FOLT. This explains the inconsistencies that arise in using GT for electromagnetic problems.

14.19 Alternate Form of Position Four-Vector

As mentioned in Section 14.5, L appears larger than L_0 and T appears smaller than T_0 in Figure 14.4b, since the rotation angle ψ is complex. In some books, for example [1], alternative definitions of the position four-vector and its norm are used as given below:

$$x_\mu = (x_1, x_2, x_3, x_4) = (x, y, z, ct) \tag{14.150}$$

$$|x_\mu| = \left(x_1^2 + x_2^2 + x_3^2 - x_4^2\right)^{1/2} = \left(x^2 + y^2 + z^2 - c^2t^2\right)^{1/2}. \tag{14.151}$$

Note the negative sign before x_4^2 in the middle term of (14.151). The norm in this case in not the distance in four-dimensional space-time in the sense of Euclidean geometry. In the previous definition of the fourth component of the position four-vector given in (14.65), the multiplier i appears before ct, making ψ complex. However, the norm, as defined by R according to (14.36), is the distance in the Euclidean sense in the four-dimensional space-time. A geometrical picture of the LT with the new definition of the four-vector and its norm given in (14.150) and (14.151), respectively, is shown in Figure 14.6.

It is called Minkowski diagram. Compare it with Figure 14.4 a. One of the problems at the end of the chapter brings out the advantages of Minkowski diagram in visualizing the concepts of length contraction and time dilation.

Given below are the changes in the form of some of the equations in the previous sections adapted to the choice of position four-vector (14.150) and the norm (14.151). The equation numbering used for these equations has the last character as n, to make it easy to compare with the corresponding equation in the previous sections:

$$J_\mu = (J, c\rho_v) \tag{14.66n}$$

$$\Box = \left(\nabla, -\frac{\partial}{\partial(ct)}\right) \tag{14.68n}$$

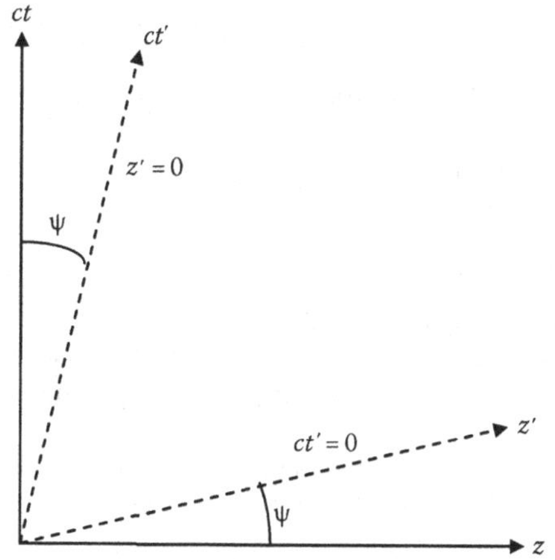

FIGURE 14.6
A geometrical picture of the LT with the new definition of the four-vector and its norm given in (14.150) and (14.151), respectively.

$$\mathbf{A}_\mu = \left(\mathbf{A}, \frac{\Phi}{c}\right) \tag{14.71n}$$

$$\mathbf{k}_\mu = (\mathbf{k}, \omega/c) \tag{14.93n}$$

$$p_\mu = (\mathbf{p}, \varepsilon/c). \tag{14.124n}$$

As far as the case of a medium moving *uniformly* with the velocity v_0, special relativity applies and the choice of (a) position four-vector as in (14.65) with the associated norm as the Euclidean distance in four dimensions or (b) position four-vector as defined in (14.150) and its norm as defined in (14.151) is a "matter of taste" [5]. The author has chosen (a).

When the medium is accelerating, general relativity applies, and there are some advantages in choosing the alternative notation given in (b). The calculations will be much facilitated by invoking additional concepts of tensors with covariant and contravariant components, the summation convention, etc. Reference [5] deals with such applications.

References

1. Lorrain, P., Corson, D. P., and Lorrain, F., *Electromagnetic Fields and Waves*, 3rd edn., W.H. Freeman and Company, New York, 1988.
2. Schmitt, R., *Electromagnetics Explained*, Newnes, and Imprint of Elsevier Science, Boston, MA, 2002.
3. Jackson, J. D., *Classical Electrodynamics*, John Wiley, New York, 1962.
4. Kong, J. U., *Electromagnetics Wave Theory*, EMW Publishing, Cambridge, MA, 2002.
5. Van Bladel, J., *Relativity and Engineering*, Springer-Verlag, New York, 1984.
6. Chawla, B. R. and Unz, H., *Electromagnetic Waves in Moving Magneto-Plasmas*, The University Press of Kansas, Lawrence/Manhattan/Wichita, 1971.
7. Shrivastava, R. K., The interaction of electromagnetic waves with relativistically moving bounded plasmas, Doctoral thesis, Birla Institute of Technology, Ranchi, India, 1975.
8. Prasad, R. C., Transient and frequency response of a bounded plasma, Doctoral thesis, Birla Institute of Technology, Ranchi, India, 1976.
9. Prasad, R., Wave propagation in plasma waveguides and experimental simulation, Doctoral thesis, Birla Institute of Technology, Ranchi, India, 1979.
10. Papas, C. H., *Theory of Electromagnetic Wave Propagation*, McGraw-Hill, New York, 1965.
11. Kalluri, D., Waveguide modes of a warm drifting unaxial electron plasma, *Proc. IEEE*, 58(2), 278–280, 1970.
12. Kalluri, D. and Shrivastava, R. K., Electromagnetic wave interaction with moving bounded plasmas, *J. Appl. Phys.*, 44(10), 4518–4521, 1973.
13. Kalluri, D. and Shrivastava, R. K., Reflection and transmission of electromagnetic waves obliquely incident on a relativistically moving isotropic plasma slab, *J. Appl. Phys.*, 44(5), 2440–2442, 1973.
14. Kalluri, D. and Shrivastava, R. K., Reflection and transmission of electromagnetic waves obliquely incident on a relativistically moving unaxial plasma slab, *IEEE Trans. Antennas Propag.*, AP-21(1), 63–70, 1973.
15. Kalluri, D. and Shrivastava, R. K., On reflection and transmission of electromagnetic waves obliquely incident on a relativistically moving unaxial and isotropic plasma slab, *IEEE Trans. Plasma Sci.*, PS-2, 206–210, 1974.

16. Kalluri, D. and Prasad, R., Comments on characteristics of waveguides filled with homogeneous lossy anisotropic drifting plasma, *Int. J. Electron. (England)*, 38(4), 573–574, 1975.
17. Kalluri, D. and Prasad, R., Waveguide modes of a warm unaxial lossy drifting electron plasma, *Int. J. Electron. (England)*, 39(6), 637–646, 1975.
18. Kalluri, D. and Shrivastava, R. K., Brewster angle for a plasma medium moving at relativistic speed, *J. Appl. Phys.*, 46(3), 1408–1409, 1975.
19. Kalluri, D., Waveguide modes of a warm drifting unaxial electron plasma for large drifting velocities, *IEEE Trans. Antenna Propag.*, AP-23(5), 745, 1975.
20. Kalluri, D. and Prasad, R. C., Interaction of electromagnetic waves with bounded plasmas moving perpendicular to the plane of incidence, *Appl. Phys.*, 48(2), 587–591, 1977.
21. Kalluri, D. and Prasad, R., Waveguide modes of a warm unaxial lossy drifting electron plasma for large drift velocities, *IEEE Trans. Plasma Sci.*, PS-6(4), 593–598, 1978.
22. Kalluri, D. and Shrivastava, R. K., On total reflection of electromagnetic waves from moving plasmas, *J. Appl. Phys.*, 49(12), 6169–6170, 1978.
23. Kalluri, D. and Shrivastava, R. K., Radiation pressure due to plane electromagnetic waves obliquely incident on moving media, *J. Appl. Phys.*, 49(6), 3584–3586, 1978.
24. Kalluri, D. and Prasad, R. C., Reflection of an impulsive plane wave by a plasma half-space moving perpendicular to the plane of incidence, *J. Appl, Phys.*, 49(5), 2696–2699, 1978.
25. Kalluri, D. and Prasad, R. C., On impulse response of plasma half-space moving normal to the interface-II: Results, *1979 IEEE International Conference on Plasma Science*, Montreal, Quebec, Canada, Conference Record-Abstracts, pp. 154–155, June 4–6, 1979.
26. Kalluri, D. and Prasad, R. C., On impulse response of plasma half-space moving normal to the interface-I: Formulation of the solution, *1979 IEEE International Conference on Plasma Science*, Montreal, Quebec, Canada, Conference Record-Abstracts, p. 155, June 4–6, 1979.
27. Shrivastava, R. K. and Kalluri, D., On brewster angles for perpendicularly polarized electromagnetic waves interacting with a moving dielectric half-space, *Proceedings of the Symposium on "Topics in Applied Physics"*, Calcutta University, Kolkata, India, February 1979.
28. Kalluri, D. and Prasad, R., Waveguide modes of a warm transversely drifting electron plasma with strong transverse magnetic field, *IEEE Trans. Plasma Sci.*, PS-7(1), 6–9, 1979.
29. Kalluri, D. and Prasad, R., Waveguide modes of a warm isotropic drifting electron plasma, *J. Appl. Phys.*, 50(4), 2675–2677, 1979.
30. Prasad, R. C. and Kalluri, D., Impulse response of a lossy magnetoplasma half-space moving along the magnetic field, *1980 IEEE International Conference on Plasma Science*, Madison, WI, Conference Record-Abstracts, pp. 11–12, May 1921, 1980.

Part III

Appendices

Appendix 1A: Vector Formulas and Coordinate Systems

1A.1 Vector Transformations

This appendix closely follows the development of the materials given in [1], which is a standard development used in many textbooks. The three coordinate systems, (a) rectangular, (b) cylindrical, and (c) spherical, are shown in Figure 1A.1.

1A.1.1 Rectangular to Cylindrical (and Cylindrical to Rectangular) Transformation

Referring to Figure 1A.1b, the coordinate transformation from rectangular (x, y, z) to cylindrical (ρ, ϕ, z) coordinates is given by

$$\begin{aligned} x &= \rho\cos\phi, \\ y &= \rho\sin\phi, \\ z &= z. \end{aligned} \tag{1A.1}$$

In rectangular coordinates, a vector **A** is written as

$$\mathbf{A} = \hat{\mathbf{x}}A_x + \hat{\mathbf{y}}A_y + \hat{\mathbf{z}}A_z, \tag{1A.2}$$

where $\hat{\mathbf{x}}, \hat{\mathbf{y}}$, and $\hat{\mathbf{z}}$ are the unit vectors and A_x, A_y, and A_z are the components of the vector **A** in the rectangular coordinate system. We can also write **A** as

$$\mathbf{A} = \hat{\boldsymbol{\rho}}A_\rho + \hat{\boldsymbol{\varphi}}A_\phi + \hat{\mathbf{z}}A_z, \tag{1A.3}$$

where $\hat{\boldsymbol{\rho}}$, $\hat{\boldsymbol{\varphi}}$, and $\hat{\mathbf{z}}$ are the unit vectors and A_ρ, A_ϕ, and A_z are the vector components in the cylindrical coordinate system. It can be shown that

$$\begin{aligned} \hat{x} &= \hat{\boldsymbol{\rho}}\cos\phi - \hat{\boldsymbol{\varphi}}\sin\phi, \\ \hat{y} &= \hat{\boldsymbol{\rho}}\sin\phi + \hat{\boldsymbol{\varphi}}\cos\phi, \\ \hat{\mathbf{z}} &= \hat{\mathbf{z}}, \end{aligned} \tag{1A.4}$$

and therefore,

$$\begin{aligned} \mathbf{A} &= \left(\hat{\boldsymbol{\rho}}\cos\phi - \hat{\boldsymbol{\varphi}}\sin\phi\right)A_x + \left(\hat{\boldsymbol{\rho}}\sin\phi - \hat{\boldsymbol{\varphi}}\cos\phi\right)A_y + \hat{\mathbf{z}}A_z, \\ \mathbf{A} &= \hat{\boldsymbol{\rho}}\left(A_x\cos\phi + A_y\sin\phi\right) - \hat{\boldsymbol{\varphi}}\left(A_x\sin\phi + A_y\cos\phi\right) + \hat{\mathbf{z}}A_z, \end{aligned} \tag{1A.5}$$

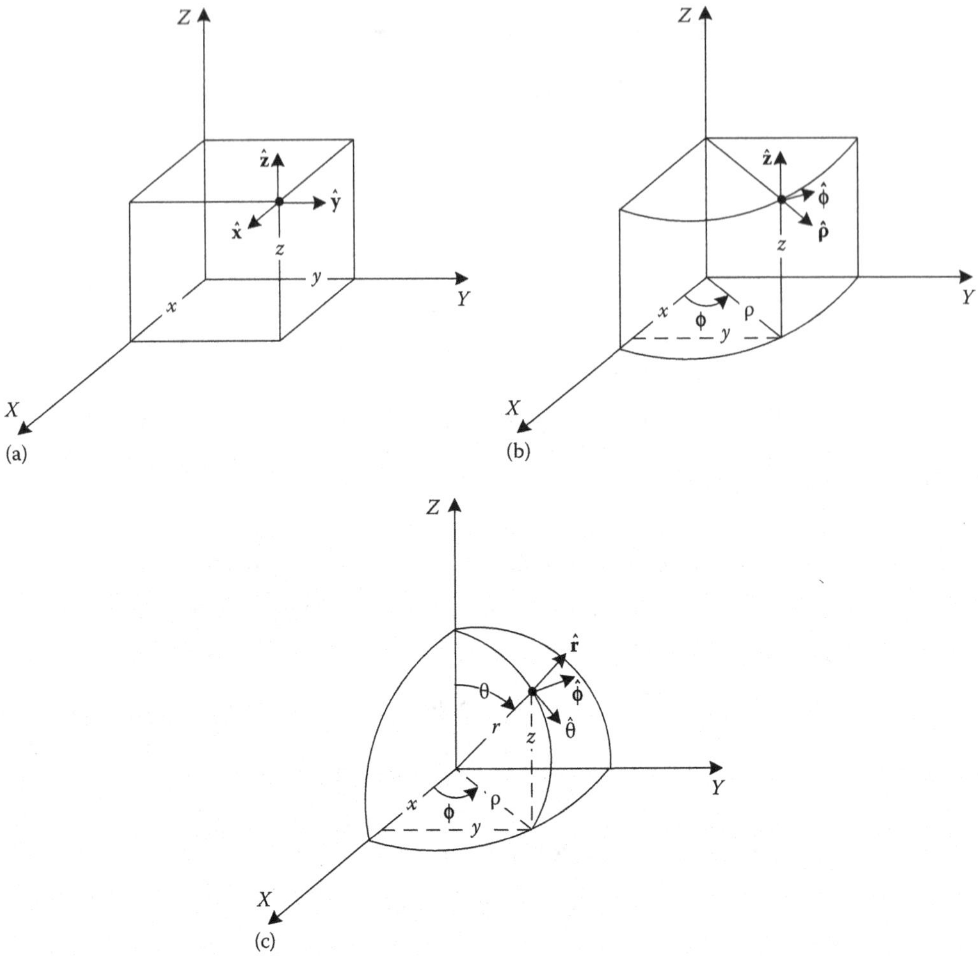

FIGURE 1A.1
(a) Rectangular, (b) cylindrical, and (c) spherical coordinate systems.

thus

$$
\begin{aligned}
A_\rho &= A_x \cos\phi + A_y \sin\phi, \\
A_\phi &= -A_x \sin\phi + A_y \cos\phi, \\
A_z &= A_z.
\end{aligned}
\tag{1A.6}
$$

This can be expressed in the matrix form as

$$
\begin{bmatrix} A_\rho \\ A_\phi \\ A_z \end{bmatrix} = \begin{bmatrix} \cos\phi & \sin\phi & 0 \\ -\sin\phi & \cos\phi & 0 \\ 0 & 0 & 1 \end{bmatrix} \begin{bmatrix} A_x \\ A_y \\ A_z \end{bmatrix} = \left[A\right]_{\text{rc}} \begin{bmatrix} A_x \\ A_y \\ A_z \end{bmatrix}, \tag{1A.7}
$$

where

$$[A]_{rc} = \begin{bmatrix} \cos\phi & \sin\phi & 0 \\ -\sin\phi & \cos\phi & 0 \\ 0 & 0 & 1 \end{bmatrix} \tag{1A.8}$$

is the transformation matrix for rectangular to cylindrical components. Since $[A]_{rc}$ is an orthonormal matrix (its inverse is equal to its transpose), the transformation matrix from cylindrical to rectangular components can be written as

$$[A]_{cr} = [A]_{rc}^{-1} = [A]_{rc}^{T} = \begin{bmatrix} \cos\phi & -\sin\phi & 0 \\ \sin\phi & \cos\phi & 0 \\ 0 & 0 & 1 \end{bmatrix}, \tag{1A.9}$$

$$\begin{bmatrix} A_x \\ A_y \\ A_z \end{bmatrix} = \begin{bmatrix} \cos\phi & -\sin\phi & 0 \\ \sin\phi & \cos\phi & 0 \\ 0 & 0 & 1 \end{bmatrix} \begin{bmatrix} A_\rho \\ A_\phi \\ A_z \end{bmatrix}, \tag{1A.10}$$

$$\begin{aligned} A_x &= A_\rho \cos\phi - A_\phi \sin\phi, \\ A_y &= A_\rho \sin\phi + A_\phi \cos\phi, \\ A_z &= A_z. \end{aligned} \tag{1A.11}$$

1A.1.2 Cylindrical to Spherical (and Spherical to Cylindrical) Transformation

From Figure 1A.1c, it can be seen that the cylindrical and spherical coordinates are related by

$$\begin{aligned} \rho &= r\sin\theta, \\ z &= r\cos\theta. \end{aligned} \tag{1A.12}$$

In a manner similar to the previous section, it can be shown that

$$\begin{aligned} A_r &= A_\rho \sin\theta + A_z \cos\theta, \\ A_\theta &= A_\rho \cos\theta - A_z \sin\theta, \end{aligned} \tag{1A.13}$$

therefore

$$A_\phi = A_\phi,$$

or in matrix notation

$$\begin{bmatrix} A_r \\ A_\theta \\ A_\phi \end{bmatrix} = \begin{bmatrix} \sin\theta & 0 & \cos\theta \\ \cos\theta & 0 & -\sin\theta \\ 0 & 1 & 0 \end{bmatrix} \begin{bmatrix} A_\rho \\ A_\phi \\ A_z \end{bmatrix}, \tag{1A.14}$$

$$[A]_{cs} = \begin{bmatrix} \sin\theta & 0 & \cos\theta \\ \cos\theta & 0 & -\sin\theta \\ 0 & 1 & 0 \end{bmatrix}. \tag{1A.15}$$

The $[A]_{cs}$ matrix is orthonormal and so its inverse is given by

$$[A]_{sc} = [A]_{cs}^{-1} = [A]_{cs}^{T} = \begin{bmatrix} \sin\theta & \cos\theta & 0 \\ 0 & 0 & 1 \\ \cos\theta & -\sin\theta & 0 \end{bmatrix} \tag{1A.16}$$

and the spherical to cylindrical transformation is accomplished by

$$\begin{bmatrix} A_\rho \\ A_\phi \\ A_z \end{bmatrix} = \begin{bmatrix} \sin\theta & \cos\theta & 0 \\ 0 & 0 & 1 \\ \cos\theta & -\sin\theta & 0 \end{bmatrix} \begin{bmatrix} A_r \\ A_\theta \\ A_\phi \end{bmatrix}, \tag{1A.17}$$

$$\begin{aligned} A_\rho &= A_r \sin\theta + A_\theta \cos\theta, \\ A_\phi &= A_\phi, \\ A_z &= A_r \cos\theta - A_\theta \sin\theta. \end{aligned} \tag{1A.18}$$

1A.1.3 Rectangular to Spherical (and Spherical to Rectangular) Transformation

From Figure 1A.1c, it can be seen that the rectangular and spherical coordinates are related by

$$x = r\sin\theta\cos\phi, \tag{1A.19}$$

or

$$\begin{aligned} y &= r\sin\theta\sin\phi, \\ x &= r\cos\theta, \end{aligned}$$

and the spherical and rectangular components by

$$\begin{aligned} A_r &= A_x \sin\theta\cos\phi + A_y \sin\theta\sin\phi + A_z \cos\theta, \\ A_\theta &= A_x \cos\theta\cos\phi + A_y \cos\theta\sin\phi - A_z \sin\theta, \\ A_\phi &= -A_x \sin\phi + A_y \cos\phi. \end{aligned} \tag{1A.20}$$

In the matrix form,

$$\begin{bmatrix} A_r \\ A_\theta \\ A_\phi \end{bmatrix} = \begin{bmatrix} \sin\theta\cos\phi & \sin\theta\sin\phi & \cos\theta \\ \cos\theta\cos\phi & \cos\theta\sin\phi & -\sin\theta \\ -\sin\phi & \cos\phi & 0 \end{bmatrix} \begin{bmatrix} A_x \\ A_y \\ A_z \end{bmatrix}, \tag{1A.21}$$

$$[A]_{rs} = \begin{bmatrix} \sin\theta\cos\phi & \sin\theta\sin\phi & \cos\theta \\ \cos\theta\cos\phi & \cos\theta\sin\phi & -\sin\theta \\ -\sin\phi & \cos\phi & 0 \end{bmatrix}. \tag{1A.22}$$

The $[A]_{rs}$ matrix is orthonormal and so its inverse is given by

$$[A]_{sr} = [A]_{rs}^{-1} = [A]_{rs}^{T} = \begin{bmatrix} \sin\theta\cos\phi & \cos\theta\cos\phi & -\sin\phi \\ \sin\theta\sin\phi & \cos\theta\sin\phi & \cos\phi \\ \cos\theta & -\sin\theta & 0 \end{bmatrix} \tag{1A.23}$$

and the spherical to rectangular transformation is accomplished by

$$\begin{bmatrix} A_x \\ A_y \\ A_z \end{bmatrix} = \begin{bmatrix} \sin\theta\cos\phi & \cos\theta\cos\phi & -\sin\phi \\ \sin\theta\sin\phi & \cos\theta\sin\phi & \cos\phi \\ \cos\theta & -\sin\theta & 0 \end{bmatrix} \begin{bmatrix} A_r \\ A_\theta \\ A_\phi \end{bmatrix}, \tag{1A.24}$$

$$\begin{aligned} A_x &= A_r \sin\theta\cos\phi + A_\theta \cos\theta\cos\phi - A_\phi \sin\phi, \\ A_y &= A_r \sin\theta\sin\phi + A_\theta \cos\theta\sin\phi + A_\phi \cos\phi, \\ A_z &= A_r \cos\theta - A_\theta \sin\theta. \end{aligned} \tag{1A.25}$$

1A.2 Vector Differential Operators

The differential operators normally include gradient of a scalar ($\nabla\psi$), divergence of a vector ($\nabla\cdot\mathbf{A}$), curl of a vector ($\nabla\times\mathbf{A}$), Laplacian of a scalar ($\nabla^2\psi$), and Laplacian of a vector ($\nabla^2\mathbf{A}$). These will be shown in rectangular, cylindrical, and spherical coordinates as given below.

1A.2.1 Rectangular Coordinates

$$\nabla\psi = \hat{\mathbf{x}}\frac{\partial\psi}{\partial x} + \hat{\mathbf{y}}\frac{\partial\psi}{\partial y} + \hat{\mathbf{z}}\frac{\partial\psi}{\partial z}, \tag{1A.26}$$

$$\nabla\cdot\mathbf{A} = \frac{\partial A_x}{\partial x} + \frac{\partial A_y}{\partial y} + \frac{\partial A_z}{\partial z}, \tag{1A.27}$$

$$\nabla\times\mathbf{A} = \begin{vmatrix} \hat{\mathbf{x}} & \hat{\mathbf{y}} & \hat{\mathbf{z}} \\ \dfrac{\partial}{\partial x} & \dfrac{\partial}{\partial y} & \dfrac{\partial}{\partial z} \\ A_x & A_y & A_z \end{vmatrix} = \hat{\mathbf{x}}\left(\frac{\partial A_z}{\partial y} - \frac{\partial A_y}{\partial z}\right) + \hat{\mathbf{y}}\left(\frac{\partial A_x}{\partial z} - \frac{\partial A_z}{\partial x}\right) + \hat{\mathbf{z}}\left(\frac{\partial A_y}{\partial x} - \frac{\partial A_x}{\partial y}\right), \tag{1A.28}$$

$$\nabla\cdot\nabla\psi = \nabla^2\psi = \frac{\partial^2\psi}{\partial x^2} + \frac{\partial^2\psi}{\partial y^2} + \frac{\partial^2\psi}{\partial z^2}, \tag{1A.29}$$

$$\nabla^2\mathbf{A} = \hat{\mathbf{x}}\nabla^2 A_x + \hat{\mathbf{y}}\nabla^2 A_y + \hat{\mathbf{z}}\nabla^2 A_z. \tag{1A.30}$$

1A.2.2 Cylindrical Coordinates

$$\nabla\psi = \hat{\rho}\frac{\partial\psi}{\partial\rho} + \hat{\varphi}\frac{1}{\rho}\frac{\partial\psi}{\partial\phi} + \hat{\mathbf{z}}\frac{\partial\psi}{\partial z}, \tag{1A.31}$$

$$\nabla\cdot\mathbf{A} = \frac{1}{\rho}\frac{\partial}{\partial\rho}(\rho A_\rho) + \frac{1}{\rho}\frac{\partial A_\phi}{\partial\phi} + \frac{\partial A_z}{\partial z}, \tag{1A.32}$$

$$\nabla\times\mathbf{A} = \hat{\rho}\left(\frac{1}{\rho}\frac{\partial A_z}{\partial\phi} - \frac{\partial A_\phi}{\partial z}\right) + \hat{\varphi}\left(\frac{\partial A_\rho}{\partial z} - \frac{\partial A_z}{\partial\rho}\right) + \hat{\mathbf{z}}\left(\frac{1}{\rho}\frac{\partial(\rho A_\phi)}{\partial\rho} - \frac{1}{\rho}\frac{\partial A_\rho}{\partial\phi}\right), \tag{1A.33}$$

$$\nabla^2\psi = \frac{1}{\rho}\frac{\partial}{\partial\rho}\left(\rho\frac{\partial\psi}{\partial\rho}\right) + \frac{1}{\rho^2}\frac{\partial^2\psi}{\partial\phi^2} + \frac{\partial^2\psi}{\partial z^2}, \tag{1A.34}$$

$$\nabla^2\mathbf{A} = \nabla(\nabla\cdot\mathbf{A}) - \nabla\times\nabla\times\mathbf{A}, \tag{1A.35}$$

$$\begin{aligned}\nabla^2\mathbf{A} = {} & \hat{\rho}\left(\frac{\partial^2 A_\rho}{\partial\rho^2} + \frac{1}{\rho}\frac{\partial A_\rho}{\partial\rho} - \frac{A_\rho}{\rho^2} + \frac{1}{\rho^2}\frac{\partial^2 A_\rho}{\partial\phi^2} - \frac{2}{\rho^2}\frac{\partial A_\phi}{\partial\phi} + \frac{\partial^2 A_\rho}{\partial z^2}\right)\\ & + \hat{\varphi}\left(\frac{\partial^2 A_\phi}{\partial\rho^2} + \frac{1}{\rho}\frac{\partial A_\phi}{\partial\rho} - \frac{A_\phi}{\rho^2} + \frac{1}{\rho^2}\frac{\partial^2 A_\phi}{\partial\phi^2} + \frac{2}{\rho^2}\frac{\partial A_\rho}{\partial\phi} + \frac{\partial^2 A_\phi}{\partial z^2}\right)\\ & + \hat{\mathbf{z}}\left(\frac{\partial^2 A_z}{\partial\rho^2} + \frac{1}{\rho}\frac{\partial A_z}{\partial\rho} + \frac{1}{\rho^2}\frac{\partial^2 A_z}{\partial\phi^2} + \frac{\partial^2 A_z}{\partial z^2}\right).\end{aligned} \tag{1A.36}$$

1A.2.3 Spherical Coordinates

$$\nabla\psi = \hat{\mathbf{r}}\frac{\partial\psi}{\partial r} + \hat{\theta}\frac{1}{r}\frac{\partial\psi}{\partial\theta} + \hat{\varphi}\frac{1}{r\sin\theta}\frac{\partial\psi}{\partial\phi}, \tag{1A.37}$$

$$\nabla\cdot\mathbf{A} = \frac{1}{r^2}\frac{\partial}{\partial r}(r^2 A_r) + \frac{1}{r\sin\theta}\frac{\partial}{\partial\theta}(\sin\theta A_\theta) + \frac{1}{r\sin\theta}\frac{\partial A_\phi}{\partial\phi}, \tag{1A.38}$$

$$\nabla\times\mathbf{A} = \frac{\hat{\mathbf{r}}}{r\sin\theta}\left(\frac{\partial}{\partial\theta}(\sin\theta A_\phi) - \frac{\partial A_\theta}{\partial\phi}\right) + \frac{\hat{\theta}}{r}\left(\frac{1}{\sin\theta}\frac{\partial A_r}{\partial\phi} - \frac{\partial}{\partial r}(rA_\phi)\right) + \frac{\hat{\varphi}}{r}\left(\frac{\partial}{\partial r}(rA_\theta) - \frac{\partial A_r}{\partial\theta}\right), \tag{1A.39}$$

$$\nabla^2\psi = \frac{1}{r^2}\frac{\partial}{\partial r}\left(r^2\frac{\partial\psi}{\partial r}\right) + \frac{1}{r^2\sin\theta}\frac{\partial}{\partial\theta}\left(\sin\theta\frac{\partial\psi}{\partial\theta}\right) + \frac{1}{r^2\sin^2\theta}\frac{\partial^2\psi}{\partial\phi^2}, \tag{1A.40}$$

$$\nabla^2\mathbf{A} = \nabla(\nabla\cdot\mathbf{A}) - \nabla\times\nabla\times\mathbf{A}, \tag{1A.41}$$

$$\begin{aligned}\nabla^2\mathbf{A} = {} & \hat{\mathbf{r}}\left(\frac{\partial^2 A_r}{\partial r^2} + \frac{2}{r}\frac{\partial A_r}{\partial r} - \frac{2}{r^2}A_r + \frac{1}{r^2}\frac{\partial^2 A_r}{\partial\theta^2} + \frac{\cot\theta}{r^2}\frac{\partial A_r}{\partial\theta} + \frac{1}{r^2\sin^2\theta}\frac{\partial^2 A_r}{\partial\phi^2} - \frac{2}{r^2}\frac{\partial A_\theta}{\partial\theta} - \frac{2\cot\theta}{r^2}A_\theta - \frac{2}{r^2\sin\theta}\frac{\partial A_\phi}{\partial\phi}\right)\\ & + \hat{\theta}\left(\frac{\partial^2 A_\theta}{\partial r^2} + \frac{2}{r}\frac{\partial A_\theta}{\partial r} - \frac{A_\theta}{r^2\sin^2\theta} + \frac{1}{r^2}\frac{\partial^2 A_\theta}{\partial\theta^2} + \frac{\cot\theta}{r^2}\frac{\partial A_\theta}{\partial\theta} + \frac{1}{r^2\sin^2\theta}\frac{\partial^2 A_\theta}{\partial\phi^2} + \frac{2}{r^2}\frac{\partial A_r}{\partial\theta} - \frac{2\cot\theta}{r^2\sin\theta}\frac{\partial A_\phi}{\partial\phi}\right)\\ & + \hat{\varphi}\left(\frac{\partial^2 A_\phi}{\partial r^2} + \frac{2}{r}\frac{\partial A_\phi}{\partial r} - \frac{1}{r^2\sin^2\theta}A_\phi + \frac{1}{r^2}\frac{\partial^2 A_\phi}{\partial\theta^2} + \frac{\cot\theta}{r^2}\frac{\partial A_\phi}{\partial\theta} + \frac{1}{r^2\sin^2\theta}\frac{\partial^2 A_\phi}{\partial\phi^2} + \frac{2}{r^2\sin\theta}\frac{\partial A_r}{\partial\phi} + \frac{2\cot\theta}{r^2\sin\theta}\frac{\partial A_\theta}{\partial\phi}\right).\end{aligned} \tag{1A.42}$$

1A.3 Vector Identities

1A.3.1 Addition and Multiplication

$$\mathbf{A}\cdot\mathbf{A} = |\mathbf{A}|^2 = A^2, \tag{1A.43}$$

$$\mathbf{A}\cdot\mathbf{A}^* = |\mathbf{A}|^2 = A^2, \tag{1A.44}$$

$$\mathbf{A}+\mathbf{B} = \mathbf{B}+\mathbf{A}, \tag{1A.45}$$

$$\mathbf{A}\cdot\mathbf{B} = \mathbf{B}\cdot\mathbf{A}, \tag{1A.46}$$

$$\mathbf{A}\times\mathbf{B} = -\mathbf{B}\times\mathbf{A}, \tag{1A.47}$$

$$(\mathbf{A}+\mathbf{B})\cdot\mathbf{C} = \mathbf{A}\cdot\mathbf{C}+\mathbf{B}\cdot\mathbf{C}, \tag{1A.48}$$

$$(\mathbf{A}+\mathbf{B})\times\mathbf{C} = \mathbf{A}\times\mathbf{C}+\mathbf{B}\times\mathbf{C}, \tag{1A.49}$$

$$\mathbf{A}\cdot\mathbf{B}\times\mathbf{C} = \mathbf{B}\cdot\mathbf{C}\times\mathbf{A} = \mathbf{C}\cdot\mathbf{A}\times\mathbf{B}, \tag{1A.50}$$

$$\mathbf{A}\times(\mathbf{B}\times\mathbf{C}) = (\mathbf{A}\cdot\mathbf{C})\mathbf{B}-(\mathbf{A}\cdot\mathbf{B})\mathbf{C}, \tag{1A.51}$$

$$\begin{aligned}(\mathbf{A}\times\mathbf{B})\cdot(\mathbf{C}\times\mathbf{D}) &= \mathbf{A}\cdot\mathbf{B}\times(\mathbf{C}\times\mathbf{D})\\ &= \mathbf{A}\cdot(\mathbf{B}\cdot\mathbf{D}\mathbf{C}-\mathbf{B}\cdot\mathbf{C}\mathbf{D})\\ &= (\mathbf{A}\cdot\mathbf{C})(\mathbf{B}\cdot\mathbf{D})-(\mathbf{A}\cdot\mathbf{D})(\mathbf{B}\cdot\mathbf{C}),\end{aligned} \tag{1A.52}$$

$$(\mathbf{A}\times\mathbf{B})\times(\mathbf{C}\times\mathbf{D}) = (\mathbf{A}\times\mathbf{B}\cdot\mathbf{D})\mathbf{C}-(\mathbf{A}\times\mathbf{B}\cdot\mathbf{C})\mathbf{D}. \tag{1A.53}$$

1A.3.2 Differentiation

$$\nabla\cdot(\nabla\times\mathbf{A}) = 0, \tag{1A.54}$$

$$\nabla\times\nabla\psi = 0, \tag{1A.55}$$

$$\nabla(\phi+\psi) = \nabla\phi+\nabla\psi, \tag{1A.56}$$

$$\nabla(\phi\psi) = \phi\nabla\psi+\psi\nabla\phi, \tag{1A.57}$$

$$\nabla\cdot(\mathbf{A}+\mathbf{B}) = \nabla\cdot\mathbf{A}+\nabla\cdot\mathbf{B}, \tag{1A.58}$$

$$\nabla\times(\mathbf{A}+\mathbf{B}) = \nabla\times\mathbf{A}+\nabla\times\mathbf{B}, \tag{1A.59}$$

$$\nabla\cdot(\psi\mathbf{A}) = \mathbf{A}\cdot\nabla\psi+\psi\nabla\cdot\mathbf{A}, \tag{1A.60}$$

$$\nabla\times(\psi\mathbf{A}) = \nabla\psi\times\mathbf{A}+\psi\nabla\times\mathbf{A}, \tag{1A.61}$$

$$\nabla(\mathbf{A}\cdot\mathbf{B}) = (\mathbf{A}\cdot\nabla)\mathbf{B} + (\mathbf{B}\cdot\nabla)\mathbf{A} + \mathbf{A}\times(\nabla\times\mathbf{B}) + \mathbf{B}\times(\nabla\times\mathbf{A}), \tag{1A.62}$$

$$\nabla\cdot(\mathbf{A}\times\mathbf{B}) = \mathbf{B}\cdot\nabla\times\mathbf{A} - \mathbf{A}\cdot\nabla\times\mathbf{B}, \tag{1A.63}$$

$$\nabla\times(\mathbf{A}\times\mathbf{B}) = \mathbf{A}\nabla\cdot\mathbf{B} - \mathbf{B}\nabla\cdot\mathbf{A} + (\mathbf{B}\cdot\nabla)\mathbf{A} - (\mathbf{A}\cdot\nabla)\mathbf{B}, \tag{1A.64}$$

$$\nabla\times\nabla\times\mathbf{A} = \nabla(\nabla\cdot\mathbf{A}) - \nabla^2\mathbf{A}. \tag{1A.65}$$

1A.3.3 Integration

$$\oint_C A\cdot \mathrm{dl} = \iint_S (\nabla\times\mathbf{A})\mathrm{ds} \quad \text{(Stokes' theorem)}, \tag{1A.66}$$

$$\oiint_S \mathbf{A}\cdot \mathrm{ds} = \iiint_V (\nabla\cdot\mathbf{A})\mathrm{d}v \quad \text{(Divergence theorem)}, \tag{1A.67}$$

$$\oiint_S (\hat{\mathbf{n}}\times\mathbf{A})\mathrm{d}s = \iiint_V (\nabla\times\mathbf{A})\mathrm{d}v, \tag{1A.68}$$

$$\oiint_S \psi \mathrm{ds} = \iiint_V \nabla\psi \mathrm{d}v, \tag{1A.69}$$

$$\oint_C \psi \mathrm{dl} = \iint_S \hat{\mathbf{n}}\times\nabla\psi \mathrm{ds}. \tag{1A.70}$$

Reference

1. Balanis, C. A., *Advanced Engineering Electromagnetics*, Wiley, New York, 1989.

Appendix 1B: Retarded Potentials and Review of Potentials for the Static Cases

1B.1 Electrostatics

The basic equation is

$$\nabla \times \mathbf{E} = \mathbf{0}. \tag{1B.1}$$

It is known that the curl of the gradient of any scalar is zero, that is,

$$\nabla \times [\text{gradient of any scalar}] \equiv 0. \tag{1B.2}$$

Therefore, **E** can be expressed as the gradient of a scalar ψ (electrostatic potential).

$$\mathbf{E} = -\nabla\psi, \tag{1B.3}$$

where ψ satisfies Poisson's equation

$$\nabla^2\psi = -\frac{\rho_v}{\varepsilon}. \tag{1B.4}$$

The solution of Equation 1B.4 at an arbitrary point P is given by

$$\psi(\mathbf{r}) = \iiint_{v'} \frac{\rho_v(\mathbf{r}')}{4\pi\varepsilon|\mathbf{r}-\mathbf{r}'|} dv', \tag{1B.5}$$

where the volume charge density ρ_v exists over volume v' as shown in Figure 1B.1.

1B.2 Magnetostatics

The basic equation is

$$\nabla \cdot \mathbf{B} = 0. \tag{1B.6}$$

It is known that the divergence of the curl of any vector is zero, that is,

$$\nabla \cdot [\text{curl of any vector}] \equiv \mathbf{0}. \tag{1B.7}$$

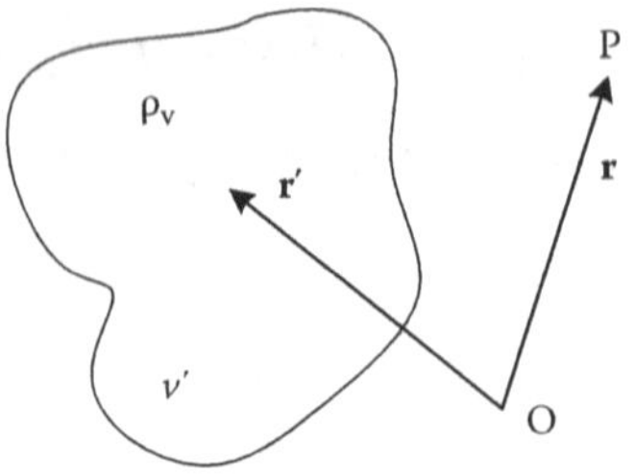

FIGURE 1B.1
Electrostatic geometry.

Therefore, **B** can be expressed as the curl of a vector **A** as

$$\mathbf{B} = \nabla \times \mathbf{A}. \tag{1B.8}$$

The magnetic vector potential **A** satisfies the vector Poisson's equation

$$\nabla^2 \mathbf{A} = -\mu \mathbf{J}. \tag{1B.9}$$

In deriving Equation 1B.9, we use Ampere's law $\nabla \times \mathbf{H} = \mathbf{J}$ and choose $\nabla \cdot \mathbf{A} = 0$. The solution of Equation 1B.9 at an arbitrary point P is given by

$$\mathbf{A}(\mathbf{r}) = \iiint_{v'} \frac{\mu \mathbf{J}(\mathbf{r}')}{4\pi|\mathbf{r}-\mathbf{r}'|} dv' \tag{1B.10}$$

and

$$\nabla \cdot \mathbf{A} = 0. \tag{1B.11}$$

1B.3 Time-Varying Case

We know

$$\nabla \cdot \mathbf{B} = 0 \tag{1B.12}$$

and

$$\mathbf{B} = \nabla \times \mathbf{A}, \tag{1B.13}$$

but

$$\nabla \times \mathbf{E} = -\frac{\partial \mathbf{B}}{\partial t}. \tag{1B.14}$$

From Equations 1B.13 and 1B.14, we obtain

$$\nabla \times \mathbf{E} + \frac{\partial}{\partial \mathbf{t}}(\nabla \times \mathbf{A}) = 0, \tag{1B.15}$$

$$\nabla \times \left[\mathbf{E} + \frac{\partial \mathbf{A}}{\partial t} \right] = 0. \tag{1B.16}$$

Therefore, from Equation 1B.2, $\mathbf{E} + (\partial \mathbf{A}/\partial t)$ can be expressed as the gradient of a scalar ψ (electric potential)

$$\mathbf{E} = -\nabla \psi - \frac{\partial \mathbf{A}}{\partial t}. \tag{1B.17}$$

Now, we shall derive the equations satisfied by **A** and ψ and we would like to have these equations without the **E** and **H** fields in them. We want the equations for the time varying case corresponding to Equations 1B.4 and 1B.9. Let us start with

$$\nabla \cdot \mathbf{D} = \rho_v. \tag{1B.18}$$

Using Equations 1B.18, 1B.17, and the relation $\mathbf{D} = \varepsilon \mathbf{E}$, we obtain

$$\nabla \cdot \varepsilon \mathbf{E} = \varepsilon \nabla \cdot \mathbf{E} = \varepsilon \nabla \cdot \left[-\nabla \psi - \frac{\partial \mathbf{A}}{\partial t} \right] = \rho_v, \tag{1B.19}$$

$$\nabla^2 \psi + \frac{\partial}{\partial t} \nabla \cdot \mathbf{A} = -\frac{\rho_v}{\varepsilon}. \tag{1B.20}$$

Similarly, start from the equation

$$\nabla \times \mathbf{H} = \mathbf{J} + \frac{\partial \mathbf{D}}{\partial t}, \tag{1B.21}$$

$$\nabla \times \mathbf{B} = \mu \mathbf{J} + \mu \frac{\partial \mathbf{D}}{\partial t} = \mu \mathbf{J} + \varepsilon \mu \frac{\partial \mathbf{E}}{\partial \mathbf{t}} = \mu \mathbf{J} + \varepsilon \mu \frac{\partial}{\partial t}\left(-\nabla \psi - \frac{\partial \mathbf{A}}{\partial t} \right). \tag{1B.22}$$

Since

$$\nabla \times \mathbf{B} = \nabla \times (\nabla \times \mathbf{A}) = \nabla(\nabla \cdot \mathbf{A}) - \nabla^2 \mathbf{A}, \tag{1B.23}$$

$$\nabla(\nabla \cdot \mathbf{A}) - \nabla^2 \mathbf{A} = \mu \mathbf{J} - \varepsilon \mu \frac{\partial^2 \mathbf{A}}{\partial t^2} - \varepsilon \mu \nabla \left(\frac{\partial \psi}{\partial t} \right), \tag{1B.24}$$

$$\nabla^2 \mathbf{A} - \varepsilon \mu \frac{\partial^2 \mathbf{A}}{\partial t^2} - \nabla \left(\nabla \cdot \mathbf{A} + \varepsilon \mu \frac{\partial \psi}{\partial t} \right) = -\mu \mathbf{J}. \tag{1B.25}$$

Equations 1B.20 and 1B.25 are coupled, that is, in each of these equations, both variables ψ and **A** appear. In the static case, we had uncoupled equations for ψ and **A** (Equations 1B.4

and 1B.9). A vector is uniquely defined only if its curl and also its divergence are specified everywhere. In the static case, **A** is made unique by specifying its curl as **B**, that is, $\nabla \times \mathbf{A} = \mathbf{B}$, and its divergence as zero, that is, $\nabla \cdot \mathbf{A} = 0$. In the time-varying case, we have specified curl of **A** as **B**, but we have not specified its divergence. It is seen from Equation 1B.25 that if we specify (Lorentz condition)

$$\nabla \cdot \mathbf{A} + \varepsilon\mu \frac{\partial \psi}{\partial t} = 0, \tag{1B.26}$$

then Equation 1B.25 becomes uncoupled (ψ does not appear in the equation):

$$\nabla^2 \mathbf{A} - \varepsilon\mu \frac{\partial^2 \mathbf{A}}{\partial t^2} = -\mu \mathbf{J}. \tag{1B.27}$$

Furthermore, if we substitute Equation 1B.26 into Equation 1B.20 to eliminate **A**, then we obtain

$$\nabla^2 \psi + \frac{\partial}{\partial t}\left(-\varepsilon\mu \frac{\partial \psi}{\partial t}\right) = -\frac{\rho_v}{\varepsilon} \tag{1B.28}$$

that is,

$$\nabla^2 \psi - \varepsilon\mu \frac{\partial^2 \psi}{\partial t^2} = -\frac{\rho_v}{\varepsilon}, \tag{1B.29}$$

which is also an uncoupled equation.

Equations 1B.27 and 1B.29 are called wave equations. Although Poisson's equation governs the static cases, time-varying phenomena are governed by the wave equation.

In free space, $1/\varepsilon_0\mu_0 = (3 \times 10^8)^2 = c^2$, where c is the velocity of light. Note that c is large but finite. Equation 1B.29 for free space is

$$\nabla^2 \psi - \frac{1}{c^2}\frac{\partial^2 \psi}{\partial t^2} = -\frac{\rho_v}{\varepsilon}. \tag{1B.30}$$

This tends to Equation 1B.4 if $1/c^2 \to 0$, that is, $c \to \infty$. The implication of this statement will be explained later. Consider a charge Q at point A as shown in Figure 1B.2. A comparison between the static and dynamic cases is summarized in Table 1B.1.

For a volume charge source of density ρ_v, the solution is

$$\psi(\mathbf{r}, t) = \iiint_{v'} \frac{\rho_v(t - R_{AP}/c)}{4\pi\varepsilon R_{AP}} dv' = \iiint_{v'} \frac{[\rho_v]}{4\pi\varepsilon R_{AP}} dv', \tag{1B.31}$$

where

$$[\rho_v] = \rho_v\left(t - \frac{R_{AP}}{c}\right).$$

is the charge density at the retarded time.

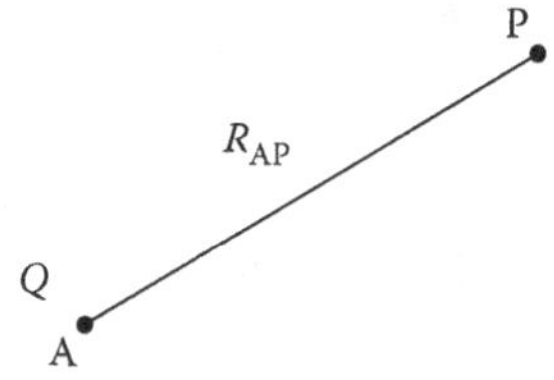

FIGURE 1B.2
Point charge at point A for static and dynamic case comparison.

TABLE 1B.1
Comparison of Static and Dynamic Cases for a Point Charge

Static Case	Dynamic Case
$\nabla^2\psi = -\frac{\rho_v}{\varepsilon_0}$	$\nabla^2\psi - \frac{1}{c^2}\frac{\partial^2\psi}{\partial t^2} = -\frac{\rho_v}{\varepsilon_0}$
For a static point change Q, $\psi = \frac{Q}{4\pi\varepsilon_0 R_{AP}}$	For a time-varying point charge $Q(t)$, $\psi(t) \neq \frac{Q(t)}{4\pi\varepsilon_0 R_{AP}}$ $\psi(t) = \frac{Q(t - R_{AP}/c)}{4\pi\varepsilon_0 R_{AP}}$
may be written if we assume that the effect is simultaneous with cause. The effect at any point is instantaneously felt ($c \to \infty$)	If we assume that the cause and effect are not instantaneously related, the potential at P is due to a charge at a previous time. The time of retardation is the time of travel for the propagation of the effect

Example 1B.1

Let a current filament be short (length = ℓ) and carry a current $I = I_0 \cos(\omega t)$ as shown in Figure 1B.3.

Then

$$\mathbf{A}(\mathbf{r},t) = \mu \iiint_{v'} \frac{[I]}{4\pi r} dv' = \mu_0 \int \frac{[I]}{4\pi r} dl'. \tag{1B.32}$$

For $I(t) = I_0 \cos \omega t$,

$$[I] = I_0 \cos\omega\left(t - \frac{r}{c}\right) = \mathrm{Re}\left(I_0 e^{j\omega t} e^{-j\omega r/c}\right), \tag{1B.33}$$

$$\mathbf{A}(\mathrm{r}, t) = \mathrm{Re}\left[\mu_0 I_0 \int \frac{e^{j\omega t} e^{-j\omega r/c}}{4\pi r}\right] dz'\hat{z}, \tag{1B.34}$$

$$\mathbf{A}(\mathrm{r}, t) = \mathrm{Re}\left[\mu_0 I_0 e^{j\omega t} \int \frac{e^{-j\omega r/c}}{4\pi r} dz'\right]\hat{z}. \tag{1B.35}$$

If $\ell \ll \lambda$,

$$A_z(r, t) \approx \mathrm{Re}\left[\frac{\mu_0 I_0 \ell e^{j\omega t}}{4\pi r} e^{-j\omega r/c}\right]. \tag{1B.36}$$

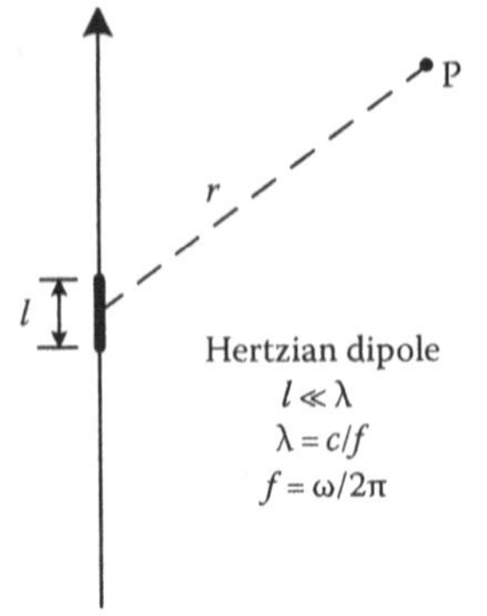

FIGURE 1B.3
Hertzian dipole.

The phasor $\tilde{A}_z$ is given by

$$\tilde{A}_z(r) = \frac{\mu_0 I_0 l}{4\pi r} e^{-j\omega r/c}.$$

1B.4 One-Dimensional Solution for the Wave Equation

In explaining the retardation effect, we assumed the solution of

$$\nabla^2 \psi - \frac{1}{c^2}\frac{\partial^2 \psi}{\partial t^2} = 0, \tag{1B.37}$$

is a wave propagating with the velocity c. Let us solve Equation 1B.37 for the one-dimensional case, $\psi = \psi(z, t)$, that is,

$$\frac{\partial^2 \psi}{\partial z^2} - \frac{1}{c^2}\frac{\partial^2 \psi}{\partial t^2} = 0. \tag{1B.38}$$

Let the Laplace transform of ψ be F and it may be written as

$$\mathbf{L}[\psi] = \int_0^\infty \psi e^{-st} dt = F. \tag{1B.39}$$

Transforming Equation 1B.37, we obtain

$$\frac{d^2 F(z,s)}{dz^2} - \frac{1}{c^2}\left[s^2 F - s\psi(z,0) - \psi'(z,0)\right] = 0. \tag{1B.40}$$

If the initial conditions are zero, then

$$\frac{d^2 F}{dz^2} - \frac{s^2}{c^2} F = 0. \tag{1B.41}$$

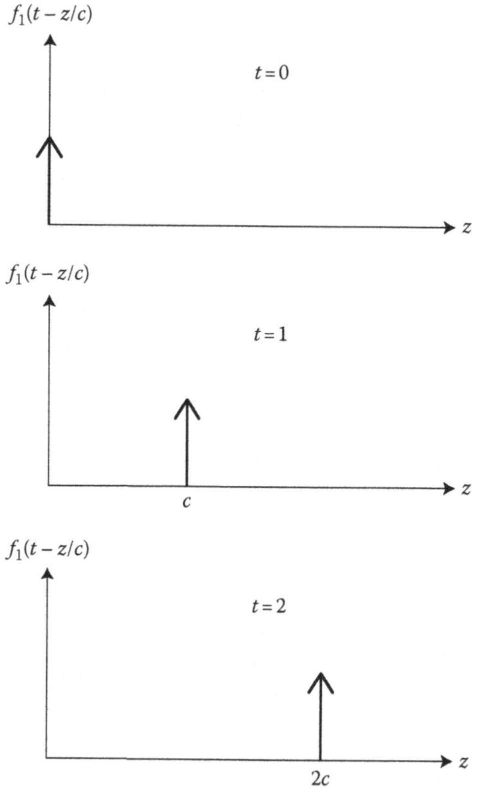

FIGURE 1B.4
$f_1(t - z/c)$ is the pulse $f_1(t)$ moving with a velocity $v = c$ in the +z-direction.

Then

$$F = F_1(s)e^{(-s/c)z} + F_2(s)e^{(+s/c)z}. \tag{1B.42}$$

If $f_1(t) = \mathcal{L}^{-1}\{F_1(s)\}$ and $f_2(t) = \mathcal{L}^{-1}\{F_2(s)\}$, then by using shifting theorem, that is, $\mathcal{L}[f(t-a)] = e^{-as}F(s)$, we get

$$\psi(z,t) = f_1\left(t - \frac{z}{c}\right) + f_2\left(t + \frac{z}{c}\right). \tag{1B.43}$$

Equation 1B.43 is the solution to Equation 1B.38.

Consider the case where $f_1(t) = \delta(t)$. Then $f_1(t - z/c)$ is the pulse $f_1(t)$ moving with a velocity c in the +z direction as depicted in Figure 1B.4.

Appendix 1C: Poynting Theorem

Poynting theorem is nothing but a statement of conservation of energy. Since the theorem can be derived from Maxwell's equations, it shows that Maxwell's equations are consistent with the more general principle of energy conservation.

From Equation 1.1,

$$\mathbf{H}\cdot\nabla\times\mathbf{E} = -\mathbf{H}\cdot\frac{\partial\mathbf{B}}{\partial t}, \tag{1C.1}$$

and from Equation 1.2,

$$\mathbf{E}\cdot\nabla\times\mathbf{H} = \mathbf{E}\cdot\mathbf{J} + \mathbf{E}\cdot\frac{\partial\mathbf{D}}{\partial t}. \tag{1C.2}$$

Subtracting Equation 1C.2 from Equation 1C.1 and making use of the vector identity

$$\mathbf{H}\cdot\nabla\times\mathbf{E} - \mathbf{E}\cdot\nabla\times\mathbf{H} = \nabla\cdot(\mathbf{E}\times\mathbf{H}), \tag{1C.3}$$

we obtain

$$\nabla\cdot(\mathbf{E}\times\mathbf{H}) = -\mathbf{H}\cdot\frac{\partial\mathbf{B}}{\partial t} - \mathbf{E}\cdot\frac{\partial\mathbf{D}}{\partial t} - \mathbf{E}\cdot\mathbf{J}. \tag{1C.4}$$

Integrating Equation 1C.4 over a volume V and using the divergence theorem (see Appendix 1A),

$$\oint\!\!\!\oint_s \mathbf{S}\cdot\mathrm{d}\mathbf{s} + \iiint_V \left[\mathbf{E}\cdot\frac{\partial\mathbf{D}}{\partial t} + \mathbf{H}\cdot\frac{\partial\mathbf{B}}{\partial t}\right]\mathrm{d}V + \iiint_V (\mathbf{E}\cdot\mathbf{J})\,\mathrm{d}V = 0, \tag{1C.5}$$

where s is the closed surface bounding the volume V. The second term on the LHS of Equation 1C.5 can be written as

$$\frac{\partial}{\partial t}\iiint_V \left(\frac{1}{2}\mathbf{E}\cdot\mathbf{D} + \frac{1}{2}\mathbf{B}\cdot\mathbf{H}\right)\mathrm{d}v \tag{1C.6}$$

for the case of an isotropic simple medium, since

$$\frac{\partial}{\partial t}(\mathbf{E}\cdot\mathbf{D}) = \mathbf{E}\cdot\frac{\partial\mathbf{D}}{\partial t} + \frac{\partial\mathbf{E}}{\partial t}\cdot\mathbf{D} = 2\mathbf{E}\cdot\frac{\partial\mathbf{D}}{\partial t}, \tag{1C.7}$$

$$\frac{\partial}{\partial t}(\mathbf{H}\cdot\mathbf{B}) = \mathbf{H}\cdot\frac{\partial\mathbf{B}}{\partial t} + \frac{\partial\mathbf{H}}{\partial t}\cdot\mathbf{B} = 2\mathbf{H}\cdot\frac{\partial\mathbf{B}}{\partial t}. \tag{1C.8}$$

Therefore, Equation 1C.6 can be rewritten as

$$\oiint_S \mathbf{S}\cdot d\mathbf{s} + \frac{\partial}{\partial t}\iiint_V \left(\frac{1}{2}\mathbf{E}\cdot\mathbf{D} + \frac{1}{2}\mathbf{B}\cdot\mathbf{H}\right)dv + \iiint_V (\mathbf{E}\cdot\mathbf{J})dV = 0. \tag{1C.9}$$

Let us interpret Equation 1C.9 for a sourceless but lossy simple medium. If **S** represents the electromagnetic power density, then the first term gives the net power leaving the closed surface **S**. The second term is the time rate of increase of the stored electric and magnetic energy. The third term is the ohmic power loss due to the conversion of electromagnetic energy to thermal energy. The time rate of decrease in the stored energy provides for the power leaving the surface plus the inevitable power loss due to the lossy medium.

Appendix 1D: Low-Frequency Approximation of Maxwell's Equations R, L, C, and Memristor M

1D.1 Introduction

Circuit theory, used extensively in low-frequency electrical engineering, is a low-frequency approximation of Maxwell's equations. It is formulated in terms of the two-terminal circuit elements such as resistance (R), inductance (L), and capacitance (C). Fano et al. [1] discussed circuit theory as a quasistatic approximation in terms of the time-rate parameter α, which was used to expand the electromagnetic fields in power series. The classical elements R, L, and C were shown to be obtained by certain combinations of the zeroth- and first-order solutions. Chua [2] pointed out that a new circuit element M, called memristor (memory + resistor), could be proposed based on the relationship between the charge $q(t)$ and the magnetic flux $\Phi_m(t)$. Such a two-terminal element realized recently [3] has interesting properties and applications.

1D.2 Time-Rate Parameter

In preparing for generating quasistatic solutions, let us first consider the family of solutions generated by changing the timescale of the excitation. As an example, suppose the excitation is $\rho(x, y, z, t)$, consider the family of solutions obtained by the excitation $\rho(x, y, z, \alpha t)$, where α is called the time-rate parameter. We can rewrite Maxwell's equations in terms of the variable

$$\tau = \alpha t. \tag{1D.1}$$

Then

$$\frac{\partial \bar{B}}{\partial t} = \frac{\partial \bar{B}}{\partial \tau}\frac{\partial \tau}{\partial t} = \alpha \frac{\partial \bar{B}}{\partial \tau}. \tag{1D.2}$$

Maxwell's equations (1.1) through (1.4) can now be written as

$$\bar{\nabla} \times \bar{E}(\bar{r}, \tau) = -\alpha \frac{\partial \bar{B}}{\partial \tau}(\bar{r}, \tau), \tag{1D.3}$$

$$\bar{\nabla} \times \bar{H}(\bar{r}, \tau) = \bar{J} + \alpha \frac{\partial \bar{D}}{\partial \tau}(\bar{r}, \tau), \tag{1D.4}$$

$$\bar{\nabla} \cdot \bar{D}(\bar{r}, \tau) = \rho_v(\bar{r}, \tau), \tag{1D.5}$$

$$\bar{\nabla} \cdot \bar{B}(\bar{r}, \tau) = 0, \tag{1D.6}$$

and Equation 1.5 becomes

$$\bar{\nabla} \cdot \bar{J}(\bar{r}, \tau) = -\alpha \frac{\partial \rho_v}{\partial \tau}(\bar{r}, \tau). \tag{1D.7}$$

Since the fields are functions of $\bar{r}$, τ, and the parameter α, they can be expressed in power series in terms of α as

$$\bar{E}(\bar{r}, \tau, \alpha) = \bar{E}_0(\bar{r}, \tau) + \alpha \bar{E}_1(\bar{r}, \tau) + \alpha^2 \bar{E}_2(\bar{r}, \tau) + \cdots, \tag{1D.8a}$$

where

$$\bar{E}_k(\bar{r}, \tau) = \frac{1}{k!}\left[\frac{\partial^k \bar{E}(\bar{r}, \tau, \alpha)}{\partial \alpha^k}\right]_{\alpha=0}, \quad k = 1, 2, \ldots, \tag{1D.8b}$$

$$\bar{\nabla} \times \bar{E} = \bar{\nabla} \times \bar{E}_0 + \alpha \bar{\nabla} \times \bar{E}_1 + \alpha^2 \bar{\nabla} \times \bar{E}_2 + \cdots.$$

If we use similar power-series expressions and use in Equations 1D.3 through 1D.7 and collect terms to the same power of α, for instance, then we obtain

$$\bar{\nabla} \times \bar{E}_0 + \alpha\left(\bar{\nabla} \times \bar{E}_1 + \frac{\partial \bar{B}_0}{\partial \tau}\right) + \alpha^2\left(\bar{\nabla} \times \bar{E}_2 + \frac{\partial \bar{B}_1}{\partial \tau}\right) + \cdots = 0. \tag{1D.9}$$

Similar equations can be obtained, based on Equations 1D.4 through 1D.9.

Equation 1D.9 must be satisfied, for all values of α. This can be met only if the coefficients of all the powers of α are separately equal to zero. Thus, we obtain the series of equations

$$\bar{\nabla} \times \bar{E}_0 = 0, \tag{1D.10}$$

$$\bar{\nabla} \times \bar{H}_0 = \bar{J}_0, \tag{1D.11}$$

$$\bar{\nabla} \cdot \bar{J}_0 = 0, \tag{1D.12}$$

$$\bar{\nabla} \times \bar{E}_1 = -\frac{\partial \bar{B}_0}{\partial \tau}, \tag{1D.13}$$

$$\bar{\nabla} \times \bar{H}_1 = \bar{J}_1 + \frac{\partial \bar{D}_0}{\partial \tau}, \tag{1D.14}$$

$$\bar{\nabla} \cdot \bar{J}_1 = -\frac{\partial \rho_{v0}}{\partial \tau}. \tag{1D.15}$$

We can generate higher-order sets of equations such as Equations 1D.13 through 1D.15. Note, however, from Equations 1D.5 and 1D.6, we obtain

$$\bar{\nabla}\cdot\bar{D}_k = \rho_{vk} \tag{1D.16}$$

$$\bar{\nabla}\cdot\bar{B}_k = 0, \quad k = 0,1,2,\ldots. \tag{1D.17}$$

1D.3 Circuit Parameters *R*, *L*, and *C*

The circuit element R is identified as the zeroth-order solution of Equations 1D.10 and 1D.11. For a resistor R, the first-order electric and magnetic fields are negligible compared with the zeroth-order fields. Its characterization as a relationship between the zeroth-order electric and magnetic fields leads to the memoryless instantaneous relationship of a pure resistor. Its gross parameter representation is Ohm's law given by

$$V(t) = RI(t). \tag{1D.18}$$

If only the first-order magnetic field is negligible but the first-order electric field is not negligible, the system can be identified as a resistor in series with an inductor [2]. Its gross parameter representation is

$$V(t) = RI(t) + L\frac{dI}{dt}. \tag{1D.19}$$

On the other hand, if only the first-order electric field is negligible but the first-order magnetic field is not negligible, the system can be identified as a resistor R (conductance $G = 1/R$) in parallel with a capacitor C. The gross parameter representation in this case is

$$I(t) = GV(t) + C\frac{dV}{dt}. \tag{1D.20}$$

Equations 1D.19 and 1D.20 can be obtained and R, G, L, and C are calculated by using the integral form of the relevant Maxwell's equations and the constitutive relations for the material:

$$\bar{J} = \sigma\bar{E}, \tag{1D.21}$$

$$\bar{B} = \mu\bar{H}, \tag{1D.22}$$

$$\bar{D} = \varepsilon\bar{E}. \tag{1D.23}$$

1D.4 Memristor

Fano et al. [1] dismissed the case where the first-order fields $\bar{E}_1$ and $\bar{H}_1$ are not negligible saying that such a case is not relevant to circuit theory. Chua [2] suggested that such a case under approximate conditions leads to a two-terminal device where the charge $q(t)$

(the surface integral of the first-order electric flux density vector $\bar{D}_1$) is related to the flux $\Phi_m(t)$ (surface integral of the first-order $\bar{B}_1$ field). The approximate conditions are: (i) both zeroth-order fields are negligible compared to the first-order fields; and (ii) the material of the device is nonlinear.

The nonlinear relationships are

$$\bar{J}_1 = \mathcal{J}(\bar{E}_1), \tag{1D.24}$$

$$\bar{B}_1 = \mathcal{B}(\bar{H}_1), \tag{1D.25}$$

$$\bar{D}_1 = \mathcal{D}(\bar{E}_1), \tag{1D.26}$$

$$\bar{\nabla} \times \bar{H}_1 = \mathcal{J}(\bar{E}_1), \tag{1D.27}$$

$$\bar{E}_1 = f(\bar{H}_1), \tag{1D.28}$$

leading to

$$\bar{D}_1 = g(\bar{B}_1). \tag{1D.29}$$

Equation 1D.29 gives the instantaneous (memoryless) relationship between $\bar{D}_1$ and $\bar{B}_1$.

A physical memristor device is essentially an ac device; otherwise, the dc electromagnetic fields will give rise to nonnegligible zeroth-order fields.

In terms of gross variables and parameters we write

$$V(t) = RI(t) = f_R(I), \tag{1D.30}$$

$$q(t) = CV(t) = f_C(V), \tag{1D.31}$$

$$\Phi_m(t) = LI(t) = f_L(I). \tag{1D.32}$$

In the case of a memristor

$$\Phi_m(t) = f_M(q). \tag{1D.33}$$

For the case of a charge-controlled memristor

$$\frac{d\Phi_m}{dt} = \frac{df_M}{dq}\frac{dq}{dt}, \tag{1D.34}$$

$$V(t) = \frac{df_M}{dq} i(t), \tag{1D.35}$$

where

$$\frac{df_M}{dq} = M(q) \tag{1D.36}$$

is the incremental memristance (in the units of ohms).

The mathematical model for the HP memristor [3,4] which is based on its fabrication process is given by

$$M(q) = R_{\mathrm{OFF}}\left(1 - \frac{R_{\mathrm{ON}}}{\beta} q(t)\right),$$

where

$$\beta = \frac{D^2}{\mu_D} \quad (\mathrm{Wb}).$$

Here, μ_D is the average drift mobility and D is the film (titanium dioxide) thickness. R_{OFF} and R_{ON} are the "OFF" and "ON" states of the resistance. The device is a TiO_2 junction where one side is doped with positive ions and the other side is undoped. The width w of the doped region depends on the charge passing through the device. The state equation describing process can be written as

$$\frac{1}{D}\frac{dw(t)}{dt} = \frac{R_{\mathrm{ON}}}{\beta} i(t),$$

$$\frac{w(t)}{D} = \frac{w(t_0)}{D} + \frac{q(t)}{Q_{\mathrm{D}}},$$

where

$$Q_{\mathrm{D}} = \frac{\beta}{R_{\mathrm{ON}}} = i \times t.$$

Figure 1D.1 gives the symbol proposed in [1] and q versus Φ_{m} curve for the memristor.

Memristor is a passive two-terminal device. It cannot store energy, when $V(t) = 0$ and $I(t) = 0$. A memristor acts as a linear resistor when its frequency goes toward infinity and as a nonlinear resistor at low frequencies.

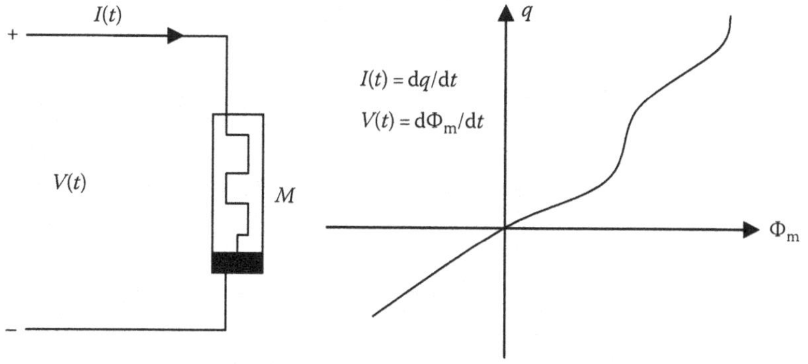

FIGURE 1D.1
Memristor and its Φ_{m} versus q curve.

It has a regime of operation with an approximately linear charge–resistance relationship as the time integral of the current stays within certain bounds. Applications are in nanoelectronic memories, computer logic, and neuromorphic computer architectures.

In an ideal description, the element remembers the amount of current passing through it in the past. When the current stops, the component retains the last resistance that it had. When the flow of charge starts again, the resistance of the circuit will be what it was when it was last active.

In summary, unlike the resistor R whose value depends on the ratio of instantaneous voltage to current, in the case of memristor the resistance depends on the ratio of the integral of the input voltage to the integral of the input current.

Since $\int V\,dt = \Phi_m$ and $\int I\,dt = q$, it is the functional relationship between Φ_m and q that determines the properties of the memristor. If this relationship is linear, it is the same as the resistor R.

This is an active field of research with many potential applications and developments.

References

1. Fano, R. M., Chu, L. N., and Adler, R. B., *Electromagnetic Fields, Energy, and Forces*, Wiley, New York, 1960.
2. Chua, L. O., Memristor, the missing circuit element, *IEEE Trans. Circ. Theory*, CT-18(5), 507–519, 1971.
3. Strukov, D. B., Snider, G. S., Steward, D. R., and Williams, R. S., The missing memristor found, *Nature*, 453(7191), 80–83, 2008.
4. Kavehei, O., Iqbal, A., Kim, Y. S., Eshraghiam, K., Al-Sarawi, S. F., and Abbott, D., The fourth element: Characteristics, modeling and electromagnetic theory of the memristor, *Proc. R. Soc. A*, 466(2120), 2175–2202, 2010.

Appendix 2A: AC Resistance of a Round Wire When the Skin Depth δ Is Comparable to the Radius a of the Wire

We have shown in Section 2.2 that the parameters α and β are given by δ^{-1} in a good conductor. The approximations made in arriving at this result assumed that the loss tangent $T = \sigma_c/\omega\varepsilon$ is large. Another way of stating the approximation is to say that the displacement current density $j\omega\varepsilon\tilde{\mathbf{E}}$ is neglected in comparison with the conduction current density $\sigma_c\tilde{\mathbf{E}}$. The propagation constant $\gamma = jk$ may be obtained in this case by neglecting the first term on the RHS of (2.4)

$$\gamma^2 = -k^2 = j\omega\mu\sigma = 2\pi f\mu\sigma e^{j\pi/2}. \tag{2A.1}$$

The wave equation for the current density $\tilde{\mathbf{J}} = \sigma_c\tilde{\mathbf{E}}$ is given by

$$\nabla^2\tilde{\mathbf{J}} - j\omega\mu\sigma_c\tilde{\mathbf{J}} = 0. \tag{2A.2}$$

For the cylindrical one-dimensional problem under consideration in this section

$$\tilde{\mathbf{J}} = \hat{z}J(\rho), \tag{2A.3}$$

and from Equation 2A.2, we obtain

$$\frac{d^2J}{d\rho^2} + \frac{1}{\rho}\frac{dJ}{d\rho} - j\omega\mu\sigma_c J = 0. \tag{2A.4}$$

This function is the Bessel equation considered in Section 2.15 and since the current-density is finite at the origin, we reject the function Y_0 and choose J_0:

$$J = AJ_0(T\rho), \tag{2A.5}$$

where

$$T = \sqrt{-j\omega\mu\sigma_c} = j^{-1/2}\frac{\sqrt{2}}{\delta}. \tag{2A.6}$$

Since T is complex, $J_0(T\rho)$ is also complex and has a real part and an imaginary part. The following special functions called Ber(x) and Bei(x) are defined and tabulated in many mathematical tables:

$$\mathrm{Ber}(x) = \mathrm{Re}\left[J_0\left(\mathrm{j}^{-1/2}x\right)\right], \tag{2A.7}$$

$$\mathrm{Bei}(x) = \mathrm{Im}\left[J_0\left(\mathrm{j}^{-1/2}x\right)\right]. \tag{2A.8}$$

In terms of the current density $\tilde{J}_a$ at the surface of the conductor, the current density in the wire may be written as

$$\tilde{J}(\rho) = \tilde{J}_a \frac{\mathrm{Ber}\left(\sqrt{2}\rho/\delta\right) + \mathrm{Bei}\left(\sqrt{2}\rho/\delta\right)}{\mathrm{Ber}\left(\sqrt{2}a/\delta\right) + \mathrm{Bei}\left(\sqrt{2}a/\delta\right)}, \quad 0 < \rho < a. \tag{2A.9}$$

Figure 2A.1 shows $|J(\rho)/J_a|$ versus the radius for two values of a/δ. The solid lines show the results based on Equation 2A.9. The broken lines are shown for comparison and are obtained assuming a parallel plane formula

$$\frac{\tilde{J}(\rho)}{\tilde{J}_a} = \mathrm{e}^{-(a-\rho)/\delta}, \quad a \gg \delta, \tag{2A.10}$$

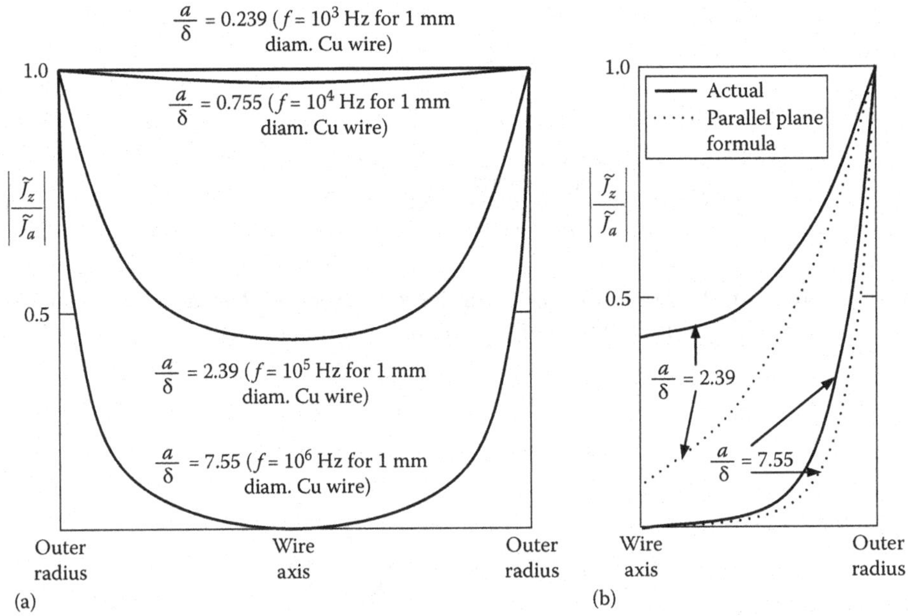

FIGURE 2A.1
(a) Current distribution in a cylindrical wire for different frequencies. (b) Actual and approximate (parallel plate formula) distribution in cylindrical wire. (From Ramo, S., Whinnery, J.R., and Van Duzer, T., *Fields and Waves in Communication Electronics*. p. 297. 1967. Copyright Wiley-VCH Verlag GmbH&Co. KGaA. Reproduced with permission.)

which is a good approximation when $a \gg \delta$. As expected, the solid and the broken curves are close for $a/\delta = 7.55$ and differ considerably for $a/\delta = 2.39$.

The magnetic field $\tilde{H}$ may be obtained from Equation 1.40 and it has only the ϕ component

$$\tilde{H}_\phi = \frac{1}{j\omega\mu}\frac{d\tilde{E}_z}{d\rho} \tag{2A.11}$$

and

$$\tilde{E}_z = \frac{\tilde{J}}{\sigma_c} = \frac{A}{\sigma_c}J_0(T\rho) = \frac{\tilde{J}_a}{\sigma_c}\frac{J_0(T\rho)}{J_0(Ta)}. \tag{2A.12}$$

Hence,

$$\tilde{H}_\phi = \frac{\tilde{J}_a}{\sigma_c}\frac{T}{j\omega\mu}\frac{J_0'(T\rho)}{J_0(Ta)} = \frac{-\tilde{J}_a}{T}\frac{J_0'(T\rho)}{J_0(Ta)}. \tag{2A.13}$$

From Ampere's law,

$$2\pi a\tilde{H}_\phi\Big|_{\rho=a} = -2\pi a\frac{\tilde{J}_a}{T}\frac{J_0'(Ta)}{J_0(Ta)} = \tilde{I}. \tag{2A.14}$$

The internal impedance per unit length is

$$Z_i' = \frac{\tilde{E}_z\Big|_{\rho=a}}{\tilde{I}} = \frac{\tilde{J}_a/\sigma_c}{\tilde{I}} = -\frac{TJ_0(Ta)}{2\pi a\sigma_c J_0'(Ta)}. \tag{2A.15}$$

Equation 2A.15 may be expressed in terms of Ber and Bei functions defined by Equations 2A.7 and 2A.8, respectively, giving

$$R' = \mathrm{Re}\left[Z_i'\right] = \frac{R_s}{\sqrt{2}\pi a}\left[\frac{\mathrm{Ber}q\,\mathrm{Ber}'q - \mathrm{Bei}q\,\mathrm{Bei}'q}{(\mathrm{Ber}'q)^2 + (\mathrm{Bei}'q)^2}\right](\Omega/\mathrm{m}), \tag{2A.16}$$

$$\omega L_i' = \frac{R_s}{\sqrt{2}\pi a}\left[\frac{\mathrm{Ber}q\,\mathrm{Ber}'q + \mathrm{Bei}q\,\mathrm{Bei}'q}{(\mathrm{Ber}'q)^2 + (\mathrm{Bei}'q)^2}\right](\Omega/\mathrm{m}), \tag{2A.17}$$

where

$$q = \frac{\sqrt{2}a}{\delta} \tag{2A.18}$$

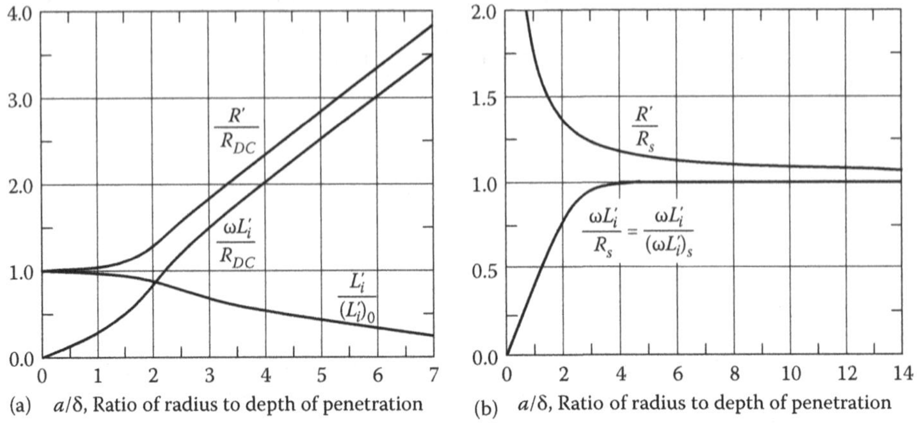

FIGURE 2A.2
(a) Solid wire skin effect quantities compared with d.c. values. (b) Solid wire skin effect quantities compared with values from high-frequency formulas. (From Ramo, S., Whinnery, J. R., and Van Duzer, T., *Fields and Waves in Communication Electronics*. p. 297. 1967. Copyright Wiley-VCH Verlag GmbH&Co. KGaA. Reproduced with permission [1].)

and

$$R_s = \frac{1}{\sigma_c \delta} = \sqrt{\frac{\pi f \mu}{\sigma_c}}. \tag{2A.19}$$

Figure 2A.1 shows the current distribution for different frequencies.

Figure 2A.2 shows R'/R_s and $\omega L_i'/R_s$ versus a/δ. As expected, they reach the horizontal line 1 as a/δ tends to infinity.

Reference

1. Ramo, S., Whinnery, J. R., and Van Duzer, T., *Fields and Waves in Communication Electronics*, Wiley, New York, 1967.

Appendix 2B: Transmission Lines: Power Calculation

2B.1 Transmission Lines: Power Calculation

A transmission line of length d is shown in Figure 2B.1a. The equivalent circuit is shown in Figure 2B.1b.

The input impedance into the transmission line of length d with load Z_L is given by

$$Z_{in} = Z_0 \frac{Z_L + jZ_0 \tan(\beta d)}{Z_0 + jZ_0 \tan(\beta d)} = Z_0 \frac{\left(1+\Gamma_0 e^{-j2\beta d}\right)}{\left(1-\Gamma_0 e^{-j2\beta d}\right)}. \tag{2B.1}$$

The voltage at the input of the transmission line can be found as

$$\begin{aligned}\tilde{V}(d) &= \frac{\tilde{V}_g Z_{in}}{Z_g + Z_{in}} \\ &= \tilde{V}_0^+ e^{j\beta d} + \tilde{V}_0^- e^{-j\beta d} = \tilde{V}_0^+ e^{j\beta d}\left[1+\Gamma_0 e^{-j2\beta d}\right],\end{aligned} \tag{2B.2}$$

where $\tilde{V}_0^+$ is the voltage of the positive traveling wave and $\tilde{V}_0^-$ is the voltage of the reflected wave at the load end. Rearranging, we obtain

$$\tilde{V}_0^+ = \frac{\tilde{V}_g Z_{in}}{\left(Z_g + Z_{in}\right)e^{j\beta d}\left[1+\Gamma_0 e^{-j2\beta d}\right]}, \tag{2B.3}$$

$$\tilde{V}_0^+ = \tilde{V}_g \frac{Z_{in}}{\left(Z_g + Z_{in}\right)\left(e^{j\beta d} + \Gamma_0 e^{-j\beta d}\right)}. \tag{2B.4}$$

In the above,

$$\Gamma_0 = \frac{Z_L - Z_0}{Z_L + Z_0}. \tag{2B.5}$$

The average incident power is given as

$$P_{av}^{i} = \mathrm{Re}\left[\frac{1}{2}\tilde{V}_0^+\left(\tilde{I}_0^+\right)^*\right] = \frac{1}{2}\mathrm{Re}\left[\tilde{V}_0^+ \frac{\left(\tilde{V}_0^+\right)^*}{Z_0}\right],$$

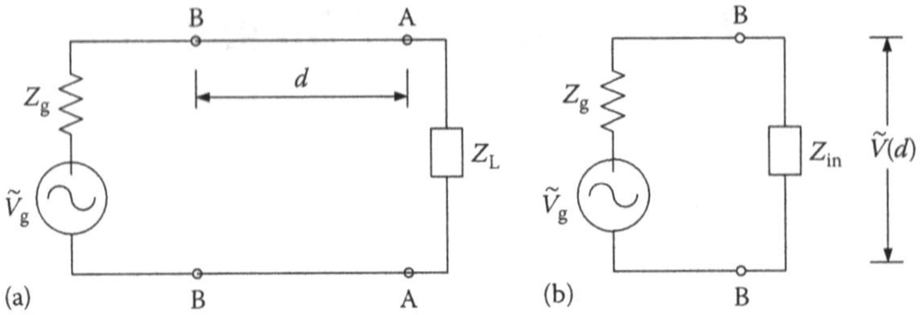

FIGURE 2B.1
Transmission line: (a) length d and (b) equivalent input impedance.

$$P_{av}^{i} = \frac{1}{2}\frac{\left|\tilde{V}_0^+\right|^2}{Z_0}. \tag{2B.6}$$

The average reflected power is given by

$$P_{av}^{r} = \text{Re}\left[\frac{1}{2}\tilde{V}_0^-\left(\tilde{I}_0^-\right)^*\right] = -\frac{1}{2}\text{Re}\left[\tilde{V}_0^-\frac{\left(\tilde{V}_0^-\right)^*}{Z_0}\right],$$

$$P_{av}^{r} = -\frac{1}{2}\frac{\left|\Gamma_0\right|^2\left|\tilde{V}_0^+\right|^2}{Z_0}. \tag{2B.7}$$

The total power consumed by the load is

$$P_{av}^{tot} = P_{av}^{i} + P_{av}^{r} = \frac{1}{2}\frac{\left|\tilde{V}_0^+\right|^2}{Z_0}\left[1 - \left|\Gamma_0\right|^2\right],$$

$$P_{av}^{tot} = P_{av}^{i}\left[1 - \left|\Gamma_0\right|^2\right]. \tag{2B.8}$$

This is shown in Figure 2B.2.

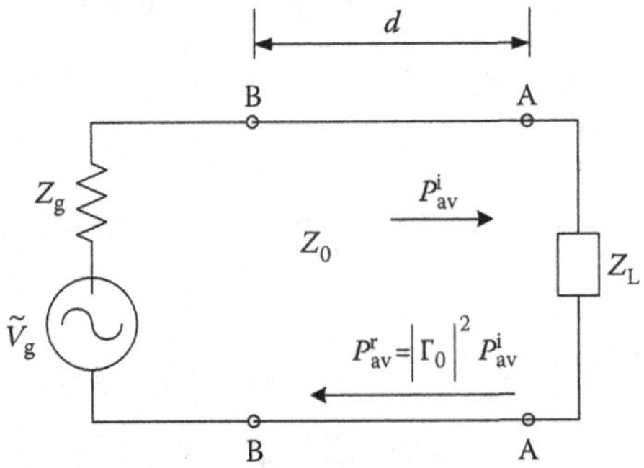

FIGURE 2B.2
Transmission line circuit showing incident and reflected power.

2B.2 Transmission Lines: Special Case $Z_g = Z_0$

Consider the special case of the transmission line circuit in Figure 2B.1a with $Z_g = Z_0$, the characteristic impedance of the transmission line. It will be shown that $\tilde{V}_0^+$ is independent of Z_L if $Z_g = Z_0$. When $Z_g = Z_0$, from Equation 2B.1, we obtain

$$Z_g + Z_{in} = Z_0 + Z_0 \frac{\left(1+\Gamma_0 e^{-j2\beta d}\right)}{\left(1-\Gamma_0 e^{-j2\beta d}\right)}. \tag{2B.9}$$

Substituting Equations 2B.1 and 2B.9 in the expression for $\tilde{V}_0^+$ given in Equation 2B.3:

$$\tilde{V}_0^+ = \frac{\tilde{V}_g Z_0 \left(1+\Gamma_0 e^{-j2\beta d}\right)/\left(1-\Gamma_0 e^{-j2\beta d}\right)}{Z_0\left[1+\left(1+\Gamma_0 e^{-j2\beta d}\right)/\left(1-\Gamma_0 e^{-j2\beta d}\right)\right] e^{j\beta d}\left[1+\Gamma_0 e^{-j2\beta d}\right]}. \tag{2B.10}$$

Rearranging,

$$\tilde{V}_0^+ = \frac{\tilde{V}_g}{2} e^{-j\beta d}, \tag{2B.11}$$

which holds whenever $Z_g = Z_0$ regardless of the value of Z_L.

Appendix 2C: Introduction to the Smith Chart

The Smith chart is a graphical aid or nomogram [1] designed as an aid in radio frequency engineering to assist in solving problems with transmission lines and matching circuits. The reflection coefficient is the relationship based on which the Smith chart is constructed:

$$\Gamma = \frac{Z_L - Z_0}{Z_L + Z_0}. \tag{2C.1}$$

The x-axis is the real part of the reflection coefficient (Γ_r) and the y-axis is the imaginary part of the reflection coefficient (Γ_i), where

$$\Gamma = \Gamma_r + j\Gamma_i. \tag{2C.2}$$

The impedance plotted on the chart will be normalized with respect to the characteristic impedance Z_0:

$$z_L = r + jx = \frac{Z_L}{Z_0} = \frac{R_L + jX_L}{Z_0}. \tag{2C.3}$$

The reflection coefficient can be represented as

$$\Gamma = \frac{z_L - 1}{z_L + 1} \tag{2C.4}$$

and the load impedance as

$$z_L = \frac{1+\Gamma}{1-\Gamma}. \tag{2C.5}$$

In polar form, $|\Gamma|$ and ϕ are, respectively the magnitude and angle of Γ as shown in Figure 2C.1.

Using Equations 2C.2 and 2C.3, and rearranging to get the real and imaginary parts of z_L, it can be seen that

$$r = \frac{1 - \Gamma_r^2 - \Gamma_i^2}{(1-\Gamma_r)^2 + \Gamma_i^2}, \tag{2C.6}$$

$$x = \frac{2\Gamma_i}{(1-\Gamma_r)^2 + \Gamma_i^2}. \tag{2C.7}$$

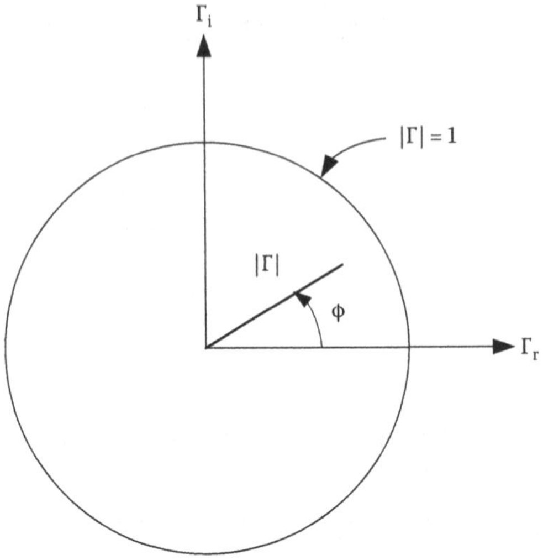

FIGURE 2C.1
Polar coordinates of the Smith chart.

After some further algebraic manipulation, it is found that

$$\left(\Gamma_r - \frac{r}{1+r}\right)^2 + \Gamma_i^2 = \left(\frac{1}{1+r}\right)^2, \tag{2C.8}$$

$$\left(\Gamma_r - 1\right)^2 + \left(\Gamma_i - \frac{1}{x}\right)^2 = \left(\frac{1}{x}\right)^2. \tag{2C.9}$$

Equation 2C.8 represents a family of circles defined by the resistance r. These circles are centered at the point $\Gamma_r = r/(1 + r)$, $\Gamma_i = 0$ (which is on the horizontal-axis), and have a radius equal to $1/(1 + r)$. When $r = 0$, this is a circle centered at the origin with radius = 1 (the unit circle). On the other hand, if $r = \infty$, the circle center is at (1, 0) with a radius of zero. Thus, for $r = \infty$, Equation 2C.8 represents a point on the Γ real axis at the far-right edge of the unit circle. If $r = 1$, Equation 2C.8 gives a circle centered at (½, 0) with a radius of ½. These circles along with those corresponding to r = ½ and 2 are shown in Figure 2C.2. Note that all the circles are centered on the positive Γ_r-axis and go through the point $\Gamma_r = 1$ and $\Gamma_i = 0$.

Equation 2C.9 also represents a family of circles, in this case centered at $\Gamma_r = 1$ and $\Gamma_i = 1/x$ with a radius of $1/x$. When $x = \pm\infty$, this represents a circle centered at (1, 0) with a radius of zero which is again the point at the right edge of the unit circle. If $x = 1$, the circle is centered at $\Gamma = 1 + j1$ with a radius of unity. Since $|\Gamma|$ is always less than or equal to 1, only the portion of these circles within the unit circle are shown (Figure 2C.2). Figure 2C.3 shows a number of these circular sections for various values of x. Note that the circle representing $x = 0$ is along the horizontal-axis as the center of the circle moves toward ∞ along the $\Gamma_r = 1$ axis and the radius becomes infinite. As can be seen in Figure 2C.4, both sets of circles are plotted on the Smith chart. If Z_L is given, one may divide by Z_0 and find the intersection of the appropriate r and x circles and determine Γ from the plot.

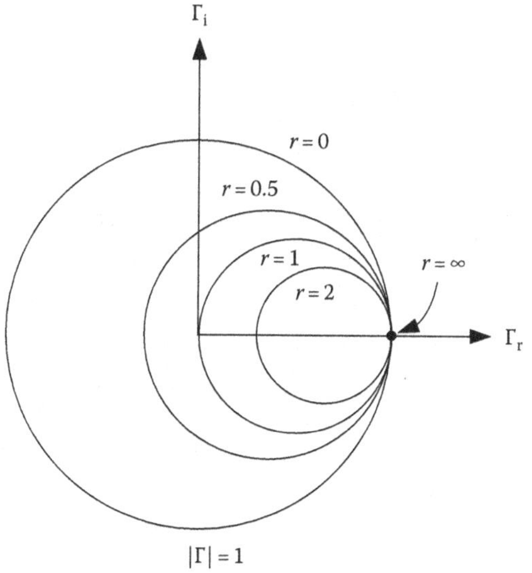

FIGURE 2C.2
Family of circles on the Smith chart corresponding to reflection coefficient for fixed, normalized load resistances.

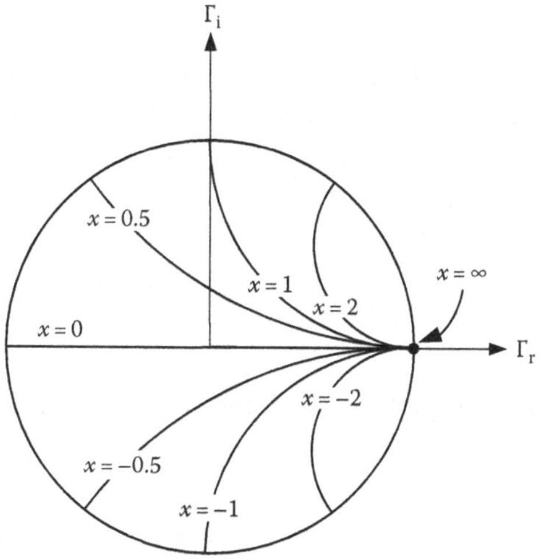

FIGURE 2C.3
Family of circles on the Smith chart corresponding to reflection coefficient for fixed, normalized load reactance.

To avoid unreadable clutter on the graph, the concentric circles for $|\Gamma|$ and the radial lines for ϕ are not shown on the Smith chart. But rather graduated line segments at the bottom of the chart and circumference labels of angle around the outer circle are included as shown in Figure 2C.4. As an example, consider the point $Z_L = 50 + j100\ \Omega$ on a 50-Ω line. Then $z_L = Z_L/50 = 1 + j2$, so that $r = 1$ and $x = 2$. This point is plotted as point P in Figure 2C.4 where it can be seen that $|\Gamma| \approx 0.7$ and $\phi = 45°$.

The final scale on the Smith chart, which is used to calculate the distance along a transmission line, is added on the circumference of the plot. The scale is in wavelength units

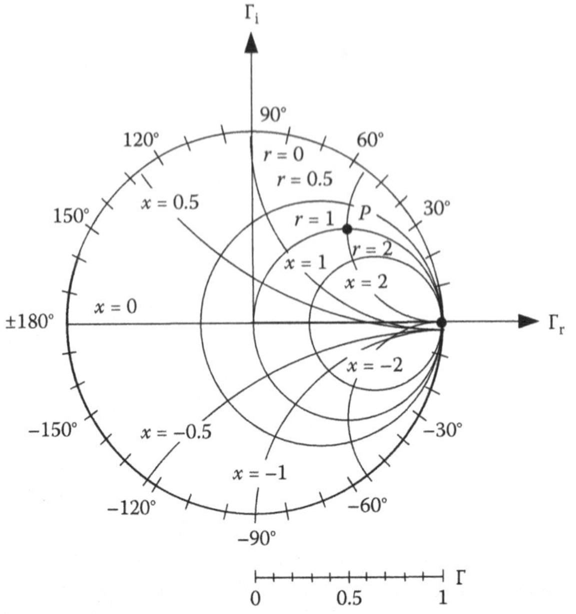

FIGURE 2C.4
Smith chart constant resistance (r) and reactance (x) contours.

with the zero point taken arbitrarily to the left. It is known that the voltage and the current at any point along a transmission line is given, respectively, by

$$\tilde{V} = \tilde{V}_0^+ \left(e^{-j\beta z} + \Gamma e^{j\beta z} \right) \tag{2C.10}$$

and

$$\tilde{I} = \frac{\tilde{V}_0^+}{Z_0} \left(e^{-j\beta z} - \Gamma e^{j\beta z} \right). \tag{2C.11}$$

Dividing Equation 2C.10 by Equation 2C.11 and normalizing gives the normalized input impedance:

$$z_{in} = \frac{\tilde{V}}{Z_0 \tilde{I}} = \frac{e^{-j\beta z} + \Gamma e^{j\beta z}}{e^{-j\beta z} - \Gamma e^{j\beta z}}. \tag{2C.12}$$

Replacing z by $-d$ and dividing the numerator and denominator by $e^{j\beta d}$, we obtain the general equation relating normalized input impedance, reflection coefficient, and line length (d):

$$z_{in} = \frac{1 + \Gamma e^{-j2\beta d}}{1 - \Gamma e^{-j2\beta d}} = \frac{1 + |\Gamma| e^{j(\phi - 2\beta d)}}{1 - |\Gamma| e^{j(\phi - 2\beta d)}}. \tag{2C.13}$$

Equation 2C.13 shows that the input impedance at any point $z = -d$ can be obtained by replacing Γ, the reflection coefficient of the load, by $\Gamma e^{-j2\beta d}$. This corresponds to a decrease in the angle of Γ by $2\beta d$ radians as we go from the load to the line input. Only the angle of

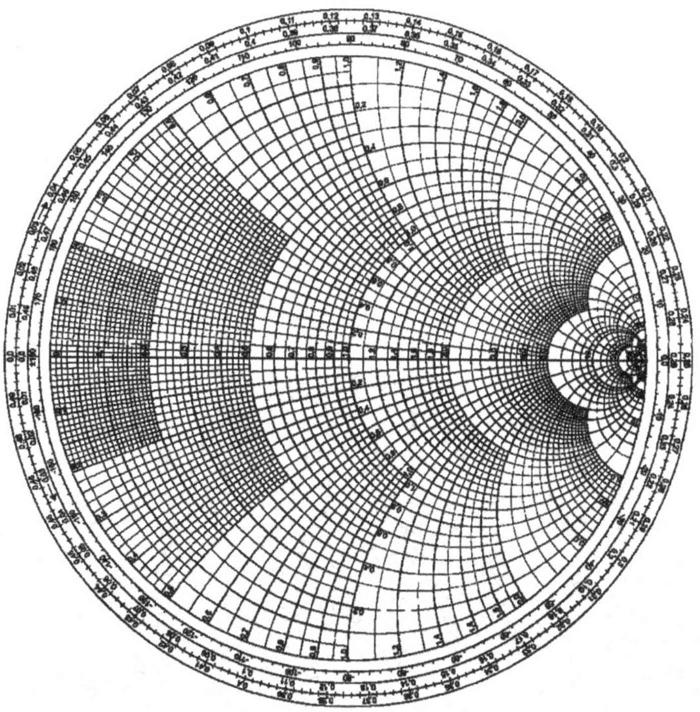

FIGURE 2C.5
The Smith chart (graduated line segments at the bottom are not shown here).

Γ changes, whereas the magnitude stays constant, thus the movement is along a constant radius arc. As we move from the load z_L to the input impedance z_{in}, we move toward the generator by a distance d on the transmission line and a clockwise angle $2\beta d$ on the Smith chart. One complete rotation around the Smith chart corresponds to a phase change of π radians or when d changes a half wavelength. The Smith chart is thus marked with a scale showing a change of 0.5λ for one circumnavigation of the unit circle. Two scales are normally given, one showing an increase in distance for clockwise movement and the other an increase for counterclockwise rotation. These two scales can be seen on the Smith chart of Figure 2C.5. One scale represents wavelengths toward generator (wtg) and increases for clockwise rotation. The other represents wavelengths toward load (wtl) and increases for counterclockwise rotation.

Example 2C.1

The scenario shown in Figure 2C.6 is considered in this example and Figure 2C.7 shows the Smith chart analysis for this example.

1. Load point A: $z_L = Z_L/Z_0 = 2 - j0.5$. $|\Gamma| = 0.37$, $\phi = -17°$. Point A is 0.274 wtg.
2. Point B, $d = 0.6\lambda$, 0.374 wtg

$$z(0.6\lambda) = 0.77 - j0.65(\text{point B})$$

$$Z(0.6\lambda) = 50(0.77 - j0.65) = 38.5 - j32.5\Omega$$

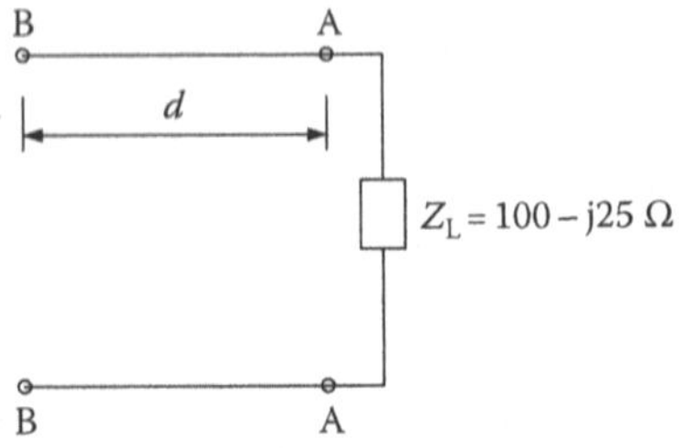

FIGURE 2C.6
Smith chart example ($Z_L = 100 - j25\ \Omega$, $d = 0.6\lambda$).

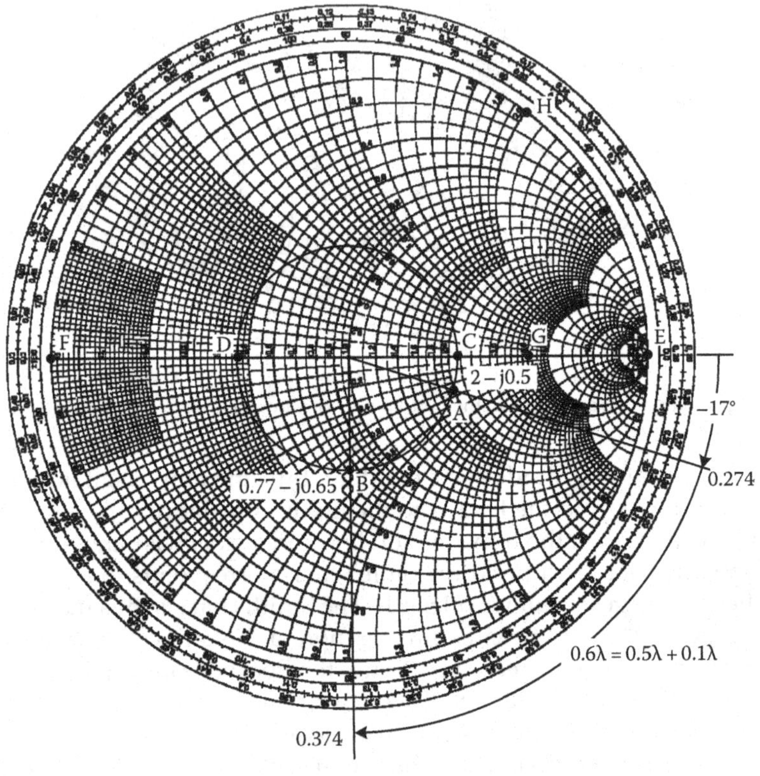

FIGURE 2C.7
Smith chart example A = load 100 – j25Ω, B = input impedance at $d = 0.6\lambda$ line, C = point of voltage maximum, D = point of voltage minimum, E = open-circuit point, F = short-circuit point, G = load 4 + j0, H = load 0 + j2.

3. $|z(d)|_{max}$ is at point C and is equal to 2.2 = s (voltage standing wave ratio). This is also the point of voltage maximum. Voltage is maximum when $d = [(0.5 - 0.274) + 0.25]\lambda = 0.476\lambda$.
4. $|z(d)|_{min}$ is at point D and is equal to 1/2.2 = 0.45. This is also the point of voltage minimum. Voltage is minimum when $d = (0.5 - 0.274)\lambda = 0.226\lambda$.
5. Point E shows the open-circuit point.
6. Point F shows the short-circuit point.
7. Point G shows the point $z = 4 + j0$.
8. Point H shows the point $z = 0 + j2$.
9. E, F, G, H are not related to the example in Figure 2C.6.

2C.1 Reflection of Uniform Plane Waves

Note that the characters with tilde (~) represent a phasor quantity. A superscript plus sign indicates a wave propagating in the +z-direction, whereas a superscript minus sign represents a wave propagating in the −z-direction.

2C.1.1 Free Space: Perfect Conductor Interface

Let us consider the following problem depicted in Figure 2C.8.

The space $z > 0$ is a perfect conductor and the space $z < 0$ is a free space. A time-harmonic electromagnetic field propagating in the +z-direction in the free space is normally incident on the interface from the left. Let the electric field of this wave be in the x-direction only, that is,

$$\tilde{\mathbf{E}}^{I}(z) = \tilde{E}_0^{+} e^{-j\beta z}\hat{\mathbf{x}}, \tag{2C.14}$$

where $\beta = \omega/c = \omega\sqrt{\mu_0\varepsilon_0}$ and $\omega = 2\pi f$. The magnetic field of this incident wave is

$$\tilde{\mathbf{H}}^{I}(z) = \frac{\tilde{E}_0^{+}}{\eta_0} e^{-j\beta z}\hat{\mathbf{y}}, \tag{2C.15}$$

where $\eta_0 = \sqrt{\mu_0/\varepsilon_0} = 120\pi$. The superscript I stands for incident wave. The electric field at $z = 0$ of this incident wave is $\tilde{E}_0^{+}$. But the boundary condition at $z = 0$ requires the electric field should be zero. Such a boundary condition can be satisfied if we assume that a reflected wave exists in the free space (propagating in the −z-direction) with the electric and magnetic fields as follows:

$$\tilde{\mathbf{E}}^{R}(z) = \tilde{E}_0^{-} e^{+j\beta z}\hat{\mathbf{x}}, \tag{2C.16}$$

$$\tilde{\mathbf{H}}^{R}(z) = -\frac{\tilde{E}_0^{-}}{\eta_0} e^{+j\beta z}\hat{\mathbf{y}}, \tag{2C.17}$$

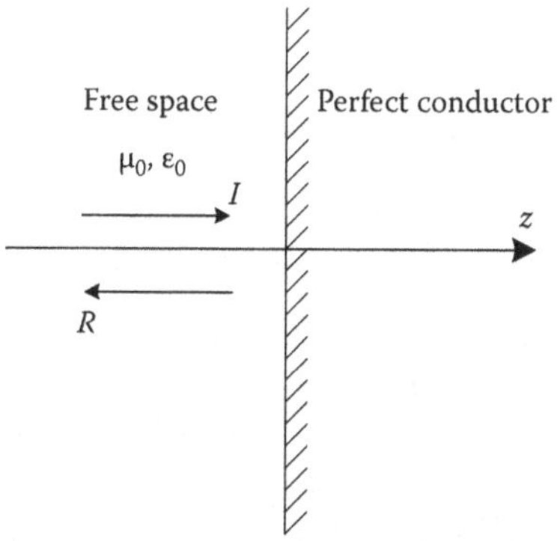

FIGURE 2C.8
Reflection of uniform plane wave at a free space to perfect conductor boundary.

and

$$\tilde{E}_0^+ + \tilde{E}_0^- = 0, \tag{2C.18}$$

so

$$\Gamma_0 = \frac{\tilde{E}_0^-}{\tilde{E}_0^+} = -1. \tag{2C.19}$$

2C.2 Comparison with a Transmission Line Problem

A comparison of a normal plane wave reflection (from a free space to a perfect electric conductor) to the transmission line analogy is shown in Figure 2C.9 and summarized in Table 2C.1.

2C.2.1 Dielectric–Dielectric Interface

Let us now consider the problem of reflection of a plane wave at the interface of two dielectrics as depicted in Figure 2C.10.

Let the incident wave be traveling in medium 1 in the positive z-direction and then for $z < 0$:

$$\tilde{\mathbf{E}}^{\mathrm{I}}(z) = \tilde{E}_0^+ e^{-j\beta_1 z}\hat{\mathbf{x}}, \tag{2C.20}$$

$$\tilde{\mathbf{H}}^{\mathrm{I}}(z) = \frac{\tilde{E}_0^+}{\eta_1} e^{-j\beta_1 z}, \ \hat{\mathbf{y}} = \tilde{H}_0^+ e^{-j\beta_1 z}\hat{\mathbf{y}}, \tag{2C.21}$$

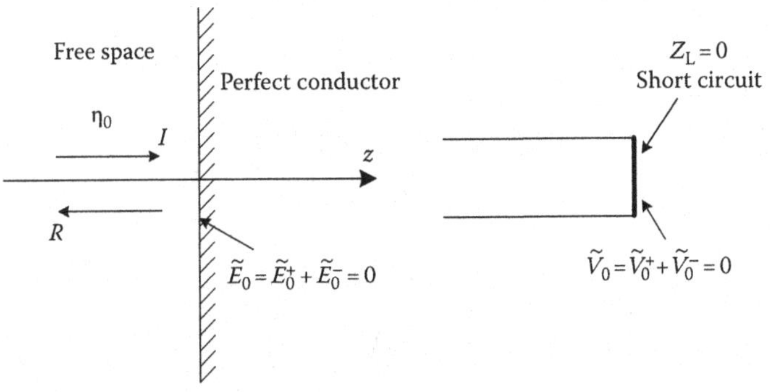

FIGURE 2C.9
Comparison of normal plane wave reflection with transmission line.

TABLE 2C.1

Plane Wave Reflection Comparison with the Transmission Line Analogy

Plane Wave	**Transmission Line**
$\Gamma_0 = \dfrac{\tilde{E}_0^-}{\tilde{E}_0^+} = -1$	$\Gamma_0 = \dfrac{Z_L - Z_0}{Z_L + Z_0} = -1$
E_x	V
H_y	I
$\eta_0 = \sqrt{\mu_0/\varepsilon_0} = 120\pi$	$Z_0 = \sqrt{L/C}$
$\beta = \omega\sqrt{\mu_0\varepsilon_0}$	$\beta = \omega\sqrt{LC}$

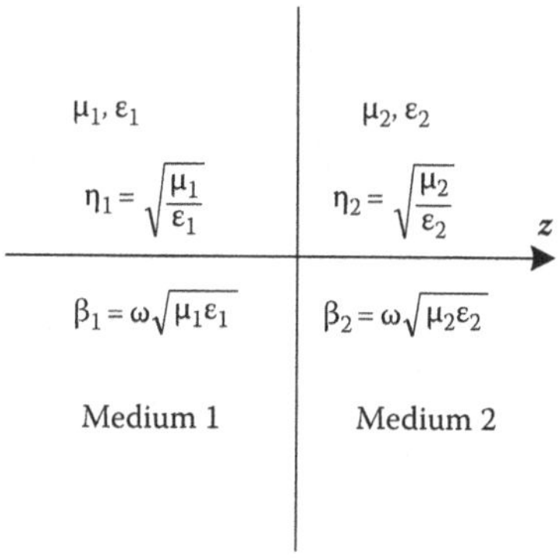

FIGURE 2C.10
Boundary between two dielectrics.

Let this give rise to a reflected wave,

$$\tilde{\mathbf{E}}^{\mathrm{R}}(z) = \tilde{E}_0^- e^{+j\beta_1 z}\hat{\mathbf{x}}, \tag{2C.22}$$

$$\tilde{\mathbf{H}}^{\mathrm{R}}(z) = -\frac{\tilde{E}_0^-}{\eta_1} e^{+j\beta_1 z}\hat{\mathbf{y}} = \tilde{H}_0^- e^{+j\beta_1 z}\hat{\mathbf{y}}, \tag{2C.23}$$

and a transmitted wave:

$$\tilde{\mathbf{E}}^{\mathrm{T}}(z) = \tilde{E}_0^{(2)} e^{-j\beta_2 z}\hat{\mathbf{x}}, \tag{2C.24}$$

$$\tilde{\mathbf{H}}^{\mathrm{T}}(z) = \frac{\tilde{E}_0^{(2)}}{\eta_2} e^{-j\beta_2 z}\hat{\mathbf{y}} = \tilde{H}_0^{(2)} e^{-j\beta_2 z}\hat{\mathbf{y}}. \tag{2C.25}$$

The problem is to determine the reflection coefficient $\Gamma_0 = \tilde{E}_0^-/\tilde{E}_0^+$ and the transmission coefficient $T_0 = \tilde{E}_0^{(2)}/\tilde{E}_0^+$. The boundary conditions give us the required number of equations to solve for Γ_0 and T_0. The boundary conditions at a dielectric–dielectric interface are continuity of tangential components of $\bar{E}$ and $\bar{H}$:

$$\tilde{E}_0^+ + \tilde{E}_0^- = \tilde{E}_0^{(2)}, \tag{2C.26}$$

$$\tilde{H}_0^+ + \tilde{H}_0^- = \tilde{H}_0^{(2)}, \tag{2C.27}$$

or

$$\frac{\tilde{E}_0^+}{\eta_1} - \frac{\tilde{E}_0^-}{\eta_1} = \frac{\tilde{E}_0^{(2)}}{\eta_2}. \tag{2C.28}$$

Using Equations 2C.26 and 2C.28, we obtain

$$\Gamma_0 = \frac{\tilde{E}_0^-}{\tilde{E}_0^+} = \frac{\eta_2 - \eta_1}{\eta_2 + \eta_1} \tag{2C.29}$$

and

$$T_0 = \frac{\tilde{E}_0^{(2)}}{\tilde{E}_0^+} = \frac{2\eta_2}{\eta_2 + \eta_1}. \tag{2C.30}$$

2C.3 Comparison with a Transmission Line Problem

Figure 2C.11a shows the normal incidence from the left of a plane wave on a dielectric–dielectric interface from dielectric 1 on the left to dielectric 2 on the right. Figure 2C.11b and c shows the transmission line analogy and the equivalent transmission line circuit, respectively. The transmission line 2 extends to infinity. On this line, the wave propagates in the +z-direction and there is no reflected wave. The reflection coefficient at $z = \infty$ is zero

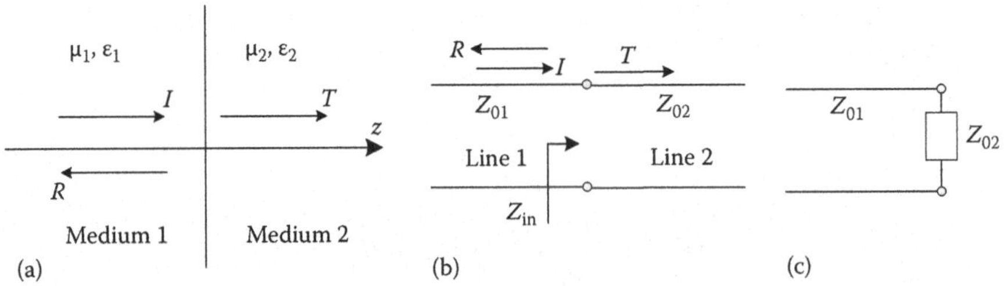

FIGURE 2C.11
Comparison of reflection at a dielectric–dielectric boundary to a transmission line: (a) dielectric interface, (b) transmission line interface, and (c) equivalent load on transmission line.

and $\Gamma_\infty = 0$, and therefore $Z_{in} = Z_{02}$ (see Figure 2C.11c). The reflection coefficient may now be obtained as follows:

$$\Gamma_0 = \frac{Z_{02} - Z_{01}}{Z_{02} + Z_{01}} = \frac{\eta_2 - \eta_1}{\eta_2 + \eta_1}. \tag{2C.31}$$

The purpose of developing the analogy with the transmission lines is to exploit the tools available for transmission line problems for solving the analogous boundary value problems involving plane waves. The Smith chart is one of them. Let us illustrate through a few examples.

Example 2C.2

A 4-GHz uniform plane wave is normally incident from region 1 ($z < 0$, $\varepsilon_1 = 5$, $\mu_1 = 1$, $\sigma_1 = 0$) toward region 2 ($z > 0$, $\varepsilon_2 = 2$, $\mu_2 = 10$, $\sigma_2 = 0$) (Figure 2C.12). Find s (Voltage Standing Wave Ratio) for regions 1 and 2 and the intrinsic impedance (η_{in}) at $z = -0.6$ cm.

For region 2, there is no reflected wave, $\Gamma_\infty = 0$, and $s = (1 + \Gamma)/(1 - \Gamma) = 1$. To determine various quantities in region 1, let us draw an equivalent transmission line (Figure 2C.13). The load is purely resistive and $Z_L = R_L > Z_0$. Therefore, s in region 1 is $Z_L/Z_0 = 5$. To determine η_{in} at -0.6 cm, we make use of the equivalent transmission line of Figure 2C.14 and the Smith chart shown in Figure 2C.15. In this case,

$$\beta_1 = \omega\sqrt{\mu_1\varepsilon_1} = \frac{2\pi \times 4 \times 10^9}{3 \times 10^8}\sqrt{5}, \tag{2C.32}$$

$$\lambda_1 = \frac{2\pi}{\beta_1} = \frac{2\pi \times 3 \times 10^8}{2\pi \times 4 \times 10^9\sqrt{5}} = \frac{0.075 \text{ m}}{\sqrt{5}} = \frac{7.5 \text{ cm}}{\sqrt{5}}, \tag{2C.33}$$

$$\frac{d}{\lambda_1} = \frac{0.6\sqrt{5}}{7.5} = 0.1789. \tag{2C.34}$$

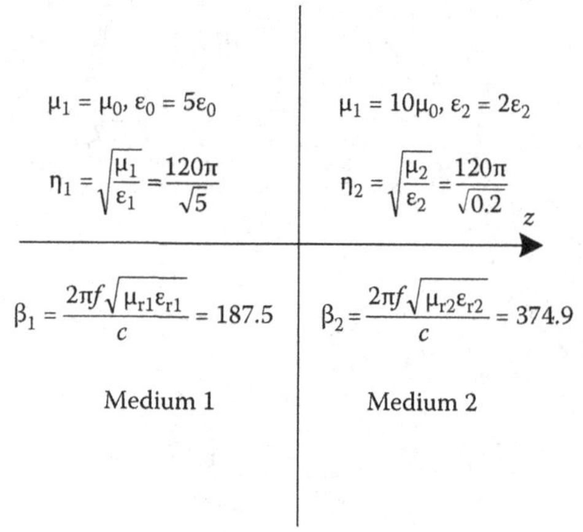

FIGURE 2C.12
Problem geometry for Example 2C.1.

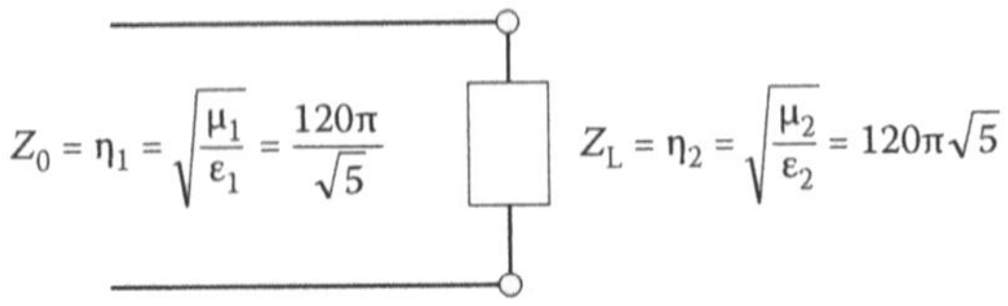

FIGURE 2C.13
Equivalent transmission line circuit for Example 2C.1.

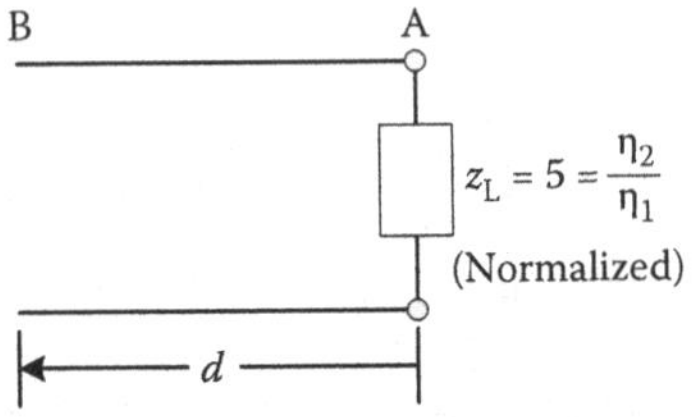

FIGURE 2C.14
Equivalent transmission line circuit with line length *d*.

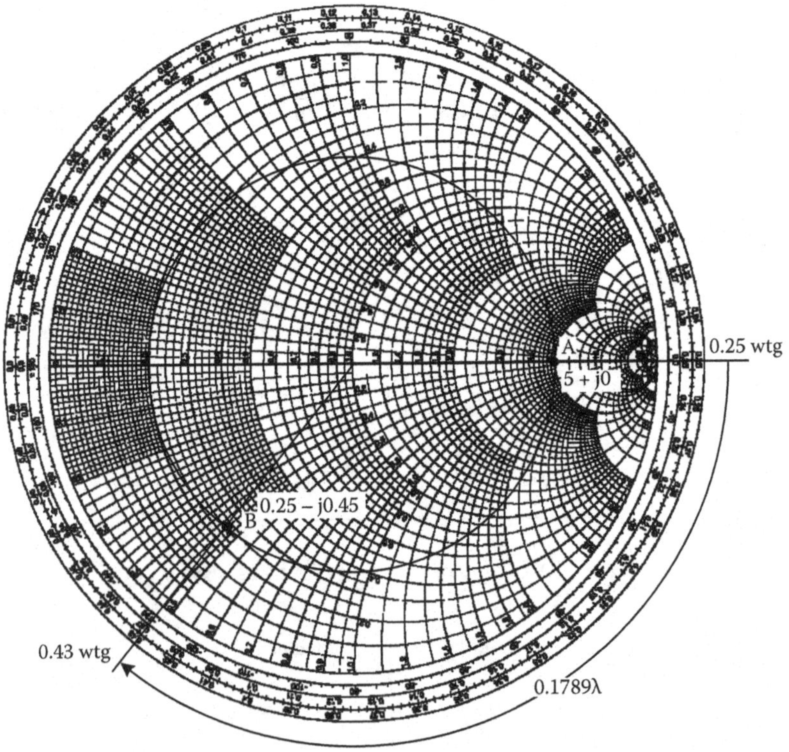

FIGURE 2C.15
Smith chart utilized to solve Example A2.2 dielectric interface problem.

Load point A is located at $r = 5$, $x = 0$ on the Smith chart (0.25 wtg). To reach B on the $|\Gamma|$ circle, move 0.1789 wtg from A on the $|\Gamma|$ circle, that is, go to (0.25 + 0.1789) wtg = 0.4289 wtg. From the Smith chart read $z_B = 0.25 - j0.45$. Therefore,

$$\eta_{in}\left(z = -0.6 \text{ cm}\right) = \frac{120\pi}{\sqrt{5}}\left(0.25 - j0.45\right) = 42.15 - j75.87 = 86.8\angle -60.9°. \quad (2C.35)$$

Example 2C.3

Design a radome. A radome is a dielectric material that covers an antenna and protects it from the weather. The thickness of the material should be such that there are no reflections (see Figure 2C.16).

For a transmission line analogy (see Figure 2C.17), it is obvious that $\eta_{input} = \eta_0$ if $d = \lambda/2$. If $\eta_{input} = \eta_0$, there is a perfect match at interface 1 and there will be no reflections at interface 1.

Consider a numerical example, where a uniform plane wave with $\lambda = 3$ cm in the free space is normally incident on a fiberglass ($\varepsilon_r = 4.9$, $\sigma = 0$) radome. (a) What thickness of fiberglass will produce no reflections? (b) What percentage of the incident energy

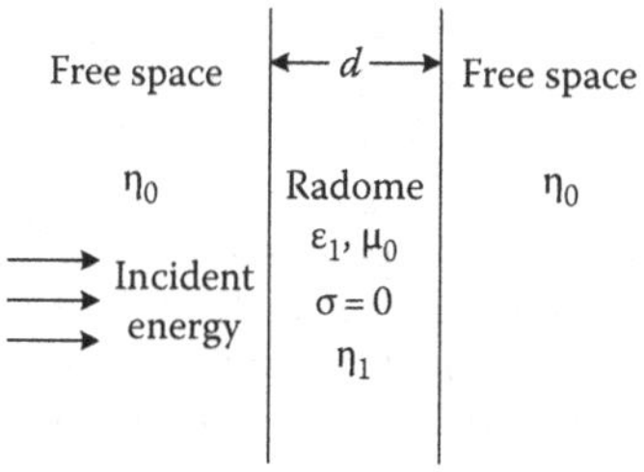

FIGURE 2C.16
Radome analysis geometry.

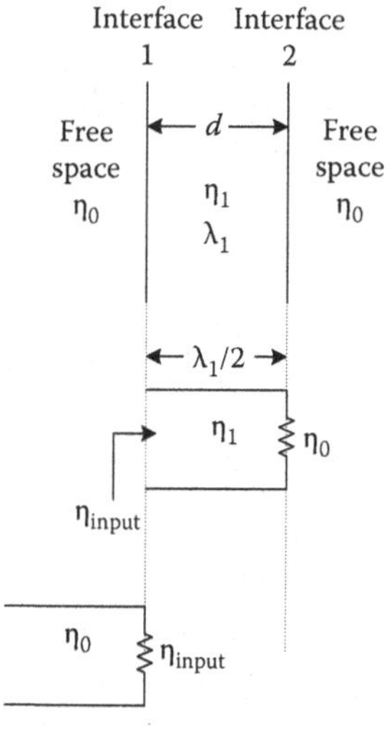

FIGURE 2C.17
Radome equivalent transmission line circuit.

will be transmitted through the fiberglass if the frequency of the incident wave is decreased by 10%?

In this problem, $\varepsilon_1 = 4.9\,\varepsilon_0$, $f = c/\lambda = 3 \times 10^8/3 \times 10^{-2} = 10$ GHz.

$$\beta_1 = \omega\sqrt{\mu_1\varepsilon_1} = \frac{2\pi \times 10 \times 10^9}{3 \times 10^8}\sqrt{4.9} \text{ and } \lambda_1 = \frac{2\pi}{\beta_1} = 1.355 \text{ cm.}$$

a. $d = \lambda_1/2 = 0.678$ cm for no reflections.
b. If $f_{new} = 0.9f$, $\lambda_0(\text{new}) = c/f_{new} = \lambda_0/0.9$.

$$\beta_1(\text{new}) = \omega_{new}\sqrt{\mu_1\varepsilon_1} = 2\pi f_{new}\sqrt{\mu_1\varepsilon_1} = 2\pi(0.9)f\sqrt{\mu_1\varepsilon_1} = 0.9\beta_1 \text{ and } \lambda_1(\text{new}) = \frac{2\pi}{\beta_1(\text{new})} = \frac{\lambda_1}{0.9}.$$

Since the physical width d is the same,

$$\frac{d}{\lambda_1(\text{new})} = \frac{d(0.9)}{\lambda_1} = \frac{(\lambda_1/2)(0.9)}{\lambda_1} = 0.45.$$

Here η_{input} may now be obtained by solving the following transmission line problem (Figure 2C.18). To use the Smith chart (Figure 2C.19), let us use normalized values:

$$Z_0 = \eta_1 = \sqrt{\frac{\mu_1}{\varepsilon_1}} = \frac{120\pi}{\sqrt{4.9}} = \frac{\eta_0}{\sqrt{4.9}}.$$

But $Z_L = \eta_0$, so $z_L = \sqrt{4.9} = 2.214$. This is point A on the Smith chart ($r = 2.214$, $x = 0$). Now draw the $|\Gamma|$ circle and move 0.45 wtg to reach B on the $|\Gamma|$ circle. Read $z_B = 1.6 + j0.85$. $\eta_{input} = \eta_1\,(1.6 + j0.85)$. The equivalent transmission line load $= \eta_1\,(1.6 + j0.85)/\eta_0 = 0.723 + j0.384$. This normalized load of $z_L = 0.723 + j0.384$ is marked as point C on the Smith chart. To get the reflection coefficient magnitude measure OC = 0.26, $|\Gamma|^2 = 0.0676$. The transmitted energy $= 1 - |\Gamma|^2 = 0.9324 = 93.24\%$.

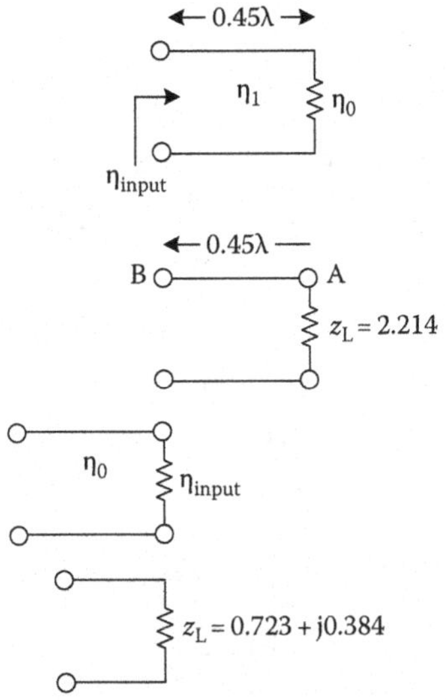

FIGURE 2C.18
Radome equivalent transmission line circuit.

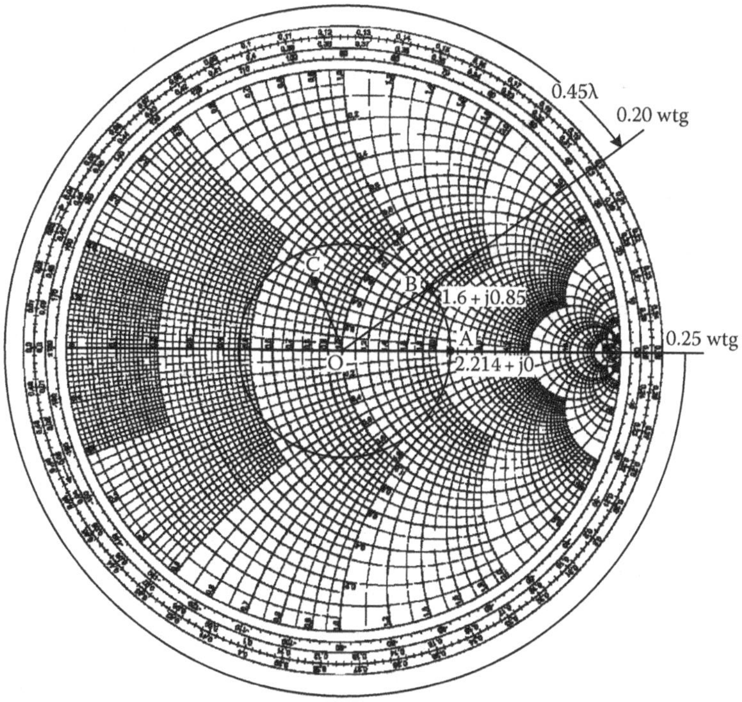

FIGURE 2C.19
Smith chart used in solution of radome problem (Example 2C.1).

2C.4 Example of Use of the Smith Chart for Oblique Incidence [2]

The analogy between plane wave reflection problems for oblique incidence and transmission lines is summarized in Table 2C.2.

Consider the example shown in Figure 2C.20 where we have oblique incidence at an angle of 40° with perpendicular polarization from free space onto two dielectric layers situated within free space at a frequency of 10 GHz.

The incident wave freespace wavelength is

$$\lambda_0 = \frac{c}{f} = \frac{3 \times 10^8}{10 \times 10^9} = 3 \text{ cm}. \tag{2C.36}$$

From Snell's law,

$$\theta_2 = \sin^{-1}\left(\frac{\sin(40^\circ)}{\sqrt{2.54}}\right) = 23.79^\circ, \tag{2C.37}$$

$$\theta_3 = \sin^{-1}\left(\frac{\sin(40^\circ)}{\sqrt{4}}\right) = 18.73^\circ, \tag{2C.38}$$

TABLE 2C.2

A Summary of Plane Wave Reflection (Oblique Incidence) Comparison with the Transmission Line Analogy

Parallelly Polarized Plane Wave	Transmission Line
$E_x = E\cos\theta_m$	V
$H_y = H$	I
$Z_{0m} = \eta_m \cos\theta_m = \sqrt{\frac{\mu_m}{\varepsilon_m}}\cos\theta_m$	$Z_0 = \sqrt{L/C}$
$q_m = k_m\cos\theta_m = \omega\sqrt{\mu_m\varepsilon_m}\cos\theta_m$	$q = \omega\sqrt{LC} = \beta$
Perpendicularly Polarized Plane Wave	**Transmission Line**
$E_y = E$	V
$\lvert H_x\rvert = H\cos\theta_m$	I
$Z_{0m} = \frac{\eta_m}{\cos\theta_m} = \sqrt{\frac{\mu_m}{\varepsilon_m}}\frac{1}{\cos\theta_m}$	$Z_0 = \sqrt{L/C}$
$q_m = k_m\cos\theta_m = \omega\sqrt{\mu_m\varepsilon_m}\cos\theta_m$	$q = \omega\sqrt{LC} = \beta$

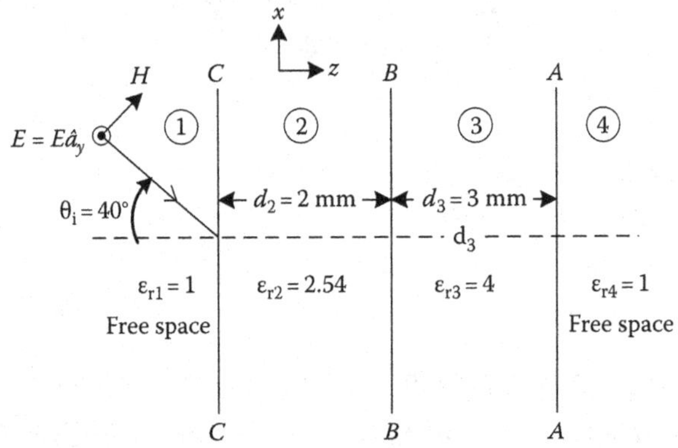

FIGURE 2C.20
Example of use of the Smith chart for oblique incidence.

$$\theta_4 = \sin^{-1}\left(\frac{\sin(40°)}{\sqrt{1}}\right) = 40°. \tag{2C.39}$$

For perpendicular polarization,

$$Z_{01} = \frac{120\pi}{\cos(40°)} = 492\Omega,$$
$$Z_{02} = \frac{120\pi/\sqrt{2.54}}{\cos(\theta_2)} = 258.5\Omega, \tag{2C.40}$$
$$Z_{03} = \frac{120\pi/\sqrt{4}}{\cos(\theta_3)} = 199\Omega.$$

The length in wavelengths of each region

$$\lambda_{mz} = \frac{2\pi}{q_m} = \frac{2\pi}{k_m \cos\theta_m} = \frac{2\pi}{\omega\sqrt{\mu_m \varepsilon_m} \cos\theta_m}, \tag{2C.41}$$

$$\lambda_{mz} = \frac{2\pi}{\omega\sqrt{\mu_0 \varepsilon_0}\sqrt{\varepsilon_r} \cos\theta_m} = \frac{\lambda_0}{\sqrt{\varepsilon_r} \cos\theta_m}, \tag{2C.42}$$

$$\frac{d_2}{\lambda_{2z}} = \frac{0.2\sqrt{2.54}\cos(23.79)}{3} = 0.097, \tag{2C.43}$$

and therefore,

$$\frac{d_3}{\lambda_{3z}} = \frac{0.3\sqrt{4}\cos(18.73)}{3} = 0.189.$$

Now starting at interface AA, we have the equivalent transmission line shown in Figure 2C.21a that is point A on the Smith chart of Figure 2C.22:

$$Z_A = Z_{L3} = \frac{120\pi}{\cos(40°)} = 492\Omega, \tag{2C.44}$$

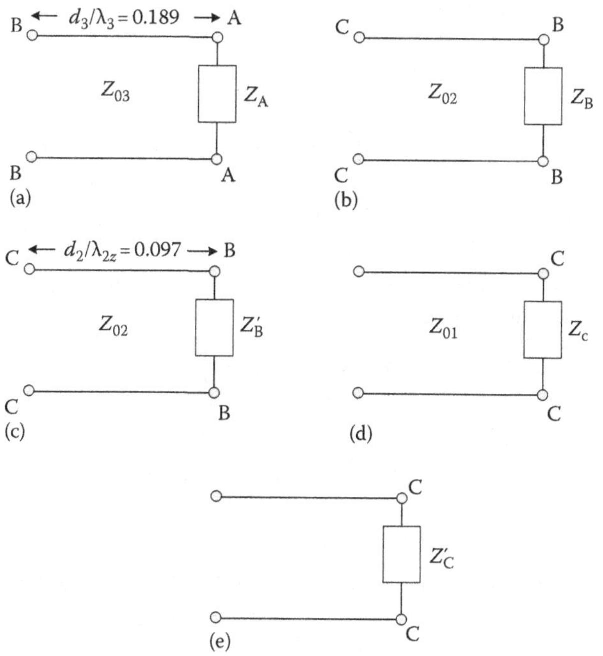

FIGURE 2C.21
Equivalent transmission lines for example problem. (a) Transmission line for 3. (b) Transmission line for 2. (c) Renormalization with Z_{02}. (d) Transmission line 1. (e) Renormalization with Z_{01}.

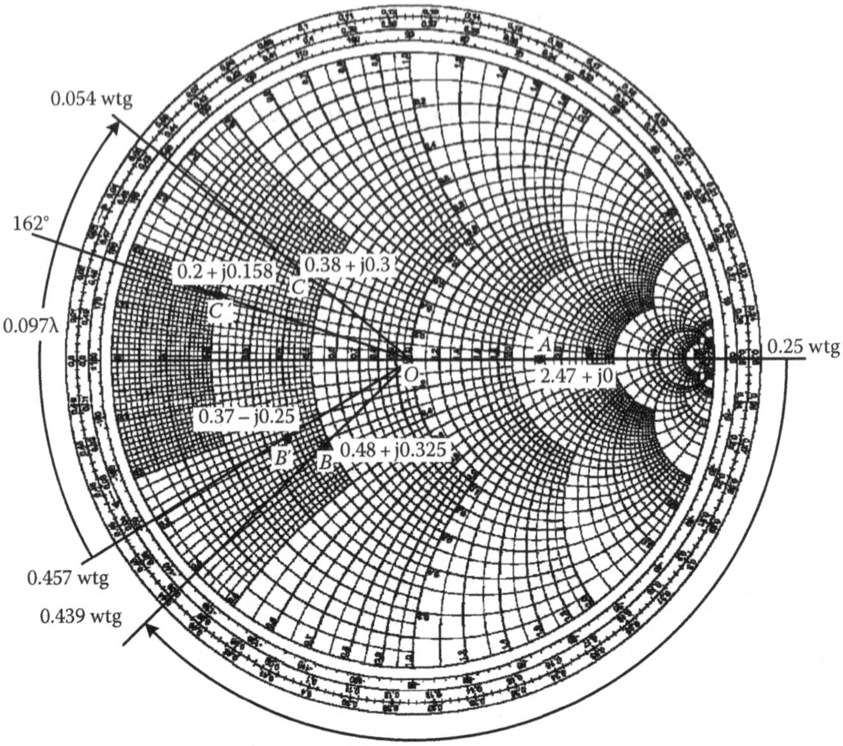

FIGURE 2C.22
Smith chart used in the solution of oblique plane wave reflection from multiple layers.

where

$$Z_{03} = 199\Omega,$$

$$z_A = \frac{492}{199} = 2.47.$$

Also shown in the Smith chart of Figure 2C.22, moving from point A toward generator 0.189 wavelengths, we read (point B):

$$z_B = 0.48 - j0.325, \tag{2C.45}$$

and therefore,

$$Z_B = (0.48 - j0.325)199, \tag{2C.46}$$

which can be seen as point B on the Smith chart and the transmission line of Figure 2C.21b.

Renormalizing,

$$z_{B'} = (0.48 - j0.325)\frac{199}{258.5} = 0.37 - j0.25. \tag{2C.47}$$

This is represented by point B′ on the Smith chart and the transmission line of Figure 2C.21c. Now moving 0.097 wavelengths toward the generator, we have the transmission line analogy shown in Figure 2C.21d and point C on the Smith chart:

$$z_C = 0.38 + j0.3, \tag{2C.48}$$

and therefore

$$Z_C = (0.38 + j0.3)258.5. \tag{2C.49}$$

Renormalizing, we get

$$z_{C'} = (0.38 + j0.3)\frac{258.5}{492} = 0.2 + j0.158. \tag{2C.50}$$

This is represented by point C′ on the Smith chart and the equivalent transmission line of Figure 2C.21e. The reflection coefficient can be read from the Smith chart as

$$\Gamma_s = 0.66\angle 162^\circ \tag{2C.51}$$

and the power reflection coefficient is given as

$$|\Gamma_s|^2 = 0.435. \tag{2C.52}$$

References

1. Hayt, W. H., Jr., *Engineering Electromagnetics*, 5th edition, McGraw-Hill, New York, 1989.
2. Ramo, S., Whinnery, J. R., and Van Duzer, T., *Fields and Waves in Communication Electronics*, Wiley, New York, 1967.

Appendix 2D: Nonuniform Transmission Lines

2D.1 Nonuniform Transmission Lines

A nonuniform transmission line can represent many physical phenomena in an inhomogeneous medium whose medium properties are functions of position. A one-dimensional equivalent ideal transmission line will have the per unit values of series inductance L' and the shunt capacitance C' as functions of one spatial coordinate.

Let

$$L' = L'(z), \tag{2D.1a}$$

$$C' = C'(z). \tag{2D.1b}$$

If ε and μ are homogenous in the transverse plane, then

$$\frac{1}{\sqrt{L'C'}} = \frac{1}{\sqrt{\varepsilon\mu}} = u_{\mathrm{p}}, \tag{2D.2}$$

where u_{p} is the phase velocity.

The first-order coupled differential equations are

$$-\frac{\partial V}{\partial z} = L'(z)\frac{\partial I}{\partial t}, \tag{2D.3a}$$

$$-\frac{\partial I}{\partial z} = C'(z)\frac{\partial V}{\partial t}. \tag{2D.3b}$$

In the phasor form,

$$-\frac{\partial \tilde{V}}{\partial z} = L'(z)\mathrm{j}\omega\tilde{I}, \tag{2D.4a}$$

$$-\frac{\partial \tilde{I}}{\partial z} = C'(z)\mathrm{j}\omega\tilde{V}. \tag{2D.4b}$$

From Equations 2D.3a and 2D.3b, we can obtain

$$\frac{\partial^2 V}{\partial z^2} - \frac{1}{L'(z)}\frac{\partial L'}{\partial z}\frac{\partial V}{\partial z} - L'(z)C'(z)\frac{\partial^2 V}{\partial t^2} = 0. \tag{2D.5a}$$

The equation for I can be obtained similarly as

$$\frac{\partial^2 I}{\partial z^2} - \frac{1}{C'(z)}\frac{\partial C'}{\partial z}\frac{\partial I}{\partial z} - L'(z)C'(z)\frac{\partial^2 I}{\partial t^2} = 0. \tag{2D.5b}$$

The phasor form of Equations 2D.5a and 2D.5b are

$$\frac{\partial^2 \tilde{V}}{\partial z^2} - \frac{1}{L'(z)}\frac{\partial L'}{\partial z}\frac{\partial \tilde{V}}{\partial z} + \omega^2 L'(z)C'(z)\tilde{V} = 0, \tag{2D.6a}$$

$$\frac{\partial^2 \tilde{I}}{\partial z^2} - \frac{1}{C'(z)}\frac{\partial C'}{\partial z}\frac{\partial \tilde{I}}{\partial z} + \omega^2 L'(z)C'(z)\tilde{I} = 0. \tag{2D.6b}$$

Analytical solutions can be obtained for an exponential transmission line and this aspect is discussed in the following section.

2D.2 Exponential Transmission Line

In this section, we define an exponential transmission line where both the inductance and capacitance are exponential functions as shown below.

Let

$$L' = L_0' e^{qz}, \tag{2D.7a}$$

$$C' = C_0' e^{-qz}. \tag{2D.7b}$$

Note that

$$L'C' = L_0' C_0' = \varepsilon\mu. \tag{2D.7c}$$

Equations 2D.5a and 2D.5b assume the form of

$$\frac{\partial^2 V}{\partial z^2} - q\frac{\partial V}{\partial z} - L_0' C_0' \frac{\partial^2 V}{\partial t^2} = 0, \tag{2D.8}$$

$$\frac{\partial^2 I}{\partial z^2} + q\frac{\partial I}{\partial z} - L_0' C_0' \frac{\partial^2 I}{\partial t^2} = 0. \tag{2D.9}$$

For harmonic variation in space and time, let

$$V(z,t) = V_0 e^{j(\omega t - k_v z)}. \tag{2D.10}$$

From Equations 2D.8 and 2D.10, we obtain

$$\begin{aligned}&(-jk_v)^2 - q(-jk_v) - L_0'C_0'(j\omega)^2 = 0,\\&-k_v^2 + jk_v q + \omega^2 L_0'C_0' = 0,\\&\omega^2 L_0'C_0' = k_v^2 - jk_v q.\end{aligned} \tag{2D.11}$$

For a real ω, k_v will be complex, so

$$k_v = \beta_v - j\alpha_v. \tag{2D.12}$$

By substituting Equation 2D.12 into Equation 2D.11, we obtain

$$\begin{aligned}&\omega^2 L_0'C_0' = (\beta_v - j\alpha_v)^2 - j(\beta_v - j\alpha_v)q,\\&\omega^2 L_0'C_0' = \beta_v^2 - \alpha_v^2 - \alpha_v q,\end{aligned} \tag{2D.13a}$$

$$-\beta_v(2\alpha_v + q) = 0. \tag{2D.13b}$$

Thus, we have

$$\alpha_v = \frac{-q}{2}. \tag{2D.14}$$

Substituting Equation 2D.14 into Equation 2D.13a, we obtain

$$\begin{aligned}\beta_v^2 - \frac{q^2}{4} + \frac{q^2}{2} &= \omega^2 L_0'C_0',\\ \beta_v &= \pm\beta,\end{aligned} \tag{2D.15a}$$

where

$$\beta = \sqrt{\omega^2 L_0'C_0' - \frac{q^2}{4}}. \tag{2D.15b}$$

Thus $k_v = \pm\beta - j\alpha_v = \pm\beta + j(q/2)$,

$$e^{-jk_v z} = e^{\pm j\beta z} e^{\frac{q}{2}z}. \tag{2D.16}$$

Let

$$\omega_c = \frac{q}{2\sqrt{L_0'C_0'}}, \tag{2D.17a}$$

$$\beta_0 = \omega\sqrt{L_0'C_0'}, \tag{2D.17b}$$

then

$$\beta_v = \beta_0\sqrt{1-\frac{\omega_c^2}{\omega^2}}, \tag{2D.18}$$

and

$$e^{-jk_v z} = e^{\left[\pm j\beta_0\sqrt{1-\frac{\omega_c^2}{\omega^2}}+\frac{q}{2}\right]z}. \tag{2D.19}$$

Therefore, the time-harmonic solution for an exponential transmission line described by Equation 2D.8 is given as

$$\tilde{V}(z) = V_0^+ e^{(q/2)z} e^{-j\beta z} + V_0^- e^{(q/2)z} e^{+j\beta z}. \tag{2D.20}$$

The first term on the RHS is the positive traveling wave and the second term is the negative traveling wave given by

$$\tilde{V}^+(z) = V_0^+ e^{(q/2)z} e^{-j\beta z} \left(\text{positive traveling wave}\right), \tag{2D.21}$$

$$\tilde{V}^-(z) = V_0^- e^{(q/2)z} e^{+j\beta z} \left(\text{negative traveling wave}\right). \tag{2D.22}$$

The solution for the currents can be obtained in a similar fashion by solving Equation 2D.9 and it can be shown that

$$\tilde{I}(z) = I_0^+ e^{-(q/2)z} e^{-j\beta z} + I_0^- e^{-(q/2)z} e^{+j\beta z}, \tag{2D.23a}$$

$$\tilde{I}^+(z) = I_0^+ e^{-(q/2)z} e^{-j\beta z} \left(\text{positive traveling wave}\right), \tag{2D.23b}$$

$$\tilde{I}^-(z) = I_0^- e^{-(q/2)z} e^{+j\beta z} \left(\text{negative traveling wave}\right). \tag{2D.23c}$$

The relationship between $\tilde{V}_0^+$ and $\tilde{I}_0^+$, and $\tilde{V}_0^-$ and $\tilde{I}_0^-$ can be obtained from Equations 2D.4a, 2D.21, and 2D.7a.

$$\begin{aligned}
\tilde{I}^+(z) &= -\frac{1}{j\omega L'(z)}\frac{\partial \tilde{V}^+(z)}{\partial z},\\
\tilde{I}^+(z) &= -\frac{1}{j\omega L'(z)}\left(\frac{q}{2} - j\beta\right) V_0^+ e^{(q/2)z} e^{-j\beta z},\\
\tilde{I}^+(z) &= -\frac{q/2 - j\beta}{j\omega L_0' e^{qz}} V_0^+ e^{(q/2)z} e^{-j\beta z},\\
\tilde{I}^+(z) &= -\frac{V_0^+\left(q/2 - j\beta\right)}{j\omega L_0'} e^{-(q/2)z} e^{-j\beta z},\\
\tilde{I}^+(z) &= \tilde{I}_0^+ e^{-(q/2)z} e^{-j\beta z},\\
Z_0^+ &= \frac{V_0^+}{I_0^+} = \frac{-j\omega L_0'}{q/2 - j\beta} = \frac{\omega L_0'}{\beta + j(q/2)}.
\end{aligned} \tag{2D.24}$$

Similarly,

$$Z_0^- = -\frac{V_0^-}{I_0^-} = \frac{\omega L_0'}{\beta - j(q/2)}. \tag{2D.25}$$

One should note that in a nonuniform transmission line, the characteristic impedances for the oppositely traveling waves are not the same:

$$Z_0^+ \neq Z_0^-. \tag{2D.26}$$

Furthermore,

$$\begin{aligned} Z^+(z) &= \frac{V^+(z)}{I^+(z)} = \frac{V_0^+ e^{(q/2)z} e^{-j\beta z}}{I_0^+ e^{-(q/2)z} e^{-j\beta z}}, \\ Z^+(z) &= Z_0^+ e^{qz}. \end{aligned} \tag{2D.27}$$

Similarly,

$$Z^-(z) = Z_0^- e^{qz}. \tag{2D.28}$$

Substituting for β from Equation 2D.18 Equation 2D.24 can be written as

$$\begin{aligned} Z_0^+ &= \frac{\omega L_0'}{\beta + j(q/2)} = \frac{\omega L_0'}{\beta_0\left[1-\omega_c^2/\omega^2\right] + j\omega_c\sqrt{L_0'C_0'}}, \\ Z_0^+ &= \frac{\omega L_0'}{\beta_0\left\{\left[1-\omega_c^2/\omega^2\right]^{1/2} + j(\omega_c/\omega)\right\}}, \\ Z_0^+ &= \sqrt{\frac{L_0'}{C_0'}}\frac{\left[\left(1-\omega_c^2/\omega^2\right)^{1/2} - j(\omega_c/\omega)\right]}{1-\omega_c^2/\omega^2 + \omega_c^2/\omega^2}, \\ Z_0^+ &= Z_0\left[\left(1-\frac{\omega_c^2}{\omega^2}\right)^{1/2} - j\frac{\omega_c}{\omega}\right]. \end{aligned} \tag{2D.29a}$$

Similarly, we can show that

$$Z_0^- = Z_0\left[\left(1-\frac{\omega_c^2}{\omega^2}\right)^{1/2} + j\frac{\omega_c}{\omega}\right], \tag{2D.29b}$$

where Z_0 is the nominal characteristic impedance given by

$$Z_0 = \sqrt{\frac{L_0'}{C_0'}}. \tag{2D.30}$$

For a uniform line $q = \omega_c = 0$, all the formulas of a uniform line can be obtained from

$$Z_0^+ = Z_0^- = \sqrt{\frac{L'}{C'}}\left(\text{uniform line } \omega_c = 0\right). \tag{2D.31}$$

2D.3 The Input Impedance

$$Z(z)=\frac{\tilde{V}_0^{+}e^{(q/2)z}e^{-j\beta z}\left[1+\Gamma_{0V}e^{j2\beta z}\right]}{\tilde{I}_0^{+}e^{-(q/2)z}e^{-j\beta z}\left[1+\Gamma_{0I}e^{j2\beta z}\right]},$$
$$Z(z)=Z_0^{+}e^{qz}\frac{\left[1+\Gamma_{0V}e^{j2\beta z}\right]}{\left[1+\Gamma_{0I}e^{j2\beta z}\right]}, \tag{2D.32}$$

where

$$\Gamma_{0V}=\frac{V_0^{-}}{V_0^{+}}, \tag{2D.33}$$

$$\Gamma_{0I}=\frac{I_0^{-}}{I_0^{+}}. \tag{2D.34}$$

Note that

$$\Gamma_{0V}=\frac{V_0^{-}}{V_0^{+}}=-\frac{Z_0^{-}I_0^{-}}{Z_0^{+}I_0^{+}},$$
$$\Gamma_{0V}=-\frac{Z_0^{-}}{Z_0^{+}}\Gamma_{0I}, \tag{2D.35}$$
$$\Gamma_{0I}=-\frac{Z_0^{+}}{Z_0^{-}}\Gamma_{0V}.$$

2D.4 Arbitrary Load at $z = 0$

$$Z_L=\frac{V_L}{I_L}=\frac{V_0^{+}+V_0^{-}}{I_0^{+}+I_0^{-}},$$
$$Z_L=\frac{V_0^{+}}{I_0^{+}}\left(\frac{1+\Gamma_{0V}}{1+\Gamma_{0I}}\right),$$
$$Z_L=Z_0^{+}\left(\frac{1+\Gamma_{0V}}{1+\Gamma_{0I}}\right),$$
$$Z_L=Z_0^{+}\left[\frac{1+\Gamma_{0V}}{1+\left(Z_0^{+}/Z_0^{-}\right)\Gamma_{0V}}\right], \tag{2D.36}$$
$$Z_L=Z_0^{+}Z_0^{-}\left[\frac{1+\Gamma_{0V}}{Z_0^{-}-Z_0^{+}\Gamma_{0V}}\right],$$
$$Z_L\left(Z_0^{-}-Z_0^{+}\Gamma_{0V}\right)=Z_0^{+}Z_0^{-}\left(1+\Gamma_{0V}\right),$$
$$\Gamma_{0V}\left[Z_0^{+}Z_0^{+}+Z_LZ_0^{+}\right]=Z_LZ_0^{-}-Z_0^{+}Z_0^{-},$$
$$\Gamma_{0V}=\frac{Z_0^{-}\left[Z_L-Z_0^{+}\right]}{Z_0^{+}\left[Z_L-Z_0^{-}\right]}.$$

Similarly,

$$\Gamma_{0I} = -\frac{\left[Z_L - Z_0^+\right]}{\left[Z_L - Z_0^-\right]}. \tag{2D.37}$$

From Equation 2D.32, the input impedance at $z = -h$ is given by

$$\begin{aligned}
Z(h) &= Z_0^+ e^{-qh}\left[\frac{1+\Gamma_{0V}e^{-j2\beta h}}{1+\Gamma_{0I}e^{-j2\beta h}}\right],\\
Z(h) &= Z_0^+ e^{-qh}\left[\frac{e^{j\beta h}+\Gamma_{0V}e^{-j\beta h}}{e^{j\beta h}+\Gamma_{0I}e^{-j\beta h}}\right],\\
Z(h) &= \frac{Z_0^+ e^{-qh}\left(Z_L+Z_0^-\right)}{Z_0^+\left(Z_L+Z_0^-\right)}\,\frac{Z_0^+\left(Z_L+Z_0^-\right)e^{j\beta h}+Z_0^-\left(Z_L+Z_0^+\right)e^{-j\beta h}}{\left(Z_L+Z_0^-\right)e^{j\beta h}-\left(Z_L+Z_0^+\right)e^{-j2\beta h}}.
\end{aligned} \tag{2D.38}$$

The above can be written as

$$\begin{aligned}
Z(h) &= e^{-qh}\frac{N}{D},\\
N &= \cos\beta h\left[Z_0^+\left(Z_L+Z_0^-\right)+Z_0^-\left(Z_L-Z_0^+\right)\right]+j\sin\beta h\left[Z_0^+\left(Z_L+Z_0^-\right)-Z_0^-\left(Z_L-Z_0^+\right)\right],\\
N &= \cos\beta h\left[Z_L\left(Z_0^++Z_0^-\right)\right]+j\sin\beta h\left[Z_0^+Z_L-Z_0^-Z_L+2Z_0^+Z_0^-\right],\\
D &= \cos\beta h\left[Z_L+Z_0^-+Z_L+Z_0^+\right]+j\sin\beta h\left[Z_L+Z_0^-+Z_L-Z_0^+\right],\\
D &= \cos\beta h\left[Z_0^++Z_0^-\right]+j\sin\beta h\left[2Z_L+Z_0^--Z_0^+\right],\\
Z(h) &= e^{-qh}\frac{Z_L\left(Z_0^++Z_0^-\right)}{\left(Z_0^++Z_0^-\right)}\left\{\frac{1+j\tan\beta h\left[\left(Z_0^+Z_L-Z_0^-Z_L+2Z_0^+Z_0^-\right)/Z_L\left(Z_0^++Z_0^-\right)\right]}{1+j\tan\beta h\left[\left(2Z_L+Z_0^--Z_0^+\right)/\left(Z_0^++Z_0^-\right)\right]}\right\},\\
Z(h) &= e^{-qh}\frac{\left(Z_0^++Z_0^-\right)}{2}\left\{\frac{Z_L+j\tan\beta h\left[\left(Z_0^+Z_L-Z_0^-Z_L+2Z_0^+Z_0^-\right)/\left(Z_0^++Z_0^-\right)\right]}{\left(Z_0^++Z_0^-\right)/2+j\tan\beta h\left[\left(2Z_L+Z_0^--Z_0^+\right)/2\right]}\right\}.
\end{aligned} \tag{2D.39}$$

Appendix 4A: Calculation of Losses in a Good Conductor at High Frequencies: Surface Resistance R_S

From Equation 2.14, the surface resistance R_S is written as

$$R_S = \sqrt{\frac{\omega\mu}{2\sigma}} = \sqrt{\frac{\pi f \mu}{\sigma}} = \frac{1}{\sigma\delta}.$$

The surface resistance is the AC resistance of a rectangular, good conductor material of 1 m long and 1 m wide, at high frequencies. In Figure 2.2, if $l = 1$ and $b = 1$, from Equation 2.38, $R_{AC} = 1/\sigma\delta$. R_S is given in the unit of Ω per square, but it increases with l and decreases as b increases. As l increases, the voltage increases ($\tilde{E}$ is in the z-direction). As b increases, the current $\tilde{I}$ increases. In computing fields in the dielectric medium at the interface of a dielectric–conductor medium, if we make the idealizing assumption that the conductor is a PEC, the current in the conductor is idealized as a surface current $\tilde{K}$ on the interface. Thus, if $b = 1$, the current is K and the power loss per meter square is

$$\frac{1}{2} I^2 R_{AC} = \frac{1}{2} K^2 R_S.$$

From the boundary condition at the PEC we obtain

$$|\tilde{K}| = |\tilde{H}|.$$

Thus, the power loss per meter square is

$$S_R = \frac{1}{2} R_S |\tilde{H}|^2 \left(\mathrm{W/m^2}\right).$$

The total power loss over a surface S is given by

$$W_L = \iint_S \frac{1}{2} R_S |\tilde{H}|^2 \, ds.$$

This is the basis of Equation 4.14.

Appendix 6A: On Restricted Fourier Series Expansion

A periodic function expressed by

$$f(t) = f(t+T), \tag{6A.1}$$

where T is the period, may be expressed in Fourier series as

$$f(t) = a_0 + \sum_{n=1}^{\infty} a_n \cos(n\omega_0 t) + \sum_{n=1}^{\infty} b_n \sin(n\omega_0 t). \tag{6A.2}$$

In Equation 6A.2,

$$\omega_0 = \frac{2\pi}{T}, \tag{6A.3}$$

$$a_0 = \frac{1}{T} \int_{-T/2}^{T/2} f(t)\mathrm{d}t, \tag{6A.4}$$

$$a_n = \frac{2}{T} \int_{-T/2}^{T/2} f(t)\cos(n\omega_0 t)\mathrm{d}t, \tag{6A.5}$$

$$b_n = \frac{2}{T} \int_{-T/2}^{T/2} f(t)\sin(n\omega_0 t)\mathrm{d}t. \tag{6A.6}$$

In trying to solve the Laplace equation in rectangular coordinates with specified boundary conditions, we need Fourier series expansion of an arbitrary function defined over an interval. The boundary conditions may require use of sine terms-only, odd harmonics-only, and so on. The following example [1] illustrates how the function will look in its entire period and what the period is.

Example 6A.1

A function $x(t)$ defined over a finite range $0 < t < t_1$ is shown in Figure 6A.1.

Sketch the various possible periodic continuations of the function $x(t)$ such that the Fourier series will have

A. odd harmonics only.
B. sine terms only.
C. cosines and odd harmonics only.
D. sines and odd harmonics only.

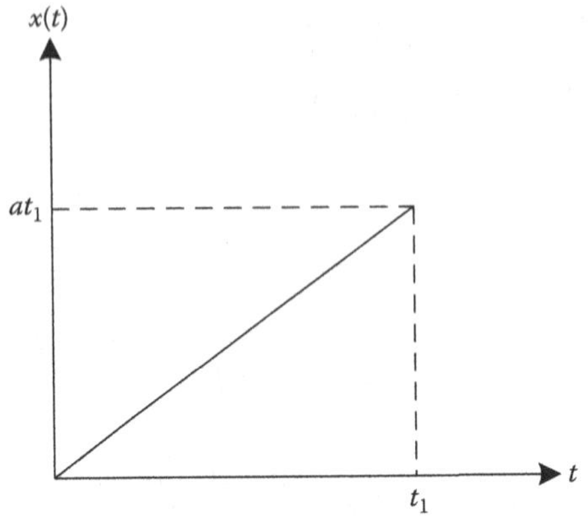

FIGURE 6A.1
A function $x(t)$ defined over the interval $0 < t < t_1$.

The period and continuation of the function in the period are adjusted so as to satisfy the requirements specified.

Solution

A. For odd harmonics-only, rotational symmetry is required, that is,

$$x(t) = -x\left(t \mp \frac{T}{2}\right). \tag{6A.7}$$

The periodic continuation is as shown in Figure 6A.2 with period $T = 2t_1$.

B. For sine terms-only, odd symmetry is required, that is,

$$x(t) = -x(-t). \tag{6A.8}$$

The periodic continuation is as shown in Figure 6A.3 with period $T = 2t_1$ sketched from $-t_1$ to t_1.

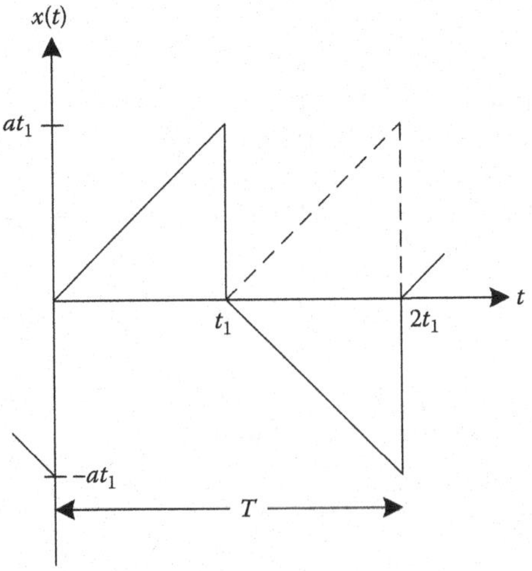

FIGURE 6A.2
$x(t)$ with odd harmonics-only. The dashed curve is $x(t - T/2)$.

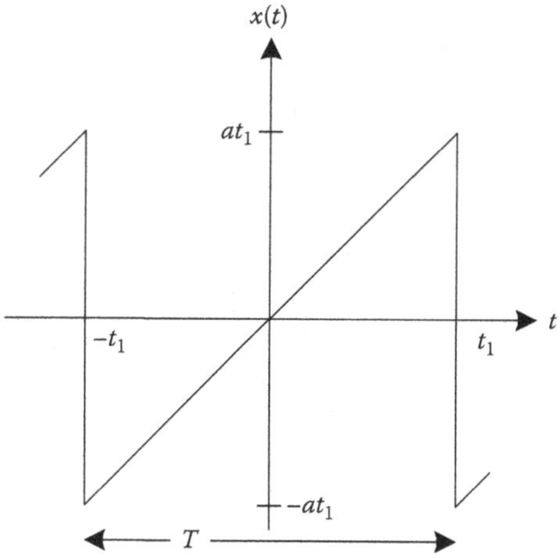

FIGURE 6A.3
$x(t)$ with sine terms-only.

C. For cosines and odd harmonics-only, even symmetry and rotational symmetry are required, that is,

$$x(t) = x(-t),$$
$$x(t) = -x\left(t \mp \frac{T}{2}\right). \tag{6A.9}$$

The periodic continuation is as shown in Figure 6A.4 with period $T = 4t_1$ sketched from $-t_1$ to $3t_1$.

D. For sines and odd harmonics-only, odd symmetry and rotational symmetry are required. That is,

$$x(t) = -x(-t),$$
$$x(t) = -x\left(t \mp \frac{T}{2}\right). \tag{6A.10}$$

The periodic continuation is as shown in Figure 6A.5 with period $T = 4t_1$ sketched from $-t_1$ to $3t_1$.

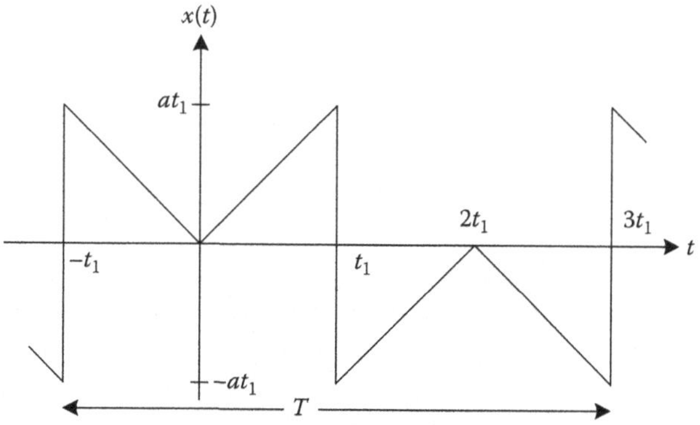

FIGURE 6A.4
$x(t)$ with cosine terms and odd harmonics-only.

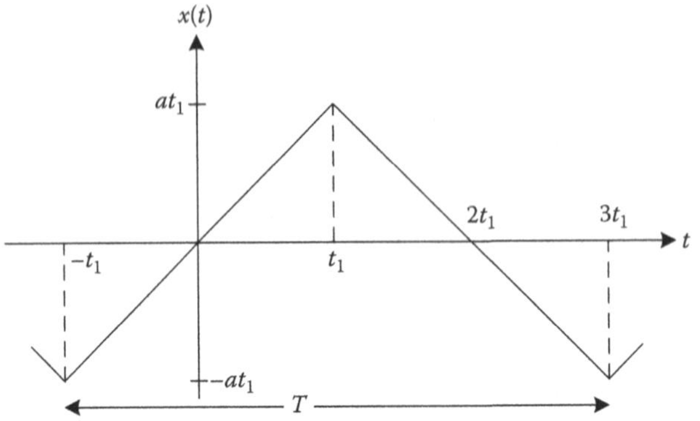

FIGURE 6A.5
$x(t)$ with sine terms and odd harmonics-only.

As an example of the effect of the periodic continuation on the evaluation of the Fourier coefficients, let us take up further the B part for which the period is $2t_1$. Therefore,

$$b_n = \frac{2}{2t_1}\int_{-t_1}^{t_1} f(t)\sin(n\omega_0 t)\mathrm{d}t. \tag{6A.11}$$

where $\omega_0 = 2\pi/2t_1 = \pi/t_1$. Since $f(t)$ is odd and $\sin(n\omega_0 t)$ is odd, the integrand $f(t)\sin(n\omega_0 t)$ is even for all integral values of n, and therefore,

$$\begin{aligned} b_n &= \frac{2}{2t_1}(2)\int_0^{t_1} f(t)\sin(n\omega_0 t)\mathrm{d}t \\ &= \frac{2}{t_1}\int_0^{t_1} f(t)\sin(n\omega_0 t)\mathrm{d}t, \end{aligned} \tag{6A.12}$$

$$b_n = \frac{2}{half\ period}\int_0^{half\ period} f(t)\sin(n\omega_0 t)\mathrm{d}t. \tag{6A.13}$$

Reference

1. Guillemin, E. A., *The Mathematics of Circuit Analysis*, Wiley, New York, 1949.

Appendix 7A: Two- and Three-Dimensional Green's Functions

7A.1 Introduction

We discussed Green's function briefly (Section 7.2). The one-dimensional Green's function of the Laplace equation, with Dirichlet boundary conditions, is the solution of the differential equation

$$\frac{d^2G}{dx^2} = -\delta(x - x'), \quad 0 < x < L, \tag{7A.1a}$$

subject to the boundary conditions

$$G = 0, \quad x = 0, \tag{7A.1b}$$

$$G = 0, \quad x = L, \tag{7A.1c}$$

and is shown to be

$$G = G_1 = x\frac{L - x'}{L}, \quad 0 < x < x', \tag{7A.2a}$$

$$G = G_2 = \frac{x'(L - x)}{L}, \quad x' < x < L. \tag{7A.2b}$$

For an arbitrary input $f(x)$, the response $y(x)$ can be obtained from a superposition integral

$$y(x) = \int_0^L f(x')G(x,x')dx'. \tag{7A.3}$$

The differential equation for y is

$$\frac{d^2y}{dx^2} = -f(x), \quad 0 < x < L, \tag{7A.4a}$$

and the boundary conditions are

$$y = 0, \quad x = 0, \tag{7A.4b}$$

$$y = 0, \quad x = L. \tag{7A.4c}$$

Equation 7A.3 for y is the solution to Equation 7A.4, where G is given by Equation 7A.2. Equation 7A.3 may be written more explicitly, by considering the solution in the two domains $0 \le x \le x'$ and $x' \le x \le L$:

$$y(x) = \int_0^L \frac{x(L-x')}{L} f(x')\mathrm{d}x', \quad 0 \le x \le x', \tag{7A.5a}$$

$$y(x) = \int_0^L \frac{x'(L-x)}{L} f(x')\mathrm{d}x', \quad x' \le x \le L. \tag{7A.5b}$$

Note: Some books define the Green's function using a positive impulse for the input strength:

$$\frac{\mathrm{d}^2 G}{\mathrm{d}x^2} = \delta(x - x'). \tag{7A.6}$$

Then the solution will be of opposite sign to Equations 7A.2a and 7A.2b :

$$G = G_1 = x\frac{x' - L}{L}, \quad 0 < x < x', \tag{7A.7a}$$

$$G = G_2 = x'\frac{x - L}{L}, \quad x' < x < L. \tag{7A.7b}$$

We also considered the one-dimensional Helmholtz equation, with Dirichlet boundary conditions,

$$\frac{\mathrm{d}^2 G}{\mathrm{d}x^2} + k^2 G = -\delta(x - x'), \tag{7A.8a}$$

$$G = 0, \quad x = 0, \tag{7A.8b}$$

$$G = 0, \quad x = L, \tag{7A.8c}$$

$$G = G_1 = \frac{1}{k}\frac{\sin kx \sin k(L - x')}{\sin kL}, \quad 0 < x < x', \tag{7A.9a}$$

$$G = G_2 = \frac{1}{k}\frac{\sin kx' \sin k(L - x)}{\sin kL}, \quad x' < x < L. \tag{7A.9b}$$

7A.2 Alternate Form: Infinite Series

The procedure used in Section 7.2 is to solve the homogeneous equation in the two domains excluding the point $x = x'$ where the impulse is applied. Using the boundary conditions, and the source conditions, the undetermined constants were determined. The result is Green's function in closed form. An alternative form will now be determined using eigenfunction expansion.

Let us solve Equation 7A.6 using this technique. The solution will be written as an infinite series

$$G(x,x') = \sum_{n=1}^{\infty} a_n(x') \sin\frac{n\pi x}{L}. \tag{7A.10}$$

Note that each of the terms on the RHS of Equation 7A.10 satisfies the boundary conditions of $G = 0$ at $x = 0$ or L. From Equation 7A.10, we obtain

$$\frac{d^2G}{dx^2} = \sum \left(-\frac{n\pi}{L}\right)^2 a_n(x') \sin\frac{n\pi x}{L}. \tag{7A.11}$$

By substituting Equation 7A.11 into Equation 7A.6, multiplying by $\sin m\pi x/L$, and then integrating from 0 to L, we obtain

$$\begin{aligned} -\sum \left(\frac{n\pi}{L}\right)^2 a_n(x') \int_0^L \sin\frac{n\pi x}{L} \sin\frac{m\pi x}{L} dx \\ = \int_0^L \delta(x - x') \sin\frac{m\pi x}{L} dx. \end{aligned} \tag{7A.12}$$

From the orthogonality property,

$$\begin{aligned} \int_0^L \sin\frac{n\pi x}{L} \sin\frac{m\pi x}{L} dx &= 0, \quad m \neq n \\ &= \frac{L}{2}, \quad m = n. \end{aligned} \tag{7A.13}$$

Thus, from Equation 7A.12, we get

$$-\left(\frac{m\pi}{L}\right)^2 a_m(x') \frac{L}{2} = \sin\frac{m\pi x'}{L},$$

$$a_m(x') = -\frac{2L}{m^2\pi^2} \sin\frac{m\pi x'}{L}.$$

Thus, we have

$$G(x,x') = -\frac{2L}{\pi^2} \sum_{n=1}^{\infty} \frac{1}{n^2} \sin\frac{n\pi x'}{L} \sin\frac{n\pi x}{L}. \tag{7A.14}$$

Equation 7A.14 is an alternative form to Equations 7A.7a and 7A.7b, although Equation 7A.14 appears as an infinite series. The solution of

$$\frac{d^2y}{dx^2} = f(x) \tag{7A.15}$$

can be written as

$$y(x) = -\frac{2L}{\pi^2}\sum_{n=1}^{\infty}\frac{1}{n^2}\sin\frac{n\pi x}{L}\int_0^L \sin\frac{n\pi x'}{L} f(x')dx'. \tag{7A.16}$$

The solution converges due to the term $1/n^2$ in Equation 7A.16.

7A.3 Sturm–Liouville Operator

A generalization of the series method for a one-dimensional problem with a general second-order differential equation is formulated in terms of a Sturm–Liouville operator L:

$$\left[L + \lambda r(x)\right]y = f(x), \tag{7A.17a}$$

where

$$L = \left\{\frac{d}{dx}\left[p(x)\frac{d}{dx}\right] - q(x)\right\}. \tag{7A.17b}$$

Let ψ_n be a complete set of orthonormal eigenfunctions for the L operator, that is,

$$\left[L + \lambda_n r(x)\right]\Psi_n(x) = 0 \tag{7A.18}$$

subject to the same boundary condition as the original problem of Equation 7A.17a. If $G(x, x')$ is Green's function

$$\left[L + \lambda r(x)\right]G(x,x') = \delta(x - x') \tag{7A.19}$$

subject to the same boundary condition as the original problem, then

$$G(x,x') = \sum_n \frac{\Psi_n(x')\Psi_n(x)}{\lambda - \lambda_n}. \tag{7A.20}$$

We will show two examples [1] of series form of Green's function for two particular problems obtained from Equation 7A.20.

Example 1

$$\frac{d^2G}{dx^2} = \delta(x - x'). \tag{7A.6}$$

Equation 7A.6 is a particular case of Equation 7A.19 where

$$\lambda = 0, \quad p(x) = 1, \quad q(x) = 0, \quad r(x) = 1, \quad L = \frac{d^2}{dx^2}.$$

The eigenfunctions are obtained from

$$\left(L + \lambda_n r(x)\right)\psi(x) = \left[\frac{d^2}{dx^2} + \lambda_n\right]\psi_n = 0$$

subject to the boundary conditions

$$\psi_n = 0, \quad x = 0 \text{ or } L.$$

Thus,

$$\psi_n = \sqrt{\frac{2}{L}}\sin\frac{n\pi x}{L} \text{ (orthonormal)}, \quad n = 1, 2, \ldots, \infty,$$

$$\lambda_n = \left(\frac{n\pi}{L}\right)^2,$$

$$G(x, x') = \sum_n \frac{\sqrt{\frac{2}{L}}\sin\frac{n\pi x'}{L}\sqrt{\frac{2}{L}}\sin\frac{n\pi x}{L}}{0 - \left(\frac{n\pi}{L}\right)^2},$$

which is the same as Equation 7A.14. The second example is Green's function of Helmholtz equation (in series form)

Example 2

$$\frac{d^2G}{dx^2} + \beta^2 G = \delta(x - x'), \tag{7A.21a}$$

$$G(0) = G(L) = 0. \tag{7A.21b}$$

Equation 7A.21a is a particular case of Equation 7A.19, where

$$\lambda = \beta^2, \quad p(x) = 1, \quad q(x) = 0, \quad r(x) = 1, \quad L = \frac{d^2}{dx^2}.$$

The orthonormal equations are

$$\psi_n = \sqrt{\frac{2}{L}}\sin\beta_n x = \sqrt{\frac{2}{L}}\sin\frac{n\pi x}{L},$$

$$\lambda_n = \beta_n^2 = \left(\frac{n\pi}{L}\right)^2.$$

Hence,

$$G(x,x') = \frac{2}{L}\sum_{n=1,2,\ldots}^{\infty} \frac{\sin(n\pi x'/L)\sin n\pi x/L}{\beta^2 - (n\pi/L)^2}. \quad (7A.22)$$

To complete this section, here we will write down Green's function in closed form for the Sturm–Liouville problem:

$$G_1 = G(x,x') = \frac{h_2(x')h_1(x)}{p(x')w(x')}, \quad 0 < x < x', \quad (7A.23a)$$

$$G_2 = G(x,x') = \frac{h_1(x')h_2(x)}{p(x')w(x')}, \quad x' < x < L, \quad (7A.23b)$$

where $h_1(x)$ is the solution form of the homogeneous equation, in the interval $0 < x < x'$, and $h_2(x)$ is the solution form of the homogeneous equation, in the interval $x' < x < a$:

$$G_1(x) = A_1 h_1(x), \quad (7A.23c)$$

$$G_2(x) = A_2 h_2(x), \quad (7A.23d)$$

and $w(x')$ is the Wronskian of h_1 and h_2 at $x = x'$:

$$w(x') = h_1(x')h_2'(x') - h_2(x')h_1'(x'). \quad (7A.23e)$$

The closed solution of Green's function given in Equation 7A.9a and 7A.9b can be obtained from the general solution given in (7A.23) by noting

$$p(x) = 1, \quad h_1(x) = \sin kx, \quad h_2(x) = \sin k(L-x)$$

and the Wronskian $w(x')$ is obtained as

$$\begin{aligned} w(x') &= h_1(x')h_2'(x') - h_2(x')h_1'(x') \\ &= (\sin kx)(-k)\cos k(L-x) - \sin k(L-x)(k)\cos kx \\ &= -[\sin kx \cos k(L-x) + \cos kx \sin k(L-x)] \\ &= -[\sin k(x+L-x)] \\ &= -k\sin kL, \end{aligned}$$

$$G = G_1 = \frac{\sin k(L-x')\sin kx}{-k\sin kL}, \quad 0 < x < x', \quad (7A.24a)$$

$$G = G_2 = \frac{\sin kx' \sin k(L-x)}{-k\sin kL}, \quad x' < x < L. \quad (7A.24b)$$

The difference in sign between Equations 7A.24 and 7A.9 is because Equation 7A.9 is the response when the impulse is of strength −1.

7A.4 Two-Dimensional Green's Function in Rectangular Coordinates

7A.4.1 Laplace Equation: Formulation and Closed Form Solution

$$\frac{d^2G}{dx^2}+\frac{d^2G}{dy^2}=\delta(x-x')\delta(y-y'), \tag{7A.25a}$$

$$\begin{aligned} x&=0 \text{ or } a, \quad G=0,\\ y&=0 \text{ or } b, \quad G=0. \end{aligned} \tag{7A.25b}$$

Let us first write

$$G(x,y,x',y')=\sum_{n=1,2,\ldots}^{\infty} g_m(y,x',y')\sin\frac{m\pi x}{a}. \tag{7A.26}$$

Here, we formulated such that the boundary conditions at the walls $x = 0$ or a are satisfied.

By substituting Equation 7A.26 into Equation 7A.25a, we obtain

$$\sum_{m=1}^{\infty}\left[-\left(\frac{m\pi}{a}\right)^2 g_m(y;x',y')\sin\frac{m\pi x}{a}+\sin\frac{m\pi x}{a}\frac{d^2 g_m(y;x',y')}{dy^2}\right]$$
$$=\delta(x-x')\delta(y-y'). \tag{7A.27}$$

Multiply both sides of Equation 7A.27 by sin $n\pi x/a$ and integrate with respect to x from 0 to a.

From orthogonality property, the LHS is

$$\frac{a}{2}\left[-\left(\frac{n\pi}{a}\right)^2 g_n(y;x',y')+\frac{d^2}{dy^2}g_n(y;x',y')\right]$$

and the RHS is

$$\int_0^a \delta(x-x')\delta(y-y')\sin\frac{n\pi x}{a}dx=\sin\frac{n\pi x'}{a}\delta(y-y').$$

Thus, the differential equation in y is obtained as

$$\frac{d^2}{dy^2}g_m(y;x',y')-\left(\frac{m\pi}{a}\right)^2 g_m(y;x',y')=\frac{a}{2}\sin\frac{m\pi x'}{a}\delta(y-y'). \tag{7A.28}$$

Equation 7A.28 can now be solved by using the recipe of Equations 7A.23a and 7A.23b.

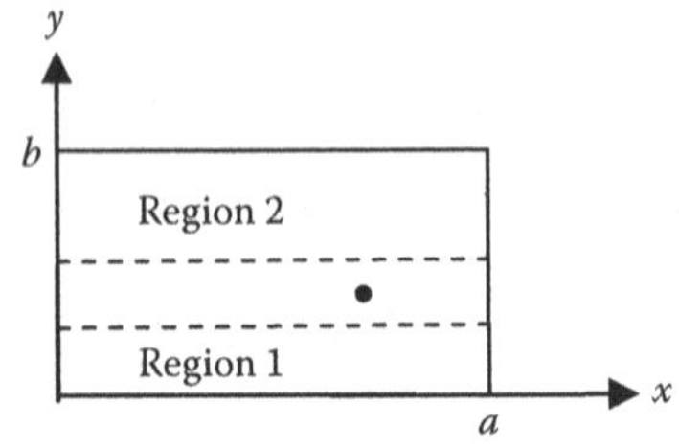

FIGURE 7A.1
Two-dimensional rectangular area divided into two regions: region 1, $0 < y < y' - \varepsilon$; region 2, $y' + \varepsilon < y < b$.

The homogeneous form of Equation 7A.28 is

$$\frac{d^2 g_m}{dy^2}(y;x',y') - \left(\frac{m\pi}{a}\right)^2 g_m(y;x',y') = 0 \tag{7A.29}$$

subject to the boundary condition $g_m = 0$, $y = 0$ or b (Figure 7A.1):

$$A_1 h_1(x) \to A_m h_m^{(1)}(y) = A_m(x',y')\sinh\frac{m\pi y}{a}, \quad \text{region 1,} \quad 0 < y < y', \tag{7A.30a}$$

$$A_2 h_2(x) \to B_m h_m^{(2)}(y) = B_m(x',y')\sinh\frac{m\pi(b-y)}{a}, \quad \text{region 2,} \quad y' < y < b. \tag{7A.30b}$$

Now the Wronskian from Equation 7A.23e, after simplification, becomes

$$w(y;x',y') = -\frac{m\pi}{a}\sinh\frac{m\pi b}{a}. \tag{7A.30c}$$

From Equation 7A.23a, we obtain

$$g_m^{(1)}(y;x',y') = \frac{(2/a)\sin(m\pi x'/a)\left(h_m^{(2)}\big|_{y'}\right)\left(h_m^{(1)}\big|_{y}\right)}{-(m\pi/a)\sinh(m\pi b/a)}$$

$$= \frac{-(2/m\pi)\sin(m\pi x'/a)\sinh(m\pi/a)(b-y')\sinh(m\pi y/b)}{\sinh(m\pi b/a)}, \quad 0 < y < y' \tag{7A.31a}$$

From Equation 7A.23b, we obtain

$$g_m^{(2)}(y;x',y') = \frac{-(2/m\pi)\sin(m\pi y'/a)\sinh\left[(m\pi/a)(b-y)\right]}{\sinh(m\pi b/a)}, \quad y' < y < b. \tag{7A.31b}$$

Now the complete Green's function is given as

$$G(x,y;x',y') = -\frac{2}{\pi}\sum_{m=1}^{\infty}\frac{\sin(m\pi x'/a)\sinh\left[(m\pi/a)(b-y)\right]}{m\sinh(m\pi b/a)}$$

$$\times \sin\frac{m\pi x}{a}\sinh\frac{m\pi y}{a}, \quad \text{for } 0 \le x \le a,\ 0 \le y \le y'. \tag{7A.32a}$$

$$= -\frac{2}{\pi}\sum_{m=1}^{\infty}\frac{\sin(m\pi x/a)\sinh\left[(m\pi/a)(b-y)\right]}{m\sinh(m\pi b/a)}$$
$$\times \sin\frac{m\pi x'}{a}\sinh\frac{m\pi y'}{a} \quad \text{for } 0\le x\le a,\ y'\le y\le b. \tag{7A.32b}$$

We can develop an alternative form of Green's function by first satisfying the boundary conditions at the walls $y = 0$ or b, that is, instead of Equation 7A.2 start with

$$G(x, y; x', y') = \sum_{n=1,2,\ldots}^{\infty} g_n(x; x', y')\sin\frac{n\pi y}{b} \tag{7A.33}$$

and develop the solution. The result is given in Equations 14.83a and b of Balanis [1].

7A.4.2 Laplace Equation: Series Form (Bilateral) Solution

For series solution of Green's function, we need the orthonormal eigenfunction ψ_{mn} for this two-dimensional problem.

These are the solutions of (for Laplace equation)

$$\frac{\partial^2\psi_{mn}}{\partial x^2}+\frac{\partial^2\psi_{mn}}{\partial y^2}+\lambda_{mn}\psi_{mn}=0 \tag{7A.34a}$$

subject to boundary condition

$$\begin{aligned}\psi_{mn} &= 0, \quad x = 0 \text{ or } a,\\ &= 0, \quad y = 0 \text{ or } b.\end{aligned} \tag{7A.34b}$$

Of course, we know that the orthogonal eigenfunctions that satisfy the boundary conditions (Equation 7A.34b) are

$$\psi_{mn} = A_{mn}\sin\frac{m\pi x}{a}\sin\frac{n\pi y}{b}.$$

From orthogonality property,

$$1 = \int_{x=0}^{a}\int_{y=0}^{b}(\psi_{mn})^2\,dxdy = \frac{ab}{4}A_{mn}^2.$$

Thus, $A_{mn} = 2/\sqrt{ab}$,

$$\psi_{mn} = \frac{2}{\sqrt{ab}}\sin\frac{m\pi x}{a}\sin\frac{n\pi y}{b}, \tag{7A.35a}$$

$$\lambda_{mn} = \left[\left(\frac{m\pi}{a}\right)^2+\left(\frac{n\pi}{b}\right)^2\right]^{1/2}. \tag{7A.35b}$$

Using the bilinear form (7A.20), we get

$$G(x,y;x',y') = \sum_m \sum_n \frac{\psi_{mn}(x',y')\psi_{mn}(x,y)}{\lambda - \lambda_{mn}}$$
$$= \frac{4}{ab}\sum_{m=1}^{\infty}\sum_{n=1}^{\infty} \frac{\sin(m\pi x'/a)\sin(n\pi y'/b)}{-\left[(m\pi/a)^2 + (n\pi/b)^2\right]} \sin\frac{m\pi x}{a}\sin\frac{n\pi y}{b}. \quad (7A.36)$$

Note that $\lambda = 0$, for the Laplace equation.
One can use Equation 7A.36 to solve Poisson's equation:

$$\frac{\partial^2 V}{\partial x^2} + \frac{\partial^2 V}{\partial y^2} = q(x,y). \quad (7A.37a)$$

By using Equation 7A.36 and the superposition integral, we obtain

$$V(x,y) = \frac{-4}{ab}\sum_{m=1}^{\infty}\sum_{n=1}^{\infty} \frac{\sin(m\pi x/a)\sin(n\pi y/b)}{(m\pi/a)^2 + (n\pi/b)^2}$$
$$\times \int_{y'=0}^{b}\int_{x'=0}^{a} q(x',y')\sin(m\pi x'/a)\sin(n\pi y'/b)\,dx'dy'. \quad (7A.37b)$$

7A.4.3 Helmholtz Equation (Series Form)

Green's function with homogeneous, Dirichlet boundary condition is given by

$$\frac{\partial^2 G}{\partial x^2} + \frac{\partial^2 G}{\partial y^2} + \beta^2 G = \delta(x-x')\delta(y-y') \quad (7A.38a)$$

with boundary conditions

$$G = 0, \quad x = 0 \text{ or } a, \quad y = 0 \text{ or } b. \quad (7A.38b)$$

For this case, $\lambda = \beta^2$:

$$\psi_{mn} = \frac{2}{\sqrt{ab}}\sin\frac{m\pi x}{a}\sin\frac{n\pi y}{b} \quad (7A.39)$$

and Equation 7A.36, gets modified as

$$G(x,y;x',y') = \frac{4}{ab}\sum_{m=1}^{\infty}\sum_{n=1}^{\infty} \frac{\sin(m\pi x'/a)\sin(n\pi y'/b)}{\beta^2 - \left[(m\pi/a)^2 + (n\pi/b)^2\right]^2} \times \sin\frac{m\pi x}{a}\sin\frac{n\pi y}{b}. \quad (7A.40)$$

7A.5 Generalized Green's Function Method

Till now, we derived Green's function that satisfied homogenous Dirichlet boundary conditions; in all the examples, the potential also satisfied the homogenous Dirichlet boundary condition.

Let us now investigate whether we can use a more general Green's function that has an impulse source but not necessarily satisfying the Dirichlet homogenous boundary condition. The scalar Helmholtz equation is

$$\nabla^2\Phi(r)+\beta^2\Phi(r)=f(r) \tag{7A.41}$$

and Green's function for the problem is $G(r, r')$ satisfying the equation

$$\nabla^2 G(r,r')+\beta^2 G(r,r')=\delta(r-r'). \tag{7A.42}$$

Green's first and second identities are given below.

Green's first identity

$$\oiint_s \Phi\frac{\partial\psi}{\partial n}\mathrm{d}s=\iiint_V \Phi\nabla^2\psi\,\mathrm{d}v+\iiint_V\left(\bar{\nabla}\Phi\cdot\bar{\nabla}\psi\right)\mathrm{d}v. \tag{7A.43}$$

In the above, s is a closed surface bounding a volume V, Φ and ψ are two scalar functions, and $\hat{n}$ is the unit vector normal to the surface.

Green's second identity

$$\oiint_s\left[\Phi\frac{\partial\psi}{\partial n}-\psi\frac{\partial\Phi}{\partial n}\right]\mathrm{d}s=\iiint_V\left[\Phi\nabla^2\psi-\psi\nabla^2\Phi\right]\mathrm{d}v. \tag{7A.44}$$

Multiply Equation 7A.41 with $G(\bar{r},\bar{r}')$ and Equation 7A.42 with $\Phi(\bar{r})$, we obtain

$$G\nabla^2\Phi+\beta^2\Phi G=fG, \tag{7A.45}$$

$$\Phi\nabla^2 G+\beta^2\Phi G=\Phi\delta(\bar{r},\bar{r}'). \tag{7A.46}$$

By subtracting Equation 7A.45 from Equation 7A.46 and integrating over the volume V, we obtain

$$\iiint_V \Phi(\bar{r}')\delta(\bar{r},\bar{r}')\mathrm{d}v-\iiint_V fG\mathrm{d}v=\iiint_V\left[\Phi\nabla^2 G-G\nabla^2\Phi\right]\mathrm{d}v. \tag{7A.47}$$

Applying Equation 7A.44 to the RHS of Equation 7A.47 and also evaluating the first term on the LHS of Equation 7A.47, we get

$$\Phi(\bar{r}')=\iiint_V f(\bar{r})G(\bar{r},\bar{r}')\mathrm{d}v+\oiint_s\left[\Phi\frac{\partial G}{\partial n}-G\frac{\partial\Phi}{\partial n}\right]\mathrm{d}s. \tag{7A.48a}$$

Since $\bar{r}'$ is an arbitrary point in V and $\bar{r}'$ is a dummy variable, $G(\bar{r},\bar{r}')=G(\bar{r}',\bar{r})$.

We can write (7A.48) as (exchanging $\bar{r}$ and $\bar{r}'$)

$$\Phi(\bar{r}) = \iiint_V f(\bar{r}')G(\bar{r},\bar{r}')dv + \oint\!\!\oint_s \left[\Phi(\bar{r}')\frac{\partial G}{\partial n} - G\frac{\partial \Phi(\bar{r}')}{\partial n}\right]ds. \tag{7A.48b}$$

In the above, the differentiations are with respect to primed (′) coordinates.

If we have homogeneous Dirichlet boundary conditions, satisfied by Φ as well as G on s, the second integral (surface integral) becomes zero, and hence

$$\Phi(\bar{r}) = \iiint_V f(\bar{r}')G(\bar{r},\bar{r}')dv. \tag{7A.49}$$

This is exactly the superposition integral we used in all the previous discussions. In all those discussions, we had the homogeneous Dirichlet boundary condition satisfaction both by Φ and G. Equation 7A.48 is the modified superposition integral for the general case. To use Equation 7A.48, we need to know Φ, $\partial\Phi/\partial n$, G, and $\partial G/\partial n$ on the closed boundary s.

7A.6 Three-Dimensional Green's Function and Green's Dyadic

Three-dimensional Green's function in free space for the scalar Helmholtz equation

$$\nabla^2 G + k^2 G = -\delta(\bar{r} - \bar{r}') \tag{7A.50}$$

is given by

$$G(\bar{r},\bar{r}') = \frac{1}{4\pi}\frac{e^{-jk|\bar{r}-\bar{r}'|}}{|\bar{r}-\bar{r}'|}. \tag{7A.51}$$

By making $k = 0$ in Equations 7A.50 and 7A.51, we obtain the Laplace equation and the associated Green's function.

Green's function $\bar{\Gamma}(\bar{r},\bar{r}')$, called Green's dyadic, satisfies the equation

$$\left[\bar{\nabla}\times\bar{\nabla}\times - k^2\right]\bar{\Gamma}(\bar{r},\bar{r}') = \bar{\mathbf{u}}\delta(\bar{r}-\bar{r}'), \tag{7A.52}$$

where $\bar{\mathbf{u}}$ is a unit dyadic given as

$$\bar{\mathbf{u}} = \sum_{m=1}^{3}\sum_{n=1}^{3}\hat{e}_m\hat{e}_n\delta_{mn}. \tag{7A.53}$$

In Equation 7A.53, $\hat{e}_m$, in Cartesian coordinates are $\hat{x}, \hat{y}, \hat{z}$ for $m = 1, 2, 3$, respectively. δ_{mn} is the Kronecker delta and given by

$$\delta_{mn} = \begin{cases} 1, & m = n, \\ 0, & m \neq n. \end{cases} \tag{7A.54}$$

A good account of dyads and their properties are given in [2,3]. In this connection, one can also define the double gradient $\bar{\nabla}\bar{\nabla}$:

$$\bar{\nabla}\bar{\nabla} = \sum_{m=1}^{3}\sum_{n=1}^{3} e_m e_n \frac{\partial}{\partial x_m}\frac{\partial}{\partial x_n}. \tag{7A.55}$$

From the properties of dyadic operations [2,3], we can show that dyadic Green's function

$$\bar{\Gamma}\left(\bar{r},\bar{r}'\right) = \left(\bar{\mathbf{u}} + \frac{1}{k^2}\bar{\nabla}\bar{\nabla}\right) G\left(\bar{r},\bar{r}'\right). \tag{7A.56}$$

Equation 7A.56 relates the scalar Green's function G given by Equation 7A.51 with dyadic Green's function given by Equation 7A.56.

Thus, the equation for the electric field

$$\bar{\nabla} \times \bar{\nabla} \times \bar{E} - k^2\bar{E} = -j\omega\mu\bar{J} \tag{7A.57}$$

can be solved as a superposition integral in terms of dyadic Green's function $\bar{\Gamma}$:

$$\bar{E}\left(\bar{r}\right) = -j\omega\mu\bar{\Gamma}\left(\bar{r},\bar{r}'\right) \cdot \bar{J}\left(\bar{r}'\right) dv'. \tag{7A.58}$$

References

1. Balanis, C. A., *Advanced Engineering Electromagnetics*, Wiley, New York, 1989.
2. Papas, C. H., *Theory of Electromagnetic Wave Propagation*, McGraw-Hill, New York, 1964.
3. Van Bladel, J., *Electromagnetic Fields*, Hemisphere Publishing, New York, 1985.

Appendix 8A: Wave Propagation in Chiral Media

It can be shown that the R and L waves are the normal modes of propagation in a chiral medium.

A mode is said to be a normal mode of propagation if the state of polarization of the wave is unaltered as it propagates. In a sourceless chiral medium, Maxwell's equations are

$$\nabla \times \mathbf{E} = -\frac{\partial \mathbf{B}}{\partial t}, \tag{8A.1}$$

$$\nabla \times \mathbf{H} = \frac{\partial \mathbf{D}}{\partial t}, \tag{8A.2}$$

$$\nabla \cdot \mathbf{B} = 0, \tag{8A.3}$$

$$\nabla \cdot \mathbf{D} = 0. \tag{8A.4}$$

Let us investigate the propagation of an R wave in such a medium. Let

$$\mathbf{E} = \left(\hat{\mathbf{x}} - \mathrm{j}\hat{\mathbf{y}}\right) E_0 \mathrm{e}^{\mathrm{j}(\omega t - k_c z)}, \tag{8A.5}$$

$$\mathbf{D} = \left(\hat{\mathbf{x}} - \mathrm{j}\hat{\mathbf{y}}\right) D_0 \mathrm{e}^{\mathrm{j}(\omega t - k_c z)}, \tag{8A.6}$$

$$\mathbf{B} = \left(\mathrm{j}\hat{\mathbf{x}} + \hat{\mathbf{y}}\right) B_0 \mathrm{e}^{\mathrm{j}(\omega t - k_c z)}, \tag{8A.7}$$

$$\mathbf{H} = \left(\mathrm{j}\hat{\mathbf{x}} + \hat{\mathbf{y}}\right) H_0 \mathrm{e}^{\mathrm{j}(\omega t - k_c z)}, \tag{8A.8}$$

where k_c is the wave number in the chiral medium. From Equation 8A.1, we obtain

$$\begin{vmatrix} \hat{\mathbf{x}} & \hat{\mathbf{y}} & \hat{\mathbf{z}} \\ 0 & 0 & -\mathrm{j}k_c \\ E_0 & -\mathrm{j}E_0 & 0 \end{vmatrix} = -\mathrm{j}\omega B_0,$$

which leads to

$$k_c E_0 = \omega B_0. \tag{8A.9}$$

From Equation 8A.2, similarly we get

$$k_c H_0 = \omega D_0. \tag{8A.10}$$

From the constitutive relations for the chiral medium given by Equations 8.174 and 8.175, on substitution of Equation 8A.5 into Equation 8A.8, we obtain

$$D_0 = \varepsilon E_0 + \xi_c B_0, \tag{8A.11}$$

$$H_0 = -\xi_c E_0 + \frac{B_0}{\mu}. \tag{8A.12}$$

From Equations 8A.10 through 8A.12, we can obtain a relation between E_0 and B_0:

$$(-k_c\xi_c - \varepsilon\omega)E_0 + \left(\frac{k_c}{\mu} - \omega\xi_c\right)B_0 = 0. \tag{8A.13}$$

Equations 8A.13 and 8A.9 may be arranged in the matrix form as

$$\begin{bmatrix} -k_c\xi_c - \omega\varepsilon & \dfrac{k_c}{\mu} - \omega\xi_c \\ k_c & -\omega \end{bmatrix} \begin{bmatrix} E_0 \\ B_0 \end{bmatrix} = 0. \tag{8A.14}$$

This set of homogeneous equations has a nontrivial solution only when the determinant of the matrix is zero, giving rise to the following equation:

$$k_c^2 - 2\omega\mu\xi_c k_c - k^2 = 0. \tag{8A.15}$$

The wave number k_{cR} of this R wave is thus given by

$$k_{cR} = \omega\mu\xi_c + \left[k^2 + (\omega\mu\xi_c)^2\right]^{1/2}. \tag{8A.16}$$

It can be further shown that the wave impedance is

$$\eta_c = \frac{E_0}{H_0} = \left[\frac{\mu}{\varepsilon + \mu\xi_c^2}\right]^{1/2}. \tag{8A.17}$$

For an L wave, the wave number is given by

$$k_{cL} = -\omega\mu\xi_c + \left[k^2 + (\omega\mu\xi_c)^2\right]^{1/2} \tag{8A.18}$$

and the wave (characteristic) impedance is still given by Equation 8A.17.

Appendix 11A discusses Faraday rotation in a magnetoplasma and compares it with the natural rotation in a chiral medium.

Appendix 8B: Left-Handed Materials and Transmission Line Analogies

8B.1 Introduction

The material [1] discussed in this section is different from the left-handed chiral material discussed in Section 8.10. The qualifier "left-handed" is explained below.

Section 2.1 discussed the propagation of the uniform plane wave in a sourceless isotropic medium. It is shown that the unit vectors in the directions of the electric field **E**, magnetic field **H**, and the direction of propagation **k** are mutually orthogonal in the right-handed sense:

$$\hat{\mathbf{E}} \times \hat{\mathbf{H}} = \hat{\mathbf{k}}.$$

Such a material is considered a right-handed material, in contrast to an artificial material called the "left-handed material" [1–8]. In this artificial material, $\hat{\mathbf{E}}$, $\hat{\mathbf{H}}$, and $\hat{\mathbf{k}}$ are mutually orthogonal in the left-handed sense. The direction of wave propagation is in the opposite sense to that in a right-handed material. Some authors prefer to designate the "left-handed material" as a "negative phase velocity" (NPV) medium.

Such a property arises in this artificial medium with negative permittivity ε and negative permeability μ. In that sense, some authors prefer to call it "double-negative" (DNG) medium. Some authors call it by yet another name "Veselago" medium, in honor of the scientist who discussed the properties of such a medium in a seminal paper [2] in 1968. Caloz and Itoh [8] explain metamaterials through the engineering approach, using the analogy with the transmission lines. The write-up in this section closely follows the approach of Caloz and Itoh [8].

Consider a transmission line, whose circuit representation of a differential length of a transmission line is given in Figure 8B.1.

The reason for using the subscripts R and L will become clear later on.

Following the approach of Section 2.7, we obtain the partial differential equation (PDE) for the voltage and the current:

$$-\frac{\partial V(z,t)}{\partial z} = L_R' \frac{\partial I(z,t)}{\partial t} + \frac{1}{C_L'} \int I(z,t)\,\mathrm{d}t, \tag{8B.1}$$

$$-\frac{\partial I(z,t)}{\partial z} = C_R' \frac{\partial V(z,t)}{\partial t} + \frac{1}{L_L'} \int V(z,t)\,\mathrm{d}t. \tag{8B.2}$$

The fourth-order PDE for the voltage variable is obtained by eliminating I, from Equations 8B.1 and 8B.2. The operator technique of replacing $\partial/\partial z$ by p and $\partial/\partial t$ by s will

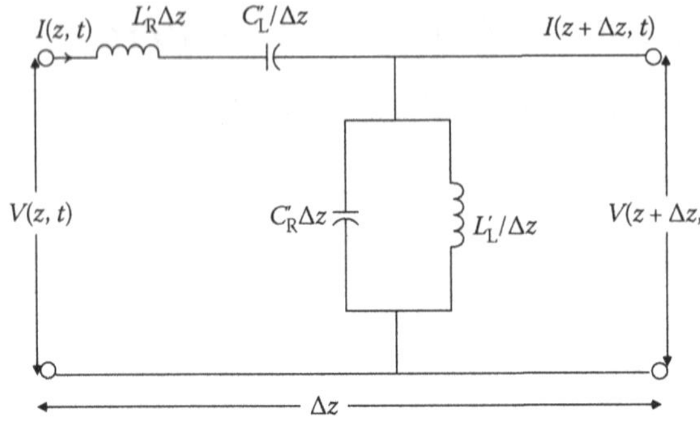

FIGURE 8B.1
Circuit representation of an artificial transmission line.

be useful in solving Equations 8B.1 and 8B.2 simultaneously. Note that $\int f dt$ will be replaced by f/s, where f stands for I or V. Thus, from Equation 8B.2, we get

$$I = -\left(C'_R s + \frac{1}{L'_L s}\right)\frac{V}{p}. \tag{8B.3}$$

In the operator form, Equation 8B.1 becomes

$$pV = -\left(L'_R s + \frac{1}{C'_L s}\right)I. \tag{8B.4}$$

Substituting Equation 8B.3 into Equation 8B.4 to eliminate I and simplifying, we get

$$p^2 s^2 V - L'_R C'_R s^4 V - \left(\frac{L'_R}{L'_L} + \frac{C'_R}{C'_L}\right)s^2 V - \frac{1}{C'_L L'_L}V = 0. \tag{8B.5}$$

Reverting back to the derivative form, Equation 8B.5 will yield the PDE for the voltage:

$$\frac{\partial^4 V(z,t)}{\partial z^2 \partial t^2} - L'_R C'_R \frac{\partial^4 V(z,t)}{\partial t^4} - \left(\frac{L'_R}{L'_L} + \frac{C'_R}{C'_L}\right)\frac{\partial^2 V(z,t)}{\partial t^2} - \frac{1}{C'_L L'_L}V = 0. \tag{8B.6}$$

For harmonic variation in space and time,

$$V(z,t) = V_0 e^{j(\omega t - kz)}. \tag{8B.7}$$

We obtain the following dispersion relation:

$$\omega^2 k^2 - L'_R C'_R \omega^4 + \left(\frac{L'_R}{L'_L} + \frac{C'_R}{C'_L}\right)\omega^2 - \frac{1}{C'_L L'_L} = 0. \tag{8B.8}$$

The ω–k diagram is obtained by tracing the curve given by Equation 8B.8 in the ω–k plane. The intersections with the ω-axis are obtained by substituting $k = 0$ in Equation 8B.8.

$$\omega^4 - \left(\frac{1}{L'_L C'_R} + \frac{1}{L'_R C'_L}\right)\omega^2 - \frac{1}{L'_L C'_L L'_R C'_R} = 0 \quad (k = 0). \tag{8B.9}$$

Let

$$\omega_{se} = \frac{1}{\sqrt{L'_R C'_L}}(\text{rad/s}), \tag{8B.10a}$$

$$\omega_{sh} = \frac{1}{\sqrt{L'_L C'_R}}(\text{rad/s}), \tag{8B.10b}$$

$$\omega'_L = \frac{1}{\sqrt{L'_L C'_L}}(\text{rad/ms}), \tag{8B.10c}$$

$$\omega'_R = \frac{1}{\sqrt{L'_R C'_R}}(\text{radm/s}) \tag{8B.10d}$$

noting that

$$\omega_{se}\omega_{sh} = \omega'_L \omega'_R, \tag{8B.10e}$$

we can write Equation 8B.9 in the form

$$\left(\omega^2 - \omega_{sh}^2\right)\left(\omega^2 - \omega_{se}^2\right) = 0. \tag{8B.11}$$

Thus, the intersection points on the ω-axis are $\omega = \pm\omega_{sh}$ and $\omega = \pm\omega_{se}$. The complete sketch of the ω–k diagram is given in Figure 8B.2, assuming that $\omega_{se} < \omega_{sh}$.

The subscripts LH and RH and the superscripts + and – in Figure 8B.2 are explained next.

The superscript + denotes a positive group velocity (positive slope of the ω–k curve) which in turn means energy flow in the positive z-direction. The superscript – denotes a negative group velocity which in turn means energy flow in the negative z-direction. The subscript RH denotes that the phase velocity and the group velocity have the same sign, the usual property of isotropic regular materials called right-handed materials. The subscript LH indicates opposite sign for the phase velocity and the group velocity, the property of the unusual materials called left-handed materials. In the LH case, the direction of power flow and the phase propagation are opposite to each other.

If $\omega_{se} > \omega_{sh}$, then the ω–k diagram is still given by Figure 8B.2, with ω_{se} and ω_{sh} interchanged. If $\omega_{se} = \omega_{sh}$, called balanced model, then the ω–k diagram degenerates into a simpler diagram shown in Figure 8B.3.

The pure right-handed (PRH) system is obtained by making $C'_L = \infty$ and $L'_L = \infty$, in which case the transmission line is the same as the ideal transmission line discussed in Sections 2.7 and 2.8. The ω–k diagram is shown dotted in Figure 8B.3 and marked as k^{PRH}. The dual case of $L'_R = 0$ and $C'_R = 0$ reduces Figure 8B.1 to Figure 8B.4, which is the transmission line analogy for the pure left-handed (PLH) material. Its dispersion relation is shown dotted and marked as k^{PLH} in Figure 8B.3.

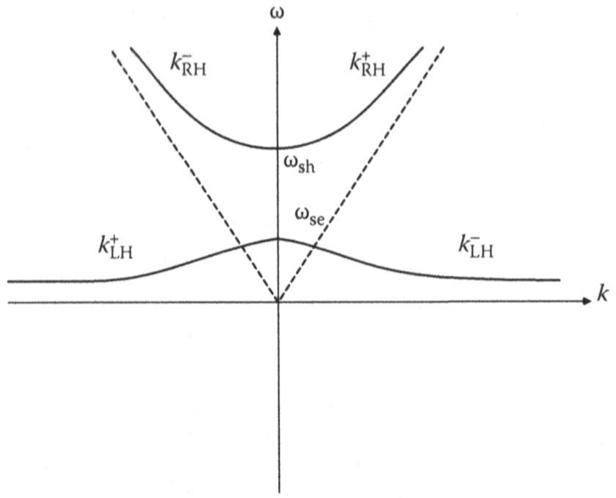

FIGURE 8B.2
ω–*k* diagram of the artificial transmission line given in Figure 9.6.

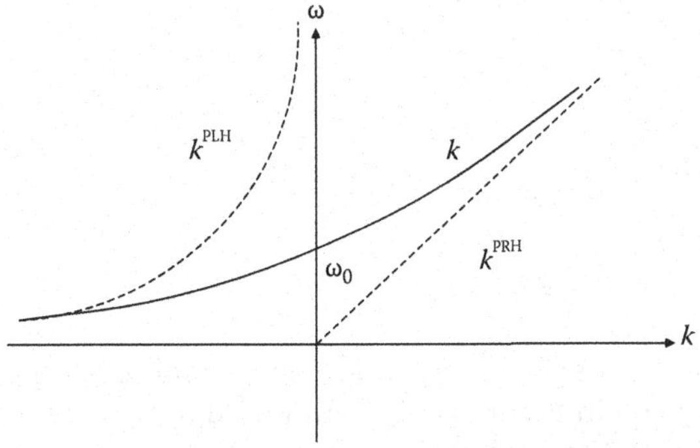

FIGURE 8B.3
ω–*k* diagram for a balanced transmission line model of Figure 9.6. $\omega_0 = \omega_{se} = \omega_{sh}$. The dotted curves are for PLH system and for PRH system.

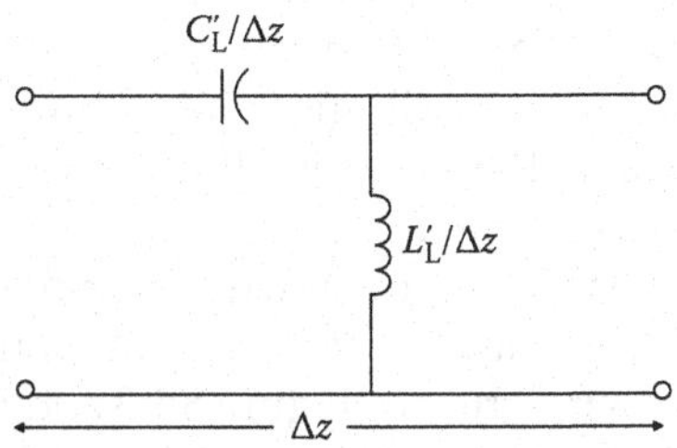

FIGURE 8B.4
Transmission line analogy to a PLH material.

Example 8B.1

In the equivalent circuit of a transmission line, for a length Δz, shown in Figure 8B.1, assume that

$$L_L' = C_L' = L_R' = C_R' = 1.$$

a. Sketch the ω–k diagram.
b. If $\omega = 0.8$, then determine (i) phase velocity V_p, (ii) group velocity V_g, and (iii) wavelength λ.
c. Repeat (b), if $\omega = 1$.
d. Repeat (b), if $\omega = 1.1$.

Solution

a. From Equation 8B.8, the dispersion relation is given by

$$\omega^2 k^2 + 2\omega^2 - 1 - \omega^4 = 0 \tag{8B.12}$$

and from Equations 8B.10a and 8B.10b,

$$\omega_{se} = 1. \tag{8B.13}$$

$$\omega_{sh} = 1. \tag{8B.14}$$

This is a balanced transmission line.
From Equation 8B.12,

$$\begin{aligned} \omega^2 k^2 &= \omega^4 - 2\omega^2 + 1 = \left(\omega^2 - 1\right)^2, \\ k^2 &= \left(\frac{\omega^2 - 1}{\omega}\right)^2 = \left(\omega - \frac{1}{\omega}\right)^2, \\ k &= \pm\left(\omega - \frac{1}{\omega}\right), \\ V_p &= \frac{\omega}{k} = \pm\frac{\omega}{\omega - 1/\omega} = \pm\frac{\omega^2}{\omega^2 - 1}, \end{aligned} \tag{8B.15}$$

$$\begin{aligned} \frac{dk}{d\omega} &= \pm\frac{d}{d\omega}\left(\omega - \frac{1}{\omega}\right) = \pm\frac{\omega^2 + 1}{\omega^2}, \\ V_g &= \frac{d\omega}{dk} = \pm\frac{\omega^2}{\omega^2 + 1}. \end{aligned} \tag{8B.16}$$

The bottom sign in the expressions for V_p and V_g correspond to negative values of ω. The sketch of the ω–k diagram is shown by the solid line in Figure 8B.3, where $\omega_0 = 1$.

b. From Equations 8B.15 and 8B.16, we have

i. $\omega = 0.8, V_p = \left.\dfrac{\omega^2}{\omega^2 - 1}\right|_{\omega=0.8} = -1.78,$

ii. $V_g = \left.\dfrac{\omega^2}{\omega^2 + 1}\right|_{\omega=0.8} = 0.39,$

iii. $\lambda = \dfrac{2\pi}{|k|} = \left.\dfrac{2\pi}{\omega - 1/\omega}\right|_{\omega=0.8} = 1.78.$

c. $\omega = 1.0$:

$$V_p = \infty, \quad V_g = \frac{1}{2}, \quad \lambda = \infty.$$

d. $\omega = 1.1$:

$$V_p = 5.76, \quad V_g = 0.548, \quad \lambda = 32.91.$$

From this example, we see that the balanced transmission line behaves like a PLH material for $\omega < \omega_0$ and as a PRH material for $\omega > \omega_0$. In the neighborhood of $\omega \approx \omega_0$, the balanced transmission line differs considerably from PLH as well as PRH.

8B.2 Electromagnetic Properties of a Left-Handed Material

Let us reverse the transmission line analogy to obtain the equivalent electromagnetic parameters ε_r and μ_r of a medium which has the same form as Equations 8B.1 and 8B.2. We will consider a TEM harmonic wave with $\hat{\mathbf{E}} = \hat{y}$ and $\hat{\mathbf{H}} = -\hat{x}$. From Maxwell's equations, we obtain

$$\frac{d\tilde{E}_y}{dz} = -j\omega\mu H_x, \tag{8B.17}$$

$$\frac{d\tilde{H}_x}{dz} = -j\omega\varepsilon E_y, \tag{8B.18}$$

whereas Equations 8B.1 and 8B.2 yield for the phasors $\tilde{V}$ and $\tilde{I}$,

$$-\frac{\partial\tilde{V}}{\partial z} = \left(j\omega L_R' + \frac{1}{j\omega C_L'}\right)\tilde{I}, \tag{8B.19}$$

$$-\frac{\partial\tilde{I}}{\partial z} = \left(j\omega C_R' + \frac{1}{j\omega L_L'}\right)\tilde{V}. \tag{8B.20}$$

Comparing Equation 8B.17 with Equation 8B.19, with $\tilde{E}_y$ playing the role of $\tilde{V}$ and $\tilde{H}_x$ playing the role of $\tilde{I}$, we obtain the equivalence

$$\mu(\omega) = L_R' + \frac{1}{\omega^2 C_L'}, \tag{8B.21}$$

$$\varepsilon(\omega) = C_R' + \frac{1}{\omega^2 L_L'}. \tag{8B.22}$$

Note that the medium is dispersive since μ as well as ε are highly dispersive.

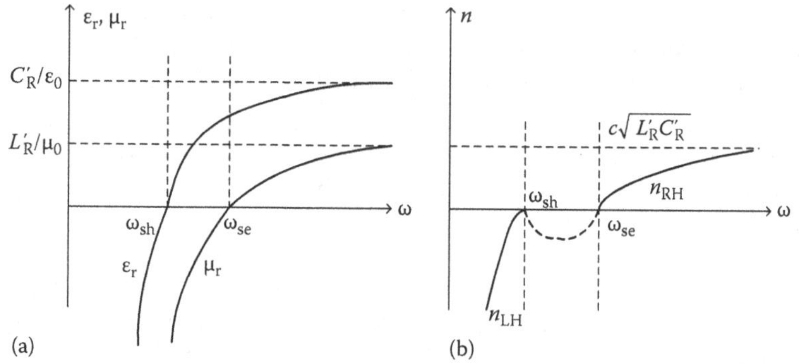

FIGURE 8B.5
Variation with frequency of (a) ε_r and μ_r and (b) refractive index n for the case $\omega_{sh} < \omega_{se}$. (Reproduced from Caloz, C. and Itoh, T. *Electromagnetic Metamaterials*, p. 68. Copyright Wiley. With permission.)

The relative permittivity ε_r and the relative permeability μ_r are given by

$$\mu_r = \frac{L'_R}{\mu_0}\left(1 - \frac{\omega_{se}^2}{\omega^2}\right), \tag{8B.23}$$

$$\varepsilon_r = \frac{C'_R}{\varepsilon_0}\left(1 - \frac{\omega_{sh}^2}{\omega^2}\right), \tag{8B.24}$$

where ω_{se} and ω_{sh} are given by Equations 8B.10a and 8B.10b, respectively.

Figure 8B.5a sketches the variation of ε_r and μ_r with frequency ω and Figure 8B.5b sketches the refractive index n defined by

$$n = \sqrt{\mu_r \varepsilon_r}. \tag{8B.25}$$

We note the following interesting features revealed by Figure 8B.5.

1. The variation of ε_r is similar to the variation of ε_p given by Equation 8B.8 for simple metals of Section 8.1, when collisions are neglected. The role of ω_p is played by ω_{sh}. ε_r is negative for $\omega < \omega_{sh}$. Perhaps metals excited by electromagnetic wave of $\omega < \omega_{sh}$ can be the material with negative ε_r. Metals have high electron density and hence the plasma frequency is in the ultraviolet frequency range. At microwave frequency, ε_r is highly negative and the metal behaves as a good conductor dominated by collisions and we usually treat it as a good conductor of conductivity given by Equation 8.27, rather than a plasma in the cutoff region (see Figure 8.7). Pendry et al. [4,5] used an interesting calculation method to compute the plasma frequency of a metallic system consisting of thin wires.
2. So far, we have not discussed any material which has the variation of μ_r given in Figure 8B.5. Pendry et al. [6] suggested the possibility of constructing such a material in 1999, using a split ring structure in a square array.
3. From Figure 8B.5 of this section, we note when $\omega < \omega_{sh}$ and $\omega < \omega_{se}$, ε_r as well as μ_r are negative and the refractive index n is also negative. In this case, the material is

left-handed. Smith et al. [3] constructed such artificial materials and showed experimentally that such a material exhibited the properties of a left-handed material. The experiments were conducted in the microwave range.

4. In the frequency range $\omega_{sh} < \omega < \omega_{se}$, ε_r is positive but μ_r is negative and the refractive index is imaginary and the wave is evanescent. This frequency domain is a stop band due to a negative μ_r and hence the band gap is called the magnetic gap. In the case $\omega_{se} < \omega < \omega_{sh}$, the gap is called the electric gap.

A negative value for ε_r (μ_r) does not imply that the stored electric energy (magnetic energy) is negative. As we have noted, the material is dispersive and the stored energy for a dispersive medium is not given by $(1/4)\varepsilon_0\varepsilon_r E^2((1/4)\mu_0\mu_r H^2)$ but by [1,9]

$$\langle W_e \rangle = \frac{1}{4}\varepsilon_0 \frac{\partial(\omega\varepsilon_r)}{\partial\omega} E^2, \tag{8B.26}$$

$$\langle W_m \rangle = \frac{1}{4}\mu_0 \frac{\partial(\omega\mu_r)}{\partial\omega} H^2. \tag{8B.27}$$

Substituting Equations 8B.23 and 8B.24 into Equations 8B.26 and 8B.27, respectively, we obtain

$$\langle W_e \rangle = \frac{1}{4}\varepsilon_0 \left(C_R' + \frac{1}{\omega^2 L_L'} \right) E^2 > 0, \tag{8B.28}$$

$$\langle W_m \rangle = \frac{1}{4}\mu_0 \left(L_R' + \frac{1}{\omega^2 C_L'} \right) H^2 > 0. \tag{8B.29}$$

The refractive index $n = \sqrt{\mu_r \varepsilon_r}$ is given by

$$n = \frac{c}{\omega_R'} \left(1 - \frac{\omega_{se}^2}{\omega^2} \right)^{1/2} \left(1 - \frac{\omega_{sh}^2}{\omega^2} \right)^{1/2}. \tag{8B.30}$$

Equation 8B.25 can be written in an alternate form as

$$n = \frac{sc}{\omega_R'} \left| \left(1 - \frac{\omega_{se}^2}{\omega^2} \right) \right|^{1/2} \left| \left(1 - \frac{\omega_{sh}^2}{\omega^2} \right) \right|^{1/2}, \tag{8B.31}$$

where

$$\begin{aligned} &s = 1, \quad \omega > \omega_{se} \quad \text{and} \quad \omega > \omega_{se}, \\ &s = -1, \quad \omega > \omega_{se} \quad \text{and} \quad \omega > \omega_{se}, \\ &s = -j \quad \text{otherwise}. \end{aligned} \tag{8B.32}$$

Equation 8B.31 is graphed in Figure 8B.5. The broken line is for the evanescent wave. n_{LH} is negative and hence the alternative name for the LH material is NPV material.

8B.3 Boundary Conditions, Reflection, and Transmission

The boundary conditions (Equations 1.14 through 1.17) are valid even if one or both of the media are LHM, since these are derived from Maxwell's equation. If ρ_s and **K** are zero, these are

$$D_{1n} = D_{2n}, \tag{8B.33}$$

$$B_{1n} = B_{2n}, \tag{8B.34}$$

$$E_{1t} = E_{2t}, \tag{8B.35}$$

$$H_{1t} = H_{2t}, \tag{8B.36}$$

where the subscripts n stands for the normal component and t stands for the tangential component. For the specific case of the medium 2 LHM and the medium 1 RHM, since μ_2 and ε_2 are negative, we have

$$E_{1n} = -\frac{|\varepsilon_2|}{\varepsilon_1} E_{2n}, \tag{8B.37}$$

$$H_{1n} = -\frac{|\mu_2|}{\mu_1} H_{2n}. \tag{8B.38}$$

A more general case can be considered by considering the sign of the electromagnetic parameters separately and writing

$$s_1|\varepsilon_1|E_{1n} = s_2|\varepsilon_2|E_{2n}, \tag{8B.39}$$

$$s_1|\mu_1|H_{1n} = s_2|\mu_2|H_{2n}, \tag{8B.40}$$

where s_1 and s_2 are +1 or −1 depending on whether the medium is RH or LH, respectively. Figure 8B.6 compares the transmission through the interface of the two media.

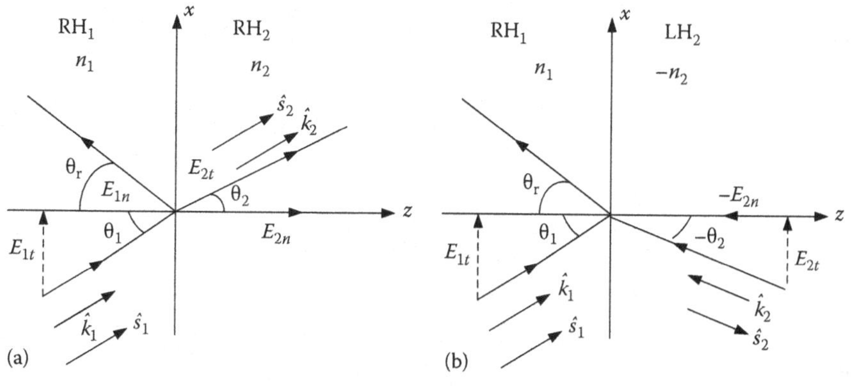

FIGURE 8B.6
Transmission through the interface: (a) both media are RH and (b) media 2 is LH with refractive index $-n_2$.

The first part of Snell's law

$$\theta_i = \theta_1 = \theta_r \tag{8B.41}$$

is the same for Figure 8B.6a and b, but the second part of the Snell's law for RH–LH interface is given by

$$n_1 \sin\theta_i = -n_2 \sin\theta_t. \tag{8B.42}$$

Thus, θ_t for the case of RH–LH interface is $(-\theta_2)$ as shown in Figure 8B.6b. We can thus state Snell's law in the general form as

$$s_1 |n_1| \sin\theta_i = s_1 |n_1| \sin\theta_r = s_2 |n_2| \sin\theta_t. \tag{8B.43}$$

Figure 8B.7 suggests that a "flat lens" consisting of RH, LH, and RH can focus and refocus if the three media have the same magnitude for n.

If $|n_L| \neq |n_R|$, then a pure focal point is not formed and the focal point gets diffused [1]. Moreover, the intrinsic impedances also satisfy the requirement

$$\eta_1 = \sqrt{\frac{\mu_1}{\varepsilon_1}} = \eta_2 = \sqrt{\frac{\mu_2}{\varepsilon_2}}, \tag{8B.44}$$

There will be no reflection. This can happen only if

$$\varepsilon_1 = |\varepsilon_2|, \tag{8B.45}$$

$$\mu_1 = |\mu_2|. \tag{8B.46}$$

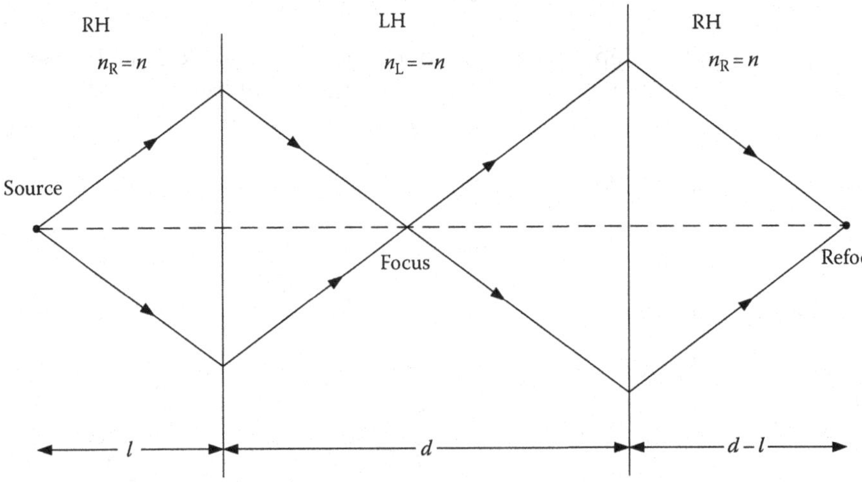

FIGURE 8B.7
Two symmetrical rays from the source is focused and refocused if the refractive indices of the three media have the same magnitude of n. (Reproduced from Caloz, C. and Itoh, T. 2006. *Electromagnetic Metamaterials*, p. 47. Copyright Wiley. With permission.)

This is obviously difficult to achieve, except perhaps for over a very narrow band of frequencies, even approximately, for a flat lens.

Pendry [10] discussed the phenomenon that LH materials can overcome diffraction limit of the conventional optics. He showed theoretically that an LH slab can overcome the diffraction limit (the resolution is limited by the wavelength) by choosing

$$\varepsilon_r = \mu_r = -1.$$

Caloz and Itoh [1] discuss the enhancement of the evanescent wave by the LH slab.

8B.4 Artificial Left-Handed Materials

Pendry et al. [4] showed that a long, thin, metallic wire array with a period p much less than the wavelength can produce a material with negative ε and positive μ. This artificial plasmonic type of structure has

$$\omega_{pe} = \sqrt{\frac{2\pi c^2}{p^2 \ln p/a}}, \quad p \ll \lambda, \tag{8B.47}$$

$$\nu = \frac{\varepsilon_0 \left(p\omega_{pe}/a\right)^2}{\pi\sigma}. \tag{8B.48}$$

In the above, p is the cell size, ω_{pe} the plasma frequency, given by Equation 8.22 for a real plasma, a the radius of the metallic thin wire of conductivity σ, and ν the collision frequency. The wire length l is taken as infinity ($a \ll l$) (Figure 8B.8).

Equations 8B.47 and 8B.48 are derived, assuming the electric field is in the z-direction, the wires are thin and long ($a \ll l$), $p \ll \lambda$, and the radius $a \ll p \ll l$. The restriction $p \ll \lambda$ is equivalent to saying that the differential model used in Figure 8B.1 ($\Delta z \sim p$) can be used. The experimental results for the structure with the following data are

$$a = 1.0 \times 10^{-6} \text{ m}, \tag{8B.49}$$

$$p = 5 \text{ mm}, \tag{8B.50}$$

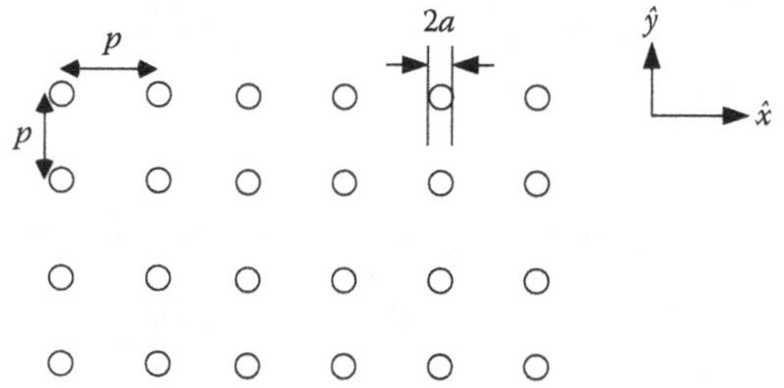

FIGURE 8B.8
Cross section of a thin-wire plasmonic structure.

The wire material, aluminum ($N = 1.806 \times 10^{29}/\mathrm{m}^3$), showed a plasma frequency f_{pe} = 8.3 GHz. Had we used (Equation 8.22) with N given above, we would have got a plasma frequency in an ultraviolet frequency band. Pendry showed that the reduction of the plasma frequency of the structure is due to the constraint that electrons are forced to move along the thin wires. Since only a part of the space is filled with metal, the effective electron density N_{eff} is given by

$$N_{eff} = N\frac{\pi a^2}{p^2}. \tag{8B.51}$$

The electrons in the magnetic field created by the current in the wire experience additional momentum. The effect of this additional momentum can be accounted for by considering an effective value m_{eff} for the mass of the electron given by

$$m_{eff} = \frac{\mu_0 q^2 \pi a^2 N}{2\pi}\ln\frac{p}{r}. \tag{8B.52}$$

The plasma frequency, using the effective values for the electron density and the mass of the electron, is given by

$$\omega_{pe} = \frac{N_{eff}^2 q^2}{m_{eff}\varepsilon_0}. \tag{8B.53}$$

Substituting Equations 8B.51 and 8B.52 into Equation 8B.53, we get Equation 8B.47. For the data of Equations 8B.49 and 8B.50, the plasma frequency turns out to be

$$f_{pe} = 8.2\,\mathrm{GHz}, \tag{8B.54}$$

close to the experimental value.

Equations 8B.47 and 8B.48 can also be obtained by considering the array as a "periodically spaced wire gratings or periodically spaced loaded transmission line" [1].

The permeability of the wire structure made of nonmagnetic conducting material such as copper or aluminum will be μ_0. The assumption that $l \gg \lambda$ means that the wires are excited at frequencies much less than the first resonance.

Pendry et al. [6] discussed a negative $-\mu$ artificial element, which is a metal split–ring resonator (SRR) shown in Figure 8B.9; Figure 8B.9a shows one SRR element and Figure 8B.9b shows the array structure.

The structure is excited by a magnetic field perpendicular to the plane of the ring. Such a structure exhibits a plasmonic-type permeability frequency function given by (ignoring losses) [1,6]

$$\mu_r(w) = 1 - \frac{Fw^2}{w^2 - w_{om}^2}, \tag{8B.55}$$

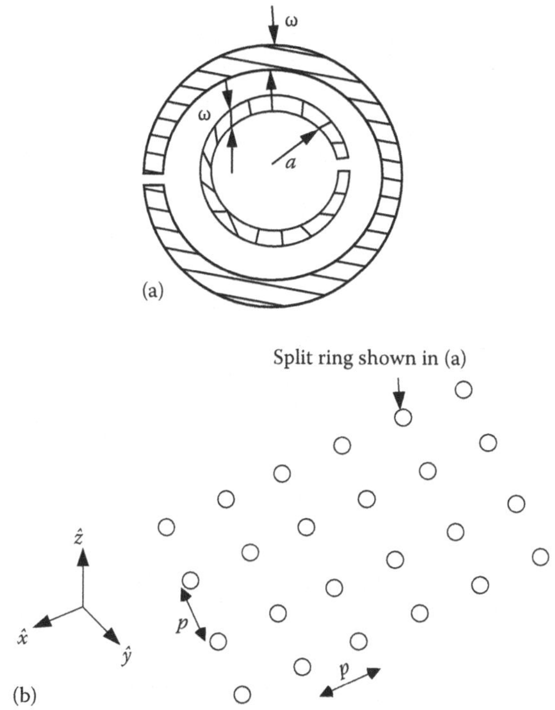

FIGURE 8B.9
Negative −μ resonant structure: (a) SRR element and (b) SRR structure.

where

$$F = \pi\left(\frac{a}{p}\right)^2, \tag{8B.56}$$

$$w_{\text{om}} = c\sqrt{\frac{3p}{\pi\ln\left(2wa^3/8\right)}}. \tag{8B.57}$$

In the above equations, p is the spatial period, a the inner radius of the smaller ring, w the width of the rings, and δ the radial spacing between the rings, as marked in Figure 8B.9a. From Equation 8B.55, we note that $\mu_r > 1$, for $1 < w < w_{\text{om}}$, has resonance ($\mu_r = \infty$) at $w = w_{\text{om}}$, negative in the frequency domain $w_{\text{om}} < w < w_{\text{pm}}$, where

$$w_{\text{pm}} = \frac{w_{\text{om}}}{\sqrt{1-F}}. \tag{8B.58}$$

First experimental LH material based on the combination of the thin-wire structure with the SRR was constructed by Smith et al. [3]. Caloz and Itoh [1, p. 8, Figure 1.4a] show a three-dimensional sketch of such a structure and remark that it is mono-dimensionally LH, since only "one direction" is allowed for the doublet (**E**, **H**): $\varepsilon_{xx} < 0$, $0 < \omega < \omega_{\text{pe}}$; $\mu_{xx} < 0$, $\omega_{\text{om}} < \omega < \omega_{\text{pm}}$ (Figure 8B.10).

A planar LH structure in microstrip technology consisting of series interdigital capacitors and shunt stub inductors was investigated [11,12] and summarized in [1].

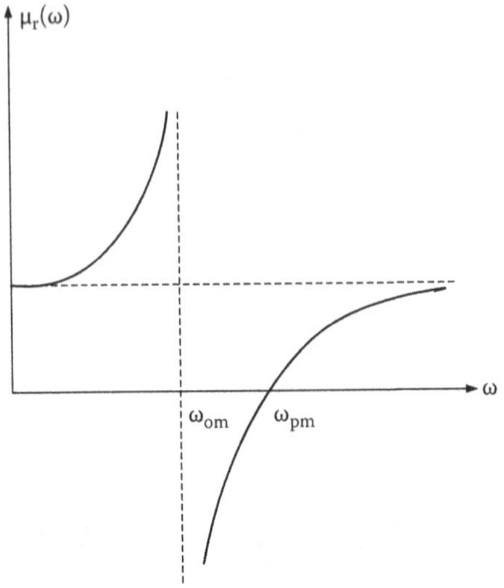

FIGURE 8B.10
Sketch of μ_r vs. w for split-ring resonate structure.

References

1. Caloz, C. and Itoh, T., *Electromagnetic Metamaterials*, Wiley, Hoboken, NJ, 2006.
2. Veselago, V., The electrodynamics of substances with simultaneously negative values of ε and μ, *Soviet Phys. Uspekhi*, 10(4), 509–514, 1968.
3. Smith, D. R., Padilla, W. J., Vier, D. C., Nemat-Nasser, S. C., and Schultz, S., Composite medium with simultaneously negative permeability and permittivity, *Phys. Rev. Lett.*, 84(18), 4184–4187, 2000.
4. Pendry, J. B., Holden, A. J., Stewart, W. J., and Youngs, I., Extremely low frequency plasmons in metallic mesostructure, *Phys. Rev. Lett.*, 76(25), 4773–4776, 1996.
5. Pendry, J. B., Holden, A. J., Robbins, D. J., and Stewart, W. J., Low frequency plasmons in thin-wire structure, *J. Phys. Condens. Matter*, 10, 4785–4809, 1998.
6. Pendry, J. B., Holden, A. J., Robbins, D. J., and Stewart, W. J., Magnetism from conductors and enhanced nonlinear phenomena, *IEEE Trans. Microw. Theory. Tech.*, 47(11), 2075–2084, 1999.
7. Smith, D. R., Vier, D. C., Kroll, N., and Schultz, S., Direct calculation of permeability and permittivity for a left-handed metamaterial, *App. Phys. Lett.*, 77(14), 2246–2248, 2000.
8. Sheley, R. A., Smith, D. R., and Schultz, S., Experimental verification of a negative index of refraction, *Science*, 292, 77–79, 2001.
9. Papas, C. H., *Theory of Electromagnetic Wave Propagation*, McGraw-Hill, New York, 1965.
10. Pendry, J. B., Negative refraction makes a perfect lens, *Phys. Rev. Lett.*, 85(18), 3966–3969, 2000.
11. Caloz, C. and Itoh, T., Application of the transmission line theory of left-handed (LH) materials to the realization of a microstrip LH transmission line, in *Proceedings of IEEE-AP-S USNC/URSI National Radio Science Meeting*, Vol. 2, San Antonio, TX, June 2002, pp. 412–415.
12. Caloz, C. and Itoh, T., Transmission line approach of left-handed (LH) structures and microstrip realization of a low-loss broadband LH filter, *IEEE Trans. Antennas Propag.*, 52(5), 1159–1166, 2004.

Appendix 9A: Backscatter from a Plasma Plume due to Excitation of Surface Waves*

Dikshitulu K. Kalluri

9A.1 Introduction

Dr. Keith Groves, my focal point at PL/GP for the Air Force Office of Scientific Research Summer Faculty Research Program (1994), suggested me to investigate the following problem: Experiments conducted by Air Force Laboratories showed that considerable unexpected electromagnetic backscatter from a plasma plume is occurring in a certain intermediate frequency band.

Figures 9A.1 and 9A.2 show the electron density and collision frequency numbers in the plasma plume. The plume is a cylindrical inhomogeneous lossy plasma column. The maximum electron density N_{0m} on the axis ($r = 0$) is 3E13 (#/cm^3) corresponding to angular plasma frequency (note: for convenience, the computer notation Exx = 10^{XX} will be used occasionally):

$$\omega_{pm} = \sqrt{\frac{N_{0m}e^2}{m\varepsilon_0}} = 3\times10^{11}\,(\text{rad}/\text{s}) \tag{9A.1}$$

and $f_{pm} = \omega_{pm}/2\pi = 47.75$ GHz. Here m and e are the mass and absolute value of the charge of an electron, respectively, and $\varepsilon_0 = 8.854\text{E}{-12}$ (F/m) is the permittivity of the free space. The collision frequency (ν) also varies with r ranging in value from 6E11 to 1E11 (rad/s). Scattering of an electromagnetic wave, in the frequency range of $f = 50$ MHz to 10 GHz, by such an inhomogeneous and lossy plasma plume is the object of this investigation.

9A.2 Problem Classification Based on Plasma Parameters

1. Plasma radius a is of the order of 0.5 m; the normalized value $a/\lambda_{pm} = af_{pm}/c = 83.3$. In this sense, the column may be classified as thick.

   Here c is the velocity of light = 3×10^8 m/s and λ_{pm} is the free space wavelength of the corresponding plasma frequency.
2. The normalized value $a/\lambda = af/c$ varies from 8.3×10^{-2} to 16.67 as f varies from 50 MHz to 10 GHz, respectively.

* Reprinted from Final Report Summer Faculty Research Program, *Air Force Office of Scientific Research*, Report Number A669853, 3B, 1994. With permission.

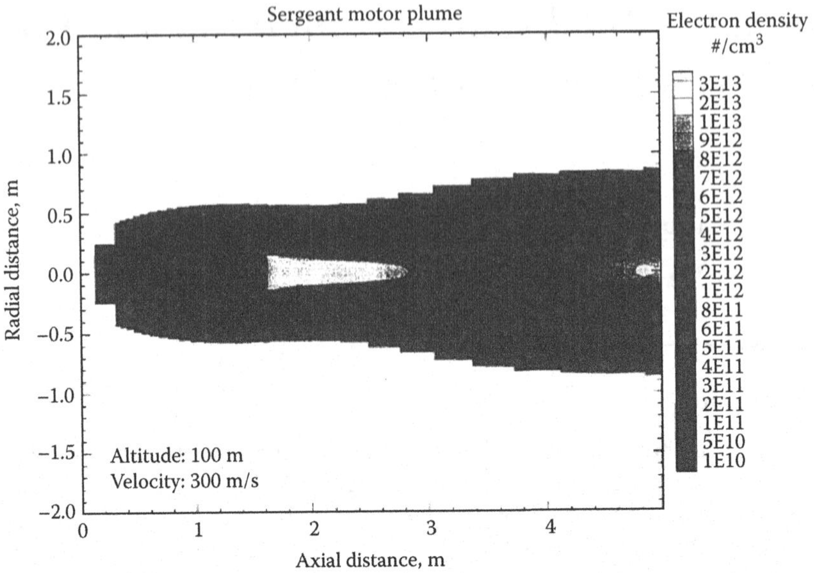

FIGURE 9A.1
Electron density of the plasma plume. (Reprinted from Final Report Summer Faculty Research Program, *Air Force Office of Scientific Research*, Report Number A669853, 38, 1994. With permission.)

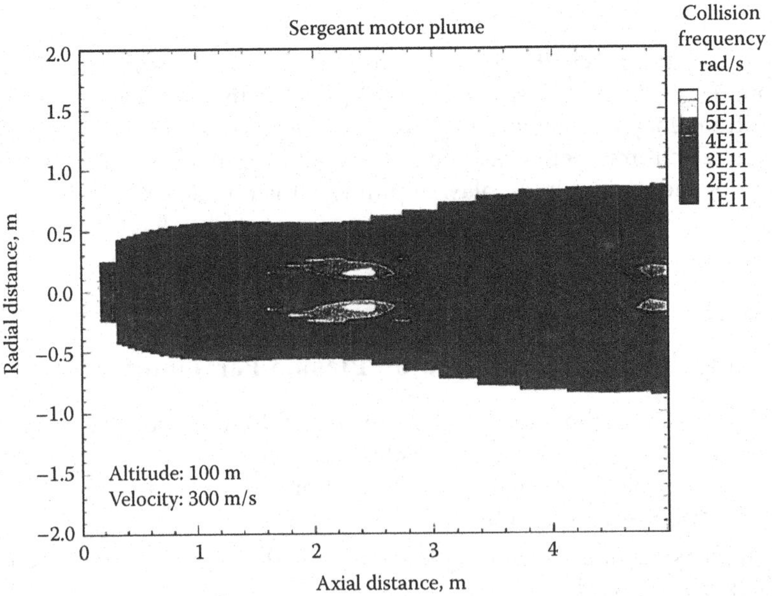

FIGURE 9A.2
Collision frequency of the plasma plume. (Reprinted from Final Report Summer Faculty Research Program, *Air Force Office of Scientific Research*, Report Number A669853, 38, 1994. With permission.)

3. The collision frequency (ν) is quite high and is of the order of plasma frequency on the axis. The plasma has to be classified as highly collisional.
4. The outer layer around $r = a$ has a turbulent character.
5. Apart from the outer layer, the rest of the column behaves like an overdense plasma (good conductor), for the frequency range 50 MHz to 10 GHz.
6. The fact that $r > a$ is free space and $r < a$, overdense plasma that behaves like a conductor suggests that the column is capable of supporting TM surface waves. The outer edge near $r = a$, being turbulent, behaves like a rough surface [1] for $f < f_p$.

The author of this report (henceforth will be referred to as the author) is strongly influenced by his experience with the study of surface plasmons at optical frequencies and wondered whether their excitation in this case will result in backscatter. However, before launching a full investigation into this aspect, the author wanted to understand the absorption of the high-frequency electromagnetic radiation by the highly collisional inhomogeneous plasma. This aspect is discussed in the following section.

9A.3 Absorption of TM Wave

The problem of the absorption of the electromagnetic wave by the highly collisional plasma column is studied by modeling the column as an inhomogeneous lossy plasma slab of width $d = 2a$. Figure 9A.3 shows this model, where the inhomogeneity is mathematically modeled, for illustrative purposes, by a trigonometric function given below:

$$\omega_p(z) = \omega_{pm}\left(\frac{\sin \pi z}{d}\right)^{m/2}. \tag{9A.2}$$

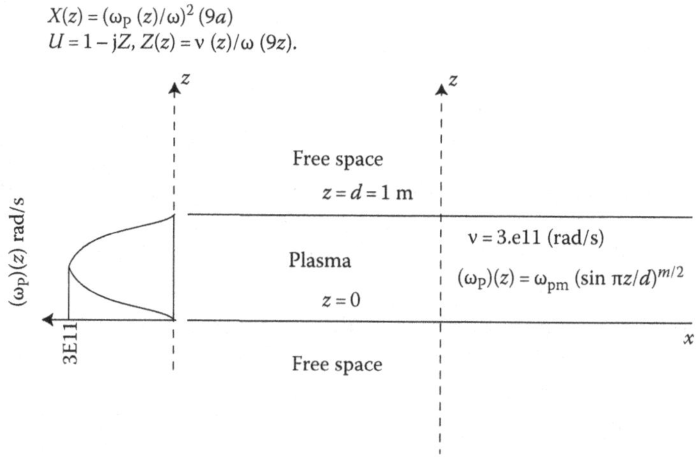

FIGURE 9A.3
Mathematical model for the inhomogeneity. (Reprinted from Final Report Summer Faculty Research Program, *Air Force Office of Scientific Research*, Report Number A669853, 38, 1994. With permission.)

By changing the value of m, the profile may be altered to fit the experimental profile approximately.

In the region $0 < z < d$, the equations satisfied by the time-harmonic fields are given below:

$$\bar{\nabla} \times \bar{E} = -j\omega\mu_0\bar{H}, \tag{9A.3}$$

$$\bar{\nabla} \times \bar{H} = -N_0 e\bar{v} + j\omega\varepsilon_0\bar{E}, \tag{9A.4}$$

$$j\omega m\bar{v} = -e\bar{E} - m\nu\bar{v}. \tag{9A.5}$$

Here E, H, and v are the electric, the magnetic, and the velocity fields, respectively. The other symbols have the usual meaning. Assuming the field quantities vary as

$$F(x, z, t) = F(z)e^{j(\omega t - Sx)}. \tag{9A.6}$$

The first-order-coupled differential equations satisfied by the state variables E_x and $(\eta_0 H_y)$ of a TM wave are obtained:

$$-\frac{1}{jk_0}\frac{dE_x}{dz} = \frac{C^2 - X/U}{1 - X/U}(\eta_0 H_y), \tag{9A.7}$$

$$-\frac{1}{jk_0}\frac{d(\eta_0 H_y)}{dz} = (1 - X/U)E_x. \tag{9A.8}$$

Here $S = \sin\theta$, $C = \cos\theta$, θ is the angle of incidence (angle with z-axis), $k_0 = \omega/c$, η_0 is the characteristic impedance of free space $= (\mu_0/\varepsilon_0)^{1/2} = 120\pi$ and the symbols X and U (notation used in magnetoionic theory) are

$$X(z) = \left(\frac{\omega_p(z)}{\omega}\right)^2, \tag{9A.9a}$$

$$U = 1 - jZ, \quad Z(z) = \frac{\nu(z)}{\omega}. \tag{9A.9b}$$

The following numerical method is devised to obtain the absorption coefficient.

1. Assume a suitable arbitrary complex value K_X for E_X at $z = d$, that is, $E_X(d) = K_X = E_X^T$. It follows that $(\eta_0 H_y)$ at $z = d$ is K_X/C.
2. Starting with these initial values, solve, numerically the first-order-coupled differential equations (9A.7) and (9A.8) by integrating downwards and obtain $E_X(0) = a_X$ and $\eta_0 H_y(0) = b_y$.
3. Calculate the reflection and transmission coefficients:

$$R_{11} - \frac{E_x^R}{E_x^I} = \frac{a_x - Cb_y}{a_x + Cb_y}, \tag{9A.10}$$

$$|T_{11}| = \left|\frac{E_x^{\mathrm{T}}}{E_x^{\mathrm{I}}}\right| = \left|\frac{2K_x}{a_x + Cb_y}\right|. \tag{9A.11}$$

Here the superscripts I, R, and T refer to incident, reflected, and transmitted fields, respectively. The subscript 11 refers to the parallel (TM) polarization of the incident wave.

4. Calculate the absorption coefficient:

$$A = 1 - \rho - \tau = 1 - |R_{11}|^2 - |T_{11}|^2, \tag{9A.12}$$

where ρ and τ are power reflection and transmission coefficients.

Figure 9A.4a through c are graphs for ρ, τ, and A versus f for an angle of incidence of 60°, respectively. Here ν is assumed constant and is equal to 3E11 (rad/s). From these graphs, it is clear that absorption increases with frequency. This result may be qualitatively explained by noting that the depth of penetration of the source wave increases with frequency since the value of z at which $\omega_p(z) = \omega$ increases with increase of f. In passing, it may be noted that a pseudo-Brewster angle exists for the TM case under consideration and shown in Figure 9A.5.

In conclusion, a theory based on only specular reflection and associated physics assumed above does not perhaps explain the increased backscatter in an intermediate frequency range. This led us to consider incorporating new physics into our model.

Turbulence present in the outer layers perhaps plays some part. Since for the frequency range under consideration, the plasma is overdense, the turbulent layer is modeled as a rough surface. The inner layers behave like a gaseous conductor. These thoughts led the author to consider the aspect of surface wave excitation.

9A.4 Review of Surface Waves

A surface wave [2,3] propagates along the surface $x = 0$ (see Figure 9A.6), with a phase velocity $V_{\mathrm{ph}} = \omega/k_z$. Its fields attenuate for $|x| > 0$. An interface between two dielectrics can support such a wave provided their dielectric constants are of opposite sign and $\varepsilon_{r1} < -\varepsilon_{r2}$. If medium 2 is the free space and medium 1 is a plasma whose plasma frequency is such that $\varepsilon_{r1} = \left(1 - \omega_p^2/\omega^2\right) < -1$, the conditions for the support of a surface wave at the interface are satisfied. Such a surface wave is called surface plasmon. For optical frequencies, medium 1 is a metal which behaves like a plasma with negative dielectric constant.

However, an electromagnetic wave in free space incident at any angle on the interface cannot excite the surface wave since the dispersion characteristic of the surface wave (k_z vs. ω) lies to the right of the light line (see Figure 9A.7b for an illustrative graph in which β is the real part of k_z). In optics, the required additional Δk_z is obtained by using a ATR coupler or a grating coupler or a rough surface.

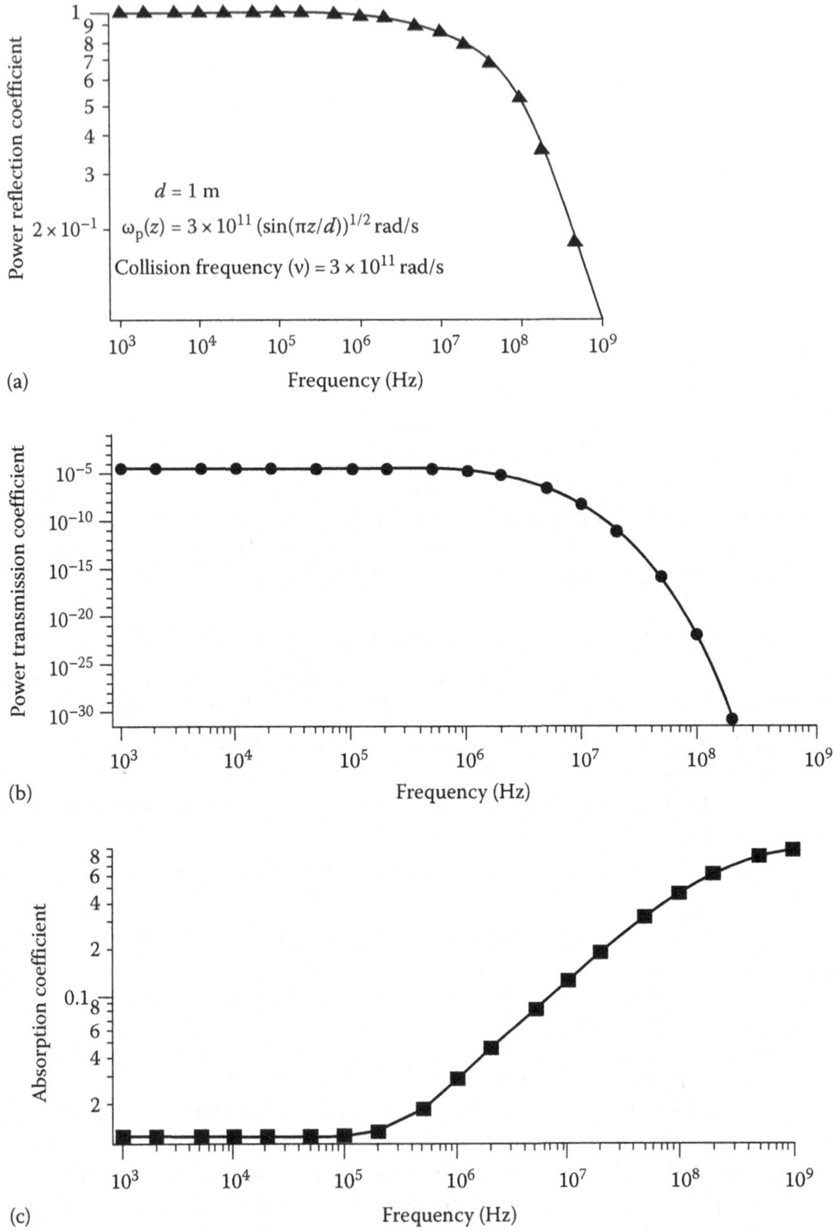

FIGURE 9A.4
Frequency dependence of various power coefficients: (a) reflection, (b) transmission, and (c) absorption angle of incidence is 60°. (Reprinted from Final Report Summer Faculty Research Program, *Air Force Office of Scientific Research*, Report Number A669853, 38, 1994. With permission.)

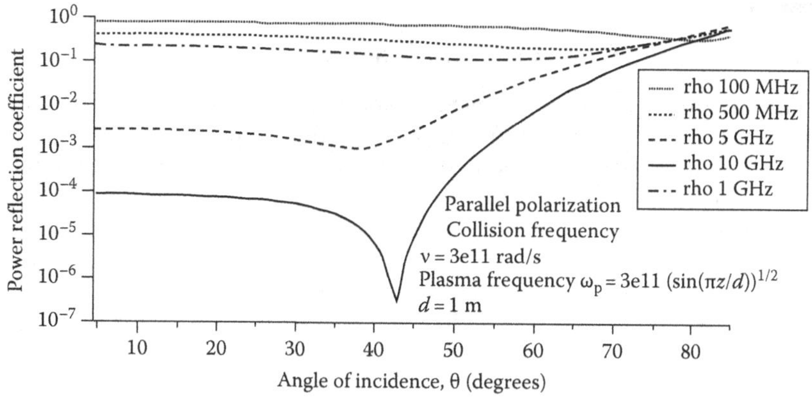

FIGURE 9A.5
Power reflection coefficient vs. angle of incidence. (Reprinted from Final Report Summer Faculty Research Program, *Air Force Office of Scientific Research*, Report Number A669853, 38, 1994. With permission.)

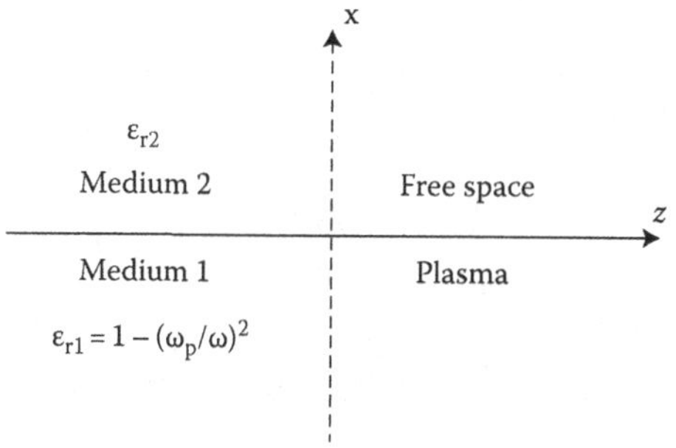

FIGURE 9A.6
Free space-plasma interface. (Reprinted from Final Report Summer Faculty Research Program, *Air Force Office of Scientific Research*, Report Number A669853, 38, 1994. With permission.)

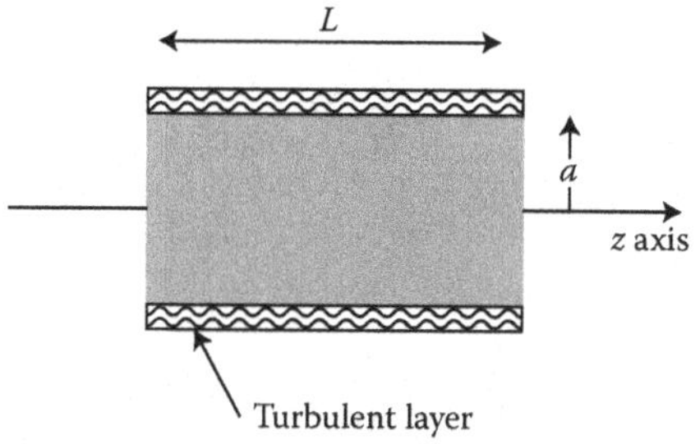

FIGURE 9A.7
Plasma with a turbulent layer. (Reprinted from Final Report Summer Faculty Research Program, *Air Force Office of Scientific Research*, Report Number A669853, 38, 1994. With permission.)

9A.5 Excitation of Surface Waves on Plasma Plume

From the material presented in Sections 9A.2 and 9A.4, it is clear that conditions exist for the excitation of a surface wave on the surface of the plasma plume. In this section, the equations needed to obtain the dispersion relation [3,4] of the surface plasmons are discussed (refer to Figure 9A.7).

Plasma region ($0 < r < a$):

$$\varepsilon_{\mathrm{p}}(r) = 1 - \frac{\omega_{\mathrm{p}}^2(r)}{\omega(\omega - \mathrm{j}\nu)}. \tag{9A.13}$$

Let

$$E_z^{\mathrm{p}}(r, \phi, z, t) = AG(r)\mathrm{e}^{\mathrm{j}(\omega t - k_z z)}, \tag{9A.14}$$

where

$$-k_{\mathrm{tp}}^2(r) + k_z^2 = k_0^2 \varepsilon_{\mathrm{p}}(r), \quad k_0 = \frac{\omega}{c} \tag{9A.15}$$

and $G(r)$ satisfies the differential equation:

$$\frac{\mathrm{d}^2 G}{\mathrm{d}r^2} + \left[\frac{1}{r} + \frac{k_z^2}{k_{\mathrm{tp}}^2(r)} \frac{\mathrm{d}\varepsilon_{\mathrm{p}}/\mathrm{d}r}{\varepsilon_{\mathrm{p}}(r)}\right]\frac{\mathrm{d}G}{\mathrm{d}r} - k_{\mathrm{tp}}^2(r)G = 0. \tag{9A.16}$$

Other field components of this TM wave can be expressed in terms of the z-component of the electric field.

$$E_r^{\mathrm{p}} = -\frac{\mathrm{j}}{k_{\mathrm{t0}}^2} k_z \frac{\partial E_z^{\mathrm{p}}}{\partial r}, \tag{9A.17a}$$

$$H_\phi^{\mathrm{p}} = -\frac{\mathrm{j}}{k_{\mathrm{t0}}^2} \omega \varepsilon_0 \varepsilon_{\mathrm{p}} \frac{\partial E_z^{\mathrm{p}}}{\partial r}, \tag{9A.17b}$$

$$Z_{\mathrm{TM}}^z = \frac{E_r}{H_\phi} = \frac{k_z}{\omega \varepsilon_0 \varepsilon_p}. \tag{9A.17c}$$

Free space region ($a < r < \infty$):

$$\frac{\mathrm{d}^2 E_z^0}{\mathrm{d}r^2} + \frac{1}{r}\frac{\mathrm{d}E_z^0}{\mathrm{d}r} - k_{\mathrm{t0}}^2 E_z^0 = 0, \tag{9A.18}$$

$$-k_{\mathrm{t0}}^2(r) + k_z^2 = k_0^2, \tag{9A.19}$$

$$E_z^0(r, \phi, z, t) = BK_0(k_{\mathrm{t0}} r)\mathrm{e}^{\mathrm{j}(\omega t - k_z z)}, \tag{9A.20}$$

$$E_r^0(r,\phi,z,t) = \mathrm{j}B\frac{k_z}{k_{t0}}K_1(k_{t0}r)\mathrm{e}^{\mathrm{j}(\omega t - k_z z)}, \tag{9A.21a}$$

$$H_\phi^0(r,\phi,z,t) = \frac{\mathrm{j}B}{k_{t0}}\omega\varepsilon_0 K_1(k_{t0}r)\mathrm{e}^{\mathrm{j}(\omega t - k_z z)}. \tag{9A.21b}$$

Here K_0 and K_1 are the modified Bessel functions of the second kind [5]. From the boundary conditions of continuity of E_z and H_ϕ at $r = a$, the following dispersion relation is obtained:

$$\frac{k_{tp}^2(a)}{\varepsilon_p(a)}\frac{G(a)}{G'(a)} + k_{t0}\frac{K_0(k_{t0}a)}{K_1(k_{t0}a)} = 0. \tag{9A.22}$$

In these equations while ω is real, k_z and several other quantities are complex and the complex mode is to be used for computations. In particular, let us denote

$$k_z = \beta - j\alpha, \tag{9A.23}$$

where β the phase constant and α the attenuation constant are real. The propagation velocity of the surface wave is given by

$$V_{ph} = \frac{\omega}{\beta}. \tag{9A.24}$$

9A.6 Numerical Method for Obtaining the Dispersion Relation

The above equations are used to compute the complex value of k_z for a given real value of ω. Numerical mode of the software Mathematica is used. The method consists of iterating between the two steps outlined below till convergence is obtained:

Step 1: Starting with a guessed value for k_z, the differential equation (9A.16) for G is numerically solved, with the initial conditions $G(0) = 0$ and $G'(0) = 0$. The singularity at $r = 0$ has to be handled with care. Thus, the values of $G(a)$ and $G'(a)$ are obtained.

Step 2: The dispersion relation which is a nonlinear algebraic equation is now solved for k_z.

Since the cylinder is thick, it is possible that for some values of ω, both $G(a)$ and $G'(a)$ are large though their ratio is not large. To cover this aspect, an alternative numerical method is also developed. The second-order differential equation for G is converted into a first-order nonlinear differential equation by defining a new variable $Y = G'/G$, which has the initial condition $Y = 0$:

$$\frac{dY}{dr} + \left[\frac{1}{r} + \frac{k_z^2}{k_{tp}^2(r)}\frac{d\varepsilon_p/dr}{\varepsilon_p(r)}\right]Y + Y^2 - k_{tp}^2(r) = 0. \tag{9A.25}$$

Once again the singularity at $r = 0$ has to be handled with care.

9A.7 Back Scatter

The surface traveling wave on the plasma plume is analogous to a current wave along thin-wire antennas [6] in the end-fire mode. It travels with a velocity close to that of light. The far-radiated field

$$E_r = -Ajk_0L\sin\theta_1 \frac{\sin\left[(1/2\mathrm{p})k_0L(1-p\cos\theta_1)\right]}{(1/2\mathrm{p})k_0L(1-p\cos\theta_1)}, \tag{9A.26}$$

where A is a constant with respect to θ_1 and L, θ_1 the angle of the point of observation away from the axis of the plume, and p the propagation velocity divided by c. p is close but slightly less than 1. The surface wave traveling a distance L along the axis is reflected when it encounters a spatial discontinuity in the properties of the medium at $z = L$. It is the reflected wave that gives rise to the backscatter. Figure 9A.8 gives a qualitative explanation [6] for the backscatter. For p close to 1, the first maximum in the pattern [5] occurs at

$$\theta_1 = 49.35\sqrt{\frac{\lambda}{L}}\,(\mathrm{deg}). \tag{9A.27}$$

9A.8 Sample Calculation

The electron density is assumed to be varying radially as a Bessel function of the first kind of zero order. Consequently, the square of the plasma frequency is given by

$$\omega_{\mathrm{p}}^2(r) = \omega_{\mathrm{pm}}^2 J_0\left(\mu\frac{r}{a}\right). \tag{9A.28}$$

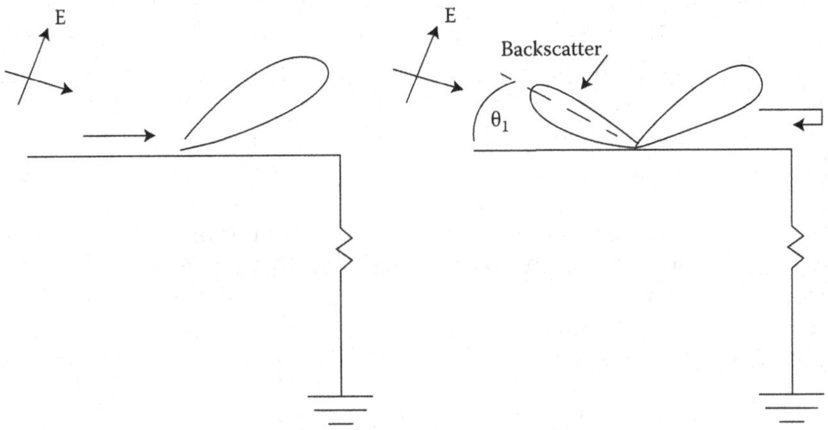

FIGURE 9A.8
Reflected wave giving rise to the back scatter. (Reprinted from Final Report Summer Faculty Research Program, *Air Force Office of Scientific Research*, Report Number A669853, 38, 1994. With permission.)

If $\mu = 0$, then we have the homogeneous case and if $\mu = 2.405$ the plasma density at the edge $r = a$ is zero. For the sample calculation, we chose $\mu = 2.1$. Thus,

$$\omega_p(a) = \omega_{pm}\sqrt{J_0(2.1)} = 0.4082\omega_{pm}. \tag{9A.29}$$

Choosing further ω_{pm} = 3E11 (rad/s), plasma frequency $f_p(a)$ = 19.51 GHz.

For the purpose of illustration, let the turbulent layer be described by $0.4 < r < 0.5$, thus $a = 0.4$ m. The collision frequency ν is taken to be 1.E11 (rad/m) which is the value in the outer layers. Using these parameters α versus f and β versus f are obtained numerically as described in Section 9A.6. Figure 9A.9 shows these results. The value of $L = 1$ m is assumed where from Figure 9A.2, there appears to be a discontinuity in the properties of the medium. For an illustrative value of f = 600 MHz, λ = 0.5 m and from Equation 9A.27, $\theta_1 = 34.9°$. The line to the left of the dispersion curve marked as surface wave on plasma column is the light line for an EM wave in free space incident at 34.9° with the axis. This curve is a straight line described by the equation:

$$k_{z0} = k_0 \cos 34.9°. \tag{9A.30}$$

The horizontal line marked as $\Delta k_z = 2.29$ is proportional to the z-component of the momentum (in the language of photonics) that is supplied by the turbulent rough surface.

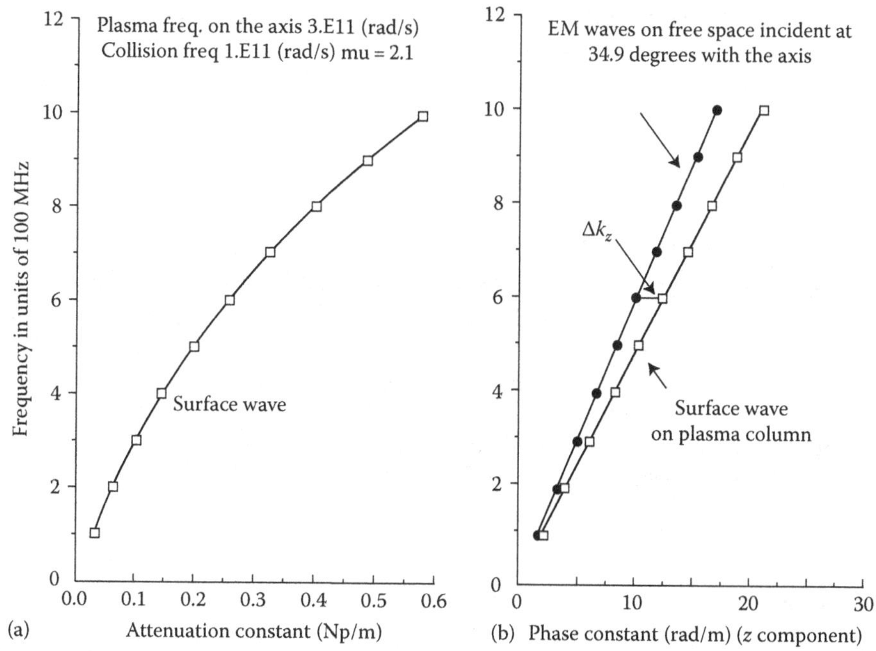

FIGURE 9A.9
Sample calculation: (a) attenuation constant and (b) phase constant. (Reprinted from Final Report Summer Faculty Research Program, *Air Force Office of Scientific Research*, Report Number A669853, 38, 1994. With permission.)

Superposition of the spectrum of plasma turbulence will lead to the frequency band of the surface waves that can be excited. Figure 9A.9a permits us to calculate the attenuation suffered by the surface wave in traveling a distance L before it reaches the point of discontinuity. At 600 MHz, $\alpha = 0.26119$ and $\exp(-2\alpha L) = 59.3\%$. Thus, 60% of power is still in tact for backscattering. All the numbers were obtained assuming p is nearly equal to 1 and this may be verified by calculating ω/β at $f = 600$ MHz. This gives a propagation velocity of 2.9767e8 m/s confirming $p \approx 1$.

9A.9 Results and Conclusions

Based on the results obtained so far, the following observations can be made with reference to the various parameters.

Effect of frequency: At low frequencies, the radiation field given in Equation 9A.26, which is inversely proportional to λ, is weak. At high frequencies, the attenuation constant is high and the surface wave is dissipated before it reaches the discontinuity and therefore the backscattering is weak.

Effect of L: As L increases, the radiation field increases as seen from Equation 9A.26 but the aspect angle θ_1 decreases, Δk_z decreases facilitating easier excitation of the surface waves. There is no change in α or β. However, the wave attenuation increases because of the increase in L. Perhaps, the center of the frequency band of significant backscatter moves toward higher frequencies as L increases.

Effect of polarization: The model assumed TM waves, since surface plasmons can be excited only when the wave electric field has a z-component.

9A.10 Scope for Future Work

A theory based on the excitation of surface waves on a plasma plume is constructed to offer a plausible explanation for increased backscatter in an intermediate frequency band. Sample calculations are made based on a simple model and approximate data. The theory will be improved and more accurate calculations will be made based on some more experimental data. Spectrum of the turbulent plasma will be incorporated into the model when the experimental data on this aspect becomes available. The author intends to submit summer research extension proposal to complete this work.

References

1. Hochstin, A. R. and Martens, C. P., *Proceedings of Symposium on Turbulence of Fluids and Plasmas*, Polytechnic Press, New York, 1968, p. 187.
2. Raether, H., *Surface Plasmons*, Springer, Berlin, Germany, 1988.

3. Boardman, A. D. (Ed.), *Electromagnetic Surface Modes*, Wiley, New York, 1982.
4. Zethoff, M. and Kortshagen, U., *J. Phys. D: Appl. Phys.*, 25, 187, 1992.
5. Abramowitz, M. and Stegun, I. A. (Eds.), *Handbook of Mathematical Functions*, Dover Publications, New York, 1965.
6. Knott, E. F., Shaeffer, J. F., and Tuley, M. T., *Radar Cross Section*, Artech House, Boston, MA, 1985, p. 148.

Appendix 10A: Thin Film Reflection Properties of a Warm Isotropic Plasma Slab between Two Half-Space Dielectric Media*

T. Markos, Constantine, and Dikshitulu K. Kalluri

10A.1 Introduction

Thin metal films will play an important role in next-generation integrated optoelectronics. Exploiting photon–electron coupling and the propagation of electron–plasma waves (plasmonics) shows great promise in building nanosized sensors. Modeling the reflection properties of thin metal films integrated with dielectrics provides valuable insights into device response to waves in the optical spectrum in terms of frequency, film thickness, and angle of incidence. Optical imaging devices using surface plasmons, at nanometer scale resolution, are being fabricated that has enhanced photocurrent, responsivity, and bandwidth [1]. Thin-film gratings are being fabricated using complex nanoimprinting methods for next-generation lithographic technology, where dielectrics/ferroelectrics/metals are integrated into single substrates to build tunable surface plasmon resonance filters [2]. Thin films fabricated as photonic crystal back-reflectors with, enhanced optical absorption, have potential for light harvesting in high-efficiency solar cell applications [3]. Understanding the behavior of waves as they interact with metal/dielectric films is crucial for building novel sensing devices.

Within an order of magnitude of the plasma frequency of a metal, some interesting phenomena occur as oscillations of the power reflection coefficient caused by the internal interference of fields within the slab [4]. The nature of the oscillations, such as the frequency and amplitude are a function of the slab thickness, obliqueness of the incident wave, indices of refraction of the bounding materials, and the spatial dispersion of the metal itself. This chapter describes the physical model of the thin film and the behavior of the power reflection coefficient. The warm plasma model will be used based on the hydrodynamic approximation that will show oscillations in the power reflection coefficient. Based on film thickness, the effect of power tunneling caused by the interaction of evanescent waves will be shown in the region of total internal reflection.

The wave equation for a warm, compressible, anisotropic magnetoplasma was derived by Unz [5]. Prasad [6] derived the power reflection coefficient using state–space methods for a warm plasma slab in the free space and described the maxima/minima points as well as the characteristic roots q. Kalluri and Prasad [7] described the thin film reflection properties of a cold plasma slab in the free space for the isotropic case and described the oscillations internal to the slab where the maxima/minima were derived. The papers from Kalluri

* The authors are with the Electromagnetics and Complex Media Research Laboratory, University of Massachusetts Lowell. The article is a reprint of the report UML-EM & CM-2010-1. Reprinted with permission from the director of the laboratory, Dikshitulu K. Kalluri.

and Prasad consider $n_I = n_T = 1$. The power reflection coefficient in this chapter is derived from detailed equations starting from Maxwell's equations and the hydrodynamic approximation with consideration of different input/output materials. The input transverse wave is obliquely incident on the slab interface, which reproduces itself inside the plasma, and, in addition, supporting longitudinal acoustic waves from the interface, with different wave numbers.

Special cases of the power reflection coefficient are presented for different refractive indices in frequency space. The overall power reflection coefficient is periodic with decreasing periodicity. When the acoustic mode velocity approaches zero, the plasma becomes cold and incompressible. The longitudinal characteristic roots are dependent on the parameter $\delta = a/c$, where a is the acoustic velocity of the electron plasma wave. In certain metals the Fermi velocity [8] plays the role of the acoustic velocity. When a is comparable to c, that is, $\delta \approx 0.1$–0.01, the plasma is spatially dispersive and is considered to be warm. A small percentage change in the velocity of the longitudinal component gives rise to a dynamic response of the power reflection coefficient, of note, which is periodic with a slight aperiodicity. In addition, the oscillations within the warm plasma reflection coefficient tend to ride atop the cold plasma component.

10A.2 Formulation of the Problem

To describe the electromagnetic waves inside the thin film, the metal is modeled as a lossy (collisions) homogeneous, isotropic (no static $\vec{B}_0$) warm plasma slab of thickness d. The plasma slab of n_p is sandwiched between two different semiinfinite dielectric media of n_I (input medium) and n_T (output medium) (Figure 10A.1).

A parallel (TM or p) wave E^I is incident on the slab at an angle θ^I with reflected wave at angle $\theta^R = \theta^I$ and transmitted wave at angle θ^T all on the plane of incidence.

10A.2.1 Waves Outside the Slab

For the incident, reflected, and transmitted (I, R, and T) waves outside the plasma slab, the electric and magnetic fields obliquely incident can be described as

$$\vec{E}^{I,R,T} = \vec{E}_0^{I,R,T}\psi^{I,R,T}, \quad \vec{H}^{I,R,T} = \vec{H}_0^{I,R,T}\psi^{I,R,T}, \tag{10A.1}$$

where $\vec{E}_0^{I,R,T}$ and $\vec{H}_0^{I,R,T}$ are the complex electric and magnetic amplitudes given by

$$\vec{E}_0^{I,R,T} = \hat{x}E_x^{I,R,T} + \hat{y}E_y^{I,R,T} + \hat{z}E_z^{I,R,T}, \quad \vec{H}_0^{I,R,T} = \hat{x}H_x^{I,R,T} + \hat{y}H_y^{I,R,T} + \hat{z}H_z^{I,R,T}. \tag{10A.2}$$

$\psi^{I,R,T}$ is the phase reference of the waveform at $\vec{r} = 0$,

$$\psi^{I,R,T} = e^{-j\vec{k}_{I,R,T}\cdot\vec{r}}. \tag{10A.3}$$

From Figure 10A.1, the geometry of the wave vector can be described as

$$\vec{k}_I = k_0 n_I\left(\hat{x}S + \hat{z}C\right), \quad \vec{k}_R = k_0 n_I\left(\hat{x}S - \hat{z}C\right), \quad \vec{k}_T = k_0 n_T\left(\hat{x}S^T + \hat{z}C^T\right), \tag{10A.4}$$

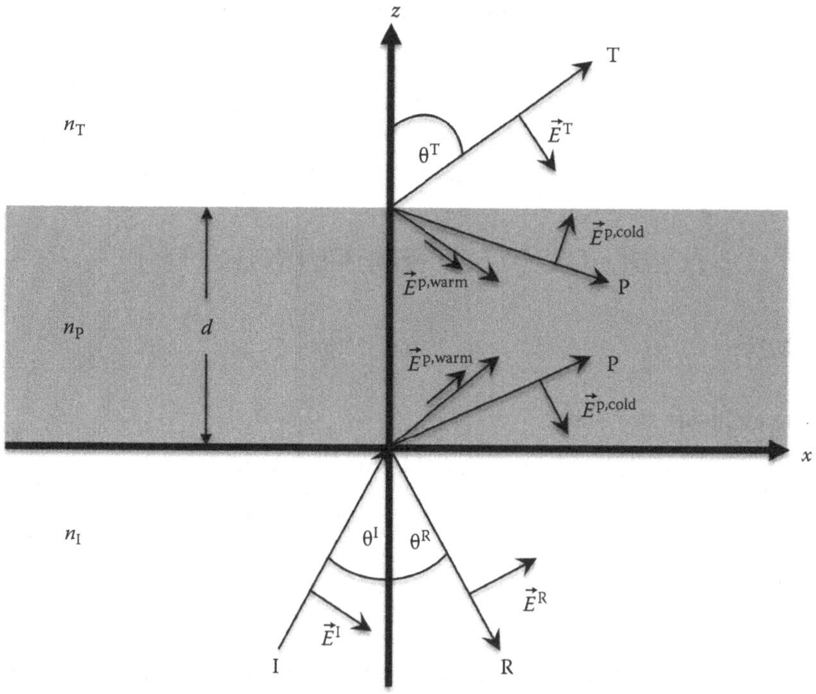

FIGURE 10A.1
Oblique incidence of a (TM or *p*) wave on a warm plasma slab and reflection/transmission waves. I, incident; R, reflected; T, transmitted; P, plasma.

where $S^T = k_I/k_T S$ and $C^T = \sqrt{1-(k_I/k_T S)^2}$. S^T and C^T describe the characteristics of the dielectric in the output of the system. Note, $S = \sin(\theta^I)$, $C = \cos(\theta^I)$, $S^T = \sin(\theta^T)$, and $C^T = \cos(\theta^T)$. If $k_I = k_T$, then $S^T = S$ and $C^T = C$. Since the position vector is $\vec{r} = \hat{x}x + \hat{y}y + \hat{z}z$, the phase factor of the waves outside the plasma slab become

$$\psi^I = e^{-jk_0 n_I (xS+zC)}, \quad \psi^R = e^{-jk_0 n_I (xS-zC)}, \quad \psi^T = e^{-jk_0 n_I (xS+zC')}, \tag{10A.5}$$

where $C' = \sqrt{(k_T/k_I)^2 - S^2} = (k_T k_I)C^T$. From ψ^T, $k_{T,x}^2 + k_{T,z}^2 = k_T^2$, where $k_T \omega/c\sqrt{\varepsilon_T} = \omega/cn_T = k_0 n_T$. The angle-dependent parameter, C', in ψ^T (Equation 10A.5) was derived using Snell's law. Snell's law implies that the boundary conditions (or phase match) has been satisfied as is evident in the *x*-component of (Equation 10A.5) and also by the fact that $\theta^I = \theta^R$. The *x*-components of the phase factor for the incident, reflected, and transmitted waves are the same to satisfy the boundary conditions, and hence, for the TM wave the tangential components of $\vec{E}$ and $\vec{H}$ are continuous across the boundary in both space and time.

For a plane wave propagation [9], the electric and magnetic fields can be described as

$$\vec{k}_{I,R,T} \cdot \vec{E}^{I,R,T} = 0, \quad \vec{H}^{I,R,T} = \frac{1}{\omega\mu_0} \vec{k}_{I,R,T} \times \vec{E}^{I,R,T}, \tag{10A.6}$$

where μ_0 is the free-space permeability. Using Equation 10A.1 in Equation 10A.6, we obtain

$$SE_x^{\mathrm{I}} = -CE_z^{\mathrm{I}}, \quad SE_x^{\mathrm{R}} = CE_z^{\mathrm{R}}, \quad SE_x^{\mathrm{T}} = -C'E_z^{\mathrm{T}}, \tag{10A.7a}$$

$$S\eta^{\mathrm{I}}H_y^{\mathrm{I}} = -E_z^{\mathrm{I}}, \quad S\eta^{\mathrm{I}}H_y^{\mathrm{R}} = -E_z^{\mathrm{R}}, \quad S\eta^{\mathrm{I}}H_y^{\mathrm{T}} = -E_z^{\mathrm{T}}\left(C'^2 + S^2\right), \tag{10A.7b}$$

$$C\eta^{\mathrm{I}}H_y^{\mathrm{I}} = E_x^{\mathrm{I}}, \quad C\eta^{\mathrm{I}}H_y^{\mathrm{R}} = -E_x^{\mathrm{R}}, \quad C'\eta^{\mathrm{I}}H_y^{\mathrm{T}}E_x^{\mathrm{T}}\left(C'^2 + S^2\right), \tag{10A.7c}$$

where $\eta^{\mathrm{I}} = \eta_0/n_{\mathrm{I}}$.

10A.2.2 Waves Inside the Slab

The basic equations for a warm plasma are Maxwell's equations:

$$\vec{\nabla} \times \vec{E}^{\mathrm{P}} = -j\omega\mu\vec{H}^{\mathrm{P}}, \tag{10A.8a}$$

$$\vec{\nabla} \times \vec{H}^{\mathrm{P}} = j\omega\varepsilon\vec{E}^{\mathrm{P}} - eN_0\vec{u}. \tag{10A.8b}$$

Using the hydrodynamic approximation for a compressible electron gas the conservation of momentum [5] (isotropic case)

$$mN_0\left(j\omega + \nu\right)\vec{u} = -\vec{\nabla}p - eN_0\vec{E}^{\mathrm{P}}. \tag{10A.9a}$$

The conservation of energy is

$$p = \gamma KT_0N. \tag{10A.9b}$$

The equation of state is

$$\frac{p}{p_0} = \frac{N}{N_0} + \frac{T}{T_0}, \quad p_0 = KN_0T_0, \tag{10A.9c}$$

where N, N_0 are the small and large signal number density (m^{-3}), p, p_0 the small and large signal pressure (kg/m s^2), T, T_0 the small and large signal temperature (K), °K the Boltzman constant, γ the specific heat in constant pressure/specific heat in constant volume, m the electron mass (kg), $\vec{u}$ the velocity vector (m/s).

The formulation given above is based on the linear theory. Unz showed that Equations 10A.8 and 10A.9 reduce to the wave equation

$$U\left(1 - n_{\mathrm{P}}^2 - \frac{X}{U}\right)\vec{E}^{\mathrm{P}} + \left(U - \delta^2\right)\left(\vec{n}_{\mathrm{P}} \cdot \vec{E}^{\mathrm{P}}\right)\vec{n} = 0, \tag{10A.10}$$

where

$$U = 1 - jZ, \tag{10A.11}$$

and Z is described in terms of the electron collision frequency ν and transmission frequency ω:

$$Z = \frac{\nu}{\omega}. \tag{10A.12}$$

The term

$$X = \left(\frac{\omega_P}{\omega}\right)^2 \tag{10A.13}$$

is inversely proportional to the normalized frequency squared, $X = 1/\Omega^2$, where $\Omega = \omega/\omega_p$. Within the film four modes exist: Transmitted/reflected transverse waves and transmitted/reflected longitudinal waves. The boundary at $z = 0$, d serves to act as the source of an acoustic wave mode only supported within the slab medium. Within the slab, a TM wave and an acoustic wave with velocity, pressure, and electric field intensity, both satisfy the wave equation (10A.10). In this chapter, the dimensionless parameter n_p can be described in terms of the input medium, input angle, and plasma characteristic root parameter, q':

$$\vec{n}_P = n_I\left(S\hat{x} + q'\hat{z}\right) \quad \text{or} \quad n_P^2 = n_I^2\left(S^2 + q'^2\right), \tag{10A.14}$$

where q' will be determined. Note: q' is not the derivative of q but rather distinguishes between a $n_I > 1$ dependent solution of the dispersion equation and a $n_I = 1$ solution. For a warm plasma, the velocity of the longitudinal component is significant and in certain metals is approximately two to three orders of magnitude less of the speed of light c. The electric and magnetic field intensities inside the plasma can be described as

$$\vec{E}^P = \vec{E}_0^P \psi^P, \quad \vec{H}^P = \vec{H}_0^P \psi^P. \tag{10A.15}$$

The electric and magnetic fields in Equation 10A.15 have complex field amplitudes as

$$\vec{E}_0^P = \hat{x}E_x^P + \hat{y}E_y^P + \hat{z}E_z^P, \quad \vec{H}_0^P = \hat{x}H_x^P + \hat{y}H_y^P + \hat{z}H_z^P. \tag{10A.16}$$

The phase of the fields within the plasma slab is defined as

$$\psi^P = e^{-jk_0 n_I (xS + zq')}. \tag{10A.17}$$

When using Equations 10A.15 and 10A.16 in Equation 10A.10, one has for oblique incidence on the slab interface:

$$U\left(1 - n_P^2 - \frac{X}{U}\right)E_x^P + \left(U - \delta^2\right)n_I^2\left(S^2 E_x^P + Sq'E_z^P\right) = 0, \tag{10A.18a}$$

$$U\left(1 - n_P^2 - \frac{X}{U}\right)E_y^P = 0, \tag{10A.18b}$$

$$U\left(1 - n_P^2 - \frac{X}{U}\right)E_z^P + \left(U - \delta^2\right)n_I^2\left(Sq'E_x^P + q'^2 E_z^P\right) = 0. \tag{10A.18c}$$

Equations 10A.18a through 10A.18c can be rearranged in a matrix form as

$$[F][E]=0, \tag{10A.19}$$

which is

$$[F][E]=\begin{bmatrix} F_{11} & F_{12} & F_{13} \\ F_{21} & F_{22} & F_{23} \\ F_{31} & F_{32} & F_{33} \end{bmatrix}\begin{bmatrix} E_x \\ E_y \\ E_z \end{bmatrix}. \tag{10A.20}$$

The components of Equation 10A.20 come from Equations 10A.18a through 10A.18c and are as follows:

$$F_{11}=U\left(1-n_{\mathrm{p}}^2-\frac{X}{U}\right)+\left(U-\delta^2\right)n_{\mathrm{I}}^2S^2, \tag{10A.21a}$$

$$F_{12}=0, \tag{10A.21b}$$

$$F_{13}=\left(U-\delta^2\right)n_{\mathrm{I}}^2Sq', \tag{10A.21c}$$

$$F_{21}=0, \tag{10A.21d}$$

$$F_{22}=U\left(1-n_{\mathrm{p}}^2-\frac{X}{U}\right), \tag{10A.21e}$$

$$F_{23}=0, \tag{10A.21f}$$

$$F_{31}=\left(U-\delta^2\right)n_{\mathrm{I}}^2Sq', \tag{10A.21g}$$

$$F_{32}=0, \tag{10A.21h}$$

$$F_{33}=U\left(1-n_{\mathrm{p}}^2-\frac{X}{U}\right)+\left(U-\delta^2\right)n_{\mathrm{I}}^2q'^2. \tag{10A.21i}$$

Setting determinant of $[F]$ to zero yields the dispersion relation $F_{11}F_{33}-F_{13}F_{31}=0$. The characteristic roots for the cold and warm plasma models are derived directly from the dispersion relation

$$q'_{1,2}=\pm\left(\frac{1}{n_{\mathrm{I}}^2}\left(1-\frac{X}{U}\right)-S^2\right)^{1/2},\quad q'_{3,4}=\pm\left(\frac{1}{\delta^2n_{\mathrm{I}}^2}(U-X)-S^2\right)^{1/2}. \tag{10A.22}$$

Subscript 1 indicates an upward-going wave and subscript 2 indicates a reflected downward-going wave of the transverse type, similarly the subscripts 3 and 4 of the longitudinal type. Using Equation 10A.8 to find the magnetic field intensity yields

$$\begin{vmatrix} \hat{x} & \hat{y} & \hat{z} \\ \partial/\partial x & \partial/\partial y & \partial/\partial z \\ E_x^{\mathrm{P}} & E_y^{\mathrm{P}} & E_z^{\mathrm{P}} \end{vmatrix}=\begin{vmatrix} \hat{x} & \hat{y} & \hat{z} \\ -jk_0n_{\mathrm{I}}S & 0 & -jk_0n_{\mathrm{I}}q' \\ E_x^{\mathrm{P}} & E_y^{\mathrm{P}} & E_z^{\mathrm{P}} \end{vmatrix}. \tag{10A.23}$$

Since the incident wave is a TM wave, only the y-component matters, and hence

$$\hat{y}\left(\vec{\nabla}\times\vec{E}^{\mathrm{P}}\right)=-\hat{y}\left(-\mathrm{j}k_0n_{\mathrm{I}}SE_z^{\mathrm{P}}+\mathrm{j}k_0n_{\mathrm{I}}q'E_x^{\mathrm{P}}\right). \tag{10A.24}$$

From dispersion relation and Equation 10A.24, the ratio of the magnetic to electric tangential components becomes

$$\frac{H_y^{\mathrm{P}}}{E_x^{\mathrm{P}}}=\frac{U-X-\delta^2n_{\mathrm{P}}^2}{\eta_0n_{\mathrm{I}}q'\left(U-\delta^2\right)}. \tag{10A.25}$$

The warm plasma electric field-dependent electron velocity field is given by

$$\vec{u}=\frac{1}{\beta_0}\left(\vec{E}-\frac{1}{k_0^2}\vec{\nabla}\times\vec{\nabla}\times\vec{E}^{\mathrm{P}}\right), \tag{10A.26}$$

where $\beta_0=eN_0/\mathrm{j}\omega\varepsilon_0=\omega mX/\mathrm{j}e$. Using Equations 10A.21a and 10A.21c, Equation 10A.26 becomes

$$\frac{u_z}{E_x}=\frac{1}{\beta_0}\left(n_{\mathrm{I}}^2q'S-\frac{U\left(1-n_{\mathrm{P}}^2-X/U\right)+\left(U-\delta^2\right)n_{\mathrm{I}}^2S^2}{\left(U-\delta^2\right)n_{\mathrm{I}}^2Sq'}\left(1-n_{\mathrm{I}}^2S^2\right)\right). \tag{10A.27}$$

10A.2.3 Power Reflection Coefficient for a Warm Isotropic Plasma Slab between Two Different Dielectric Media (i.e., n_{I} and n_{T})

In considering the case where the plasma medium is warm, a longitudinal electric field must be taken into account. The power reflection coefficient is derived for a warm plasma sandwiched between two different dielectric media n_{I} and n_{T}. Using Equations 10A.25 and 10A.27 with the following equations for a collisionless plasma, $U = 1$, we have

$$\eta_{y1}=\frac{\eta_0n_{\mathrm{I}}H_{y1}^{P}}{E_{x1}^{\mathrm{P}}}=\frac{1-X}{q_1},\quad \eta_{y3}=\frac{\eta_0n_{\mathrm{I}}H_{y3}^{\mathrm{P}}}{E_{x3}^{\mathrm{P}}}=0, \tag{10A.28a}$$

$$\beta_{z1}=\beta_0\frac{u_{z1}^{\mathrm{P}}}{E_{x1}^{\mathrm{P}}}=-\mathrm{j}\frac{S}{q_1},\quad \beta_{z3}=\beta_0\frac{u_{z3}^{\mathrm{P}}}{E_{x3}^{\mathrm{P}}}=\mathrm{j}\frac{q_3}{SX}, \tag{10A.28b}$$

$$\eta_{y2}=-\eta_{y1},\quad \eta_{y4}=\eta_{y3},\quad \beta_{z2}=-\beta_{z1},\quad \beta_{z4}=-\beta_{z3}. \tag{10A.28c}$$

We use boundary conditions where the tangential components of $\vec{E}$ and $\vec{H}$ are continuous at $z = 0$ and d to generate four equations. For time-harmonic waves, tangential boundary conditions are used as they are derived by Maxwell's curl equations. Normal boundary conditions (from divergence equations) can be derived from the curl equations. Hence, normal boundary conditions are not independent but are assured to be satisfied when the tangential boundary conditions are satisfied [10]. Two more equations are generated from the boundary conditions to account for the z component of the electron velocity to be zero at the interfaces since there is no electron flow within the dielectrics. Writing the system of equations from the boundary conditions, Equations 10A.1 through 10A.7 and (10A.28):

$$\sum_{i=1}^{4} E_{xi}^{\mathrm{P}} = E_x^{\mathrm{I}} + E_x^{\mathrm{R}}, \tag{10A.29a}$$

$$\sum_{i=1}^{4} H_{yi}^{\mathrm{P}} = H_y^{\mathrm{I}} + H_y^{\mathrm{R}}, \tag{10A.29b}$$

$$\sum_{i=1}^{4} u_{zi}^{\mathrm{P}} = 0, \tag{10A.29c}$$

$$\sum_{i=1}^{4} E_{xi}^{\mathrm{P}} \mathrm{e}^{-\mathrm{j}k_0 n_{\mathrm{I}} q_i' d} = E_x^{\mathrm{T}} \mathrm{e}^{-\mathrm{j}k_0 n_{\mathrm{I}} C' d}, \tag{10A.29d}$$

$$\sum_{i=1}^{4} H_{yi}^{\mathrm{P}} \mathrm{e}^{-\mathrm{j}k_0 n_{\mathrm{I}} q_i' d} = H_y^{\mathrm{T}} \mathrm{e}^{-\mathrm{j}k_0 n_{\mathrm{I}} C' d}, \tag{10A.29e}$$

$$\sum_{i=1}^{4} u_{zi}^{\mathrm{P}} \mathrm{e}^{-\mathrm{j}k_0 n_{\mathrm{I}} q_i' d} = 0. \tag{10A.29f}$$

The longitudinal component of the waveform inside the plasma is comprised of the electric field only, since η_{y3} is equal to zero. Putting Equation 10A.29 in a matrix form gives

$$\begin{bmatrix} 1 & 1 & 1 & 1 & -1 & 0 \\ \dfrac{c\eta_{y1}}{n_{\mathrm{I}}^2} & -\dfrac{c\eta_{y1}}{n_{\mathrm{I}}^2} & 0 & 0 & 1 & 0 \\ \beta_{z1} & -\beta_{z1} & \beta_{z3} & -\beta_{z3} & 0 & 0 \\ \lambda_1' & \lambda_2' & \lambda_3' & \lambda_4' & 0 & -1 \\ \dfrac{C\lambda_1'\eta_{y1}}{n_{\mathrm{I}}^2\left(C'^2+S^2\right)} & \dfrac{C\lambda_2'\eta_{y1}}{n_{\mathrm{I}}^2\left(C'^2+S^2\right)} & 0 & 0 & 0 & -1 \\ \lambda_1'\beta_{z1} & -\lambda_2'\beta_{z1} & \lambda_3'\beta_{z3} & -\lambda_4'\beta_{z4} & 0 & 0 \end{bmatrix} \begin{bmatrix} E_{x1}^{\mathrm{P}} \\ E_{x2}^{\mathrm{P}} \\ E_{x3}^{\mathrm{P}} \\ E_{x4}^{\mathrm{P}} \\ E_x^{\mathrm{R}} \\ E_x^{\mathrm{T}'} \end{bmatrix} = \begin{bmatrix} 1 \\ 1 \\ 0 \\ 0 \\ 0 \\ 0 \end{bmatrix} E_x^{\mathrm{I}}, \tag{10A.30}$$

where $\lambda_i' = \mathrm{e}^{-\mathrm{j}k_0 n_{\mathrm{I}} q_i' d}$, $E_x^{\mathrm{T}'} = E_x^{\mathrm{T}} \mathrm{e}^{-\mathrm{j}k_0 n_{\mathrm{I}} C' d}$, and $H_y^{\mathrm{T}'} = H_y^{\mathrm{T}} \mathrm{e}^{-\mathrm{j}k_0 n_{\mathrm{I}} C' d}$.

The complex reflection coefficient can be solved by the ratio of the reflected and incident electric fields

$$_{\parallel}R_{\parallel} = \frac{E_x^{\mathrm{R}}}{E_x^{\mathrm{I}}}. \tag{10A.31}$$

The symbol $\|$ implies that the incident and reflected waves are p-waves on the plane of incidence. After substantial simplification of Equation 10A.31, ${}_{\|}R_{\|}$ becomes

$$\begin{aligned}
{}_{\|}R_{\|} = -\Big[& \beta_{z1}\beta_{z3}\Big[\left(\lambda_4'+\lambda_3'\right)\left(\lambda_2'-\lambda_1'\right)\eta_{y1}n_I^2\left(C'-C\left(C'^2+S^2\right)\right) \\
& -\left(\lambda_1'\lambda_2'+\lambda_3'\lambda_4'\right)4n_I^4\left(C'^2+S^2\right)+\left(\lambda_4'+\lambda_3'\right)\left(\lambda_1'+\lambda_2'\right)2n_I^4\left(C'^2+S^2\right)\Big] \\
& +\beta_{z1}^2\left(\lambda_4'-\lambda_3'\right)\left(\lambda_1'-\lambda_2'\right)n_I^4\left(C'^2+S^2\right) \\
& +\beta_{z3}^2\Big[\left(\lambda_4'-\lambda_3'\right)\left(\lambda_1'+\lambda_2'\right)\eta_{y1}n_I^2\left(C\left(C'^2+S^2\right)-C'\right) \\
& +\left(\lambda_4'-\lambda_3'\right)\left(\lambda_1'-\lambda_2'\right)\left(n_I^4\left(C'^2+S^2\right)-CC'\eta_{y1}^2\right)\Big]\Big]/\Delta
\end{aligned} \tag{10A.32}$$

where

$$\begin{aligned}
\Delta = & \beta_{z1}\beta_{z3}\Big[\left(\lambda_4'+\lambda_3'\right)\left(\lambda_1'-\lambda_2'\right)\eta_{y1}n_I^2\left(C'-C\left(C'^2+S^2\right)\right) \\
& +\left(\lambda_1'\lambda_2'+\lambda_3'\lambda_4'\right)4n_I^4\left(C'^2+S^2\right)-\left(\lambda_4'+\lambda_3'\right)\left(\lambda_1'+\lambda_2'\right)2n_I^4\left(C'^2+S^2\right)\Big] \\
& +\beta_{z1}^2\left(\lambda_3'-\lambda_4'\right)\left(\lambda_1'-\lambda_2'\right)n_I^4\left(C'^2+S^2\right) \\
& +\beta_{z3}^2\Big[\left(\lambda_4'-\lambda_3'\right)\left(\lambda_1'+\lambda_2'\right)\eta_{y1}n_I^2\left(C\left(C'^2+S^2\right)-C'\right) \\
& -\left(\lambda_4'-\lambda_3'\right)\left(\lambda_1'-\lambda_2'\right)\left(n_I^4\left(C'^2+S^2\right)-CC'\eta_{y1}^2\right)\Big].
\end{aligned} \tag{10A.33}$$

The power reflection coefficient is defined by the magnitude squared of ${}_{\|}R_{\|}$

$$\rho_{\|} = \left|{}_{\|}R_{\|}\right|^2. \tag{10A.34}$$

A closed form of Equation 10A.34 is not given here because of its high complexity; it can either be studied using computer code or from approximation. The refractive indices n_I and n_T cause the angle-dependent parameter C' to become zero at the critical angle. The real and imaginary parts of the complex reflection coefficient change depending on whether the incident angle is less than or greater than the critical angle. Equation 10A.34 is much dependent on n_I and n_T such that the choice of material will dramatically change the reflection properties of the composite material. Figure 10A.2 shows the power reflection coefficient for a normalized film thickness of $d_p = d/\lambda_p = 0.2$, where λ_p is the plasma wavelength, $\lambda_p = 2\pi c/\omega_p$, an input angle of $\theta^I = 60°$ and materials $n_I = 1.1$ and $n_T = 1.3$ with $\delta = 0.1$. The power reflection coefficient changes from a monotonic to a complex oscillatory behavior that occurs between zero and one with a periodicity that decreases with increasing frequency.

10A.3 Different Regions of the Power Reflection Coefficient

A method in understanding the reflection properties is by examining that the dispersion relation, when solved, leads to the characteristic roots of the plasma medium. For the isotropic case, there are four roots in total, two for an upward- and downward-going set of

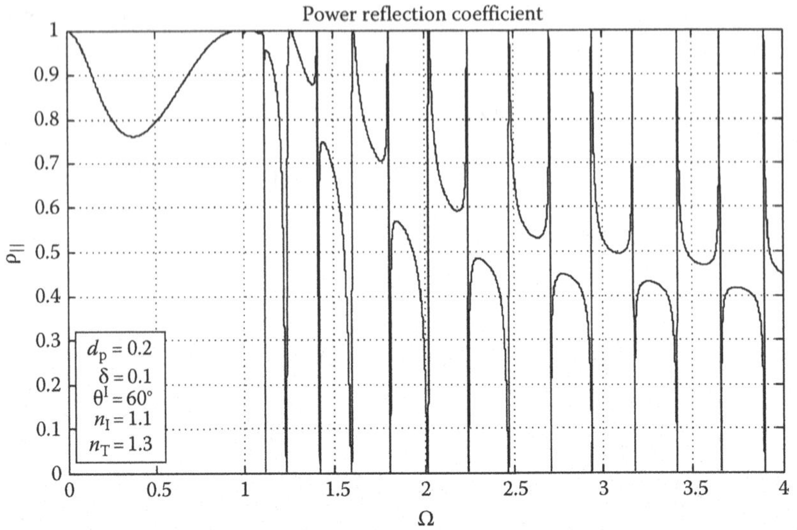

FIGURE 10A.2
Power reflection coefficient for $d_P = 0.2$, $\delta = 0.1$, $\theta^I = 60°$, $n_I = 1.1$, and $n_T = 1.3$.

transverse modes, $q'_{1,2}$ and similarly two for a set of longitudinal (acoustic) modes, $q'_{3,4}$. The transverse modes can be considered as the cold plasma modes and the longitudinal modes can be considered as the warm plasma modes. For the lossless case $U = 1$, the characteristic roots will be either real or imaginary but not complex. The zeros of the roots, considered here as critical points, are dependent on a number of parameters. The critical points are also the points where the roots change from real to imaginary and vice versa. Equation 10A.22 shows that as Ω approaches zero the characteristic roots approaches infinity. In addition, as Ω approaches infinity, the roots asymptotically reach levels dependent on the incident angle. The critical points divide $\rho_{||}$ into three main regions (see Figure 10A.3): The first region, $0 < \Omega < [1 - (\delta n_I S)^2]^{-1/2}$: both q'_1 and q'_3 are imaginary causing total reflection of the incident power for thick materials. Note, in $\rho_{||}$ there is another phenomenon known as power tunneling that will be described to explain a dip seen in this first region for thin materials. The first region is also where attenuated total reflection (ATR) can occur by careful choice of index material and incident angle. The second region $[1 - (\delta n_I S)^2]^{-1/2} < \Omega < [1 - (n_I S)^2]^{-1/2}$: q'_1 is still imaginary but q'_3 is real allowing the longitudinal (acoustic) plasma wave to propagate through the plasma. In this region, $\rho_{||}$ takes on an oscillatory behavior as a result of the internal reflections of the longitudinal component. For the case of $n_I = n_T = 1$, an approximation to $\rho_{||}$ is made to show that the oscillatory behavior is from the longitudinal mode only. Deriving a cosecant term that determines the zeros $\rho_{||}$ (approximate) shows a q'_3 only dependence. The frequency of the oscillations can change dramatically with small changes in film thickness indicating a large sensitivity of the film properties. The third region, $[1 - (n_I S)^2]^{-1/2} < \Omega < \infty$: both q'_1 and q'_3 are real and the transverse and longitudinal waves are propagating through the slab. Oscillations of a different complexity occur in $\rho_{||}$ for this third region.

Varying the incident angle or material properties n_I can dramatically change the zeros of q'_1 and q'_3 (see Figure 10A.3). This sets the critical points for $\rho_{||}$ which changes its dynamic. The critical points bound the regions of $\rho_{||}$ and can be described as

$$\Omega_{\text{Zero,Cold}} = \left[1 - (n_I S)^2\right]^{-1/2}, \quad \Omega_{\text{Zero,Warm}} = \left[1 - (\delta n_I S)^2\right]^{-1/2}, \tag{10A.35a}$$

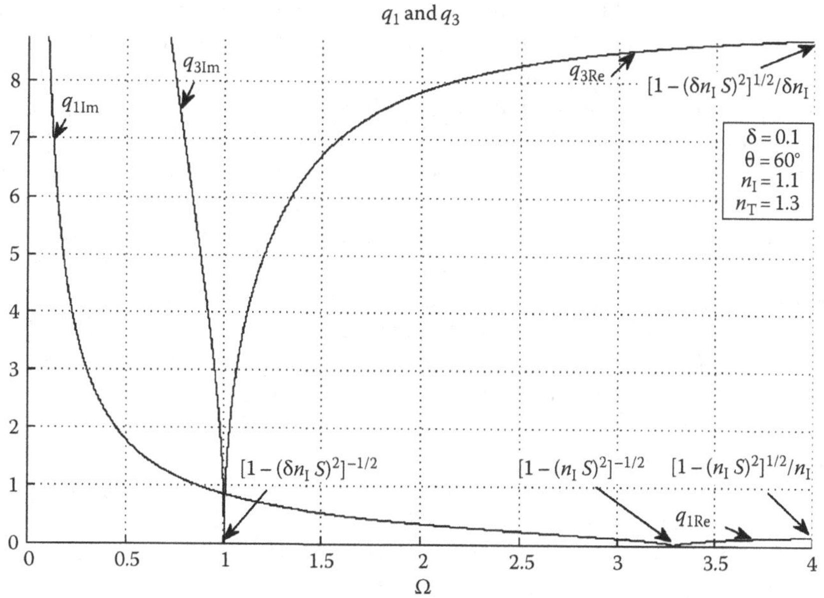

FIGURE 10A.3
Characteristic roots for $\delta = 0.1$ and $\theta = 60°$. The critical points are shown for refractive index $n_I = 1.1$. Both q_1' and q_3' do not depend on n_T.

$$q_1'(\Omega \to \infty) = \frac{\left[1-(n_I S)^2\right]^{1/2}}{n_I}, \quad q_3'(\Omega \to \infty) = \frac{\left[1-(\delta n_I S)^2\right]^{1/2}}{\delta n_I}. \tag{10A.35b}$$

The following sections describe in detail the behavior of $\rho_{\parallel}$ within the regions bounded by the critical points of q'. Equation 10A.32 is analyzed for a few special cases.

10A.3.1 Power Reflection Coefficient for Region 1, $0 < \Omega < [1 - (\delta n_I S)^2]^{-1/2}$

Based on Figure 10A.3, the respective characteristic roots are

$$\begin{bmatrix} jq_1 \\ jq_3 \end{bmatrix}. \tag{10A.36}$$

All of the λ_i' terms will be real, and hence, Equation 10A.32 will consist of hyperbolic sines and hyperbolic cosines only. As a result, there is no propagation of either transverse or longitudinal modes and the waves are evanescent. The power reflection coefficient will have no oscillations and will stay at unity for this region. The following section will show that as the film gets sufficiently thin, power tunneling will take effect and $\rho_{\parallel}$ will not always be unity, but will allow some power to propagate, or tunnel, through the slab.

10A.3.1.1 Region 1 and the Effects of Power Tunneling

For thick slabs, the waves are evanescent and the decaying fields will not interact. When the slab becomes sufficiently thin power tunneling will occur. The tunneling represents itself as a peak in region 1 of the transmission coefficient $\tau_{\parallel}$. Kalluri [11] gave an expression at the peak which is a transcendental equation related to the phase term of oscillation in $\rho_{\parallel}$. Tuning Ω and measuring the peak in $\tau_{\parallel}$ for a known incident angle gives Ω_{max} that can lead to determining the slab thickness or the plasma frequency.

Figure 10A.4 shows the effect of power tunneling for normalized thicknesses of $d_p = 0.05$, 0.1, 0.2, 1, and $n_I = n_T = 1$. The power transmission coefficient, $\tau_{\parallel} = 1 - \rho_{\parallel}$, is also shown. For thin films at $\lambda_p/20$, the slab is almost completely transmissive. For $d_P = 1$, the intensity of light impinging on the warm plasma slab is completely reflected away from the interface. However, as the thickness of the slab decreases, the cross-interaction between evanescent forward-going electric and backward-going magnetic (and vice versa) fields create a real propagating power in a plane parallel to the interface that emerges at $z = d$. This explains the dip seen in this first region. For the air–plasma–air cold plasma case, an experimental measurement of the maximum power transfer could yield information about the material thickness with the use of a transcendental equation. As the thickness of the slab increases, this phenomenon decreases as the decaying fields within the slab become negligibly small as they reach the opposite interface. Figure 10A.5 shows results for dielectric–plasma–dielectric cold and warm plasma case, $n_I = 1.46$, $n_T = 1.48$. For all thicknesses, there is a noticeable difference in the tunneling effect between the cold plasma model ($\delta = 0$) and the warm plasma case ($\delta = 0.1$) and there are differences when different dielectrics are introduced.

Looking at both Figures 10A.4 and 10A.5, the largest difference between warm and cold is seen in $\rho_{\parallel}$ at $d_p = 0.05$ near the plasma frequency. For all thicknesses, more light transmits through the slab for the cold plasma model. In the measurement of optical constants, the

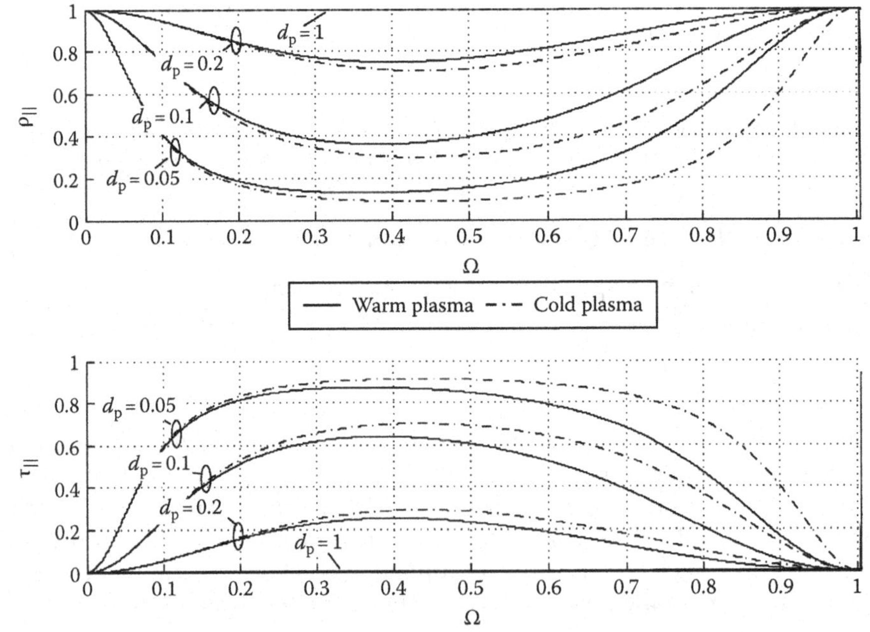

FIGURE 10A.4
Region 1, $0 < \Omega < [1 - (\delta n_I S)^2]^{-1/2}$. This shows the power reflection coefficient at different film thickness: $d_p = 0.05$, 0.1, 0.2, and 1. Refractive indices are $n_I = n_T = 1$ and the warm plasma plots have $\delta = 0.1$ with $\theta = 60°$. The bottom plot is the transmission coefficient, $\tau_{\parallel} = 1 - \rho_{\parallel}$.

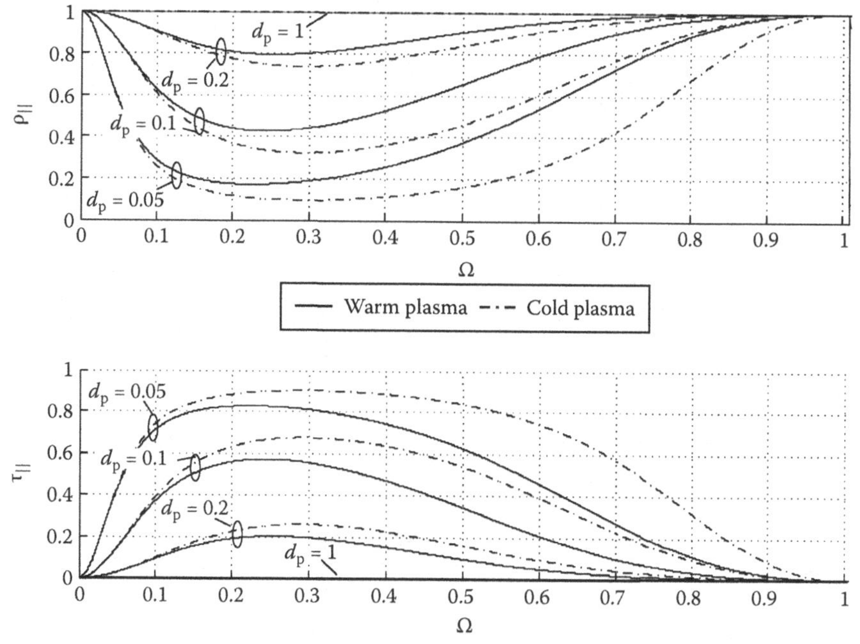

FIGURE 10A.5
Region 1, $0 < \Omega < [1 - (\delta n_I S)^2]^{-1/2}$. Similar to Figure 10A.4 but with $n_I = 1.46$, $n_T = 1.48$.

dip seen for $0 \le \Omega \le 1$ can provide valuable information about the material n_p with knowledge of n_I and n_T by an appropriate correction factor to the transcendental equation. In addition, the figures show that at $\Omega = 0, 1$, $\rho_{\|} = 1$. From Equation 10A.32 as $\Omega \to 0$ ($\omega = 0$), $X \to \infty$ then as shown in Figure 10A.3, $q_1 \to \infty$ and $q_3 \to \infty$. Subsequently, η_{y1}, $\beta_{z1} \to 0$, β_{z3}, $\lambda_i' \to \infty$ and $_{\|}R_{\|} \to 1$. When $\Omega = 1$ ($\omega = \omega_p$), then $X \to 1$, $\eta_{y1} \to 0$, β_{z1}, $\beta_{z3} \to -1$ and λ_i' will all be equal, then $_{\|}R_{\|} \to 1$. For both $\Omega = 0, 1$, $\rho_{\|}$ is forced to unity, independent of any n_I or n_T creating a confluence point for any choice of material.

10A.3.1.2 Region 1 and Surface Plasmons

If one considers the electron plasma wave velocity a to become several orders of magnitudes slower than $c = 3 \times 10^8$ m/s, then from Equations 10A.22 and 10A.28b, $\delta \to 0$, $q_3 \to \infty$, $\beta_{z3} \to \infty$ and the plasma is cold. Also, from Equation 10A.32, only the terms multiplied by β_{z3}^2 survive and one can derive the power reflection coefficient for a cold plasma between two different dielectrics,

$$\rho_{\|}^2 = \left| \frac{\left(1 - C\eta_{y1}'/n_I^2\right)\left(1 - C'\eta_{y2}'/n_I^2\left(C'^2 + S^2\right)\right)\lambda_2' - \left(1 - C\eta_{y2}'/n_I^2\right)\left(1 - C'\eta_{y1}'/n_I^2\left(C'^2 + S^2\right)\right)\lambda_1'}{\left(1 + C\eta_{y1}'/n_I^2\right)\left(1 - C'\eta_{y2}'/n_I^2\left(C'^2 + S^2\right)\right)\lambda_2' - \left(1 + C\eta_{y2}'/n_I^2\right)\left(1 - C'\eta_{y1}'/n_I^2\left(C'^2 + S^2\right)\right)\lambda_1'} \right|. \tag{10A.37}$$

Plotting Equation 10A.37, Figure 10A.6, using the material chosen by Lekner [12] and Otto [13] the input dielectric was glass of $n_I = 1.9018$, silver film of $n_p = 0.055 + j3.28$, which

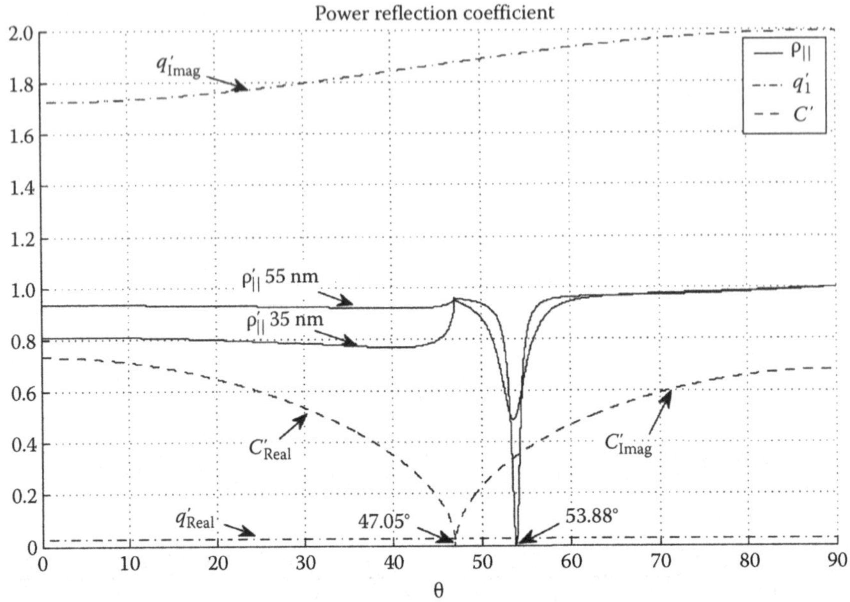

FIGURE 10A.6
The power reflection coefficient at two different film thicknesses versus angle of incidence in degrees: $d = 35$ and 55 nm. Refractive indices are $n_I = 1.9018$, $n_T = 1.392$, and $n_p = 0.055 + j3.28$.

is based on the plasma frequency, and lithium fluoride of $n_T = 1.392$ as the output medium. Collisions were taken into account as is indicated by the real part of n_p.

The result in Figure 10A.6 compares quite well with Lekner's result. From the results of Equation 10A.37, the collision term of n_p causes a nonzero real part of q'. Since this is the case of ATR, the frequency corresponding to the wavelength of $\lambda = 546.1$ nm is below the plasma frequency and in the region of total reflection. Figure 10A.6 also shows the C' term, which can be shown to be zero at the critical angle $C' = \sqrt{(n_T/n_I)^2 - S^2(\theta = 47.05°)} = 0$ causing $\rho_{\parallel}$ to transition downward to zero as photons couple to electrons exciting a surface plasmon at the surface plasmon angle of 53.9°. The critical angle (47.05° for this case) is the point where C' turns from a purely real number to a purely imaginary number. This zero point is the angle at which total internal reflection would occur and the wave would travel along the boundary if no film was present between n_I and n_T. The critical angle is also the point where the analytical expressions for the real and imaginary parts of $_{\parallel}R_{\parallel}$ (not shown in this chapter) change, hence, $\mathrm{Re}\{_{\parallel}R_{\parallel}\}_{\theta^I<\theta^C} \neq \mathrm{Re}\{_{\parallel}R_{\parallel}\}_{\theta^I>\theta^C}$ and $\mathrm{Im}\{_{\parallel}R_{\parallel}\}_{\theta^I<\theta^C} \neq \mathrm{Im}\{_{\parallel}R_{\parallel}\}_{\theta^I>\theta^C}$. The surface plasmon resonance is the effect of a metal film present sandwiched between n_I and n_T. In a sense, the insertion of the metal film induces an equivalent Brewster angle for an incident p-wave that would not exist for the half-space materials only. From Averitt [14], the angle where the power reflection coefficient is totally suppressed, and for the real part of ε^P the surface plasmon resonance can be calculated as

$$\theta_{SP} = \arcsin\left(\frac{\sqrt{\varepsilon^T\varepsilon^P/\left(\varepsilon^T+\varepsilon^P\right)}}{\sqrt{\varepsilon^I}}\right) = 53.9°. \tag{10A.38}$$

10A.3.2 Power Reflection Coefficient for Region 2, $[1 - (\delta n_I S)^2]^{-1/2}\ \Omega < [1 - (n_I S)^2]^{-1/2}$

Based on Figure 10A.3, the respective characteristic roots are

$$\begin{bmatrix} jq_1 \\ q_3 \end{bmatrix}. \tag{10A.39}$$

The power reflection coefficient takes on an oscillatory dynamic. The purely real q_3 implies that the longitudinal (acoustic) mode will propagate through the film. The λ_i' terms in Equation 10A.32 containing q_3 will be complex implying propagation of power through the slab. As can be seen in Figure 10A.2, the oscillations in this region are dramatic as a result of the spatially dispersive plasma medium. The warm plasma component is an acoustic mode of electrostatic field intensity with amplitude and velocity a, it is longitudinal, and is a solution to Maxwell's equations. When a is small compared to c, there is no wave propagation and the plasma becomes cold, then only plasma oscillations exist at ω_p. The λ_i' terms containing q_1 will be real and the transverse modes will be evanescent.

10A.3.2.1 Region 2 and an Approximation of $\rho_{\parallel}$ for $n_I = n_T = 1$

Since, $\sin(k_0 jq_1 d) \equiv j\sinh(k_0 q_1 d)$ and $\cos(k_0 jq_1 d) \equiv \cosh(k_0 q_1 d)$,

$$\rho_{\parallel} = \frac{1}{1+\begin{bmatrix} 2Cq_3(1-X)\left[q_1q_3\sin(k_0q_3d)+S^2X\sinh(k_0q_1d)\right]/ \\ 2q_1q_3S^2X\left[1-\cosh(k_0q_1d)\cos(k_0q_3d)+\left[-q_1^2q_3^2-C^2q_3^2(1-X)^2+S^4X^2\right]\sinh(k_0q_1d)\sin(k_0q_3d)\right] \end{bmatrix}^2}. \tag{10A.40}$$

The power reflection coefficient in this region can be approximated and put into a simpler form and can be written as

$$\rho_{\parallel} = \frac{1}{1+\left[A_0\left[q_1q_3\sin(k_0q_3d)+S^2X\sinh(k_0q_1d)\right]/\left(A_2+A_3\cos(k_0q_3d)+A_4\sin(k_0q_3d)\right)\right]^2}, \tag{10A.41}$$

where

$$A_0 = 2Cq_3(1-X), \tag{10A.42a}$$

$$A_1 = \left[-q_1^2q_3^2 - C^2q_3^2(1-X)^2 + S^4X^2\right], \tag{10A.42b}$$

$$A_2 = 2q_1q_3S^2X, \tag{10A.42c}$$

$$A_3 = -A_2\cosh(k_0q_1d), \tag{10A.42d}$$

$$A_4 = A_1\sinh(k_0q_1d). \tag{10A.42e}$$

The A_2 term does not contribute much to Equation 10A.41 when compared to A_3 and A_4. The A_3 and A_4 terms are large and negative for most of the region of interest, as such, A_2 can be neglected. An approximation to Equation 10A.41 (designated as ρ_a) becomes

$$\rho_a = \frac{1}{1+\left[A_0\left[q_1q_3\sin\left(k_0q_3d\right)+S^2X\sinh\left(k_0q_1d\right)\right]/\left(A_3\cos\left(k_0q_3d\right)+A_4\sin\left(k_0q_3d\right)\right)\right]^2}, \tag{10A.43}$$

which can be reduced to

$$\rho_a = \frac{1}{1+\left[A_0\left[q_1q_3\sin\left(k_0q_3d\right)+S^2X\sinh\left(k_0q_1d\right)\right]/\sqrt{A_3^2+A_4^2}\right]^2 \times\csc^2\left(k_0q_3d+\tan^{-1}\left(-A_4/A_3\right)\pm\pi/2\right)}. \tag{10A.44}$$

Equation 10A.44 is in the same form as the cold plasma slab result reported by Kalluri and Prasad [7]. The plots for $\rho_{\parallel}$ and ρ_a are given in Figure 10A.7 for $\delta = 0.3$ with film thickness of $d_p = 1$. Region 2 is approximately $\{1.04 < \Omega < 2\}$. The approximation of $\rho_{\parallel}$ is very good until $\Omega \approx 1.86$. Values of Ω in the region $\sim 1.86 < \Omega < 2$ is where $A_2 \approx A_3 \approx A_4$ and the suppression of the A_2 term is no longer valid in Equation 10A.43. The oscillations have a periodicity that decreases as Ω increases, which is caused by the Ω-dependent $\csc^2$ term. Although there is an oscillatory term in the bracketed term $[\sin(\ )]^2$, the zeros

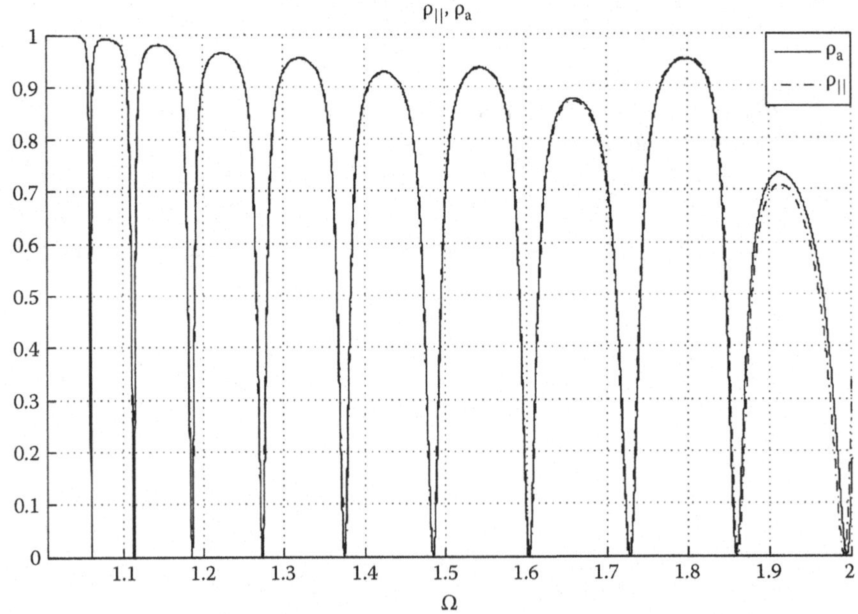

FIGURE 10A.7
Comparison of Equations 10A.40 and 10A.44 and their approximation. Refractive indices are $n_I = n_T = 1$, $\delta = 0.3$, input incident angle of $\theta = 60°$, and film thickness is $d_p = 1$.

of ρ_a are not determined by this term but are completely dependent on the phase of the $\csc^2$ term, and hence

$$k_0 q_3 d + \tan^{-1}\left(-\frac{A_4}{A_3}\right) \pm \frac{\pi}{2} = m\pi, \quad m = 0,1,2,\ldots. \tag{10A.45}$$

At $m = 0,1,2, \ldots$, the $\csc^2$ term goes to infinity defining discrete zeros for ρ_a and consequently for $\rho_{\|}$. The peaks seen in Figure 10A.7 are at values of Ω where $d\rho_a/d\Omega = 0$. If one allows Equation 10A.44 to become

$$\rho_a = \frac{1}{1 + A_5^2/\left(A_3^2 + A_4^2\right)\csc^2\left(A_6\right)}, \tag{10A.46}$$

where $A_5 = A_0(q_1 q_3 \sin(k_0 q_3 d) + S^2 X \sinh(k_0 q_1 d))$ and $A_6 = k_0 q_3 d + \tan^{-1}(-A_4/A_3) \pm \pi/2$.

Then from Equation 10A.46, the equation to find the peaks of ρ_a is a transcendental equation,

$$\tan\left(A_6\right) = f\frac{dA_6}{d\Omega}, \tag{10A.47}$$

where

$$f = \frac{\left(A_3^2 + A_4^2\right)}{\left(A_3^2 + A_4^2\right)\left(1/A_5\right)dA_5/d\Omega - \left(A_3 dA_3/d\Omega + A_4 dA_4/d\Omega\right)}, \tag{10A.48}$$

and

$$\frac{dA_6}{d\Omega} = k_0 q_3 d\left(\frac{1}{q_3}\frac{dq_3}{d\Omega} + \frac{1}{\Omega}\right) + \frac{d}{d\Omega}\tan^{-1}\left(-\frac{A_4}{A_3}\right). \tag{10A.49}$$

10A.3.2.2 Region 2 and Film Thickness

Figure 10A.8 shows $\rho_{\|}$ (Equation 10A.34) for different slab thicknesses $d_p = 0.05, 0.1, 0.2, 0.5, 0.7$, and 1. The index of refraction of the dielectrics are $n_I = n_T = 1$, with $\delta = 0.1$ and an input incident angle of $\theta = 60°$. The range for this region is $1.0037 < \Omega < 2$. As the thickness increases, the number of oscillations increases. The amplitude of the oscillations spans the entire range between zero and one. Since $n_I = n_T = 1$, the points where $\rho_{\|} = 1$ are when $q_1 q_3 \sin(k_0 q_3 d) = -S^2 X \sinh(k_0 q_1 d)$. The points where $\rho_{\|} = 0$ are when the phase of the $\csc^2$ is equal to $m\pi$.

For the case where $n_I = 1.4$, $n_T = 1.45$, the range of Ω has no second critical point, and hence $1.0074 < \Omega < j1.458$. This implies that q_1' stays imaginary for all Ω as shown in Figure 10A.9. The refractive indices for the input and output mediums have the effect of tuning the second critical point of q_1' which changes the behavior of the oscillations in $\rho_{\|}$. Shown in Figure 10A.10, as the thickness increases, sharp notches in the frequency band occur.

Power Reflection Coefficient for $([1 - (n_I S)^2]^{-1/2} < \Omega < \infty)$. Region 3, where both q_1' and q_3' are real will be discussed in a future paper.

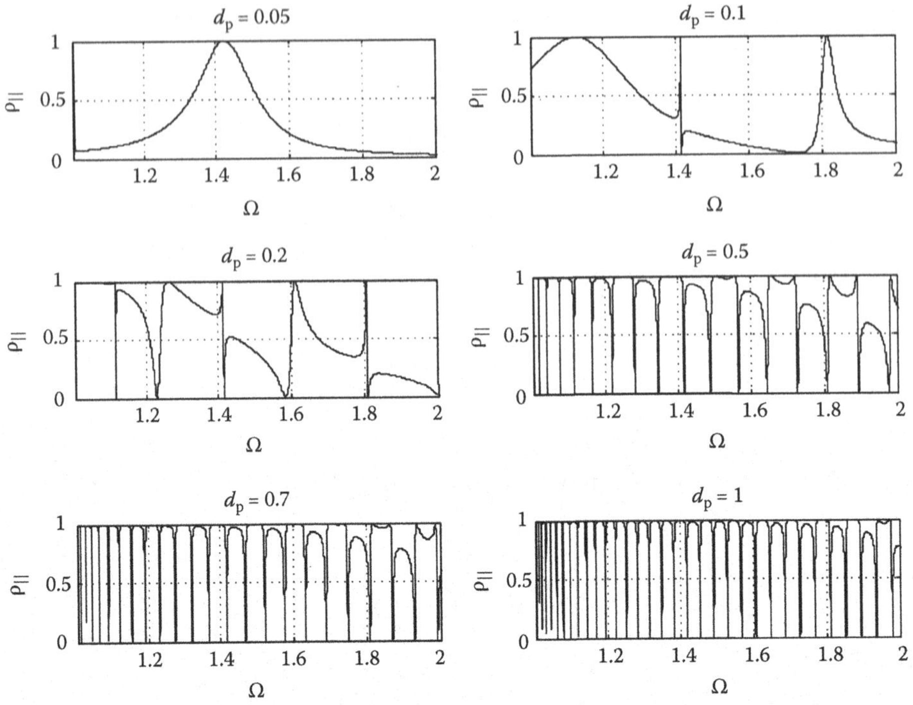

FIGURE 10A.8
Plots of the power reflection coefficient for different film thicknesses. Refractive indices are $n_I = n_T = 1$, $\delta = 0.1$, input incident angle of $\theta = 60°$, and film thicknesses of $d_p = 0.05, 0.1, 0.2, 0.5, 0.7$, and 1.

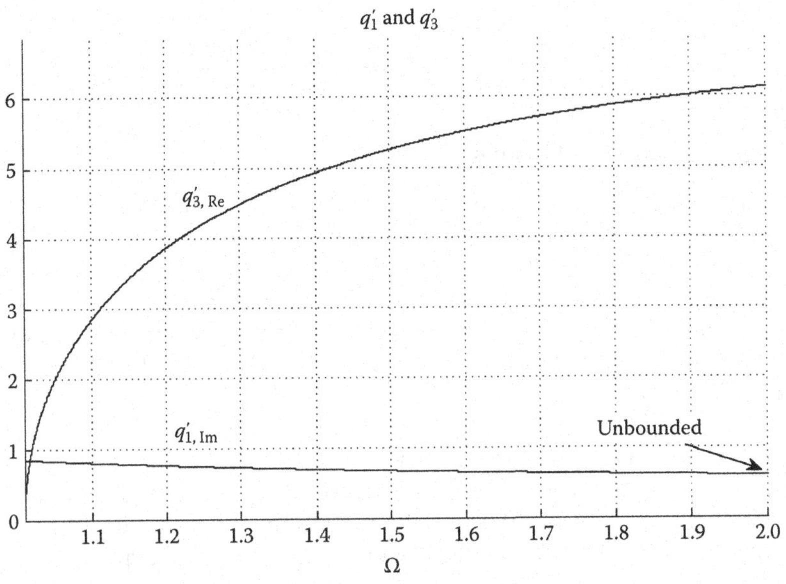

FIGURE 10A.9
q_1' and q_3' for $n_I = 1.4$, $n_T = 1.45$ in the frequency range of $1.0074 < \Omega < \infty$.

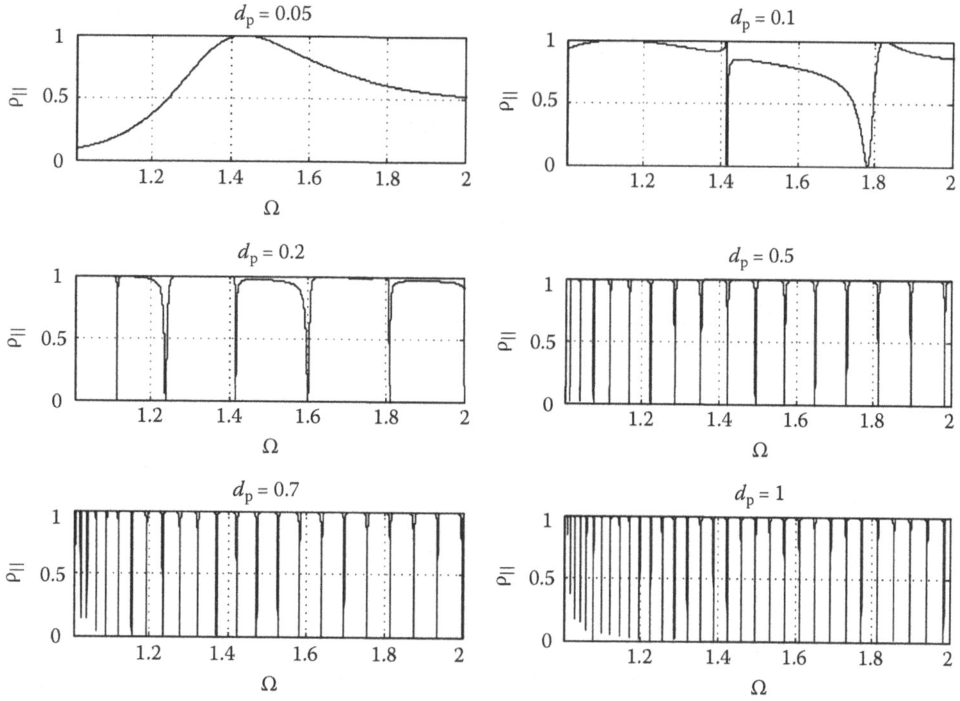

FIGURE 10A.10
Plots of the power reflection coefficient for different film thicknesses. Refractive indices are $n_I = 1.4$, $n_T = 1.45$, $\delta = 0.1$, input incident angle of $\theta = 60°$, and film thicknesses of $d_p = 0.05, 0.1, 0.2, 0.5, 0.7$, and 1.

10A.3.3 Power Reflection Coefficient for a Sodium Thin Film

Using the material properties of Na from Forstmann and Gerhardts [8]: The plasma frequency is $\omega_p = 8.2 \times 10^{15}$ rad/s, with a Fermi velocity of $V_F = 9.87 \times 10^7$ cm/s resulting in $\delta = V_F/c = 3.29 \times 10^{-3}$. Using a film thickness of $d = 10$ nm and an incident angle of $\theta = 60°$, Figure 10A.11 shows the power reflection coefficient for the warm (top) and cold (bottom) plasma model. Only the first two regions are shown since the third region is unbounded based on the input and output materials. The first region ($0 < \Omega < 1$) shows the effect of power tunneling as seen as the large dip. The plasma wavelength of $\lambda_p = 229$ nm is large when compared with the 10 nm film thickness. The oscillations in the second region are very clear as a result of the Fermi velocity of the metal film. The warm plasma oscillations ride atop of the cold plasma result. Similar to previous results, the periodicity of the oscillations caused by real q_3 and imaginary q_1 decrease as the Ω increases.

10A.4 Conclusion

Reflection properties for a thin metal slab bounded by two different dielectrics have been presented with emphasis on the power reflection coefficient and its corresponding characteristic roots. The film was modeled as a warm plasma that supports a longitudinal electromagnetic acoustic mode that is a solution to Maxwell's equations. The oscillations in the

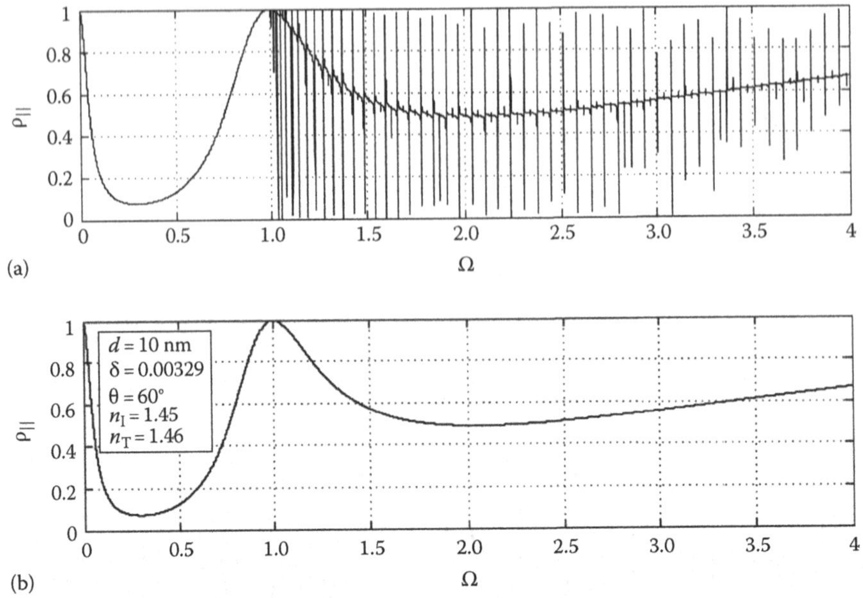

FIGURE 10A.11
Plots of the warm and cold power reflection coefficients for a sodium (Na) material thin film with thickness of $d = 10$ nm. Refractive indices are $n_I = 1.45$, $n_T = 1.46$, the plasma frequency for sodium is $\omega_p = 8.2 \times 10^{15}$ rad/s, and the Fermi velocity is $V_F = 9.87 \times 10^7$ cm/s resulting in $\delta = V_F/c = 3.29 \times 10^{-3}$. The input incident angle is $\theta = 60°$. Power reflection coefficient (a) $\delta = 0.00329$, and (b) $\delta = 0$.

power reflection coefficient follow the trend of the cold plasma component. The cold plasma case was presented for different dielectric materials and the model shows surface plasmon excitation.

The choice of the material has an impact on the characteristics of the energy propagating through the plasma material. The characteristic roots are dependent on the index of refraction of the input material, the velocity of the longitudinal plasma wave, and the angle of incidence.

Choice of the refractive index of the input material such that $n_I S = 1$, makes q_1 purely imaginary for all frequencies. The sines and cosines, which are functions of q_1 wave number become hyperbolic, creating power reflection coefficient similar to comb function of slightly decreasing periodicity. Hence, the refractive index of the input medium has the effect of tuning the reflection properties of the thin film.

Markos and Kalluri [15] have recently published an updated version of the material in this chapter.

References

1. Shackleford, J. A., Grote, R., Currie, M., Spanier, J. E., and Nabet, B., Integrated plasmonic lens photodetector, *Appl. Phys. Lett.*, 94, 083501-1–083501-3, 2009.
2. Chen, H. L., Hsieh, K. C., Lin, C. H., and Chen, S. H., Using direct nanoimprinting of ferroelectric films to prepare devices exhibiting bi-directionally tunable surface plasmon resonances, *Nanotechnology*, 19, 1–6, 435304, 2008.

3. Curtin, B., Biswas, R., and Dalal, V., Photonic crystal based back reflectors for light management and enhanced absorption in amorphous silicon solar cells, *Appl. Phys. Lett.*, 95, 083501-1–083501-3, 231102, 2009.
4. Stratton, J. A., *Electromagnetic Theory*, IEEE Press, Piscataway, NJ, p. 513, 2007.
5. Unz, H., Oblique wave propagation in a compressible general magnetoplasma, *Appl. Sci. Res.*, 16, 105–120, 1965.
6. Prasad, R. C., Transient and frequency response of a bounded plasma, PhD thesis, Birla Institute of Technology, Ranchi, India, 1976.
7. Kalluri, D. K. and Prasad, R. C., Thin-film reflection properties of isotropic and uniaxial plasma slabs, *Appl. Sci. Res.*, 27, 415–424, 1973.
8. Forstmann, F. and Gerhardts, R. R., *Metal Optics near the Plasma Frequency*, Springer, Berlin, Germany, pp. 9–24, 1986.
9. Pozar, D. M., *Microwave Engineering*, Wiley, New York, pp. 21–24, 1998.
10. Ramo, S., Whinnery, J. R., and Van Duzer, T., *Fields and Waves in Communication Electronics*, 3rd edition, Wiley, New York, p. 147, 1994.
11. Kalluri, D. K., *Electromagnetics of Time Varying Complex Media: Frequency and Polarization Transformer*, 2nd edition, CRC Press, New York, pp. 26–27, 2010.
12. Lekner, J., *Theory of Reflection, Developments in Electromagnetic Theory and Applications*, Academic Publishers, Dordrecht, the Netherlands, pp. 169–170, 1987.
13. Otto, A., Spectroscopy of surface polaritons by attenuated total reflection, in: Seraphin, B. O. (Ed.), *Optical Properties of Solids: New Developments*, North-Holland, Amsterdam, the Netherlands, p. 703, 1976.
14. Averitt, R. D., Plasmonics and metamaterials at terahertz and mid-infrared frequencies, *IEEE LEOS Plasmonics Workshop*, MIT Lincoln Laboratory, Lexington, MA, October 30, 2007.
15. Markos, C. T. and Kalluri, D., Thin film reflection properties of a warm isotropic plasma slab between two half-space dielectric media, *J. Infrared Millim. Terahertz Waves*, 32, 914–934, 2011.

*Appendix 10B: First-Order Coupled Differential Equations for Waves in Inhomogeneous Warm Magnetoplasmas**

Dikshitulu K. Kalluri and H. Unz

The first-order coupled differential equations for waves in *temperate* inhomogeneous magnetoplasmas are well known [1]. The extension of this method for the case of waves in inhomogeneous *warm* magnetoplasmas is presented here. In this form, they are well suited for solution on a digital computer by the Runge–Kutta method. The plasma is assumed to be neutral and in equilibrium, and the motion of the ions is neglected. Taking $\bar{E}_0$, $\bar{H}_0$, p_0, N_0, and $T_0\left(\bar{u}_0=0\right)$ as the stationary values of the plasma which are given functions of position, and $\bar{E}_1$, $\bar{H}_1$, p_1, N_1, T_1, and $\bar{u}_1$ as the harmonic time-varying ($e^{i\omega t}$) small components of the wave in the plasma, and defining the plasma parameters

$$X(\bar{r})=\frac{e^2N_0(\bar{r})}{\omega^2\varepsilon m},\quad Y(\bar{r})=\frac{|e|\mu}{m\omega}\bar{H}_0(\bar{r}),\quad U(\bar{r})=1-\mathrm{i}\frac{\nu(\bar{r})}{\omega},$$

$$\delta(\bar{r})=\frac{a^2(\bar{r})}{c^2}=\frac{\mu\varepsilon\gamma KT_0(\bar{r})}{m},$$

where $a(\bar{r})$ is the acoustic velocity in the electron gas, by taking a new set of dependent variables

$$\bar{E}=\bar{E}_1,\quad \bar{H}=\left[\frac{\mu}{\varepsilon}\right]^{\frac{1}{2}}\bar{H}_1,\quad \bar{u}=\frac{\omega m}{e}\bar{u}_1,\quad p=\frac{e}{\omega m}\left[\frac{\mu}{\varepsilon}\right]^{\frac{1}{2}}p_1, \tag{10B.1}$$

one obtains [2] for small-signal theory, where $k_0=\omega(\mu\varepsilon)^{1/2}=\omega/c$:

$$\nabla\times\bar{E}=-\mathrm{i}k_0\bar{H}, \tag{10B.2a}$$

$$\nabla\times\bar{H}=\mathrm{i}k_0\bar{E}-k_0X\bar{u}, \tag{10B.2b}$$

$$-\mathrm{i}UX\bar{u}=\frac{1}{k_0}\nabla p+X\bar{E}+\frac{\delta}{\gamma k_0^2}\frac{\nabla p_0}{p_0}\left(\nabla\cdot\bar{E}\right)+X\bar{u}\times\bar{Y}, \tag{10B.2c}$$

$$-\nabla\cdot\bar{u}=\frac{\mathrm{i}k_0}{\delta X}p+\bar{u}\cdot\frac{\nabla p_0}{\gamma p_0}. \tag{10B.2d}$$

* Reprinted with permission from Kalluri, D. and Unz, H., The first-order coupled differential equations for waves in inhomogeneous warm magnetoplasmas, *Proc. IEEE*, 55(9), 1620–1621, © September 1967 IEEE.

For the case of a free-space wave of arbitrary polarization with direction cosines S_1, S_2, and C, obliquely incident on the plasma medium, one [1] obtain the following:

$$\frac{\partial}{\partial x} = -ik_0 S_1, \quad \frac{\partial}{\partial y} = -ik_0 S_2. \tag{10B.3}$$

Assuming the plasma parameters vary in the z-direction only such that $N_0(z)$, $T_0(z)$, $v(z)$, and $H_0(z)$, and substituting Equation 10B.3 into Equation 10B.2, one obtains 10 equations with 10 unknowns $\bar{E}$, $\bar{H}$, $\bar{u}$, and p as functions of z only. Eliminating E_z, H_z, u_x, and u_y from these equations, one obtains the following six first-order linear coupled differential equations with six unknowns, where primes denote derivatives with respect to z:

$$\frac{-1}{ik_0}E_x' = S_1 S_2 H_x + \left(1 - S_1^2\right)H_y - iS_1 X u_z, \tag{10B.4a}$$

$$\frac{-1}{ik_0}E_y' = -\left(1 - S_2^2\right)H_x - S_1 S_2 H_y - iS_2 X u_z, \tag{10B.4b}$$

$$\begin{aligned}\frac{-1}{ik_0}H_x' = &-\left(S_1 S_2 + \frac{iXY_z}{U^2 - Y_z^2}\right)E_x + \left(S_1^2 - 1 + \frac{UX}{U^2 - Y_z^2}\right)E_y \\ &+\left(\frac{UXY_x + iXY_yY_z}{U^2 - Y_z^2}\right)u_z - \left(\frac{S_1 Y_z + iUS_2}{U^2 - Y_z^2}\right)p,\end{aligned} \tag{10B.4c}$$

$$\begin{aligned}\frac{-1}{ik_0}H_y' = &\left(1 - S_2^2 - \frac{UX}{U^2 - Y_z^2}\right)E_x + \left(S_1 S_2 - \frac{iXY_z}{U^2 - Y_z^2}\right)E_y \\ &+\left(\frac{UXY_y - iXY_xY_z}{U^2 - Y_z^2}\right)u_z + \left(\frac{iS_1 U - S_2 Y_z}{U^2 - Y_z^2}\right)p,\end{aligned} \tag{10B.4d}$$

$$\begin{aligned}\frac{-1}{ik_0}u_z' = &-\left(\frac{S_2 Y_z + iUS_1}{U^2 - Y_z^2}\right)E_x + \left(\frac{S_1 Y_z - US_2}{U^2 - Y_z^2}\right)E_y \\ &+\left(\frac{S_1 Y_x Y_z + S_2 Y_y Y_z + iUS_1 Y_y - iUS_2 Y_x}{U^2 - Y_z^2} - \frac{i}{\gamma k_0}\frac{p_0'}{p_0}\right)u_z \\ &+\frac{1}{X}\left(\frac{1}{\delta} - \frac{US_1^2 + US_2^2}{U^2 - Y_z^2}\right)p,\end{aligned} \tag{10B.4e}$$

$$\begin{aligned}\frac{-1}{ik_0}p' = &\left(\frac{UXY_y + iXY_xY_z}{U^2 - Y_z^2}\right)E_x + \left(\frac{iXY_yY_z - UXY_x}{U^2 - Y_z^2}\right)E_y \\ &- iS_2 X H_x + iS_1 X H_y \\ &+\left\{UX - X^2 - UX\frac{Y_x^2 + Y_y^2}{U^2 - Y_z^2} + \frac{\delta X}{k_0^2 \gamma^2}\left(\frac{p_0'}{p_0}\right) - \frac{X'\delta}{\gamma k_0^2}\frac{p_0'}{p_0}\right\}u_z \\ &+\left(\frac{S_2 Y_y Y_z + S_1 Y_x Y_z + iUS_2 Y_x - iUS_1 Y_y}{U^2 - Y_z^2} + \frac{i}{k_0 \gamma}\frac{p_0'}{p_0}\right)p.\end{aligned} \tag{10B.4f}$$

Equations 10B.4 may be written in the matrix form as

$$\frac{-1}{ik_0}\frac{d}{dz}[e]=[T][e], \tag{10B.5}$$

where $[e]$ is a (6×1) column matrix with the elements E_x, E_y, H_x, H_y, u_z, and p.

For the particular case of a homogeneous plasma, the elements of the $[T]$ matrix are constants. The characteristic equation of the coupled differential equation (10B.5) obtained by setting

$$\frac{d}{dz}=-ik_0 q \tag{10B.6a}$$

is given by the determinant

$$\det([T]-q[I])=0. \tag{10B.6b}$$

Equation 10B.6b is a sixth-order algebraic equation and it is found to be in agreement with an earlier result by Unz [3].

From the above definitions, one has [2]

$$\frac{\nabla p_0}{p_0}=\frac{\nabla X}{X}+\frac{\nabla \delta}{\delta}. \tag{10B.7}$$

In a recent communication, Burman [4] has suggested that $\nabla p_0 = 0$. By substituting $p_0' = 0$ into Equation 10B.4, one obtains our corresponding result.

Acknowledgment

This work was supported in part by the National Science Foundation.

References

1. Budden, K. G., *Radio Waves in the Ionosphere*, Cambridge University Press, Cambridge, MA, 1961.
2. Unz, H., Wave propagation in inhomogeneous gyrotropic warm plasmas, *Am. J. Phys.*, 35, 505–508, 1967.
3. Unz, H., Oblique wave propagation in a compressible general magneto-plasma, *Appl. Sci. Res.*, 16, 105–120, 1966.
4. Burman, R., Coupled wave equations for propagation in generally inhomogeneous compressible magnetoplasmas, *Proc. IEEE (Lett.)*, 55, 723–724, 1967.
5. Kalluri, D. and Unz, H., The first-order coupled differential equations for waves in inhomogeneous warm magnetoplasmas, *Proc. IEEE*, 55(9), 1620–1621, September 1967.

Appendix 10C: Waveguide Modes of a Warm Drifting Uniaxial Electron Plasma*

Dikshitulu K. Kalluri

10C.1 Introduction

The guided wave propagation in an anisotropic plasma medium is investigated by many authors. The results for a cold stationary anisotropic plasma are well summarized by Allis et al. [1]. The guided slow-wave propagation in drifting cold anisotropic plasma is considered by Trivelpiece [2]. Mode theory of a waveguide filled with stationary warm uniaxial plasma has been recently reported by Tuan [3]. Electron stream interactions with a cold plasma in a waveguide are considered by Smullin and Chorney [4] and Briggs [5].

In this appendix, the guided wave propagation in a drifting warm plasma in an infinite static magnetic field is investigated with emphasis on the phase characteristics of the waves. The motivation for such a study was provided by an investigation by Articolo [6] on modeling the drifting warm anisotropic plasma medium as a dielectric medium with temporal and spatial dispersions under the assumptions that (1) the neutral molecules and positive ions are cold, (2) the electrons are warm with an average temperature and have a preferred drift superimposed on their thermal motion, and (3) the drift and thermal velocities are nonrelativistic. The additional assumptions made here are that (1) the drift velocity υ_0 is along the *z*-axis, (2) the static magnetic field of infinite strength is also along the *z*-axis, and (3) collisional effects are neglected.

10C.2 Basic Equations

Let all field quantities vary as

$$F_1(x, y, z, t) = F(x, y)e^{j\omega t - \gamma z}. \tag{10C.1}$$

The Maxwell equations are

$$\bar{\nabla} \times \bar{H} = j\omega\varepsilon_0\bar{\bar{\varepsilon}} \cdot \bar{E}, \tag{10C.2}$$

$$\bar{\nabla} \times \bar{H} = -j\omega\mu_0\bar{H}. \tag{10C.3}$$

* Reprinted with permission from Kalluri, D., Waveguide modes of a warm drifting uniaxial electron plasma, *Proc. IEEE*, 58(2), 278–280, February 1970 IEEE.

For the case of an infinite static magnetic field and the drift velocity υ_0 both lying along the positive z-axis, it can be shown [6] that

$$\bar{\varepsilon}(\omega,\gamma) = \hat{x}\hat{x} + \hat{y}\hat{y} + \varepsilon_3 \hat{z}\hat{z}, \tag{10C.4}$$

$$\varepsilon_3(\omega,\gamma) = 1 - \frac{\omega_p^2}{(\omega + j\gamma v_0)^2 + a^2\gamma^2}, \tag{10C.5}$$

where a is the acoustic velocity indicative of the average temperature of the plasma and ω_p the plasma frequency. From Equations 10C.1 through 10C.3, the expressions for the transverse field $\bar{E}_t$ and $\bar{H}_t$ in terms of E_z and H_z and the wave equations for H_z can be derived. These are the same as for the stationary case derived by Tuan [3] and need not be repeated here. From the wave equation for H_z, it may also be concluded that the TE modes are unaffected by the presence of the plasma.

From the wave equation for E_z, we have

$$\left[\nabla_t^2 + (\gamma^2 + k_0^2)\varepsilon_3\right]E_z = 0, \tag{10C.6}$$

where $k_0^2 = \omega^2\mu_0\varepsilon_0$ and ∇_t is the transverse part of the ∇ operator. The dispersion relation for TM modes can be shown to be

$$(\gamma^2 + k_0^2)\varepsilon_3 = \left(\frac{\omega_c}{c}\right)^2, \tag{10C.7}$$

where ω_c is the cutoff frequency of the empty waveguide and c is the velocity of light. Letting $\gamma = j\beta$, where β is the propagation constant, Equation 10C.7 may be written as

$$(\omega^2 - \beta^2c^2)\left[1 - \frac{\omega_p^2}{(\omega - v_0\beta)^2 - a^2\beta^2}\right] = \omega_c^2. \tag{10C.8}$$

The discussion of the ω–β diagram (Figure 10C.1) is facilitated by using the following normalized variables:

$$\Omega = \frac{\omega}{\omega_p}, \quad B = \frac{\beta c}{\omega_p}, \quad \Omega_c = \frac{\omega_c}{\omega_p}, \quad \alpha = \frac{v_0}{c}, \quad \delta = \frac{a}{c}. \tag{10C.9}$$

Substituting Equation 10C.9 into Equation 10C.8, we obtain

$$(\Omega^2 - B^2)\left[(\Omega - \alpha B)^2 - \delta^2B^2 - 1\right] = \Omega_c^2\left[(\Omega - \alpha B)^2 - \delta^2B^2\right]. \tag{10C.10}$$

This is a fourth-degree polynomial equation in Ω, B. The radius vector to each point on the diagram is proportional to the phase velocity $[v_\phi = c(\Omega/B)]$ and the slope of the tangent

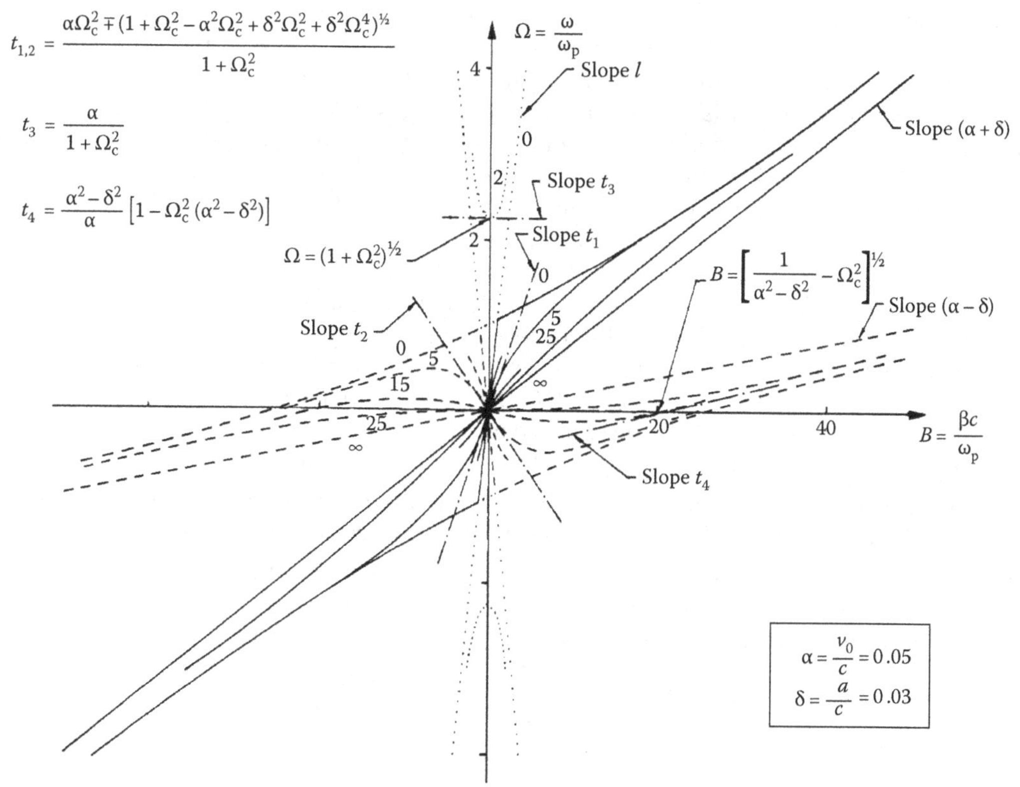

FIGURE 10C.1
The ω–β diagram normalized to the plasma frequency, for guided waves in a warm uniaxial drifting electron plasma for the case of the drift velocity v_0 greater than the acoustic velocity a. The numbers on the curves are the values of Ω_c, and the mode cutoff frequency of the empty guide ω_c normalized to the plasma frequency. (Reprinted with permission from Kalluri, D., Waveguide modes of a warm drifting uniaxial electron plasma, *Proc. IEEE*, 58(2), 278–280, © February 1970 IEEE.)

at each point is proportional to the group velocity ($v_g = c\, d\Omega/dB$). The Ω–B diagram has a double point at the origin and the tangents at this point are given by

$$t_{1,2} = \frac{\alpha\Omega_c^2 \mp \sqrt{1+\Omega_c^2-\alpha^2\Omega_c^2+\delta^2\Omega_c^4}}{1+\Omega_c^2}. \tag{10C.11}$$

The diagram cuts the Ω-axis at $\Omega = \mp\left(1+\Omega_c^2\right)^{1/2}$ and the tangent to the curve at this point is given by

$$t_3 = \frac{\alpha}{1+\Omega_c^2}. \tag{10C.12}$$

It cuts the B-axis at $B = \left\{\left[1/\left(\alpha^2-\delta^2\right)\right]-\Omega_c^2\right\}^{1/2}$ and the tangent at this point is given by

$$t_4 = \frac{\alpha^2-\delta^2}{\alpha}\left[1-\Omega_c^2\left(\alpha^2-\delta^2\right)\right]. \tag{10C.13}$$

The general characteristics of the Ω–B diagram for a given Ω_c may be obtained by considering the two extreme cases of $\Omega_c = 0$ (one-dimensional case) and $\Omega_c = \infty$. In the former case, the fourth-degree curve degenerates into two straight lines $\Omega = \pm B$ and the hyperbola $(\Omega - \alpha B)^2 - \delta^2 B^2 - 1 = 0$. In the latter case ($\Omega_c = \infty$), the curve degenerates into two straight lines $\Omega = (\alpha \mp \delta)B$. The branches of the fourth-degree curve for any other Ω_c will be boxed by the compartments formed by the branches of the curves for the two extreme cases. Figures 10C.1 and 10C.2 illustrate this point. In these diagrams, one may identify [2,6] the "fast wave" whose phase velocity $v_{\phi 1} > c(\alpha - \delta)$, the "slow wave" with phase velocity $v_{\phi 2} < c(\alpha - \delta)$, and the "waveguide wave." In Figure 10C.1, the Ω–B diagram is shown for $\alpha = 0.05$ and $\delta = 0.03$, the parameter being Ω_c. For $\omega < \omega_p$, one of these curves (e.g., the dashed curve marked $\Omega_c = 5$) shows the existence of the backward waves (with positive group velocity and negative phase velocity) as well as a stationary wave ($v_g = 0$). However, these disappear for $\Omega_c \geq (\alpha^2 - \delta^2)^{-1/2}$ (see, e.g., the dashed curve marked $\Omega_c = 25$ of Figure 10C.1). Figure 10C.2 shows the Ω–B diagram for $\alpha < \delta$ ($\alpha = 0.03$, $\delta = 0.05$). In Figure 10C.3, the phase characteristics of the waveguide wave for $\alpha = 0.05$, $\delta = 0.03$, and $\Omega_c = 0.1$ are exhibited in great detail. It can be seen from this diagram that while the low-frequency cutoff for the forward waves of the waveguide wave is given by $\left(1 + \Omega_c^2\right)^{1/2}$ (point P, Figure 10C.3), there exist backward waves for a small band of frequencies (branch AP) below the cutoff frequency $\left(1 + \Omega_c^2\right)^{1/2}$. This band extends in width as Ω_c decreases with a maximum value of $\alpha/(1 + \alpha)$. The transition of the waveguide wave from the forward wave type to the backward wave type takes place at a group velocity $v_g = ct_3$.

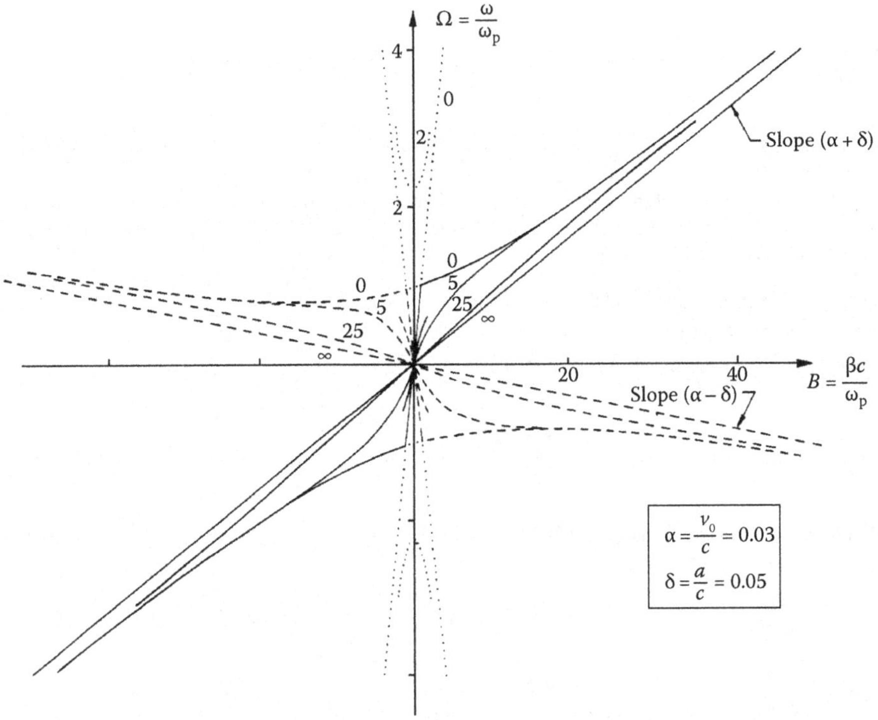

FIGURE 10C.2
The ω–β diagram normalized to the plasma frequency ω_p, for guided waves in a warm uniaxial drifting electron plasma for the case of the drift velocity v_0 less than the acoustic velocity a. The numbers on the curves are as given in Figure 10C.1. (Reprinted with permission from Kalluri, D., Waveguide modes of a warm drifting uniaxial electron plasma, *Proc. IEEE*, 58(2), 278–280, © February 1970 IEEE.)

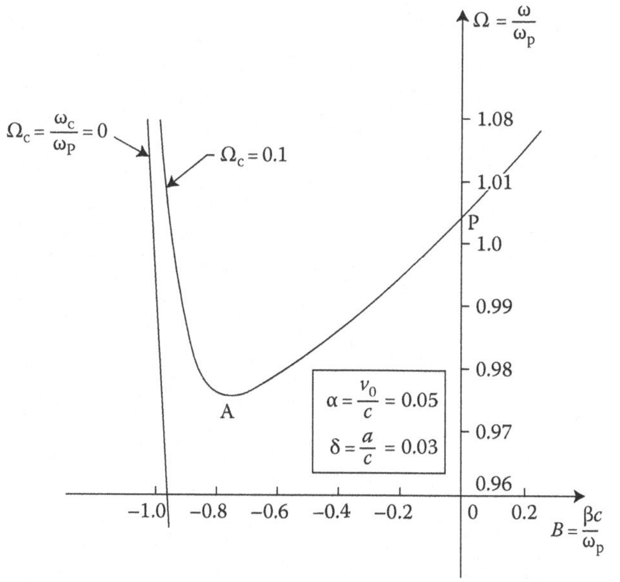

FIGURE 10C.3
Phase characteristics of the "waveguide wave" of a warm uniaxial drifting electron plasma near the point $\Omega = [1 + (\omega_c/\omega_p)^2]^{1/2}$, where ω_c is the mode cutoff frequency of the empty guide. (Reprinted with permission from Kalluri, D., Waveguide modes of a warm drifting uniaxial electron plasma, *Proc. IEEE*, 58(2), 278–280, © February 1970 IEEE [7].)

References

1. Allis, W. P., Buchbaum, S. J., and Bers, A., *Waves in Anisotropic Plasmas*, MIT Press, Cambridge, MA, 1963.
2. Trivelpiece, A. W., *Slow-Wave Propagation in Plasma Waveguides*, San Francisco Press, San Francisco, CA, 1967.
3. Tuan, H. S., Mode theory of waveguide filled with warm uniaxial plasma, *IEEE Trans. Microw. Theory Tech.*, MTT-17, 134–137, 1969.
4. Smullin, L. D. and Chorney, P., Propagation in ion loaded waveguide, *Proceedings of the Symposium on Electronic Waveguides*, Polytechnic Press, Brooklyn, NY, 1958.
5. Briggs, R. J., *Electron–Stream Interaction with Plasmas*, MIT Press, Cambridge, MA, 1964.
6. Articolo, G. A., Derivation of an equivalent dielectric tensor for a warm, drifting, lossy, electron plasma, *J. Appl. Phys.*, 40, 1896–1902, 1969.
7. Kalluri, D., Waveguide modes of a warm drifting uniaxial electron plasma, *Proc. IEEE*, 58(2), 278–280, February 1970.

Appendix 11A: Faraday Rotation versus Natural Rotation

Equation 11.31 gives the expression for the angle ψ of the linearly polarized wave as it propagates in the magnetoplasma medium. In this appendix, we show how that expression is obtained. We also show an application of this Faraday rotation in measuring the total electron content in the ionosphere [1].

Such a rotation also occurs in chiral medium called natural rotation or optical activity discussed in Section 8.10. However, there is an important difference between Faraday rotation and natural rotation. This aspect is also explained (see Figure 11A.1).

11A.1 Faraday Rotation

At $z = 0$, consider the electric field of a linearly polarized wave propagating in the z-direction to be

$$\tilde{\tilde{E}}\Big|_{z=0} = \hat{x}E_0 e^{-j\beta z}\Big|_{z=0} = \hat{x}E_0. \tag{11A.1}$$

This is an x-polarized (linear) wave. This can be written as a superposition of R and L waves:

$$\tilde{\tilde{E}}\,|_{z=0} = \frac{E_0}{2}(\hat{x} - j\hat{y}) + \frac{E_0}{2}(\hat{x} + j\hat{y}). \tag{11A.2}$$

Since the wave is propagating in the $+z$-direction, the wave at $z = d$ can be written as

$$\tilde{\tilde{E}}\Big|_{z=d} = \frac{E_0}{2}(\hat{x} - j\hat{y})e^{-jk_{pR}d} + \frac{E_0}{2}(\hat{x} + j\hat{y})e^{-jk_{pL}d}, \tag{11A.3}$$

where

$$k_{pR} = k_0\sqrt{\varepsilon_{pR}}, \tag{11A.4}$$

$$k_{pL} = k_0\sqrt{\varepsilon_{pL}}, \tag{11A.5}$$

$$\varepsilon_{pR} = 1 - \frac{\omega_p^2}{\omega(\omega - \omega_b)}, \tag{11A.6}$$

$$\varepsilon_{pL} = 1 - \frac{\omega_p^2}{\omega(\omega + \omega_b)}. \tag{11A.7}$$

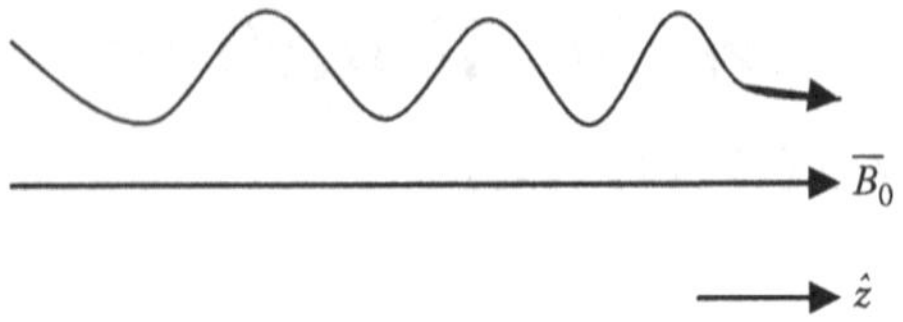

FIGURE 11A.1
Definition of longitudinal propagation.

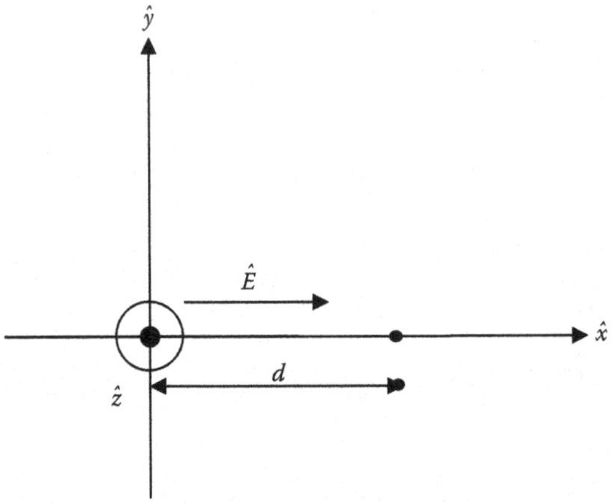

FIGURE 11A.2
The electric field at $z = 0$.

Equation 11A.3 may be written as (see Figure 11A.2)

$$\tilde{\tilde{E}} = E_0 e^{-j[k_{pR}+k_{pL}]d} \left\{ \hat{x}\cos\left[\frac{(k_{pR}-k_{pL})d}{2}\right] - \hat{y}\sin\left[\frac{(k_{pR}-k_{pL})d}{2}\right]\right\}, \tag{11A.8}$$

which is a linearly polarized wave, but it is no longer x-polarized. The electric field traces a straight line but the line at an angle ψ_F is given by (see Figure 11A.3)

$$\psi_F = \tan^{-1}\frac{E_y}{E_x} = -\tan^{-1}\left\{\frac{\tan[k_{pR}-k_{pL}]d}{2}\right\},$$
$$\psi_F = \frac{(k_{pL}-k_{pR})d}{2}. \tag{11A.9}$$

From Equations 11A.4 through 11A.7 and 11A.9, and using the definition of ω_p and ω_b, we obtain

$$\psi_F = \frac{q^3 B_0}{2m^2 \varepsilon_0 \omega^2 c} N_0 d, \tag{11A.10}$$

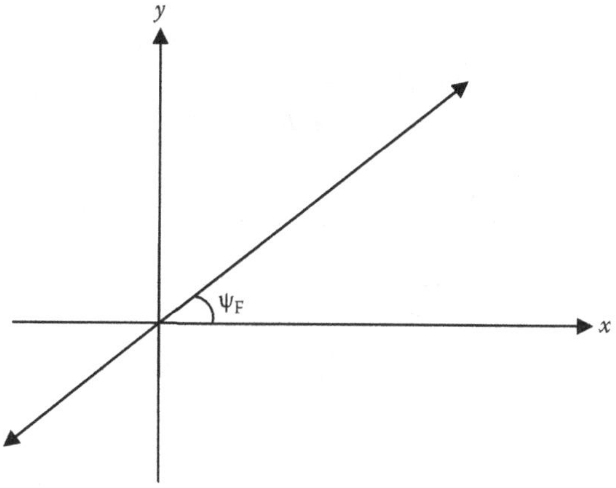

FIGURE 11A.3
The electric field at $z = d$.

where q and m are the absolute value of the charge and mass of the electron, respectively, and N_0 is the electron density. If the electron density varies with distance and if B_0 is approximately constant, the net rotation angle is given by

$$\psi_F = \frac{q^3 B_0}{2m^2 \varepsilon_0 \omega^2 c} \int N_0(z) \mathrm{d}z,$$

where the integration is along the propagation path. Measurements of the total electron content in 1 m^2 column are made using satellite–earth transmissions, by measuring ψ_F.

11A.2 Natural Rotation and Comparison

Note that the chiral media also cause rotation of linearly polarized wave. In calculating the angle $\psi = \psi_c$, we use k_{cL} and k_{cR} instead of k_{pR} and k_{pL}. See Appendix 8A.

However, there is one important difference in the rotation angle for a "round-trip" calculation, for the two cases.

The calculation for a return trip can be made by observing (Figure 11A.4).

Let us first consider Faraday rotation.

For a negatively going wave, the direction of B_0 and the direction of propagation are 180° apart. The wave numbers k_{pR} and k_{pL} can be obtained by replacing ω_b by $(-\omega_b)$ in Equations 11A.6 and 11A.7. This is tantamount to replacing

$$k_{pL} \rightarrow k_{pR}$$

and

$$k_{pR} \rightarrow k_{pL}$$

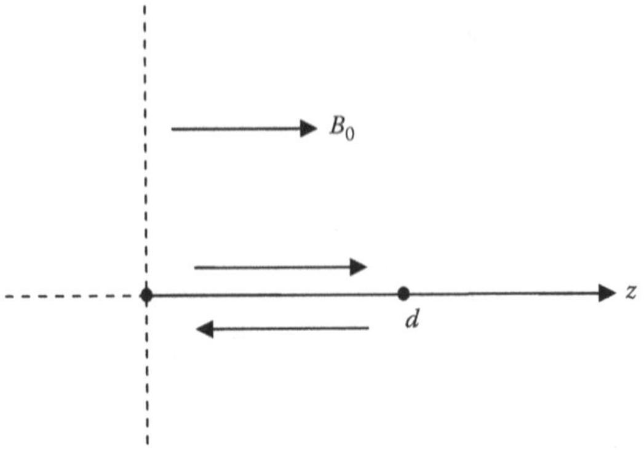

FIGURE 11A.4
Natural rotation.

in Equation 11A.9. Also, since the difference between the z-coordinate of the starting point and the end point is $-d$, we have to replace $d \rightarrow -d$ in Equation 11A.9.

Thus, the value of ψ_F remains unchanged. The rotation angle for a round trip is thus twice that of a one-way trip.

On the other hand, for a chiral media, k_{cL} and k_{cR} are unchanged but d will be replaced by $-d$. Thus, the rotation angle for the return trip is the same as for the trip except for the sign. The round-trip rotation angle is zero.

Reference

1. Inan, S. I. and Inan, S. A., *Electromagnetic Waves*, Prentice Hall, Upper Saddle River, NJ, 2000.

Appendix 11B: Ferrites and Permeability Tensor

In Chapter 11, we discussed the anisotropic properties of a magnetized plasma by modeling it as an anisotropic dielectric whose permittivity is a tensor. Ferrites in the presence of an external static magnetic field behave like an anisotropic magnetic material whose permeability is a tensor. Such material has many applications in microwave engineering. Here we briefly discuss the permeability tensor [1].

Ferrites are magnetic primarily because of the magnetic dipole moment $\bar{m}$ created due to the electron spin. In the presence of a magnetic field $\bar{H}$, torque $\bar{T}$ exerted on the electron is given by

$$\bar{T} = \mu_0 \bar{m} \times \bar{H}. \tag{11B.1}$$

The equation of angular motion of mechanics is given by

$$\frac{d\bar{L}}{dt} = \bar{T}, \tag{11B.2}$$

where $\bar{L}$ is the angular momentum.

The motion of the electron is gyroscopic around $\bar{H}$. The difference between a mechanical gyroscope and the spinning electron motion is that in the case of electrons the torque is of magnetic origin rather than mechanical. The dipole moment to be used in Equation 11B.1 is given by

$$\bar{m} = \gamma_m \bar{L}, \tag{11B.3}$$

where

$$\gamma_m = \frac{q}{m_e}, \tag{11B.4}$$

m_e being the mass of the electron, γ_m the gyromagnetic ratio ($=1.76 \times 10^{11}$ rad/sT).

From Equations 11B.1 through 11B.3, we obtain

$$\frac{d\bar{m}}{dt} = \gamma_m \mu_0 \bar{m} \times \bar{H}. \tag{11B.5}$$

Equation 11B.5 describes (Larmor) precession motion with the frequency ω_0 (Larmor frequency) given by

$$\omega_0 = \mu_0 \gamma_m \left|\bar{H}\right|. \tag{11B.6}$$

If there are N electrons per unit volume with each electron creating magnetic dipole moment $\bar{m}$ due to its spin, the magnetization vector $\bar{M}$ is given by

$$\bar{M} = N\bar{m}. \tag{11B.7}$$

Equation 11B.5 can now be written as

$$\frac{d\bar{M}}{dt} = \gamma_m \mu_0 \bar{M} \times \bar{H}. \tag{11B.8}$$

For harmonic variation of $\bar{M}$ with frequency ω,

$$\tilde{\bar{M}} = \frac{\gamma_m \mu_0}{j\omega} \tilde{\bar{M}} \times \bar{H}. \tag{11B.9}$$

Equation 11B.9 contains the origin of the anisotropy of the permeability. Let us assume that the external magnetic field is

$$\bar{H}_0 = \hat{z} H_0 \tag{11B.10}$$

and the small-signal H_1 in the z-direction is neglected. From Equation 11B.9, one can obtain the relation between the small-signal value of $\bar{M}$ and the small-signal values of H_x and H_y as follows:

$$\bar{B} = \mu_0 \left(\bar{H}_1 + \bar{M}_1 \right). \tag{11B.11}$$

We can now write

$$\bar{B} = \bar{\mu}_{eff} \cdot \bar{H}, \tag{11B.12}$$

where

$$\bar{\mu}_{eff} = \begin{bmatrix} \mu_{eff}^{11} & \mu_{eff}^{12} & 0 \\ \mu_{eff}^{21} & \mu_{eff}^{22} & 0 \\ 0 & 0 & \mu_0 \end{bmatrix}. \tag{11B.13}$$

The elements of the matrix are:

$$\mu_{eff}^{11} = \mu_0 \left[1 + \frac{\omega_0 \omega_M}{\omega_0^2 - \omega^2} \right] = \mu_{eff}^{22}, \tag{11B.14a}$$

$$\mu_{eff}^{12} = \left(\mu_{eff}^{21} \right)^* = j\mu_0 \frac{\omega_M \omega}{\left(\omega_0^2 - \omega^2 \right)}, \tag{11B.14b}$$

$$\omega_M = \mu_0 \gamma_m M_0. \tag{11B.14c}$$

If μ_r of the ferrite material is known, then

$$M_0 = (\mu_r - 1)H_0. \tag{11B.14d}$$

Equation 11B.13 looks similar to the dielectric tensor $\bar{\mathbf{K}}$ discussed in Chapter 11. One can thus expect phenomena of Faraday rotation. Sohoo [2] discusses a number of microwave applications.

References

1. Inan, S. I. and Inan, S. A., *Electromagnetic Waves*, Prentice-Hall, Upper Saddle River, NJ, 2000.
2. Sohoo, R. F., *Microwave Magnetics*, Harper & Row, New York, 1985.

Appendix 11C: Thin Film Reflection Properties of a Warm Magnetoplasma Slab: Coupling of Electromagnetic Wave with Electron Plasma Wave

Dikshitulu K. Kalluri and Constantine T. Markos

11C.1 Introduction

The study of interaction of electromagnetic waves with warm magnetoplasma continues to attract the attention of researchers in electromagnetics [1]. Earlier research in the 1960s and 1970s was motivated by its possible application (1) in opening the passband below plasma frequency for the problem of radio reentry blackout [1] and (2) RF heating of fusion plasmas [2,3]. There is a resurgence of research in this area due to new applications. The broad area of plasmonics, which utilizes the coupling of electromagnetic waves with an electron plasma wave, has applications in (1) HEMT [4–6] and in the general area of (2) plasma wave electronics [7–10]. The important features of the interaction can be obtained by studying the reflection and transmission properties of a warm magnetoplasma slab.

Numerical results for a general problem of oblique incidence of an electromagnetic plane wave on a warm, thin magnetoplasma slab [2,11] with different input and output media can be obtained by formulating it as a boundary value problem. Reference 1 gives a detailed study of this aspect.

To interpret and explain the results—by unraveling the effects of various parameters given in Table 11C.1 on the (a) behavior near resonances, (b) behavior due to triply refractive magnetoplasma medium, (c) excitation of evanescent and/or propagating positive-going or negative-going waves in various frequency bands, (d) tunneling of power through a thin slab in the stop bands [12–16], and (e) excitation of surface plasmons [13,17]—requires simpler formulations that bring out each of these consequences into focus. Characterizing reflection properties for a magnetoplasma slab can be achieved by careful choice of various parameters in Table 11C.1.

Consider first the simpler case of isotropic, warm plasma slab. For normal incidence, it is obvious that electron plasma wave (which requires electric field in z-direction) will not be excited in the slab. For oblique incidence, the electromagnetic as well as electron plasma wave will be excited in the slab, and the results for this case are discussed in [13] and Appendix 10A.

For normal incidence of R wave (right circular polarization) or L wave (left circular polarization), when the static magnetic field is in the z-direction, the electron plasma wave will not be excited [18], and the results for *cold* or *warm* magnetoplasma slab are the same. For normal incidence of x- or y-polarized wave, when the static magnetic field is in the z-direction, the results of cold and warm plasma slab are the same since the linear polarization in x- or y-direction can be considered as a superposition of R and L waves. For normal

TABLE 11C.1

Parameters of Interest for a Warm Magnetoplasma Slab

Symbol	Description	Units
d	Slab ($0 < z < d$) thickness	m
θ^{I}	Angle of incidence	rad
ω	Incident wave frequency (and its polarization)	rad/m
n_i, n_t	Input/output refractive indices	1
T	Temperature	K
ω_p	Plasma frequency	rad/m
ω_b	Electron gyrofrequency	rad/m

incidence of y-polarized wave, when the static magnetic field is in the y-direction, the results are the same as those of isotropic plasma slab.

For normal incidence of x-polarized wave, when the static magnetic field is in the y-direction, the results are very different for the *cold* and *warm* plasmas. The incident electromagnetic wave excites in the plasma an extraordinary wave (called X wave) [12] with the results for warm and cold cases very different near upper hybrid frequency ω_{uh}. For the warm plasma case, the resonance (refractive index infinity) for the cold case will be replaced by a finite value for the refractive index. In the neighborhood of the upper hybrid frequency, the characteristic of the waves in the magnetoplasma will change from a basically electromagnetic wave to that of an electron plasma wave [17]. In this appendix, we formulate and solve the associated electromagnetic boundary value problem to study this aspect of coupling of the waves and the effect of various parameters on the coupling. Moreover, the role of the evanescent waves on the tunneling of power through a thin slab will be explored.

11C.2 Formulation of the Problem

The power reflection coefficient in this chapter is derived from detailed equations starting with Maxwell's equations and the hydrodynamic approximation. The wave equation for a warm, compressible, anisotropic magnetoplasma was derived by Unz [19].

To describe the electromagnetic waves inside the slab, the medium is modeled as a lossless (collisionless), homogeneous, anisotropic ($\vec{B}_0$) warm plasma of thickness d. The plasma slab of n_p is bounded by two half-space media $n = 1$. The incident wave is x-polarized, equivalent to normal incidence ($\theta^I = 0$) of a parallel-polarized (TM or p) wave (see Figure 11C.1). In addition, there is a magnetostatic field normal to the plane of incidence, or $\vec{B}_y = \hat{y}B_0$.

11C.2.1 Waves Outside the Slab

For the incident, reflected, and transmitted (I, R, and T) waves outside the plasma slab, the incident wave can be described as follows:

$$\vec{E}^{I,T} = \hat{x}E_{I,T}e^{-jk_0z}, \quad \vec{H}^{I,T} = \hat{y}E_{I,T}/\eta_0\, e^{-jk_0z}, \tag{11C.1a}$$

$$\vec{E}^{R} = \hat{x}E_{R}e^{jk_0z}, \quad \vec{H}^{R} = -\hat{y}E_{R}/\eta_0\, e^{jk_0z}, \tag{11C.1b}$$

where $\eta_0 = k_0/\omega\mu_0$.

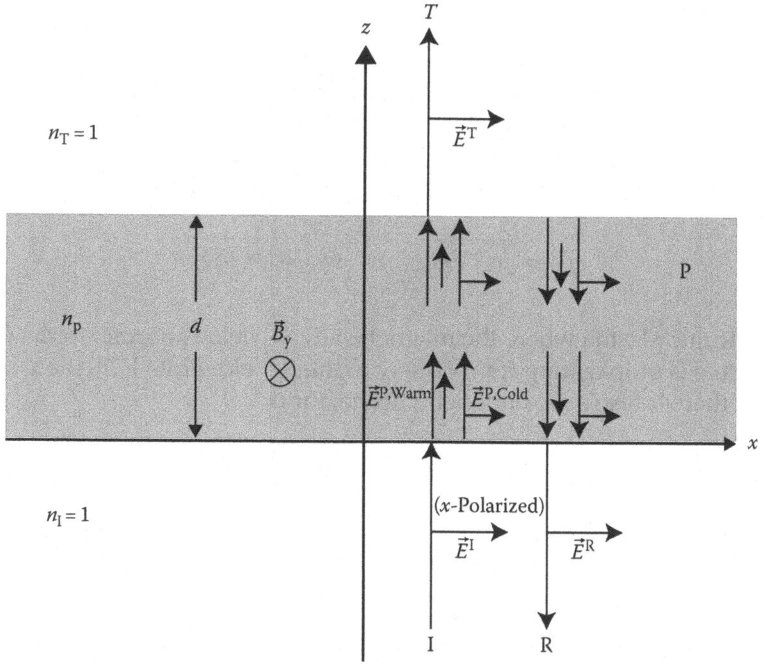

FIGURE 11C.1
Normal incidence of an x-polarized wave on a warm, anisotropic plasma slab. Shown are the reflection and transmission waves. I, incident; R, reflected; T, transmitted, P, plasma.

11C.2.2 Waves Inside the Slab (Cold Plasma)

Using the exponential factor,

$$\psi_P = e^{-jk_0 z q_i}, \tag{11C.2}$$

where $k_p = k_0 q_i = k_0\sqrt{\varepsilon_{px}}$, for $i = 1, 2$. Note: For normal incidence, $q_i = n_{p,i}$. From [12], the dielectric constant for an X wave in an anisotropic plasma is

$$\varepsilon_{p,x} = q_i^2 = n_p^2 = 1 - \frac{\omega_p^2/\omega^2}{1 - \dfrac{\omega_b^2}{\omega^2 - \omega_p^2}} = \frac{(1-X)^2 - Y^2}{1 - X - Y^2}. \tag{11C.3}$$

Equation 11C.3 agrees with [15] for normal incidence dispersion relation when normalizing ω and ω_b by the plasma frequency ω_p: $\Omega = \omega/\omega_p$, $\Omega_b = \omega_b/\omega_p$ and by allowing $X = \Omega^{-2}$ with $Y = \omega/\omega_b$. It also agrees with Appleton–Hartree formula for a wave at normal incidence on a collisionless cold plasma as described by Budden [20]. The dielectric constant can also be written in terms of cutoff and upper hybrid frequencies:

$$\omega_{c1,c2} = \mp\frac{\omega_b}{2} + \sqrt{\left(\frac{\omega_b}{2}\right)^2 + \omega_p^2}. \tag{11C.4}$$

When (11C.4) is normalized,

$$\Omega_{c1,c2} = \mp\frac{\Omega_b}{2} + \sqrt{\left(\frac{\Omega_b}{2}\right)^2 + 1}. \quad (11C.5)$$

The upper hybrid frequency, and its normalized file are given by

$$\omega_{uh} = \sqrt{\omega_p^2 + \omega_b^2} \quad \text{or} \quad \Omega_{uh} = \sqrt{1 + \Omega_b^2}. \quad (11C.6)$$

For an anisotropic plasma where the magnetostatic B_0 field is normal to the plane of incidence, the X wave is propagating transversely to the B_0 field. From [12], the x-component is preserved, but there is also a z-component; hence, for $i=1,2$,

$$\vec{E}_{p,i} = \left(\hat{x}E_{p,x,i} + \hat{z}E_{p,z,i}\right)e^{-jk_0 q_i z}, \quad (11C.7a)$$

$$\vec{H}_{p,i} = \hat{y}\frac{E_{p,x,i}}{\eta_{p,i}}e^{-jk_0 q_i z}, \quad (11C.7b)$$

where $\eta_{p,i}=\eta_0/q_i$. From Equations 11C.1 and 11C.7 and using tangential boundary conditions, for $z=0$ and d, one can derive the power reflection coefficient:

$$\rho = \left|\frac{E_R}{E_I}\right|^2 = \frac{1}{1 + \left[\frac{2q_1}{1-q_1^2}\csc\left(2\pi q_1 d_p \Omega\right)\right]^2}, \quad (11C.8)$$

where d_p is the normalized film thickness and $d_p = d/\lambda_p$, λ_p is the normalized plasma wavelength.

11C.2.3 Waves Inside the Slab (Warm Plasma)

The symbols used for the parameters of *warm plasma* medium are given in Table 11C.2. The basic equations for a warm plasma are Maxwell's equations:

$$\vec{\nabla} \times \vec{E}^P = -j\omega\mu\vec{H}^P, \quad (11C.9a)$$

$$\vec{\nabla} \times \vec{H}^P = j\omega\varepsilon\vec{E}^P - eN_0\vec{u}. \quad (11C.9b)$$

Using the hydrodynamic approximation for a compressible electron gas, the conservation of momentum [19] is given as follows:

$$mN_0\left(j\omega + \nu\right)\vec{u} = -\vec{\nabla}p - eN_0\vec{E}^P - e\mu N_0\vec{u} \times \vec{H}_0^P, \quad (11C.10a)$$

The conservation of energy is given by

$$p = \gamma K T_0 N. \quad (11C.10b)$$

TABLE 11C.2

Warm Plasma Parameters

Symbol	Description	Units
N, N_0	Small and large signal number density	m^{-3}
p, p_0	Small and large signal pressure	$kg/m\ s^2$
T, T_0	Small and large signal electron thermodynamic temperature	K
K	Boltzmann's constant	
γ	C_p/C_v Specific heat in constant pressure/specific heat in constant volume	1
m	Electron mass	kg
$\vec{u}$	Velocity vector	m/s

Varying device geometries can be justified using compressible electron gas models.

The equation of state is given by

$$\frac{p}{p_0} = \frac{N}{N_0} + \frac{T}{T_0}; \quad p_0 = KN_0T_0. \tag{11C.10c}$$

The formulation given above is based on linear theory. Unz showed that Equations 11C.9 and 11C.10 reduce to the wave equation

$$\begin{aligned} &U\left(1-n_p^2-\frac{X}{U}\right)\vec{E}^P+\left(U-\delta^2\right)\left(\vec{n}_p\cdot\vec{E}^P\right)\vec{n}_p \\ &-\left[\left(1-n_p^2\right)\vec{E}^P+\left(\vec{n}_p\cdot\vec{E}^P\right)\vec{n}_p\right]\times j\vec{Y}=0, \end{aligned} \tag{11C.11}$$

$$\vec{Y} = e\mu\vec{H}_0/m\omega, \tag{11C.12a}$$

In the above

$$U = 1 - jZ, \tag{11C.12b}$$

where Z is given in terms of the electron collision frequency ν and the transmission frequency ω:

$$Z = \frac{\nu}{\omega}, \tag{11C.13a}$$

Also e is the absolute value of the charge and m is the mass of the electron. The normalized velocity variable

$$\delta = \frac{a}{c} = \frac{\sqrt{\gamma T_0\left(C_p - C_v\right)}}{c}, \tag{11C.13b}$$

where a is the acoustic velocity of the electron plasma wave and n_p is the plasma refractive index. The electric and magnetic field intensities inside the plasma can be described as

$$\vec{E}^P = \vec{E}^P_{x,y,z}\psi^P; \quad \vec{H}^P = \vec{H}^P_{x,y,z}\psi^P. \tag{11C.14}$$

The phase of the fields within the plasma slab is defined as

$$\psi^{\mathrm{P}} = e^{-jk_0qz}. \tag{11C.15}$$

For normal incidence, $n_{\mathrm{p}} = q$ and $\vec{n}_{\mathrm{p}} = \hat{z}q$. When using (11C.14) and (11C.15) in (11C.10a), one has for normal incidence on the slab interface the dispersion relation

$$\left\{q^2 - (1-X)\right\} \times \left\{ \begin{array}{c} \delta^2 q^4 - \left[\left(1 - X - Y^2\right) + \delta^2(1-X)\right]q^2 \\ +\left[(1-X)^2 - Y^2\right] \end{array} \right\} = 0. \tag{11C.16}$$

The first term is the dispersion relation for an O wave for an electric field within the plasma in the y-direction. The second term in the {} is the dispersion relation for the X wave. Although solvable, the solution in the second term cannot be decoupled between cold and warm plasmas; however, for a solution of $q_{1,2}|_{\delta\to 0}$, Equation 11C.3 would result and $q_{3,4}|_{\delta\to 0}$ has no physical significance.

We use boundary conditions where the tangential components of $\vec{E}$ and $\vec{H}$ are continuous at $z = 0$ and d to generate four equations. For time-harmonic waves, tangential boundary conditions are used as they are derived by Maxwell's curl equations. Normal boundary conditions (from divergence equations) can be derived from the curl equations. Hence, normal boundary conditions are not independent but are assured to be satisfied when the tangential boundary conditions are satisfied [21]. Using a scalar electron pressure, hydrodynamic approach, it is assumed that the electron velocity normal to the plasma/dielectric boundary is zero [11,21,22], which results in two more boundary equations. However, we do not assume electron pressure to be zero at the boundaries. Writing the system of equations from the boundary conditions, solving and one has the power reflection coefficient for a warm magnetoplasma,

$$\rho = \left|\frac{E_{\mathrm{R}}}{E_{\mathrm{I}}}\right|^2 = \frac{1}{1 + \left[\dfrac{\xi_1}{\xi_2 + \xi_3}\right]^2}, \tag{11C.17a}$$

where

$$\begin{aligned}\xi_1 = 8\left(q_1^2 - q_3^2\right)[(1-X)\left(q_1\sin\phi_3 - q_3\sin\phi_1\right) \\ + q_1q_3\left(q_1\sin\phi_1 - q_3\sin\phi_3\right)],\end{aligned} \tag{11C.17b}$$

$$\begin{aligned}\xi_2 = 4\sin\phi_1\sin\phi_3[\left(q_1^2 + q_3^2\right)\left((1-X)^2 + q_1^2q_3^2\right) \\ -\left(2q_1^2q_3^2(1-2X) + q_1^4 + q_3^4\right)],\end{aligned} \tag{11C.17c}$$

$$\xi_3 = 8q_1q_3\left(\cos\phi_1\cos\phi_3 - 1\right)\left(1 - q_1^2 - X\right)\left(1 - q_3^2 - X\right). \tag{11C.17d}$$

In (11C.17b) and (11C.17d), $\phi_1 = 2\pi d_{\mathrm{p}}\Omega q_1$ and $\phi_3 = 2\pi d_{\mathrm{p}}\Omega q_3$. Note how Equation 11C.17a compares with Equation 11C.8 and yet considerably more complex as a result of the warm plasma medium even when it is modeled by hydrodynamic approximation.

11C.3 Characteristic Roots and the Power Reflection Coefficient

The hydrodynamic approximation takes into account that electron plasma wave has a velocity high enough where the plasma is considered compressible. When the acoustic mode velocity approaches zero, the plasma becomes cold and incompressible. The longitudinal characteristic roots are dependent on the parameter $\delta = a/c$, where a is the acoustic velocity of the electron plasma wave. In certain metals, the Fermi velocity [17] plays the role of the acoustic velocity. When a is comparable to c, that is, $\delta \approx 0.1-0.01$, the plasma is spatially dispersive and is considered to be warm. A small percentage change in the velocity of the longitudinal component gives rise to a dynamic response of the power reflection coefficient. In addition, the oscillations (or resonances) within the warm plasma reflection coefficient tend to ride atop the cold plasma component.

A method to understand the reflection properties is to examine the dispersion relation, which when solved leads to the characteristic roots of the plasma medium. Note that we consider here the lossless case, $U=1$, where the characteristic roots are either real or imaginary but not complex. For the anisotropic case, X wave only, there are four roots to consider, $q_{1,2}$ and $q_{3,4}$, which create four distinct regions that affect ρ. In our previous paper [13], the regions of the power reflection coefficient were defined by the cutoff frequencies (zeros of q). In this paper, the regions are defined by the cutoff and upper hybrid frequencies.

Figure 11C.2 shows the results of Equations 11C.5 (cold) and 11C.17a (warm) power reflection coefficients. The bottom smaller plots separate the two for clarity. There is a considerable difference between the cold and warm plasma reflection coefficients. This is a

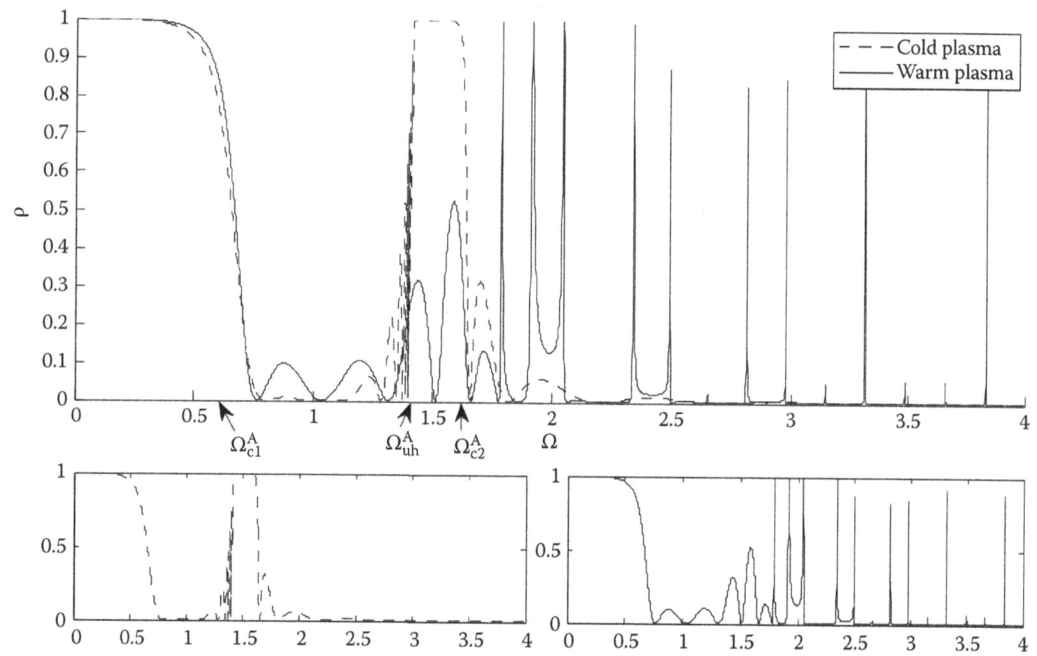

FIGURE 11C.2
Power reflection coefficient for $d_P=1$, $\delta=0.1$, $\theta^I=0°$, $\Omega_b=0.5$, and $n_I=n_T=1$.

result of an E_z component that excites longitudinal propagation (acoustic mode), which is a solution to the wave equation.

Figures 11C.3 and 11C.4 show the characteristic roots for the cold and warm plasma models. From Equation 11C.3, a resonance would occur when $1-X-Y^2=0$ for the cold plasma, that is, $q_i \to \infty$ [18] determines that the resonance was located at $X=1-Y^2$. This produced Equation 11C.8, which was determined here as the upper hybrid frequency. The resonance is located in the region where the approximations to warm plasma dispersion equations fail [18]. For the warm plasma, the characteristic root q_i will not go to infinity when $1-X-Y^2=0$ and remove the resonance in the cold plasma model. To ascertain the exact value of q_i at the resonance region from the second term of Equation 11C.16 yields the characteristic root at the upper hybrid frequency

$$q_{i,uh}^2 = \frac{\delta^2 Y_{uh}^2 \pm \sqrt{\delta^4 Y_{uh}^4 + 4\delta^2 Y_{uh}^2 X_{uh}}}{2\delta^2}, \tag{11C.18}$$

where $Y_{uh}=\Omega_b/\Omega_{uh}$ and $X_{uh}=\Omega_{uh}^{-2}$. The negative sign gives the value for q_i^2 and leads to an evanescent wave. Equation 11C.18 is equivalent to the "true point" that Tanenbaum [18] describes. For a warm plasma, at the upper hybrid frequency, there exists a true point that replaces a resonance as shown in Figure 11C.3, in magnetoionic plasma theory. Hence, there is a transition from a transverse electromagnetic wave to a longitudinal electroacoustic wave.

To qualitatively understand the dynamics inside the plasma, one can look at how the characteristic roots change in frequency space. Fundamental changes in q_i occur at the cutoff points $\Omega_{C1,C2}$ and the upper hybrid frequency Ω_{uh} (Equation 11C.5).

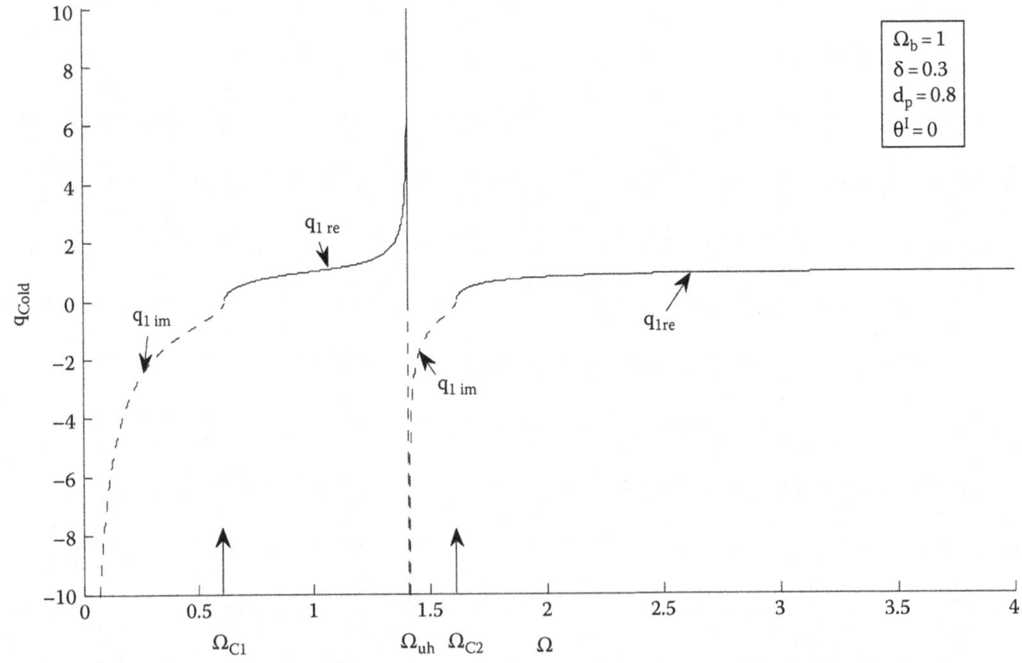

FIGURE 11C.3
Cold plasma characteristic roots for $\theta^I=0°$. The resonance is at the upper hybrid frequency $\Omega_{uh}=1.414$. The cutoff frequencies are $\Omega_{C1}=0.618$ and $\Omega_{C2}=1.618$; dotted, imaginary; solid, real.

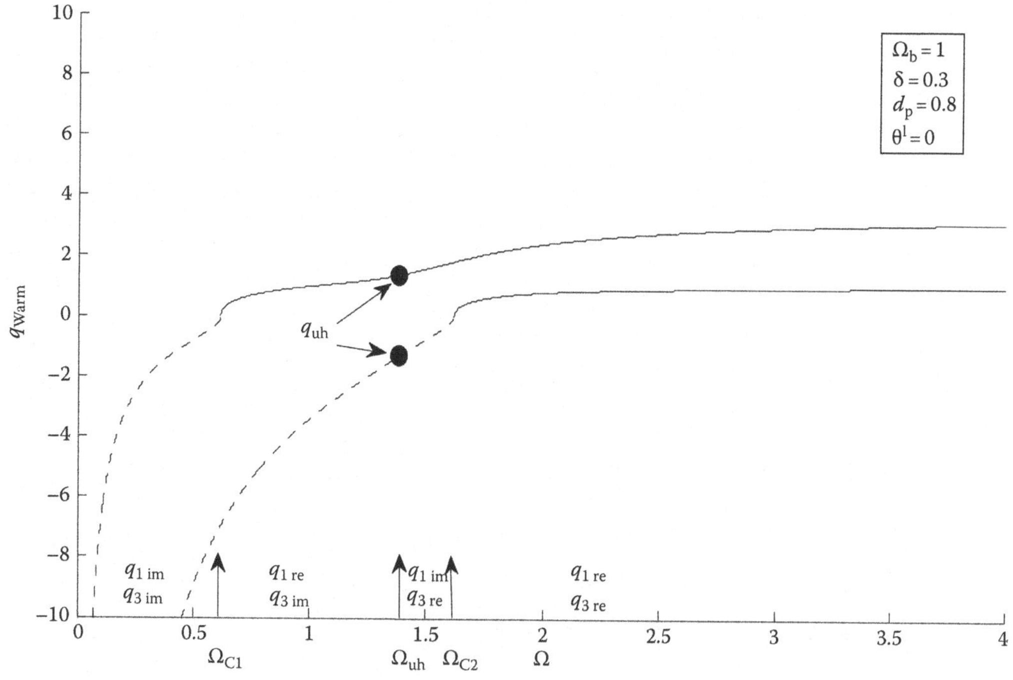

FIGURE 11C.4
Warm plasma characteristic roots; dotted, imaginary; solid, real.

11C.3.1 Regions of the Power Reflection Coefficient

a. For $0<\Omega<\Omega_{C1}$: Both q_1 and q_3 are imaginary, noted as dashed lines in Figures 11C.3 and 11C.4. This means that both the electromagnetic and electroacoustic waves do not propagate and are evanescent. The power reflection coefficient is close to unity as all the energy is reflected back (Figure 11C.5). A unity value of ρ means $\xi_1 \approx 0$, and by looking at Figure 11C.4 as $\Omega \to 0$, both q_1 and q_3 are very large and negative and can be considered approximately equal. Both terms in ξ_1 can be shown to be zero if $q_1 = q_3$. However, the reflection coefficient starts on a downward path as $\Omega \to \Omega_{c1}$. This downward path also shows the deviation between the cold and warm results. The deviation results as Ω approaches the first cutoff frequency and q_1 can no longer be considered equal to q_3; therefore, $\xi_1 \neq 0$.

b. For $\Omega_{C1}<\Omega<\Omega_{C2}$: There are two subregions to consider, as the characteristic roots cross over the upper hybrid frequency and wave coupling occurs.

   i. For $\Omega_{C1}<\Omega<\Omega_{uh}$: From Figure 11C.4, q_1 is real and q_3 is imaginary, and the electromagnetic wave now propagates through the slab, while the electroacoustic wave remains evanescent. As a result, from Equation 11C.7a, $E_{p,z,i}$ is small. The electromagnetic propagation presents itself by observing that the power reflection coefficient for both the cold and warm results starts to oscillate. The cold plasma result oscillates at increasing rates before a sharp rise at $\Omega = \Omega_{uh} = 1.414$, then stops oscillating (Figure 11C.6). The warm plasma also oscillates albeit at a slow rate almost following the cold plasma result until the cold plasma result oscillates faster on its own.

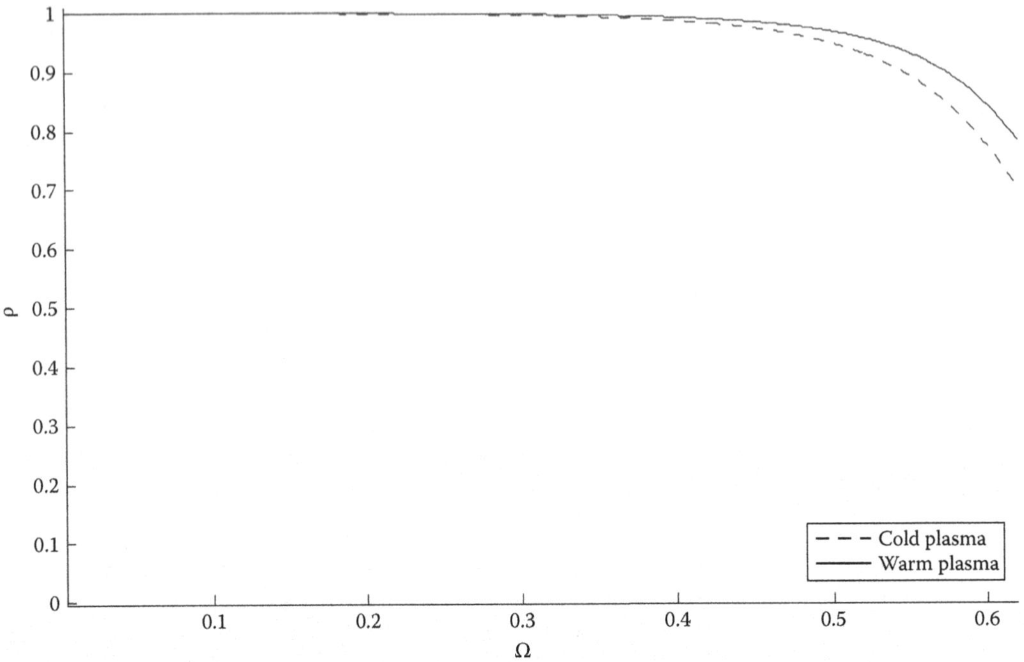

FIGURE 11C.5
The power reflection coefficient ρ for $0<\Omega_{C1}$.

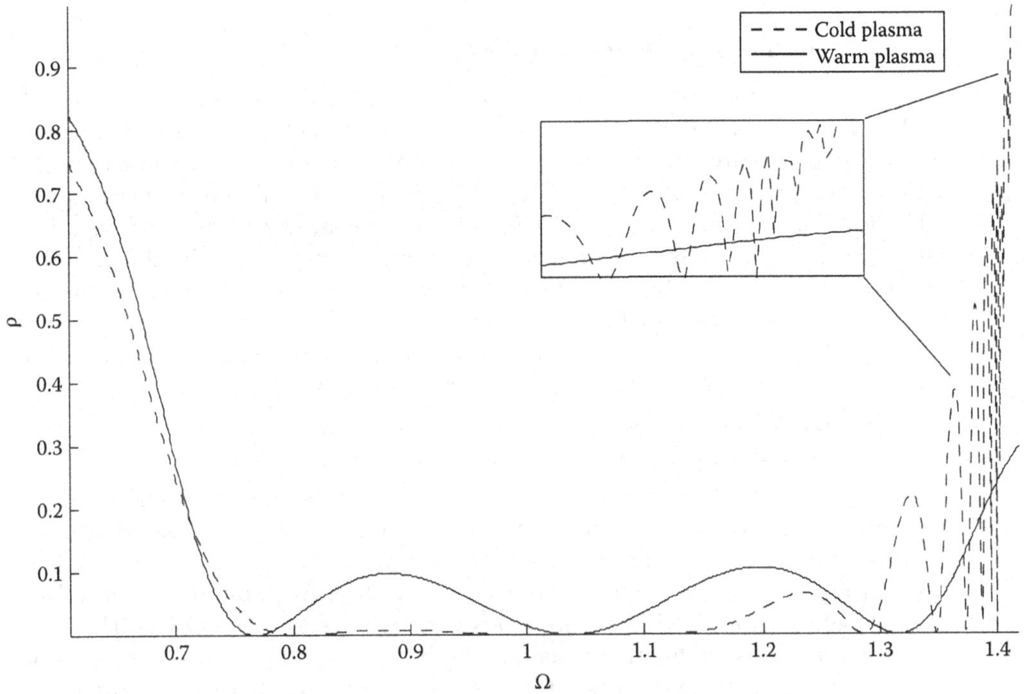

FIGURE 11C.6
The power reflection coefficient ρ for $\Omega_{C1}<\Omega<\Omega_{uh}$.

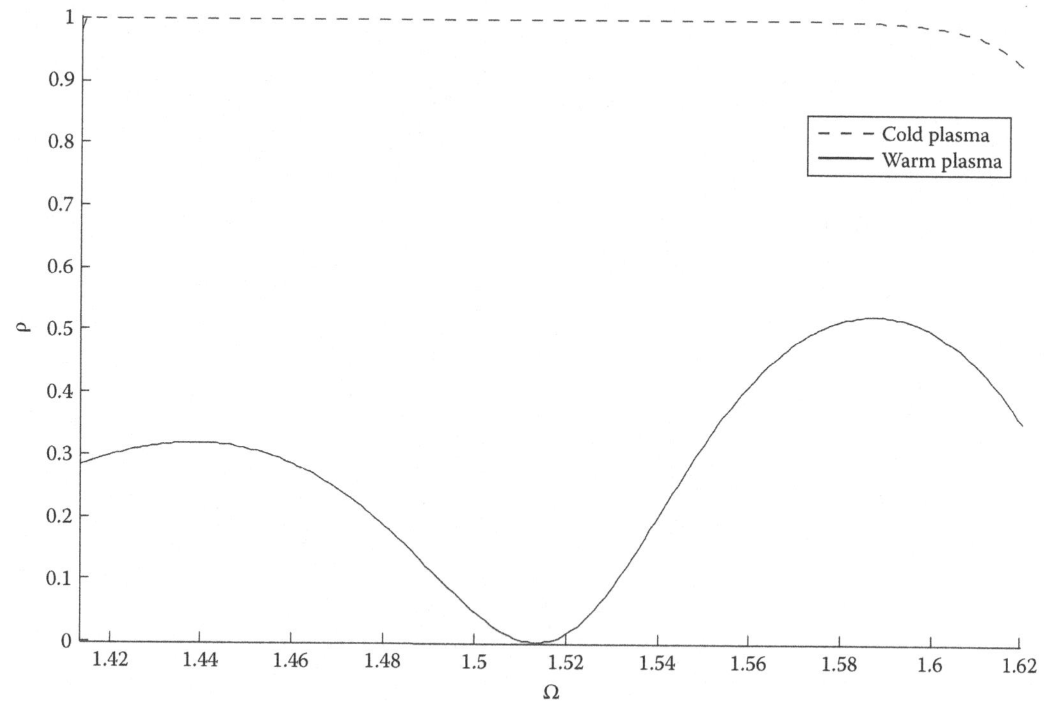

FIGURE 11C.7
The power reflection coefficient ρ for $\Omega_{uh} < \Omega < \Omega_{C2}$.

ii. For $\Omega_{uh} < \Omega < \Omega_{C2}$: From Figure 11C.4, q_1 is imaginary and q_3 is real, and the electroacoustic mode now propagates through the slab, while the electromagnetic wave remains evanescent and $E_{p,z,i}$ becomes large. This is possible as a result of the coupling of the electroacoustic wave to the electromagnetic wave at the true points; hence, $\Omega = \Omega_{uh}$. The cold plasma power reflection coefficient remains constant reflecting the electromagnetic wave in this region, but the warm plasma part continues to oscillate by propagating the electroacoustic wave (Figure 11C.7).

c. For $\Omega_{C2} < \Omega < \infty$: Both q_1 and q_3 are real, shown as solid lines in Figures 11C.3 and 11C.4. Both the electromagnetic and electroacoustic waves are propagating through the slab. For the power reflection coefficient, the warm plasma resonances ride atop of the smooth, cold plasma result, which could be qualitatively described as Tonks–Dattner resonances (Figure 11C.8) [13].

11C.3.2 Other Cases of the Power Reflection Coefficient

Variations of the power reflection coefficient are shown with different magnetic field strengths as well as different δ's and film thicknesses. Three cases are chosen and described in Table 11C.3 to show how the power reflection coefficient changes with variations in the physical parameters in the table and, more importantly, how the cold and warm plasma power reflection coefficients change for the same parameters.

For case A, the film thickness is same as the plasma wavelength, λ_p. In the low frequency range the characteristic roots for both cold and warm cases are very large in the negative

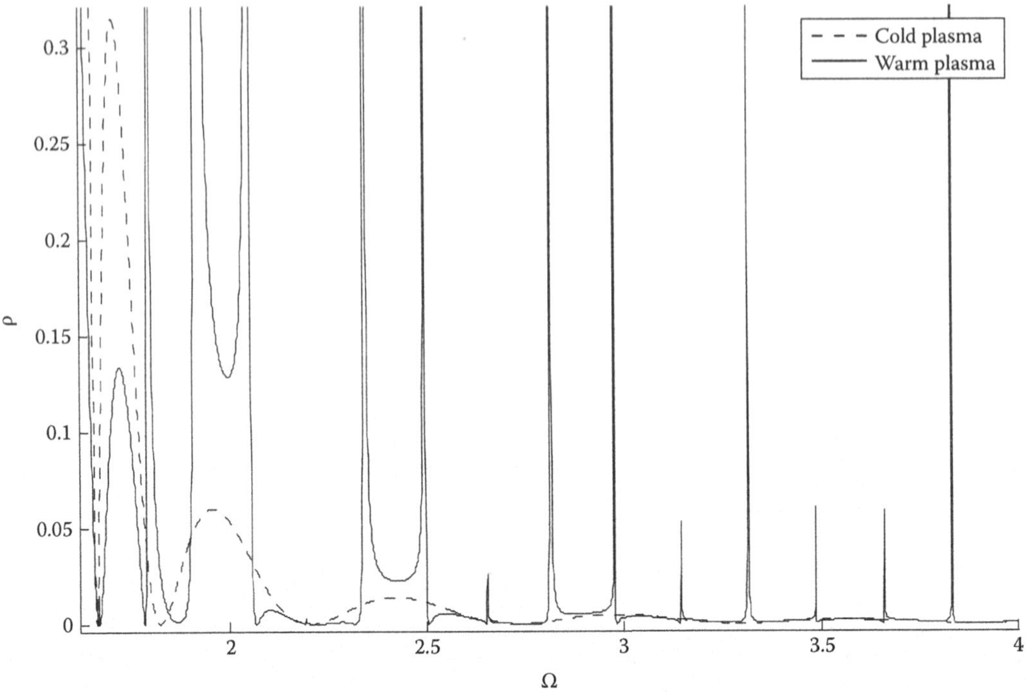

FIGURE 11C.8
The power reflection coefficient ρ for $\Omega_{C2} < \Omega < \infty$.

TABLE 11C.3
Different Cases A, B, and C of Parameters

Parameter	A	B	C
Ω_b	0.5	1.0	2.0
d_p	1.0	0.1	0.1
δ	0.1	0.3	0.05
Ω_{uh}	1.118	1.414	2.236
Ω_{C1}	0.781	0.618	0.414
Ω_{C2}	1.281	1.618	2.414

direction and are negative imaginary (Figure 11C.9 and 11C.10). Both the cold and warm power reflection coefficients follow each other fairly well with $\rho \approx 1$, and no energy propagates through the slab (Figure 11C.11). As $\Omega \to \Omega_{C1}$, the characteristic roots are small and ρ starts to trend downward and is approximately 0.9 at Ω_{C1}. For $\Omega_b = 0.5$, there is a cold plasma resonance at the upper hybrid frequency $\Omega_{uh} = 1.118$ (Figure 11C.9). Between Ω_{C1} and Ω_{uh}, the cold plasma characteristic root becomes real and transverse waves will propagate through the slab. The power reflection coefficient for both cold and warm models will continue to track each other and oscillate.

However, as $\Omega \to \Omega_{uh}$, the cold plasma power reflection coefficient will oscillate at increasing rates until it reaches the upper hybrid frequency, then steps up to 1, where as the warm plasma coefficient will continue to oscillate but increases its rate slowly through the upper

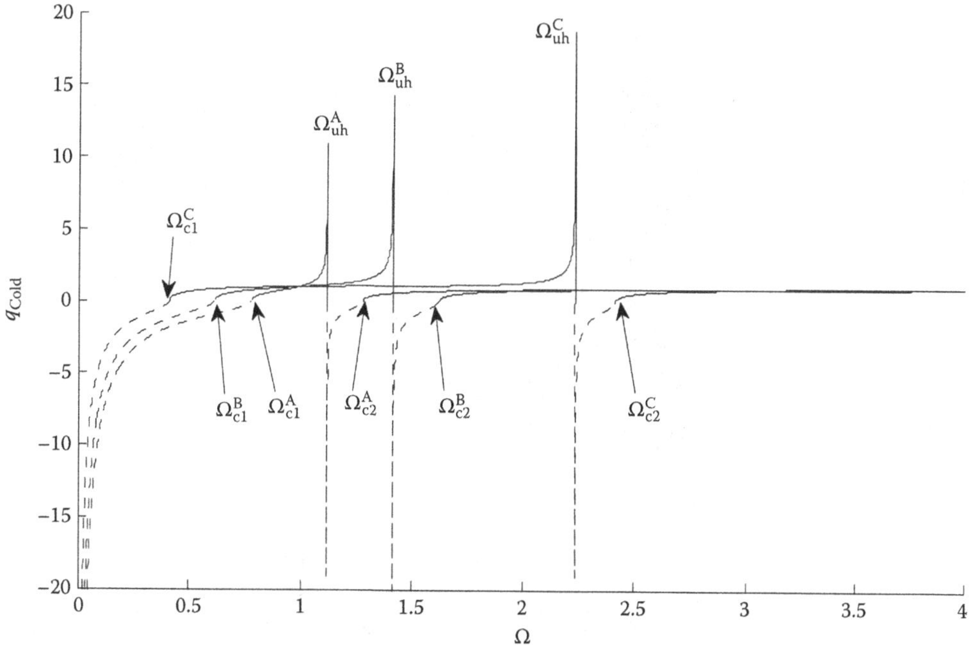

FIGURE 11C.9
Cold plasma model; characteristic roots for cases A, B, and C. The junction between the dotted and solid lines are marked with the cutoff or resonant frequencies.

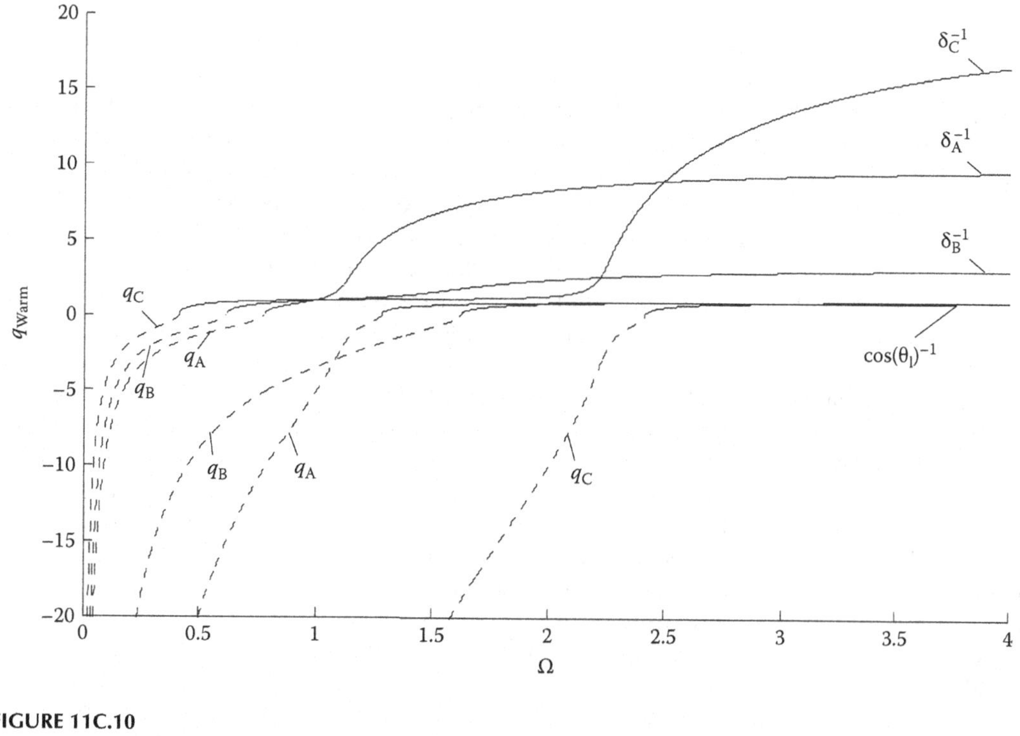

FIGURE 11C.10
Warm plasma model; characteristic roots for cases A, B, and C.

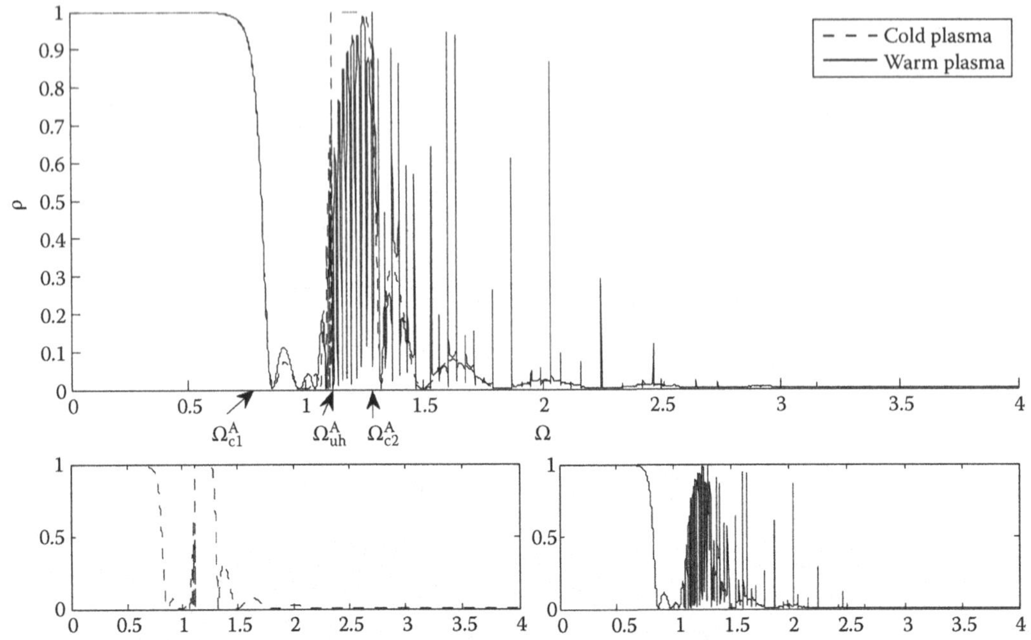

FIGURE 11C.11
Case A, $\Omega_b = 0.5$, $d_p = 1.0$, and $\delta = 0.1$.

hybrid frequency. At Ω_{uh}, the transition from transverse waves to longitudinal waves and vice versa takes place, and the characteristic roots resolve the resonance with true points at 2.0252 and i1.9752. The cold characteristic root now becomes imaginary and the warm is real; thus, electrostatic waves only propagate. The cold power reflection coefficient rises sharply from 0 to 1 after oscillating. The warm power reflection coefficient doesn't necessarily track the cold plasma part but oscillates under the portion of the cold plasma that looks like a step function. At the second cutoff frequency, Ω_{C2}, both characteristic roots are real and both transverse and longitudinal waves propagate through the slab. The cold plasma power reflection coefficient dips down to zero, then slowly oscillates, while the warm counterpart continues to oscillate, yet it tracks the cold component and its oscillations tend to ride atop of the cold plasma results similar to the Tonks–Dattner resonances [22] counterpart. When $\Omega \to \infty$, the cold characteristic root tends to 1 and the warm tends to δ^{-1}.

For case B (Figure 11C.12), the normalized electron gyrofrequency, proportional to the strength of the static magnetic field, $\Omega_b = 1$. From Figures 11C.8 and 11C.9, we see that increasing the magnetostatic field shifts the resonance of the cold plasma characteristic root as well as the warm plasma true points to the right. The film thickness is one-tenth of the plasma wavelength and $\delta = 0.3$. The power reflection coefficient takes on a different behavior when compared to case A. The behavior of the characteristic roots is similar in all cases, and they will not be repeated. The power reflection coefficient between 0 and Ω_{C1} tends downward with clear difference between the cold and warm models. When $\Omega \to \Omega_{uh}$, the ρ of the cold model tends upward with a sharp dip to 0 just before $\Omega_{uh} = 1.414$, then upwards again, while the warm model continues toward 0, and then hits two resonances at 2.07 and 3.31.

For case C, the normalized B field strength is $\Omega_b = 2$, and although there are distinct differences between cold and warm power reflection coefficients (Figure 11C.13) from 0 to

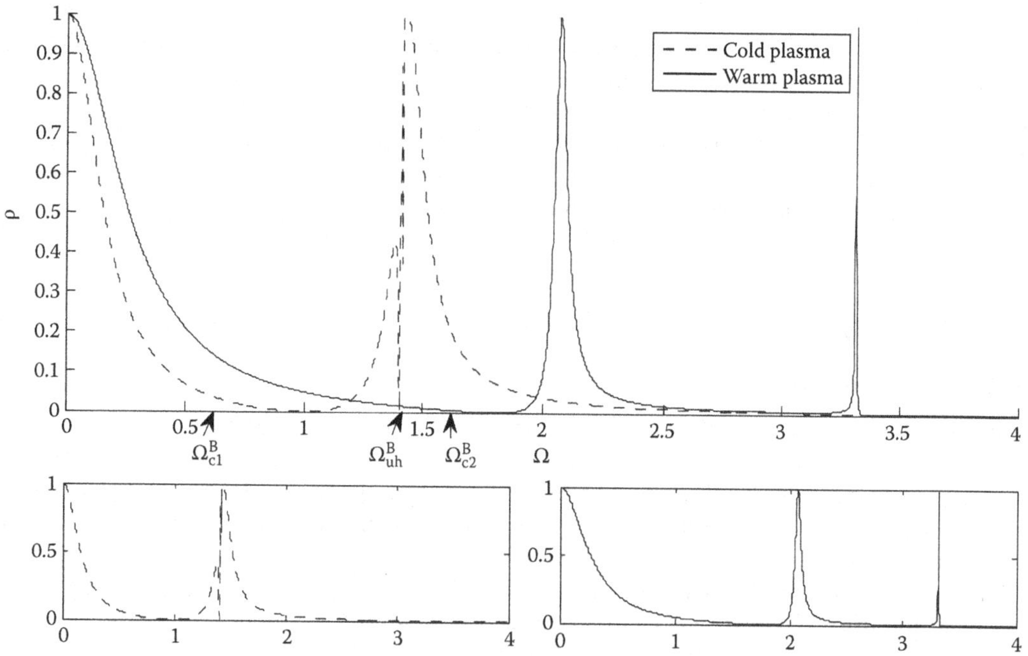

FIGURE 11C.12
Case B, $\Omega_b = 1.0$, $d_p = 0.1$, and $\delta = 0.3$.

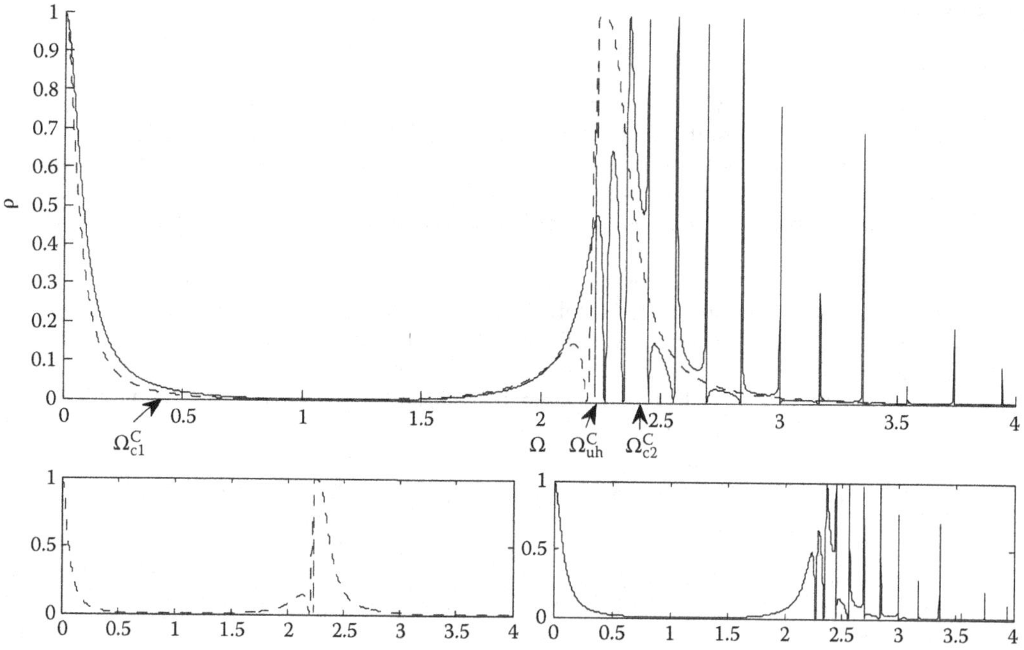

FIGURE 11C.13
Case C, $\Omega_b = 2.0$, $d_p = 0.1$, and $\delta = 0.05$.

Ω_{C1}, they both trend down to $\rho=0$ until the warm characteristic root becomes appreciably small. "As the warm plasma characteristic root approaches 0 or $\Omega \to \Omega_{uh}$, the cold and warm plasma power reflection coefficients trend upward but start to oscillate. The cold plasma power reflection coefficient rises up to and the warm plasma counterpart continues to oscillate until Ω reaches Ω_{C2}. Beyond the second cutoff frequency, ρ for the cold plasma monotonically decreases to 0 while the ρ for the warm plasma oscillates atop it". These corrections are not shown on the corrected page proofs file.

11C.4 Conclusion

Reflection properties for a thin metal slab with an incident X wave have been presented with emphasis on the power reflection coefficient and its corresponding characteristic roots. The film was modeled as a cold and warm plasma that supports a longitudinal electromagnetic acoustic mode, which is a solution to Maxwell's equations. For the X wave, the cold and warm plasma results differ greatly where there would be no difference for (R–L–O–) waves. The characteristic roots show that the warm plasma resolves the resonance at the upper hybrid frequency at a true point, and in frequency space, a transition occurs where a transverse wave is coupled to a longitudinal wave and vice versa. The power reflection coefficient changes behavior as a result of the wave coupling and frequency cutoff points.

References

1. Markos, C. T., Thin film reflection properties of warm anisotropic plasma slabs, PhD thesis, University of Massachusetts Lowell, Lowell, MA, 2012.
2. Kalluri, D. K., Numerical solutions of electromagnetic waves in inhomogeneous magnetoplasma slabs, PhD thesis, University of Kansas, Lawrence, KS, 1968.
3. Miyamoto, K., *Plasma Physics for Nuclear Fusion*, Revised edition, The MIT Press, Cambridge, MA, 1987.
4. Shur, M. S. and Lu, J. Q., Terahertz sources and detectors using two-dimensional electronic fluid in high electron-mobility transistors, *IEEE Trans. Microw. Theory Tech.*, 48(4), 750–756, 2000.
5. Kushwaha, M. S. and Vasilopoulos, P., Resonant response of a FET to an AC signal: Influence of magnetic field, device length, and temperature, *IEEE Trans. Electron Dev.*, 51(5), 803–813, 2004.
6. Muravjov, A. V., Veksler, D. B., Hu, X., Gaska, R., Pala, N., Saxena, H., Peale, R. E., and Shur, M. S., Resonant terahertz absorption by plasmons in grating-gate GaN HEMT structures, *Proc. SPIE*, 7311, 73110D-1–73110D-7, 2009.
7. Shur, M., AlGaN/GaN plasmonic terahertz electronic devices, 2014, *J. Phys.: Conf. Ser.* 486 012025. http://iopscience.iop.org/1742-6596/486/1/012025. Accessed on August 16, 2017.
8. Fourkal, E., Velshev, I., Ma, C.-M., and Smolyakov, A., Evanescent wave interference and the total transparency of a warm high-density plasma slab, *Phys. Plasmas*, 13, 092113 (2006); doi: http://dx.doi.org/10.1063/1.2354574. Accessed on August 17, 2017.
9. Smolyakov, A. I., Fourkal, E., Krasheninnnikov, S. I., and Sternberg, N., Resonant modes and resonant transmission in multi-layer structures, *Prog. Electromagn. Res.*, 107, 293–314, 2010.

10. Yesil, A. and Aydogdu, M., The effect of the electron sound speed on wave propagation in the ionospheric plasma, *Acte Geophys.*, 59(2), 398–406, April 2011.
11. Chawla, B. R., Kalluri, D., and Unz, H., Propagation of oblique electromagnetic waves through a plasma slab with normal magnetostatic field, *Radio Sci.*, 2, 869–879, 1967.
12. Kalluri, D., *Electromagnetic Waves, Materials, and Computation with MATLAB*, CRC Press, Boca Raton, FL, 2011.
13. Markos, C. T. and Kalluri, D., Thin film reflection properties of a warm isotropic plasma slab between two half-space dielectric media, *J. Infrared Millim. Terahertz Waves*, 32(7), 914–934, 2011.
14. Kalluri, D. K. and Prasad, R. C., Thin-film reflection properties of isotropic and uni-axial plasma slabs, *Appl. Sci. Res.*, 27, 415–424, 1973.
15. Kalluri, D. and Prasad, R. C., Thin-film reflection properties of an anisotropic plasma slab immersed in a static magnetic field normal to the plane of incidence, *Int. J. Electron.*, 35(6), 801–816, 1973.
16. Prasad, R. C., Transient and frequency response of a bounded plasma, PhD thesis, Birla Institute of Technology, Ranchi, India, 1976.
17. Forstmann, F. and Gerhardts, R. R., *Metal Optics Near the Plasma Frequency*, Springer-Verlag, Berlin, Germany, pp. 9–24, 1986.
18. Tanenbaum, B. S., *Plasma Physics*, McGraw-Hill Book Company, New York, NY, pp. 122–129, 1967.
19. Unz, H., Oblique wave propagation in a compressible general magnetoplasma, *Appl. Sci. Res.*, 16, 105–120, 1965.
20. Budden, K. G., *Radio Waves in the Ionosphere: The Mathematical Theory of the Reflection of Radio Waves from Stratified Ionised Layers*, Cambridge University Press, New York, NY, p. 144, 1966.
21. Crawford, F. W., Internal resonances of a discharge column, *J. Appl. Phys.*, 35, 1365–1369, 1964.
22. Vandenplas, P. E., *Electron Waves and Resonances in Bounded Plasmas*, Interscience Publishers, New York, NY, p. 29, 1968.

Appendix 13A: Maxwell Stress Tensor and Electromagnetic Momentum Density

From the electromagnetic force equation (13.105), we can write the equation for the force density f as follows:

$$f = [\rho E + J \times B]. \tag{13A.1}$$

Assume the medium is free space. We can eliminate the volume charge density ρ and the current density J by using Maxwell's equations, and after using some vector identities, we can write [1]

$$f = [\rho E + J \times B] = \nabla \cdot \bar{T} - \varepsilon_0 \mu_0 \frac{\partial S}{\partial t}, \tag{13A.2}$$

where the tensor $\bar{T}$, called the Maxwell stress tensor, in free space medium is given by [1–3]

$$T_{ij} = \varepsilon_0 \left(E_i E_j - \frac{1}{2} \delta_{ij} E^2 \right) + \frac{1}{\mu_0} \left(B_i B_j - \frac{1}{2} \delta_{ij} B^2 \right). \tag{13A.3}$$

The first term on the right side of (13A.3) can be called the electric stress tensor $[T_e]$. If written out in matrix form in Cartesian coordinates,

$$[T_e] = \varepsilon_0 \begin{bmatrix} \frac{1}{2}\left(E_x^2 - E_y^2 - E_z^2\right) & E_x E_y & E_x E_z \\ E_x E_y & \frac{1}{2}\left(E_y^2 - E_z^2 - E_x^2\right) & E_y E_z \\ E_x E_z & E_y E_z & \frac{1}{2}\left(E_z^2 - E_x^2 - E_y^2\right) \end{bmatrix}. \tag{13A.4}$$

The second term on the right side (13A.3) can be similarly written out. The term S in (13A.2) is the usual Poynting vector in free space given by (1.26).

The force F on a volume V bounded by a closed surface s in free space medium containing charges and currents can be obtained by integrating (13A.2). After using an identity similar to the vector divergence theorem involving a tensor of second order [3, p. 99] $\bar{T}$,

$$\iiint_V \nabla \cdot \bar{T} \mathrm{d}V = \oint\!\!\oint_s \bar{T} \cdot \mathrm{d}s, \tag{13A.5}$$

we can write

$$\frac{\mathrm{d}p}{\mathrm{d}t} = F = -\varepsilon_0 \mu_0 \frac{\mathrm{d}}{\mathrm{d}t} \iiint_V S \mathrm{d}V + \oint\!\!\oint_s \bar{T} \cdot \mathrm{d}s. \tag{13A.6}$$

Equation 13A.6 is the statement of conservation of momentum, parallel to that of Poynting theorem for conservation of energy given by (1.27). Thus, one interprets

$$g = \mu_0 \varepsilon_0 S \tag{13A.7}$$

as the electromagnetic momentum density contained in the fields. Let us denote the electromagnetic momentum in the volume V by **G**. Substituting (13A.7) in (13A.6),

$$\frac{d(p+G)}{dt} = \oiint_s \bar{T} \cdot ds. \tag{13A.8}$$

Equation 13A.8 in words states that the time rate of increase of the total momentum (the sum of the mechanical momentum and electromagnetic momentum) is equal to the momentum carried *inward* [3, p. 104] of the volume V through its bounding surface s. Proper interpretation of the abbreviated tensor divergence theorem stated in (13A.5) is given in [3, p. 104] to explain the *inward* word.

The Maxwell stress tensor can be used to calculate the radiation pressure due to a plane electromagnetic wave obliquely incident on a conductor or dielectric half-space moving with relativistic speed (see Appendix 14B).

References

1. Griffths, D. J., *Introduction to Electrodynamics*, Prentice Hall, Englewood Cliffs, NJ, 1981.
2. Panofsky, W. K. H. and Phillips, M., *Classical Electricity and Magnetism*, 2nd edition, Dover Publications, Mineola, NY, 2005.
3. Stratton, J. A., *Electromagnetic Theory*, McGraw-Hill, New York, 1941.

Appendix 13B: Electric and Magnetic Forces and Newton's Third Law

Newton's third law, as stated in words in many books, is that when one body exerts a force on a second body, the second body simultaneously exerts a force equal in magnitude and opposite in direction on the first body. In this appendix, we will examine it from the viewpoint of electric and magnetic forces [1–3].

It is easily shown to be valid in electrostatics by considering the vector form of Coulomb's law between stationary charges Q_1 and Q_2. See Figure 13B.1 for an example of the satisfaction of Newton's third law for the electric force.

Let us assume that Q_1 is located at A and Q_2 is located at B. The electric force at A on Q_1 exerted by the electric field of the source Q_2 at B is given by

$$\boldsymbol{F}_{eA} = Q_1 \frac{Q_2}{4\pi\varepsilon_0 R^2}\hat{a}_{BA}. \tag{13B.1}$$

The electric force at B on Q_2 exerted by the electric field of the source Q_1 at A is given by

$$\boldsymbol{F}_{eB} = Q_2 \frac{Q_1}{4\pi\varepsilon_0 R^2}\hat{a}_{AB}. \tag{13B.2}$$

We will next calculate the force between two differential current elements to show that Newton's third law is violated for the case of the magnetic force. Let the first element $I_1\Delta x\hat{x}$ be located at the point $A(R, 0)$ on the x-axis and the second element $I_2\Delta y\hat{y}$ be located at $B(0, R)$. We wish to calculate the magnetic force $\Delta\boldsymbol{F}_{mA}$ experienced by the element at A. We use the formula derived from the Lorentz force equation (1.6):

$$\Delta\boldsymbol{F}_m = \rho_V\Delta V\ \boldsymbol{v}\times\boldsymbol{B} = \boldsymbol{J}\Delta V\times\boldsymbol{B} = \boldsymbol{K}\Delta s\times\boldsymbol{B} = I\Delta\boldsymbol{L}\times\boldsymbol{B}, \tag{13B.3}$$

where the left-hand side is the differential vector magnetic force at the location of the differential element experiencing the force and B is the external B field at the same location (There should be no confusion in the use of the symbol B. When used as a subscript, it denotes the location of the point on the y-axis.) It then follows that

$$\Delta\boldsymbol{F}_{mA} = I_1\Delta x\hat{x}\times\boldsymbol{B}_{AB}, \tag{13B.4}$$

where $\boldsymbol{B}_{AB}$ is the vector B field at A produced by the current element at B. Assuming that the source currents are steady, from the Biot–Savart law, one can write

$$\boldsymbol{B}_{AB} = \mu_0 \frac{I_2\Delta y\hat{y}}{4\pi\left(R\sqrt{2}\right)^2}\times\hat{a}_{BA}. \tag{13B.4}$$

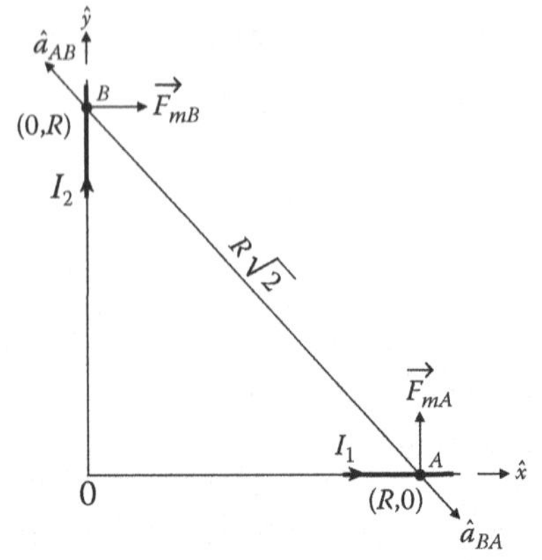

FIGURE 13B.1
Geometry to check the validity of Newton's third law for the electric and magnetic forces.

From Figure 13B.1,

$$\hat{a}_{BA} = \frac{1}{\sqrt{2}}(\hat{x} - \hat{y}). \tag{13B.5}$$

Thus,

$$\boldsymbol{B}_{AB} = \mu_0 \frac{I_2 \Delta y}{8\sqrt{2}\pi R^2}(-\hat{z}). \tag{13B.6}$$

From (13B.6) and (13B.4),

$$\Delta \boldsymbol{F}_{mA} = \mu_0 \frac{I_1 I_2 \Delta x \Delta y}{8\sqrt{2}\pi R^2}\hat{y}. \tag{13B.7}$$

The computation of $\Delta \boldsymbol{F}_{mB}$ proceeds on similar lines and results in

$$\Delta \boldsymbol{F}_{mB} = \mu_0 \frac{I_1 I_2 \Delta x \Delta y}{8\sqrt{2}\pi R^2}\hat{x}. \tag{13B.8}$$

From (13B.7) and (13B.8), we conclude that the two forces, while equal in magnitude, are not opposite in direction, thus violating Newton's third law. In fact, they are perpendicular to each other as shown in Figure 13B.1. A few remarks should be made with regard to the example of the force between two differential current elements discussed above. It is artificial since a differential filamentary current element is used as a help in computing, and any physical filamentary source will be a complete circuit obeying the continuity equation. For example, had we formulated the question as to the force between two parallel, infinitely long filaments, carrying currents in the opposite directions, the repulsive force between them is equal and opposite [3].

A subtle question is whether Newton's third law will be violated if the charges in the example move with a velocity v [1]. The Lorentz force equation is still valid, but we need to address the method of computing the electromagnetic fields of a moving point charge. This is discussed in detail in Appendix 14G. However, we can note that a moving charge gives rise to time-varying electric field and magnetic field and, as per Appendix 13A, gives rise to an electromagnetic momentum. The problem is thus a three-body problem [2] rather than a two-body problem, and Newton's third law is applicable to a two-body problem only. In any case for nonrelativistic velocities, the magnetic force is much less than the Coulomb force.

References

1. Griffiths, D. J., *Introduction to Electrodynamics*, Prentice-Hall, Englewood Cliffs, NJ, 1981.
2. Popovic, B. D., *Introductory Engineering Electromagnetics*, Addison_Wesley, Reading, MA, 1971.
3. Hayt, W. H., Jr., *Engineering Electromagnetics*, 5th edition, McGraw-Hill, New York, 1989.

Appendix 13C: Frequency and Polarization Transformer (10–1000 GHz): Interaction of a Whistler Wave with a Collapsing Plasma in a Cavity*

Dikshitulu K. Kalluri and Robert Kevin Lade

13C.1 Introduction

The electric power industry extensively uses voltage transformers of high transformation ratio. If such devices for transforming frequencies with high transformation ratio are available, one can use off-the-shelf inexpensive sources and transform the frequency to the level desired by the application. It is well known that the wave number k of an electromagnetic wave is conserved in a time-varying but space-invariant medium. This property can be used to construct a frequency transformer using a collapsing magnetoplasma medium in a cavity. The time variation of the relative permittivity is easily produced in a magnetoplasma medium by altering the ionization level. Representative literature on the interaction of an electromagnetic wave with a time-varying medium pertinent to the topic under discussion is listed as references [1–13].

The frequency transformer system [1,14] can be represented by a black box (see Figure 13C.1) where the input is an electromagnetic wave of frequency ω_0 (in radians per second) and amplitude A_0 (of the electric or the magnetic field) and the output is an electromagnetic wave of frequency ω_n and amplitude A_n.

The k-conservation property can be used to show [1–5]

$$\omega_n(t)n(t) = \omega_0 n_0, \tag{13C.1}$$

where n is the refractive index. The relative permittivity $\varepsilon_p = n^2$ of a right-hand circularly (RHC) polarized wave (R wave) whose direction of propagation coincides with the direction of the background static magnetic field (called longitudinal propagation) is given by [10]

$$\varepsilon_p = n^2 = 1 - \frac{\omega_p^2}{\omega(\omega - \omega_b)}, \tag{13C.2}$$

where $\omega = 2\pi f$ is the operating frequency, $\omega = 2\pi f_p$ is the plasma frequency, that is,

$$\omega_p = \left(\frac{N_0 q_e^2}{m_e \varepsilon_0}\right)^{1/2} \tag{13C.3}$$

* C IEEE. Reprinted from *IEEE Trans. Plasma Science*, 40 (11), pp. 3070–3078, November 2012. With permission. The on-line version of the original paper has some of the figures in color.

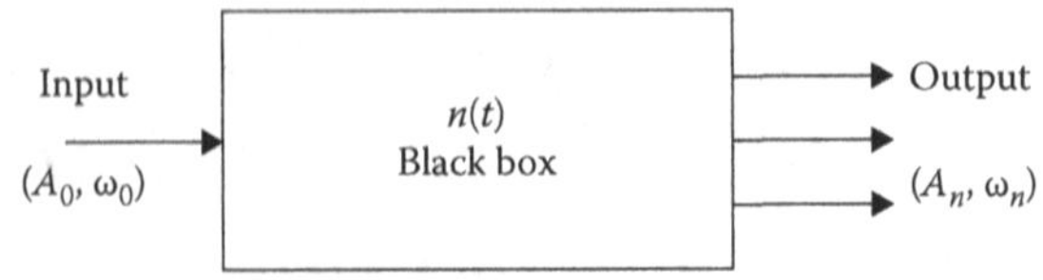

FIGURE 13C.1
Black box description of frequency transformation showing single input wave and multiple output waves.

and $\omega_b = 2\pi f_b$ is the electron cyclotron frequency

$$\omega_b = \frac{q_e \mathbf{B}_0}{m_e} = \omega_b \hat{\mathbf{B}}_0, \tag{13C.4}$$

where $\mathbf{B}_0$ is the background static magnetic field and $\hat{\mathbf{B}}_0$ is a unit vector in the direction of $\mathbf{B}_0$. In a previous equation (see (13C.3)), N_0 is the electron density (plasma density); q_e and m_e are the absolute value of the charge and the mass of the electron, respectively; and ε_0 is the permittivity of free space. The radio approximation (ions are immobile) is used in obtaining (13C.2) [10].

In a decaying plasma, ω_p, and therefore n, is a decaying function of time. Substituting (13C.2) in (13C.1) results in

$$n^2\omega^2 = \omega^2 \left[1 - \frac{\omega_p^2(t)}{\omega(\omega - \omega_b)}\right] = n_0^2\omega_0^2. \tag{13C.5}$$

Equation 13C.5 is a cubic in ω

$$\omega^3 - \omega_b\omega^2 - \left(n_0^2\omega_0^2 + \omega_p^2(t)\right)\omega + n_0^2\omega_0^2\omega_b = 0. \tag{13C.6}$$

Thus, the output in Figure 13C.1 consists of three waves of different frequencies. If the input wave has a frequency $\omega_0 < \omega_b$ and close to resonance and $\omega_p^2/\omega_0(\omega_b - \omega_0) \gg 1$, ε_p can be quite large and n_0 is approximately given by

$$n_0 \approx \frac{\omega_{p0}}{\sqrt{\omega_0(\omega_b - \omega_0)}}, \tag{13C.7}$$

where $\omega_{p0} = \omega_p(0)$.

A large value for n_0 can be also obtained by choosing an extraordinary wave (X wave) with the source wave frequency ω_0 close to the upper hybrid resonance. However, in such a case, the effect of temperature (warm plasma) resolves the resonance into a lesser and finite n_0. When the source wave is an R wave, the electromagnetic wave and the electron plasma wave are uncoupled and the resonance at ω_b remains [15].

One can also argue that a large frequency upshift can be obtained by simply switching the medium from free space to a very dense plasma of $f_p = 1000$ GHz [3]

$$f = \sqrt{f_0^2 + f_p^2} = \sqrt{10^2 + 1000^2} \approx 1000 \text{ GHz} \tag{13C.8}$$

rather than using a more complicated process of a decaying magnetoplasma. However, it can be shown that, in such a case, the signal of frequency f is very weak and most of the energy will go into a wiggler magnetic field [12]. The case under consideration in this appendix has the upshifted output frequency signal with the amplitude of the electric field greater than that of the input wave.

If $f_0 = 10$ GHz, $f_{p0} = 30$ GHz, and $f_b = 10.009$ GHz, then $n_0 \approx 100$. If the R input wave is a standing wave in a cavity filled with magnetoplasma with the above given parameters and $n(\infty) = 1$ in Figure 13C.1, $f(\infty) = (100)(10) = 1000$ GHz. Thus, the principle of frequency transformation in a time-varying magnetoplasma medium can be used to construct a frequency transformer to upshift the frequency of 10–1000 GHz (1 THz) radiation. A 10 GHz X-band source is a fairly inexpensive off-the-shelf item, but the present-day terahertz sources are either very expensive and of large size or of very small output [16,17].

13C.2 Longitudinal Propagation

Consider the general case of a cold lossy anisotropic magnetoplasma medium. In general, as an electromagnetic wave propagates in a plasma, in the presence of a static magnetic field, the polarization state is constantly changing. However, there are specific directions of the static magnetic field relative to the direction of wave propagation where the state of polarization is unaltered. These normal modes of propagation occur when the static magnetic field is along the direction of propagation (longitudinal modes) or normal to the direction of propagation (transverse modes). Waves with left (L wave) or right (R wave) circular polarization are the normal longitudinal modes, and the ordinary (O wave) and extraordinary (X wave) waves are the normal modes for transverse propagation.

Longitudinal R wave propagation will be considered in this appendix. It can be shown that, for R wave propagation, the magnetized plasma can be modeled as a dielectric with a dielectric constant given by (13C.2), which is plotted as a function of frequency in Figure 13C.2. This results in propagation that can be divided into three distinct frequency regions. The first region (labeled C) is given by $\omega < \omega_b$ where the dielectric constant is

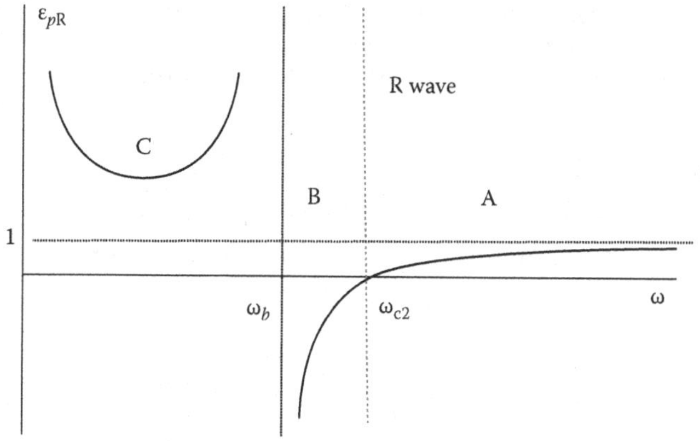

FIGURE 13C.2
Relative permittivity for longitudinal R wave propagation.

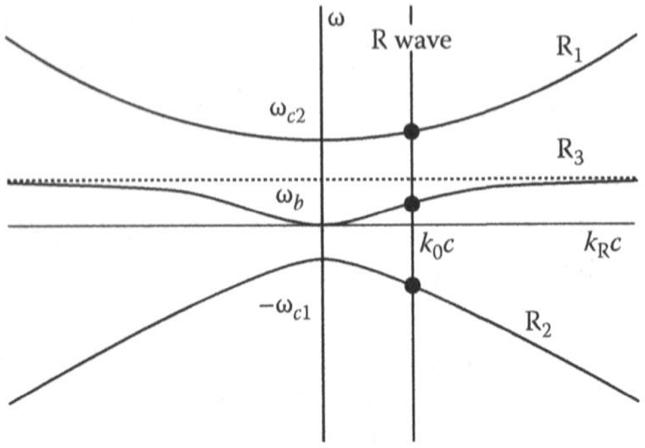

FIGURE 13C.3
ω–k diagram for longitudinal R wave propagation.

greater than unity (potentially much greater than unity as ω approaches zero or ω_b). This mode of propagation is called the whistler mode in the literature on ionospheric physics and the helicon mode in the literature on solid-state plasma, and it is designated here as the R3 mode. The second frequency region is defined for $\omega_b < \omega < \omega_{C2}$, where ω_{C2} is the waveguide mode cutoff frequency given by (13C.9) with the positive sign [1]:

$$\omega_{C1,C2} = \mp\frac{\omega_b}{2} + \sqrt{\left(\frac{\omega_b}{2}\right)^2 + \omega_p^2}. \tag{13C.9}$$

In this region (labeled B), the dielectric constant is negative, the waves are evanescent, and no propagation is supported. The third distinct frequency region (labeled A) is where $\omega > \omega_{c2}$ and ε_p is between zero and one. This mode is referred to as the waveguide mode of propagation and will be denoted as the R_1 mode.

The ω–k diagram for R wave propagation [1] is shown in Figure 13C.3. If a vertical line is drawn through the diagram at a fixed value of the wave number k, it is shown that propagation is supported at three frequencies, which are denoted by modes R_1, R_2, and R_3. R_1 and R_3 are the forward propagating waveguide mode and the whistler mode, respectively, as aforementioned. A third mode, that is, R_2, is supported at a negative frequency with $|\omega| > \omega_{c1}$, where ω_{c1} is the waveguide mode cutoff frequency given by (13C.9) with the negative sign. The negative frequency of the R_2 wave simply represents an RHC polarized wave propagating in a direction opposite to the R_1 and R_3 waves (backward propagating wave).

13C.2.1 Basic Field Equations

The electric field $\mathbf{E}(\mathbf{r}, t)$ and the magnetic field $\mathbf{H}(\mathbf{r}, t)$ and the velocity field of the electrons $\mathbf{v}(\mathbf{r}, t)$ in a lossy magnetized Lorentz plasma are governed by Maxwell's equations and the momentum equation

$$\nabla \times \mathbf{E} = -\mu_0 \frac{\partial \mathbf{H}}{\partial t} \tag{13C.10}$$

$$\nabla \times \mathbf{H} = \varepsilon_0 \frac{\partial \mathbf{E}}{\partial t} + \mathbf{J} \tag{13C.11}$$

$$m_e \frac{d\mathbf{v}(t)}{dt} + m_e v \mathbf{v}(t) = -q_e \mathbf{E} - q_e \mathbf{v}(t) \times \mathbf{B}_0, \tag{13C.12}$$

where μ_0 and ε_0 are the permeability and the permittivity of free space, respectively, and **J** is the current density vector.

13C.2.2 Constitutive Relation for a Collapsing Magnetoplasma

The constitutive relationship between the current density vector **J** and the electric field in the plasma will depend on the ionization (or deionization) process [18–21] that creates (or decays) the plasma. The constitutive equation for the case of a building-up plasma will be different from that of the collapsing plasma [1,22]. This appendix considers the case of collapsing plasma. It is initially assumed that a plasma is present with electron density N_0 and initial plasma frequency given by (13C.3). The plasma is in the presence of a static magnetic field $\mathbf{B}_0$, and the electron gyrofrequency is given by (13C.4).

The decrease in the electron density is assumed to be due to the capture and sudden removal of $\Delta N(t)$ electrons; the velocities of all the remaining electrons are unaffected by this capture and have the same instantaneous velocity $\mathbf{v}(t)$. The model has to ensure the continuity of $\mathbf{v}(t)$ for the remaining electrons. The momentum equation for the electron (13C.12) and the equation relating current density $\mathbf{J}(t)$ and electron velocity $\mathbf{v}(t)$

$$\mathbf{J}(t) = -qN(t)\mathbf{v}(t) \tag{13C.13}$$

lead to the constitutive relation between $\mathbf{J}(t)$ and electric field $\mathbf{E}(t)$. Based on this model, the result is

$$\frac{d\mathbf{J}}{dt} + \left[v - \frac{\left[\omega_p^2(t)\right]'}{\omega_p^2(t)} \right] \mathbf{J} = \varepsilon_0 \omega_p^2(t) \mathbf{E} - \mathbf{J} \times \boldsymbol{\omega}_b, \tag{13C.14}$$

where $\left[\omega_p^2(z,t)\right]'$ is the time derivative of $\omega_p^2(z,t)$. Equation 13C.14 can be shown to be equivalent to the constitutive equation developed and verified in [22] for a collapsing plasma. Even if it is assumed that the plasma is loss free ($v = 0$), the term involving $\left[\omega_p^2(z,t)\right]'$ ensures that current density **J** ultimately disappears when the plasma collapses completely. The parameter v in (13C.14) represents the energy dissipated as heat, whereas the $\left[\omega_p^2(t)\right]'/\omega_p^2(t)$ term represents the energy loss due to the $\Delta N(t)$ electrons lost in the plasma decay process. Neglecting this term can lead to errors, as shown in [22].

13C.2.3 Profile of the Collapsing Plasma

One of the powerful features of the finite-difference time-domain (FDTD) approach is the ability to generate arbitrary plasma decay profiles. In the analysis that follows, an exponentially decaying profile is assumed, as shown in Figure 13C.4. Here, the b/T factor defines the decay parameter where $T = 2\pi/\omega_0$ is the period of the wave in the initial magnetoplasma medium. The rate of plasma collapse can be controlled by varying

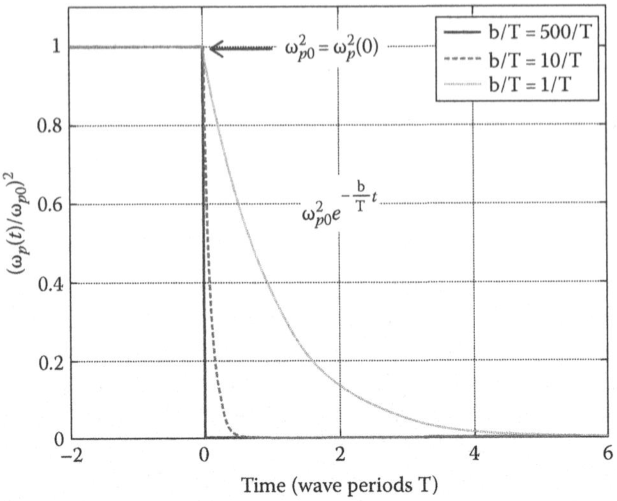

FIGURE 13C.4
Exponential plasma decay profile.

the b/T parameter. For example (see Figure 13C.4), setting b/T = 500/T provides a step plasma collapse and is used to represent the sudden collapse case. Using a value of b/T = 1, on the other hand, provides a relatively slow decay of the plasma. The general results are not strongly dependent on the exact shape of the plasma collapse (but more on the rate of the collapse), and the exponential profile is used for analytical simplicity.

13C.2.4 Output Frequencies for a Collapsing Magnetoplasma

As shown by (13C.6), the output in Figure 13C.1 consists of three waves of different frequencies. After the complete plasma collapse, $\omega_p(\infty) = 0$ and the roots of (13C.6) become

$$\omega_1 = \omega_0 n_0 \qquad \omega_2 = -\omega_0 n_0 \qquad \omega_3 = \omega_b, \tag{13C.15}$$

where a negative value indicates a reflected wave traveling in a direction opposite to that of the source wave. Since the resulting medium after complete plasma collapse is free space, the mode at ω_b will disappear as the plasma collapses. The normalized output frequency, $\log(\omega_{out}/\omega_0) = \log(n_0)$, for various values of ω_{p0}/ω_0 and ω_b/ω_0 is shown in Figure 13C.5.

Three regions are identified in the plot with letters A, B, and C (C_1 and C_2). These regions correspond to those shown in Figure 13C.2 and are defined by the initial mode of propagation in the plasma. Region B represents a region where $\omega_b < \omega_0 < \omega_{c2}$ such that ω_0 is between ω_b and the R wave waveguide mode cutoff frequency. Here, the dielectric constant is less than zero and no initial propagation is supported. Region A is defined by the region where $\omega_0 > \omega_{c2}$, $0 < \varepsilon_p < 1$, and the input wave is the waveguide mode R_1. Here, the output wave(s) will be at the same frequency or less than that of the incident wave (since $\omega_{out} = \pm n_0\omega_0$). Finally, region C is given by the condition $\omega_0 < \omega_b$, which means that the initial wave is a whistler wave (R_3 mode). Region C is further divided into subregions C_1 and C_2 where the boundary between the two regions is given by $\omega_{out} = \omega_b$. Region C_1 is defined where

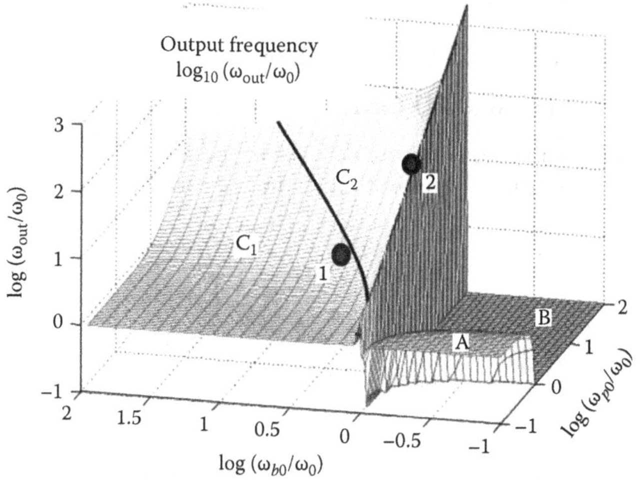

FIGURE 13C.5
Normalized output frequency (log scale) versus ω_{p0} and ω_b.

the output frequency is less than ω_b, and thus, the output wave remains in the whistler mode throughout the plasma collapse. Region C_2 (the area of real interest) is defined where $\omega_{out} > \omega_b$, and the incident whistler wave is converted to the waveguide mode(s) R_1 and R_2. Large upward frequency shifts are possible in region C_2 where the index of refraction, and thus output frequency, increases as ω_{p0} grows and as ω_b approaches ω_0. Frequency shifts of 100:1 or more can be seen. These regions are labeled and summarized in Figure 13C.6 for the ω_{p0} versus ω_b plane.

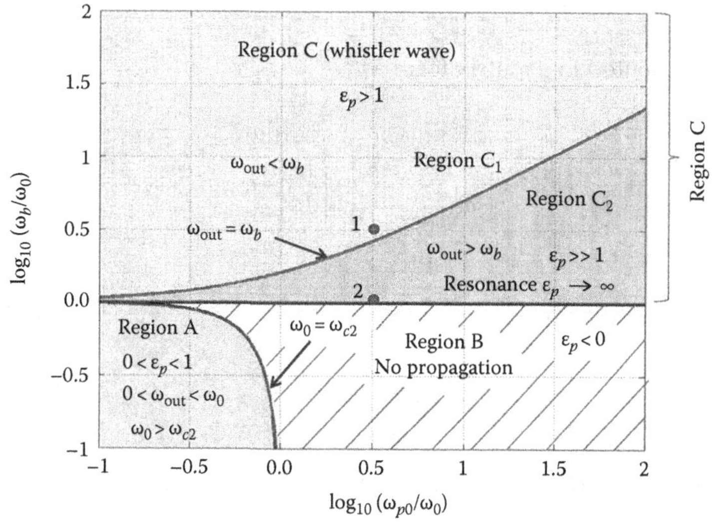

FIGURE 13C.6
Definition of various regions of the ω_{p0} and ω_b plane.

13C.3 One-Dimensional FDTD Simulation

13C.3.1 Collapsing Plasma in a 1-D Cavity

Consider the case of a 1-D cavity where it is assumed that all parameters vary only in the z-direction and the static magnetic field is in the z-direction. Equation 13C.11 then reduces to

$$\frac{\partial E_x}{\partial t} = -\frac{1}{\varepsilon_0}\frac{\partial H_y}{\partial z} - \frac{1}{\varepsilon_0}J_x \tag{13C.16}$$

$$\frac{\partial E_y}{\partial t} = \frac{1}{\varepsilon_0}\frac{\partial H_x}{\partial z} - \frac{1}{\varepsilon_0}J_y \tag{13C.17}$$

and (13C.10) similarly reduces to

$$\frac{\partial H_x}{\partial t} = \frac{1}{\mu_0}\frac{\partial E_y}{\partial z} \tag{13C.18}$$

$$\frac{\partial H_x}{\partial t} = \frac{1}{\mu_0}\frac{\partial E_y}{\partial z}. \tag{13C.19}$$

Finally, (13C.14) results in the following two coupled equations:

$$\frac{\partial J_x}{\partial t} + \left[v - \left(\omega_p^2\right)'/\omega_p^2\right]J_x = \varepsilon_0\omega_p^2 E_x - \omega_b J_y \tag{13C.20}$$

$$\frac{\partial J_y}{\partial t} + \left[v - \left(\omega_p^2\right)'/\omega_p^2\right]J_y = \varepsilon_0\omega_p^2 E_y + \omega_b J_x, \tag{13C.21}$$

which can be represented in a matrix form as

$$\begin{bmatrix} \dfrac{\partial J_x}{\partial t} \\ \dfrac{\partial J_y}{\partial t} \end{bmatrix} = \begin{bmatrix} -v_1 & -\omega_b \\ \omega_b & -v_1 \end{bmatrix} + \begin{bmatrix} J_x \\ J_y \end{bmatrix} + \varepsilon_0\omega_p^2 \begin{bmatrix} E_x \\ E_y \end{bmatrix}, \tag{13C.22}$$

where

$$v_1 = v - \left(\omega_p^2\right)'/\omega_p^2 \tag{13C.23a}$$

and for the exponential profile

$$v_1 = v + \frac{b}{T}. \tag{13C.23b}$$

The FDTD grid is chosen such that E (E_x, E_y) and J (J_x, J_y) lie on the same space coordinate but will be offset by a 1/2 time step [1]. H (H_x, H_y) is at the same time step as J and offset spatially by a 1/2 step. Using the central difference formula for (13C.16) through (13C.19), the following FDTD equations can be generated:

$$E_x\Big|_k^{n+1} = E_x\Big|_k^{n} - \frac{\Delta t}{\varepsilon_0 \Delta_z}\left(H_y\Big|_{k+(1/2)}^{n+(1/2)} - H_y\Big|_{k-(1/2)}^{n+(1/2)}\right) - \frac{\Delta t}{\varepsilon_0} J_x\Big|_k^{n+(1/2)} \tag{13C.24}$$

$$E_y\Big|_k^{n+1} = E_y\Big|_k^{n} + \frac{\Delta t}{\varepsilon_0 \Delta_z}\left(H_x\Big|_{k+(1/2)}^{n+(1/2)} - H_x\Big|_{k-(1/2)}^{n+(1/2)}\right) - \frac{\Delta t}{\varepsilon_0} J_y\Big|_k^{n+(1/2)} \tag{13C.25}$$

$$H_x\Big|_{k+(1/2)}^{n+(1/2)} = H_x\Big|_{k+(1/2)}^{n-(1/2)} + \frac{\Delta t}{\mu_0 \Delta z}\left(E_y\Big|_{k+1}^{n} - E_y\Big|_k^{n}\right) \tag{13C.26}$$

$$H_y\Big|_{k+(1/2)}^{n+(1/2)} = H_y\Big|_{k+(1/2)}^{n-(1/2)} - \frac{\Delta t}{\mu_0 \Delta z}\left(E_x\Big|_{k+1}^{n} - E_x\Big|_k^{n}\right). \tag{13C.27}$$

13C.3.2 Exponential Time Stepping

The central difference formula and spatial averaging can be used to obtain the FDTD equation for (13C.22). However, for a large ω_b, highly lossy media (large v), or rapidly decaying plasma, the time rate of amplitude change of J during the time step is very rapid that the usual time stepping becomes inaccurate. To handle this difficulty, (13C.22) is analytically solved over one time step by treating it as a first-order differential equation for [J], assuming all other terms to be constant during the time interval.

Using the Laplace transform technique on the matrix equation of (13C.22), a new FDTD equation can be now expressed as

$$\begin{bmatrix} J_x\Big|_k^{n+\frac{1}{2}} \\ J_x\Big|_k^{n+\frac{1}{2}} \end{bmatrix} = \exp\left(-v_1\Big|_k^n \Delta t\right)\begin{bmatrix} c_{11} & c_{12} \\ c_{21} & c_{22} \end{bmatrix}\begin{bmatrix} J_x\Big|_k^{n-\frac{1}{2}} \\ J_x\Big|_k^{n-\frac{1}{2}} \end{bmatrix} + \frac{\varepsilon_0 \omega_p^2\Big|_k^n \exp\left(-v_1\Big|_k^n \Delta t\right)}{\left(v_1\Big|_k^n\right)^2 + \omega_b^2}\begin{bmatrix} k_{11} & k_{12} \\ k_{21} & k_{22} \end{bmatrix}\begin{bmatrix} E_x \\ E_y \end{bmatrix} \tag{13C.28}$$

$$\mathrm{C} = \begin{bmatrix} c_{11} & c_{12} \\ c_{21} & c_{22} \end{bmatrix} = \begin{bmatrix} \cos(\omega_b \Delta t) & -\sin(\omega_b \Delta t) \\ \sin(\omega_b \Delta t) & \cos(\omega_b \Delta t) \end{bmatrix} \tag{13C.29}$$

$$\begin{aligned} \mathbf{K} &= \begin{bmatrix} k_{11} & k_{12} \\ k_{21} & k_{22} \end{bmatrix} \\ k_{11} &= v_1\Big|_k^n\left[\exp\left(v_1\Big|_k^n \Delta t\right) - c_{11}\right] + \omega_b c_{21} \\ k_{12} &= -\exp\left(v_1\Big|_k^n \Delta t\right)\omega_b + \omega_b c_{11} + v_1\Big|_k^n c_{21} \\ k_{21} &= \exp\left(v_1\Big|_k^n \Delta t\right)\omega_b - \omega_b c_{11} - v_1\Big|_k^n c_{21} \\ k_{22} &= v_1\Big|_k^n\left[\exp\left(v_1\Big|_k^n \Delta t\right) - c_{11}\right] + \omega_b c_{21}. \end{aligned} \tag{13C.30}$$

This is called exponential time stepping [23], and it maintains the second-order accuracy. Details of a similar derivation can be found in [24].

13C.4 Results for a Collapsing Magnetoplasma Medium in a Cavity

The input wave is a standing R wave in a 1-D cavity with perfect electric conductor (PEC) plates with a separation d given by

$$d = \frac{m\pi c}{n_0 \omega_0}, \tag{13C.31}$$

where m is the harmonic number of the standing wave. The use of a cavity is desirable since it is much easier to control a time-varying magnetoplasma medium and the background magnetic field over a small distance between plates than in an unbounded volume.

The x- and y-components of the electric field of the input standing wave are assumed to be

$$E_x^- = E_0 \sin\frac{m\pi z}{d}\cos\left(\omega_0 t + \phi_0\right) \tag{13C.32}$$

$$E_y^- = E_0 \sin\frac{m\pi z}{d}\sin\left(\omega_0 t + \phi_0\right), \tag{13C.33}$$

where ϕ_0 is the switching angle (phase at the instant of switching $t = 0$). The FDTD equations (13C.24) through (13C.30) are used to calculate the values of E_x, E_y, H_x, H_y, J_x, and J_y in a leapfrog manner over time. For each of the following scenarios, parameters $\omega_0 = 10$ Grad/s, $v = 0$ (no loss), $\phi_0 = 90°$, and $m = 1$ are used. The values of ω_b and ω_{p0} were varied from $0.1\omega_0$ to $100\omega_0$. For this analysis, the space and time step sizes (Δt and Δz) were varied as needed (based on the plasma parameters) to ensure stability and adequate accuracy of the algorithm [23]. The electric field was sampled at the center of the cavity. A successive reduction method was used [24] to obtain the amplitude and frequencies from the time series of the fields after the plasma time variation was completely finished.

13C.4.1 Sudden Collapse (FDTD)

In this section, the switching speed is taken as very fast with (b/T = 500/T) and the plasma (ω_p) decays to 1% of the initial value at approximately T/100 or 0.006 ns. The FDTD results for the normalized amplitude of the x-component of the E field (E_x/E_0) are plotted on the top of Figure 13C.7a versus normalized ω_b and ω_{p0}. Note that logarithmic scales are used for all axes. Similar results for the y-component of the E field (E_y/E_0) are shown in the lower plot in Figure 13C.7a. In the region labeled as B, there is no output wave for E_x or E_y as expected since no initial propagation is supported. In region A, where the initial wave mode is the waveguide mode, the output wave has E_x and E_y components equal to those of the initial wave. From Figure 13C.5, the frequency of the output wave is also nearly the same as that of the input wave. Thus, in region A, the output standing wave that results after plasma collapse is very similar to the input standing wave.

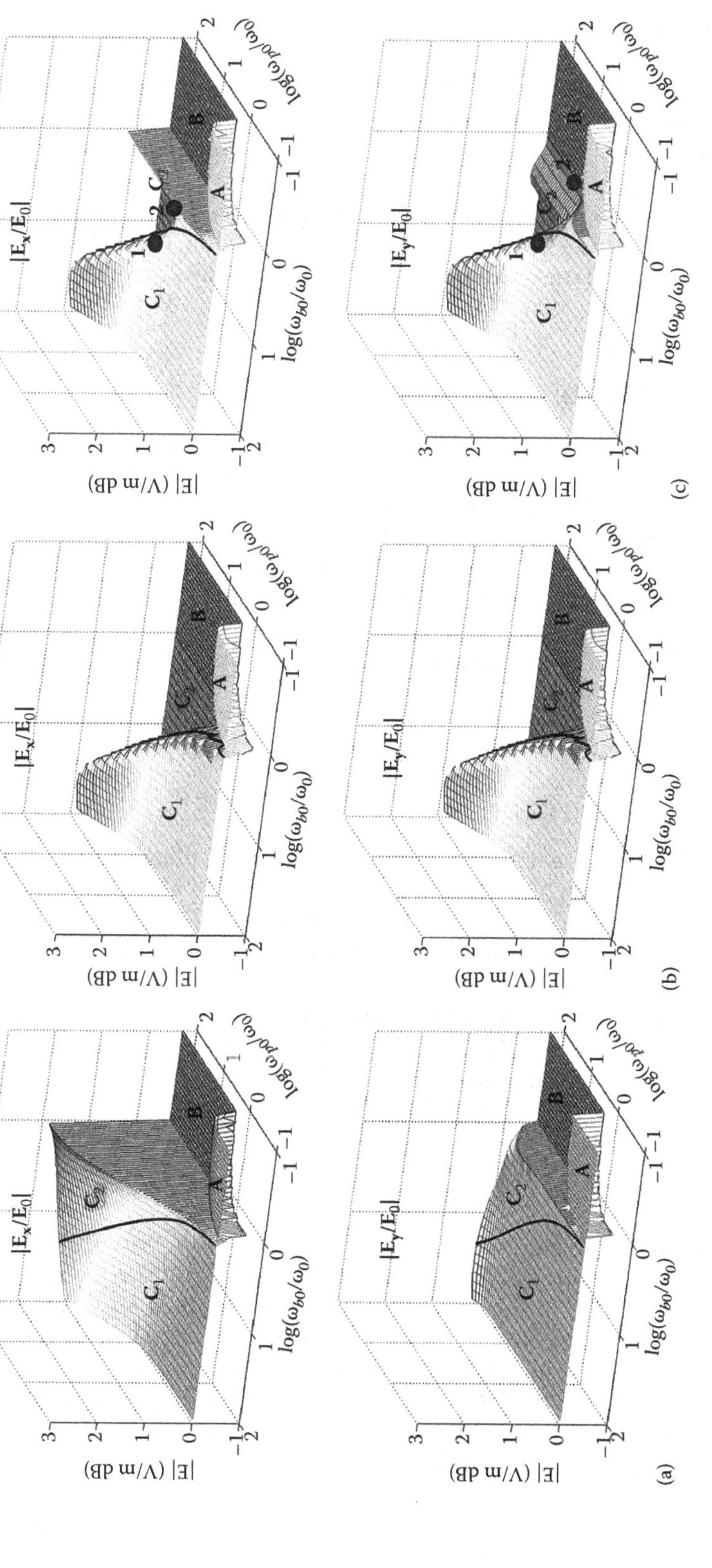

FIGURE 13C.7
Normalized amplitude of (top) the *x*-component and (bottom) the *y*-component of output standing wave E field for (a) sudden ($b = 500$), (b) slow ($b = 1$), and (c) moderate ($b = 10$) plasma decay rates.

In part of region C_1 (see Figures 13C.5 and 13C.7a), the results are similar to those of A in that the output standing wave is very similar in amplitude, frequency, and polarization to the input wave. However, as the C_1 region approaches the C_1/C_2 boundary, the x-component of the output wave starts to dominate and the amplitude and frequency start to increase over those of the incident wave. In the C_1 region, the incident whistler mode remains a whistler mode as the plasma collapses and only moderate frequency shifts are observed.

Region C_2, on the other hand, represents the region where the initial whistler mode transforms to the waveguide mode(s). In this region, it is shown that the x-component of the E field dominates and large amplitude and frequency shifts are observed for this sudden switching case.

13C.4.2 Slow Decay (FDTD)

The analysis of the previous section was repeated here but with a much slower plasma decay (b/T = 1/T), where the plasma (ω_p) decays to 1% of the initial value at approximately 4.6T or 2.9 ns. The results of this section are shown in Figure 13C.7b.

The behaviors of regions A and B are seen to be the same as the previous section and are indeed independent of the switching speed. However, for this slow switching case, the behavior of region C is much different. Over all of region C_1, the E_x and E_y components have equal amplitudes with values equal to or moderately greater than the incident wave. Region C_2 is shown to produce no output. The slow collapsing case is not sudden enough to produce the transition from the whistler to waveguide modes with the associated large frequency shifts.

13C.4.3 Moderate Decay (FDTD)

The analysis was again repeated with a moderate rate of plasma decay (b/T = 10/T). With b/T = 10/T, the plasma (ω_p) decays to 1% of the initial value at approximately 0.5T or 0.3 ns. The results in this section, as shown in Figure 13C.7c, are seen to be in between that of the sudden and slow decay cases in Figure 13C.7a and b. Region C_1 produces equal amplitude E_x and E_y components with moderate amplitude and frequency shifts in some areas. Much of region C_2 still has no output response except very near the $\omega_0 = \omega_b$ resonance region where the E_x component is starting to show significant amplitude. Here, there are large frequency shifts (see Figure 13C.5) and the output is linearly polarized.

13C.4.4 Polarization of the Output Waves

The polarization of the output standing wave is really a function of the superposition of four possible waves in the cavity after plasma collapse. Before the plasma collapse, a standing wave is present in the cavity, which is the result of an R wave traveling in both directions in the cavity. Note that the left traveling wave (−z-direction) is really a left-hand circularly (LHC) polarized wave (due to the PEC wall reflection) traveling in an opposite direction to the background magnetic field. Due to symmetry, this results in propagation properties that are the same as the R wave. In general, upon plasma collapse, each traveling wave will result in a forward and a backward (reflected) propagating wave all at the same output frequency $\omega_{out} = n_0\omega_0$. The forward and reflected waves will have the same polarization as their originating wave. Thus, after collapse, there may be an RHC and an LHC polarized wave moving in each direction. The output wave polarization will thus depend on the relative amplitude and phases of these four wave components.

For all of region C_1 for slow switching speeds and much of region C_1 for all switching speeds, the plasma collapse does not excite appreciable amplitude in the reflected wave for each initial traveling wave in the cavity. Thus, in these cases, the cavity (after collapse) contains two waves that have amplitude and polarization properties similar to the initial waves. The output standing wave will therefore be circularly polarized with equal E_x and E_y components.

For fast switching, some of region C_1 (near the C_1/C_2 boundary) and all of C_2, the output will consist of four waves with nearly equal amplitude in the cavity after collapse. The resultant standing wave is the superposition of these four waves (an RHC and an LHC wave moving in each direction). The end result can easily be shown to be a linearly polarized standing wave. In addition, the slant of this polarization is directly dependent on the switching angle ϕ_0. For the above analysis, $\phi_0 = 90°$ was used, which resulted in an x-polarized linear standing wave in these regions. If $\phi_0 = 0°$ is used, the output wave will be a y-polarized linear standing wave.

13C.5 Whistler-to-Whistler Mode Frequency Transformation (Region C_1)

As a specific example of a case in region C_1, let b/T = 10/T, $f_0 = 10$ GHz, $f_{p0} = 30$ GHz, $f_b = 30$ GHz, $k_v = v/\omega_{p0} = 0.0$, and $\phi_0 = 90°$. Let $m = 1$, which results in a plate separation of $d = 6.39$ mm. The profile of the plasma collapse is shown in Figure 13C.4, and this point on the ω_{p0} versus ω_b plane is labeled as point 1 in Figures 13C.6 and 13C.7c. Figure 13C.8 shows the FDTD results for this set of parameters. Figure 13C.8a and b gives the E field x- and y-components as a function of time, respectively, and Figure 13C.8c shows the polarization (E_x vs. E_y) over time. This is an example from region C_1 where the whistler mode remains a whistler mode as the plasma collapses and the polarization remains nearly circular. From Figure 13C.8, the output frequency is slightly greater than the input frequency, the output amplitude for E_x and E_y is close to the input amplitude, and the output polarization is nearly circular.

A more detailed view of the evolution of the frequencies and amplitudes of the three modes (for each of the two initial waves that form the standing wave) as a function of $\omega_p(t)$ is presented in Figure 13C.9a and b, respectively. The collapse of the plasma occurs right to left in these plots where initially the whistler mode (at frequency ω_0) has a normalized amplitude of unity and the other modes are not excited (amplitude = 0). As the plasma collapses (right to left), the whistler mode frequency shifts from ω_0 to $n_0\omega_0$, which is below ω_b and thus still a whistler wave. The R_1 mode shifts frequency to ω_b, and the R_2 mode shifts to $-n_0\omega_0$, neither of which gains any amplitude. The R_3 mode (that is left for each of the initial waves) combines to produce a circularly polarized standing wave similar to the input wave, E_x and E_y of which are shown in Figure 13C.8a and b. Figure 13C.10 gives a spectrogram showing the evolution of the whistler mode to the final output wave.

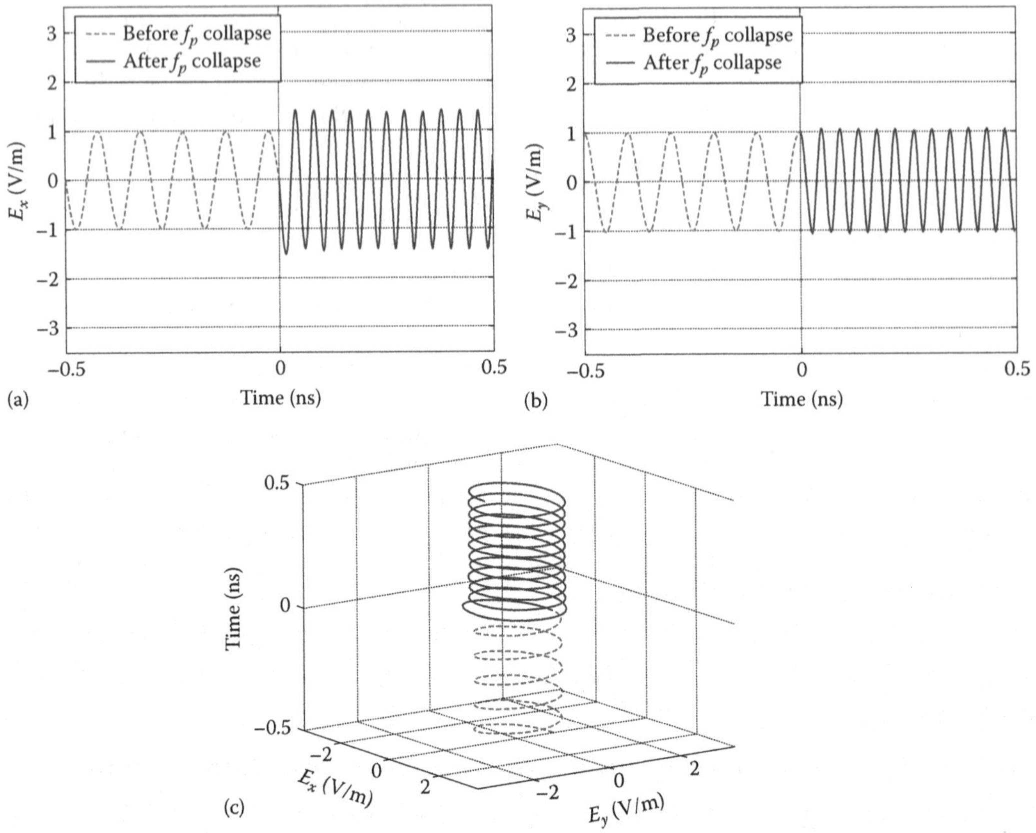

FIGURE 13C.8
FDTD results for collapsing plasma in a cavity (region C_1) with $b = 10$, $f_0 = 10$ GHz, $f_{p0} = 30$ GHz, $f_p = 30$ GHz, $k_v = v/\omega_{p0} = 0.0$, and $\phi_0 = 90°$ (switching angle) for (a) E_x versus time, (b) E_y versus time, and (c) E_x versus E_y over time (z-axis).

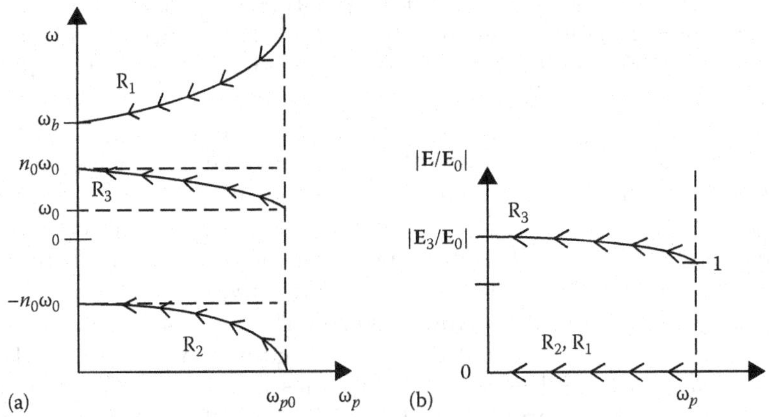

FIGURE 13C.9
Sketch of the variation of the (a) frequencies and (b) amplitudes of the three modes as the plasma collapses for the case $n_0\omega_0 < (C_1)$.

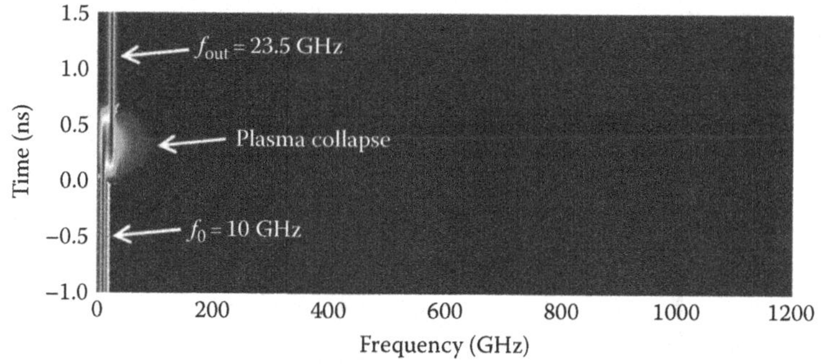

FIGURE 13C.10
Spectrogram for the x-component of the E field (C_1 case).

13C.6 Whistler-to-Waveguide Mode Frequency Transformation (Region C_2)

As a specific example of a case in region C_2, let b/T = 10/T, f_0 = 10 GHz, f_{p0} = 30 GHz, f_b = 10.009 GHz, $k_v = v/\omega_{p0} = 0.0$, and $\phi_0 = 90°$. Let $m = 1$, which results in a plate separation of d = 0.15 mm. It is possible to maintain the required values over this small plate separation, for the plasma step profile in space and the background static magnetic field, to a high degree of accuracy. The effects of any small inaccuracies are discussed in [14]. The profile of the plasma collapse is shown in Figure 13C.4, and this point on the ω_{p0} versus ω_b plane is labeled as point 2 in Figures 13C.6 and 13C.7c. Figure 13C.11 shows the FDTD analysis results for this set of parameters where (a) gives the E field x-component and (b) gives the E field y-component as a function of time. Figure 13C.11c shows the polarization (E_x vs. E_y) over time. This is an example from region C_2 where the whistler mode converts to the waveguide mode, the polarization becomes linear, and there is a large upward frequency shift.

13C.6.1 Puzzle and the Solution

The result presented as FDTD solution of this ideal problem has one puzzling aspect: If the frequency of the wave changes from 10 to 1000 GHz, it will go through the cyclotron resonance at $f = f_b$ = 10.009 GHz, should there not be a strong absorption, when the plasma is assumed lossy. The simulation results do not show any such strong absorption even for v/ω_p = 0.1. A more detailed picture of the variation of the amplitudes and frequencies of the three modes with the decaying plasma parameter $\omega_p(t)$ will clarify the results. Figure 13C.12 shows the variation of (a) the frequency and (b) the amplitude of the electric field of the three modes labeled as R_1, R_2, and R_3 (for each of the initial components of the standing wave). As ω_p decreases and becomes zero, ω_1 becomes $n_0\omega_0$, ω_2 becomes $-n_0\omega_0$, and ω_3 becomes ω_b.

The puzzle is now solved. At the instant of switching off the plasma, that is, at t = 0+, the three modes are excited, the first mode has a frequency higher than $n_0\omega_0$, but the amplitude of the electric field is zero; the amplitude builds up as ω_p further decreases, reaching a final value E_f. Mode 2 is a negative-going wave but otherwise has similar characteristics as mode 1 and builds up as the plasma collapses. Mode 3 is a positive-going wave with an initial frequency ω_0 and amplitude E_0 of the source R wave. However, as plasma decays, ω_3

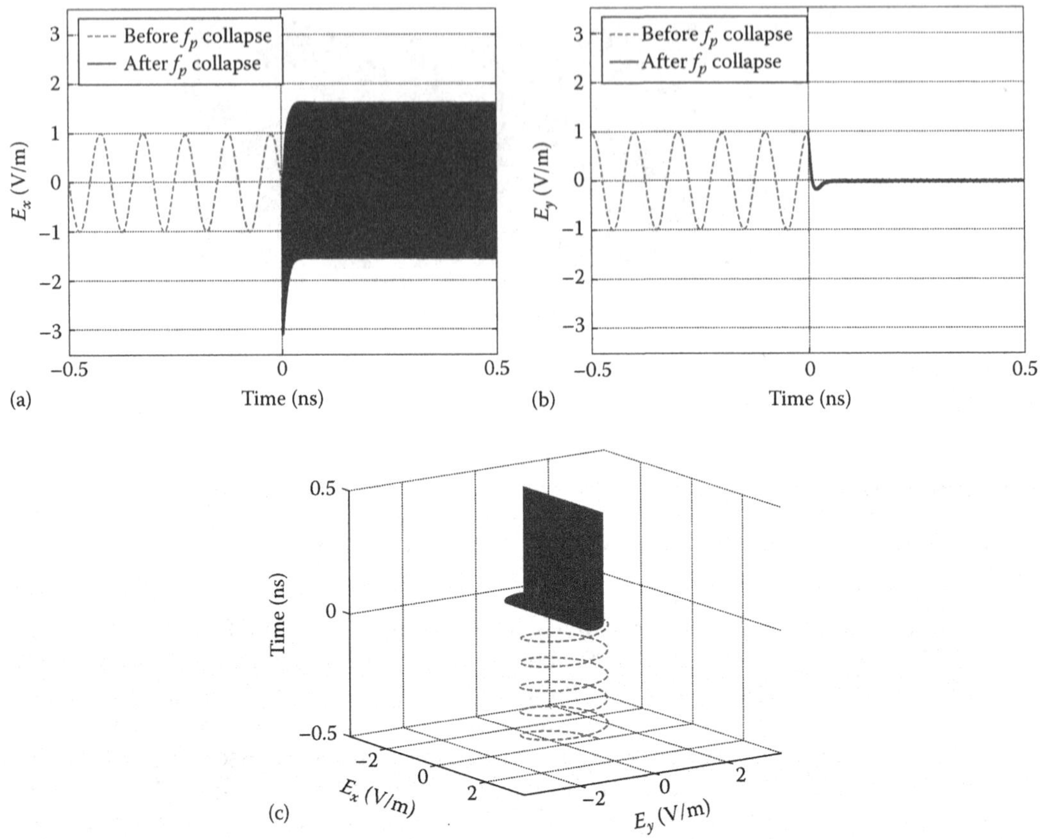

FIGURE 13C.11
FDTD results for collapsing plasma in a cavity (region C_2) with $b = 10$, $f_0 = 10$ GHz, $f_{p0} = 30$ GHz, $f_b = 10.009$ GHz, $k_v = v/\omega_{p0} = 0.0$, and $\phi_0 = 90°$ (switching angle) for (a) E_x versus time, (b) E_y versus time, and (c) E_x versus E_y over time (z-axis).

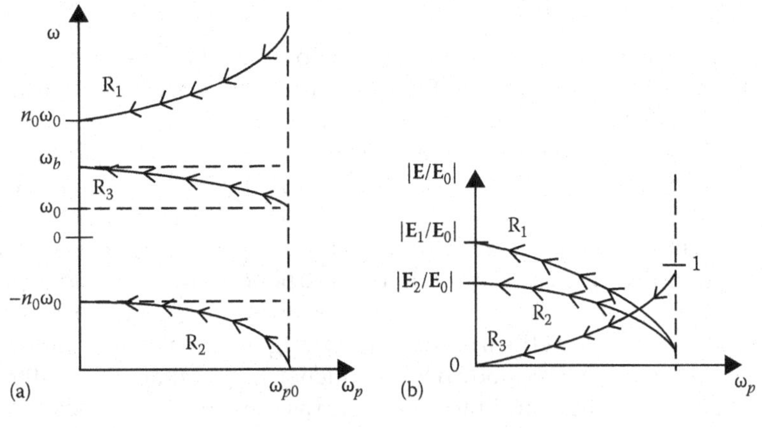

FIGURE 13C.12
Sketch of the variation of the (a) frequencies and (b) amplitudes of the three modes as the plasma collapses for the case $n_0\omega_0 > \omega_b$ (C_2).

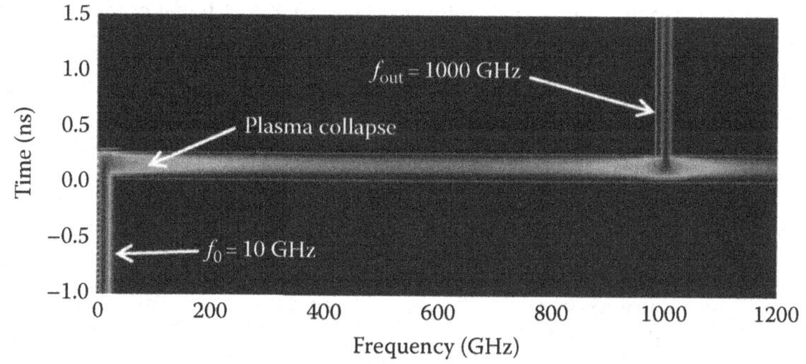

FIGURE 13C.13
Spectrogram for the x-component of the E field (C_2 case).

increases from ω_0 to ω_b and E_3 amplitude decreases from E_0 to 0. Even if we assume no losses in the plasma medium, this mode dies. In the presence of collisions, it is absorbed at a faster rate. The absorption in this case serves a useful purpose in removing the unwanted mode 3. Figure 13C.13 gives a spectrogram of the evolution of the whistler mode to the final output waveguide mode with the large frequency transformation to 1 THz.

The presence of a cavity introduces the switching angle ϕ_0 as a controlling parameter, controlling the relative amplitudes and phases of the x- and y-components of the electric fields of the output wave. The polarization of the output wave can be altered by changing the switching angle. Figure 13C.11 shows that the output wave is linearly x-polarized for $\phi_0 = 90°$. However, if the switching angle is 0°, the output wave will be y-polarized. Thus, the case under study for the chosen parameters (see Figure 13C.11) gives a frequency and polarization transformer, transforming a 10 GHz input wave to a 1000 GHz output wave with a change of polarization from right circularly polarized to x-polarized.

13C.7 Conclusion

The concepts discussed in this appendix make it possible to construct a cost-effective (inexpensive) frequency transformer of high transformation ratio. One can buy an off-the-shelf inexpensive power source in a low frequency band and use this frequency transformer to produce a radiation source in a frequency band not easily available (e.g., 1 THz).

Two more interesting aspects of this appendix for ongoing investigation are (1) reducing the ohmic losses due to real conductors at terahertz frequencies and (2) development of a technique to couple the terahertz radiation to devices outside the cavity.

References

1. Kalluri, D. K., *Electromagnetics of Time-Varying Complex Media: Frequency and Polarization Transformer*, 2nd edition, CRC Press, Boca Raton, FL, 2010.
2. Auld, B. A., Collins, J. H., and Zapp, H. R., Signal processing in a non-periodically time-varying magnetoelastic medium, *Proc. IEEE*, 56(3), 258–272, March 1968.

3. Jiang, C. L., Wave propagation and dipole radiation in a suddenly created plasma, *IEEE Trans. Antennas Propag.*, AP-23(1), 83–90, January 1975.
4. Felsen, B. L. and Whitman, G. M., Wave propagation in time-varying media, *IEEE Trans. Antennas Propag.*, AP-18(2), 242–253, March 1970.
5. Fante, R. L., Transmission of electromagnetic waves into time-varying media, *IEEE Trans. Antennas Propag.*, AP-19(3), 417–424, May 1971.
6. Joshi, C. J., Clayton, C. E., Marsh, K., Hopkins, D. B., Sessler, A., and Whittum, D., Demonstration of the frequency upshifting of microwave radiation by rapid plasma creation, *IEEE Trans. Plasma Sci.*, 18(5), 814–818, October 1990.
7. Kuo, S. P., Frequency up-conversion of microwave pulse in a rapidly growing plasma, *Phys. Rev. Lett.*, 65(8), 1000–1003, 1990.
8. Yablonovitch, E., Spectral broadening in the light transmitted through a rapidly growing plasma, *Phys. Rev. Lett.*, 31(14), 877–879, October 1973.
9. Kuo, S. P., Bivolaru, D., Orlick, L., Alexeff, I., and Kalluri, D. K., A transmission line filled with fast switched periodic plasma as a wideband frequency transformer, *IEEE Trans. Plasma Sci.*, 29(2), 365–370, April 2001.
10. Booker, H. G., *Cold Plasma Waves*, Kluwer, Hingham, MA, 1984.
11. Kalluri, D. K., Frequency upshifting with power intensification of a whistler wave by a collapsing plasma medium, *J. Appl. Phys.*, 79(8), 3895–3899, April 1996.
12. Ehsan, M. M. and Kalluri, D. K., Plasma induced wiggler magnetic field in a cavity, *Int. J. Infrared Millim. Waves*, 24(8), 1215–1234, August 2003.
13. Mendonca, J. T. and Oliveria e Silva, L., Mode coupling theory of flash ionization in a cavity, *IEEE Trans. Plasma Sci.*, 24(1), 147–151, February 1996.
14. Kalluri, D. K., Eker, S., and Ehsan, M. M., Frequency transformation of a whistler wave by a collapsing plasma medium in a cavity: FDTD solution, *IEEE Trans. Antennas Propag.*, 57(7), 1921–1930, July 2009.
15. Tanenbaum, B. S., *Plasma Physics*, McGraw-Hill, New York, p. 124, 1967.
16. Federici, J. F., Gary, D., Barat, R., and Michalopoulou, Z.-H., T-rays vs. terrorists, *IEEE Spectr.*, 44(7), 47–52, July 2007.
17. Smart, A. G., Room-temperature source delivers record-power terahertz beam, *Phys. Today*, 64(2), 13–15, February 2011.
18. Baños, Jr., A., Mori, W. B., and Dawson, J. M., Computation of the electric and magnetic fields induced in a plasma created by ionization lasting a finite interval of time, *IEEE Trans. Plasma Sci.*, 21(1), 57–69, February 1993.
19. Kalluri, D. K., Goteti, V. R., and Sessler, A. M., WKB solution for wave propagation in a time-varying magnetoplasma medium: Longitudinal propagation, *IEEE Trans. Plasma Sci.*, 21(1), 70–76, February 1993.
20. Kalluri, D. K. and Goteti, V. R., WKB solution for wave propagation in a decaying plasma medium, *J. Appl. Phys.*, 66(8), 3472–3475, 1989.
21. Stepanov, N. S., Dielectric constant of unsteady plasma, *Sov. Radio-Phys. Quantum Electron.*, 19(7), 683–689, July 1976.
22. Lade, R. K., Lee, J. H., and Kalluri, D. K., Frequency transformer: Appropriate and different models for a building-up and collapsing magnetoplasma medium, *J. Infrared Millim. Terahertz Waves*, 32(7), 960–972, July 2010.
23. Taflove, A., *Computational Electrodynamics: The Finite-Difference Time-Domain Method*, Artech House, Norwood, MA, 1995.
24. Lee, H. J., Kalluri, D. K., and Nigg, G. C., FDTD simulation of electromagnetic wave transformation in a dynamic magnetized plasma, *Int. J. Infrared Millim. Waves*, 21(8), 1223–1253, August 2000.

Further Reading

Ehsan, M. M. and Kalluri, D. K., Plasma induced wiggler magnetic field in a cavity II—FDTD method for a switched lossy plasma, *Int. J. Infrared Millim. Waves*, 24(10), 1655–1676, October 2003.
Heald, M. A. and Wharton, C. B., *Plasma Diagnostics with Microwaves*, Wiley, New York, 1965.
Kalluri, D. K., Conversion of a whistler wave into a controllable helical wiggler magnetic field, *J. Appl. Phys.*, 79(9), 6770–6774, May 1996.
Kalluri, D. K., *Electromagnetics of Complex Media*, CRC Press, Boca Raton, FL, 1999.
Lai, C. H., Katsouleas, T. C., Mori, W. B., and Whittum, D., Frequency upshifting by an ionization front in a magnetized plasma, *IEEE Trans. Plasma Sci.*, 21(1), 45–52, February 1993.
Lampe, M. and Ott, E., Interaction of electromagnetic waves with a moving ionization front, *Phys. Fluids*, 21(1), 42–54, January 1978.
Lee, J. H., Finite difference time domain simulation of electromagnetic wave transformation in a dynamic, inhomogeneous, bounded, and magnetized plasma, PhD dissertation, Department of Electrical Engineering, University of Massachusetts at Lowell, Lowell, MA, 1998.
Lee, J. H. and Kalluri, D. K., Three-dimensional FDTD simulation of electromagnetic wave transformation in a dynamic inhomogeneous magnetized plasma, *IEEE Trans. Antennas Propag.*, 47(7), 1146–1151, July 1999.
Yee, K. S., Numerical solution of initial boundary value problems involving Maxwell's equations in isotropic media, *IEEE Trans. Antennas Propag.*, AP-14(3), 302–307, May 1966.

Appendix 14A: Electromagnetic Wave Interaction with Moving Bounded Plasmas*

Dikshitulu K. Kalluri and R.K. Shrivastava

14A.1 Introduction

Recent studies on electromagnetic wave interaction with bounded relativistic plasmas have generally neglected *oblique* incidence and the results have been limited to the case of normal incidence. In a previous article [1], the reflection and transmission of electromagnetic waves obliquely incident on a relativistically moving uniaxial plasma slab was studied, bringing out the nature of oscillations in the reflection coefficient against the normalized slab velocity. The problem was solved in the rest system of the slab and it was seen that as the slab *recedes* from the incident wave, the angle of reflection as viewed from the observer's frame turns out to be 90° for some critical slab velocity, thus explaining the zero reflected power at this velocity. This is in agreement with Yeh's finding [2] for the dielectric half-space moving normal to the interface. However, his observation that beyond this critical point the reflected wave is evanescent seems to be incorrect.

The purpose of this chapter was to unfold the interesting effects that occur beyond the above critical slab velocity. The conclusion drawn is very general and has relevance whenever a plane wave is *obliquely* incident on a medium moving *normal* to the interface.

The maxima and minima of the power reflection coefficient as a function of the normalized slab velocity are also shown for electromagnetic waves obliquely incident on an isotropic plasma slab. Both parallel and perpendicular polarizations of the wave are considered.

14A.2 Basic Equations

Let the primed quantities be measured in Σ' which is at rest with respect to the moving plasma slab and the corresponding unprimed quantities with respect to the observer's frame Σ. Assume that the field quantities in plasma have an exponential variation of the form

$$\Psi'_{p} = \exp\left\{\left[\omega' t' - k'_0\left(S'x' + q'z'\right)\right]\right\}, \tag{14A.1}$$

where

$$\omega' = \left[\gamma(1 - C\beta)\right]\omega = p\omega, \tag{14A.2}$$

$$S' = \sin\theta' = \frac{S}{p}. \tag{14A.3}$$

Using Maxwell's equations in plasma, one may obtain q', the root of the dispersion equation, as

$$q' = \pm\left(\epsilon - S'^2\right)^{1/2} = \pm\left(\frac{\left(p^2 - S^2\right)\Omega^2 - 1}{p^2\Omega^2}\right)^{1/2}. \tag{14A.4}$$

In the above equations, $S = \sin\theta$, $C = \cos\theta$ (θ being the angle of incidence), $\Omega = \omega/\omega_p$, and $\varepsilon' = 1(\omega_p/\omega')^2\, 1 - (\omega_p/\omega')^2$.

Other symbols have their usual meaning [1]. The parameter p has the significance that it transforms the incident wave frequency to the rest system of the plasma. The relations between the primed and unprimed quantities are as given in Reference 1.

The power reflection and transmission coefficients may be put in the form [1]:

$$\rho = \frac{\alpha}{1 + B\mathrm{cosec}^2 A}, \quad \tau = \frac{1}{1 + (1/B)\sin^2 A} \tag{14A.5a}$$

in the real range of q' and

$$\rho = \frac{\alpha}{1 + B\mathrm{cosech}^2 |A|}, \quad \tau = \frac{1}{1 + (1/B)\sinh^2 A} \tag{14A.5b}$$

in the imaginary range of q', where α, A, and B for the two polarizations are given by the following.

Parallel polarization:

$$\alpha_{\|} = \beta'\left(\frac{C - \beta}{C + \beta\beta'}\right)^2 = \frac{\left(1 + \beta^2 - 2\beta C\right)\left(1 + \beta^2 - 2\beta/C\right)}{\left(1 - \beta^2\right)^2}, \tag{14A.6a}$$

$$A_{\|} = k_0' q' d' = 2\pi\left[\left(p^2 - S^2\right)\Omega^2 - 1\right]^{1/2} d_p', \tag{14A.6b}$$

$$B_{\|} = \left(\frac{2q'T'C'}{q'^2 - T'^2 C'^2}\right)^2$$

$$= \frac{4p^4\Omega^2\left(p^2 - S^2\right)\left(p^2\Omega^2 - 1\right)\left(\overline{p^2 - S^2\Omega^2 - 1}\right)}{\left[\left(p^2 \quad S^2\right)\left(2p^2\Omega^2 - 1\right) - p^4\Omega^2\right]^2}. \tag{14A.6c}$$

Perpendicular polarization:

$$\alpha_{\perp} = \frac{1}{\beta'}\left(\frac{1-\beta C}{1+\beta C/\beta'}\right)^2 = \frac{\left(1+\beta^2-2\beta C\right)\left(1+\beta-2\beta/C\right)}{\left(1-\beta^2\right)^2}, \tag{14A.7a}$$

$$A_{\perp} = 2\pi\left[\left(p^2-S^2\right)\Omega^2-1\right]^{1/2} d_p', \tag{14A.7b}$$

$$B_{\perp} = \left(\frac{2q'C'}{q'^2-C'^2}\right)^2$$
$$= 4\Omega^2\left(p^2-S^2\right)\left(\overline{p^2-S^2\Omega^2}-1\right). \tag{14A.7c}$$

In the above expressions,

$$C' = \cos\theta' = \frac{C-\beta}{1-C\beta}, \tag{14A.7d}$$

$$\beta' = \frac{1+\beta^2-2\beta C}{1+\beta^2-2\beta/C}, \tag{14A.7e}$$

and $d_p' = d'/\lambda_p$, λ_p being the free-space wavelength corresponding to plasma frequency and d' is the thickness of the plasma slab.

14A.3 Numerical Results and Discussion

The power reflection coefficient versus normalized wave frequency (Ω) curves are similar in shape to those given in Reference 3 and hence omitted in this chapter.

The development of the power reflection coefficient as a function of the normalized slab velocity (β) follows the procedure outlined in Reference 1 and the equations giving the maxima and minima in the oscillatory range are listed in Section 14A.3.1.

Although the two types of wave polarization lead to the same result for the limiting case of normal incidence [4,5], it is not so for oblique incidence and numerical results are therefore given for both polarizations taking $C = 0.5$, $d_p = 1.0$, and $\Omega = 2.0$ (Figures 14A.1 and 14A.2). It is seen that the oscillations decay faster for parallel polarization than for perpendicular polarization. The curve beyond $\beta = (1 - S)/C$ is shown dashed to bring out the significant point that the reflected wave in this range propagates *toward* the slab. This is explained as follows.

The factor α in Equation 14A.5 has the physical significance that it is the reflected power from a moving mirror and the reflected power for the slab is given by multiplying α by an appropriate term which takes into account that plasma is partly conducting and that the parallel boundaries cause interference of the internally reflected waves. The plot of α versus β is given in Figure 14A.3 (along with the variation of θ_r and θ' with β) where the negative value of α beyond $\beta = (1 - S)/C$ is shown dashed. α is zero at $\beta = (1 - S)/C$ since the

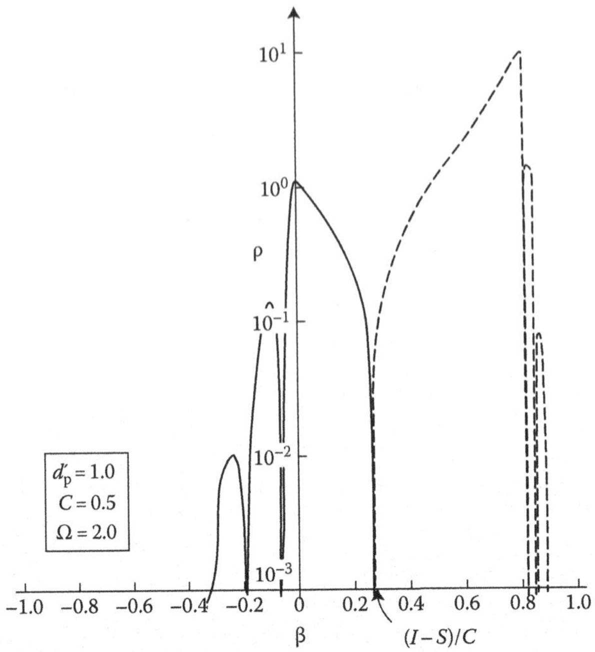

FIGURE 14A.1
Power reflection coefficient ρ versus normalized slab velocity β for parallel polarized wave. The curve beyond the critical slab velocity [$\beta = (1 - S)/C$] represents −ρ and is shown as dashed lines. The significance is that for $p > (1 - S)/C$ the reflected wave propagates along the positive z-axis. (Reprinted with permission from Kalluri, D. and Shrivastava, R.K., Electromagnetic wave interaction with moving bounded plasmas, *J. Appl. Phys.*, 44, 4518–4521, 1973. Copyright 1973, American Institute of Physics.)

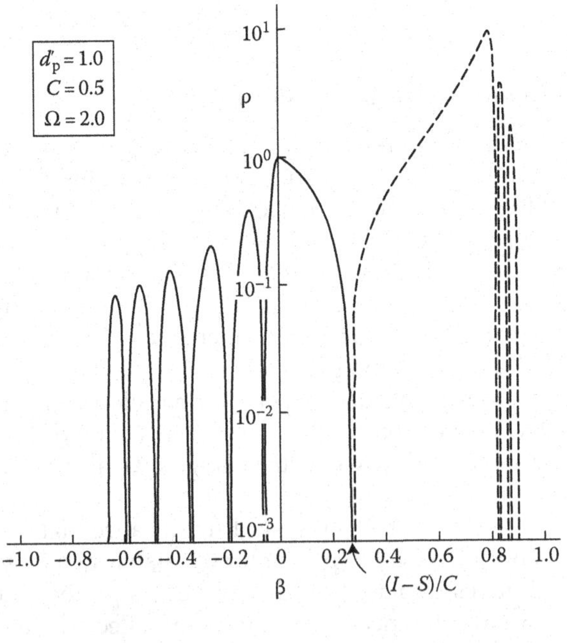

FIGURE 14A.2
ρ versus β for perpendicularly polarized wave. The slow oscillations when compared to fast-decaying oscillations in the corresponding curve for parallel polarization is to be noted.

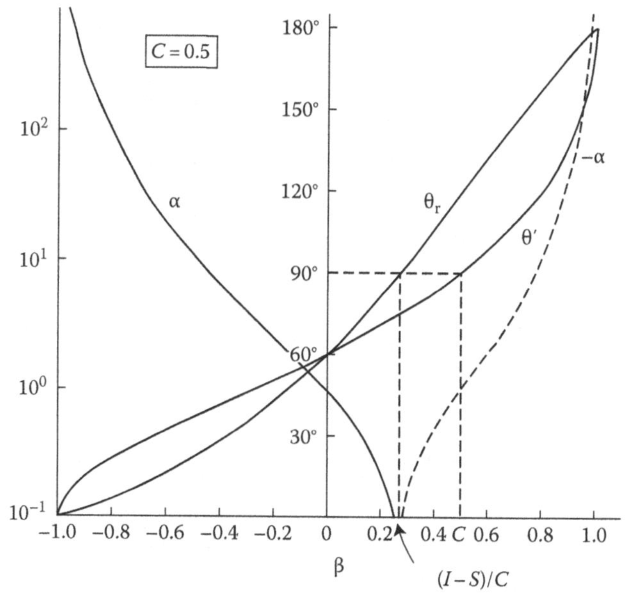

FIGURE 14A.3
θ′, the angle of incidence in Σ′, and θ_r, the angle of reflection as seen in Σ, plotted as functions of β. α versus β is also shown. α is the factor that represents power reflected from a moving mirror.

angle of reflection $\theta_r[\cos^{-1}(C/\beta')]$ as seen in Σ is 90° at this value of β. It is seen that in the range $(1 - S)/C < \beta < C$, $\theta_r > 90°$ and α is thus negative. This means that in this range of β, the reflected wave actually travels *toward* the plasma medium. It is further interesting to note that α continues to represent the reflected power of a moving mirror for $C < \beta < 1$. However, in this range, while the incident wave from below never catches the moving mirror, the moving mirror can interact with an existing plane wave above the mirror. The wave is, no doubt, moving away from the mirror, but since the velocity of the mirror is greater than the normal component of the phase velocity of the incident wave, the mirror, so to speak, will be impinging on the free-space wave.

The interaction of the incident wave with the plasma slab in the entire range $-1 < \beta < 1$ is explained qualitatively in Figures 14A.4 and 14A.5.

Figure 14A.4a depicts the situation in the range $-1 < \beta < (1 - S)/C$. The incident wave (I wave) and the reflected wave (R wave) are below the slab, the reflected wave moving away from the slab. The angle of reflection θ_r continuously increases and becomes 90° at $\beta = (1 - S)/C$ as noted earlier (Figure 14A.4b). The transmitted wave (T wave) is above the slab. The situation seen in Σ′ is also depicted. At this β, the reflected wave appears to travel parallel to the slab in Σ frame and the normal component of the Poynting's vector associated with this wave in zero.

In the range $(1 - S)/C < \beta < C$, the R wave is still below the slab since the normal component of its phase velocity is less than the slab velocity, but the wave now propagates *toward* the slab. The observation made by Yeh[2] that since $\theta_r > 90°$, the reflected wave is evanescent is thus in error.

At $\beta = C$, the normal component of the phase velocity of the incident wave is equal to the slab velocity, and the wave moves just parallel to the slab interface as seen in Σ′ (Figure 14A.5a). Thus, there is no interaction between the wave and the plasma medium at $\beta = C$.

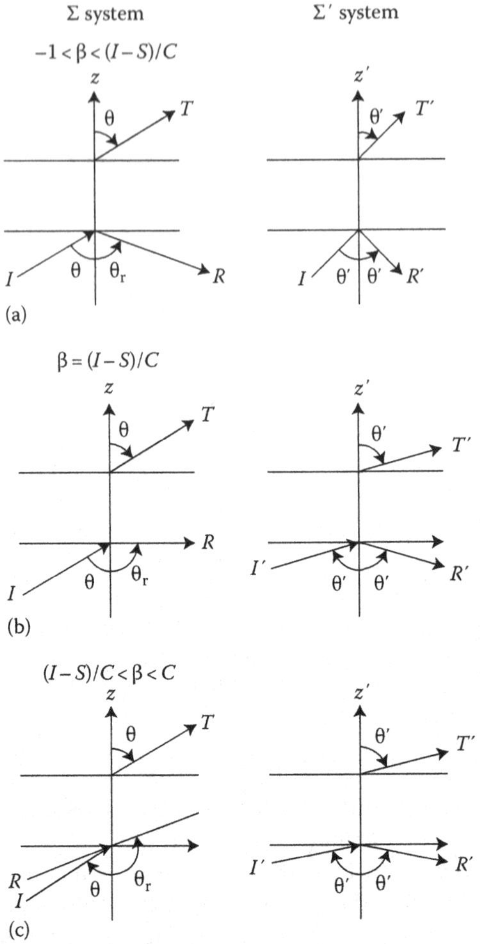

FIGURE 14A.4
Geometries explaining the interaction of the wave with the plasma slab qualitatively for the range $-1 < \beta < C$. The orientation of the various free-space waves in both Σ and Σ' is clearly depicted.

For $C < \beta < 1$, the situation in Σ and Σ' is shown in Figure 14A.5b. The I and R waves are above the slab, moving away from the slab. The T wave, of course, is below the slab, moving toward it.

As is known, the reflected power (ρ, power in the R wave) and transmitted power (τ, power in the T wave) add up to the incident power only at $\beta = 0$ and, for other values of β, the conservation of power requires that the excess of the time-averaged power of the generated waves over the incident power must come from the rate of decrease of stored energy and the mechanical power to be put in to maintain the motion against the radiation pressure. Using a detailed energy balance analysis on the lines indicated by Daly and Gruenberg [6], the authors have checked the assumed orientation of the various free-space waves (as given in Figures 14A.4 and 14A.5) for both the lower and upper interfaces of a *dielectric* slab and this is a sufficient check since the given wave orientations do not depend on the nature of the medium considered.

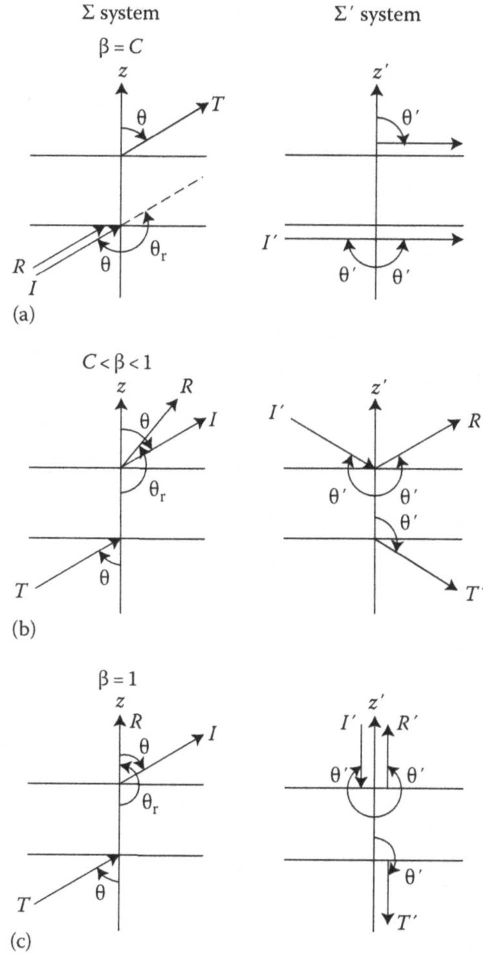

FIGURE 14A.5
Geometries explaining the interaction of the wave with the plasma slab for $C \leq \beta \leq 1$.

A similar analysis for the plasma slab, requiring suitable modifications* in the expression for the stored energy and the stress tensor, is currently in progress and will be reported later.

14A.3.1 Appendix

The minima in ρ/α occur for p and is given by

$$p_{\min,n} = \left[\left(S^2 + \frac{1}{\Omega^2}\right) + \frac{n^2}{16\Omega^2 {d'_p}^2}\right]^{1/2}_{(n \text{ even})}. \tag{14A.8}$$

* Reprinted with permission from *J. Appl. Phys.*, 44(10), 1973, 4518–4521 [7].

The maxima in ρ/α occur for p satisfying the transcendental equation $\tan A = f_p A$, where

$$\left(f_{\mathrm{P}}\right)_{\parallel} = \frac{K_1 K_2}{K_1\left[K_2 + K_3 + K_5\left(p^2 p^2 \Omega^2 - 1 + 2p^2\Omega^2 \overline{p^2 - S^2}\right)\right] - p^2 K_3 K_4},$$

$$\begin{aligned}
K_1 &= 2\left(p^2 - S^2\right)p^2\Omega^2 - p^4\Omega^2 - \left(p^2 - S^2\right),\\
K_2 &= p^2\Omega^2\left(p^2 - S^2\right)\left(p^2\Omega^2 - 1\right),\\
K_3 &= 2\left(p^2 - S^2\right)\left(p^2\Omega^2 - 1\right)\left(\overline{p^2 - S^2\Omega^2} - 1\right),\\
K_4 &= 2\left(p^2 - S^2\right)\Omega^2 - 1,\\
K_5 &= \left(p^2 - S^2\right)\Omega^2 - 1,\\
\left(f_{\mathrm{P}}\right)_{\perp} &= \frac{\left(p^2 - S^2\right)\Omega^2 - 1}{2\left(p^2 - S^2\right)\Omega^2 - 1}.
\end{aligned} \tag{14A.9}$$

References

1. Kalluri, D. and Shrivastava, R. K., *IEEE Trans. Antennas Propag.*, AP-21, 63, 1973.
2. Yeh, C., *J. Appl. Phys.*, 38, 5194, 1967.
3. Kalluri, D. and Shrivastava, R. K., *J. Appl. Phys.* , 44(5), 2440–2442, 1973.
4. Yeh, C., *J. Appl. Phys.*, 37, 3079, 1966.
5. Yeh, C., *J. Appl. Phys.*, 38, 2871, 1967.
6. Daly, P. and Gruenberg, H., *J. Appl. Phys.*, 38, 4486, 1967.
7. Kalluri, D. and Shrivastava, R. K., Electromagnetic wave interaction with moving bounded plasmas, *J. Appl. Phys.*, 44(10), 4518–4521, 1973.

*Appendix 14B: Radiation Pressure Due to Plane Electromagnetic Waves Obliquely Incident on Moving Media**

Dikshitulu K. Kalluri and R.K. Shrivastava

In electromagnetic theory, the force on a given volume element within a dielectric is expressible in terms of the electric and magnetic fields at the surface of the volume element and the associated concept of surface stress tensor is well understood [1]. The calculation of the radiation pressure exerted on a *moving* interface between two media forms an interesting aspect of study in relativistic electrodynamics. Motivation for such a study arises in the context of explaining the excess of radiated power from a medium moving *normal* to the interface. Daly and Gruenberg [2] have considered this problem while studying the energy balance for perpendicularly polarized plane waves reflected from moving media. Later, Yeh [3] also touched upon this aspect in his investigation of the Brewster angle for a moving dielectric medium.

A detailed consideration of the radiation pressure due to a parallel or perpendicularly polarized plane electromagnetic wave obliquely incident (*xz*-plane of incidence) on a moving conducting half-space and the mechanical power (P_m) supplied to keep it uniformly moving is presented for the *entire* velocity range. The analogous study for a dielectric medium entends Yeh's work [3] by including the velocity range beyond the critical medium velocity [4] $\beta = C$ ($C = \cos\theta$, θ being the angle of incidence). The symbols used have their usual meaning [4].

Using the concept of the Maxwell stress tensor [5] and following the procedure outlined by Daly and Gruenberg [2], one may obtain the force per unit area on the interface:

$$(F_z)_{\parallel} = \epsilon_0 \left(E_x^I\right)^2 \gamma^2 \left(1 - \frac{\beta}{C}\right)^2 \tag{14B.1a}$$

and

$$(F_z)_{\perp} = \epsilon_0 \left(E_y^I\right)^2 \gamma^2 (C - \beta)^2. \tag{14B.1b}$$

The subscripts ∥ and ⊥ indicate the type of wave polarization involved.

The mechanical power supplied to the conductor to overcome this radiation pressure is then given by

$$P_m = (-F_z)\upsilon = S_z^I\left(-\frac{2\beta}{C}\gamma^2 (C-\beta)^2\right), \tag{14B.2}$$

where S_z^I is the normal component of the Poynting vector associated with the incident wave given by

$$\left(S_z^I\right)_{\parallel} = \frac{1}{2}\frac{\epsilon_0\left(E_x^I\right)^2 c}{C} \tag{14B.3a}$$

and

$$\left(S_z^I\right)_{\perp} = \frac{1}{2}\frac{\epsilon_0\left(E_y^I\right)^2 c}{C}. \tag{14B.3b}$$

It is clear from Equation 14B.1 that the field force increases with an increase in C, being maximum at normal incidence. Further, the normalized mechanical power, P_m / S_z^I, is seen to be the same for both types of wave polarization. Figure 14B.1 shows the variation of this normalized mechanical power supplied with medium velocity. It is seen that the mechanical power supplied is positive in the range $-1 < \beta < 0$. The conductor is moving *toward* the wave, and, the field force on it being one of compression, mechanical power has to be supplied to the conductor to keep it uniformly moving. This power supplied will appear in the form of radiation, that is, the reflected power is more than the incident power. For positive β, that is, conductor moving *away* from the incident wave, there are three ranges to be considered. In the first range given by $0 < \beta < (1 - S)/C$, the field does work on the conductor and so P_m is negative. The magnitude of the normalized rate of change of mechanical power per unit area increases and reaches a maximum. Due to the work done by the fields

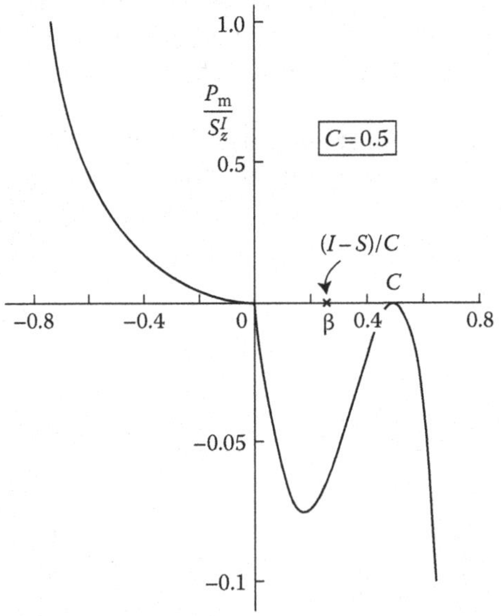

FIGURE 14B.1
Normalized mechanical power supplied $\left(P_m/S_z^I\right)$ as a function of the normalized velocity (β) for the relativistic conducting half-space. The vertical scale is not uniform. (Reprinted with permission from Kalluri, D. and Shrivastava, R.K., Radiation pressure due to plane electromagnetic waves obliquely incident on moving media, *J. Appl. Phys.*, 49, 3584–3586, 1978. Copyright 1978, American Institute of Physics.)

and the increase in the stored energy in the system, the reflected power is less than the incident power, that is, the power reflection coefficient is less than unity. The reflection coefficient eventually reduces to zero at $\beta = (1 - S)/C$. In the second range $(1 - S)/C < \beta < C$, the reflected wave propagates *toward* [4] the conductor and the reflection coefficient continues to be less than unity. At $\beta = C$, the conductor has no interaction with the wave and so P_m is zero. The last range, $C < \beta < 1$, corresponds to the problem of a conductor impinging on an existing plane wave above and the variation of P_m with β is similar to that in the range $-1 < \beta < 0$ except for the change in sign. The reason for the negative sign becomes clear from the orientation of the free-space waves as depicted in Figure 5 of Reference 4.

Assume that the fields in the dielectric medium (Σ') have an exponential variation of the form

$$\Psi'_P = \exp\left\{j\left[\omega' t' - k'_0\left(S'x' + q'z'\right)\right]\right\}. \tag{14B.4a}$$

The characteristic root q' may be determined in the usual manner by solving Maxwell's equations in Σ':

$$q' = \left(\epsilon' - S'^2\right)^{1/2} \tag{14B.4b}$$

One may calculate the radiation pressure exerted on the moving dielectric surface following Daly and Gruenberg [2]. Alternatively, the radiation pressure may be calculated in the rest frame of the moving medium (Σ') and then transformed to the laboratory system [3] (Σ) according to the transformation

$$F_z = F'_z. \tag{14B.5}$$

Thus,

$$\left(F_z\right)_{\|} = \frac{1}{2}\epsilon_0\left(E^I_x\right)^2\gamma^2\left(1 - \frac{\beta}{C}\right)^2\left[1 + \left(\frac{q' - \epsilon' C'}{q' + \epsilon' C'}\right)^2 - \frac{4\epsilon' q'^2}{\left(q' + \epsilon' C'\right)^2}\right]. \tag{14B.6}$$

Finally,

$$\left(P_m\right)_{\|} = -S^I_z\left[\beta C\left(1 - \frac{\beta}{C}\right)^2\gamma^2\left(\frac{q'^2 + \epsilon'^2 C'^2 - 2\epsilon' q'^2}{\left(q' + \epsilon' C'\right)^2}\right)\right]. \tag{14B.7}$$

For a perpendicularly polarized wave, one obtains

$$\left(P_m\right)_{\perp} = S^I_z\left[2\beta C\left(1 - \frac{\beta}{C}\right)^2\gamma^2\frac{\epsilon' - 1}{\left(q' + C'\right)^2}\right]. \tag{14B.8}$$

Our result (Equation 14B.8) for perpendicular polarization agrees with that of Daly and Gruenberg derived by another method. This may be verified by setting Q of their expression equal to $(1 - \beta C)q'$. Figure 14B.2 shows the variation of the normalized P_m with β for

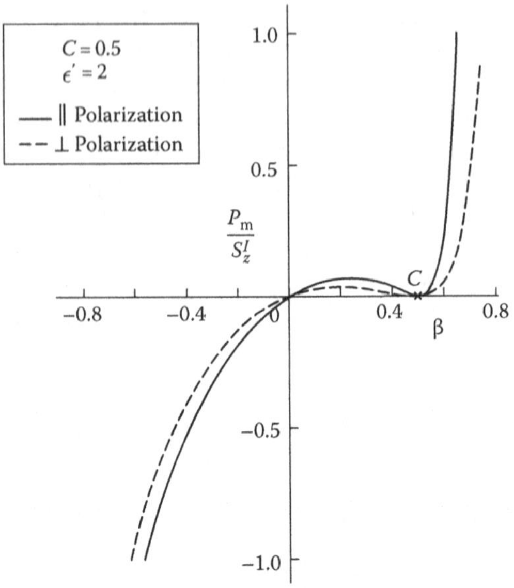

FIGURE 14B.2
Normalized mechanical power supplied versus β for a moving dielectric half-space.

$e' = 2$ for both wave polarizations. The general features of these curves are same as that for the relativistic conductor except for the change in sign. This is because the force on a conductor and that on a dielectric are in opposite directions [2]. This is depicted in Figure 14B.3 where cF / S_z^I is plotted against β. The radiation pressure pushes on a conducting surface, whereas it has a sucking effect on a dielectric. Another conclusion drawn from Figure 14B.3

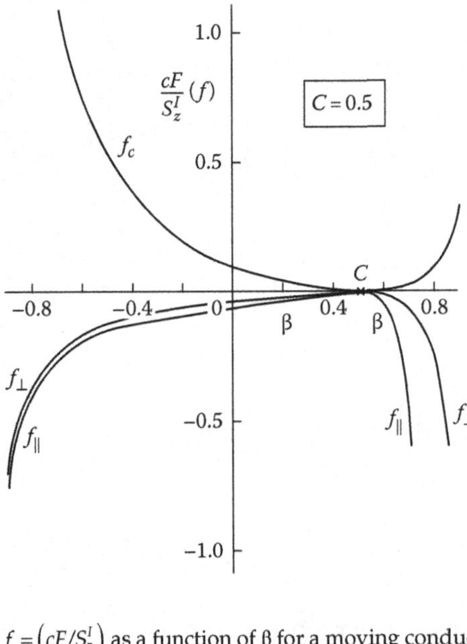

FIGURE 14B.3
Normalized radiation pressure $f = \left(cF/S_z^I\right)$ as a function of β for a moving conductor (f_c) as well as moving dielectric medium ($f_\parallel, f_\perp$). The subscripts ∥ and ⊥ refer to the type of wave polarization. The force on a conductor and that on a dielectric are in opposite directions.

is that the radiation pressure on the dielectric is less for a perpendicularly polarized wave than for the parallel polarization.

References

1. Panofsky, W. K. H. and Phillips, M., *Classical Electricity and Magnetism*, Addison-Wesley, Reading, MA, 1962.
2. Daly, P. and Gruenberg, H., *J. Appl. Phys.*, 38, 4486, 1967.
3. Yeh, C., *J. Appl. Phys.*, 38, 5194, 1967.
4. Kalluri, D. and Shrivastava, R. K., *J. Appl. Phys.*, 44, 4518, 1973.
5. Sommerfeld, A., *Electrodynamics*, Academic Press, New York, 1964.
6. Kalluri, D. and Shrivastava, R. K., Radiation pressure due to plane electromagnetic waves obliquely incident on moving media, *J. Appl. Phys.*, 49, 3584–3586, 1978.

Appendix 14C: Reflection and Transmission of Electromagnetic Waves Obliquely Incident on a Relativistically Moving Uniaxial Plasma Slab*

Dikshitulu K. Kalluri and R.K. Shrivastava

14C.1 Introduction

The electromagnetic wave interaction with relativistic plasmas has been receiving considerable attention in recent literature. A number of investigators [1–14] have studied the problem of reflection and transmission of electromagnetic waves by semi-infinite moving dielectric media and plasmas as well as dielectric slabs. Yeh [15,16] studied the reflection and transmission of electromagnetic waves by a moving isotropic plasma slab, whereas Chawla and Unz [17] considered the case of normal incidence on a moving plasma slab subjected to a finite static magnetic field normal to the slab. Kong and Cheng [18] obtained numerical results for the transmission coefficient at normal incidence when the uniaxial plasma slab is moving along the interface and the technique used was based on the concept of bianisotropy.

The purpose of this appendix is to present the solution for reflection and transmission coefficients of parallel-polarized electromagnetic waves incident on a relativistically moving anisotropic plasma slab with strong normal magnetostatic field. The emphasis is on the nature of oscillations in the reflection and transmission coefficients (due to the interference of internally reflected waves) that are discussed against the incident wave frequency as well as the velocity of the slab, the latter being the more interesting one.

The problem is solved in the rest system of the slab and relativistic transformations are then used to obtain the reflection and transmission coefficients as observed in the laboratory system.

14C.2 Formulation of Problem

The geometry of the problem is shown in Figure 14C.1. Consider a plane wave excited in the laboratory system Σ propagating in free space in the xz-plane at an angle θ to the positive z-axis. Assuming a harmonic time variation the wave takes the form

$$E^{I} = \left(\hat{x}E_x^{I} + \hat{z}E_z^{I}\right)\Psi_I,$$
$$\Psi_I = \exp\left\{j\left[\omega t - \left(k_x x + k_z z\right)\right]\right\}, \quad (14C.1)$$

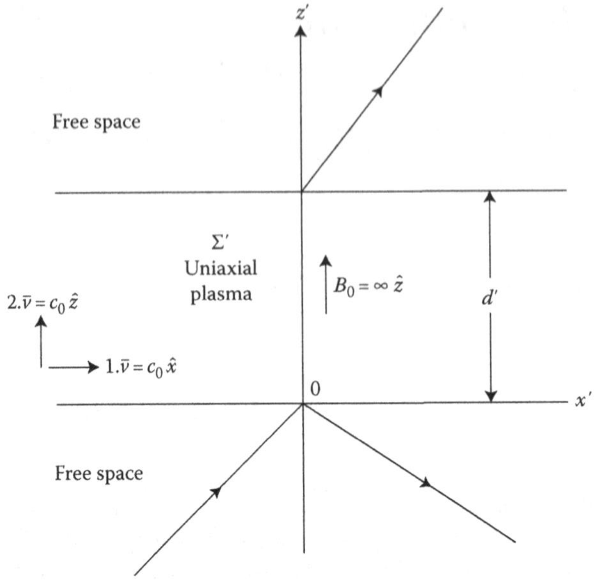

FIGURE 14C.1
Geometry of the problem. (Reprinted from Kalluri, D. and Shrivastava, R.K., Reflection and transmission of electromagnetic waves obliquely incident on a relativistically moving uniaxial plasma slab, *IEEE Trans. Antennas Propagat.*, AP-21, 63–70. © 1973 IEEE. With permission.)

where $k_x = k_0 \sin\theta = k_0 S$, $k_z = k_0 \cos\theta = k_0 C$, $k_0 = \omega/c$, and $c = 1/(\mu_0\epsilon_0)^{1/2}$ = velocity of light in free space.

This wave is transformed to the Σ' system that is stationary with respect to the moving plasma slab bounded by the planes $z' = 0$ and $z' = d'$. The infinitely strong magnetostatic field is assumed to be normal to the slab, so that the plasma medium may be modeled as a uniaxially anisotropic dielectric medium whose relative dielectric constant is a tensor:

$$\bar{\bar{\epsilon}}'_r = \hat{x}\hat{x} + \hat{y}\hat{y} + e'_z\hat{z}\hat{z}, \quad \epsilon'_z = 1 - \left(\frac{\omega_p}{\omega'}\right)^2, \tag{14C.2}$$

where ω_p is the plasma frequency and ω' the wave frequency in the Σ' system.

In the moving system Σ', the exponential variation of the field components takes the form

$$\Psi'_I = \exp\left\{j\left[\omega' t' - \left(k'_x x' + k'_z z'\right)\right]\right\}, \tag{14C.3}$$

where $k'_x = k'_0 S'$, $k'_z = k'_0 C$, and $k'_0 = \omega'/c$.

The reflected and transmitted waves, respectively, must then have the following exponential variation:

$$\Psi'_R = \exp\left\{j\left[\omega' t' - \left(k'_x x' - k'_z z'\right)\right]\right\}, \tag{14C.4}$$

$$\Psi'_T = \exp\left\{j\left[\omega' t' - \left(k'_x x' + k'_z z'\right)\right]\right\}. \tag{14C.5}$$

In the stationary system Σ, the corresponding forms for the reflected and transmitted waves are

$$\Psi_R = \exp\left\{j\left[\omega_r t - \left(k_x^r x - k_z^r z\right)\right]\right\}, \tag{14C.6}$$

$$\Psi_T = \exp\left\{j\left[\omega_t t - \left(k_x^t x - k_z^t z\right)\right]\right\}. \tag{14C.7}$$

Using Maxwell's equations in free space, one can obtain all other field components:

$$\eta_0 H_y'^{\mathrm{I,T}} = \frac{1}{C'} E_x'^{\mathrm{I,T}}, \quad \eta_0 H_y'^{\mathrm{R}} = -\frac{1}{C'} E_x'^{\mathrm{R}} \tag{14C.8}$$

in the Σ′ system, and

$$\eta_0 H_y^{\mathrm{I,T}} = \left(\frac{k_0^{\mathrm{I,T}}}{k_z^{\mathrm{I,T}}}\right) E_x^{\mathrm{I,T}}, \quad \eta_0 H_y^{\mathrm{R}} = -\left(\frac{k_0^{\mathrm{R}}}{k_z^{\mathrm{R}}}\right) E_x^{\mathrm{R}}, \tag{14C.9a}$$

$$k_0^{\mathrm{I,T},R} = \frac{\omega^{\mathrm{I,T,R}}}{c}, \tag{14C.9a}$$

in the Σ system.

Making use of the principle of phase invariance of the plane waves, the covariance of Maxwell's equations, and the Lorentz transformations, one obtains the following relations between the primed and unprimed quantities [4,11,19] for the two cases of slab motion.

Case 1: $\bar{\upsilon} = \upsilon_0 \hat{x}$.

$$\omega' = p_x \omega, \quad k_x' = \gamma k_0 (S - \beta), \quad k_z' = k_0 C, \tag{14C.10a}$$

$$\omega_r = \omega_t = \omega, \quad k_x^r = k_x^t = k_0 S, \quad k_z^r = k_z^t = k_0 C, \tag{14C.10b}$$

$$C' = \frac{C}{p_x}, \quad S' = \frac{S - \beta}{1 - S\beta}, \quad E_x'^{\mathrm{I,T,R}} = E_x^{\mathrm{I,T,R}}, \tag{14C.10c}$$

where

$$\beta = \frac{\upsilon_0}{c},$$
$$\gamma = \frac{1}{\left(1 - \beta^2\right)^{1/2}} \quad \text{and} \quad p_x = g(1 - S\beta). \tag{14C.10d}$$

Case 2: $\bar{\upsilon} = \upsilon_0 \hat{z}$.

$$\omega' = p_z \omega, \quad k_x' = k_0 S, \quad k_z' = \gamma k_0 (C - \beta), \tag{14C.11a}$$

$$\omega_r = \gamma^2\omega\left(1+\beta^2-2\beta C\right) \quad k_z^r = k_0 S, \quad k_z^r = \gamma^2 k_0\left[\left(1+\beta^2\right)C-2\beta\right], \tag{14C.11b}$$

$$\omega_t = \omega, \quad k_x^t = k_0 S, \quad k_z^t = k_0 C, \tag{14C.11c}$$

$$C = \frac{C-\beta}{1-C\beta}, \quad S = \frac{S}{p_z}, \tag{14C.11d}$$

$$E_x'^{\mathrm{I,T}} = \gamma\left(1-\frac{\beta}{C}\right)E_x^{\mathrm{I,T}}, \quad E_x'^{\mathrm{R}} = \gamma\left(1-\frac{\beta\beta'}{C}\right)E_x^{\mathrm{R}}, \tag{14C.11e}$$

where

$$pz = \gamma\left(1-C\beta\right) \quad \text{and} \quad \beta' = \frac{1+\beta^2-2\beta C}{1+\beta^2-2\beta/C}. \tag{14C.11f}$$

It can easily be seen that

$$\cos\theta_r = \frac{C}{\beta'}, \tag{14C.11g}$$

where θ_r is the angle of reflection as seen from the Σ system.

14C.3 Dispersion Relation and Other Relations

Let the waves in the plasma have the exponential variation:

$$\Psi_{\mathrm{p}}' = \exp\left\{\mathrm{j}\left[\omega' t' - \left(k_x' x' + k_0' q' z'\right)\right]\right\}. \tag{14C.12}$$

Using Maxwell's equations in the plasma, one obtains

$$\eta_0 H_y'^{\mathrm{P}} = \frac{1}{q'}E_x'^{\mathrm{P}}, \quad q' = \pm\left(1-\frac{S'^2}{\epsilon_z'}\right)^{1/2}. \tag{14C.13}$$

On substituting for S' and ϵ_z from Equations 14C.2, 14C.10, and 14C.11, the dispersion coefficient q' relevant to the two cases of slab motion may be written as

$$q_x' = \pm\left(\frac{C^2\Omega^2-1}{p_x^2\Omega^2-1}\right)^{1/2}$$

and

$$q_z' = \pm\left(\frac{\left(p_z^2 - S^2\right)\Omega^2 - 1}{p_z^2\Omega^2 - 1}\right)^{1/2}, \tag{14C.14}$$

$\Omega(= \omega/\omega_p)$ being the normalized incident wave frequency.

14C.4 Reflection and Transmission Coefficients

Making use of the usual continuity conditions of tangential electromagnetic fields at the boundaries $z' = 0$ and $z' = d'$ in the rest frame Σ' of the plasma and then transforming the field strengths in the observer's frame Σ according to Equations 14C.10c and 14C.11e, one can derive the expressions for the power reflection and transmission coefficients (ρ and τ, respectively) as observed in the Σ frame.

Case 1: $\bar{\upsilon} = \upsilon_0\hat{x}$.

$$\rho_x = \left|\frac{E_x^R}{E_x^I}\right|^2 = \frac{1}{1+\left[2q_x'C'/\left(q_x'^2 - C'^2\right)\operatorname{cosec} k_0'q_x'd'\right]^2}, \tag{14C.15a}$$

$$\tau_x = \left|\frac{E_x^T}{E_x^I}\right|^2 = \frac{1}{1+\left[\left(\left(q_x'^2 - C'^2\right)/2q_x'C'\right)\sin k_0'q_x'd'\right]^2}. \tag{14C.15b}$$

Case 2: $\bar{\upsilon} = \upsilon_0\hat{z}$.

$$\rho_z = \beta'\left|\frac{E_x^R}{E_x^I}\right|^2 = \beta'\left|\frac{C-\beta}{C+\beta\beta'}\right|^2 \frac{1}{1+\left[\left(2q_z'C'/\left(q_z'^2 - C'^2\right)\right)\operatorname{cosec} k_0'q_z'd'\right]^2}, \tag{14C.16a}$$

$$\tau_z = \left|\frac{E_x^T}{E_x^I}\right|^2 = \frac{1}{1+\left[\left(\left(q_z'^2 - C'^2\right)/2q_z'C'\right)\sin k_0'q_z'd'\right]^2}. \tag{14C.16b}$$

14C.5 Numerical Results and Discussion

Substituting for q', the reflection and transmission coefficients take the form

$$\rho = \frac{\alpha}{1+B\operatorname{cosec}^2 A}, \quad \tau = \frac{1}{1+(1/B)\sin^2 A} \tag{14C.17a}$$

in the real range of q' and

$$\rho = \frac{\alpha}{1 - B\operatorname{cosech}^2|A|}, \quad \tau = \frac{1}{1 - (1/B)\sinh^2|A|} \tag{14C.17b}$$

in the imaginary range of q', where α, A, and B for the two cases of slab motion are given by the following.

Case 1:

$$\alpha_x = 1 \tag{14C.18a}$$

$$A_x = 2\pi\Omega p_x \left(\frac{C^2\Omega^2 - 1}{p_x^2\Omega^2 - 1} \right)^{1/2} \cdot d_p', \tag{14C.18b}$$

$$B_x = \frac{4p_x^2C^2\left(p_x^2\Omega^2 - 1\right)\left(C^2\Omega^2 - 1\right)}{\left(p_x^2 - C^2\right)^2}. \tag{14C.18c}$$

Case 2:

$$\alpha_z = \frac{\left(1 + \beta^2 - 2\beta C\right)\left(1 + \beta^2 - 2\beta / C\right)}{\left(1 - \beta^2\right)^2}, \tag{14C.19a}$$

$$A_z = 2\pi\Omega p_z \left(\frac{\left(p_z^2 - S^2\right)\Omega^2 - 1}{p_z^2\Omega^2 - 1} \right)^{1/2} \cdot d_p', \tag{14C.19b}$$

$$B_z = \frac{4p_z^2\left(p_z^2 - S^2\right)\left(p_z^2\Omega^2 - 1\right)\left[\left(p_z^2 - S^2\right)\Omega^2 - 1\right]}{S^4}, \tag{14C.19c}$$

where $d_p' = d' / \lambda_p$, λ_p being the free-space wavelength corresponding to the plasma frequency.

It is thus seen that ρ is an oscillatory function in the real range of q', whereas for imaginary q' it is nonoscillatory. The reflection coefficient is now discussed as a function of the normalized wave frequency (Ω) and the slab velocity (β).

14C.5.1 Variation with Ω

Numerical results were obtained for $C = 0.5$, $d_p' = 1$, and $p = 1.1$. Figures 14C.2 and 14C.3 show the plot of q' and ρ versus Ω for the two cases of slab motion. For obtaining the maxima and minima in the oscillatory ranges [20], it should be noted that ρ_{min} is zero and this occurs for Ω satisfying the equation

$$A = k_0'q'd' = \frac{n\pi}{2} \quad \text{for } n \text{ even.} \tag{14C.20a}$$

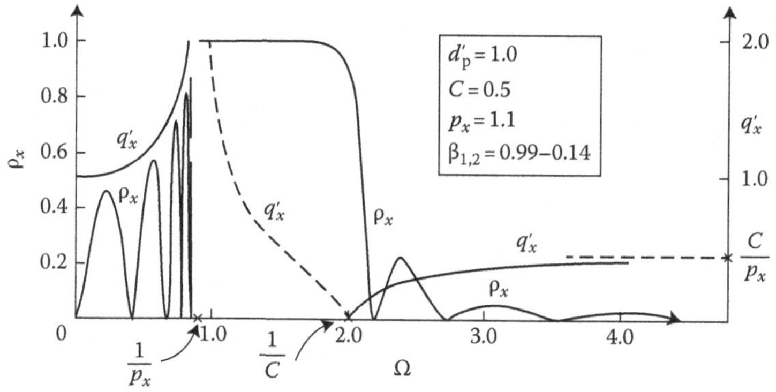

FIGURE 14C.2
Characteristic root (q'_x) and power reflection coefficient (ρ_x) as functions of normalized wave frequency for slab motion parallel to the interface. Dashed portion of the curve indicates imaginary values of q'_x. (Reprinted from Kalluri, D. and Shrivastava, R.K., Reflection and transmission of electromagnetic waves obliquely incident on a relativistically moving uniaxial plasma slab, *IEEE Trans. Antennas Propagat.*, AP-21, 63–70. © 1973 IEEE. With permission.)

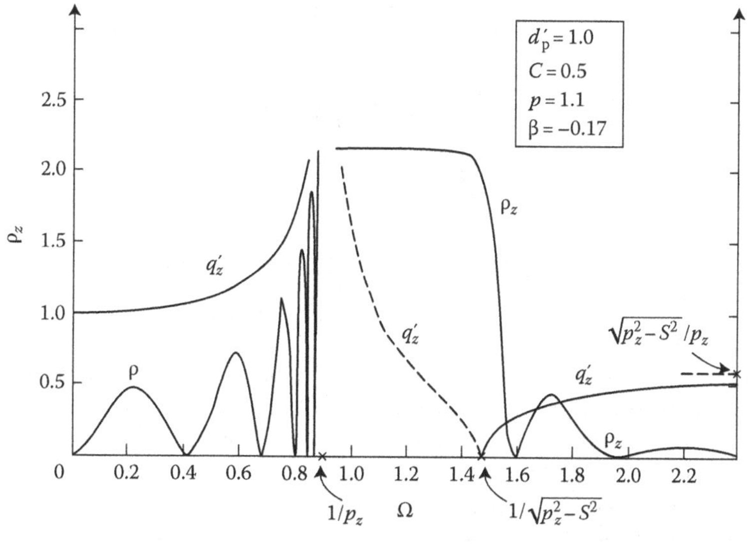

FIGURE 14C.3
Characteristic root (q'_z) and power reflection coefficient (ρ_z) as functions of normalized wave frequency for slab motion normal to the interface. Dashed portion of the curve indicates imaginary values of q'_z. (Reprinted from Kalluri, D. and Shrivastava, R.K., Reflection and transmission of electromagnetic waves obliquely incident on a relativistically moving uniaxial plasma slab, *IEEE Trans. Antennas Propagat.*, AP-21, 63–70. © 1973 IEEE. With permission.)

This is equivalent to the condition $2q'd' = (n/2)\lambda'$ (n even), which means that the optical path difference between two successive reflected waves (there being an infinite series of internal reflections inside the plasma layer) is an integral multiple of one wavelength, so that these waves have the same phase and reinforce each other and in this case the first reflected wave from the lower boundary exactly annuls the resultant of the remaining reflected waves, thus giving zero reflection [21].

Thus, one obtains, after some algebra,

$$\left(\Omega_{\min,n}\right)_x = \left[\frac{\left(n^2+16d_p'^2\right)\pm\left[\left(n^2+16d_p'^2\right)^2-\left(8Cnd_p'/p_x\right)^2\right]^{1/2}}{32C^2d_p'^2}\right]^{1/2} \tag{14C.20b}$$

and

$$\left(\Omega_{\min,n}\right)_z = \left[\frac{\left(n^2+16d_p'^2\right)\pm\left[\left(n^2+16d_p'^2\right)^2-\left(8\left(\left(p_z^2-S^2\right)/p_z^2\right)^{1/2}nd_p'\right)^2\right]^{1/2}}{32\left(p_z^2-S^2\right)d_p'^2}\right]^{1/2}. \tag{14C.20c}$$

The minus sign in the preceding equations gives the minima in the first oscillatory range, while the plus sign in the other range. There are thus an infinite number of oscillations in either range.

The maxima in ρ occur for Ω satisfying the transcendental equation

$$\tan A = f_\Omega A, \tag{14C.21a}$$

where

$$\left(f_\Omega\right)_x = \frac{2C^2\Omega^2 - p_x^2C^2\Omega^4 - 1}{\Omega^2\left[p_x^2 - 2p_x^2C^2\Omega^2 + C^2\right]} \tag{14C.21b}$$

and

$$\left(f_\Omega\right)_z = \frac{\Omega^4p_z^2\left(p_z^2-S^2\right)-2\Omega^2\left(p_z^2-S^2\right)+1}{2\Omega^4\left(p_z^2-S^2\right)-\Omega^2\left(2p_z^2-S^2\right)}. \tag{14C.21c}$$

The maximum point between the two minima of a loop was first determined using a subroutine to solve the preceding transcendental equation. Having located this maximum, each interval between the minimum and ρ maximum points was divided into a number of parts and calculated at each point. This procedure was repeated to get the second loop and so on.

It is seen from the curves in Figures 14C.2 and 14C.3 that as q' tends to infinity in its real range, the oscillations in ρ become more and more rapid and of increasing amplitude. In the other range (where q' reaches its asymptotic limit C'), the oscillations slow down progressively as Ω is increased and ultimately die down. The rapidity of oscillations is due to steep rise in q' with Ω in this range. The amplitude of oscillations is governed by the magnitude of q'. This is easily seen by noting that if we ignore the oscillations and draw a curve representing average value, this will give the ρ versus Ω curve for the semiinfinite case. For this case, reflection will be strong when the free-space fields are badly matched [22] to the plasma fields ($q' \neq C'$) and zero when the free-space fields are perfectly matched to the plasma fields ($q' = C'$).

As is known, $\rho + \tau > 1$ and the reflected wave emerges out amplified when the slab moves toward the incident wave. This is because the stored energy decreases and mechanical power has to be externally supplied to the medium against the radiation pressure exerted by the field on the medium. When the medium has a tangential velocity, the force on its surface can do no work and the time-averaged stored energy remains constant, thus giving $\rho + \tau = 1$ for parallel motion [11].

14C.5.2 Variation with β

It will be seen from the equations that the reflection and transmission coefficients are complicated functions of the slab velocity. To study these coefficients directly against β would be too involved. So, the following technique is evolved.

1. q' is first studied against p and ρ/α plotted against p, since the maxima and minima of this curve are easily obtained as before.
2. p is studied against β. The parameter p $(= \omega'/\omega)$ is significant in that it transforms the incident wave frequency to the rest system of the plasma. It is a quadratic in β and the two values of β for any p are given by

$$(\beta_{1,2})_x = \frac{S}{p_x^2 + S^2} \pm \left[\left(\frac{S}{p_x^2 + S^2}\right)^2 + \frac{p_x^2 - 1}{p_x^2 + S^2}\right]^{1/2} \tag{14C.22a}$$

and

$$(\beta_{1,2})_z = \frac{C}{p_z^2 + C^2} \pm \left[\left(\frac{C}{p_z^2 + C^2}\right)^2 + \frac{p_z^2 - 1}{p_z^2 + C^2}\right]^{1/2}. \tag{14C.22b}$$

The plot of p versus β for $C = 0.5$ is given in Figure 14C.4. It shows the variation of ω' with β.

3. ρ/α is now plotted against β by combining the preceding two steps. With reference to a different horizontal axis $(\rho/\alpha = 1)$, this represents τ.

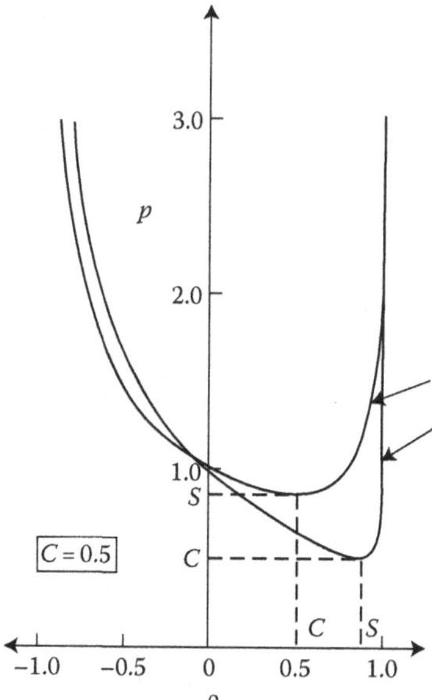

FIGURE 14C.4
Parameter p as a function of the normalized slab velocity $(\beta = v_0/c)$. p being ω'/ω, the curve shows variation of ω' with β. The subscripts x and z on p indicate the direction of slab motion. (Reprinted from Kalluri, D. and Shrivastava, R.K., Reflection and transmission of electromagnetic waves obliquely incident on a relativistically moving uniaxial plasma slab, *IEEE Trans. Antennas Propagat.*, AP-21, 63–70. © 1973 IEEE. With permission.)

4. α versus β is studied according to Equation 14C.19a and plotted for $C = 0.5$ along with the ρ/α versus β curve.
5. Finally, ρ is obtained as a function of β using steps 3 and 4.

It should be noted that since $\alpha = 1$ for parallel slab motion, the development of the ρ versus β curve in this case involves only three steps, q' versus p gives interesting results. q' is seen to vary in a different manner with p for various ranges of Ω given by $\Omega < 1/C$, $\Omega = 1/C$, and $\Omega > 1/C$ for slab motion parallel to the interface and $\Omega \leq 1/S$ and $\Omega > 1/S$ for normal motion.

14C.5.3 Slab Motion Parallel to Interface ($\bar{\upsilon} = \upsilon_0\hat{x}$)

Numerical results were obtained for $C = 0.5$, $d_p' = 1$ in the three ranges of values of aforementioned Ω. For $\Omega = 1/C$, $q' = 0$ and it can be shown that ρ_x for this case is given by

$$\rho_x = \frac{1}{1 + 1/\pi^2 d_p'^2} = 0.908.$$

For the other two cases, namely $\Omega < 1/C$ and $\Omega > 1/C$, curves were obtained with $\Omega = 1.2$ and 2.5, respectively ($1/C = 2$).

The plot of q_x' and ρ_x against p_x has been shown in Figure 14C.5 (for $\Omega < 1/C$). In the oscillatory range, $(\rho_{\min})_x$ is zero and occurs for p_x given by

$$A_x = 2\pi\Omega p_x \left(\frac{C^2\Omega^2 - 1}{p_x^2\Omega^2 - 1}\right)^{1/2} d_p' = \frac{n\pi}{2} \quad \text{for } n \text{ even.}$$

Thus, one obtains

$$\left(p_{\min,n}\right)_x = \frac{n}{\Omega} \frac{1}{\left[n^2 - 16 d_p'^2 \left(\Omega^2 C^2 - 1\right)\right]^{1/2}}. \tag{14C.23}$$

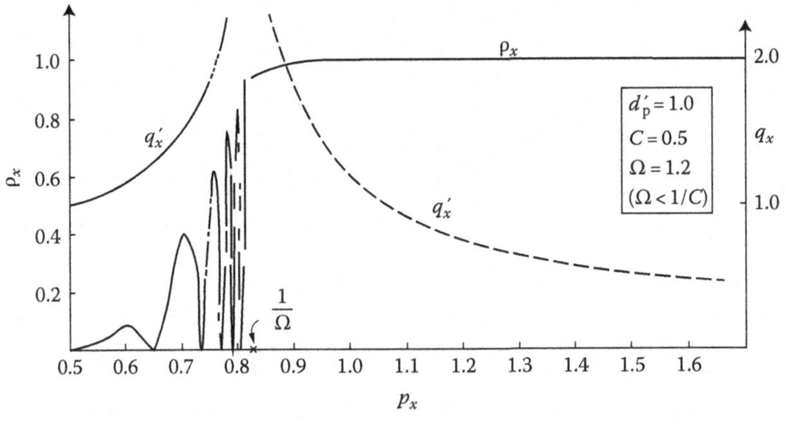

FIGURE 14C.5
q_x' and ρ_x versus p_x, for $\Omega > 1/C$, slab motion being parallel to the interface. Dashed portion of the curve indicates imaginary values of q_x'. (Reprinted from Kalluri, D. and Shrivastava, R.K., Reflection and transmission of electromagnetic waves obliquely incident on a relativistically moving uniaxial plasma slab, *IEEE Trans. Antennas Propagat.*, AP-21, 63–70. © 1973 IEEE. With permission.)

It should also be noted that there is an additional minimum point at $p_x = C$. This is readily seen by setting $B_x = \alpha$ in Equation 14C.18c. For this condition, $q' = C' = 1$ representing a perfect match for the free-space and plasma fields and the incident wave is wholly transmitted.

The maxima in ρ_x occur for p_x satisfying the transcendental equation $\tan A_x = (f_p)_x A_x$, where

$$(f_{\mathrm{P}})_x = \frac{p_x^2 - C^2}{p_x^2\left(\Omega^2C^2 - 1\right) + C^2\left(p_x^2\Omega^2 - 1\right)}. \tag{14C.24}$$

Case 1: $\Omega < 1/C$.

It should be noted in Equation 14C.23 that $n = 0$ is not a permissible even integer since this will correspond to a zero value of p_x. The lowest permissible even integer may be determined from the minimum value of p_x that equals C. This value of p_x corresponds to $n = n' = 4d_p\Omega C$. Since this value may not be an even integer, the lowest permissible even integer is the one that is greater than n'. The highest permissible value for n is infinity when $p_x = 1/\Omega$. Beyond the oscillatory region, ρ_x is monotonic.

The ρ_x versus p_x curve is finally developed into the ρ_x versus β curve (Figure 14C.6) with the help of p_x versus β curve that has been earlier explained. As there are two values of β for any p_x, one loop in the ρ_x versus p_x curve leads to two loops in the ρ_x versus β curve, the two loops being one on each side of $\beta = S$ (corresponding to the starting point $p_x = C$ in the ρ_x versus p_x curve).

It is seen from the curve that ρ_x is oscillatory in the range of β from 0.23 to 0.98 and for other values of β, ρ_x is monotonic. The rapidity of the oscillations (of increasing amplitude) increases as one approaches the end of the oscillatory region on both sides of $\beta = S$. The oscillations are even more rapid to the right of $\beta = S$ than to the left. This is because the wavelength λ' decreases much faster in the right range than the left range as can be seen from an inspection of the p_x versus β curve.

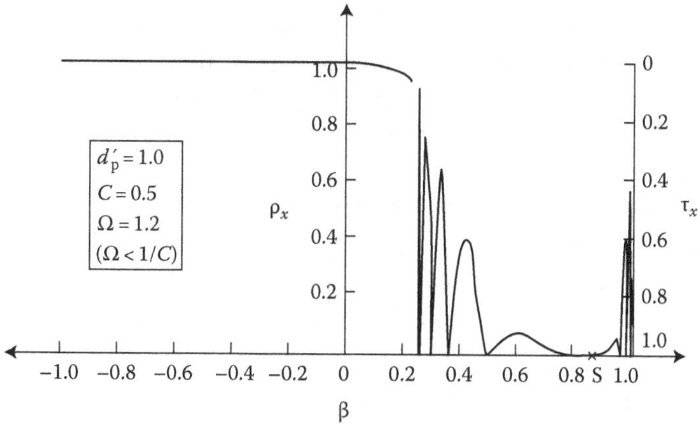

FIGURE 14C.6
ρ_x and τ_x versus β, for $\Omega > 1/C$, slab motion being parallel to the interface. (Reprinted from Kalluri, D. and Shrivastava, R.K., Reflection and transmission of electromagnetic waves obliquely incident on a relativistically moving uniaxial plasma slab, *IEEE Trans. Antennas Propagat.*, AP-21, 63–70. © 1973 IEEE. With permission.)

Case 2: $\Omega > 1/C$

Referring to Equation 14C.23 again, it should be noted that n cannot assume all even integer values. The limits on n are set from the requirement that $p_{\min}$ has to lie between C and ∞. $p_x = C$ corresponds to $n = n' = 4d_p'\Omega C$ and $p_x = \infty$ corresponds to

$$n = n'' = 4d_p'\left(\Omega^2C^2 - 1\right)^{1/2}. \quad (14C.25)$$

Noting that $n'' < n'$, the lowest permissible value for n is an even integer which is greater than or equal to n''. The highest permissible value of n is again an even integer that is less than or equal to n'.

For the chosen data of $d_p' = 1$, $C = 0.5$, and $\Omega = 2.5$, range of n as given by Equation 14C.25 is 3–5 and as only even n is allowed, the value to be used is 4. This corresponds to $p_x = 0.6$. Thus, we have only two minima in ρ_x, one at $p_x = 0.5$ and the other at $p_x = 0.6$, giving one loop in the region.

For any other data, the number of the oscillatory loops will change. For example, if $d_p' = 10$, $C = 0.5$, and $\Omega = 2.1$, we have $n' = 42$ and $n'' = 12.8$. So n'' has to be taken as 14 (the next higher even number), giving the range of n as 14–42. This leads to 14 oscillatory loops and as $n = 14$ corresponds to $p_x = 1.18$, it will be seen that all these loops will be in the range of p_x from 0.5 to 1.18, which means intense crowding of the oscillations in the region.

As p_x tends to infinity, A_x tends to $2\pi d_p'\left(\Omega^2C^2 - 1\right)^{1/2}$ and B_x tends to $4\Omega^2C^2(\Omega^2C^2 - 1)$ (refer Equation 14C.18). Thus, as p_x increases beyond the oscillatory region, ρ_x rises asymptotically to a value given by

$$\left(\rho_x\right)_{p_x\to\infty} = \frac{1}{1 + 4\Omega^2C^2\left(\Omega^2C^2 - 1\right)\operatorname{cosec}^2 2\pi d_p'\left(\Omega^2C^2 - 1\right)^{1/2}}$$

$$= 0.22 \quad \text{for chosen data of } d_p' = 1,\ C = 0.5,$$

and

$$\Omega = 2.5. \quad (14C.26)$$

Only the final plot of ρ_x versus β is shown (Figure 14C.7), from which it is seen that after some oscillation, the reflection coefficient increases to its final value on either side of $\beta = S$, the rise to the right of $\beta = S$ being extremely steep when compared to the slow rise in the left region. A physical explanation has already been given.

14C.5.4 Slab Motion Normal to Interface ($\bar{\upsilon} = \upsilon_0\hat{z}$)

Curves were obtained for $\Omega = 0.5$ ($<1/S$) and $\Omega = 2.0$ ($>1/S$) taking $d_p' = 1$ and $C = 0.5$. Figures 14C.8 and 14C.11 show the plot of q_z and ρ_z/α_z versus p_z for the two ranges of Ω. The maxima and minima in ρ_z/α_z are obtained as before and occur for p_z given by

$$\left(p_{\min,n}\right)_z = \left[\frac{\left[n^2 + 16d_p'^2\left(1 + \Omega^2S^2\right)\right] \pm \left\{\left[n^2 + 16d_p'^2\left(1 + \Omega^2S^2\right)\right]^2 - \left(8nd_p'\right)^2\right\}^{1/2}}{32\Omega^2 d_p'^2}\right]^{1/2} \quad \text{for } n \text{ even}$$

(14C.27)

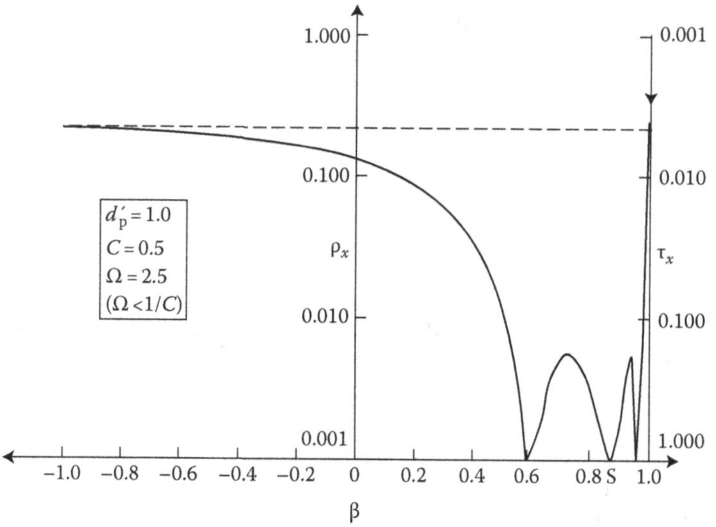

FIGURE 14C.7
ρ_x and τ_x versus β, for $\Omega > 1/C$, slab motion being parallel to the interface. (Reprinted from Kalluri, D. and Shrivastava, R.K., Reflection and transmission of electromagnetic waves obliquely incident on a relativistically moving uniaxial plasma slab, *IEEE Trans. Antennas Propagat.*, AP-21, 63–70. © 1973 IEEE. With permission.)

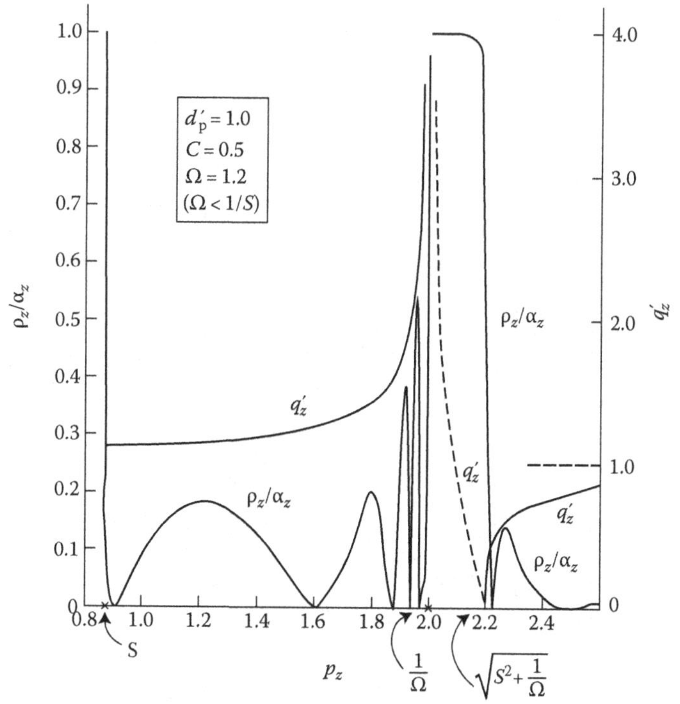

FIGURE 14C.8
q'_z and ρ_z/α_z versus p_z, for $\Omega < 1/S$, normal slab motion. Dashed portion of the curve indicates imaginary values of q'_z. (Reprinted from Kalluri, D. and Shrivastava, R.K., Reflection and transmission of electromagnetic waves obliquely incident on a relativistically moving uniaxial plasma slab, *IEEE Trans. Antennas Propagat.*, AP-21, 63–70. © 1973 IEEE. With permission.)

(the minus sign before the square-root sign is not permissible for $\Omega > 1/S$ as in this case q_z' is real in one range only, that is, $[S^2 + (1/\Omega^2)]^{1/2} < p_z < \infty$) and $\tan A_z = (f_p)_z A_z$, where

$$(f_p)_z = \frac{1/p_z^2 + \Omega^2/\left(\left(p_z^2 - S^2\right)\Omega^2 - 1\right) - \Omega^2/\left(p_z^2\Omega^2 - 1\right)}{1/p_z^2 + \Omega^2/\left(\left(p_z^2 - S^2\right)\Omega^2 - 1\right) + 1/\left(p_z^2\Omega^2 - 1\right)}. \tag{14C.28}$$

Case 1: $\Omega < 1/S$.

The minimum value of p_z is S and this corresponds to $n = n' = 4d_p'\Omega S/(1-\Omega^2 S^2)^{1/2}$. Since this value may not be an even integer, the lowest permissible even integer is the one that is greater than n'. The highest permissible value of n is, of course, infinity corresponding to $p_z = 1/\Omega$.

The development of the ρ_z versus β curve is illustrated in Figures 14C.8 through 14C.10 following the procedure described before. The significant point to be noted here is the appearance of an additional factor α_z in the expression for reflection coefficient. This is actually the reflection coefficient for a perfect

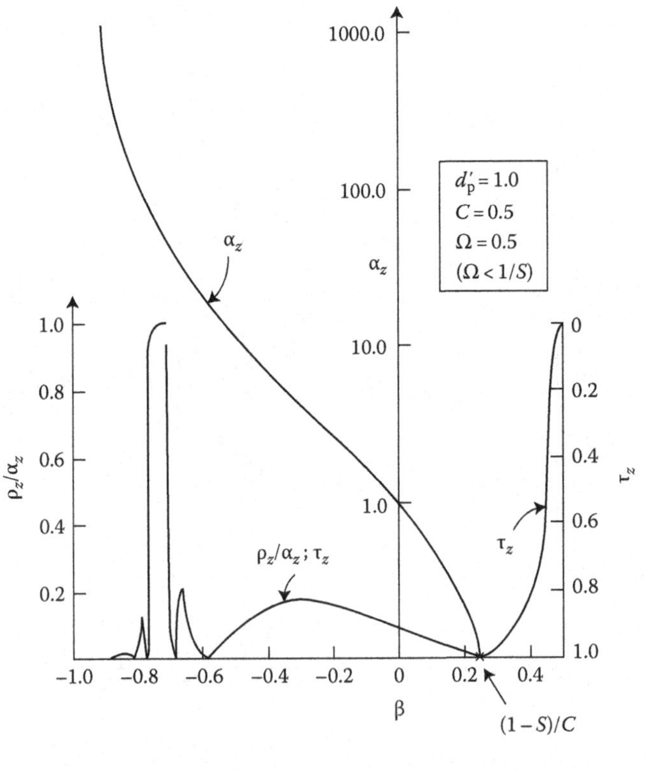

FIGURE 14C.9
ρ_z/α_z and τ_z as functions of β, for $\Omega < 1/S$, normal slab motion. The curve α_z versus β is also shown. (Reprinted from Kalluri, D. and Shrivastava, R.K., Reflection and transmission of electromagnetic waves obliquely incident on a relativistically moving uniaxial plasma slab, *IEEE Trans. Antennas Propagat.*, AP-21, 63–70. © 1973 IEEE. With permission.)

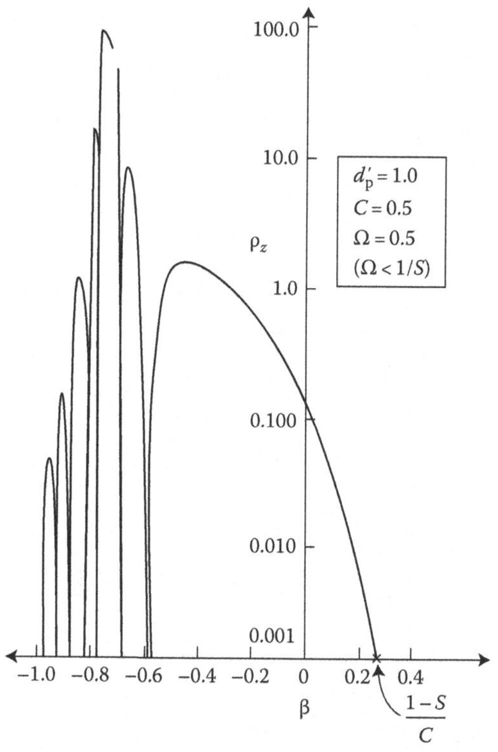

FIGURE 14C.10
ρ_z versus β, for $\Omega < 1/S$, normal slab motion. (Reprinted from Kalluri, D. and Shrivastava, R.K., Reflection and transmission of electromagnetic waves obliquely incident on a relativistically moving uniaxial plasma slab, *IEEE Trans. Antennas Propagat.*, AP-21, 63–70. © 1973 IEEE. With permission.)

moving minor [11] and the multiplier $1/(1 + B_z \operatorname{cosec}^2 A_z)$ is because of the uniaxial anisotropy of the slab.

It is seen (Figure 14C.9) that starting from infinity α_z reduces to unity as β varies from −1 to 0. Thus the reflected wave emerges out amplified in this range (physical explanation given before). In the range $0 < \beta < (1 - S)/C$, α_z reduces from 1 to 0. So, the reflection coefficient is less than unity in the above range, finally becoming zero at $\beta = (1 - S)/C$. In this range, the energy balance may be shown as

$$\begin{aligned}\text{Incident power} = &\ \text{Reflected power} + \text{transmitted power}\\ &+ \text{mechanical power supplied by the field}\\ &+ \text{stored energy change}\end{aligned}$$

Thus, when the medium recedes from the incident wave, field supplies work to the medium. At $\beta = (1 - S)/C$, the angle of reflection $\theta_R = 90°$, thus explaining the zero-reflected power (Figure 14C.11).

Oscillations in the reflection coefficient have been explained before.

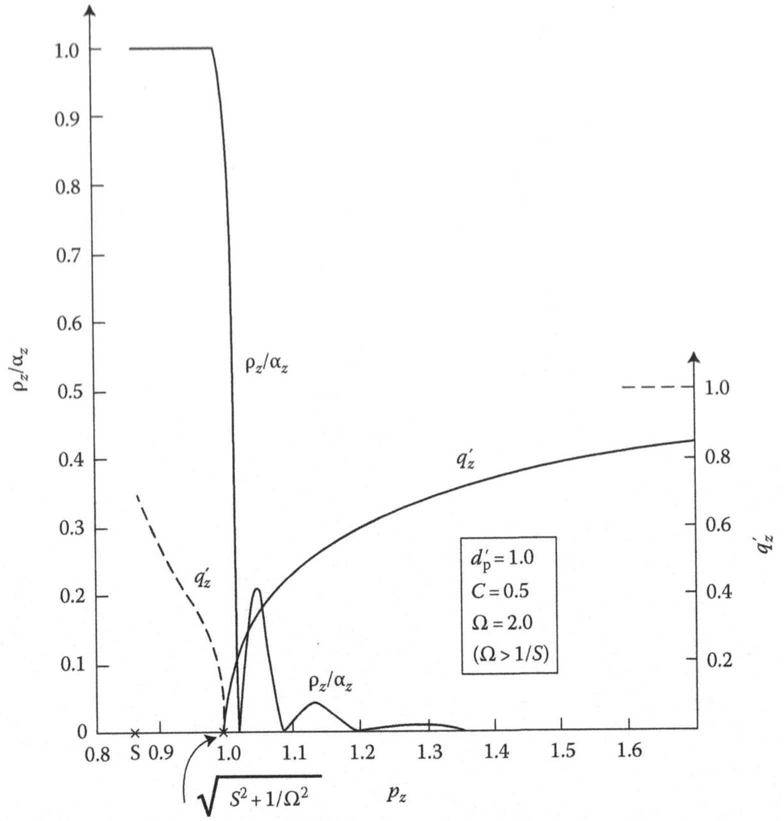

FIGURE 14C.11
q_z' and ρ_z/α_z versus p_z, for $\Omega > 1/S$, normal slab motion. (Reprinted from Kalluri, D. and Shrivastava, R.K., Reflection and transmission of electromagnetic waves obliquely incident on a relativistically moving uniaxial plasma slab, *IEEE Trans. Antennas Propagat.*, AP-21, 63–70. © 1973 IEEE. With permission.)

Case 2: $\Omega \geq 1/S$.

ρ_z versus β is shown in Figure 14C.12 for $\Omega = 2.0$ ($>1/S$) taking $d_p' = 1$ and $C = 0.5$. This may be compared with the previous curve for $\Omega < 1/S$ (Figure 14C.10) noting that the difference is because of just one oscillatory range for ρ_z in the former in contrast to the two oscillatory ranges in the latter.

The curve for $\Omega = 1/S$ is similar to the preceding and hence not given.

At β greater than or equal to C, the incident wave from below (Figure 14C.1) has no interaction with the slab since the normal component of the phase velocity of the wave is less than or equal to that of the slab. However, the results given here are applicable even when $\beta > C$ if the problem is understood to be that the slab impinges on the portion of an existing incident wave above the slab. This aspect of the problem will be reported later.

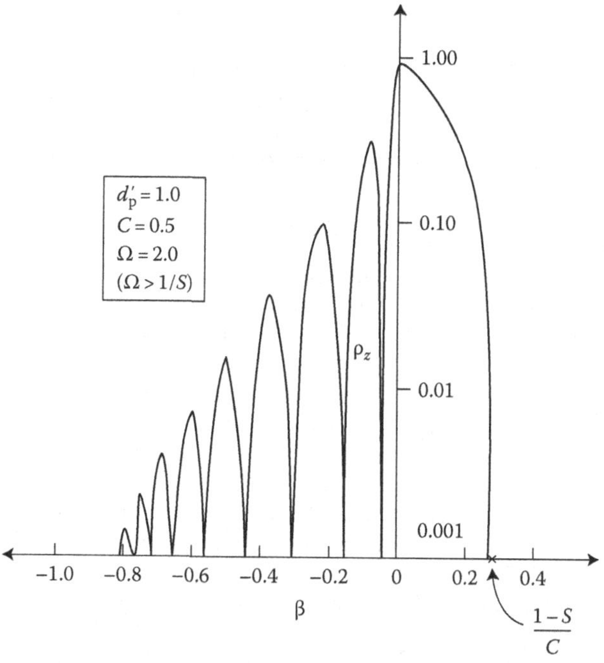

FIGURE 14C.12
ρ_z versus β, for $\Omega > 1/S$, normal slab motion. (Reprinted from Kalluri, D. and Shrivastava, R.K., Reflection and transmission of electromagnetic waves obliquely incident on a relativistically moving uniaxial plasma slab, *IEEE Trans. Antennas Propagat.*, AP-21, 63–70. © 1973 IEEE. With permission [23].)

References

1. Landecker, K., Possibility of frequency multiplication and wave amplification by means of some relativistic effect, *Phys. Rev.*, 86, 449–455, 1952.
2. Tai, C. T., Two scattering problems involving moving media, Antenna Lab., Ohio State Univ., Columbus, OH, Rep. 1691–1697, 1964.
3. Yeh, C., Reflection and transmission of electromagnetic waves by a moving dielectric medium, *J. Appl. Phys.*, 36, 3513–3517, 1965.
4. Yeh, C. and Casey, K. F., Reflection and transmission of electromagnetic waves by a moving dielectric slab, *Phys. Rev.*, 144, 665–669, 1966.
5. Yeh, C., Brewster angle for a dielectric medium moving at relativistic speed, *J. Appl. Phys.*, 38, 5194–5200, 1967.
6. Yeh, C., Reflection and transmission of electromagnetic waves by a moving dielectric slab II. Parallel polarisation, *Phys. Rev.*, 167, 875–877, 1968.
7. Pyati, V. P., Reflection and transmission of electromagnetic waves by a moving dielectric medium, *J. Appl. Phys.*, 38, 652–655, 1967.
8. Lee, S. W. and Lo, Y. T., Reflection and transmission of electromagnetic waves by a moving uniaxially anisotropic medium, *J. Appl. Phys.*, 38, 870–875, 1967.

9. Tsai, C. S. and Auld, B. A., Wave interaction with moving boundaries, *J. Appl. Phys.*, 38, 2106–2115, 1967.
10. Shiozawa, T., Hazama, K., and Kumagai, N., Reflection and transmission of electromagnetic waves by a dielectric half-space moving perpendicular to the plane of incidence, *J. Appl. Phys.*, 38, 4459–4461, 1967.
11. Daly, P. and Gruenberg, H., Energy relations for plane waves reflected from moving media, *J. Appl. Phys.*, 38, 4486–4489, 1967.
12. Shiozawa, T. and Kumagai, N., Total reflection at the interface between relatively moving media, *Proc. IEEE*, 55, 1243–1244, 1967.
13. Kong, J. A. and Cheng, D. K., Wave behavior at an interface of a semi-infinite moving anisotropic medium, *J. Appl. Phys.*, 39, 2282–2286, 1968.
14. Cheng, D. K. and Kong, J. A., Time-harmonic fields in source-free bi-anisotropic media, *J. Appl. Phys.*, 39, 5792–5796, 1968.
15. Yeh, C., Reflection and transmission of electromagnetic waves by a moving plasma medium, *J. Appl. Phys.*, 37, 3079–3082, July 1966.
16. Yeh, C., Reflection and transmission of electromagnetic waves by moving plasma medium, II. Parallel Polarization, *J. Appl. Phys.*, 38, 2871–2873, 1967.
17. Chawla, B. R. and Unz, H., Reflection and transmission of electromagnetic waves normally incident on a plasma slab moving uniformly along a magnetostatic field, *IEEE Trans. Antennas Propagat.*, AP-17, 771–777, 1969.
18. Kong, J. A. and Cheng, D. K., Reflection and transmission of electromagnetic waves by a moving uniaxially anisotropic slab, *J. Appl. Phys.*, 40, 2206–2212, 1969.
19. Chawla, B. R. and Unz, H., *Electromagnetic Waves in Moving Magneto-Plasmas*, Kansas University Press, Lawrence, KS, 1971.
20. Kalluri, D. and Prasad, R. C., Thin film reflection properties of isotropic and uniaxial plasma slabs, *Appl. Sci. Res.*, 27, 415–424, 1973.
21. Budden, K. G., *Radio Waves in the Ionosphere*, Cambridge University Press, Cambridge, U.K., p. 111, 1961.
22. French, I. P., Cloutier, G. G., and Bachynski, M. P., The absorptivity spectrum of a uniform anisotropic plasma slab, *Can. J. Phys.*, 39, 1273–1290, 1961.
23. Kalluri, D. and Shrivastava, R. K., Reflection and transmission of electromagnetic waves obliquely incident on a relativistically moving uniaxial plasma slab, *IEEE Trans. Antennas Propagat.*, AP-21, 63–70, 1973.

*Appendix 14D: Brewster Angle for a Plasma Medium Moving at a Relativistic Speed**

Dikshitulu K. Kalluri and R.K. Shrivastava

The Brewster angle for parallel-polarized electromagnetic waves incident on a relativistically moving isotropic plasma half-space is investigated. There exist two Brewster angles for certain ranges of the medium velocity irrespective of whether the medium moves parallel or normal to the interface. In the case of parallel motion, it is found that there are no Brewster angles for wave frequencies less than the plasma frequency.

Yeh [1] investigated the Brewster angle for a dielectric medium moving at a relativistic speed. Lee and Lo [2] studied, among other things, the conditions for total transmission for electromagnetic waves incident at an interface between a dielectric medium and an uniaxial plasma moving parallel to the interface. The plane of incidence, however, was confined to be parallel to the direction of motion. Later, Kong and Cheng [3] and Pyati [4] removed this restriction and considered an arbitrary orientation of the plane of incidence. In a recent paper, Chuang and Ko [5] obtained the generalized Brewster law for a dielectric medium moving parallel to its surface.

This communication reports the Brewster-angle phenomenon for a moving *plasma* half-space. The results are expected to be useful in maximizing the strength of a transmitted signal through a relativistically moving bounded plasma. The plasma medium being dispersive in nature, the Brewster angle will, in general, be a function of the incident wave frequency as well as the medium velocity. The problem in its full generality too would be involved and as such the present treatment is restricted to the case in which the plane of incidence is parallel to the direction of medium motion. The motions of the medium both parallel and perpendicular to the interface are considered.

Consider a parallel-polarized plane electromagnetic wave obliquely incident (*xz*-plane of incidence) on an isotropic plasma half-space (μ_0, ϵ') moving relativistically through the free space (μ_0, ϵ_0). The power reflection coefficient (ρ_x for the medium moving parallel to the interface, i.e., $\upsilon = \upsilon_0\hat{x}$, and ρ (for normal motion, i.e., $\upsilon = \upsilon_0\hat{z}$) can be easily obtained by the conventional theory of reflection and transmission:

$$\rho_x = \left(\frac{q' - \epsilon' C'}{q' + \epsilon' C'}\right), \tag{14D.1a}$$

$$\rho_x = \frac{\left(1+\beta^2 - 2\beta C\right)\left(1+\beta^2 - 2\beta/C\right)}{\left(1-\beta^2\right)^2}\left(\frac{q' - \epsilon' C'}{q' + \epsilon' C'}\right). \tag{14D.1b}$$

In the above, q', the root of the dispersion equation, is given by

$$q' = \left(\epsilon' - S'^2\right)^{1/2}$$

and

$$C' = \begin{cases} \dfrac{C}{\gamma(1 - S\beta)} & \text{for } \upsilon = \upsilon_0\hat{x}, \\ \dfrac{C - \beta}{1 - C\beta} & \text{for } \upsilon = \upsilon_0\hat{z}, \end{cases}$$

where $C = \cos\theta$ and $S = \sin\theta$ (θ is the angle of incidence).

The symbols and expressions used are as in References 6,7.

The power reflection coefficient is zero when $q' = \epsilon' C$. This condition really means that the ratio of the tangential electric and magnetic fields in plasma $\left(E_x'^{\,p}/H_y'^{\,p} = \eta_0 q'/\epsilon'\right)$ is the same as that in the free space $\left(E_x'/H_y' = \eta_0 C'\right)$, thus giving a perfect match of the plasma-free space system. Substituting for q', the above conditions may be recast in the form

$$C'^2 = \left(\epsilon' + 1\right)^{-1}. \tag{14D.2}$$

For a moving dielectric medium, Equation 14D.2 will lead to a quadratic equation in $S(C)$ for parallel (normal) motion which can be easily solved to give the Brewster angle [1]. However, when the moving medium is a plasma, the relative permittivity ϵ' is a function of the angle of incidence, the medium velocity and the incident wave frequency leading now to a quartic equation in S or C:

$$\upsilon = \upsilon_0\hat{x},$$

$$S^4 + a_1 S^3 + b_1 S^2 + c_1 S + d_1 = 0, \tag{14D.3}$$

where

$$a_1 = \frac{-4}{\beta\left(2 - \beta^2\right)},$$

$$b_1 = \frac{2\left(1 + \beta^2 + \beta^4\right)}{\beta^2\left(2 - \beta^3\right)} - \frac{\left(1 - \beta^2\right)^2}{\beta^2\Omega^2\left(2 - \beta^2\right)},$$

$$c_1 = \frac{-4\beta}{2 - \beta^2},$$

and

$$d_1 = \frac{2\beta^2 - 1}{\beta^2\left(2 - \beta^2\right)} + \frac{\left(1 - \beta^2\right)^2}{\beta^2\Omega^2\left(2 - \beta^2\right)},$$

$$\upsilon = \upsilon_0 \hat{z},$$

$$C^4 + a_2 C^3 + b_2 C^2 + c_2 C + d_2 = 0, \tag{14D.4}$$

where

$$a_2 = \frac{-4}{\beta\left(2-\beta^2\right)},$$

$$b_2 = \frac{2\left(1+\beta^2+\beta^4\right)}{\beta^2\left(2-\beta^2\right)} - \frac{\left(1-\beta^2\right)^2}{\beta^2\Omega^2\left(2-\beta^2\right)},$$

$$c_2 = \frac{2\left(1-\beta^2\right)}{\beta\Omega^2\left(2-\beta^2\right)} - \frac{4\beta}{2-\beta^2},$$

and

$$d_2 = \frac{2\beta^2-1}{\beta^2\left(2-\beta^2\right)} - \frac{1-\beta^2}{\Omega^2\left(2-\beta^2\right)}.$$

The variation of the Brewster angle is now examined as a function of the medium velocity with incident wave frequency as the parameter.

Case 1: Plasma half-space moving parallel to the interface $\left(\upsilon = \upsilon_0 \hat{x}\right)$.

The Brewster angle for this case is $\sin^{-1} S$, S being given by Equation 14D.3. This is an eighth-degree curve in $S - \beta$ and after tracing this complicated curve it is found that, for $\Omega < 1$, no portion of this curve exists within the physically meaningful limits of S and β ($0 < S < 1$ and $-1 < \beta < 1$) relevant to the problem. Thus, there is no Brewster angle for $\Omega < 1$. This is because, in this range, plasma supports [6] evanescent waves only.

Figure 14D.1 shows the sine of the Brewster angle as a function of the medium velocity with the parameter Ω in the range $\Omega > 1$. It is seen that the Brewster angle depends on the magnitude as well as the direction of the velocity of motion; for parallel motion (β positive) the angle increases from its stationary value of $\sin^{-1}\{[(\Omega^2 - 1)(2\Omega^2 - 1)^{-1}]^{1/2}\}$ with velocity, while for the antiparallel motion (β negative) the angle decreases with velocity. Another Brewster angle appears at $\beta = \beta_0 = [(1 - \Omega^2) + \Omega^2(\Omega^2 - 1)^{1/2}]^{1/2}$ and extends over a narrow range of medium velocity beyond β_0. For a given velocity, the Brewster angle increases with the incident wave frequency.

Case 2: Plasma half-space moving normal to the interface $\upsilon = \upsilon_0 \hat{z}$.

The Brewster angle for this case is $\cos^{-1} C$, C being obtained by the solution of Equation 14D.4. The C versus β curve was traced over the complete range of Ω and, as in parallel motion, some branches of the curve are outside the physically meaningful limits of C and β.

Figure 14D.2 incorporates the allowed branches and shows the cosine of the Brewster angle as a function of the medium velocity. The Brewster angle for the stationary case is given by $\cos^{-1}(2 - 1/\Omega^2)^{-1/2}$. When the medium moves *away* from

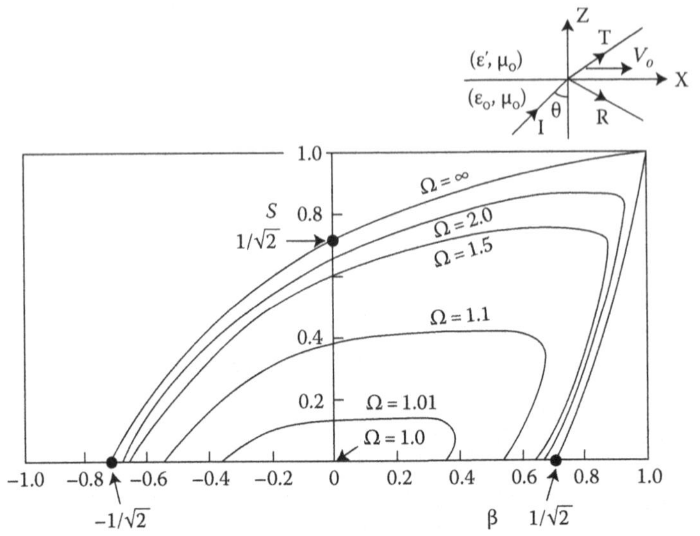

FIGURE 14D.1
Sine of the Brewster angle (S) shown as a function of the medium velocity (β) (medium moving parallel to the interface). All values of the parameter Ω are in the range $\Omega > 1$, since there are no Brewster angles for $\Omega < 1$. (Reprinted with permission from *J. Appl. Phys.*, 46(3), 1408–1409, 1975. Copyright 1975, American Institute of Physics.)

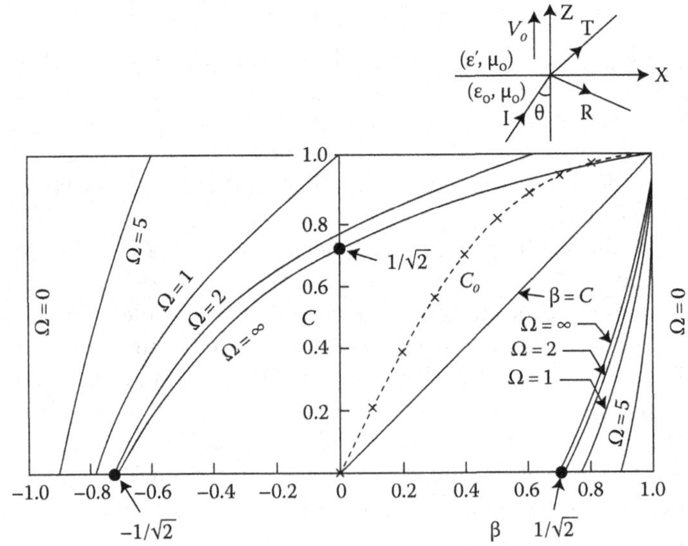

FIGURE 14D.2
Cosine of the Brewster angle (C) as a function of the medium velocity (normal motion). The dashed curve is given by $C_0 = 2\beta/(1+\beta^2)$. The branches to the right of the $\beta = C$ line correspond to the medium impinging on the incident wave. (Reprinted with permission from *J. Appl. Phys.*, 46(3), 1408–1409, 1975. Copyright 1975, American Institute of Physics.)

the incident wave (β positive), there is an *additional* angle of incidence for which the reflected wave as seen from the observer's frame grazes the interface, thus giving zero-reflected power [7]. This is given by $\beta = (1 - S)/C$, that is, $C_0 = 2\beta/(1 + \beta^2)$ and is included as a dashed curve in Figure 14D.2. The existence of C_0 is independent of the nature of the moving medium and as such this aspect checks with Yeh's conclusions for a dielectric medium [1].

There is no Brewster angle between $\beta = (1 - S)/C$ and $\beta = C$ (this is the range where the reflected wave travels toward the moving medium [7]) for $\Omega < 2\sqrt{2}$. The reason is that for this range of β and Ω, plasma supports only evanescent waves. For $\Omega < 2\sqrt{2}$, the relevant Brewster angle is given by the portion of the curve boxed by the curves C_0 and $\beta = C$. There exist two Brewster angles for large Ω, and one of them, due to the branch to the right of $\beta = C$ line, corresponds to the case of the medium impinging on the wave. It should be further noted that $\beta = (\Omega^2 - 1)/(\Omega^2 + 1)$ corresponds to $C = 1$ on the C versus β curve which means that for these medium velocities the Brewster angle is 0°.

Finally, for the medium moving *toward* the incident wave (β negative), the Brewster angle is seen to increase with velocity, becoming 90° at $-\beta_0 = -\left\{\frac{1}{2}\left[1 - 2\Omega^2 + \left(4\Omega^4 + 1\right)^{1/2}\right]\right\}^{1/2}$. For $-\beta_0 > \beta > -1$, there is no Brewster angle.

References

1. Yeh, C., *J. Appl. Phys.*, 38, 5194, 1967.
2. Lee, S. W. and Lo, Y. T., *J. Appl. Phys.*, 38, 870, 1967.
3. Kong, J. A. and Cheng, D. K., *J. Appl. Phys.*, 39, 2282, 1968.
4. Pyati, V. P., *J. Appl. Phys.*, 38, 652, 1967.
5. Chuang, C. W. and Ko, H. C., *J. Appl. Phys.*, 45, 1154, 1974.
6. Kalluri, D. and Shrivastava, R. K., *J. Appl. Phys.*, 44, 2440, 1973.
7. Kalluri, D. and Shrivastava, R. K., *J. Appl. Phys.*, 44, 4518, 1973.

*Appendix 14E: On Total Reflection of Electromagnetic Waves from Moving Plasmas**

Dikshitulu K. Kalluri and R. K. Shrivastava

When a plane electromagnetic wave is obliquely incident on a dielectric boundary, there are particular angles of incidence leading to a condition of total reflection. The critical angle is given by arcsin $(\epsilon_2/\epsilon_2)^{1/2}$, and, thus, total reflection is not possible from an optically denser medium [1]. The situation for the case of two relatively moving dielectric media was examined by Shiozawa and Kumagai [2]. If the wave traveling in free space is incident on a *plasma* medium, total reflection can take place under certain conditions; at normal incidence, waves of frequency less than the plasma frequency are totally reflected [3], and, for oblique incidence, the range of wave frequencies for total reflection is generally extended [4]. The purpose of this appendix was to investigate how this frequency range is modified by the motion of the plasma medium that may be isotropic or uniaxially anisotropic. Both parallel and perpendicular polarizations of the obliquely incident wave are considered. A few workers [5–7] have also touched upon total reflection in a limited manner in their treatment of reflection and transmission of electromagnetic waves from moving media.

Consider a plane electromagnetic wave obliquely incident (x–z-plane of incidence) on a plasma half-space moving relativistically through free space. Let the transmitted wave in the moving plasma (Σ') have the exponential variation

$$\Psi'_{\mathrm{P}} = \exp\left\{\mathrm{j}\left[\omega' t' - \left(k'_x x' + k'_0 q' z'\right)\right]\right\}, \tag{14E.1}$$

where q' is the root of the dispersion equation in Σ' and may be determined for uniaxial z [8] (infinitely strong magnetostatic field along the z-axis), uniaxial x [9], uniaxial y, or isotropic plasmas [10] in the usual manner by solving Maxwell's equations in Σ'. The notation used are the same as in Reference 8.

In the laboratory frame (Σ), the form of the transmitted wave may be written as

$$\Psi_{\mathrm{P}} = \exp\left\{\mathrm{j}\left[\omega^{\mathrm{P}} t - \left(k_x^{\mathrm{P}} x + k_z^{\mathrm{P}}\right)\right]\right\}. \tag{14E.2}$$

The wave vector k^{P} and frequency ω^{P} of the plasma wave as seen from the laboratory frame (Σ) are related to those in the rest frame (Σ') by the usual transformation formulas [11]. We consider two cases of medium motion.

One case is a plasma half-space moving parallel to the interface $(\bar{\upsilon} = \upsilon_0 \hat{x})$:

$$\omega^{P} = \gamma\left(\omega' + \upsilon_0 k_x'\right) = \omega$$

$$k_x^{P} = \gamma\left[k_x' + \left(\frac{\omega'}{c^2}\right)\upsilon_0\right] = k_0 S, \tag{14E.3}$$

$$k_z^{P} = k_0' q'.$$

For total reflection, the wave number k^P of the transmitted wave must be zero or imaginary. When k_z^P equals $k_0 q$, the conditions under which q' can become imaginary may be easily derived and are as follows:

$$\begin{aligned} &\text{a. isotropic (both polarization)} \\ &\quad \text{and uni-}y\text{ (perpendicular polarization)}, \quad \le 1/C, \\ &\text{b. uni-}x\text{ (parallel polarization)}, \quad \le 1/p_x, \\ &\text{c. uni-}z\text{ (parallel polarization)}, \quad 1/p_x \le \ \le 1/C. \end{aligned} \tag{14E.4}$$

In Equation 14E.4, Ω $(= \omega/\omega_p)$ is the normalized incident wave frequency and $p_x = \gamma\,(1 - S\beta)$. It is seen that the condition for total reflection from "isotropic" plasma is independent of the medium velocity. This conclusion also applies to the uni-y perpendicular polarization case. For a parallel-polarized wave incident on uni-y plasma or a perpendicularly polarized wave incident on uni-x or uni-z plasma, $q' = C'$, that is, the plasma has no effect and the reflected power is zero.

However, the phenomenon of total reflection for a parallel-polarized wave incident on a uni-x or uni-z plasma half-space is modified in an interesting manner by the medium velocity. Figure 14E.1 shows how the frequency range for total reflection is increased or decreased by motion. Region I, that is, area enclosed under the $\Omega = 1/p_x$ curve, gives the total reflection for the uni-x plasma, whereas, for the uni-z case, the corresponding area is

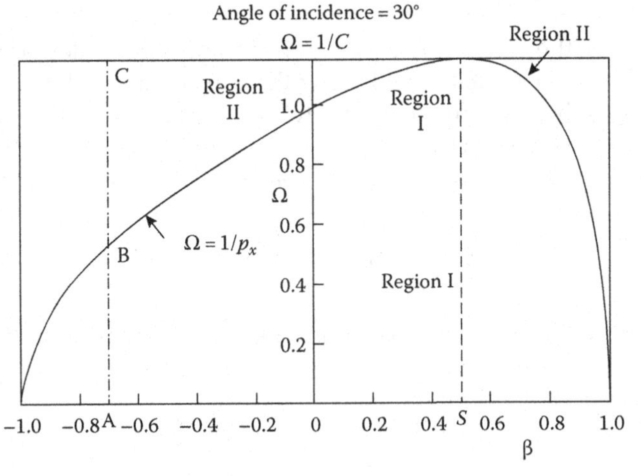

FIGURE 14E.1
Effect of the medium velocity on the frequency range for total reflection from uni-x and uni-z plasmas. (Reprinted with permission from *J. Appl. Phys.*, 49(12), 6169–6170, 1978. Copyright (1975), American Institute of Physics.)

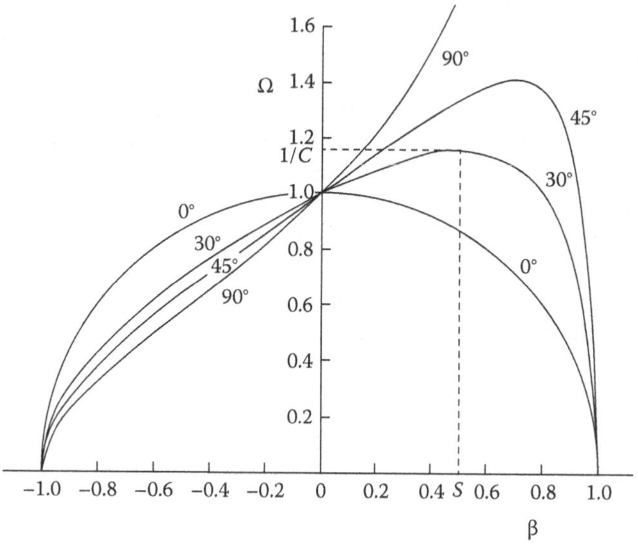

FIGURE 14E.2
Effect of the angle of incidence on the frequency range for total reflection. (Reprinted with permission from *Appl. Phys.*, 49(12), 6169–6170, 1978. Copyright (1975), American Institute of Physics.)

that enclosed between the curve $\Omega = 1/p_x$ and the line $\Omega = 1/C$ (region II). It is seen that for a given angle of incidence as β increases from −1 to S, frequency range AB (BC) is extended (narrowed down) for the uni-x (uni-z) case attaining the critical value $1/C$ (zero) at $\beta = S$. Just the reverse happens in the range $S < \beta < 1$.

Figure 14E.2 is introduced to show how the frequency range for a given medium velocity is affected by the angle of incidence. For negative β, as the angle of incidence increases, the frequency range for total reflection narrows for the uni-x case and extends for the uni-z case.

For positive β, the frequency range for the uni-x case is extended with an increase in the angle of incidence, while for the uni-z case, the range can either be increased or decreased, and it is not possible to draw any general inference concerning this, because the frequency range for total reflection in this case is given by the portion included between the curve $\Omega = 1/p_x$ and the line $\Omega = 1/C$; the latter moves up with an increase in the angle of incidence, but the former first increases and then decreases with β.

The condition $\Omega \leq 1/p_x$ for the uni-x plasma, expressed in a different form, would mean that the angle of incidence should be equal to or greater than $\arcsin(1/\beta)[1-(1/\Omega)(1-\beta^2)^{1/2}]$, and this is in agreement with the result given by Lee and Lo [5]. The corresponding relation for the uni-z case is $\arcsin[1-(1/\Omega^2)]^{1/2} \leq \theta \leq \arcsin[1-(1/\Omega)\,(1-\beta^2)^{1/2}]$.

The other case is a plasma half-space moving normal to the interface $\left(\bar{\upsilon} = \upsilon_0 \hat{z}\right)$:

$$\begin{aligned}
\omega^{P} &= \gamma\left(\omega' + \upsilon_0 k_0' q'\right) = \gamma\omega'\left(1+\beta q'\right),\\
k_x^{P} &= k_x' = k_0' S' = k_0 S, \qquad (14E.5)\\
K_z^{P} &= \gamma\left[k_0' q' + \left(\frac{\omega'}{c^2}\right)\upsilon_0\right] = \gamma k_0'\left(\beta + q'\right).
\end{aligned}$$

Unlike in the previous case, imaginary q' does *not* lead to total reflection. The reason is that though imaginary q' gives rise to the evanescent waves in Σ', the transmitted wave frequency and the wave vector as viewed from Σ are both complex, and an evanescent wave in the rest frame of the plasma appears to be a propagating wave with attenuation in the laboratory frame [12,13]. Also, the imaginary part of the complex frequency of the waves in the plasma might signify a damping of the plasma wave as a result of energy transfer to the radiation fields.

The general definition of total reflection is that situation for which there is no power transmitted into the second medium. While imaginary q' signifies this situation when the plasma is stationary, this is not necessarily the case when it is moving. Further work on this interesting aspect of the case of a plasma half-space moving normal to the interface is in progress.

References

1. Ramo, S., Whinnery, J. R., and Van Duzer, T., *Fields and Waves in Communication Electronics*, Wiley, New York, 1965.
2. Shiozawa, T. and Kumagai, N., *Proc. IEEE*, 55, 1243, 1967.
3. Budden, K. G., *Radio Waves in the Ionosphere*, Cambridge University Press, Cambridge, U.K., 1961.
4. Kalluri, D. and Prasad, R. C., *Appl. Sci. Res.*, 27, 415, 1973.
5. Lee, S. W. and Lo, Y. T., *J. Appl. Phys.*, 38, 870, 1967.
6. Pyat, V. P., *J. Appl. Phys.*, 38, 652, 1967.
7. Kong, J. A. and Cheng, D. K., *J. Appl. Phys.*, 39, 2282, 1968.
8. Kalluri, D. and Shrivastava, R. K., *IEEE Trans. Antennas Propag.*, AP-21, 63, 1973.
9. Kalluri, D. and Shrivastava, R. K., *IEEE Trans. Plasma Sci.*, PS–2, 206, 1974.
10. Kalluri, D. and Shrivastava, R. K., *J. Appl. Phys.*, 44, 2440, 1973.
11. Papas, C. H., *Theory of Electromagnetic Wave Propagation*, McGraw-Hill, New York, 1965.
12. Yeh, C., *Appl. Phys. Lett.*, 9, 184, 1966.
13. Chawla, B. R. and Unz, H., *Electromagnetic Waves in Moving Magneto-plasmas*, Regents Press of Kansas, Lawrence, KS, 1973.

Appendix 14F: Interaction of Electromagnetic Waves with Bounded Plasmas Moving Perpendicular to the Plane of Incidence*

Dikshitulu K. Kalluri and R.C. Prasad

14F.1 Introduction

In a previous paper [1], the reflection and transmission properties of isotropic and uniaxially anisotropic bounded *stationary* plasmas were considered. The results for the case when the plasma is moving perpendicular to the interface or parallel to the interface in the plane of incidence have been considered by many investigators [2–8] and are summarized by Kalluri and Shrivastava [9].

Investigation of the problem when the movement of the medium is perpendicular to the plane of incidence and the medium is a dielectric is reported in the literature [10,11]. Shiozawa et al. [10] studied the problem of reflection and transmission of an electromagnetic wave by an isotropic dielectric half-space moving perpendicular to the plane of incidence for perpendicular polarization. Pyati [11] has also considered the isotropic dielectric half-space, but the plane of incidence of the wave is arbitrary. Hence, the motion perpendicular to the plane of incidence is a particular case of his study.

When the moving medium is a plasma which is dispersive, the reflection and transmission coefficients also depend on the wave frequency and this chapter is concerned with a systematic and comprehensive investigation of this aspect for movement (along the y-axis) perpendicular to the plane of incidence (xz-plane) for *a* number of cases. The cases considered are as follows: The moving medium is (a) an isotropic plasma, (b) a uni-x ($\mathbf{B}_0 = \infty\hat{x}$) plasma, (c) a uni-$y$ ($\mathbf{B}_0 = \infty\hat{y}$) plasma, and (d) a uni-$z$ ($\mathbf{B}_0 = \infty\hat{z}$) plasma. For each of these cases, the effect of the polarization of the incident wave (parallel or perpendicular) in free space is also dealt with. Kong and Cheng [12] solved case (c); however, their results pertain to a different aspect than those discussed by the present authors.

In the literature, two methods have been used to solve such problems with moving boundaries. In the first method [12,13], the problem is solved in the laboratory frame Σ, and appropriate boundary conditions on the electric and magnetic fields as measured in Σ are imposed. In the other method [2,5], the fields are transformed to a frame of reference Σ' at rest with respect to the moving medium, and in the rest frame the problem is solved as for a stationary plasma [1,14]. To find the reflection and transmission coefficients in the laboratory frame, the fields in Σ' are transformed back to the laboratory frame. The present authors have chosen this method since, in their view, this allows the total effect to be decomposed into two distinct effects. One is the effect of the mismatch of the stationary plasma with free space as calculated in Σ', and the other is the effect of the moving boundary that is independent of the specific nature of the moving medium. As seen from Σ', the incident wave is arbitrarily polarized with an arbitrary plane of incidence. The characteristic roots and ratio of field components in the plasma in Σ' are obtained by arranging the appropriate equations in state-variable form and finding the eigenvalues and eigenvectors of the associated matrix.

The solution is outlined in Section 14F.2 mainly to introduce the notation and point out the significant steps.

14F.2 Formulation of the Problem

Let the region $z > 0$ be occupied by the plasma (linear, lossless, homogeneous) half-space and the region $z < 0$ be free space, and xz be the plane of incidence of a uniform plane wave in free space with exponential variation

$$\Psi_I = \exp\left[-jk_0\left(Sx + Cz\right) + j\omega t\right] \tag{14F.1}$$

in the laboratory frame Σ, where ω is the frequency of the incident wave, $k_0 = \omega/c$ the free-space wave number, c the velocity of light in free space, $C = \cos\theta_I$, $S = \sin\theta_I$, and θ_I is the angle which the wave normal makes with the positive z-axis. Let the movement of the medium be along the y-direction with a constant velocity v_0 and Σ' be the rest frame of the medium. The incident wave as seen from Σ' appears arbitrarily polarized with an arbitrary plane of incidence and hence

$$\Psi'_I = \exp\left[-jk'_0\left(S'_1y' + S'_2y' + C'z'\right) + j\omega' t'\right], \tag{14F.2a}$$

where

$$S'_1 = \frac{S}{\gamma},\quad S'_2 = -\beta,\quad C' = \frac{C}{\gamma},\quad \omega' = \gamma\omega,\quad \beta = \frac{\upsilon_0}{c},$$
$$\gamma = \left(1 - \beta^2\right)^{-1/2}. \tag{14F.2b}$$

In the plasma as observed in Σ', the exponential variation of the waves appears like

$$\Psi'_P = \exp\left[-jk'_0\left(S'_1x' + S'_2y' + q'z'\right) + j\omega' t'\right], \tag{14F.3}$$

where q' are the characteristic roots of the waves in the plasma and are found to be as follows for the four cases.

a. Isotropic plasma ($\mathbf{B}_0 = 0$):

$$q'_{1,3} = q'_{2,4} = \pm\left[\epsilon - \left(S_1'^2 + S_2'^2\right)\right]^{1/c2}. \tag{14F.4a}$$

b. Uni-x plasma ($\mathbf{B}_0 = \infty\hat{x}$):

$$q'_{1,3} = \pm\left[C'^2 - X'\left(1 - S_1'^2\right)\right]^{1/2}, \quad q'_{2,4} = \pm C'. \tag{14F.4b}$$

c. Uni-y plasma ($\mathbf{B}_0 = \infty\hat{y}$):

$$q'_{1,3} = \pm\left[C'^2 - X'\left(1 - S_2'^2\right)\right]^{1/2}, \quad q'_{2,4} = \pm C'. \tag{14F.4c}$$

d. Uni-z plasma ($\mathbf{B}_0 = \infty\hat{z}$):

$$q'_{1,3} = \pm\left[C'^2 - X'\left(1 - X'\right)\right]^{1/2}, \quad q'_{2,4} = \pm C', \tag{14F.4d}$$

where $\epsilon' = (1 - X') = \left[1 - (\omega_p/\omega')^2\right]$ and ω_p is the plasma frequency. The corresponding values of q (measured in Σ) can be obtained from the relation (Equation 14F.14) $q = k_z^t / k_0 = \gamma q'$.

It is seen that q_1' (or q_1) is either real or imaginary in each case, whereas q_2' (or q_2) is always real for the last three cases. For the case of an isotropic plasma, the characteristic roots are repeated. In the case of a semi-infinite plasma, there will be only two forward waves corresponding to the positive value of $q_i'(i = 1, 2)$. The q_i' in Equation 14F.4 are obtained by finding the eigenvalues [14,15] of the $\hat{T}'$ matrix when the equations, obtained from Maxwell's equations in a plasma modeled as an anisotropic dielectric [1] for the assumed field variation, are arranged in the state-variable form [14]:

$$\frac{d\hat{F}'}{dz'} = -jk_0'\hat{T}'\hat{F}', \tag{14F.5}$$

where $\hat{F}'$ is the column matrix of elements $(E'^{P}_{xi}, E'^{P}_{yi}, \eta_0 H'^{P}_{xi}, \eta_0 H'^{P}_{yi})$, $\eta_0 = (\mu_0/\epsilon_0)^{1/2}\mu_0$, and ϵ_0 are, respectively, permeability and permittivity of the free space, and $\hat{T}'$ is obtained for each of the four cases. The ratios of the field components for each mode are now obtained by finding the corresponding eigenvectors [14,15]. (The eigenvector method always gives the correct ratios, whereas the method given in Reference 16 gives an indeterminate form of the ratio for some of the cases considered here.) The unknowns are now reduced for the semi-infinite case to E'^{P}_{x1}, E'^{P}_{x2} (or E'^{P}_{y1}, E'^{P}_{y2}), E'^{R}_{x}, and E'^{R}_{y}. By matching the tangential electric and magnetic fields at the boundary, these quantities are now found in terms of E'^{I}_{x} and E'^{I}_{y}.

Let the electric field of the reflected wave in free space in Σ be

$$\mathbf{E}^{R} = \left(E^{R}_{x}\hat{x} + E^{R}_{y}\hat{y} + E^{R}_{z}z\right)\Psi_{R}, \tag{14F.6a}$$

where

$$\Psi_R = \exp\left[-j\left(k^{R}_{x}x + k^{R}_{y}y - k^{R}_{z}z\right) + j\omega^{R}t\right]. \tag{14F.6b}$$

One obtains

$$E_x^{\mathrm{R}} = \left(\frac{1}{\gamma}\right) E_x'^{\mathrm{R}} - \beta S E_y'^{\mathrm{R}}, \tag{14F.7a}$$

$$E_y^{\mathrm{R}} = E_y'^{\mathrm{R}}, \tag{14F.7b}$$

$$k_x^{\mathrm{R}} = k_0 S, \quad k_y^{\mathrm{R}} = 0, \quad k_z^{\mathrm{R}} = k_0 C, \quad \omega^{\mathrm{R}} = \omega. \tag{14F.8}$$

For the parallel-polarized incident wave $\left[\mathbf{E}^{\mathrm{I}} = \left(E_x^{\mathrm{I}}\hat{x} + E_z^{\mathrm{I}}\hat{z}\right)\Psi_{\mathrm{I}}\right]$ in Σ, one obtains in Σ':

$$E_x'^{\mathrm{I}} = \gamma E_x^{\mathrm{I}}, \tag{14F.9a}$$

$$E_y'^{\mathrm{I}} = 0. \tag{14F.9b}$$

Now the reflection coefficients ${}_{\|}R_{\|} = E_x^{\mathrm{R}}/E_x^{\mathrm{I}}$ and ${}_{\|}R_{\perp} = E_y^{\mathrm{R}}/E_x^{\mathrm{I}}$ are obtained by using Equations 14F.7 and 14F.9.

For the perpendicular-polarized incident wave $\left(\mathbf{E}^{\mathrm{I}} = E_y^{\mathrm{I}}\hat{y}\Psi_{\mathrm{I}}\right)$ in Σ, one obtains in Σ':

$$E_y'^{\mathrm{I}} = \gamma\beta S E_y^{\mathrm{I}}, \tag{14F.10a}$$

$$E_y'^{\,\mathrm{I}} = E_y^{\mathrm{I}}, \tag{14F.10b}$$

and again ${}_{\perp}R_{\|} = E_x^{\mathrm{R}}/E_y^{\mathrm{I}}$ and ${}_{\perp}R_{\perp} = E_y^{\mathrm{R}}/E_y^{\mathrm{I}}$ are obtained from Equations 14F.7 and 14F.10.

Let the electric field of the transmitted waves in the plasma in Σ be

$$\mathbf{E}_i^{\mathrm{P}} = \left(E_{xi}^{\mathrm{P}}\hat{x} + E_{yi}^{\mathrm{P}}\hat{y} + E_{zi}^{\mathrm{P}}\hat{z}\right)\Psi_{\mathrm{P}i} \quad (i = 1,2). \tag{14F.11a}$$

where

$$\Psi_{\mathrm{P}i} = \exp\left[-\mathrm{j}\left(k_x^t x + k_y^t y + k_z^t z\right) + \mathrm{j}\omega^t t\right]. \tag{14F.11b}$$

By using the Lorentz transformations and Minkowski's constitutive relation in the moving medium [17], one obtains

$$E_{xi}^{\mathrm{P}} = \left(\frac{1}{\gamma}\right) E_{xi}'^{\mathrm{P}} - \beta S E_{yi}'^{\mathrm{P}}, \tag{14F.12a}$$

$$E_{yi}^{\mathrm{P}} = E_{yi}'^{\mathrm{P}}, \tag{14F.12b}$$

$$H_{xi}^{\mathrm{P}} = \left(\frac{1}{\gamma}\right) H_{xi}'^{\mathrm{P}} - \beta S H_{yi}'^{\mathrm{P}}, \tag{14F.13a}$$

$$H_{yi}^{\mathrm{P}} = H_{yi}'^{\mathrm{P}}, \tag{14F.13b}$$

$$k_x^t = k_0 S, \quad k_y^t = 0 \quad k_z^t = k_0 \gamma q_i', \quad \omega^t = \omega. \tag{14F.14}$$

From Equations 14F.7, 14F.8, and 14F.12 through 14F.14, one concludes that for the incident E (perpendicularly polarized) or incident H (parallel-polarized) plane wave, both the reflected as well as the transmitted waves in Σ are no longer an E- or H-plane wave but a linear combination of E- and H-plane waves (except in the case of a uni-y plasma where the polarizations of the reflected and transmitted waves are the same as the incident wave) while the propagation vectors of the incident, reflected, and transmitted waves all lie in the same plane, that is, in the plane of incidence. This is in agreement with the conclusions of earlier investigators [10–12].

So far, an outline for obtaining the field components in the plasma and in free space in Σ has been described. Now the power reflection and transmission coefficients will be discussed.

14F.3 Power Reflection and Transmission Coefficients

The power reflection coefficient ρ and the power transmission coefficient τ are defined, respectively, by the relations

$$\rho = -\frac{\mathbf{n}\cdot\mathbf{S}_{\mathrm{R}}}{\mathbf{n}\cdot\mathbf{S}_{\mathrm{I}}} \tag{14F.15}$$

and

$$\tau = \frac{\mathbf{n}\cdot\mathbf{S}_{\mathrm{t}}}{\mathbf{n}\cdot\mathbf{S}_{\mathrm{I}}}, \tag{14F.16}$$

where $\mathbf{n}$ is a unit vector in the z-direction, and $\mathbf{S}_{\mathrm{I}}$, $\mathbf{S}_{\mathrm{R}}$, and $\mathbf{S}_{\mathrm{t}}$ are the time-averaged Poynting vectors of the incident, reflected, and transmitted waves, respectively. Equation 14F.15 can be reduced to the following forms [16], for the parallel and perpendicular polarization, respectively, as

$$\rho_{\parallel} = \left|{}_{\parallel}R_{\parallel}\right|^2 + C^2\left|{}_{\parallel}R_{\perp}\right|^2 \tag{14F.17a}$$

and

$$\rho_{\perp} = \left|{}_{\perp}R_{\perp}\right|^2 + \frac{1}{C^2\left|{}_{\perp}R_{\parallel}\right|^2}. \tag{14F.17b}$$

Equation 14F.16 takes the form

$$\begin{aligned}\tau = \mathrm{Re}\Big[&\left(E^{\mathrm{P}}_{x1}\Psi_{\mathrm{P}1} + E^{\mathrm{P}}_{x2}\Psi_{\mathrm{P}2}\right)\left(H^{\mathrm{P}*}_{y1}\Psi^{*}_{\mathrm{P}1} + H^{\mathrm{P}*}_{y2}\Psi^{*}_{\mathrm{P}2}\right)\\ &-\left(E^{\mathrm{P}}_{y1}\Psi_{\mathrm{P}1} + E^{\mathrm{P}}_{y2}\Psi_{\mathrm{P}2}\right)\left(H^{\mathrm{P}*}_{x1}\Psi^{*}_{\mathrm{P}1} + H^{\mathrm{P}*}_{x2}\Psi^{*}_{\mathrm{P}2}\right)\Big]\left(\mathrm{n}\cdot\mathrm{S}_{\mathrm{I}}\right)^{-1}.\end{aligned} \tag{14F.18}$$

The power reflection coefficient (ρ) and the power transmission coefficient (τ) for different cases have been obtained by using Equations 14F.17 and 14F.18, for parallel as well as perpendicular polarizations, in the ranges of real and imaginary q_1'. These are given in Table 14F.1.

TABLE 14F.1

Power Reflection Coefficient (ρ) and Power Transmission Coefficient (τ) for Parallel and Perpendicular Polarization, for Various Cases

Cases		$\rho_\parallel$	$\rho_\perp$	$\tau_\parallel$	$\tau_\perp$
Uni-x	q_1' real	$\frac{\gamma^2C^2}{\gamma^2-S^2}\left\lvert\frac{C'-q_1'}{C'+q_1'}\right\rvert^2$	$\left(\frac{\beta^2S^2}{C^2}\right)\rho_\parallel$	$\frac{\gamma^2C^2}{\gamma^2-S^2}\frac{4C'q_1'}{\left(C'+q_1'\right)^2}+\frac{\gamma^2\beta^2S^2}{\gamma^2-S^2}$	$\frac{\gamma^2\beta^2S^2}{\gamma^2-S^2}\frac{4C'q_1'}{\left(C'+q_1'\right)^2}+\frac{\gamma^2C^2}{\gamma^2-S^2}$
	q_1' imaginary	$\frac{\gamma^2C^2}{\gamma^2-S^2}$	$\left(\frac{\beta^2S^2}{C^2}\right)\rho_\parallel$	$\frac{\gamma^2\beta^2S^2}{\gamma^2-S^2}$	$\frac{\gamma^2C^2}{\gamma^2-S^2}$
	q_1' real	0	$\left\lvert\frac{C'-q_1'}{C'+q_1'}\right\rvert^2$	1	$\frac{4C'q_1'}{\left(C'+q_1'\right)^2}$
Uni-y	q_1' imaginary	0	1	1	0
	q_1' real	$\frac{\gamma^2S^2}{\gamma^2-C^2}\left\lvert\frac{C'-q_1'}{C'+q_1'}\right\rvert^2$	$\left(\frac{\beta^2C^2}{S^2}\right)\rho_\parallel$	$\frac{\gamma^2S^2}{\gamma^2-C^2}\frac{4C'q_1'}{\left(C'+q_1'\right)^2}+\frac{\gamma^2\beta^2C^2}{\gamma^2-C^2}$	$\frac{\gamma^2\beta^2C^2}{\gamma^2-C^2}\frac{4C'q_1'}{\left(C'+q_1'\right)^2}+\frac{\gamma^2S^2}{\gamma^2-C^2}$
Uni-z	q_1' imaginary	$\frac{\gamma^2S^2}{\gamma^2-C^2}$	$\left(\frac{\beta^2C^2}{S^2}\right)\rho_\parallel$	$\frac{\gamma^2\beta^2C^2}{\gamma^2-C^2}$	$\frac{\gamma^2S^2}{\gamma^2-C^2}$
	q_1' real	$\left[\left(\frac{C'-q_1'}{C'+q_1'}\right)^2-\frac{4S^2X'^2C'q_1'}{\left[\left(C'+q_1'\right)\left(\epsilon'C'+q_1'\right)\right]^2}\right]$	$\left[\left(\frac{\left(C'-q'\right)_1}{\left(C'+q_1'\right)}\right)^2-\frac{4\beta^2C^2X'^2C'q_1'}{\left[\left(C'+q_1'\right)\left(\epsilon'C'+q_1'\right)\right]^2}\right]$	$\left(\frac{4\epsilon q_1'C'}{\left(q_1'+C'\epsilon'\right)^2}-\frac{4\beta^2C^2X'^2C'q_1'}{\left[\left(C'+q_1'\right)\left(q_1'+\epsilon'C'\right)\right]^2}\right)$	$\left(\frac{4C'q_1'}{\left(C'+q_1'\right)^2}+\frac{4\beta^2C^2X'^2C'q_1'}{\left[\left(C'+q_1'\right)\left(q_1'+\epsilon'C'\right)\right]^2}\right)$
Isotropic	q_1' imaginary	1	1	0	0

Source: Reprinted with permission from *J. Appl. Phys.*, 48(2), 587–591. Copyright (1977), American Institute of Physics.

It is found that $\rho + \tau = 1$. The expressions for $\rho_\perp$ and $\tau_\perp$ for an isotropic plasma given in Table 14F.1 are in agreement with the results given by Shiozawa et al. [10], when $\rho_\perp$ and $\tau_\perp$ are deduced for a dielectric medium. The expression for $\rho_\perp$ given in Table 14F.1 for uni-*y* agrees with the result given by Kong and Cheng [12]. (But the numerical results presented by them in Figure 14F.4 of their paper, for $\theta = 20°$ and $30°$, show that the reflected power first increases and then decreases with β, whereas the reflected power is actually a maximum when the medium is stationary and decreases with the increase in the medium velocity.)

14F.4 Mechanism of Power Transmission into the Plasma

Shiozawa et al. [10] decomposed the reflected and transmitted power into a contribution due to a *E*-wave component and an *H*-wave component for the case of an isotropic dielectric when the incident wave is perpendicularly polarized and showed that the sum of the *H*-wave component of the reflected power and the transmitted power is zero and also that the sum of the *E*-wave component of the reflected power and transmitted power is 1.

It is found by our analysis that for an isotropic plasma also, the same conclusions hold good when q_1' is real. But when q_1' is imaginary, the *H*-wave component and the *E*-wave component of the transmitted power are equal in magnitude and opposite in sign and hence the net power transmitted is zero and *H*- and *E*-wave components of the reflected power add to 1.

For the uni-*y* case, for an incident *E* wave (perpendicular polarization), only *E* waves are excited in the plasma corresponding to the q_1' mode. Thus, whenever q_1' is imaginary, the transmitted power is zero. For an incident *H* wave (parallel polarization), only *H* waves are excited in the plasma corresponding to the q_2' mode and since $q'_2 = C'$ ($q_2 = C$) there is a perfect match at the interface and all the incident power gets transmitted through the plasma.

For uni-*x* and uni-*z* cases, both the modes are excited in the plasma. Even if q_1' becomes imaginary (for certain ranges of wave frequency and medium velocity) and the corresponding wave is evanescent, a certain amount of power is transmitted by the second wave, q_2' being always real. The mechanism of this power transmission is examined below.

When q_1' is real, it can be shown from Equation 14F.18 for the transmitted power that $\mathrm{Re}\left(E_{x1}^{P} H_{y2}^{P*} \Psi_{P1} \Psi_{P2}^{*} + E_{x2}^{P} H_{y1}^{P*} \Psi_{P2} \Psi_{P1}^{*}\right)$ is exactly equal to $\mathrm{Re}\left(E_{y1}^{P} H_{x2}^{P*} \Psi_{P1} \Psi_{P2}^{*} + E_{y2}^{P} H_{x1}^{P*} \Psi_{P2} \Psi_{P1}^{*}\right)$. Therefore, the net power contribution by these components is zero at any level in the plasma. Hence, the power contribution is from the components $\mathrm{Re}\left(E_{x1}^{P} H_{y1}^{P*} \Psi_{P1} \Psi_{P1}^{*} + E_{x2}^{P} H_{y2}^{P*} \Psi_{P2} \Psi_{P2}^{*}\right)$ and $\mathrm{Re}\left(E_{y1}^{P} H_{x2}^{P*} \Psi_{P1} \Psi_{P1}^{*} + E_{y2}^{P} H_{x2}^{P*} \Psi_{P2} \Psi_{P2}^{*}\right)$ and thus both the modes corresponding to q_1' and $q_2' = C'$ carry power.

When q_1' is imaginary, it can be shown from Equation 14F.18 that $\mathrm{Re}\left(E_{x1}^{P} H_{y2}^{P*} \Psi_{P1} \Psi_{P2}^{*} + E_{x2}^{P} H_{y1}^{P*} \Psi_{P2} \Psi_{P1}^{*}\right)$ is exactly equal to $\mathrm{Re}\left(E_{y1}^{P} H_{x2}^{P*} \Psi_{P1} \Psi_{P2}^{*} + E_{y2}^{P} H_{x1}^{P*} \Psi_{P2} \Psi_{P1}^{*}\right)$. Therefore, the net power contribution by these components is zero at any level in the plasma in this case also. The power contribution by $\mathrm{Re}\left(E_{x1}^{P} H_{y1}^{P*} \Psi_{P1} \Psi_{P1}^{*}\right)$ and $\mathrm{Re}\left(E_{y1}^{P} H_{x1}^{P*} \Psi_{P1} \Psi_{P1}^{*}\right)$ is zero. Therefore, the power transmitted in the plasma when q_1' is imaginary is only due to the components $\mathrm{Re}\left(E_{x2}^{P} H_{y2}^{P*} \Psi_{P2} \Psi_{P2}^{*}\right)$ and $\mathrm{Re}\left(E_{y2}^{P} H_{x2}^{P*} \Psi_{P2} \Psi_{P2}^{*}\right)$. This power is being carried by the mode in the plasma corresponding to the characteristic root $q_2' = C'$.

14F.5 Numerical Results and Discussion

The variation of ρ with the normalized wave frequency $\Omega = \omega/\omega_p$ for the cases of uni-*y*, uni-*x*, and uni-*z* plasmas has been presented for different β and $C = 0.5$ in Figures 14F.1 through 14F.3, respectively. The variation of q_1' with Ω for $\beta = 0$ and 0. 75 and $C = 0.5$ has been shown as insets for the above cases. The sign of β does not affect the final results as is evident from the physical picture and also from the relevant equations where even powers of β only appear. In the case of a uni-*y* plasma and incident *E* wave (Figure 14F.1), q_1' is imaginary for $0 < \Omega < 1/\gamma C$ and real for $1/\gamma C \leq \Omega < \infty$. For a given Ω in the range $\Omega \geq 1/C$, q_1' is real for $0 \leq \beta \leq 1$ and $\rho_\perp$ decreases with increase in β. Further, for $\Omega < 1/C$, q_1' is imaginary for $0 \leq \beta \leq \beta_1$ [$= (1 - C^2\Omega^2)^{1/2}$] and real for $\beta_1 \leq \beta_1 \leq 1$. In this range of Ω, $\rho_\perp$ is unity for $\beta \leq \beta_1$ and decreases for $\beta > \beta_1$.

In the case of a uni-*x* plasma and incident *H*-wave (Figure 14F.2), q_1' is imaginary for $0 \leq \Omega \leq \Omega_1$ [$=(1 - S_1'^2)^{1/2}/C$] and real for $\Omega < \Omega \leq \infty$. In this case, three ranges of Ω are of interest. For a given Ω in the range $\Omega \geq 1/C$, q_1' is real for all β and $\rho_\parallel$ increases with increase in β. For $\Omega \leq 1$, q_1' is imaginary for all β and $\rho_\parallel$ decreases with increase in β. For $1 < \Omega < 1/C$, q_1 is real for $0 \leq \beta \leq \beta_1$ [$= C(\Omega^2 - 1)^{1/2}/S$] and imaginary for $\beta_1 \leq \beta \leq 1$. In this range of Ω, $\rho_\parallel$ increases with $\beta \leq \beta_1$ and decreases with $\beta \leq \beta_1$. However, $\rho_\parallel$ is constant (Figure 14F.2) with Ω in the range of q_1' imaginary.

In Figure 14F.3, the case of a uni-*z* plasma has been presented. In this case, q_1' is imaginary for $1/\gamma < \Omega \leq 1/C$ and real in the ranges $0 < \Omega < 1\gamma$ and $1/C < \Omega < \infty$. Here also, three ranges of Ω are of interest. For a given Ω in the range $\Omega \leq 1$, q_1' is real for $0 < \beta \leq \beta_1$ [$= (1 - \Omega)^{1/2}$] and imaginary for $\beta_1 < \beta \leq 1$. q_1' is infinity at $\beta = \beta_1$ and zero at $\beta = 1$. In this range of Ω, $\rho_\parallel$ increases first with increase in $\beta < \beta_1$ and then decreases for $\beta > \beta_1$. In the range $1 < \Omega < 1/C$, q_1' is imaginary for all β and $\rho_\parallel$ decreases with β. For $\Omega > 1/C$, q_1' is real for all β and there is

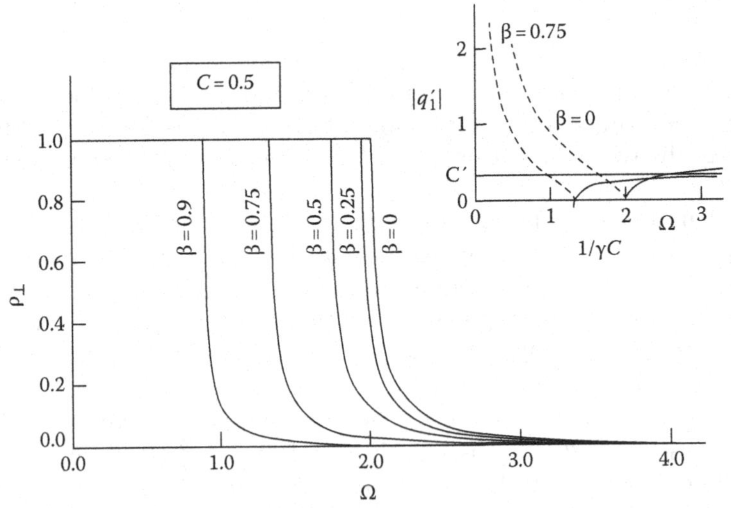

FIGURE 14F.1
Power reflection coefficient for perpendicularly polarized waves incident on a moving uni-*y* plasma with β as parameter. The variation of the characteristic root is shown in the inset (broken line, q_1' imaginary; solid line, q_1' real). (Reprinted with permission from *J. Appl. Phys.*, 48(2), 587–591. Copyright 1977, American Institute of Physics.)

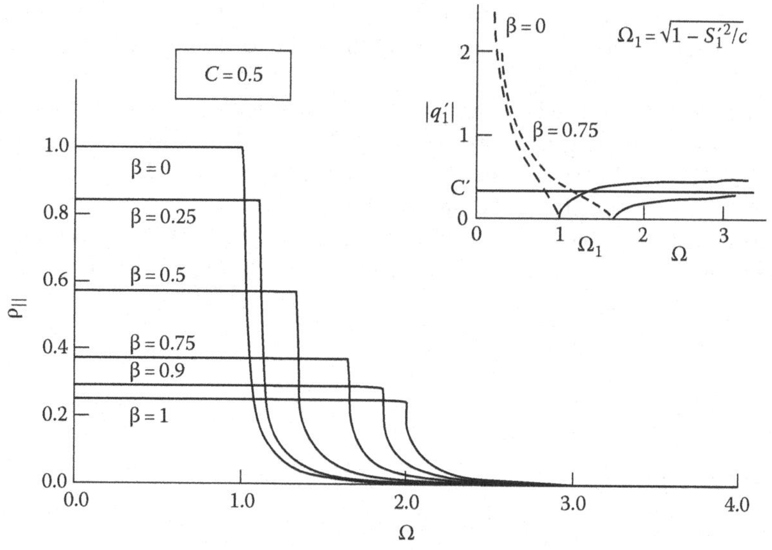

FIGURE 14F.2
Power reflection coefficient for parallel-polarized waves incident on a moving uni-x plasma with β as parameter. The variation of the characteristic root is shown in the inset (broken line, q_1' imaginary; solid line, q_1' real). (Reprinted with permission from *J. Appl. Phys.*, 48(2), 587–591. Copyright 1977, American Institute of Physics.)

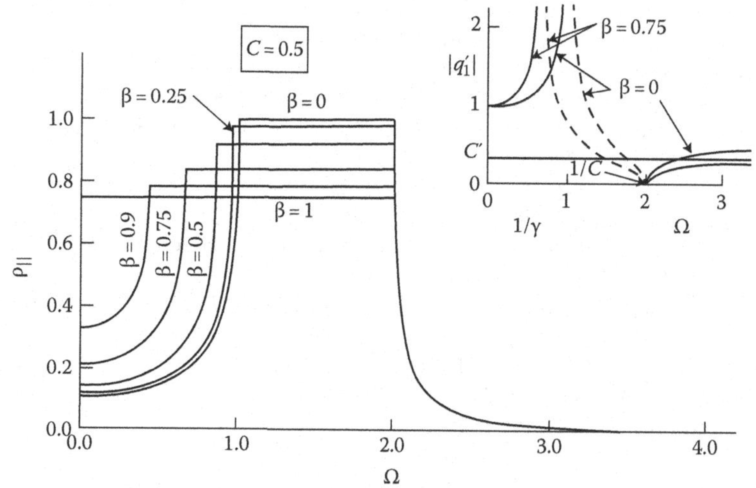

FIGURE 14F.3
Power reflection coefficient for parallel-polarized waves incident on a moving uni-z plasma with β as parameter. The variation of the characteristic root is shown in the inset (broken line, q_1' imaginary; solid line, q_1' real). (Reprinted with permission from *J. Appl. Phys.*, 48(2), 587–591. Copyright 1977, American Institute of Physics.)

an increase in $\rho_{||}$ with increase in β, but the effect is negligible and hence only one curve is shown in this range. In this case also, it is interesting to note that for q_1' imaginary, $\rho_{||}$ is constant with Ω (Figure 14F.3).

The numerical results for the isotropic plasma have not been presented here. In this case, one would see that q_1' is imaginary in the range $0 < \Omega \leq 1/C$ and real in the range $1/C < \Omega \leq \infty$ for all β. $\rho_{||}$ is unity in the range of q_1' imaginary for all β. In the range of q_1' real, $\rho_{||}$ increases as β increases, but the effect of β is small.

Expressions are also derived for the plasma slab case and computations made showed oscillations in the range of q_1' real whose nature can be qualitatively explained as before [1,18].

Acknowledgment

The authors thank the reviewer for his valuable comments.

References

1. Kalluri, D. and Prasad, R. C., *Appl. Sci. Res.*, 27, 415, 1973.
2. Yeh, C., *J. Appl. Phys.*, 37, 3079, 1966.
3. Yeh, C., *J. Appl. Phys.*, 38, 2871, 1967.
4. Lee, S. W. and Lo, Y. T., *J. Appl. Phys.*, 38, 870, 1967.
5. Chawla, B. R. and Unz, H., *IEEE Trans. Antenna Propag.*, AP-17, 771, 1969.
6. Kalluri, D. and Shrivastava, R. K., *IEEE Trans. Antenna Propag.*, AP-21, 63, 1973.
7. Kalluri, D. and Shrivastava, R. K., *J. Appl. Phys.*, 44, 2440, 1973.
8. Kalluri, D. and Shrivastava, R. K., *J. Appl. Phys.*, 44, 4518, 1973.
9. Kalluri, D. and Shrivastava, R. K., *IEEE Trans. Plasma Sci.*, PS-2, 206, 1974.
10. Shiozawa, T., Hazma, K., and Kumgai, N., *J. Appl. Phys.*, 38, 4459, 1967.
11. Pyati, V. P., *J. Appl. Phys.*, 38, 652, 1967.
12. Kong, J. A. and Cheng, D. K., *J. Appl. Phys.*, 39, 2282, 1968.
13. Cheng, D. K. and Kong, J. A., *Proc. IEEE*, 56, 248, 1968.
14. Budden, K. G., *Radio Waves in the Ionosphere*, Cambridge University Press, Cambridge, U.K., 1961.
15. Derusso, P. M., Roy, R. J., and Close, C. M., *State Variables for Engineers*, Wiley, New York, p. 232, 1965.
16. Chawla, B. R., Kalluri, D., and Unz, H., *Radio Sci.*, 2, 869, 1967.
17. Sommerfeld, A., *Electrodynamics*, Academic Press, New York, p. 287, 1964.
18. Kalluri, D. and Prasad, R. C., *Int. J. Electron.*, 35, 801, 1973.

Appendix 14G: Moving Point Charge and Lienard–Wiechert Potentials

In Appendix 1B, we wrote down the scalar wave equation (1B.30) and its solution (1B.31) in terms of the integral of the charge density at the retarded time $[\rho_V]$. For the free space medium,

$$[\rho_V] = \rho_V\left(t - \frac{R_{SP}}{c}\right). \tag{14G.1}$$

Assuming that a point charge Q is defined as an integral of the volume charge density in the limit the volume shrinks to a point, the solution for the scalar potential can be written as

$$\Phi_P(\boldsymbol{r},t) = \frac{Q\left(t - \frac{R_{SP}}{c}\right)}{4\pi\varepsilon_0 R_{SP}}, \tag{14G.2}$$

where

$$R_{SP} = |\boldsymbol{r} - \boldsymbol{r}'|. \tag{14G.3}$$

In the above, S is the source point where $Q(t)$ is permanently located and P is the field point. This cannot be an isolated, time-varying, single charge because of the conservation of charge requirement and can be one of the charges in a dipole source. A single charge is used here mainly to introduce the concept of retarded time. Even if Q is not moving, we have to use the value of the charge at a retarded time t_r:

$$t_r = \left(t - \frac{R_{SP}}{c}\right). \tag{14G.4}$$

Let us now consider a point charge moving with a velocity $\boldsymbol{v}$ along a specified trajectory as shown in Figure 14G.1

We wish to compute the scalar and vector potentials at a field point P due to this moving source. Note that the retarded time t_r is governed by the equation

$$|\boldsymbol{r} - \boldsymbol{r}'(t_r)| = R_{SP}(t_r) = c(t - t_r). \tag{14G.5}$$

This is a classical problem, solved in many textbooks [1,2], and the solution is given under Lienard–Wiechert potentials, stated below:

$$\Phi_P(\boldsymbol{r},t) = \frac{1}{4\pi\varepsilon_0}\frac{Q}{R_{SP}\left(1 - \frac{\hat{R}_{SP}\cdot\boldsymbol{v}}{c}\right)} \tag{14G.6}$$

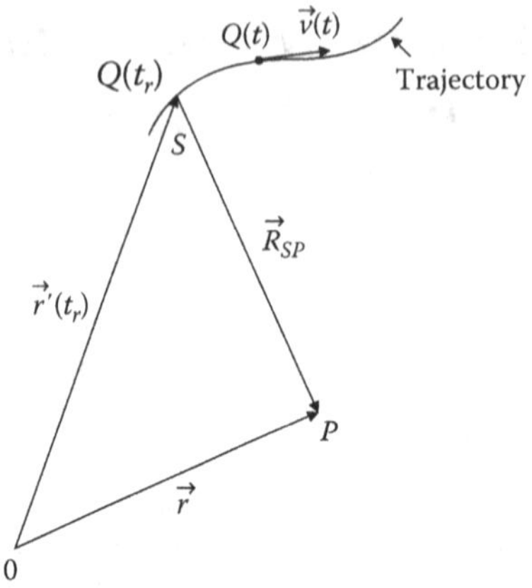

FIGURE 14G.1
Geometry for calculation of Lienard–Wiechert potentials due to a moving point charge Q. S is the retarded position of the charge. $\boldsymbol{R}_{SP}$ is the directed line segment from the retarded position to the field point.

$$\boldsymbol{A}_P(\boldsymbol{r},t) = \frac{\mu_0}{4\pi} \frac{Q\boldsymbol{v}}{R_{SP}\left(1 - \frac{\hat{R}_{SP}\cdot \boldsymbol{v}}{c}\right)} = \frac{\boldsymbol{v}}{c^2}\Phi_P(\boldsymbol{r},t). \tag{14G.7}$$

The electric and B fields are then obtained using (1.7) and (1.8):

$$\boldsymbol{E}_P(\boldsymbol{r},t) = \frac{Q}{4\pi\varepsilon_0} \frac{R_{SP}}{(\boldsymbol{R}_{SP}\cdot \boldsymbol{u})^3}\left[\boldsymbol{u}\left(c^2 - v^2\right) + \boldsymbol{R}_{SP}\times(\boldsymbol{u}\times\boldsymbol{a})\right], \tag{14G.8}$$

where

$$\boldsymbol{u} = \left(c\hat{R}_{SP} - \boldsymbol{v}\right) \tag{14G.9}$$

and $\boldsymbol{a} = \dot{\boldsymbol{v}}$ is the acceleration at the retarded time.

The B field is given by

$$\boldsymbol{B} = \frac{1}{c}\hat{R}_{SP}\times \boldsymbol{E}. \tag{14G.10}$$

The first term in (14G.8) is called the generalized Coulomb field and the second term is called the radiation field [1]. In arriving at these results, there are several subtle points to be considered as explained in detail in [1,2]:

1. Concept of the extended particle.

   You probably have come across several places in the book where we treated electron as a point charge. On the other hand, the electron in some other context is treated in nuclear physics as a distributed charge in a spherical volume of radius r_0 given by

$$r_0 = \frac{e^2}{4\pi\varepsilon_0 mc^2}, \tag{14G.11}$$

where e is the charge and m is the mass of the electron. In the relativity theory, a rigid dimension for an object is not correct in view of the Lorentz contraction of the length of a moving object. Let us consider the point charge Q as the total charge due to a volume charge density, in the sense

$$Q = \lim_{V\to 0,} \iiint_V \rho_V dV. \tag{14G.12}$$

The numerator N in the solution given in (1.23) for the retarded potential,

$$N = \iiint_V [\rho_V] dV', \tag{14G.13}$$

does not represent the total charge Q, when Q is moving with a velocity $\boldsymbol{v}$. Note that the different source points in the volume V have different retarded times t_r. Thus, Q_{extd} is the new total charge of the extended particle [1]:

$$Q_{extd} = \iiint_V \rho_V(\boldsymbol{r}', t_r) dV' = \frac{Q}{\left(1 - \frac{\hat{R}_{SP}\cdot\boldsymbol{v}}{c}\right)}. \tag{14G.14}$$

One can interpret (14G.14) as evaluating (14G.13) as though the volume element is distorted due to the Lorentz contraction in the direction of the charge motion.

2. The differentiation in space and time involved in obtaining the fields from the potentials requires careful evaluations, as given in [1].

   Let us now consider the power radiated by the moving charge. The relevant electric field in the radiation zone is the second term in (14G.8) and the radiated power density [1] is

$$\boldsymbol{S}_{rad} = \frac{1}{\mu_0 c} E_{rad}^2 \hat{R}_{SP}. \tag{14G.15}$$

The radiated power can be shown to be [1,2]

$$P_{rad} = \frac{1}{4\pi\varepsilon_0}\frac{2}{3}\frac{Q^2}{c^3}\gamma^6\left(a^2 - \left|\frac{\boldsymbol{v}}{c}\times\boldsymbol{a}\right|^2\right), \tag{14G.16a}$$

where

$$\gamma = \frac{1}{\sqrt{1 - v^2/c^2}}. \tag{14G.16b}$$

For nonrelativistic velocities , and when the second term is negligible compared to the first on the right side of (14G.16a), the radiated power is given by the Larmor formula

$$P_{rad} = \frac{1}{4\pi\varepsilon_0}\frac{2}{3}\frac{Q^2}{c^3}a^2. \tag{14G.17}$$

This will be the power radiated by the charge Q, which is oscillating (the trajectory of the particle is harmonic) at the instant its velocity is zero and the acceleration is high.

One can discuss several interesting radiation phenomena [1,2], like (1) braking radiation, or bremsstrahlung ($\boldsymbol{v}$ and $\boldsymbol{a}$ are collinear), and (2) synchrotron radiation (circular motion), based on the analysis of a point charge moving in a specified trajectory, the subject of this appendix.

References

1. Griffiths, D. J., *Introduction to Electrodynamics*, Prentice-Hall, Englewood Cliffs, NJ, 1981.
2. Panofsky, W. K. H. and Phillips, M., *Classical Electricity and Magnetism*, 2nd edition, Dover Publications, Mineola, NY, 2005.

Part IV

Chapter Problems

Problems

Chapter 1

P1.1 For a simple lossless medium, derive the wave equations for **E** and H.

P1.2 Obtain the wave Equations 1.18 and 1.19 subject to the Lorentz condition given by Equation 1.21.

P1.3 Obtain Equation 1.27.

P1.4 Show that for a time-harmonic case, Equation 1.22 reduces to Equation 1.50.

P1.5 a. Obtain the expression for the time-harmonic retarded potential **A** for a current element of small length h located on the z-axis as shown in the figure.

b. Hence obtain the electric and magnetic fields at P. Assume that the current in the filament is $I_0(A)$.

c. In the far zone, that is, for $R \gg \lambda$, $h \ll \lambda$, obtain $\mathbf{E}, \mathbf{H}, \mathbf{E} \times \mathbf{H}$, and $\langle \mathbf{S} \rangle$.

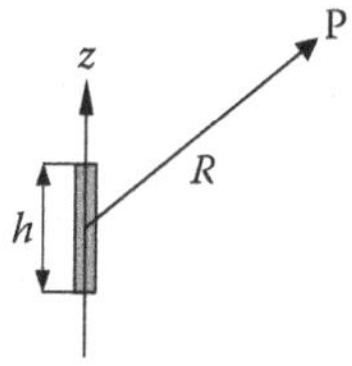

P1.6 Lecture module-week 1 is devoted to reviewing Maxwell's equations. In the undergraduate prerequisite course(s), considerable amount of time is spent in arriving at these equations based on experimental laws. In the process, vector calculus and coordinate systems are explained at length. The questions for the week make you revise the undergraduate background. You can look into the textbook you used in your undergraduate prerequisite course or search the internet in answering these two questions. The questions are:

a. Equations 1.1 through 1.4 are in differential form. Write down the corresponding equations in the integral form.

b. The integral form of Equation 1.1, Faraday's law, allows you to compute the voltage induced (emf) in a circuit. There are two components to it: (i) transformer emf and (ii) motional emf. Give an example where only (i) is induced; give an example where only (ii) is induced and give a third example where both (i) and (ii) are induced. In the third case, if your interest is in computing the total induced voltage and not in separating them into the two components, is there a simpler way of computing the total induced voltage? Illustrate the simpler way through an example.

P1.7 This question is to make you think about the concept of retarded potentials. Equations 1.22 and 1.24 are useful in answering this question. Note that Equation 1.22 is written when the source is a volume current element. For a filamentary current source, you will replace **J** dV by I d**L**.

Suppose the source is a differential filamentary current element located at the origin along the y-axis, where

$$I\,d\vec{L} = 2t\,\hat{y}\;dy.$$

Obtain the differential vector potential $d\vec{A}_P$ at the point P (5, 10, 15).

Note the following. You can use a bold letter or a letter with an arrow on the top to denote a vector. The hat on the top of y denotes a unit vector in the y-direction in the Cartesian system of coordinates (rectangular coordinate system).

P1.8 Consider a cylindrical capacitor of length ℓ. The radius of the inner cylinder is a and the radius of the outer cylinder is b. The dielectric constant of the dielectric inside the capacitor varies as (K is a constant)

$$\varepsilon_r(\rho) = K\rho^2.$$

Show that the conduction current $I(t)$ in the wire is the same as the displacement current in the capacitor.

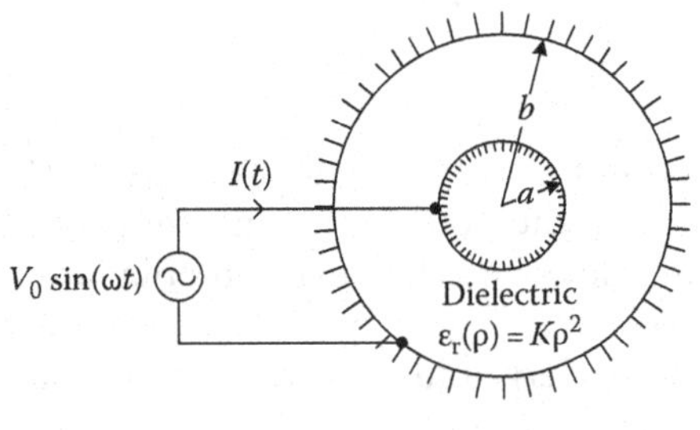

P1.9 A parallel plate capacitor is of cross-sectional area A and is filled with a dielectric material whose permittivity varies linearly from $\varepsilon = \varepsilon_1$ at one plate ($y = 0$) to $\varepsilon = \varepsilon_2$ at the other plate ($y = d$). Neglecting fringing effects, determine the capacitance.

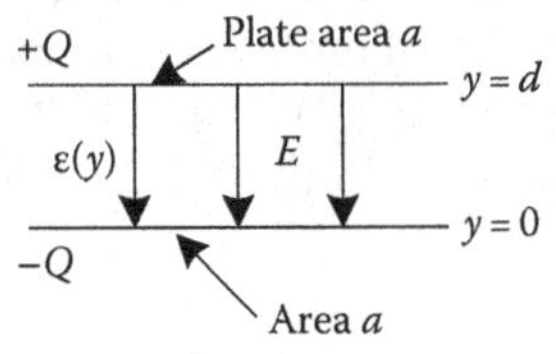

P1.10 Show that the displacement current between two concentric cylindrical conducting shells of radii r_1 and r_2, $r_2 > r_1$ is exactly the same as the conduction current in the external circuit. The applied voltage is $V = V_0 \sin \omega t$.

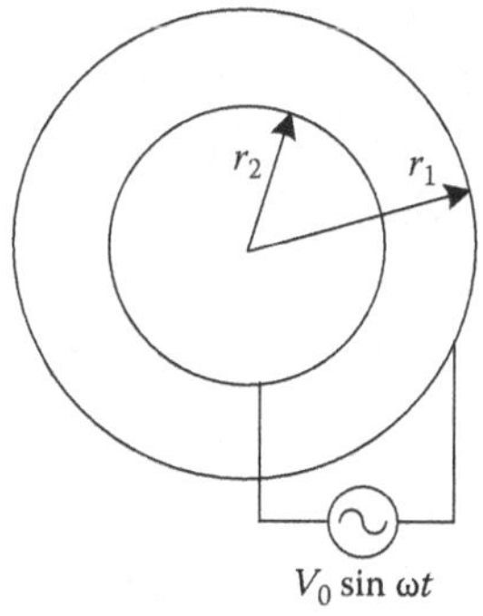

P1.11 Derive (2.42) through (2.44), based on Section 1.6.

P1.12 Determine the vector $\boldsymbol{H}$ field at a general point in free space due to a DC current I A in an infinitely long filament lying on the z-axis. Use Ampere's law and symmetry arguments.

P1.13 Suppose that an infinite sheet of current of surface current density $\hat{x}K$ A/m lying on the xy plane. Determine $\boldsymbol{H}$ at a general point P(x, y, z). Use Ampere's law and symmetry arguments. Show that the boundary condition (1.17) is satisfied by your results.

P1.14 Equation 1A.14 is the well-known divergence theorem. Less well known are some more theorems that can be derived from the divergence theorem. These are (1A.68) and (1A.69). Prove them

P1.15 Equation 1A.66 is the well-known Stoke's theorem. Less well known are some more theorems that can be derived from the Stoke's theorem. One of them is (1A.70). Prove it.

P1.16 Derive wave equation for electroquasistatics and magnetoquasistatics discussed in Section 1.5. Assuming the medium is free space and the fields vary with only z coordinate, obtain their form.

P1.17 Derive an expression for ψ, the angle between the position vectors $\boldsymbol{r}$ and $\boldsymbol{r}'$. Express it in terms the of their spherical coordinates.

Chapter 2

P2.1 Derive Equations 2.1 through 2.3.

P2.2 Assume that the time-harmonic fields in a perfect conductor are zero. Show that boundary conditions at the interface of a perfect conductor are given by Equations 2.23 or 2.24 or 2.25 or 2.26.

P2.3 The principle of induction heating is explored in this problem. A sheet of metal (good conductor) of conductivity σ_c is inserted into AC magnetic field whose value at the surface $z = \pm d$ is assumed to be $\mathbf{H}_1 = -\hat{y}H_1$.

a. Show that the AC magnetic field in the conductor is given by $H(z) = H_1(\cosh \tau z/\cosh \tau d)$ where $\tau = (1 + j)/\delta$ and $\delta = 1\sqrt{\pi f \mu \sigma_c}$. Assume that $\delta \sim d$.

b. Show that $\mathbf{J} = J\hat{x}$, where $J = \tau H_1(\sinh \tau z/\cosh \tau d)$.

c. Find the power consumed by the sheet of length (along x-axis) 1 m and width (along y-axis) 1 m. You do not have to show it for the homework, but a little further work will show that the power consumed per unit volume is maximized if $d \approx 1.125\ \delta$.

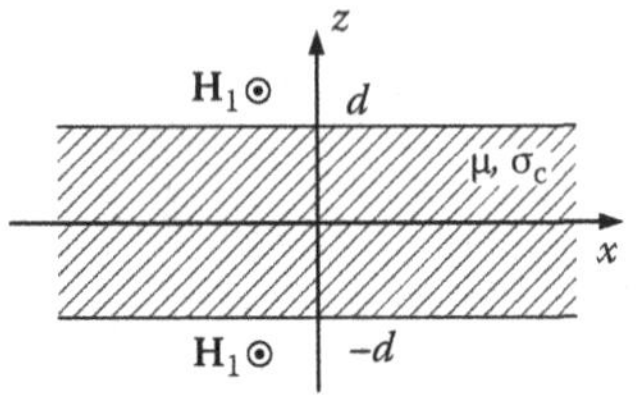

P2.4 Figure P2.4a shows the circuit equivalent of a section Δz of a transmission line.

a. Find the differential equation for the instantaneous current $I(z, t)$,

b. Assume that the current varies harmonically in space and time, that is,

$$I(z,t) = I\exp\left[\mathrm{j}(\omega t - kz)\right].$$

Find the relation between ω and k.

c. A sketch of ω and k is given in Figure P2.4b. Find the value of the cutoff frequency ω_c and the slope of the asymptotes (dotted lines).

P2.5 The electric and magnetic fields of a coaxial transmission line whose cross-section is shown in Figure P2.5 are given as follows: The line connects a source to a load.

$$\text{At } \rho = 1: \mathbf{E} = 100\,\hat{a}_\rho + 0.05\,\hat{a}_z \text{ V/m}, \quad \mathbf{H} = 2\hat{a}_\phi \text{ A/m}$$

$$\text{At } \rho = 1: \mathbf{E} = C\hat{a}_\rho + 0.02\hat{a}_z \text{ V/m}, \quad \mathbf{H} = D\hat{a}_\phi \text{ A/m}$$

a. Determine C and D.

b. Let the length of the line be 30 m. Find the source power, power lost in the line, and the power received by the load.

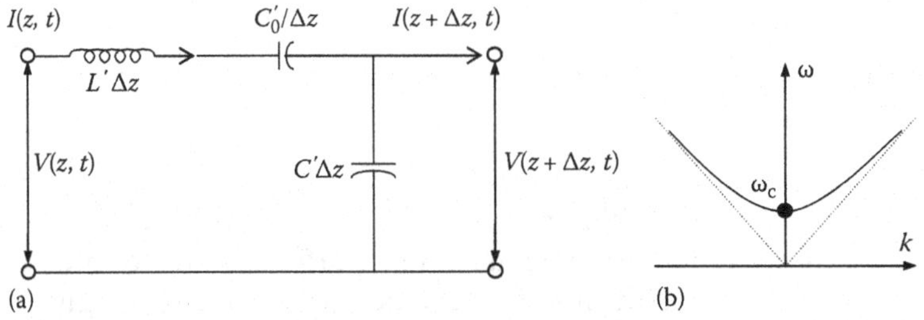

FIGURE P2.4

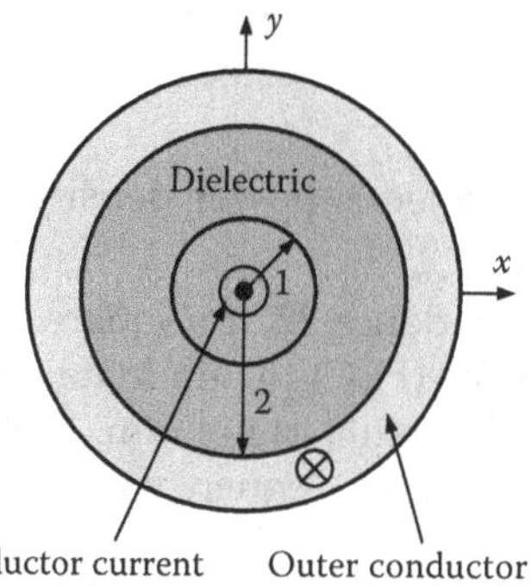

FIGURE P2.5

c. There could be a misprint in the statement: "At $\rho = 2: \mathbf{E} = C\hat{a}_\rho + 0.02\hat{a}_z$ V/m." Find the misprint and correct it. Give one or two reasons as to why you think there is a misprint. This misprint perhaps has not affected your answers to parts (a) and (b).

P2.6 A loss-free nonuniform transmission line has

$$L' = L'(z),$$
$$C' = C'(z),$$

where L' and C' are per meter values of the series inductance and parallel capacitance of the transmission line, respectively.

a. Determine the partial differential equation for the instantaneous voltage $V(z, t)$.
b. For an exponential transmission line

$$L'(z) = L_0 \exp(qz),$$
$$C'(z) = C_0 \exp(-qz),$$

assuming $V(z, t) = V_0 \exp[j(\omega t - kz)]$, determine the relation between ω and k.

P2.7 A distortionless line satisfies the condition

$$\frac{R'}{L'} = \frac{G'}{C'}.$$

Find α, β, and Z_0 for such a line.

P2.8 The magnetic field vector-phasor of a plane wave is given by

$$\tilde{\bar{H}}(x) = 2[\hat{y} + j\hat{z}]\exp(jkx). \tag{P2.1}$$

a. Use the properties of a plane wave to determine $\bar{E}(x,t)$.
b. You probably recognized from the expression for $\bar{E}(x,t)$ that it is circularly polarized. Is it L wave or R wave?

c. Had you used Equation P2.1 to determine the polarization, you would get the same answer as that for (b). Check this out by drawing the diagram like Figure 2.6b or c.

d. Determine the instantaneous power density $\bar{S}(x,t)$ and active power density $\langle \bar{S}_R(x) \rangle$.

P2.9 A conducting film of impedance 377 Ω/square is placed a quarter-wavelength in air from a plane conductor (PEC) to eliminate wave reflection at 9 GHz. Assume negligible displacement currents in the film. Plot a curve showing the fraction of incident power reflected versus frequency for frequencies 6–18 GHz.

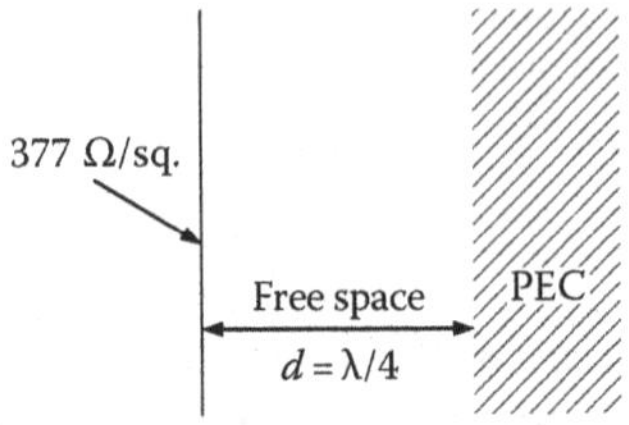

P2.10 Calculate the reflection coefficient and percent of incident energy reflected when a uniform plane wave is normally incident on a Plexiglas radome (dielectric window) of thickness 3/8″, relative permittivity $\varepsilon_r = 2.8$, with free space on both sides. $\lambda_0 = 20$ cm. Repeat for $\lambda_0 = 10$ cm. Repeat for $\lambda_0 = 3$ cm. Comment on the results obtained.

P2.11 A green ion laser beam, operating at $\lambda_0 = 5.45$ μm, is generated in vacuum, and then passes through a glass window of refractive index 1.5 into water with $n = 1.34$. Design a window to give zero reflection at the two surfaces for a wave polarized with E in the plane of incidence, that is, find $\theta_{B2} - \theta_{B1}$.

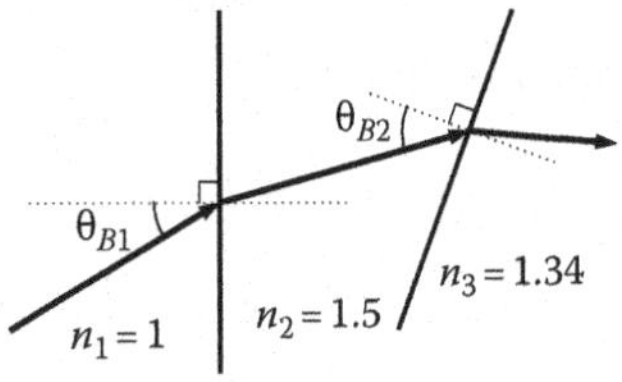

P2.12 The transmitting antenna of a ground-to-air communication system is placed at a height of 10 m above the water, as shown in the figure. For a separation of 10 km between the transmitter and the receiver, which is placed on an airborne platform, find the height h_2 above water of the receiving system so that the wave reflected by the water does not possess a parallel-polarized component. Assume that the water surface is flat and lossless [Reference 2 of Chapter 2].

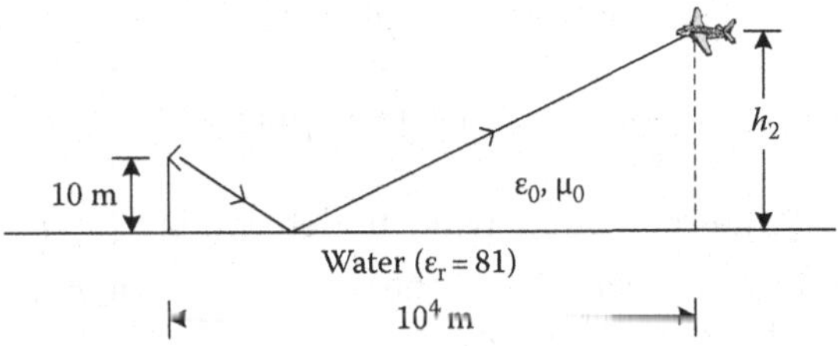

P2.13 A source $V_{gs} = 60\angle 0°$ with internal impedance 200 Ω is connected to a resistive load $Z_{L1} = 200\ \Omega$ by a two-wire lossless transmission line in air (velocity of propagation $c = 3 \times 10^8$ m/s) of length $l_1 = 10$ cm and characteristic impedance $Z_{01} = 50\ \Omega$. This load is connected to another load Z_{L2} by the two-wire transmission line of length $l_2 = 5$ cm. Let the load Z_{L2} be replaced by a short-circuit. Determine the current I_{gs} for two values of frequency (f):

a. $f = 60$ Hz. This is the frequency of the house current.

Hint: For very small values of βl, $\tan \beta l$ may be approximated by zero.

b. $f = 1.5$ GHz.

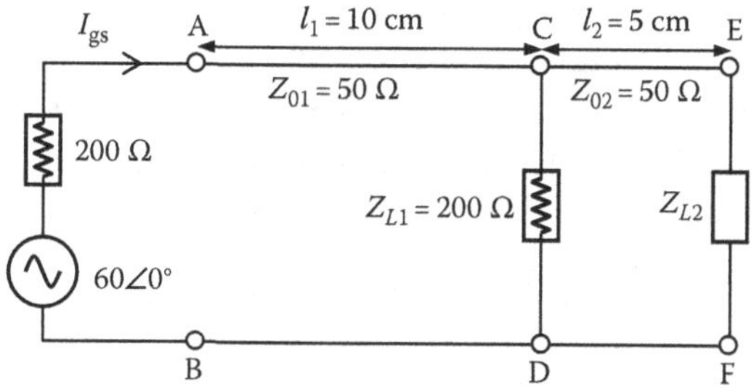

P2.14 A dielectric slab of dielectric constant ε_r is shown in the figure. A wave is incident on it from one of its ends at an oblique angle $0° < \theta_i < 90°$. Show that for $\varepsilon_r \geq 2$, the wave is contained in the slab for all angles of θ_i.

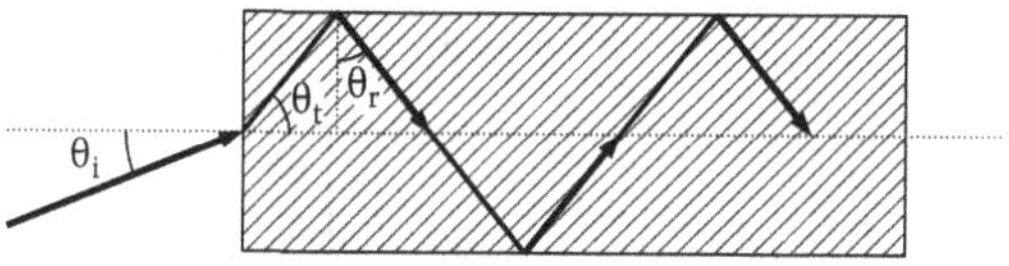

P2.15 A source that radiates isotropically is submerged at a depth d below the surface of water as shown in the figure. How far in the x-direction (both positive and negative) can an observer (on the water surface) go and still receive the radiation? Assume that the water is flat lossless with a dielectric constant of 81 [Reference 2 of Chapter 2].

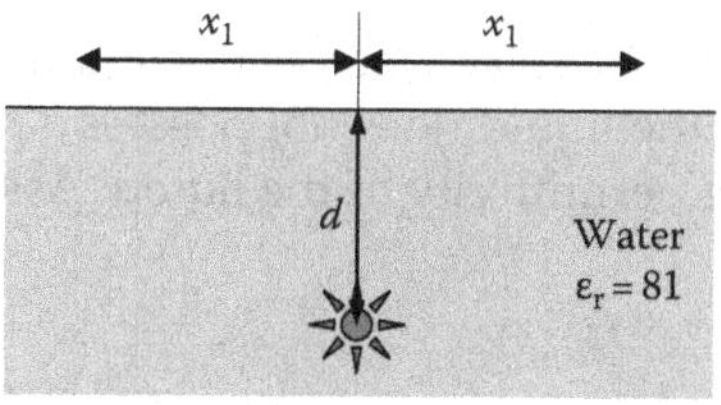

P2.16 Obtain Equations 2.112 through 2.115.

P2.17 The reflection coefficient Γ is real for a dielectric–dielectric interface for $0° < \theta_1 < \theta_c$, where θ_c is the critical angle. For $\theta_2 > \theta_c$, there is total reflection and the magnitude of the reflection coefficient is 1. Thus, in the range $\theta_c < \theta_2 < 90°$, the reflection coefficient may be represented as $\Gamma = e^{j\phi}$. Plot ϕ versus θ_1 for (a) s wave and (b) p wave. Take $n_1 = 1.5$ and $n_2 = 1.0$. In the above θ_1 is the angle of incidence.

P2.18 a. Obtain Equation 2.126.

b. Calculate the radiated power leaving the surface of a cylinder of radius ρ and length 1 m. Assume that the cylinder is in the far zone and the source is a constant harmonic current $\tilde{I}$ along a long wire on the z-axis.

P2.19 The electric field phasor of an elliptically polarized wave is given below:

$$\mathbf{E}(y) = \left[100\exp\left(\frac{j\pi}{4}\right)\hat{x} + 50\exp\left(-\frac{j\pi}{4}\right)\hat{\mathbf{z}}\right]\exp(-jk_0 y)\ (\text{V/m}).$$

Show that this field can be expressed as the sum of the fields of a left circularly polarized wave (LCP) and a right circularly polarized wave (RCP). Write the complete expression for the electric field phasor of the (a) LCP wave and (b) RCP wave.

P2.20 See below figure. Determine the electric field of the reflected wave ($\mathbf{E}^R$) when the electric field of the incident wave (assume normal incidence) is

$$\mathbf{E}^I = 2e^{j\left[2\pi\times10^{10}t - k_0 z\right]}\hat{\mathbf{x}}\ (\text{mV/m})$$

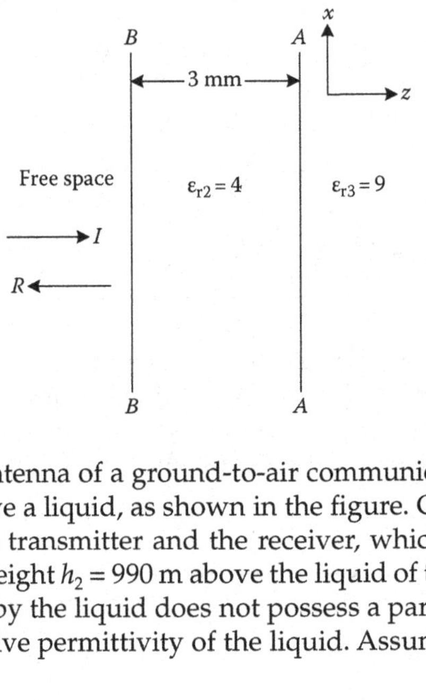

P2.21 The transmitting antenna of a ground-to-air communication system is placed at a height of 10 m above a liquid, as shown in the figure. Given that the separation of 10 km between the transmitter and the receiver, which is placed on an airborne platform, and the height h_2 = 990 m above the liquid of the receiving system so that the wave reflected by the liquid does not possess a parallel-polarized component, determine the relative permittivity of the liquid. Assume that the liquid surface is flat and lossless.

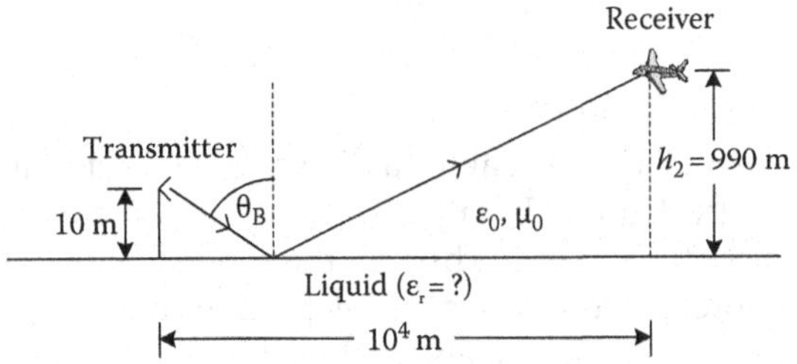

P2.22 A longitudinal section of a dielectric is shown in the figure. If $\theta_1 = 30°$ and θ_3 is the critical angle at interface 2, determine £$_r$ of the dielectric.

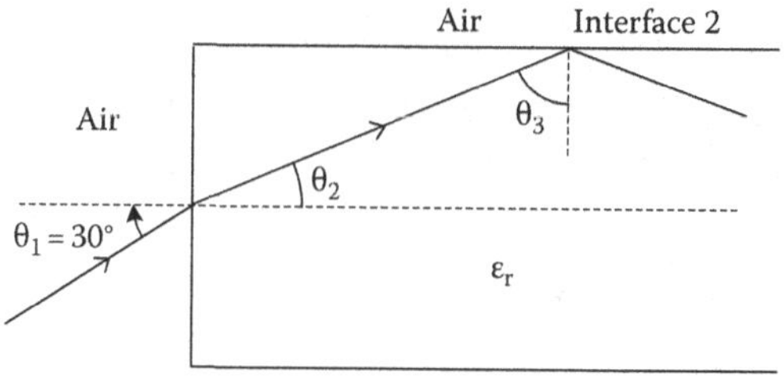

P2.23 Refer to the figure. The instantaneous electric field of the incident wave consisting of a *p* as well as a *s* wave is given by

$$E_I = (\hat{\mathbf{x}}3 - \hat{\mathbf{z}}4)\cos(\omega t - k_x x - k_z z) + 5\hat{\mathbf{y}}\sin(\omega t - k_x x - k_z z)\ (\text{V/m}).$$

a. Determine the angle of incidence.
b. Let $\varepsilon_2 = A\varepsilon_0$, $\mu_2 = \mu_0$, and the angle of incidence is the Brewster angle. Determine A. Also determine (i) Γ_s and (ii) the instantaneous electric field of the reflected wave.

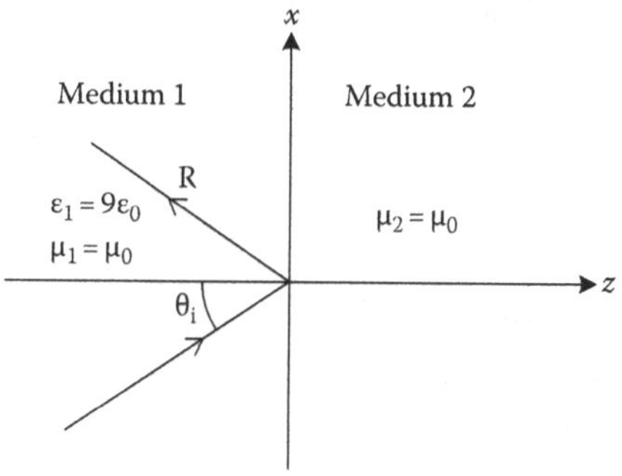

P2.24 a. A *p* wave is propagating in the ζ-direction as shown in Figure P2.24a. Its electric field is given by

$$\hat{\mathbf{E}}\text{p} = (3\hat{\mathbf{x}} - 4\hat{\mathbf{z}})\exp[-j(k_x x + k_z z)]\ (\text{V/m}).$$

Determine the angle θ. Determine $\hat{\mathbf{H}}_p$ if the medium is air.

b. The instantaneous electric field of an incident wave (see Figure P2.24b) consisting of both *p* and *s* waves is given by

$$\mathbf{E}^I = (6\hat{\mathbf{x}} - 8\hat{\mathbf{z}})\cos(\omega t - k_x x - k_z z) + 10\hat{\mathbf{y}}\sin(\omega t - k_x x - k_z z)\ (\text{V/m}).$$

1. Determine θ_1.
2. Determine θ_r.
3. If $k_x = 0.5$ rad/m, determine the frequency *f* in Hertz and wavelength λ in meters.

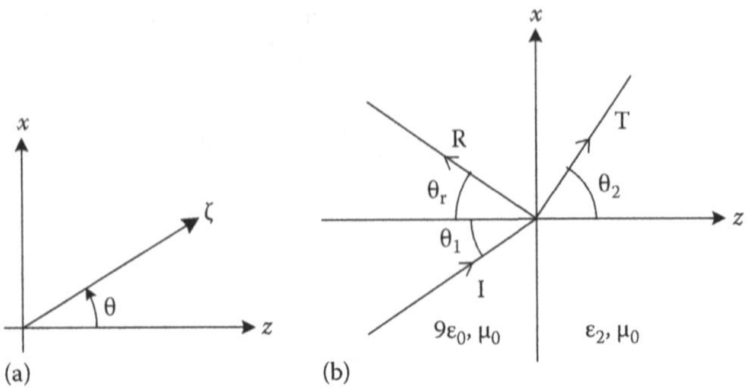

FIGURE P2.24

4. Determine the polarization of the incident wave.
5. Find the value of ε_2 for which the incident angle is equal to the critical angle.
6. The value of ε_2 for which the reflected wave is linearly polarized.
7. If $\varepsilon_2 = 16\varepsilon_0$, determine the instantaneous electric field of the reflected wave and the instantaneous electric field of the transmitted wave.
8. If $\varepsilon_2 = 25\varepsilon_0$, determine the instantaneous electric field of the reflected wave.

P2.25 The phasor electric field of a circularly polarized standing wave in free space is given by

$$\tilde{\mathbf{E}} = (\hat{\mathbf{x}} - j\hat{\mathbf{y}})E_0 \sin(kz). \quad \text{(P2.2)}$$

i. Determine
   a. The magnetic field $\tilde{\mathbf{H}}$
   b. Instantaneous power density
   c. Time-averaged power density
   d. Assuming the standing wave is linearly polarized and its phasor electric field is given by

$$\tilde{\mathbf{E}} = \hat{\mathbf{x}}E_0 \sin(kz). \quad \text{(P2.3)}$$

   Determine the quantities in (a), (b), and (c).
ii. Let the field given by Equation P2.2 exist between two perfectly conducting plates located at $z = 0$ and $z = d$. Find the relation between the angular frequency ω and d.

P2.26 The $z = 0$ plane is a sheet of current carrying a harmonic surface current density

$$\tilde{\mathbf{K}}(x,y) = \hat{\mathbf{x}}K_0 \exp\left(-j\frac{\sqrt{3}}{2}k_0 x\right) \ (\text{A/m})$$

Determine the vector phasor electric field $\tilde{\mathbf{E}}$ and magnetic field $\tilde{\mathbf{H}}$ (i) for $z < 0$, (ii) for $z > 0$, in terms of K_0.

P2.27 A *s*-wave in medium 1 is incident (plane of incidence *x*–*z* plane) on medium 2 at an angle θ with the normal to the interface, that is, the phasor electric field of the incident wave is given by

$$\mathbf{E}_S^I = \hat{\mathbf{y}} E_0 e^{\left[-jk_1(x\sin\theta + z\cos\theta)\right]} \ (\mathrm{V/m})$$

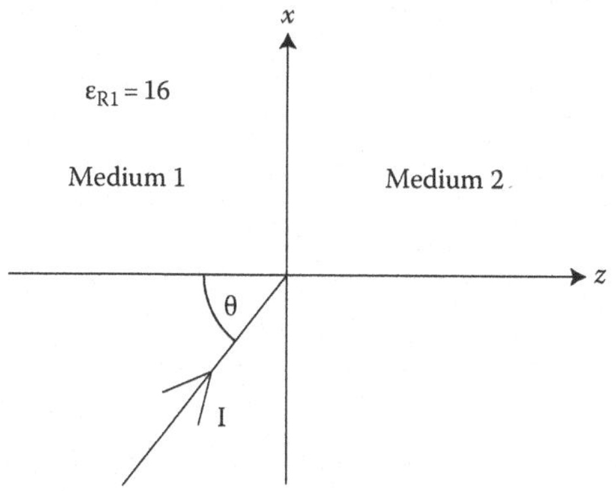

Medium 1 is a dielectric with relative permittivity $\varepsilon_{R1} = 16$ and the frequency is 10 GHz.

a. Determine k_1.

b. Let medium 2 be a perfect conductor. (i) Write down the expression for the vector magnetic field H^R of the reflected wave. (ii) Reflection coefficient Γ_S. (iii) Power reflection coefficient. (iv) Total magnetic field $H(x, y, 0^-, t)$. (v) Surface current K at the interface $z = 0$.

c. Let medium 2 be a dielectric with relative permittivity $\varepsilon_{R2} = 9$. (i) Determine the critical angle θ_c. (ii) Take $\theta = \theta_c + 5°$. The time domain expression for the *x*-component of the reflected magnetic field may be written as $H_x^R = A\cos(\omega t + bx + cz + 2\phi)$: determine ω, A, b, c, ϕ.

d. The medium 2 is the same as part (c). A *p*-wave of double the electric field strength of *s*-wave is added to the incident wave.

   i. Write down the expression for the electric field $\mathbf{E}^I$ of the incident wave consisting of both *s*- and *p*-waves in terms of E_0 and θ.

   ii. Determine the Brewster angle θ_B.

   iii. If $\theta = \theta_B$, write down the expressions for the time domain reflected electric field E^R and the time domain transmitted electric field E^T.

P2.28 Figure P2.28 shows the circuit equivalent of a section Δ*z* of a transmission line. Find the differential equation for the instantaneous voltage $V(z, t)$. Your solution should not include the current variable $I(z, t)$.

P2.29 The region $z > 0$ is a perfect conductor and the region $z < 0$ is free space. Let the magnetic field in free space be given by $\boldsymbol{H} = \hat{y}\cos(3\times10^8 t - 4x)$ (A/m).

a. Determine the surface current density $\vec{J}_s$ at $P(4,2,0)$ at $t = 5$ ns.

b. Determine the surface charge density ρ_s at $P(4,2,0)$ at $t = 5$ ns.

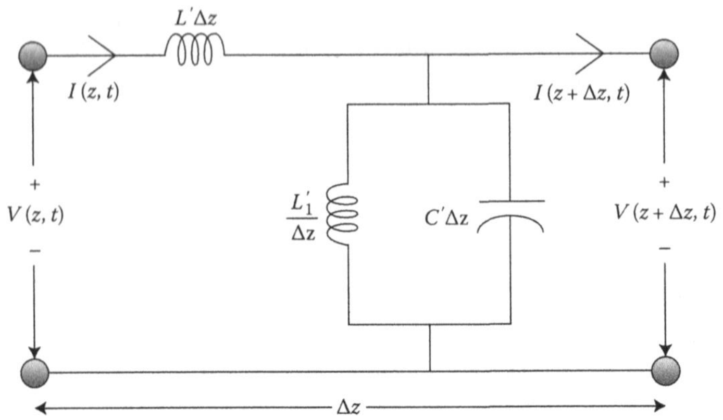

FIGURE P2.28

P2.30 The electric field intensity at the origin is $2\hat{x} - 10\hat{y} + 3\hat{z}\ (\text{V/m})$ at $t = 0$.
If the origin lies on a perfectly conducting surface while the material adjacent to the origin has $\varepsilon_R = 10$, $\mu_R = 2$, and $\sigma = 0$, find the magnitude of the surface charge density at the origin at $t = 0$.

P2.31 Wave propagation in an unbounded good conductor may be studied through a transmission line model. The below figure shows the circuit equivalent of a section Δz of a transmission line.

a. Obtain the first-order coupled differential equations for $V(z, t)$ and $I(z, t)$. Eliminate V and obtain a second-order differential equation for I.
b. Assume that the voltage varies harmonically in space and time, that is,

$$I(z,t) = I_0 \exp\left[j(\omega t - kz)\right],$$

find the relation between ω and k diagram.
c. Sketch the ω–k diagram.
d. Using your knowledge of wave propagation in a good conductor, express L' and G' of the figure in terms of μ and σ of the conductor.

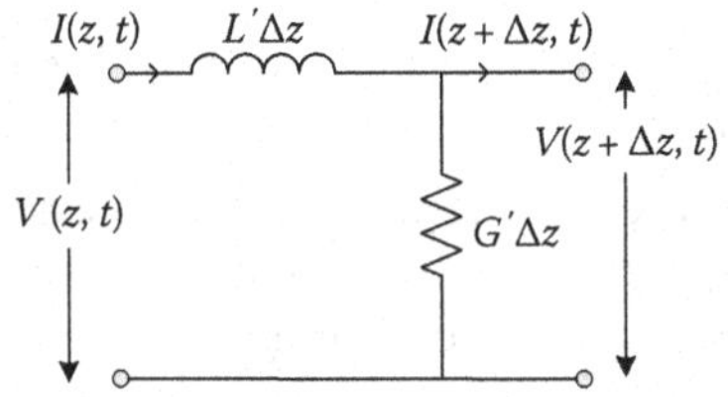

P2.32 The following figure shows two nonconducting media with a boundary at $z = 0$. The parameters of the medium and the electric field expressions in the two media are also written in the figure.

a. Find A, B, C constants in these expressions.

b. Determine the expression for the vector $\mathbf{H}_2$.

$\mu_{r2} = 20$

$\varepsilon_{r2} = 5$

$\sigma_2 = 0$

$\mathbf{E}_2 = \hat{\mathbf{x}} A \cos(Bt - Cz)(\text{V/m})$

$\mu_{r1} = 1$

$\varepsilon_{r1} = 1$

$\sigma_1 = 0$

$\mathbf{E}_1 = \hat{\mathbf{x}}[60 \cos(15 \times 10^8 t - 5z)$
$+ 20 \cos(15 \times 10^8 t + 5z)] \ (\text{V/m})$

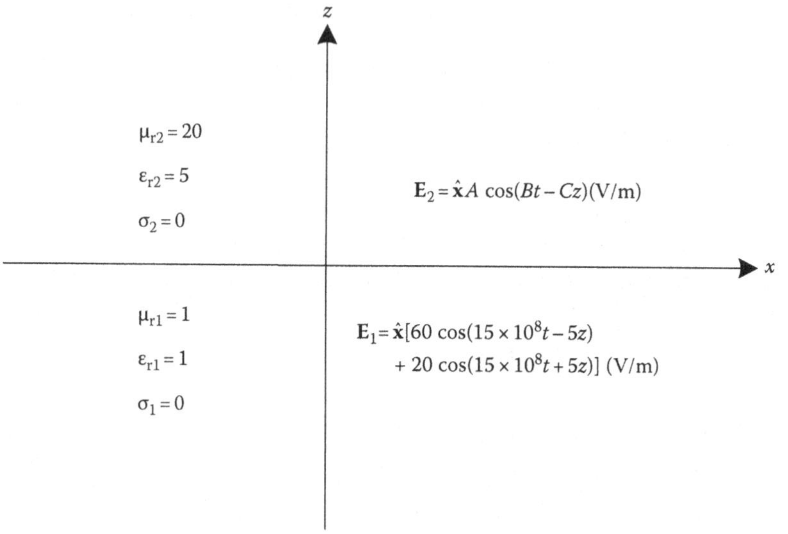

P2.33 The oblique wave propagation from an optically dense medium (medium 1) to an optically rare medium (medium 2) exhibits the phenomenon of critical angle. Let $n_1 = 1.5$ and $n_2 = 1$.

  a. Find the critical angle. If the incident wave is an *s* wave, determine the depth of penetration of the evanescent wave for an angle of incidence 5° more than the critical angle. Find the phase velocity of the evanescent wave along the interface. Express your answer in terms of the wave number in the second medium.

  b. Now consider the propagation of a *p* wave. Find the Brewster angle.

P2.34 In Section 2.4, the AC resistance R_{AC} was obtained using (2.36). Another way of obtaining R_{AC} is by calculating the impedance Z_{AC}. Such an impedance is defined as the voltage difference at the top of the conductor divided by the current entering the conductor. Show you get the same answer for R_{AC} as that of (2.38). Hint: See (8.135).

P2.35 Do Problem P2.13 when Z_{L2} is an open-circuit.

P2.36 Use (1.50) to obtain (2.121) by considering the dA_z due to differential length along the z-axis and integrating for the infinite length line.

Chapter 3

P3.1 Obtain Equations 3.29, 3.32, and 3.33.

P3.2 A standard X-band rectangular waveguide with inner dimensions of $a = 2.286$ cm (0.9″) and $b = 1.106$ cm (0.4″) is filled with lossless polystyrene ($\varepsilon_r = 2.56$). For the lowest order mode of the waveguide, determine at $f = 10$ GHz the following values:

  a. Cutoff frequency f_C (GHz).

  b. Guide wavelength λ_g (in cm).

  c. Wave impedance.

  d. Phase velocity.

P3.3 Show that $1/\lambda_g^2 = 1/\lambda^2 - 1/\lambda_c^2$.

P3.4 The $z = 0$ plane is a sheet of current carrying a harmonic surface current density

$$\mathbf{K}(x,y,t) = \hat{x}K_0 \exp(j\omega t) \text{ (A/m)}.$$

Assume the medium is free space.

a. Write the differential equation for the phasor vector magnetic potential **A**.

b. What component of the vector potential will be nonzero? On what coordinate does this component depend? Write the ordinary differential equation satisfied by this component. Find its solution valid for $z > 0$. The solution will contain an undetermined constant A_0.

c. Determine the vector phasor electric field **E** and magnetic field **H** in terms of A_0.

d. Use Amperes law to relate A_0 with K_0.

e. Calculate the radiated power per square meter.

f. What will be the values of **E** and **H** if $\omega = 0$.

g. Write down the electric and magnetic fields **E** and **H** for $z < 0$.

P3.5 Design a circular waveguide filled with a lossless dielectric medium whose relative permeability is unity. The waveguide must operate in a single dominant mode over a bandwidth of 1.5 GHz. Assume that the radius of the waveguide is 1.12 cm.

a. Find the dielectric constant of the medium that must fill the waveguide to meet the desired specifications.

b. Find the lower and upper frequencies of operation.

P3.6 Design a rectangular waveguide (determine a and b) for use in a microwave oven to operate at a frequency $f = 900$ MHz. Let $a/b = 1.8$. Choose a so that the frequency of the magnetron is 30% above the cutoff frequency of the TE_{10} mode, that is,

$$\frac{f}{(f_c)_{TE10}} = 1.3.$$

Determine the guide wavelength.

P3.7 Refer to the figure, given that the electric field $E_y = 0.5 \sin(\pi x/a)\, e^{-j\beta z}$ in a rectangular waveguide (TE_{10} mode), determines the surface charge density ρ_s on the bottom PEC plate at the point P $(a/3, 0, 0)$.

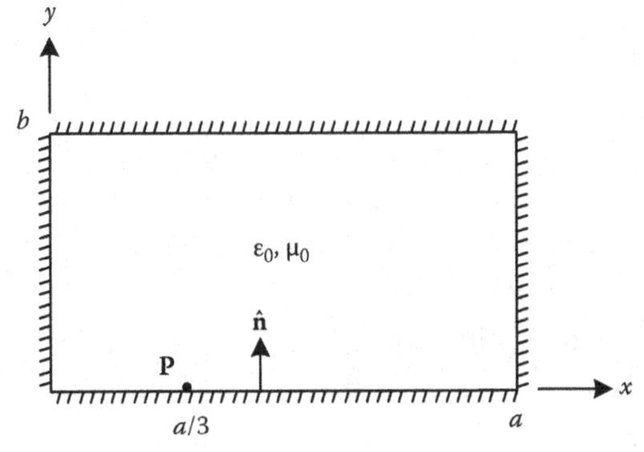

P3.8 Let a rectangular waveguide have a discontinuity in dielectric at $z = 0$, that is, ε_1 and μ_1 for $z < 0$, and ε_2 and μ_2 for $z > 0$.

a. Determine f/f_{c10} for which there is no reflection. In the above, f is the signal frequency and f_{c10} is the cutoff frequency of the TE_{10} mode.

b. Repeat for TM_{mn} mode, that is, find f/f_{cmn} for which there is no reflection (for TM_{mn} mode).

P3.9 a. Obtain the harmonic solution to the partial differential equation of Problem P2.28 by assuming $V(z, t) = V_0\, e^{j(\omega t-\beta z)}$, that is, obtain the relation between ω and β.

b. A sketch of ω–β diagram is shown in figure (a). Obtain the value of ω_c and the slope of the dotted asymptotic line.

c. Show that the transmission line in Problem P3.9 can be made to represent a TE_{mn} mode of propagation in a waveguide by a proper choice of $L', L_1', $ and C', choose $L' = \mu$, $C' = \varepsilon$. Determine the analogous entities of the waveguide for V, I, and L_1' of the transmission line.

d. Apply above results to TE_{23} mode in a rectangular waveguide shown in figure (b) and express L_1' in terms of a, b, and so on.

e. Can the same transmission line represent TM_{23} mode. Give reasons for your answer.

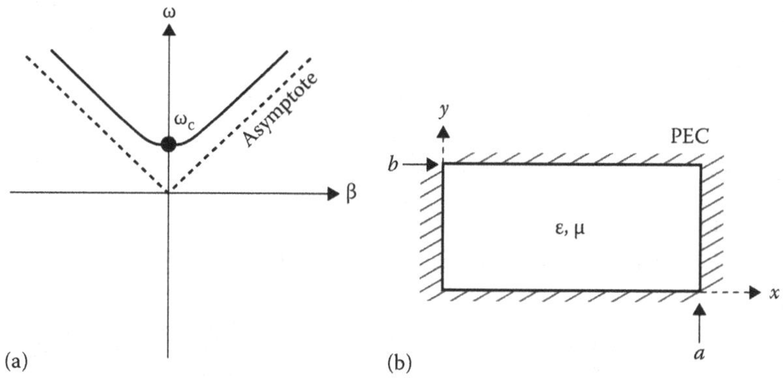

P3.10 The z-component of the magnetic field of TE_{11}^z mode of the rectangular waveguide is given below:

$$H_z = -j0.5\frac{(\pi/0.1)^2 + (\pi/0.05)^2}{2\pi\times 10^{10}}\left(3\times 10^8\right)^2 \cos(10\pi x)\cos(20\pi y)e^{-j\beta z}.$$

Determine

i. a

ii. b

iii. Cutoff frequency f_c.

iv. Assuming the signal frequency $f = 10^{10}$ Hz, determine β.

$$c = \frac{1}{\sqrt{\mu_0 \varepsilon_0}} = 3 \times 10^8 \text{m/s}.$$

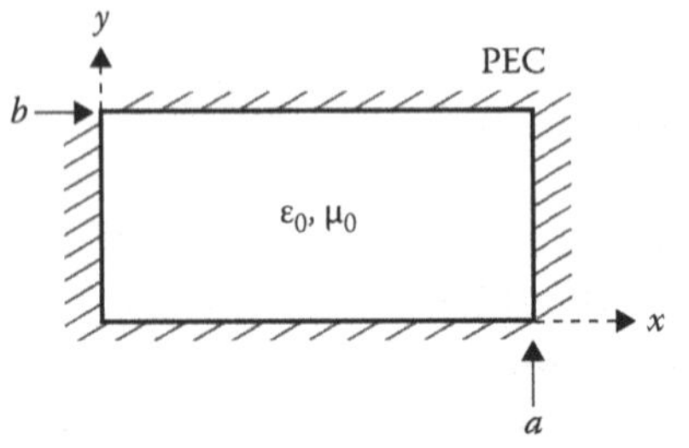

P3.11 Figure shows a conducting waveguide with cross section of a right-angled isosceles triangle.

Part A:

i. For TM modes show that the z-component of the electric field:

$$E_z = C_{mn}\left[\left(\frac{\sin m\pi x}{a}\right)\left(\frac{\sin n\pi y}{a}\right) - \left(\frac{\sin n\pi x}{a}\right)\left(\frac{\sin m\pi y}{a}\right)\right].$$

ii. What are the allowed values of m and n?

iii. Find the two lowest cutoff frequencies.

Part B:

i. For the TE modes, find H_z.

ii. Find the lowest cutoff frequency of TE modes.

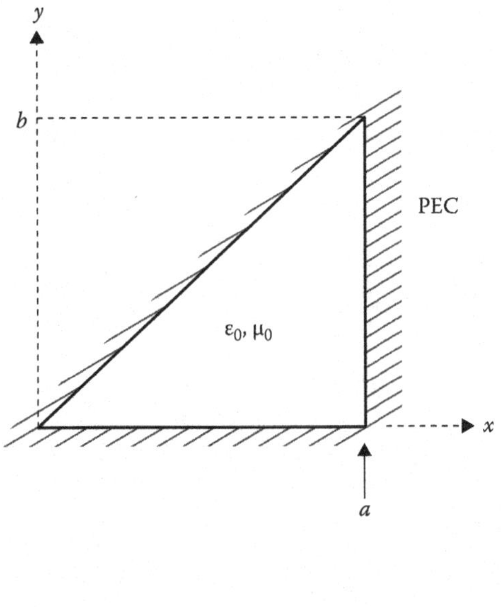

P3.12 Prove (3.54) and (3.55).

P3.13 Obtain (3.54). Several steps are needed to obtain (3.54). These steps are as follows:

(a) Show that

$$\alpha_c = -\frac{dP/dz}{2P} \approx \frac{W_L}{2P},$$

where, P is the active power carried by the guide assuming the walls of the guide are perfect conductors, W_L is the power loss per meter length.

(b) Calculate W_L using the surface resistance concept given in *Appendix 4A*.

(c) Calculate P.

(d) Simplify by expressing β and cosψ in terms of (f_c/f).

Chapter 4

P4.1 Find the three lowest resonant frequencies of the cylindrical cavity shown in the figure. Consider both TE and TM modes.

Note: After reviewing Section 3.7 on cylindrical wave guides and Sections 4.1 and 4.2 on rectangular cavities you can extend the theory to cylindrical cavities.

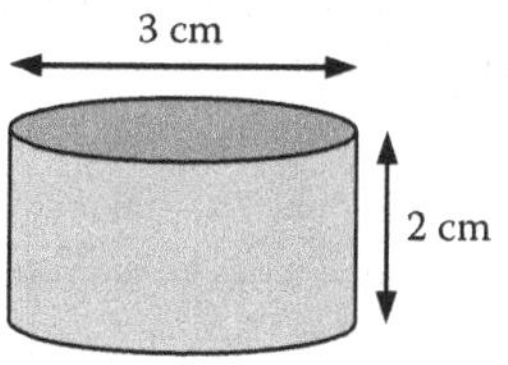

P4.2 A circular cylindrical cavity excited in the TM_{011} mode appears to resonate at a frequency of 9.375 GHz. The cavity is made of aluminum ($\sigma = 3.5 \times 10^7$ S/m) walls and is of length 2 cm.

a. Find the diameter of the cavity.
b. If the cavity is completely filled with wet snow (dielectric constant = 3.5 – j0.1), what will be the new resonance frequency?

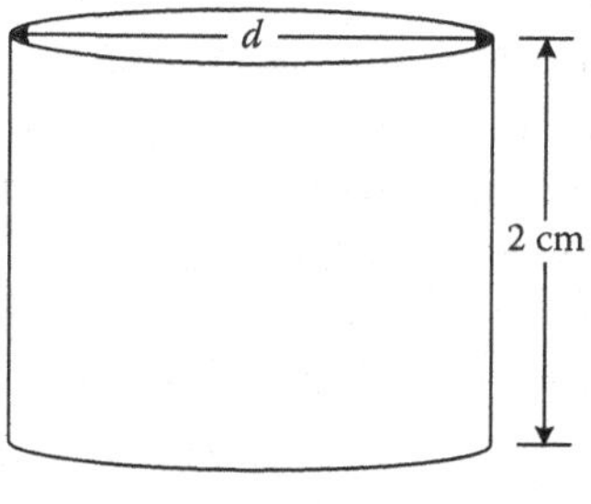

P4.3 Find the resonant frequencies of the following modes of the cylindrical cavity shown in the figure.

a. TM_{212}
b. TE_{222}

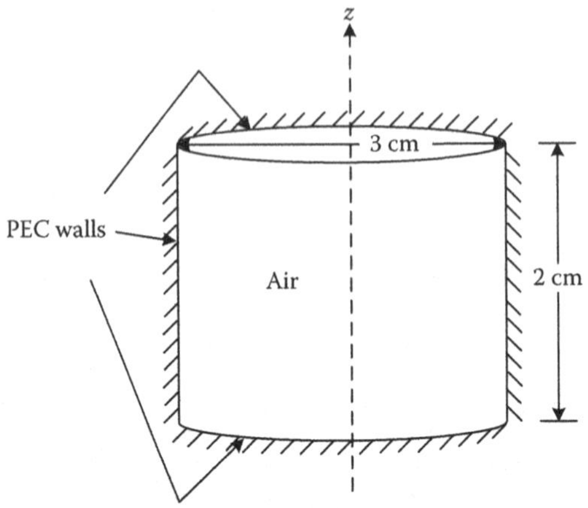

P4.4 Design a square-based cavity with height one-half (1/2) the width of the base to resonate at 1 GHz (a) when it is air-filled and (b) when it is polystyrene- filled. (c) Calculate the quality factor Q of the cavity if it is air-filled and the walls are made of copper.

P4.5 In Section 4.3, we derived the expression for the Q of a cubic cavity for the TE_{101} mode. Do the same for TM_{111} mode.

Chapter 5

P5.1 (a) Write down the expression for $\tilde{F}_e$ in (5.23) for the lowest resonant frequency mode TM^r_{101} of the spherical cavity. Use F_{101} as the mode constant.

(b) Determine the expression for the electric and magnetic fields of this resonant mode.

(c) What are the degenerate modes of this lowest resonant frequency mode? Write down the expression for $\tilde{F}_e$ of these degenerate modes.

P5.2 Determine the next lowest resonant frequency of the spherical cavity. What are its degenerate modes? Write down the expressions for $\tilde{F}_e$ of these degenerate modes.

P5.3 (a) Design a spherical cavity (determine its radius a) with a difference of 1 GHz between its lowest resonant frequency and its next lowest resonant frequency. The medium in the cavity is air.

(b) The radius a is reduced to half its size by using a nonmagnetic dielectric with the relative permittivity ε_r. Determine ε_r.

P5.4 Repeat the calculations of P5.3, if you are restricted to TE^r modes only.

P5.5 Calculate Q of a spherical cavity of radius a =10 cm for the lowest resonant frequency.

P5.6 Show that $e^{jz} = e^{jr\cos\theta} = \sum_{n=0}^{\infty} j^n (2n+1) j_n(r) P_n(\cos\theta)$ (Harrington, p. 290).

P5.7 Show that $j_n(r) = \frac{j^{-n}}{2} \int_0^{\pi} e^{jr\cos\theta} P_n(\cos\theta)\sin\theta d\theta$.

P5.8 Prove the addition theorem for spherical Hankel functions:

$$h_0^{(2)}(|\boldsymbol{r} - \boldsymbol{r}'|) = \sum_{n=0}^{\infty} (2n+1) h_n^{(2)}(r') j_n(r) P_n(\cos\xi) \quad r < r'$$

$$h_0^{(2)}(|\boldsymbol{r} - \boldsymbol{r}'|) = \sum_{n=0}^{\infty} (2n+1) h_n^{(2)}(r) j_n(r') P_n(\cos\xi) \quad r > r',$$

where ξ is the angle between $\boldsymbol{r}$ and $\boldsymbol{r}'$.

P5.9 Assume a hemispherical cavity of radius a. Determine the dominant mode and the expression for its resonant frequency. Determine the Q of this cavity (P10.23, Balanis)

P5.10 Consider a wedge consisting of two PEC plates of infinite extent, one described by $\phi = 0$ and the other by $\phi = \alpha$. The field region is defined by, $0 < r < \infty, 0 < \theta < \phi, 0 < \phi < \alpha$. (a) Describe its TE^r modes. (b) TM^r modes.

P5.11 Convert the expression in (7.87) for the vector potential $\boldsymbol{A}$ of hertzian dipole into spherical Bessel function. Determine the $\boldsymbol{E}$ and $\boldsymbol{H}$ fields from this expression in terms of spherical Bessel functions and other functions (P10.1, Balanis).

Chapter 6

P6.1 Consider a box having dimensions $a \times b \times c$ in the x-, y-, z-directions, respectively. Choose the center of the rectangular coordinate system to be at one corner of the box so that the coordinates of all points in the box are positive.

The potential on the face of the box lying on the xy-plane is given by

$$V_{0xy} = V_1 \sin\left(\pi \frac{2x}{a}\right) \sin\left(\pi \frac{3y}{b}\right).$$

The potential on the face of the box lying in the yz-plane is given by

$$V_{0yz} = V_2 \sin\left(\pi \frac{y}{b}\right) \sin\left(\pi \frac{5z}{c}\right).$$

The other four sides of the box are at zero potential. Find the potential at all points inside the box.

P6.2 Consider a rectangular prism of width a in the x-direction and b in the y-direction with all four sides at zero potential extending from $z = 0$ to $z =$ infinity. At $z = 0$, the cylinder has a cap with the following potential distribution:

$$V(x,y,0) = 0 \quad \text{for} \quad 0 < x < a/2, \ \text{all } y,$$
$$V(x,y,0) = V_0 \quad \text{for} \quad a/2 < x < a, \ \text{all } y.$$

Find $V(x, y, z)$ within the prism.

P6.3 A cylindrical capacitor is shown in the figure below. Find the potential distribution inside and outside the capacitor.

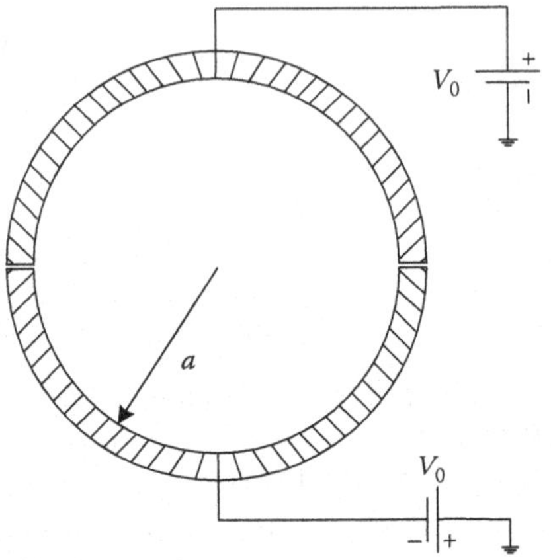

P6.4 Find a series for the potential inside the cylindrical region of figure below with end plates $z = 0$ and $z = l$ at zero potential. The cylinder has a radius of a and is in two parts: From $z = 0$ to $z = l/2$ it is at potential V_0 and $z = l/2$ to $z = l$, it is at potential $-V_0$.

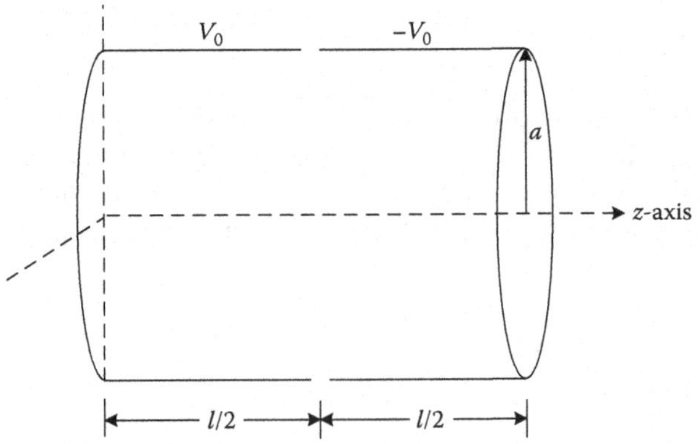

P6.5 Suppose that the end plate at $z = l$ of figure below is divided into insulated rings and connected to sources in such a way that the potential approximates a single J function of the radius given by

$$\Phi(\rho,\ell) = CJ_0\left(\frac{p_1\rho}{a}\right).$$

The end plate at $z = 0$ is at zero potential. Write the solution for $\Phi(\rho, z)$ at any point inside the cylindrical region.

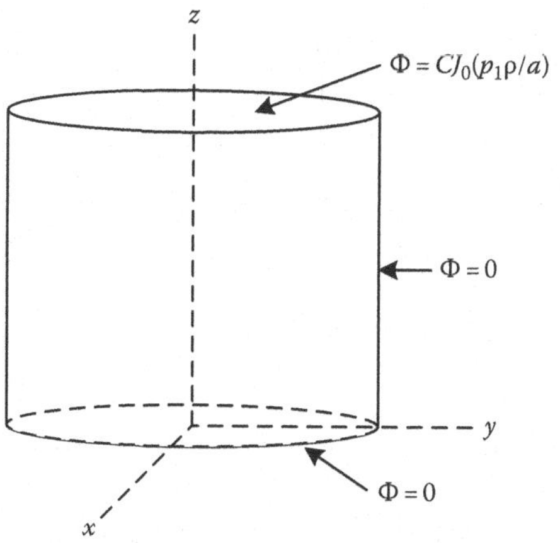

P6.6 In a region where the Laplace equation is satisfied, the potential can have neither a maximum nor a minimum. Prove this.

Consider a volume of free space bounded (enclosed) by a closed conducting surface at a potential 100 V. Can there be a point inside at 80 V? Or another point at 120 V? Can you find the voltage at the origin? Can you show that all points inside the closed surface have to be at 100 V? This is the principle of Faraday's cage.

P6.7 Find the electrostatic potential Φ at a general point $P(x, y)$ in the region $0 < x < \infty$, and $0 < y < b$. The boundary potentials are shown in the figure.

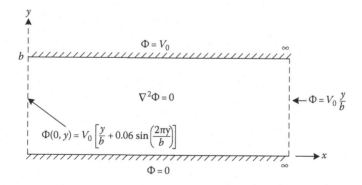

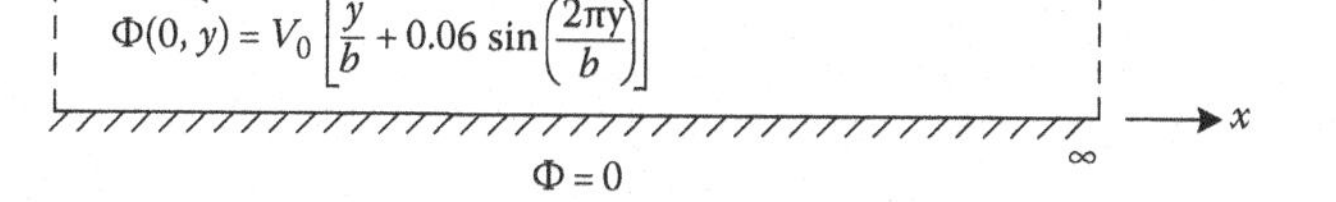

P6.8 A conducting hemisphere is placed on a flat conducting infinite sheet, as shown in the figure. Before the hemisphere was introduced, the electric field everywhere above the sheet was normal to it and equal to E_0. The radius of the sphere is a. The angle θ is measured from the normal to the sheet. The medium above the sheet is air. Assume the potential of the hemisphere and sheet to be zero. Find the potential everywhere above the sheet and the hemisphere. Find $\Phi(r, \theta, \phi)$. Note $P_1(\cos\theta) = \cos\theta$, $P_n(0) = 0$ for n odd, and $P_n(0) = (-1)^{n/2} \frac{1\cdot 3\cdot 5\cdots(n-1)}{2\cdot 4\cdot 6\cdots n}$ for n even.

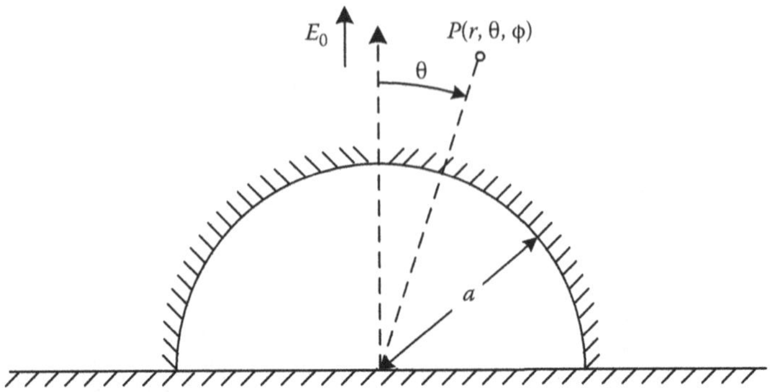

P6.9 A spherical capacitor is filled with an inhomogeneous dielectric of dielectric constant $\varepsilon_r(r) = r^3$. Determine its capacitance.

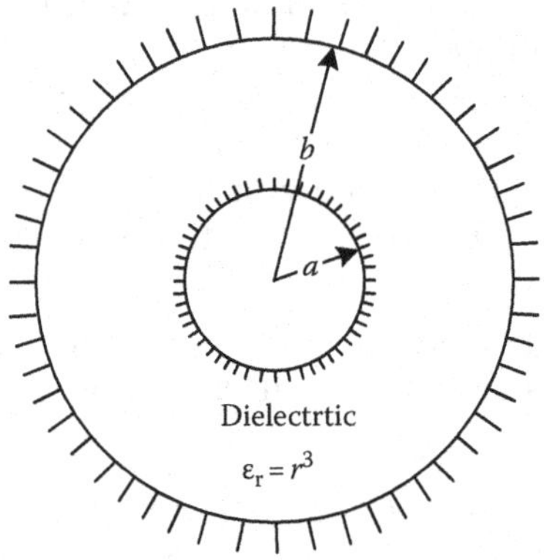

P6.10 Suppose the end plate at $z = 1$ of the figure is at a potential

$$\Phi_1 = 100 J_0\left(p_1 \frac{\rho}{a}\right)$$

and the end plate at $z = 0$ is at a potential,

$$\Phi_2 = 20 J_0\left(p_3 \frac{\rho}{a}\right).$$

Here p_1 and p_3 are the first and third zeros of the Bessel function J_0. The surface $\rho = a$ is at zero potential. Write the solution for $\Phi(\rho, z)$ at any point inside the cylindrical region.

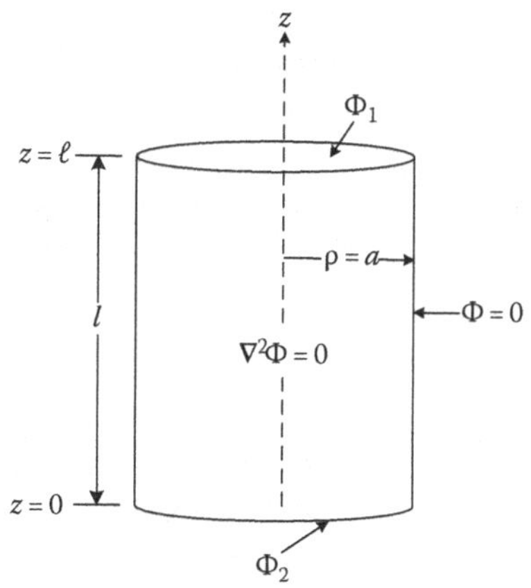

P6.11 In this two-dimensional problem, the potential on the circle $\rho = a$ varies as

$$\Phi(a,\phi) = \ln a + a^3 \cos 3\phi.$$

Determine the potential at the point $P\ (a/2,60°)$.

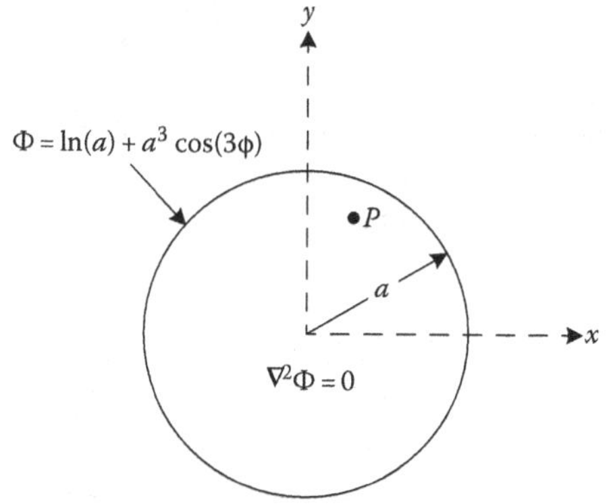

P6.12 In a circular cylindrical region $a < \rho < b$ as shown in the figure, the electrostatic potential V satisfies Laplace's equation. The boundary conditions are:

$$V(a,\phi) = 100a^2 \cos(2\phi) + \left(20/a^3\right)\cos(3\phi),$$
$$V(b,\phi) = 100b^2 \cos(2\phi) + \left(20/b^3\right)\cos(3\phi).$$

Given that $b = 2a$ and $V(3a/2, 90°) = -4$ V, find a.

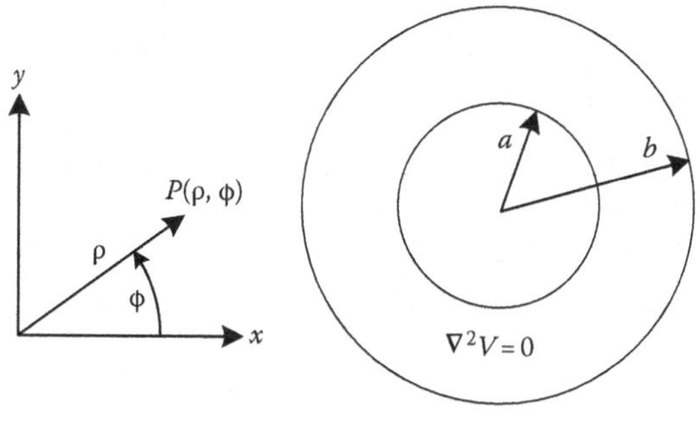

P6.13 A piece of material of conductivity σ, of circular cross section $2a$, is shown in the figure. Find the resistance R between the faces A (horizontal line) and B (vertical line). Show that the approximate value of

$$R = \frac{\pi\rho_0/2}{\sigma\pi a^2} = \frac{\rho_0}{2\sigma a^2} \text{ is valid for } \frac{a}{\rho_0} \ll 1.$$

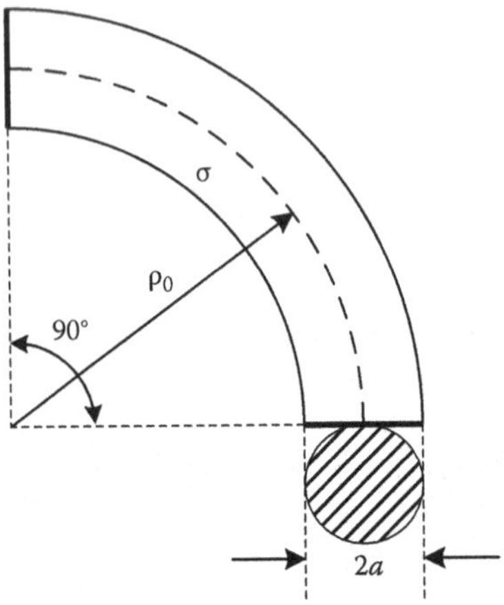

P6.14 A hemispherical electrode of very high conductivity is buried in poorly conducting earth as shown in the figure below. Determine the resistance of such a grounding system. Given the following data ($\sigma = 10^{-2}$, $I = 1000$ A), find the potential difference between A and B. If A and B represent the feet of a man would you consider the voltage dangerously high?

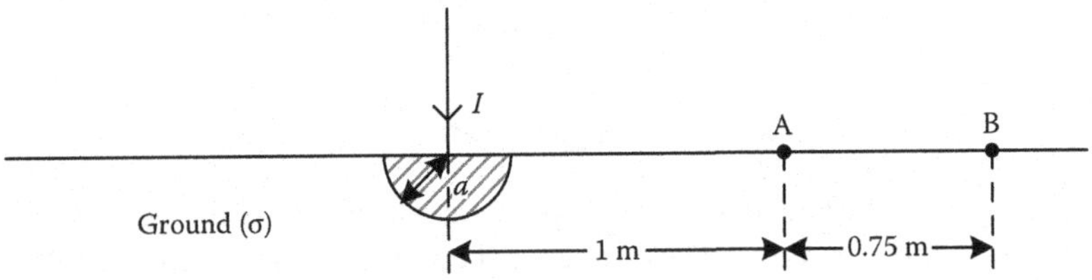

P6.15 On the cylindrical surface $\rho = a$, a surface current $\mathbf{K} = F(\phi)\hat{\mathbf{z}}$ flows. Find $\mathbf{B}$ at all points. Use the vector magnetic potential $\mathbf{A}$ to solve the problem.

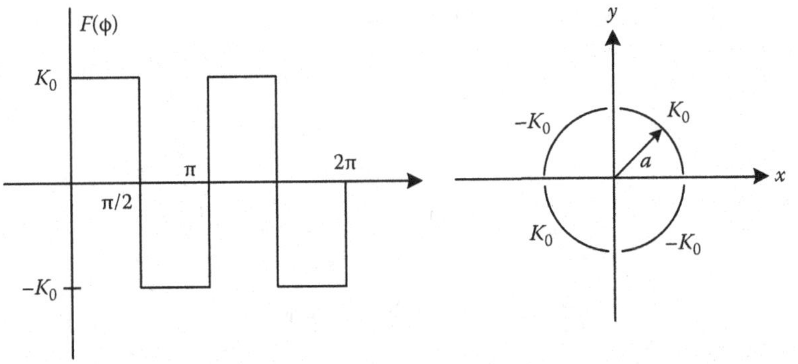

P6.16 Determine the potential in the regions:

Region 1: $0 < x < c$ and $0 < y < b$,

Region 2: $c < x < a$ and $0 < y < b$.

$x = c$ is a sheet of surface charge of constant surface charge density ρ_s (C/m^2).

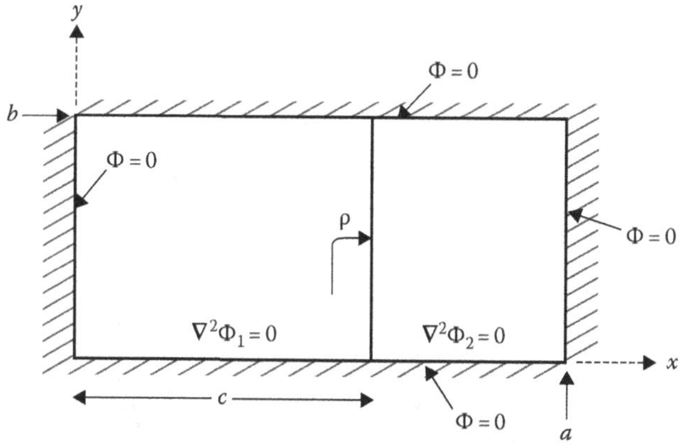

P6.17 a. In a region R bounded by a closed surface S, the potential V satisfies Laplace equation $\nabla^2 V = 0$. Show that V can have neither a maximum nor a minimum in the region R.

b. Refer to the figure below. Given that the potential V on the surface S (sphere of radius $r = 2$) is $V_s = 50 + 20 \cos\theta$. Show that we can put bounds for the potential V at a point C inside the sphere: that is, $V_{min} < V_C < V_{max}$. Determine V_{min} and V_{max}.

c. It may be shown that the solution to the Laplace equation in spherical coordinates is given by

$$V = \sum_{m=0}^{\infty} A_m r^m P_m(\cos\theta).$$

Here V is assumed to be independent of ϕ and $r < 2$. $P_m(\cos\theta)$ is the symbol for Legendre polynomials and some of these are given in Table 5.3.

Determine V_C, where C has spherical coordinates $r = 0.5$, $\theta = 30°$, and $\phi = 120$.

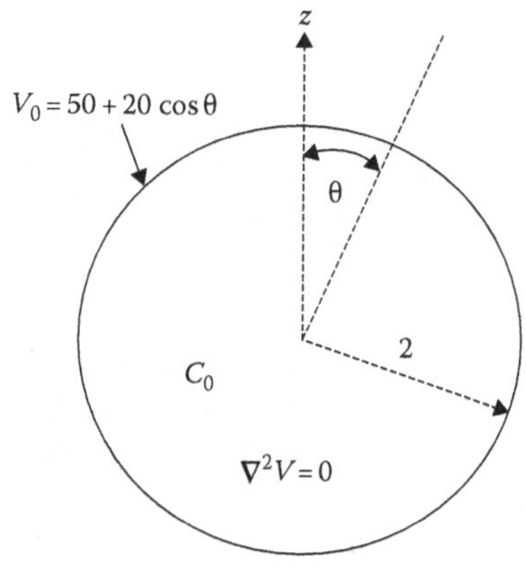

P6.18 Consider a box having dimensions $a \times b \times c$ in the x-, y-, and z-directions, respectively. The medium inside the box is free space and the potential on the faces of the box satisfy the following equations:

$$\Phi(0,y,z)=0, \quad \frac{\partial\Phi(a,y,z)}{\partial x}=0,$$
$$\Phi(x,0,z)=0, \quad \Phi(x,b,z)=0,$$
$$\Phi(x,y,0)=0, \quad \Phi(x,y,c)=V_0\left[\sin\left\{\frac{3\pi x}{2a}\right\}\right]\left[\sin\left\{\frac{4\pi y}{b}\right\}\right].$$

Find $\Phi(x, y, z)$ in the volume region: $0 < x < a, 0 < y < b, 0 < z < c$.

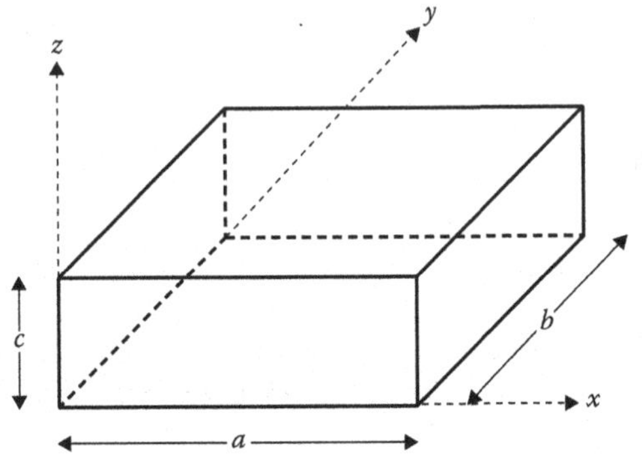

P6.19 The figure shows the cross section in the x-y plane. Determine the potential at the point $P(2, 1, 3)$. The conductors extend from $-80 < z < 80$. One conductor is at 0 V and the other is at 50 V. They are separated by a small insulating gap as shown in the figure. Use a suitable approximation.

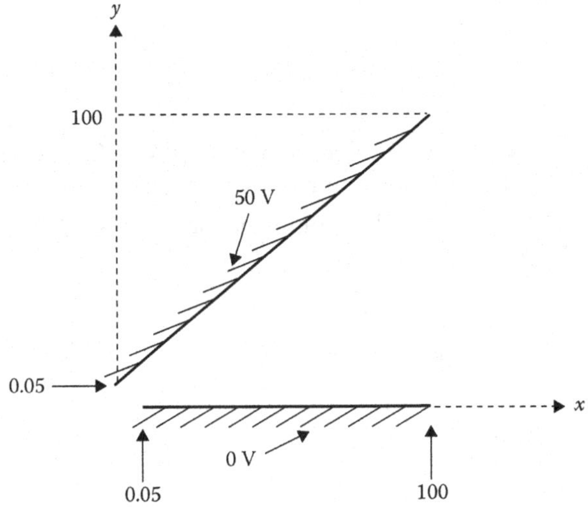

P6.20 This problem is a spherical-coordinate variation of the problem in P6.19.

A plane conductor (x–y plane) is at zero volts. The conducting surface of a cone making 45° with z axis ($\theta = 45°$, in spherical coordinates) is at 50 volts. Determine the potential at P(1,2,3); the numbers in the parenthesis are the Cartesian coordinates. Assume there is a small insulating gap at the tip of the cone and the origin.

Chapter 7

P7.1 Derive Equation 7.8.

P7.2 Show that the group velocity in an empty waveguide is given by

$$v_g = c\sqrt{1-\left(\frac{f_c}{f}\right)^2}.$$

P7.3 Derive Equation 7.30.

P7.4 Let $\tilde{\bar{G}} = \hat{z}G(\rho)$ and $\nabla^2 G + k_0^2 G = \delta(\bar{\rho}-\bar{\rho}')$.

Find the one-dimensional Green's function $G(\bar{\rho},\bar{\rho}')$.

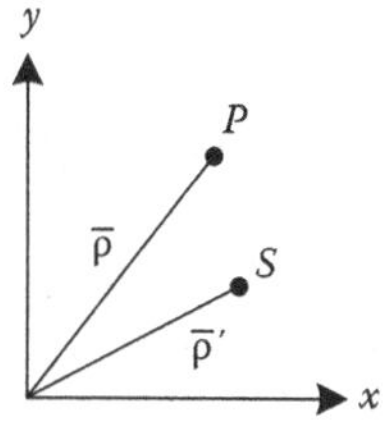

P7.5 Determine Green's function of the differential equation

$$\frac{d^2G}{dx^2} = -\delta(x-x'), 0 < x < L$$

subject to the boundary conditions

$$\left.\frac{dG}{dx}\right|_{x=0} = 0 \quad \text{and} \quad G = 0,\ x = L.$$

P7.6 Determine Green's function G for the two cases given below:

a. $\dfrac{d^3G}{dx^3} = \delta(x-x')$, the domain of x is $0 < x < L$

$$G = 0, x = 0, \left.\frac{dG}{dx}\right|_{x=0} = 0, G(L) = 0.$$

b. $\dfrac{d^2G}{dx^2} + k^2G = \delta(x-x')$, the domain of x is unbounded, that is, $-\infty < x < \infty$

P7.7 Refer to Section 7.3 and Figure P7.8.

$$\text{Given } Z_{11} - Z_{12} = j2$$
$$Z_{22} - Z_{12} = j5$$
$$Z_{12} = Z_{21} = j$$

Find the *ABCD* parameters and *S* parameter. Utilize the results to find the input reflection coefficient if the output is matched.

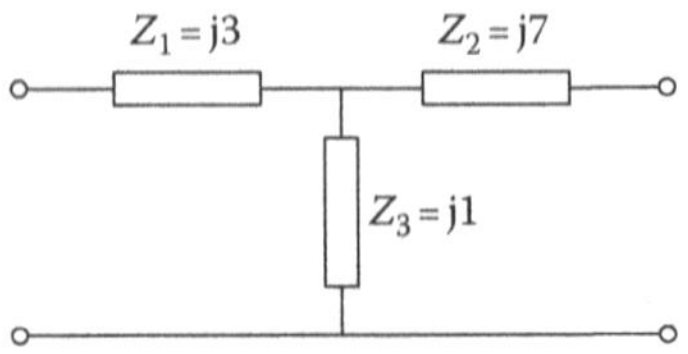

FIGURE P7.8

P7.8 Given

$$Z_1 = Z_{11} - Z_{12} = j3,$$
$$Z_2 = Z_{22} - Z_{12} = j7,$$
$$Z_3 = Z_{12} - Z_{21} = j1.$$

Find *ABCD* parameters. Utilize the results to find the input reflection coefficient if the output is matched (Figure P7.8).

P7.9 Consider the propagation of a *p* wave in a dielectric layer of width *d* as shown. Find the *ABCD* parameters (consider the positive-going and negative-going waves in the layer).

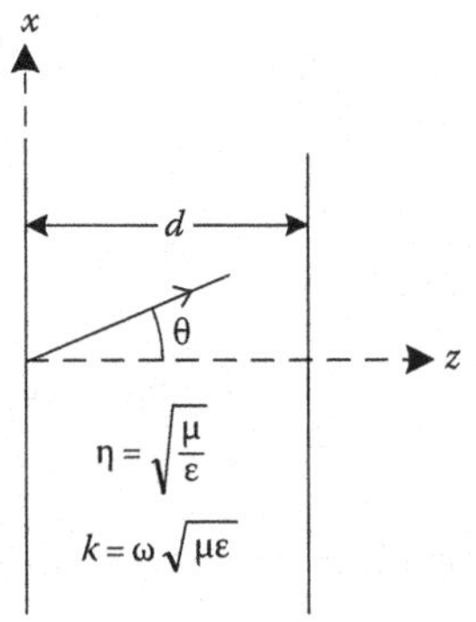

P7.10 Repeat the above problem for an *s* wave.

P7.11 Figure shows the circuit equivalent of a section Δz of a transmission line.

a. Derive the partial differential equation for the voltage $V(z, t)$.

b. Assume $V(z, t) = V_0\, e^{j(\omega t - kz)}$, find the relation between ω and k.

c. Draw the ω–k diagram and label it in terms of

$$\omega_1 = \frac{1}{\sqrt{L_1' C_1'}}, \quad \omega_2 = \frac{1}{\sqrt{L_2' C_2'}}.$$

Suppose the above transmission line is to act as a filter, rejecting the frequencies of 5–10 GHz. Choose reasonable values for L_1', L_2', C_1', and C_2' that will accomplish this goal.

For these chosen values, determine the group velocity and phase velocity when the frequency of the signal is 12 GHz.

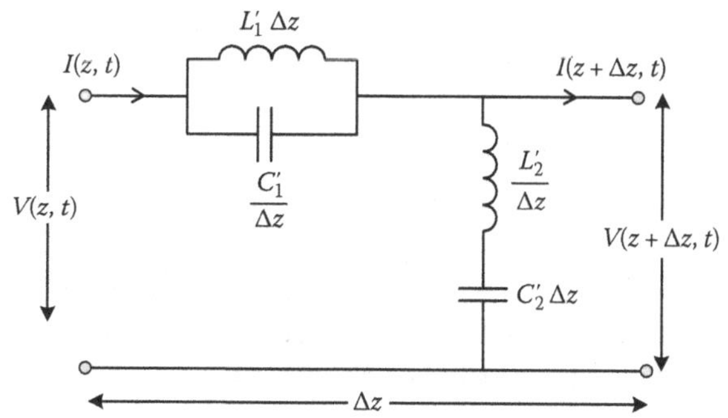

P7.12 a. Find the *ABCD* parameters of the lumped inductor L_1 of impedance Z_1 shown in the figure.

b. A lossless air transmission line of length l and wave number k has a characteristic impedance of Z_0 ohms. This line is loaded with the inductor as shown in figure b. Determine the *ABCD* parameters of this system.

c. The system shown in figure b is a unit cell of a periodic system with infinite number of cells to the left and the right as shown in figure c.

Obtain the dispersion relation of the periodic system relating the propagation constant β of the periodic system with the parameters of the unit cell.

d. Determine the first pass band of the system. Express your answer in the form $\omega_1 < \omega < \omega_2$. Find ω_1 and ω_2.

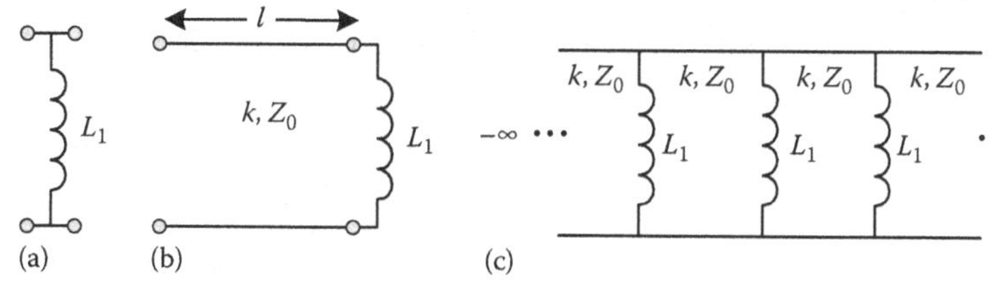

P7.13 a. Find the *ABCD* parameters of the lumped capacitor element C_1 of admittance Y_1 shown in figure a below.

b. A lossless air transmission line of length l and wave number k has a characteristic admittance of $Y_0 = 1/Z_0$. This line is loaded with the capacitor as shown in figure b. Determine the *ABCD* parameters of this system.

c. The system shown in Figure 7.13b is a unit cell of a periodic system with infinite number of cells to the left and the right as shown in Figure 7.13c. Obtain the dispersion relation of the periodic system relating the propagation constant β of the periodic system with the parameters of the unit cell.

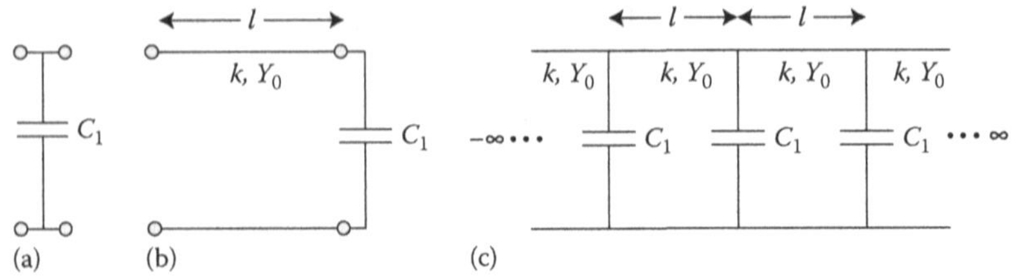

P7.14 A 1.5-m long dipole antenna is excited by a harmonic current of amplitude 8 A and frequency f = 1.5 MHz. Find the time-averaged power density radiated by the dipole at a distance of 5 km in a direction 60 degrees from the axis of the dipole.

P7.15 A half-wavelength center-fed dipole is used as an antenna at 100 MHz. The current amplitude I_0 is 10 A. The dipole lies along z-axis.

a. Determine the z component of the vector phasor $\tilde{\mathbf{E}}$ at the following points described in spherical coordinates: (i) $R = 100$, $\theta = 90$ degrees, $\phi = 90$ degrees (ii) $R = 100$, $\theta = 30$ degrees, $\phi = 90$ degrees.

b. The antenna wire is made of copper ($\sigma = 5.8 \times 10^7$ S/m) and has a radius a = 0.4 mm.

Calculate the radiation resistance and the radiation efficiency of the dipole antenna.

P7.16 A half-wavelength center-fed dipole is used as an antenna at 100 MHz. The current amplitude I_0 is 10 A. The dipole lies along z-axis.

a. Determine the z component of the vector phasor $\tilde{\mathbf{E}}$ at the point P described in spherical coordinates: $R = 100$, $\theta = 60$ degrees, $\phi = 90$ degrees.

b. The center of a vertical (parallel to z-axis) receiving antenna 3 cm long is located at the point P. Calculate the open circuit voltage $\tilde{V}_{oc}$ induced on the receiving antenna. Assume the medium in which the antennas are located is air.

P7.17 Derive the cylindrical wave transformations given by (7.124a) and (7.124b).

P7.18 Derive the cylindrical wave transformations given by (7.125a) and (7.125b).

P7.19 Derive the cylindrical wave transformations given by (7.126).

Chapter 8

P8.1 Simple metal like sodium has a plasma frequency of $\omega_p = 2 \times 10^{16}$ rad/s and collision frequency of $\nu = 2 \times 10^{13}$ rad/s. Plot α, β, n_R, and n_I as in Figure 8.6. Mark conducting, cutoff, and dielectric regions.

P8.2 Let the plasma medium shown in Figure 8.7 have a collision frequency ν (rad/s). Determine the time rate of decay of plasma oscillations given by Equation 8.41.

P8.3 Show that the group velocity in an unbounded plasma is given by

$$v_g = c\sqrt{1-\left(\frac{\omega_p}{\omega}\right)^2}.$$

P8.4 Find the cutoff frequency of a waveguide in which the medium is a plasma of frequency f_p. Let f_c be the cutoff frequency of the empty waveguide.

P8.5 The intense friction around a space vehicle re-entering the atmosphere generates a plasma sheath which is 1 m thick and is characterized by an electron density of $N_0 = 10^{13}/\text{cm}^3$, and the collision frequency $\nu = 10^{11}/\text{s}$. Calculate the power loss in decibels if the signal frequency f is equal to (a) 30 GHz, (b) 300 GHz, and (c) 350 GHz.

P8.6 The dielectric function for water is given below.

$$\varepsilon_r(\omega) = \varepsilon_\infty + \frac{\varepsilon_s - \varepsilon_\infty}{1 + j\omega t_0},$$

where ε_∞ is the infinite frequency relative permittivity, ε_s is the static relative permittivity at zero frequency, and t_0 is the relaxation time.

Determine the (a) intrinsic impedance η_C, (b) phase velocity, and (c) group velocity at frequencies:

i. 5 GHz,

ii. 20 GHz,

iii. 50 GHz,

Assume $\varepsilon_s = 81$, $\varepsilon_\infty = 1.8$, $t_0 = 9.4 \times 10^{-12}$ s.

P8.7 Calculate and plot the effective dielectric constant of a mixture of two media with $\varepsilon_{r1} = 2$ and $\varepsilon_{r2} = 3$ as a function of f, the fractional volume of inclusions of a material of dielectric constant ε_{r2}.

P8.8 A super conductor thin film of thickness d where d is comparable to the London penetration depth λ_L is excited as follows:

$$\mathbf{H}(d/2) = \hat{\mathbf{y}}H_1,$$
$$\mathbf{H}(-d/2) = \hat{\mathbf{y}}H_1.$$

Assume DC excitation and $d << w$.

a. Find the expression for **H** in the region $-d/2 < z < d/2$.

b. Find the expression for **J** the region $-d/2 < z < d/2$.

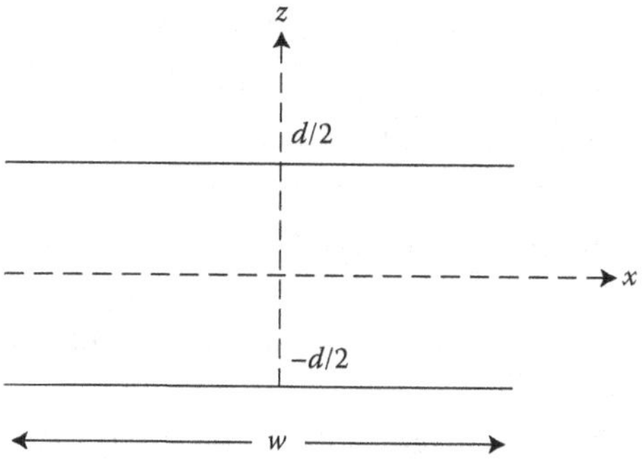

P8.9 Derive Equations 8.124 and 8.130.

P8.10 Starting from Equation 8.135, obtain Equation 8.141.

P8.11 A superconducting microstrip has the following parameters:

Normal conductivity $\sigma_n = 0.5 \times 10^6$ (S/m).

London penetration depth $\lambda_L = 0.14 \times 10^{-6}$ m.

Calculate the surface resistance R_s of the superconducting strip and compare it with the surface resistance if it were a normal conductor with the value of σ_n given above. Take the frequency $f = 10$ GHz.

P8.12 a. Refer to Figure 4.1. A rectangular cavity of square cross section ($a = b$) and height h utilizing the simple TE_{101} mode is to be designed for a millimeter-wave electron device for operation at $f_r = 0.2$ THz. Height of the cavity must be kept small (take $h = 1$ mm) to minimize electron transit time. The medium in the cavity is air. The walls are made of copper $\sigma = 5.8 \times 10^7$ s/m. Design the cavity by calculating a, also calculate the quality factor Q of the cavity.

b. Suppose the walls of the cavity are made of superconducting material with the following parameters:

Normal conductivity $\sigma_n = 5 \times 10^6$ (S/m);

London penetration depth $\lambda_L = 0.5 \times 10^{-7}$ m.

Determine the Q of the cavity.

P8.13 In the equivalent circuit of a transmission line (for a length Δz) shown in the figure, assume $L_R = C_R = L_L = C_L = 1$.

a. Sketch the ω–k diagram, for k varying from $-\infty$ to ∞.

b. If $\omega = 0.8$, determine (i) phase velocity v_p, (ii) group velocity v_g, and (iii) wavelength λ.

c. Repeat (b) if $\omega = 1$.

d. Repeat (b) if $\omega = 1.1$.

P8.14 Determine the wave number k_c and the wave impedance η_c for the case of propagation of L wave in a chiral medium.

P8.15 A linearly polarized wave can be decomposed into circularly polarized waves, for example, $\hat{x}E_0 = E_0/2(\hat{x} + j\hat{y}) + E_0/2(\hat{x} - j\hat{y})$. This kind of decomposition could be useful in solving the problem. A chiral medium whose constitutive relations are given by Equations 8.174 and 8.175 has the following parameters at 10 GHz: $\varepsilon = \varepsilon_0\varepsilon_r$, $\varepsilon_r = 2$, $\mu = \mu_0\mu_r$, $\mu_r = 1$, $\xi_c = 0.005$. The electric field of a plane wave, propagating in the positive z-direction in this medium, is linearly polarized in x-direction at $z = 0$ and is given by $\mathbf{E} = \hat{x}3\cos\left(2\pi \times 10^{10} t\right)$ V/m.

a. Determine the electric field in this medium at $z = 2 \times 10^{-2}$.

b. Determine the polarization of the wave and sketch it in the $z = 2 \times 10^{-2}$ plane like one of the sketches in Figure 2.6.

c. Determine the magnetic field $\mathbf{H}$ of the wave at $z = 2 \times 10^{-2}$.

P8.16 a. Draw the ω–k diagram for an unbounded periodic media consisting of alternating layers of chiral media. Assume the following values for the parameters of the layers in a unit cell:

First layer: $\varepsilon_1 = 2\varepsilon_0$, $\mu_1 = \mu_0$, $\xi_{c1} = 10^{-3}$, $L_1 = L_0 = 3$ cm.

Second layer: $\varepsilon_2 = 3\varepsilon_0$, $\mu_2 = \mu_0$, $\xi_{c2} = -4\xi_{c1}$, $L_2 = 2L_0$.

b. Determine the location of the center of the first stop band. Also, determine the bandwidth of this stop band.

c. Same data as above but $\xi_{c1} = 0$, that is, the layers are not chiral. Determine the first stop band and its width.

Chapter 9

P9.1 Let $\varepsilon_p(z, \omega)$ is given by

$$\varepsilon_p(z,\omega) = \varepsilon_1 \quad -\infty < z < 0, \tag{P9.1}$$

$$= \varepsilon_1 + (\varepsilon_2 - \varepsilon_1)\frac{z}{L}, \quad 0 < z < L, \tag{P9.2}$$

$$= \varepsilon_2, \quad L < z < \infty. \tag{P9.3}$$

Show that the solution of Equation 9.15 in the two regions given below may be written as

$$E = Ie^{-jk_1 z} + Re^{+jk_1 z} \quad -\infty < z < 0, \tag{P9.4}$$

$$= Te^{-jk_2 z} + Ae^{jk_2 z}, \quad L < z < \infty, \tag{P9.5}$$

where I, R, T, and A are constants and $k_1 = k_0\sqrt{\varepsilon_1}$ and $k_2 = k_0\sqrt{\varepsilon_2}$. Give physical reasons for setting $A = 0$. Note that if $I = 1$, R and T give the reflection and transmission coefficients. Assume the waves are propagating along z-axis.

P9.2 Find an analytical expression for $|R|^2$ if $L = 0$ in previous Equations P2.2 and P2.3.

P9.3 Light of wavelength 0.633 μm is incident from air on to an air–aluminum interface. Take the dielectric constant of aluminum at 0.633 μm as $\varepsilon_r = -60.56 - j24.86$.

Plot reflectivity (ρ) versus θ_i of a s wave, where θ_i is the angle of incidence.

P9.4 The geometry is given in Figure P9.4.

Find the reflectivitfy of a p-wave in free space incident on the metal at an angle of incidence of 45° when:

a. $\omega = 8.2 \times 10^{15}$ rad/s.

b. $\omega = \left(1.1\sqrt{2}\right)8.2\times10^{15}$ rad/s.

P9.5 The geometry is given in Figure P9.4.

Investigate the propagation of a surface plasmon TM on this interface at $\omega = 4.1 \times 10^{15}$ rad/s.

a. Determine the phase velocity of the surface charge wave on the interface.

b. Determine the distance, in meters, in which the z-component of the electric field in free space reaches 0.3679 ($= e^{-1}$) of its value at the interface.

c. Determine the distance, in meters, in which the z-component of the electric field in metal reaches 0.3679 ($= e^{-1}$) of its value at the interface.

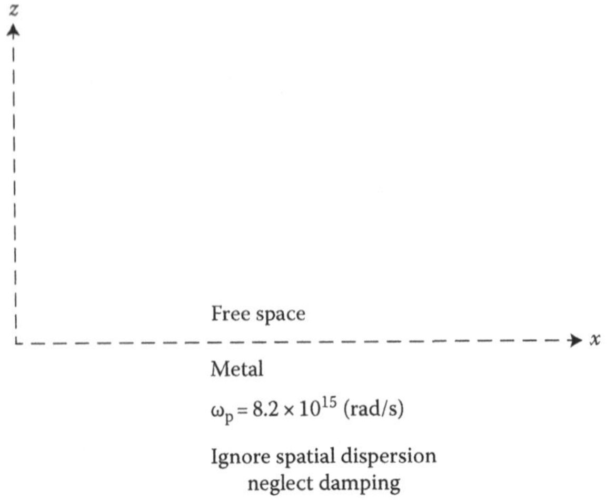

FIGURE P9.4

P9.6 The geometry is given in Figure P9.4.

Investigate the propagation of a surface plasmon TM on this interface at $\omega = 9.84 \times 10^{15}$ rad/s.

a. Determine the phase velocity of the surface charge wave on the interface.
b. Determine the distance, in meters, in which the z-component of the electric field in free space reaches 0.3679 ($= e^{-1}$) of its value at the interface.
c. Determine the distance, in meters, in which the z-component of the electric field in metal reaches 0.3679 ($= e^{-1}$) of its value at the interface.

P9.7 In this problem, we investigate the nature of the solution of Equation 9.15 in the domain $0 < z < L$.

By introducing a new variable $\xi = k_0^2 \varepsilon_P(z,\omega)$, show that Equation 9.15 may be transformed to

$$\frac{d^2E}{d\xi^2} + \frac{\xi E}{\beta^2} = 0, \tag{P9.6}$$

where

$$\beta = k_0^2 \frac{\varepsilon_2 - \varepsilon_1}{L}. \tag{P9.7}$$

Note that the solution of Equation P9.6 may be written in terms of Airy function *Ai* and *Bi*:

$$E = c_1 Ai\left[-\beta^{-2/3}\xi\right] + c_2 Bi\left[-\beta^{-2/3}\xi\right], \quad 0 < z < L. \tag{P9.8}$$

P9.8 Determine the reflection coefficient R and plot the power reflection coefficient $|R|^2$ versus L. Take $\varepsilon_1 = (4/3)^2$ and $\varepsilon_2 = 1$. These data approximate the water–air interface at optical frequencies. Assume that the dielectric function is a linear profile given by

$$\varepsilon_P(z,\omega) = \varepsilon_1 \quad -\infty < z < 0, \tag{P9.9}$$

$$= \varepsilon_1 + (\varepsilon_2 - \varepsilon_1)\frac{z}{L}, \quad 0 < z < L, \tag{P9.10}$$

$$= \varepsilon_2, \quad L < z < \infty. \tag{P9.11}$$

P9.9 Discuss the possibility of having a TE surface wave mode at a metal–vacuum interface.

P9.10 Derive the dispersion relation and draw the ω–β diagram of an unbounded periodic media consisting of alternating layers of dielectric layers. Assume the following values for the parameters of the layers in a unit cell

First layer: $\varepsilon_1 = 2\varepsilon_0$, $\mu_1 = \mu_0$, $L_1 = L_0 = 3$ cm.

Second layer: $\varepsilon_2 = 3\varepsilon_0$, $\mu_2 = \mu_0$, $L_2 = 2L_0$.

Determine the location of the center of the first stop band. Also, determine the bandwidth of this stop band.

P9.11 The cutoff frequency of a waveguide with free-space medium is 4 GHz. The medium is a loss-free plasma of plasma frequency 3 GHz. The signal frequency is 6 GHz. Determine the b. Determine the group velocity.

P9.12 Obtain (9.127) using ABCD parameter technique of Section 7.4.

Chapter 11

P11.1 Derive Equation 11.25.

P11.2 Show that Equations 11.25 and 11.28 are the same.

P11.3 Given $f = 2.8$ GHz and $B_0 = 0.3$ T, find the electron density of the plasma at which the L wave has a cutoff.

P11.4 Given $f = 10$ GHz and $B_0 = 0.3$ T, find the electron density of the plasma at which the R wave has a cutoff.

P11.5 Faraday rotation of an 8 mm wavelength microwave beam in a uniform plasma in a 0.1 T magnetic field is measured. The plane of polarization is found to be rotated by 90° after traversing 1 m of plasma. What is the electron density in the plasma?

P11.6

a. For a whistler mode, that is, $\omega << \omega_p$, $\omega << \omega_b$, show that the group velocity is given approximately by

$$v_g \approx \frac{2c\sqrt{\omega\omega_b}}{\omega_p}. \tag{P11.1}$$

b. Assume that due to lighting, signals in the frequency range of 1–10 kHz are generated. If these signals are guided from one hemisphere to the other hemisphere of the Earth along Earth's magnetic field over a path length of 5000 km, calculate the travel time of the signals as a function of frequency. Take the following representative values for the parameters: $f_p = 0.5$ MHz and $f_b = 1.5$ MHz.

P11.7 Figure 11.1 shows the variation of the dielectric constant with frequency for the case of R wave propagation. The branch between 0 and ω_b is the whistler mode. Obtain the frequency ω at which the ε_{pR} is minimum. What is the minimum value of ε_{pR} at this frequency?

P11.8 A right circularly polarized electromagnetic wave of frequency $f_0 = 60$ Hz is propagating in the z-direction in the metal potassium, at a very low temperature of a few degrees Kelvin, in the presence of a z-directed static magnetic flux density field B_0.

Determine (a) the phase velocity and (b) wavelength of the electromagnetic wave in the metal.

Assume that electron density N_0 of the metal is such that $\omega_p = 10^{16}$ rad/s.

Assume that B_0 value is such that the electron gyrofrequency $\omega_b = 10^{12}$ rad/s.

Assume that the collision frequency ν is negligible.

The calculations become simple if approximations are made based on the inequalities mentioned above. Relate the specification of a very low temperature to one of the assumptions given.

P11.9 Derive Equations 11.52 and 11.54.

P11.10 Derive Equations 11.56 through 11.58.

P11.11 An x-polarized plane wave in free space of frequency $f = 2$ GHz is incident normally on a magnetoplasma. The electron density of the plasma and the strength and direction of the static magnetic field $\mathbf{B}_0$ of the magnetoplasma are shown in the figure. Determine the power reflection coefficient.

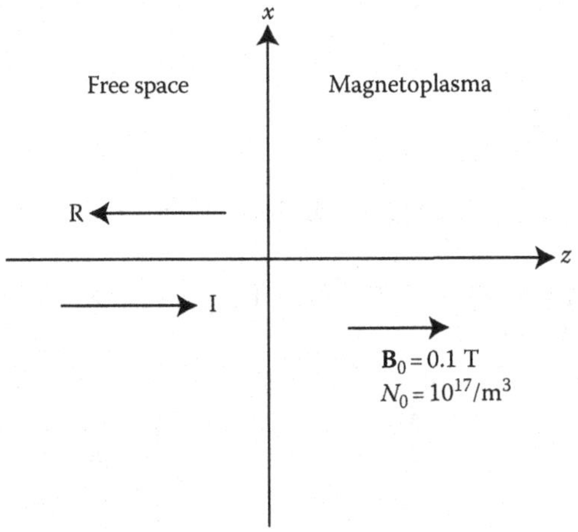

P11.12 For a whistler mode, where the electron gyrofrequency ω_b is only slightly more than the source frequency ω_0 and ω_p^2 is such that the inequality

$$\frac{\omega_p^2}{\omega_0\left(\omega_b - \omega_0\right)} << 1$$

is satisfied, show that the relative permittivity will be quite high so close to resonance. Obtain an expression for the group velocity. Calculate the group velocity if $f_b = 10.009$ GHz, $f_p = 30$ GHz, $f_0 = 10$ GHz.

P11.13 For the mode and the inequalities described in P11.12, determine the attenuation constant α, if the collision frequency is ν rad/sec. Also determine the damping distance constant z_p and the damping time constant t_p.

P11.14 For the mode and the inequalities described in P11.6, determine, α, z_p, and t_p.

P11.15 Medium 1 in which the incident plane wave is traveling in the z-direction, is the half-space $z < 0$. The electric field of the incident wave is in x-direction. The frequency $f = 6$ GHz.

Medium 2 is a magnetoplasma with the parameters $f_b = 3$ GHz and $f_p = 4$ GHz. Determine the power reflection coefficient in each of the cases given below

Case a: the static magnetic field is in x-direction

Case b: the static magnetic field is in y-direction.

Chapter 12

P12.1 Biaxial crystal mica has the principal refractive indices $n_x = 1.552$, $n_y = 1.582$, and $n_z = 1.588$. The electric field of a wave propagating in the x-direction at $x = 0$ is given by

$$\mathbf{E} = \sqrt{2}E_0\left(\hat{\mathbf{y}} + \hat{\mathbf{z}}\right)e^{j\omega t}.$$

a. Determine **E** at $x = 1$ m.

b. Determine **H** at $x = 1$ m.

P12.2 Consider rutile (positive uniaxial crystal) with $n_0 = 2.616$ and $n_e = 2.903$. Let the optic axis of the crystal be perpendicular to the paper. If $\theta_I = 30°$, determine the transmission angles θ_{to} and θ_{te} of the doubly refracted wave.

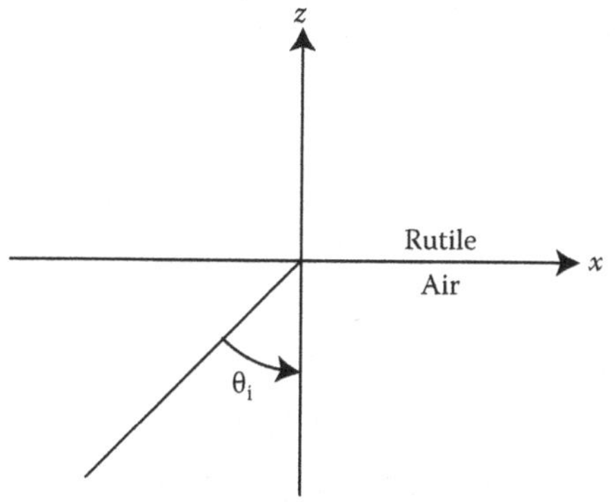

P12.3 Crystal is rutile (data as in P12.2) but the optic axis of the crystal is along the x-axis. Determine the transmission angles θ_{to} and θ_{te}.

P12.4 Crystal is rutile (data as in P12.2). The optic axis is in the x–z-plane making an angle of −30° with the x-axis as shown in the figure. Determine the transmission angles θ_{to} and θ_{te}.

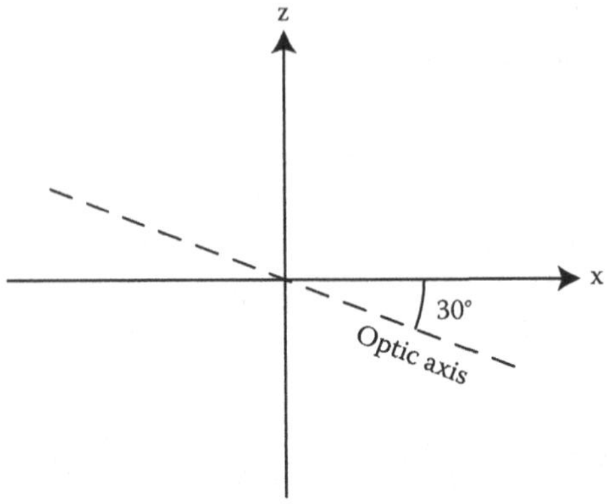

P12.5 Consider calcite (negative uniaxial crystal) with $n_0 = 1.658$ and $n_e = 1.486$. Let the optic axis of the crystal be perpendicular to the paper. If $\theta_i = 30°$, determine the transmission angles θ_{to} and θ_{te} of the doubly refracted wave. (The figure is same as the figure of Problem P12.2 except the upper medium is calcite.)

P12.6 Consider calcite (negative uniaxial crystal) with $n_0 = 1.658$ and $n_e = 1.486$. The optic axis of the crystal is along the x-axis. Determine the transmission angles θ_{to} and θ_{te}. (The figure is same as the figure of Problem P12.2 except the upper medium is calcite).

P12.7 Consider calcite (negative uniaxial crystal) with $n_0 = 1.658$ and $n_e = 1.486$. The optic axis is in the x–z-plane making an angle of −30° with the x-axis as shown in the figure above. Determine the transmission angles θ_{to} and θ_{te}.

P12.8 Consider a light beam incident on the plane boundary from the inside of a calcite crystal ($n_0 = 1.658$ and $n_e = 1.486$). Suppose the optic axis of the crystal is normal to the plane of incidence. Find the range of apex angles α so that the ordinary wave is totally reflected.

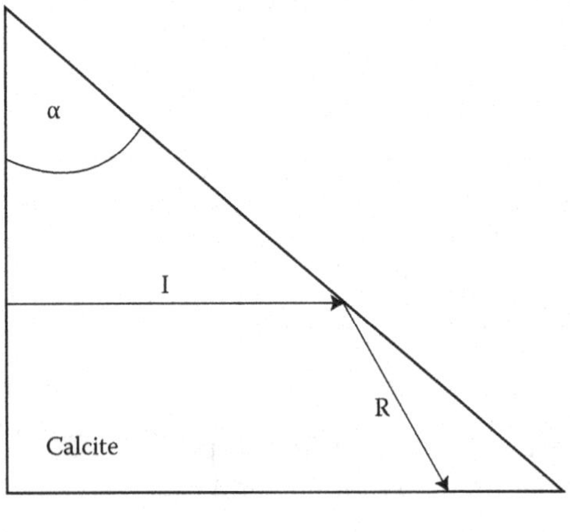

P12.9 A p wave is propagating in the x–z-plane in an uniaxial dielectric crystal (optic axis along the z-axis) whose constitutive relations are given by

$$\tilde{\mathbf{B}} = \mu_0 \tilde{\mathbf{H}},$$

$$\tilde{D}_x = \varepsilon_0 n_0^2 \tilde{E}_x,$$

$$\tilde{D}_y = \varepsilon_0 n_0^2 \tilde{E}_y,$$

$$\tilde{D}_z = \varepsilon_0 n_e^2 \tilde{E}_z.$$

For an x–z-plane of propagation shown in the figure of Problem P12.8, the exponential phase factor has the form

$$\psi = e^{-jk_0[Sx+qz]}.$$

a. Determine q in terms of n_0, n_e, and S.
b. A p wave in free space is incident on the uniaxial crystal as shown in the figure. Determine the relation between θ_t, θ_i, n_0, and n_e.
c. If $n_0 = n_e = n = 1.5$, $\theta_i = 30°$, determine θ_t.

(*Note*: The p wave has $\tilde{E}_x$, $\tilde{E}_z$, $\tilde{H}_y$ components only.)

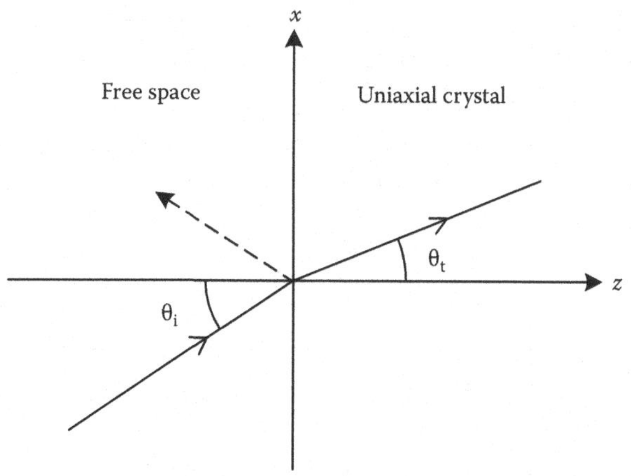

Chapter 13

P13.1 Assume the data given in Example 13.8 except that the load end is not matched and $R_L = \infty$. Sketch the load voltage.

P13.2 Give sketches of $V_S(t)$ for Example 13.8 and Problem 13.1 above.

P13.3 Suppose the voltage wave form $V(0,t)$ described below is observed at the sending end of a 50 Ω transmission line in response to a step voltage introduced by a generator with $V_g = 15$ V and an unknown series resistance R_g.

$$V(0,t) = 5 \text{ V}, \quad 0 < t < 8 \ \mu\text{s}$$

$$V(0,t) = 2.5 \text{ V}, \quad 8 \ \mu\text{s} < t < \infty.$$

The line is 1.2 km in length and its velocity of propagation 10^8 m/s, and it is terminated in a load of $Z_L = 75$ Ω.

Determine R_g and the shunt resistance R_f and the location of the fault responsible for the fault [1].

P13.4 TDR is useful in determining the inductance of a bonding wire between two microstrip interconnects. Reference 2 illustrates a suitable theory to facilitate such a measurement. Formulate the problem and explain how the inductance can be measured.

The device is called Time Domain Reflectometer. Its principle is illustrated through the example.

P13.5 Show that the solution of (13.96) and transformation back to the lab frame in z, t gives (13.86).

P13.6 Show that the x component of (13.172) can be written as (13.173).

P13.7 Show that if L and hence H is not an explicit function of t, Equation 13.184 is satisfied.

P13.8 The circuit representation of a differential length of a nonlinear transmission line with dispersion is shown in Figure 13.15. Show that the voltage of the transmission line satisfies (13.83) when the nonlinear capacitance is given by (13.82).

P13.9 Derive the dispersion equation (13.84), when $C_N' = 0$ in (13.83).

P13.10 In Section 13.7 EMP1, suppose G is a function of the voltage across the capacitor ($G = G_0 V$) and the current source has a double exponential variation given below:

$$I = I_0\left(e^{-\alpha t} - e^{-\beta t}\right).$$

Determine the voltage across the capacitor.

P13.11 Show that the LHS of (13.249) can be combined into one term given by the LHS of (13.248).

P13.12 Starting from (13.300), fill in the steps leading to (13.304).

P13.13 Show that the RMS velocity, using Maxwellian distribution, is given by (13.318).

P13.14 Derive (13.334) and prove the assertion that the force due to wave magnetic field can be neglected).

P13.15 Derive (13.336).

P13.16 Show that (13.339) reduces to (13.340) for the special case of the collision frequency ν independent of the velocity variable v.

P13.17 See Figure P13.17a. A uniform plane wave traveling in free space to the right along z-axis has an electric field, $\boldsymbol{E}^I = E_0 e^{j(\omega_0 t - k_0 z)}\hat{a}_x$.

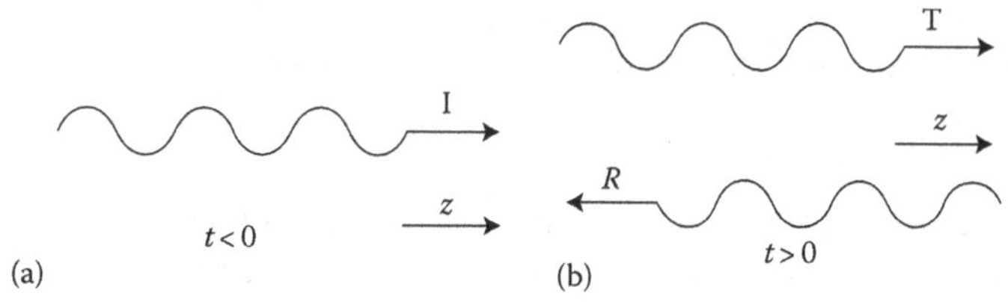

FIGURE P13.17
(a) Free space and (b) dielectric (ε_R).

See Figure P13.17b. At $t = 0$ the whole medium is suddenly turned into a dielectric of dielectric constant ε_R. Assume that this sudden switching of the medium gives rise to a reflected wave (left-going) and transmitted wave (right-going) whose electric fields are written as:

$$\mathbf{E}^R = Ae^{j(-\omega_R t - k_R z)}\hat{a}_x$$

$$\mathbf{E}^T = E_0 e^{j(\omega_T t - k_T z)}\hat{a}_x$$

(a) Determine the relation between ω_R and k_R; ω_T and k_T.

(b) Also determine ω_R, k_R, ω_T and k_T in terms of ω_0, k_0 and ε_R.

(c) Assume that the suddenly switched dielectric medium is a plasma of plasma frequency 4 GHz. If ω_0 is 3 GHz determine ω_R and ω_T.

Note: The dielectric constant of plasma is:

$$\varepsilon_R = 1 - \frac{\omega_p^2}{\omega^2}$$

where ω_p is the angular plasma frequency in rad/s and ω is the angular signal frequency in rad/s.

P13.18 Let ω_1 be the frequency of a propagating wave in a plasma medium of plasma frequency ω_{p1}. At $t = 0$, the electron density suddenly drops. Let $\omega_p = \omega_{p2}$ for $t > 0$, where $\omega_{p2} < \omega_{p1}$. Find the new frequencies and fields of all the modes generated by the switching action.

Chapter 14

P14.1 Refer to Figure 14.1. Determine the frequency of the reflected wave for normal incidence on a moving mirror. If the reflected wave frequency is $(1 + 10^{-7})$ times the incident wave frequency, determine the velocity of the mirror. Is the mirror moving towards the source of EM radiation or away from the radiation.

P14.2 Obtain (14.26).

P14.3 What is the relation between angles α and α'? What is the value of α when the velocity v tends to c [1].

P14.4 Two events occur at the same place in the laboratory at an interval of 2 s. What is the spatial distance between these two events in a moving frame with respect to which the events occur 5 s apart, and what is the relative speed of the moving and laboratory frames? [1]

P14.5 A physicist, to avoid fine for driving through a red light, argues that the red light looked to him as green from his moving vehicle. The judge fined him $ 1 for every kilometer of speed in excess of 50 km/h. What is the fine? Wavelength: Green: 0.53 microns, Red: 0.65 microns [1].

P14.6 Refer to Figure 14.4. If S′ is moving at a uniform velocity of 0.35 c in the z-direction with reference to S, determine the complex angle ψ [1].

P14.7a A linear accelerator accelerates the electrons to the full energy of 30 GeV. Calculate the mass of this electron [1].

P14.7b A proton has a kinetic energy of 450 million electron volts. Find the mass and velocity [1].

P14.8 A charged parallel plate capacitor moves with a velocity v in a direction normal to the plates. The capacitor plates have an area A and are separated by the distance s. The electric field is E′ in the direction of the capacitor movement. Find V, vector potential A, and the fields E, B with respect to a stationary reference frame [1].

P14.9 Show that Equation 14.76 is included in the four dimensional curl equation (14.77).

P14.10 Determine the power reflection coefficient for normal incidence (wave propagating along z direction) of a plane wave on a moving dielectric half space moving along z direction with a velocity $v_0 = \hat{z}v_0 = \hat{z}\beta c$. Assume $n = 2$ is the refractive index of the dielectric measured in the rest frame of the moving dielectric. Sketch ρ (power reflection coefficient for the entire range of β from −1 to +1.

Appendix 1B

P1B.1 A conducting filament extends from $z = -5$ to $z = 5$ m on the z-axis in free space and carries a current $I = 4t$ in the $\hat{\mathbf{z}}$-direction. Find and sketch $\mathbf{A}(t)$ at (0, 0, 10) for $-0.1 \le t \le 0.1$ µs.

P1B.2 Given the retarded potentials $\Phi = x - ct,\ \mathbf{A} = (x/c - t)\hat{\mathbf{x}}$, where $c = 1/\sqrt{\mu_0\varepsilon_0}$.

a. Show that $\nabla \cdot \mathbf{A} = -\mu_0\varepsilon_0(\partial\Phi/\partial t)$.

b. Find **B**, **H**, **E**, and **D**.

c. Show that these results satisfy Maxwell's equations in free space.

P1B.3 Given $\tilde{A}_z = C \sin(\pi x/a) \sin(\pi y/b) \cos(\pi z/d)$,find the corresponding electric and magnetic field phasors. Find the electric and magnetic fields as a function of time.

P1B.4 Given $\tilde{A}_\phi = CJ_1(k_c r)\, e^{-j\beta z}$, find the electric and magnetic fields.

P1B.5 Given that the vector potential in free space is given by

$$\tilde{\tilde{A}} = \hat{x}A_0 e^{-jk_0 z}$$

determine

a. Vector magnetic field phasor $\tilde{\tilde{H}}$
b. Vector electric field phasor $\tilde{\tilde{E}}$
c. Time-averaged power density at the point P (1, 2, 3).
d. A PEC plate is placed at $z = 0$. Determine the surface charge density $\tilde{\rho}_s$ on the plate.
e. A PEC plate is placed at $z = 0$. Determine the surface current density $\tilde{\tilde{J}}_s$.

Appendix 2C

P2C.1 An s-wave with free space wavelength $\lambda_0 = 3$ cm is incident on the dielectric layers as shown in the figure. Calculate the reflection and transmission coefficients. $n_1 = 1.5$, $n_2 = 2.0$, $n_3 = 2.5$, $d_1 = 1$ cm, $d_2 = 1.5$ cm, $\theta_i = 30°$

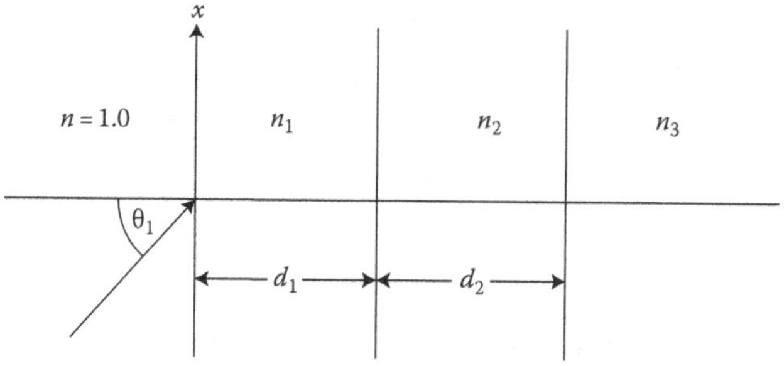

Appendix 7A

P7A.1 a. The cross section of a grounded cylindrical pipe is shown in the figure below. Show that the three-dimensional Green's function G, satisfying Dirichlet boundary conditions on the surfaces $\phi = \phi_1$, $\phi = \phi_2$ and $\rho = b$ is as given in Equation P2.1.

b. Given $\phi_1 = 30°$ and $\phi_2 = 120°$, determine S in terms of m.

c. Determine (i) k_{S1} and (ii) k_{S2}, if $b = 2$ and $m = 1$.

$$\nabla^2 G = -\delta(\boldsymbol{R} - \boldsymbol{R}'),$$

$$G(\rho,\phi,z;\rho',\phi',z') = \frac{4}{b^2[\phi_2-\phi_1]}\left\{\sum_{m=1}^{\infty}\sum_{n=1}^{\infty}\left[\frac{J_s(k_{Sn}\rho)J_s(k_{Sn}\rho')}{k_{Sn}J_{S+1}^2(k_{Sn}b)}\right]\right\} \times\left[\sin S(\phi-\phi_1)\sin S(\phi'-\phi_1)\right]\left[e^{-k_{Sn}|z-z'|}\right]. \quad \text{(P7A.1)}$$

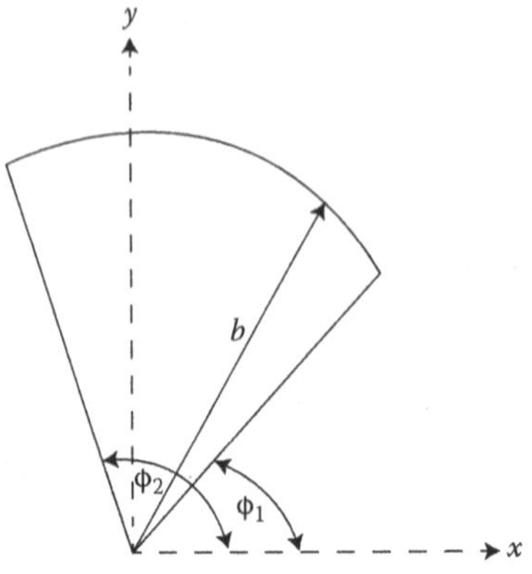

$0 < \rho < b$.

Note:

$$\int_0^c x\left[J_n(\lambda_k x)\right]^2 dx = \frac{c^2}{2}\left[J_{n+1}(\lambda_k c)\right]^2$$

where λ_k is a root of $J_n(\lambda_k c) = 0$.

Appendix 13A

P13A.1 Derive (13A.2).

P13A.2 Obtain (14B.1a) and (14B.1b).

P13A.3 Determine the pressure on the interface of two dielectrics due to electric fields with normal as well as tangential components. Express it in terms of the permittivities of the two media and the field components in medium 1.

P13A.4 Determine the pressure on the interface of two magnetic media due to magnetic fields with normal as well as tangential components. Express it in terms of the permeabilities of the two media and the field components in medium 1.

Appendix 13B

P13B.1 Use (1B.10) to derive the Biot-savart law for computing the B-field at a field point P due to a differential filamentary current element $I\Delta \boldsymbol{L}$ at the source point S.

Index

K

L

S

T

U

V

W

Printed in the United States
by Baker & Taylor Publisher Services